CONTEMPORARY PERSPECTIVES OF BIOLOGY

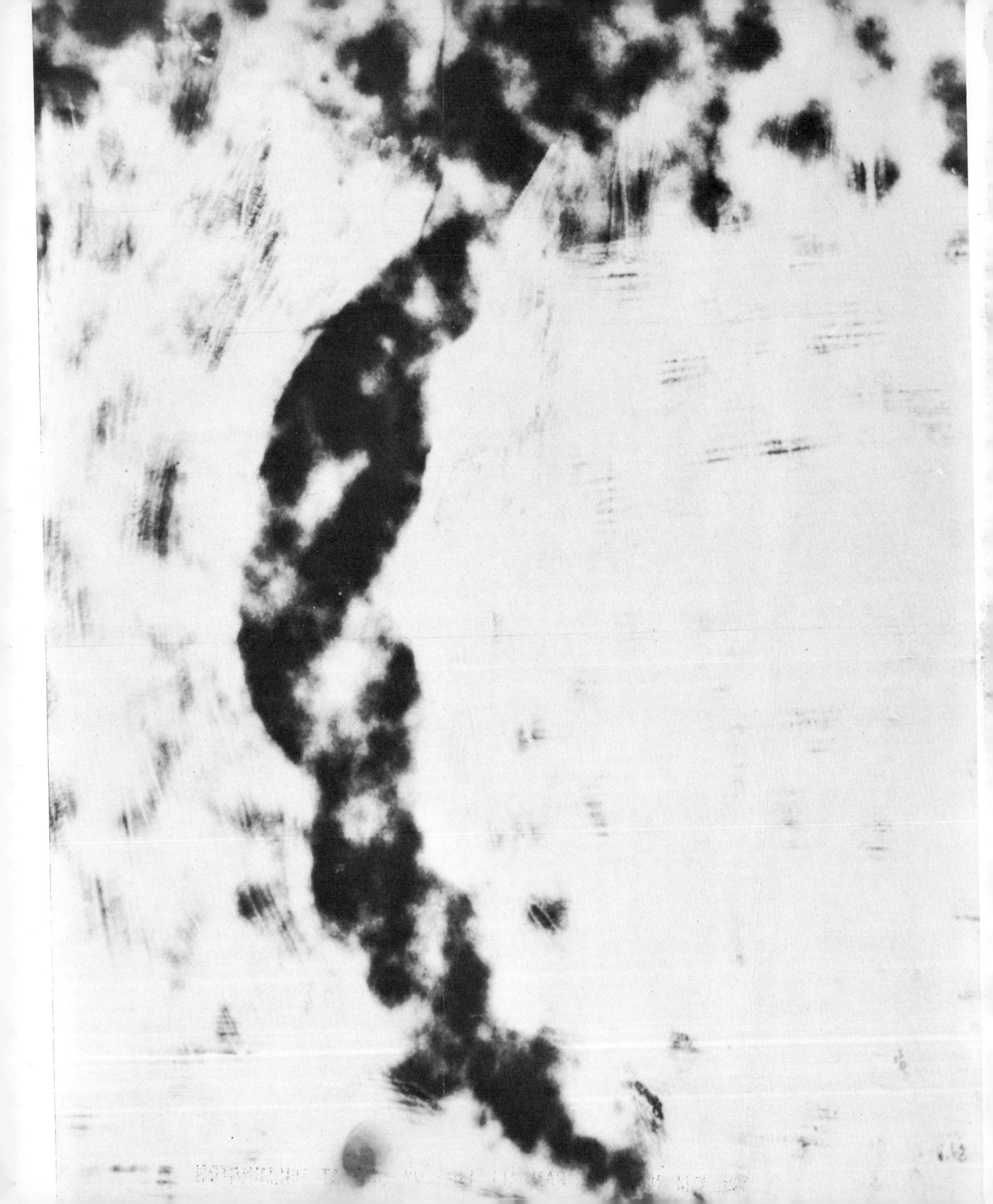

ROBERT W. KORN AND ELLEN J. KORN

CONTEMPORARY PERSPECTIVES OF BIOLOGY

John Wiley & Sons, Inc. New York, London, Sydney, Toronto

Cover etching by Rudy O. Pozzatti

Book designed by Robert Goff

Library of Congress Catalogue Card Number: 79-140178

ISBN 0-471-50376-2

Printed in the United States of America.

10 9 8 7 6 5 4 3 2 1

PREFACE

A science does not exist to be taught, but as a means of inquiry and of interpreting natural phenomena through experimentation. It is a travesty of good science to assume that the existing body of knowledge can be organized and taught without considerations of prevalent research attitudes of the field.

In biology, a course of study must be envisioned as a series of perspectives aligned as in a hall of mirrors, in which the reflections from a single mirror include partial images from others. Molecular biology, for example, is more than a new area of inquiry in the biological sciences. It is an entirely unique way of viewing biological organization. And the principles that have been found through research in molecular biology have far-reaching implications that permeate many seemingly unrelated areas, thus adding a new dimension to the study of biology.

The "perspective" view of biology leads to several organizational features in the text. Research attitudes suggest that there are two related but different perspectives of genetics: the genetic control of cellular activity, and the genetic basis of population structure, from which several principles of evolution can be derived. In keeping with the perspectives of research, these two aspects are discussed separately.

Another important organizational feature involves the source of the information available on the developmental aspects of organisms. Cellular differentiation, (once an aspect of embryology), has made great strides under the tutelage of cellular and molecular biologists, and has subsequently become a perspective of cell biology. Accordingly, the chapters that include this information ("Growth" and "Cellular Differentiation") follow the section on the cell.

We have chosen four major themes to accomplish a building organization: (1) energy, (2) the cell and cellular control, (3) evolution, and (4) the population-environment interaction. These themes are viewed many times from many different perspectives throughout the text.

We have attempted to develop some of the critical ideas in biology at a pace that can be followed by the average student. We do not consider biology so difficult, and so dull, that it can only be appreciated by students who have come to college already interested in (or committed to) the subject. A distilled science course sells short both the subject and the student. Real science is not a compilation of experimental conclusions—a series of facts—but, instead, the logical extension of ideas as manifested in experimental design or the formulation of principles. Real science can be truly appreciated, whereas distilled science is absorbed too quickly and does not lend itself to meditation.

The students of the 1960s demanded relevancy in their course material; the students of the 1970s will expect it in theirs. Yet, even as we enter the 1970s, biology courses retain the aura of pure academia—a fact that is surprising, since biological research has made advances eminently relevant to modern society. Social problems such as the biological effect of drugs, the biological basis for radiation damage, the lack of a cure for cancer, and the fate of human populations are all of intimate concern to the aware student. Is a biology course that goes into great detail about the different molecular species found in biological systems really complete without carrying the idea a few steps further to include the types of molecular species that are affected by radiation and their importance in maintaining the integrity of the individual?

To emphasize the relevancy of biology in modern society, we have interspersed "special-interest" chapters throughout the book. These chapters are placed after the regular chapters that provide the basic information necessary to handle the material. For example, before a student can fully comprehend the biological implications of radiation damage, he must have a knowledge of cell chemistry, integration of molecular species, cell form and function, and the genetic control of the cell. Therefore, the special-interest chapter "Radiation" is placed after the chapters dealing with these topics.

The basis for choosing the topics for the special-interest chapters was student curiosity and concern. In light of the interest of young adults in the NASA program, we felt that students would be receptive to a speculative study of extraterrestrial biology and would enjoy the mental exercise that goes into such a study. Moreover, we handled the material within special-interest chapters on controversial subjects in as objective a way as possible.

The inclusion of special-interest chapters has permitted an alteration of the perspectives of certain material. For instance, the presence of the special-interest chapter "Human Heredity" has made possible the inclusion of modifications of data analysis and statistical approaches, unique to human genetic studies.

We thank many people for their aid in the preparation of this manuscript:

Martin Albert for his invaluable help in the preparation of medically oriented chapters, and for his moral support during trying times; Chandran Kaimal for his help in the preparation of the discussion of the laws of thermodynamics; George Lesher for evaluating the "Chemical Background for Biology" chapter; Richard Hanson for evaluating the

"Metabolism" chapter; Mrs. Lorraine Kaimal for editing the early chapters and setting the tone for the remainder of the book; Berne Bachrach and Mrs. Gloria Hanson for proofreading several of the early chapters; Mrs. Jacqueline Bachrach for taking on the ponderous task of typing large sections of the early stages of the manuscript; and Rudy Pozzatti, for taking time from his busy schedule to create the cover for this text. We would also like to thank William Bryden, biology editor of Wiley, who was responsible for the translation of an idea into a manuscript; and Stella Kupferberg, photography editor and coordinator *par excellence* of Wiley, who was responsible for the translation of a manuscript into a textbook.

We dedicate this book to the many people who made life endurable during the years of its preparation: to our many friends for remaining our friends; to Joyce Young for creating a home for our children while their mother was busy writing; to Mrs. Lee Cosden, for the faith and $upport so necessary in those years; and to Nicholas, Jennifer, Matthew, and Jeremy for their love and patience, and for their acting as constant reminders that four college educations loomed ahead, thus adding a practical purpose to our literary attempts.

Robert W. Korn
Ellen Korn

Bellarmine College
Louisville, Ky.

CONTENTS

CONTEMPORARY PERSPECTIVES OF BIOLOGY

PART ONE: CHEMICAL ENERGY AND BIOLOGICAL SYSTEMS

The concept of chemical energy is the recurrent theme in the second through sixth chapters. The presentation of these areas from an energy perspective requires background information in physical as well as organic chemistry, which is given in the chapter on "The Chemical Background for Biology." The energetic considerations of chemical reactions are developed methodically, and only material that directly leads to those chemical principles used in discussion of biology is given.

Because of the nature of the material, this part of the text includes several mathematical references. The concept of negative and positive free energy (ΔG) is developed in the chemistry chapter, and used repeatedly in the "Metabolism" and "Photosynthesis" chapters in conjunction with discussions of energy yields. The concept of first- and zero-order kinetics is first encountered in the chapter on "The Structure and Function of Proteins," and later appears as an important feature of research in the "Photosynthesis" chapter. The operation of the Electron Transport System, which is discussed in the "Metabolism" chapter, is handled through redox equations. Thus, each mathematical development leads to important principles that are used in other contexts for the development of further principles.

The "Nutrition" chapter serves as both a summary and an extension of the material. The elemental cycles, usually discussed in an ecological perspective, are discussed in this chapter instead in order to emphasize the interaction of *enzyme systems* of different organisms, rather than the organisms themselves.

(Opposite) "Counterpoint Castle," Ibram Lassaw, Collection of the artist

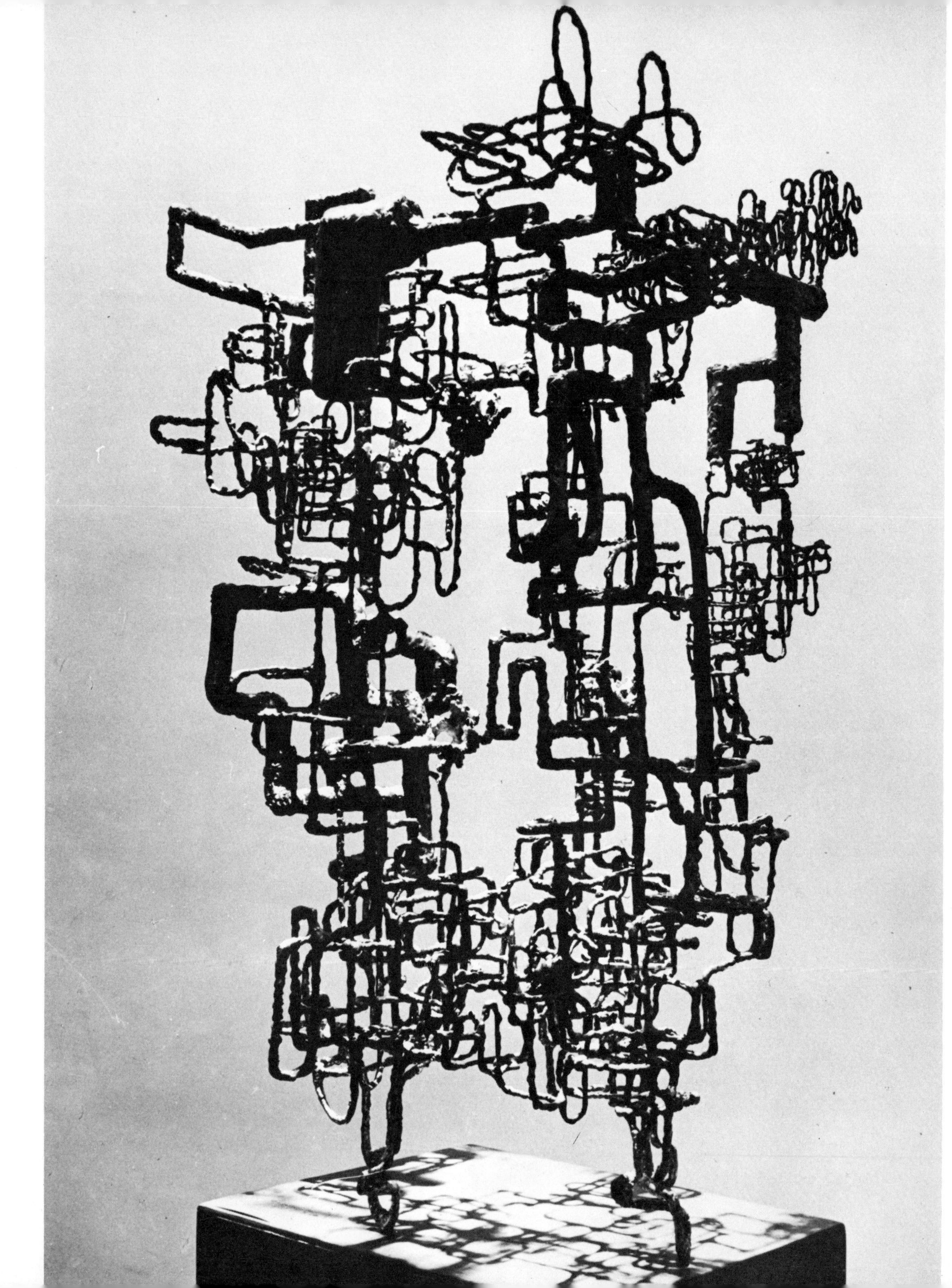

CHAPTER 1

BIOLOGY AS A SCIENCE

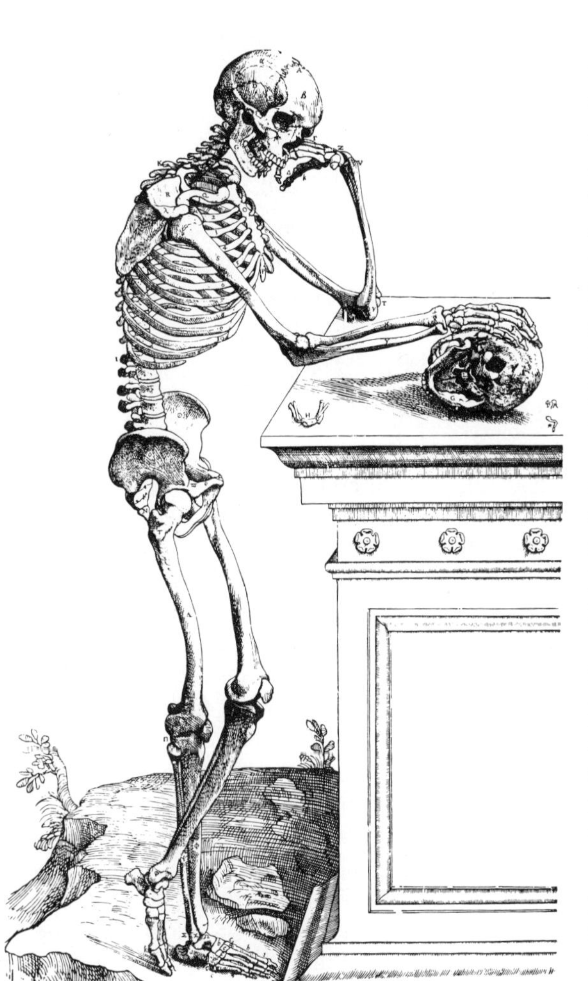

—Vesalius (1514–1565)

Man is the only creature on earth who is able to examine his total environment and draw abstract conclusions from his observations. All academic subjects are products of this examination. A student will often choose a particular course of study because he feels intellectually capable of handling the material within that area, and because he is sympathetic to the manner in which the discipline interprets the universe. A person studies psychology because, to him, the most important aspect of the universe is the human mind. Another person studies languages because he finds the translation of ideas into meaningful and descriptive words most rewarding. Still another person studies science because, to him, the most important pursuit is the systematic investigation of natural phenomena. Each academic field looks at the same universe, but from a different point of view.

The scientist, in his systematic investigation of natural phenomena, asks questions and expects answers. Some of the questions he asks require careful observations of natural phenomena, and some involve experimental manipulation of a system. Other questions may involve manipulation of data collected by others or construction of models to clarify the phenomena.

A scientist wants meaningful answers to his questions. Questions that require careful observation, and some of those that require experimental manipulation, are often answered in descriptive ways. Common language often does not suffice to explain observations or results, and scientific terms must be invented. Although a student may find scientific terminology rather formidable, it, like any language, becomes less difficult with repetitive use.

The ultimate goal of a scientist is to detect patterns, and from these patterns, to develop **rigorous** and **universal** concepts. Many times even the scientific language does not suffice to provide the degree of rigor necessary for conceptualization, and the sciences have complemented their description with the language of mathematics. Each of the major natural sciences (physics, chemistry, and biology) has undergone a metamorphosis in the course of attaining rigorous and universal concepts for natural phenomena in their domain. In the course of this metamorphosis, conceptualization changed from a purely descriptive point of view to a rigorous one, possible only through measurement and subsequent quantitative analysis.

The descriptive phase of physics ended in the seventeenth century, when Newton, Kepler, and Galileo found that phenomena of the physical world could be expressed more simply and precisely in mathematical terms. Interpretations of equations provided basic and meaningful concepts that could not have been obtained from unmeasured observations. In the late seventeenth century, chemists found that giving only verbal descriptions of chemical phenomena limited them, and they began to use suitable measurement and mathematical procedures for explaining phenomena. The transformation of biology from a purely descriptive conceptual science to a quantitatively conceptual science has been a gradual one. Measurement (the first step in quantification) was used in the seventeenth century, but did not become an integral part of biological inquiry until the nineteenth century. Universal biological concepts developed in the nineteenth century lent impetus to the quantitative approach, and subsequently developed into the hallmark of biological research in the twentieth century.

TRANSITIONS IN BIOLOGICAL THOUGHT

Most people draw a distinction between the natural sciences. Biology is the descriptive science. It is concerned with the names of bones, muscles, and the like. But in many ways, biology has come of age. One can see the "coming of age" in the general attitudes that prevailed in the history of biological studies.

The First Universal Concept in Biology

The first man who looked at himself, or a plant, or another animal, and recognized that he had a quality different from that of a stone, was the first biologist. The acknowledgment of a qualitative difference between living and nonliving things was the first biological concept. It was not a rigorous concept, but it most assuredly was a universal one.

The first biologist could not (and indeed, we still cannot) define this quality that we have chosen to give the name of "life"; it is one of those abstract concepts that is more in the realm of philosophy than of biology.

For thousands of years in cultures of Egyptian, Greek, and then Roman domination, men substantiated this first universal concept by discovering living forms. Since they were limited to the forms they could see with the naked eye, only the organisms thus discerned were given names. It soon became apparent that these living forms could be categorized. Those that exhibited motion, that had nervous response, and that varied in pigmentation were defined as *animals;* those that were sessile (incapable of movement) and green were considered *plants.* Within each of these major categories, different levels of complexity were recognized. Groups of plants and animals were described by where they were found, by how large they grew, and so on.

Although the ancients detected similarities *within* specific groups, and saw that a set of characteristics could be viewed as representative of that group, the specific groups were thought of as discrete units. No one group had any relation to any other group in terms of origin or existence, except that they all possessed the mystical quality of life. The categorization of the ancients, then, *emphasized differences* by emphasizing uniqueness among organisms.

This approach continued through the Middle Ages. During this time, the study of biological forms was man-centered. **Zoologists** tried animal breeding in attempts to get better beasts of burden *for man*. The forerunners of the present-day **botanists** extracted drugs from plants *for man's use*. More and more types of animals and plants were collected and given names, but they were examined in the most general ways.

At the end of the Middle Ages, the cumulative results of biological inquiry were little more than a conglomeration of organisms, categorized in a manner that stressed their differences. Thus, when physics and chemistry were ready for a quantitative conceptualization through the universality of natural phenomena, biology had nothing for which to seek a rigorous concept. Its only universal concept was an undefinable quality that separated the living from the nonliving.

Origin of Universal Concepts in Biology

During the seventeenth and eighteenth centuries, great studies were undertaken on the descriptive phase of biology. The emphasis changed from man-centered (what can the organism do for man?) to organism-centered (how does the organism function?). As biologists became interested in knowledge for its own sake, biology became a true scientific inquiry.

From an organism-centered scientific perspective came important indications of the universal similarity between organisms. Botanists and zoologists found that plants and animals had systems that carried out particular functions. A plant could be viewed as having a root system, a stem system, and a leaf system (see Fig. 1-1). The root system is responsible for the movement of water from the surrounding soil into the plant itself. The stem system is responsible for the transportation of water and nutrients within the plant, and for the support of the leaf system. The leaf system is the site of photosynthetic activity through which food is manufactured. Higher animals can also be thought of in terms of systems. Food that was eaten is broken down into smaller and smaller particles and finally into a liquid state by the digestive system. Harvey discovered that blood traveled through the body in such a way that it always returned to the heart, and termed the elements contributing to this phenomenon, the circulatory system. The voluntary movements of the body can be accounted for by the action of the muscles, which collectively comprised the muscle system. The collection of bones can be considered a skeletal system. The brain and spinal cord were found to be responsible for carrying nerve impulses through the

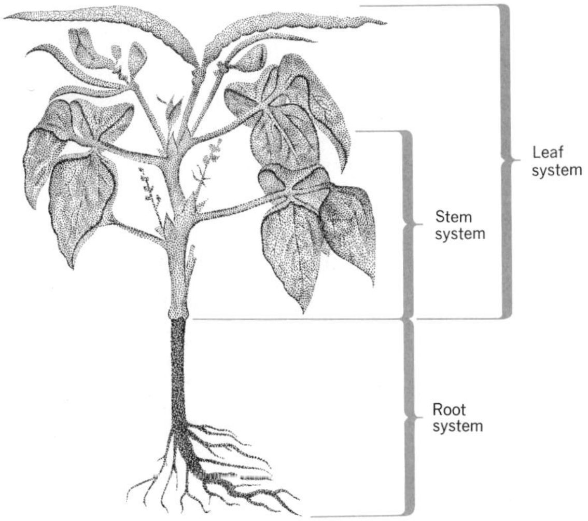

Leaf system

Stem system

Root system

Fig. 1-1 Plant systems. Higher plants have three basic systems. The root system responsible for the passage of water and minerals into the plant body. The stem system supports the plant and provides passage for the transmission of water and minerals to the leaves and of photosynthetic products from the leaves. The leaf system is the site of photosynthesis.

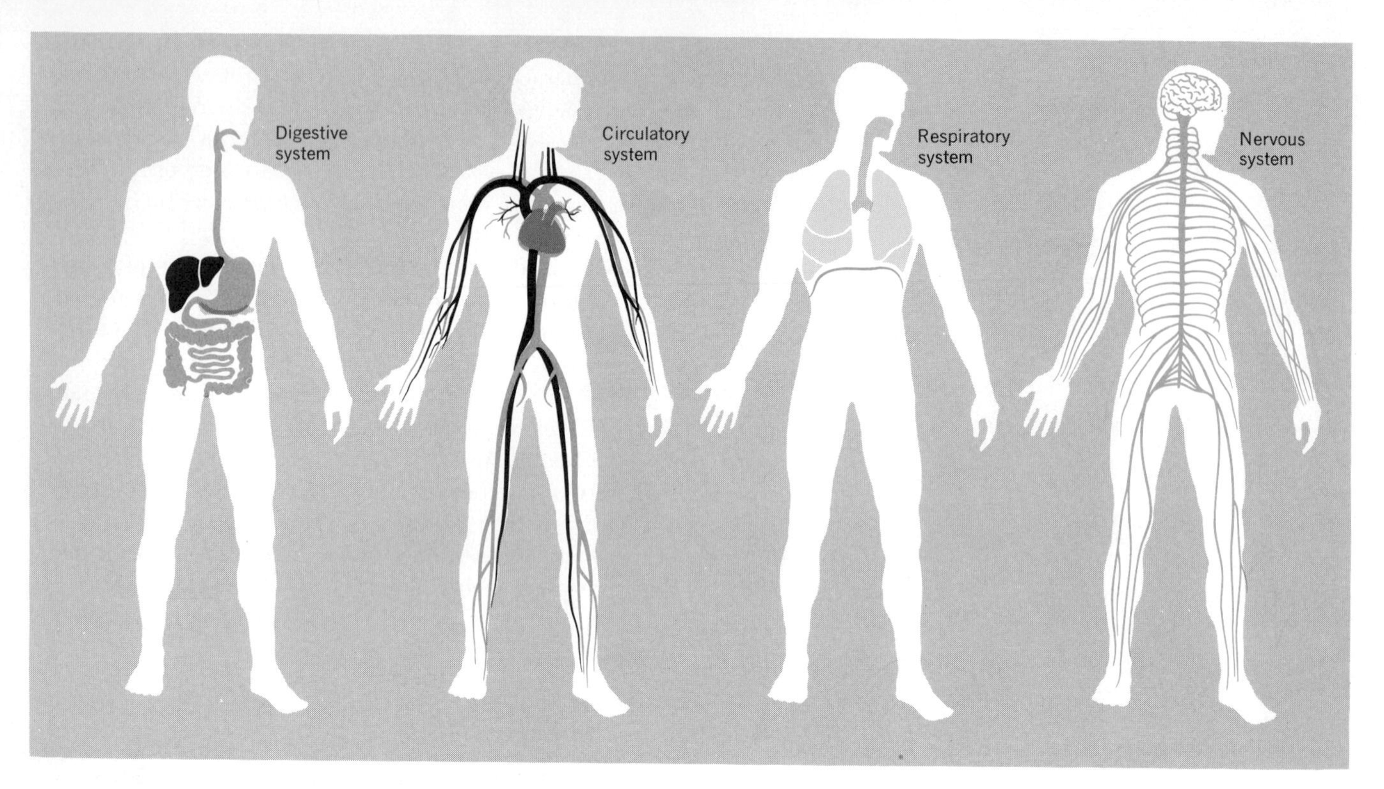

Fig. 1-2 Animal systems. Higher animals have a large number of systems, only four of which are shown here. The digestive system breaks down foodstuffs to a form usable in cellular metabolism. The circulatory system carries nutrients, wastes, and hormones to various parts of the body where they are absorbed or eliminated. The respiratory system is where the exchange of gases takes place. The nervous system provides the means of bodily response to internal and external environmental changes.

body and were collectively called the nervous system (see Fig. 1-2).

Each system was made up of separate and distinct portions, termed organs. The digestive system was made up of the **mouth,** the **esophagus** (a tubelike organ that connects the mouth and the stomach), the **stomach** (which is a large J-shaped organ where food is partially digested), etc. Concomitant with these discoveries was the growing realization of the universality of organismal function. All animals consume food. Therefore, their bodies must have a digestive system of some type. Many animal forms were dissected in the search for systems that were either structurally different or the same as those found in the higher forms.

In the seventeenth century, Anton von Leuwenhoek improved the single-lens microscope and peered into a whole new world full of unnamed organisms. But this new world was very perplexing to him and his microbiologist successors, for its inhabitants defied the type of categorization designed for larger organisms. Little green "animalcules" appeared in the drops of water (see Fig. 1-3). But the green pigment was reserved in the biological world for plant life, and plants were supposedly sessile. The arbitrarily chosen definition of plant life had to be reevaluated. Small particles of different shapes, rods, spheres, and corkscrews could be seen moving around. These were classified as bacteria, and set apart from the animals and plants. So, initially, the microscope became a tool to further divide living forms into different groups.

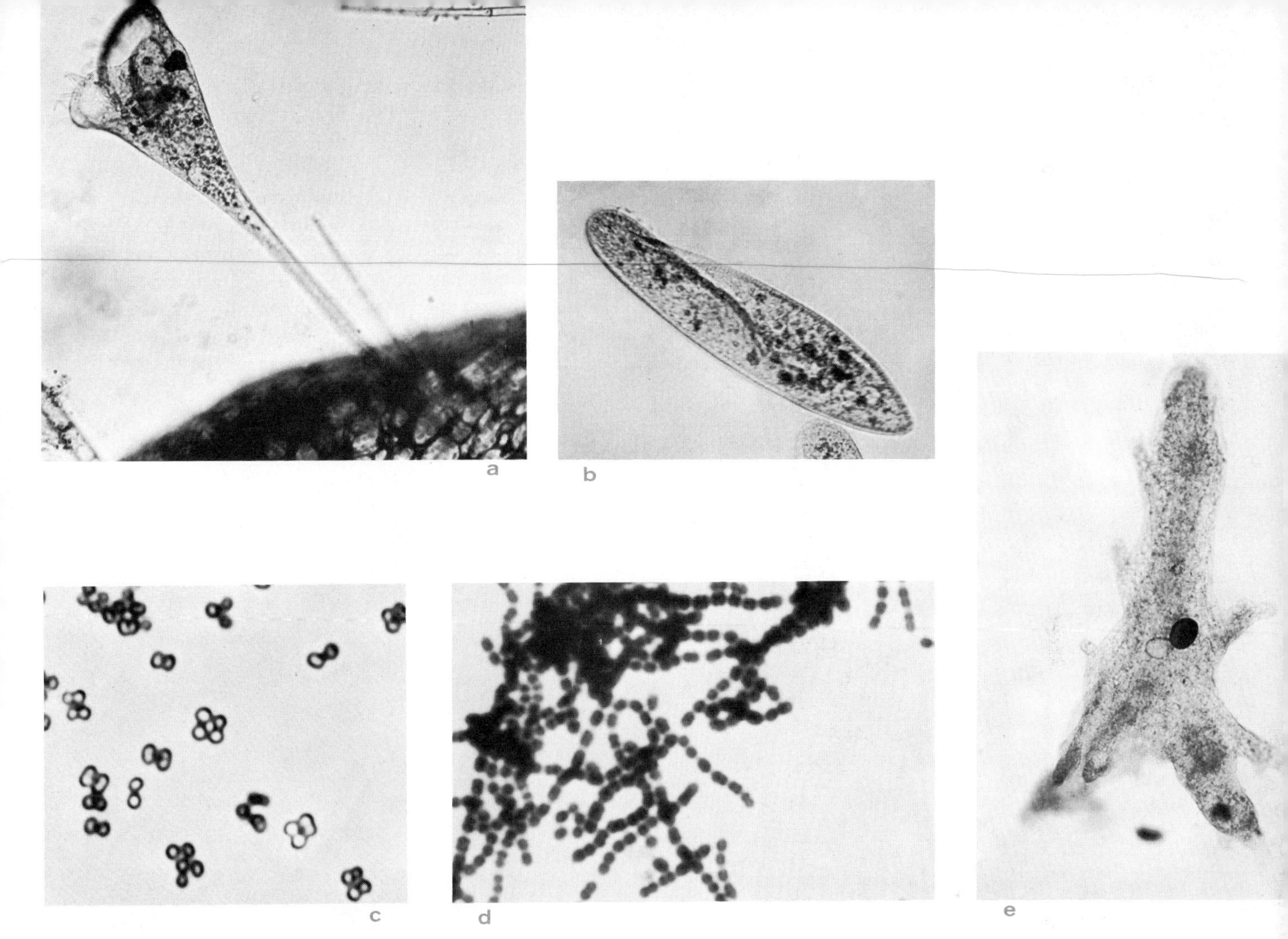

Fig. 1-3 Examples of microbes. Some of the various types of organisms discernible by microscopic examination of pond waters or bacterial cultures. (*a*) *Stentor*, an animal-like organism that lives on the substratum of shallow ponds. *Stentor* is a single-celled, highly differentiated organism, one end of which remains attached to the substratum while the other end pulsates, creating a current that brings smaller organisms into the oral cavity (75×). (*b*) *Paramecium*, a slipper-shaped ciliate also found in ponds (225×). (*c*) *Sarcina lutea* (3200×), a bacterium that characteristically grows in a quartet of cells. *Sarcina* is classified as a coccus and is often found as part of the natural flora of the mouth. (*d*) *Streptococcus pyrogenes*, also classified as a coccus, characteristically grows in chains (2700×). This organism is the cause of certain infections. (*e*) *Amoeba proteus*, found on the bottom of ponds, creates pseudopodium (false feet) by changes in the viscosity of the cytoplasm (75×). (*f*) *Volvox*, a colonial organism is also found in ponds. Note the cytoplasmic strands that connect one "cell" to another (20×).

[(*a*), Dr. William Amos; (*b*), Hugh Spencer; (*c*), Walter Dawn; (*d*), Philip Feinberg, Fellow of the New York Microscopical Society; (*e*), Lester V. Bergman; (*f*), Dr. Richard C. Starr.]

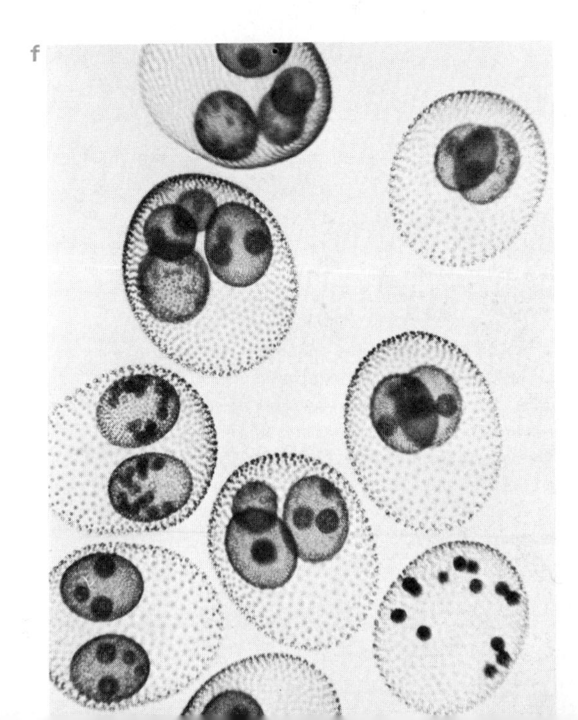

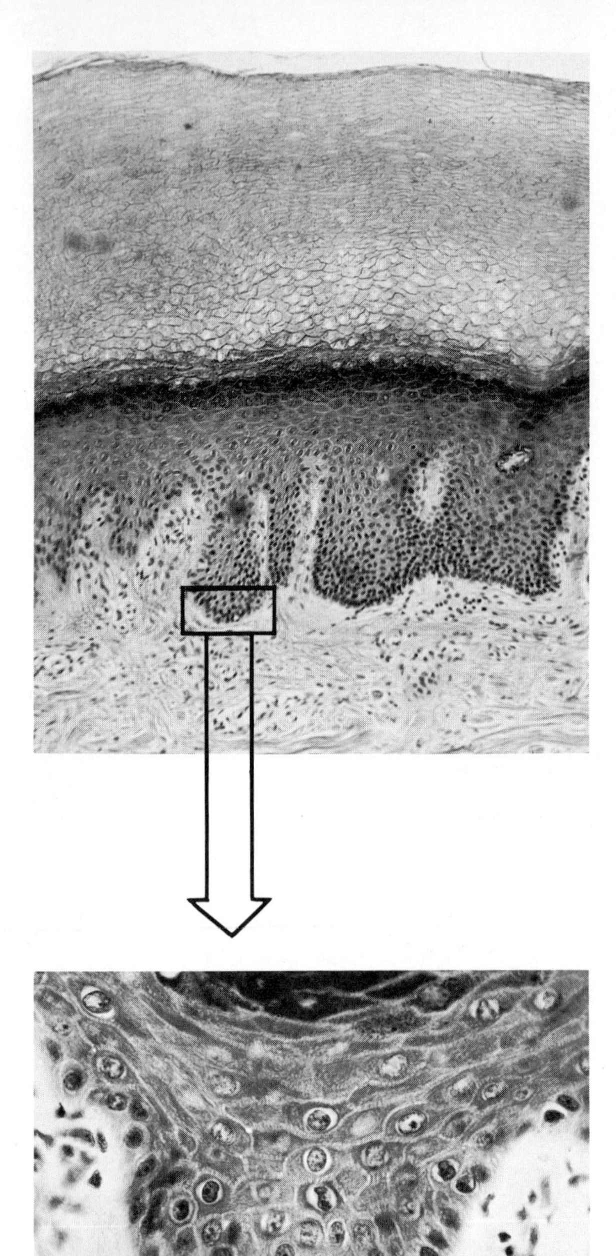

Fig. 1-4 Epidermal tissue. Different tissues are easily recognizable in some organ systems, as shown in this photograph of human epidermis. In some systems, however, only the trained eye can distinguish various types of tissue. (*Above*) A portion of human epidermis (110×). (*Below*) A higher magnification of the encircled area. Notice that the physical association of cells as well as the nature of the individual cells give the tissue its character. [Lester V. Bergman]

Methods were developed for cutting organs into very thin sections and examining them microscopically. Such examination led to the discovery that the organs were also divisible and were made up of different parts called **tissues** (see Fig. 1-4). So an organ was made up of different types of tissues: a stomach was made up of muscle tissue, gastric mucosa, connective tissue, etc. The muscle tissues found in the stomach and the intestine were very similar, but both were unlike the tissue with which they were associated in each organ. But also visible under the microscope was a lower level of organization, the **cells** (see Fig. 1-5). The tissues were made up of cells, small units that were fitted together in such a way as to give the total effect of the tissue.

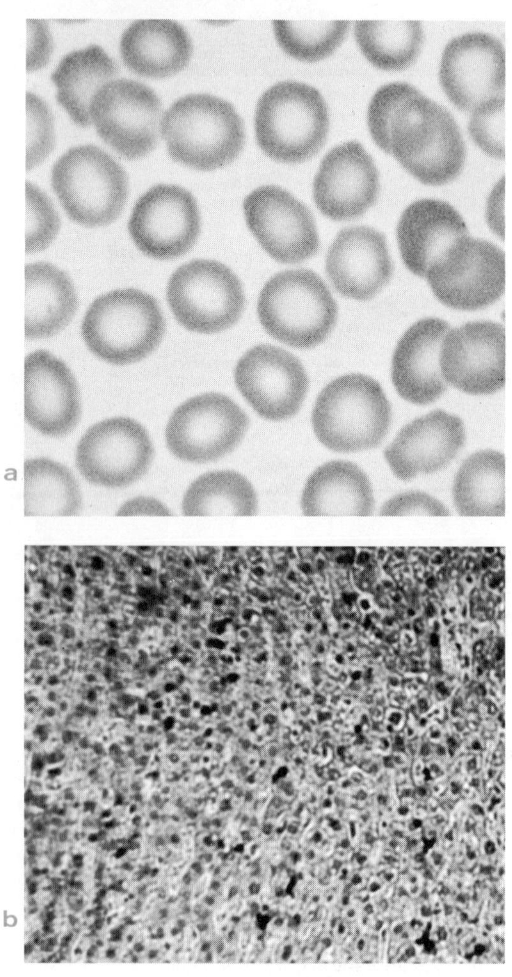

Fig. 1-5 Various types of cells discernible by microscopic examination. (*a*) Human red blood cells (1400×); these cells are donut-shaped but rest on the slide so as to appear round. These specialized cells have no nucleus, but have a high concentration of the protein hemoglobin in the cytoplasm. (*b*) Liver cells are closely packed cells with distinct nuclei (small black areas) and very dense cytoplasm characteristic of active cells (144×). (*c*) Animal epidermal cells—frog (260×). Notice the thin multi-

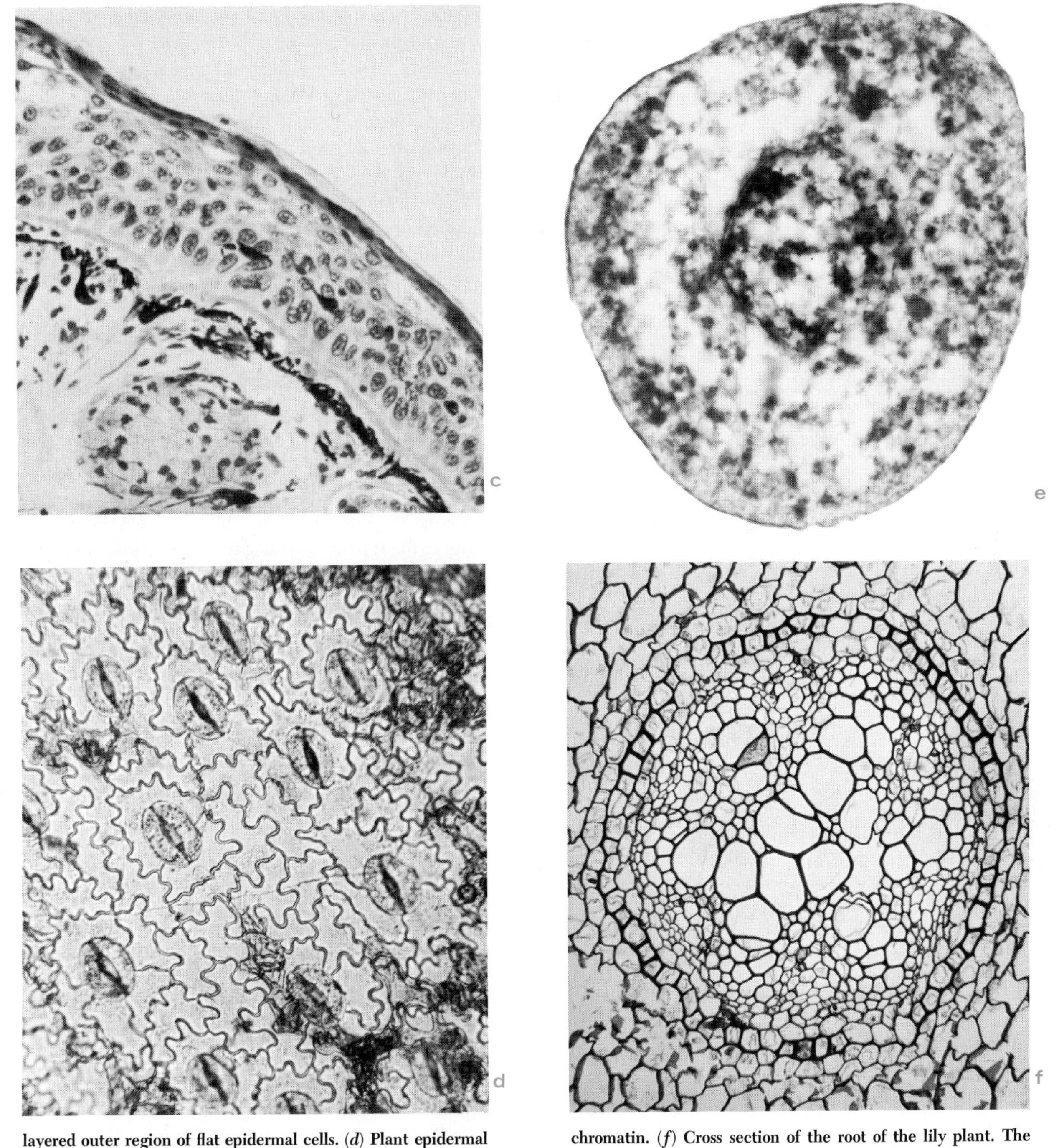

layered outer region of flat epidermal cells. (*d*) Plant epidermal cells of the leaf of a fern (200×). Notice the jig-saw-puzzlelike shapes of the epidermal cells. Of what structural value is this shape? The somewhat round structures are stomates composed of two sausage-shaped guard cells separated by a space (opening) through which gases enter and leave the leaf. (*e*) Animal egg cell—cat ovum (1650×). This a large, thin-membraned cell that has dense cytoplasm and a central nucleus with threadlike chromatin. (*f*) Cross section of the root of the lily plant. The large central cells are xylem cells that have thick, darkly stained walls and an empty lumen. A xylem cell is essentially a tube, for it is at least 1000 times greater in length than in diameter.

[(*a*) and (*c*), Philip Feinberg, Fellow of the New York Microscopical Society; (*b*), Russ Kinne-Photo Researchers; (*d*), Hugh Spencer-National Audubon Society; (*e*), Lester V. Bergman; (*f*), Russ Kinne-Photo Researchers]

It was from this cellular level that the first truly unifying principle of biology arose. As more and more organisms (both animal and plant) were examined microscopically, it became apparent that every one was made up of cells. The final statement of the **cell theory** by the botanist Schleiden and the zoologist Schwann, in 1838, was in essence the formal statement of an idea that was already present in the minds of biologists: "all living things are made up of cells." To be sure, some forms had only one cell and were termed **unicellular** organisms, while others had many cells and were called **multicellular.** But the cell is the least common denominator of living forms. Biologists were beginning to think in terms of universals. It should be emphasized, however, that this universal was achieved only through a study of particulars.

In 1859, Charles Darwin added a second unifying idea in his theory of **evolution.** His idea that all living forms have a **common origin** and that differences were the result of changes occurring over a long period of time (thousands of years) started biologists looking for the basic similarities rather than the differences among organisms.

The nineteenth-century biologist delved deeper into the cell. With the aid of better microscopes that permitted greater magnification and higher resolution, it became obvious that the cell was made up of component parts (see Fig. 1-6). It was surrounded by a **membrane,** a thin sheetlike structure that seemed to hold the cell together. The

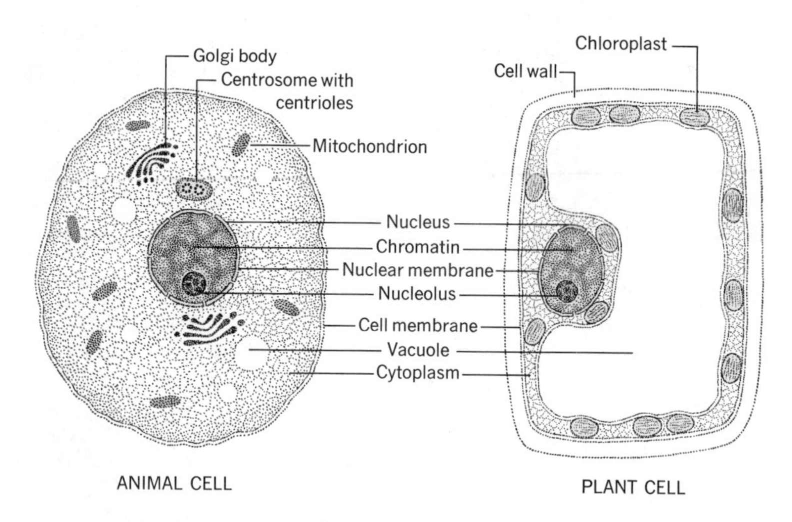

Fig. 1-6 Generalized drawing of plant and animal cells. This illustration is a conglomerate, showing some of the features common to most cells.

membrane was not rigid, and so even though all cells had membranes, they were not all the same shape. In addition to the membrane, plant cells had a rigid **cell wall** that seemed to be responsible for the rather rigid shape of these cells. Bacteria were also found to have cell walls. Within each cell was a central body called a **nucleus,** common to both plant and animal cells. Small particles within both plant and animal cells that were called **organelles** were also found, but the resolution of microscopes used at that time did not permit a true discernment of the organelles into different types. Plant cells were found to contain little packets of green material, called **chloroplasts** (chloro = green, plast =

body). The remainder of the plant cell was not pigmented. Thus it was only because the human eye could not distinguish individual chloroplasts that man saw the whole plant as being green.

Rigorous and Universal Conceptualization in Biology

The types of observations made in the nineteenth century generated the creation of a universal set of concepts in biology. If a biologist was interested in studying one activity of the nucleus, he could, by manipulative procedures, remove the nuclei from a large number of cells (all of the same type) and submit only that nuclear fraction to experimentation. At the conclusion of his experiments, he could make a statement on one aspect of nuclear activity. Other researchers, working with other cell types, would look for this aspect in their systems. If a pattern was found, a universal statement could be made. A plant physiologist could study the role of the chloroplast in plant activity by extracting the chloroplast fraction. After his experiments, he could perhaps make a statement that was valid for all plants.

From another standpoint, groups of the same organismal type (*species*) could be viewed as populations, and universal concepts could be derived from studies on the population level that would be valid regardless of the type of organisms of which a population was composed.

Chemical studies revealed that all matter was made up of molecules. Biochemists probed the chemical depths of the cell and found a universality in molecular forms. All plant walls were made up of **cellulose.** All cells contained large amounts of **protein.** Studies of cellulose could give valid conclusions for all plant cells. Studies of proteins could give valid conclusions for all cells. The stage was set for probing the chemical and ultimately the physical basis of life.

Some aspects of cellular structure defied universality. Although both plant and bacterial cells were bounded by cell walls, analysis of the cell-wall material indicated that there was a significant difference in chemical composition. Therefore, any statement based on the chemical structure of the bacterial cell wall could not be universal. One of the major problems in biology has been to distinguish between the factors that could be viewed in universal terms and the factors that could not.

But the microorganisms, such as bacteria, have many features in common with cells of higher plant and animal forms. Utilization of these features in organisms that are so easily handled under laboratory conditions has contributed greatly to the wealth of universal statements for living forms.

With the creation of a viewpoint of universality, there was a reason for handling biological material in a quantitative manner, a manner that would lead to further development and more precise definitions of rigorous and universal concepts.

To be sure, there is still observational and nonquantitative research in biology. The areas in which explosive discoveries have been made, however, are the experimental ones, in which measurement and mathematical analysis of data play an integral part. In these areas, a high level of rigor has developed. It is not the rigor of physics or chemistry, in which concepts are expressed mathematically, but rather a rigor germane to the problems of biology, in which mathematics is used to interpret measurements and provide a semimathematic or semidescriptive conclusion.

THE ART OF MEASUREMENT

Conceptualization in a science arises from the interpretation of data. The collection of data involves observation and measurement, the translation of the primary features of the phenomena under investigation into numerical values. From these measurements and their mathematical manipulations, it is possible to observe certain relationships that could not otherwise be detected. Good science demands that the data be manipulated in such a way that a *maximum amount of information is obtained from a minimum amount of data.*

An important feature in making measurements is the degree of precision with which they are made. Two scientists, Smith and Jones, measured the number of unicellular organisms in a population. Each measured the same population over a series of consecutive days, and obtained the following data, which are given in table form:

Day	1	2	3	4	5
Smith	2000	3000	6000	12000	22000
Jones	1784	3394	6138	11671	22157

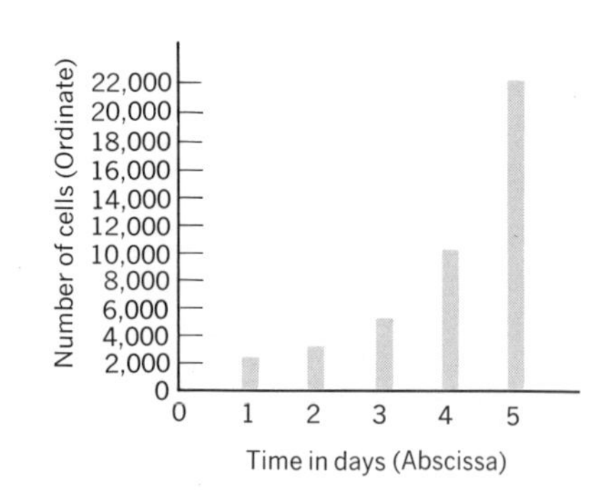

Fig. 1-7 Histogram of Jones data.

It is apparent that Jones sought exact values, while Smith was content with approximations. Smith's data could be interpreted as a pattern of growth in which the rate was initially slow, then changed to a doubling per day and later tapered off. On the other hand, Jones would conclude that the rate was constant, and that this constant rate was a little less than a two fold increase per day. Generally, both would conclude that the population grew in numbers of organisms but, in terms of the growth pattern, Jones is more likely to be correct *within the time interval under consideration.* More precise measurements often yield something that cannot be seen by approximations.

Another advantage to precise measurements is the further manipulations of the data into different forms of expression. The results of these manipulations can reveal certain aspects not readily discernible when the data are in table form. The validity of the translation of the data into different forms is dependent solely on the precision with which the data were originally collected.

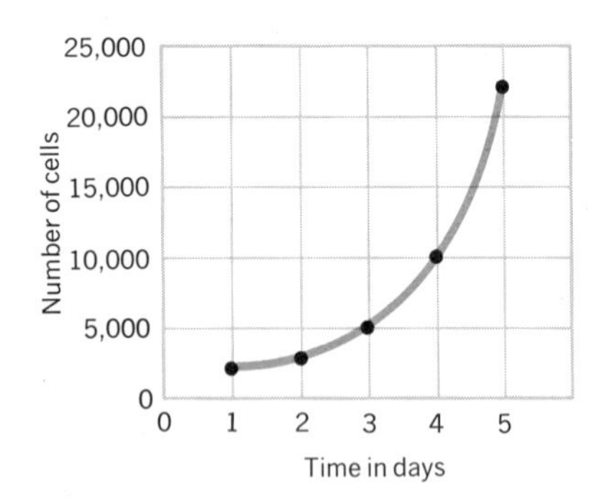

Fig. 1-8 Point graph of Jones data.

Translation of Data. It is possible to convert Jones' data into a histogram (see Fig. 1-7). It is also possible to express this data as a **point graph** (see Fig. 1-8). In this case, the point graph is more desirable because lines can be drawn between adjacent points to obtain a more **kinetic** (dynamic) picture. From the curve generated by plotting the data, it is possible to discern the growth pattern immediately. When the data are in table form, one must laboriously compare each number with the previous one.

Once the data are in graph form, Jones may further manipulate them mathematically. All curves are expressible by equations [$(x - a)^2 + (y - b)^2 = r^2$ is the general equation for a circle of radius r; $(x - a)^2 = y$

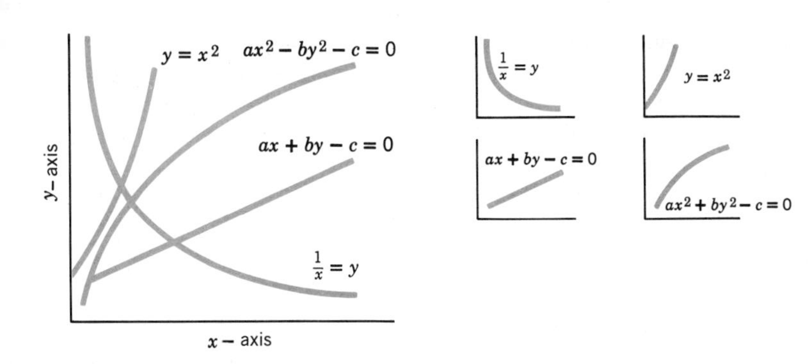

Fig. 1-9 Curves generated by various equations.

is the general equation for a parabola; etc.]. Thus, Jones may attempt to find an equation expressing a mathematical relationship between the number of organisms counted at a particular day (N_c), the initial number of organisms (1784), and the time (T) that would give the same curve that was obtained by graphing the data. The equations for certain simple curves are fairly standard, and it is a question of matching the simple curve to well-established equations (see Fig. 1-9). Examine the curves and their standard equations given in Fig. 1-9. The curve that most resembles Jones' data is $y = 2^x$. This is quite reasonable, because Jones noted that his population underwent a doubling each day. But his first count did not indicate the presence of only two unicellular organisms; it indicated 1784 organisms. The equation must reflect his starting point. Also, the coordinates in the standard curve are the x and y axis. The coordinates for a curve drawn from Jones' data must be reflective of the factors in the experiment. Therefore, x is set equal to time (T), and y is set equal to the final counts for each day (N_c). The equation now becomes

$$N_c = 1784 \times 2^T$$

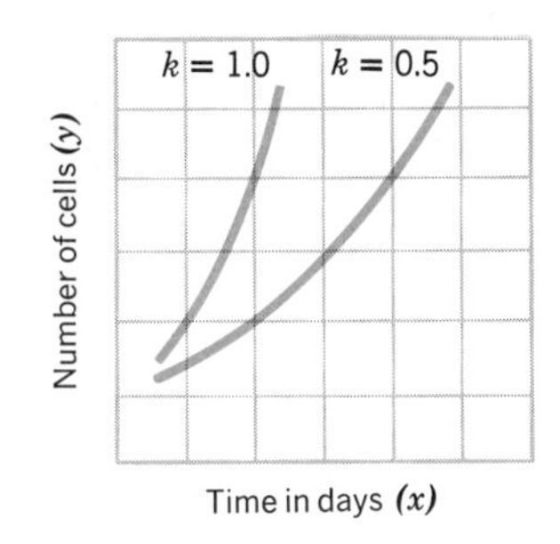

Fig. 1-10 Curve generated when $k = 0.5$.

When the data is plugged into the equation, the proper shape of the curve is produced, but it rises too sharply. Since the N_c value is fixed (the number counted on a given day), the right-hand side of the equation must be modified. This can be done by adjusting the value of T. This adjustment can be made by multiplying T by some constant k. The equation then becomes

$$N_c = 1784 \times 2^{kT}$$

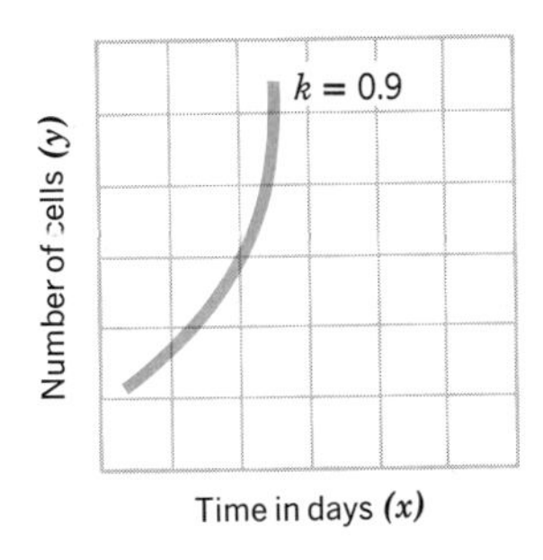

Fig. 1-11 Curve generated when $k = 0.9$.

The value of k must be less than one, for the greater the exponential value, the steeper the slope of the curve, and $k = 1$ would simply be the equation $N_c = 1784 \times 2^{1T}$ or $N_c = 1784 \times 2^T$, which is already known to be too steep. If the value $k = 0.5$ is used, the curve becomes too flat (see Fig. 1-10). So the value of k is somewhere in between 1.0 and 0.5. Trial and error reveals it to be about 0.9 (see Fig. 1-11). For then a curve is generated that is almost identical to that drawn from the data. The equation can now be written

$$N_c = 1784 \times 2^{kT} \quad \text{where} \quad k = 0.9$$

This is a much more concise analysis of Jones' data. If someone looked at the tabular form of Jones' data, he could only come to the conclusion that the population underwent an approximate doubling each day. The graph permits the same conclusion, but it is more easily discerned. The equation of the curve allows Jones to see the exact value of the increase in population rather than just a general picture.

Jones may now interpret the experiment by coupling the mathematical information with descriptive findings. Microscopic observation indicated that single-celled organisms, like those in the population that Jones counted, underwent division. A single cell divided into two cells, thereby increasing the size of the population in a doubling fashion. The graph indicates that in this particular population, 90% ($0.9 \times 100 = 90\%$) of the cells in the population were undergoing division within a 24-hour period. By appropriate calculations, Jones can calculate how long it takes for the population level to double. This can be done by setting up a proportionality, the symbol for which is α, and by solving for x.

$$\frac{90}{24} \propto \frac{100}{x}$$

$$x = 26.6$$

Therefore, every 26.6 hours, the population doubled in size.

In its present form, the curve on the graph expressing the growth pattern is rather cumbersome. It is desirable to change the curve into a straight-line relationship for reasons that will soon become apparent. The conversion of the curve to a linear form may be achieved by changing the ordinate (N_c) to an exponential form. This is a standard manipulation. The numbers

$$1 \quad 10 \quad 100 \quad 1000 \quad 10000 \quad 100000$$

are multiples of 10. When these numbers are written in exponential form, they become

$$10^0 \quad 10^1 \quad 10^2 \quad 10^3 \quad 10^4 \quad 10^5$$

and when the ordinate is denoted as being reflective of values to the base 10, the scale becomes linear:

$$0 \quad 1 \quad 2 \quad 3 \quad 4 \quad 5$$

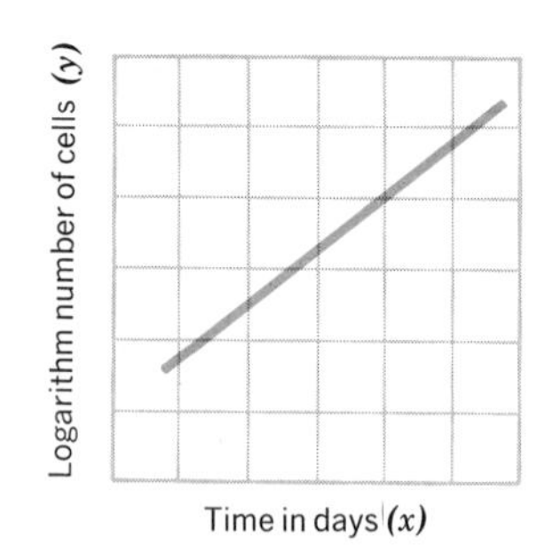

Fig. 1-12 Conversion of graph when log values for data are used.

Using this system, a number such as 87 may be expressed as 1.93. The number to the left of the decimal point indicates that the number is between 10 and 99 (an exponential value of 1) and the numbers to the right of the decimal point are obtained by looking up the value of 87 on a standard logarithmic table. When this conversion is done, for all values of the data, straight-line relationship between N_c and T is generated (see Fig. 1-12).

The straight-line form of the graph now permits further interpretation of the data. One type of interpretation may be made by

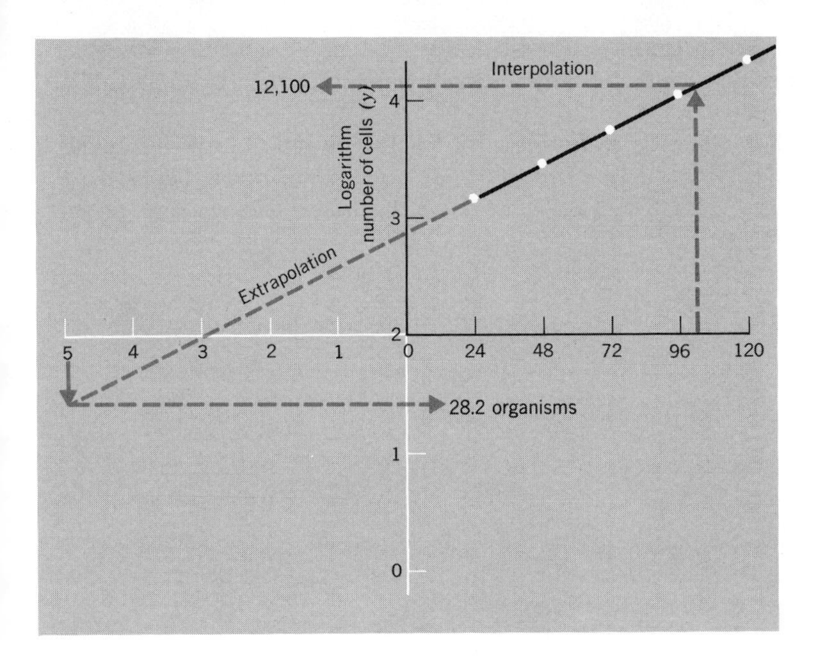

Fig. 1-13 **Extrapolation and interpolation of data from a graph.**

extrapolation. This involves extending the line into regions of the graph that had not come under direct observation. Jones may want to know what number of organisms he would have found if he had started counting the population six days before. If so, he would extend the line graph backward to a T value of -5 ($1 - 6 = -5$). The corresponding value on the ordinate would give the number of organisms Jones would have found, but it would be in an exponential form. This would have to be converted to a numerical form. From the dash line on Fig. 1-13, it may be seen that there was an exponential value of 1.45 or 28.2 organisms present in the population five days before the counting was started.

Another type of interpretation possible from a straight-line graph is **interpolation.** This involves finding a value within the range of the experiment. Suppose Jones wanted to know how many organisms he would have found after 100 hours (4.16 days). This may be directly read from the graph (see Fig. 1-13), and corresponds to an N_c value of 4.114 or 12,100 organisms. So if Jones had counted the population at 100 hours, this is the value he should have gotten. And if he repeated his experiment, starting with the same size population under the same conditions, this is the number he should obtain after 100 hours.

If one examines the straight-line graph of Jones' data carefully, it can be seen that the points do not fall directly on the straight line. This discrepancy may be significant from an experimental point of view, in which case the equation is incorrect for the data and the conclusions drawn from the equation are also incorrect. Or the discrepancy may reflect some inherent fluctuation in the system that does not alter the overall pattern and hence does not invalidate the conclusions. It is necessary to determine whether this variation is of sufficient magnitude to affect the conclusion. This is done by submitting the data to **statistical analysis.**

If Jones were really concerned with accurate measurements, he would have run series of counts on different populations. Had he done that, the values listed on p. 14 would have been the averages of his counts.

Days	1	2	3	4	5
Series A	1793	3287	6022	11248	23118
Series B	1722	3428	6348	11933	21646
Series C	1805	3367	6217	12048	22805
Series D	1733	3388	6188	11321	20099
Series E	1868	3520	5916	11805	23117
Average	1784	3394	6138	11671	22157

The average in this case is the **arithmetic mean** (m), which represents the ideal value obtained by adding all the counts and dividing by the number of counts. For example

$$\frac{1793 + 1721 + 1805 + 1733 + 1868}{5} = 1784$$

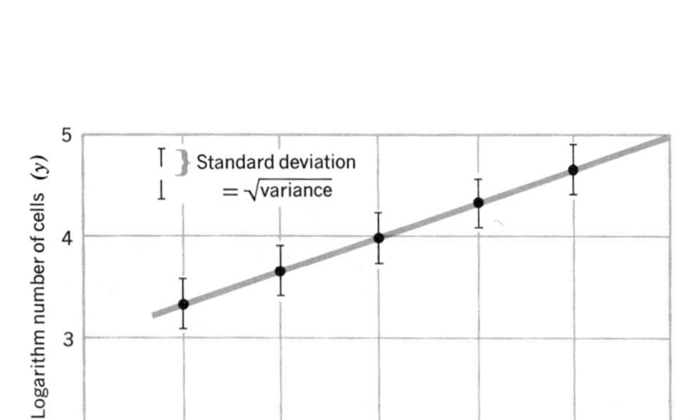

Fig. 1-14 Area included by the computed variance.

The mean value relates only half the truth, since the values from which it is derived can be any distance from it. For instance, number 1784 can be the average of 1793, 1721, 1805, 1733, and 1868; but it is also the average of 622, 4093, 1040, 322, and 2843. It is therefore important to measure the amount of dispersion or **variance** from the average. The variance is a numerical value that is obtained directly from the equation itself. It indicates the maximum amount by which a point can be off the line and still be valid for the equation. The mathematical computation of the variance is beyond the scope of this book, but the area included by the variance is shown in Fig. 1-14. Analyzing the data in terms of the variance is indicative of the validity of stating the graph as the equation, and is reflective of the validity of the conclusions drawn from that equation.

Standard Units of Measurement

If measurement is to be acceptably dressed numerically, it should be given proper labels. An elementary but nevertheless significant achievement in science was the invention of a series of standardized quantities. The metric system has been employed almost exclusively in science because it is based entirely on the number 10. It is far easier to handle multiples of 10 than the cumbersome multiples of 12 of the British-American system.

If one is to measure accurately, it is necessary to have a device with an agreed-upon standard, calibrated to the limit of the craftsman's ability. Such devices include rulers to measure distances, scales to measure weights, etc.

The standard unit of distance is the **meter,** which is about 39.4

TABLE 1-1 LINEAR UNITS OF MEASUREMENT

1 meter (m) = 39.37 inches
= 10 decimeters
= 100 centimeters
1 centimeter (cm = 10 millimeters
1 millimeter (mm) = 1,000 microns
1 micron (μ) = 1,000 millimicrons
1 millimicon (mμ) = 1 nanometer
1 nanometer = 10 Angstroms (Å)

TABLE 1-2 VOLUMETRIC UNITS OF MEASUREMENT

1 liter (1) = 1.056 quarts
= 1,000 milliliters
1 milliliter (ml) = 1,000 microliters (μl)
= 1,000 lambda (λ)

TABLE 1-3 TEMPERATURE SCALES

	Fahrenheit (F)	Centigrade (C)	Absolute or Kelvin (K)
Absolute zero	$-459.4°$	$-273°$	$0°$
Water:			
Boiling point	$212°$	$100°$	$373°$
Freezing point	$32°$	$0°$	$272°$

$°K = °C + 100$
$°C = 5/9(°F - 32)$

TABLE 1-4 MASS UNITS OF MEASUREMENT

1 kilogram (kg) = 35.27 ounces
= 2.2 pounds
= 1,000 grams
1 gram (g) = 1,000 milligrams
1 milligram (mg) = 1,000 micrograms
1 microgram (μg) = 1,000,000 picograms
1 picogram ($\mu\mu$g) = 10^{-12} grams

TABLE 1-5 ENERGY UNITS OF MEASUREMENT

1 calorie = amount of heat required to raise 1 ml of water from 14.5 to 15.5° C
1 erg (unit of work) = work required to move an object with a force of 1 dyne through a distance of 1 cm
1 joule = 10,000,000 ergs

inches. The **millimeter** (milli = 1/1000; millimeter = 1/1000 of a meter) is about the width of an ant. It is commonly used to measure any small but visible parts of organisms. Further down the scale is the **micron** (μ), which is 1/1000 of a millimeter, and far below the range of human vision. With the aid of a microscope it is possible to discern this level of distance measurement, and cells and cell components are measured in microns. Another thousandfold decrease in size brings one down to the **millimicron** level (mμ), which is used to measure the length of electromagnetic radiation. Red light is 400 mμ, blue light is 650 mμ, and ultraviolet extends over the range of 200 to 300 mμ. Another tenfold decrease (1/10,000 of a micron) is the **angstrom** (Å), used to measure the size of molecules and chemical bonds (see Table 1-1).

The standard unit of volume is the **liter,** approximately equal to a quart. One one-thousandth of a liter is a **milliliter** (ml). In the metric system, there is a relationship between distance measurement and volume that does not exist in the British-American system. One cubic centimeter (a cube, 1 cm × 1 cm × 1 cm) has a volume equal to 1 ml. Therefore, the milliliter is sometimes expressed as cm^3 or cc (cubic centimeters), and it should be recognized that these are equivalent quantities. Smaller units of volume are the **microliter** or **lambda,** one millionth of a liter. Even smaller volume measurements may be used by employing linear measures in a dimensional form, i.e. cubic millimicrons or cubic angstroms (see Table 1-2).

The temperature scale in the metric system is in terms of **centigrade** rather than the British-American Fahrenheit. The freezing point of water in the centigrade scale is **0°** and corresponds to 32° on the Fahrenheit scale. The boiling point of water is **100°** on the centigrade scale and corresponds to 212° on the Fahrenheit scale (see Table 1-3).

The weight ranges used in biology are usually in the lower half of the available spectrum. The **gram** is equal to about 0.36 ounces. The **milligram** (mg) is 1/1000 of a gram and is about the weight of the amount of pepper one puts on a fried egg. The **microgram** (μg) is 1/1,000,000 of a gram (10^{-6} gm) and is beyond the experience of everyday life. Even smaller is the **picogram** (10^{-12} gm, 1/1,000,000,000,000 gm). The gram is equal to the weight of one milliliter of water at 15.5° C. The selection of water as a standard was purely arbitrary (see Table 1-4).

Energy content has several standard units, depending on the form in which the energy is expressed. Energy expressed as heat is measured in terms of **calories.** *A calorie is defined as the amount of heat required to raise one gram of water from 14.5° C to 15.5° C.* The **kilocalorie** is the amount of heat required to raise one kilogram (1000 grams) of water from 14.5° C to 15.5° C. Mechanical energy is given in terms of joules, etc. (see Table 1-5).

As will be seen shortly, other measured units may be derived from equations employing the standard units.

Estimations. Before one measures, it is important to determine the *order of magnitude* within which the counts will be made. At times, so little is known about a particular problem that a rough estimation is necessary. Some crude calculations may be made to determine the size of the numbers (counting numbers to the left of the decimal place). If you were asked to give the number of houseflies on earth in a given year, what estimate would you give? Is it 10^5 (100,000) or 10^{10} (10,000,000,000)? What about the number of gas molecules in the lobby of Grand Central Station? Is it 10^{15} (1,000,000,000,000,000), 10^{20} (100,000,000,000,000,000,000), or 10^{30} (1,000,000,000,000,000,000,000,000,000,000)? How about the number of cells in the human body? This can be estimated by determining the volume of the body and the volume of an average cell in the body.

Let us say that an average cell has a length of 5 microns. This would mean that it has a volume of 125 μ^3.

$$(5\ \mu)^3 = 125\ \mu^3$$

An average body is about five feet nine inches tall (1.75 meters), one foot wide (0.3 meters) and one and a half feet broad (approximately 0.5 meters). Therefore, the volume of a human body in meters is

$$1.75 \times 0.3\ 0.5 = 0.26\ \text{meters}^3$$

In order to go any further with the estimation, both values must be in terms of the same standard units. Either meters^3 must be converted to microns^3 or microns^3 to meters^3. Converting meters^3 to microns^3

$$1\ \text{meter}^3 = 10^{18}\ \text{microns}^3$$

$$(1\ \text{meter} = 10^6\ \text{microns};\ 1\ \text{meter}^3 = (10^6)^3\ \text{microns}^3 = 10^{18}\ \text{microns}^3.)$$

$$0.26\ \text{meters}^3 = 0.26 \times 10^{18}\ \text{microns}^3$$

If the decimal place is moved one unit to the right, the number is increased by a factor of ten. The number may be returned to its original value by reducing the right side of the number by a factor of 10. This is done by subtracting one from 18. The number may now be expressed in the more acceptable form of

$$2.6 \times 10^{17}\ \text{microns}^3$$

By dividing the total volume of the body by the volume of a single cell, one can estimate the number of cells in the body.

$$\frac{2.6 \times 10^{17}\ \text{microns}^3/\text{body}}{1.25 \times 10^2\ \text{microns}^3/\text{cell}} = 2.1 \times 10^{15}\ \text{cells/body}$$

The division of numbers in this form is done by handling each side of the number separately. The left side of the number is a standard division of 2.625/1.25, which equals 2.1. The right side of the number is in an exponential form, and one divides exponential forms by *subtracting* the exponent in the denominator from the exponent in the numerator ($17 - 2 = 15$). The answer then becomes 2.1×10^{15}.

Notice that the units of measurement are handled as part of the equation.

$$\frac{\text{microns}^3/\text{body}}{\text{microns}^3/\text{cell}} = \frac{\cancel{\text{microns}^3}}{\text{body}} \times \frac{\text{cell}}{\cancel{\text{microns}^3}} = \frac{\text{cells}}{\text{body}}$$

Since the microns^3 term appears in both the numerator and the denominator, it may be canceled, and the answer is given in terms of cells/body. Thus the measured units are derived from equations.

One is not interested in the left side of the number, but only in the right side, the order of magnitude, the power of 10. An estimation is made allowing for an error of one order in magnitude in either direction. Therefore, the number of cells in a human body would be most likely between 10^{14} and 10^{16}.

Instrumentation and Measurement

Finally, a word or two should be mentioned about the role of instrumentation in measurement and scientific investigation. Over the decades, the combined efforts of science and technology have produced a collection of elegant devices for measuring nearly all the quantitative aspects of reality known to man. It is important to realize that machines do count better than men. You can count off seconds by saying "one-a-thousand, two-a-thousand . . . etc.", but see how competent you are over one hundred counts compared to the accuracy of a watch. Machines can count with far more precise intervals, over longer periods of time, and with complete objectivity. It is especially important in the realm of scientific inquiry to measure those qualities that one cannot detect by his own sensory capacities. Ultraviolet light, high-energy radiation, and high and low temperatures or cellular components of organisms are among the most common subjects under investigation that are beyond man's sensory limits. Through the use of instruments, the scientist has acquired a feeling for working with phenomena that cannot be detected in everyday experience.

BIOLOGY AS A MEANS OF SCIENTIFIC INQUIRY

Measurements cannot be separated from concepts, at least not in science. Measurement and theory are a totally integrated pair of processes. This integration can follow either of two patterns.

The noted physicist Bridgman felt that measurements are in some ways concepts themselves. Conversely, given the concept in accurate descriptions, anyone could collect the facts to support it. This process is called **operationalism.** It demands that the concepts include the details by which the measurements were made. On the other hand, **rationalism** makes large, discontinuous leaps from the observations to the concept.

Operationalism has become an article of faith in physics, while rationalism is very common in the social sciences and psychology. Biology rides the fence between the two. In some areas, especially those that lend themselves to a more quantitative approach, the principles of operationalism are adhered to ("growth patterns may be expressed as 2^{KT}"). More descriptive areas of biology still use the rationalistic approach ("an organism is the product of its environment and its heredity").

The Art of Scientific Inquiry

Scientific inquiry may follow either of two acceptable patterns. The **inductive** procedure involves the passage from particulars to universals. For example, "all of twenty-five different kinds of organisms examined have been found to be made up of cells; therefore, most if not all organisms are made up of cells."

The other pattern is the **deductive** procedure. It involves the passage from universals to particulars. For example, "All organisms are composed of cells; the human is an organism; therefore, the human is composed of cells." As one can see, the truth of the conclusion is dependent on the truth of the universal. The formation of the universal is from the inductive procedure. So it is the inductive procedure that forms the basis of scientific inquiry.

This inductive method was elaborated by the philosopher and scientist Francis Bacon (1561–1626) and popularized by Karl Pearson during the nineteenth century as the "scientific method." The general form of the scientific method is usually given as

1. Construction of a hypothesis: The formation of a hypothesis comes from a series of observations or from the results of other experiments. It is, in a sense, an educated guess of the conclusion that could be drawn from precise data collection. Going back to Jones' experiment of counting the number of unicellular organisms in a population, his hypothesis might have been "Since it has been observed that unicellular organs divide to form two organisms, a population of such organisms should increase by doubling their numbers within a given length of time."
2. Collection of data pertinent to the hypothesis: Jones then proceeded to construct an experiment that would permit him to collect pertinent data. He chose to do this by counting the number of organisms at daily intervals.
3. Determination of the results of the collected data: Jones determined the results of his data by converting them to a graphic form and deriving the equation for [illegible]urve.
4. Drawing the con[illegible]sion that either supports or refutes the hypothesis: Jones drew the conclusion that if the original observations on which his hypothesis was based were correct, then (since 90% of the population doubled in one day) the full population would double in 26.6 hours. Therefore it would take 26.6 hours for all the cells in the population to divide.

Notice that the conclusions are dependent on the assumption that the [illegible]othesis was correct. This is always true. If the experimental evi[illegible]does not support the original hypothesis, it is very often true that th[illegible]othesis was based on incorrect assumptions. Even if this is not the c[illegible], the scientific method is not foolproof. For example, two individuals may draw different conclusions from the same data (the human factor is always present in any scientific inquiry in the form of the scientist himself). Very often, data cannot be collected in the way one would like to have it. And there is always the problem of how one obtains the hypothesis in the first place. So the scientific method is really a grossly oversimplified means of stating the process of systematic inquiry.

It should also be mentioned that scientists do not consciously follow

Fig. 1-15 Procedure of Priestly experiment described in text.

the scientific method in their investigations. Very often, one can fit a particular experiment or series of experiments into the pattern of the scientific method in retrospect, but this does not mean that the scientist actually sat down and said "I am now going to formulate a hypothesis," etc. His very training is such that he follows the scientific method on a subconscious level.

Another approach in scientific inquiry is the **controlled** experiment. Some experiments take the form of altering a system in some way, introducing a **single variable** and studying the results of this alteration. One of the classical controlled experiments in biology was performed by Joseph Priestley. He placed a mouse under a bell jar, providing it with sufficient food. He then placed another mouse under another bell jar, again with food and, in addition, a green plant. The mouse that was alone in the bell jar died within a few hours, while the one with the plant continued to thrive (see Fig. 1-15). In doing this experiment, he had introduced a single variable—the plant in the bell jar. Both bell jars were identical, and both were sealed so that no air from the outside could get in. Both mice had sufficient food, so that it could not be the lack of food that accounted for the demise of one of the mice. Obviously, the presence of the green plant caused a change in the air within the bell jar. The bell jar with the green plant represented Priestley's **experiment;** the bell jar without the green plant represented his **control.** He could not have come to any conclusion about what the green plant contributed to the environment without having an indication of what happened in its absence.

The type of scientific inquiry in which a natural phenomenon is studied does not lend itself to controlled experimentation. Jones, for instance, was observing and measuring the natural phenomenon of growth. There is no control in Jones' experiment. If, however, Jones had wanted to change the temperature of his system, and see how this affected the growth pattern, he would have had to run two sets of populations, one series at the new temperature, and the other at the original temperature. In this case, the series run under his original temperature would have served as a control.

"Chance Favors the Prepared Mind"

Perhaps one of the greatest truisms in scientific inquiry was Pasteur's statement that "chance favors the prepared mind." The great experiments in biology, or for that matter any science, are the products of their time. Experiments utilizing microorganisms as research tools could not be made until methods of growing individual populations of microorganisms and handling them under laboratory conditions had been developed. The exploration of certain aspects of life processes could not be made until other aspects had been clarified. Yet when the time was ripe, and there were hundreds of scientists working within the area, only one or two perceived the hypothesis and the way in which the hypothesis could be tested. Only these had the "prepared mind."

But, just as the time must be ripe for the discovery, the discovery must be ripe for the times. Toward the end of the nineteenth century,

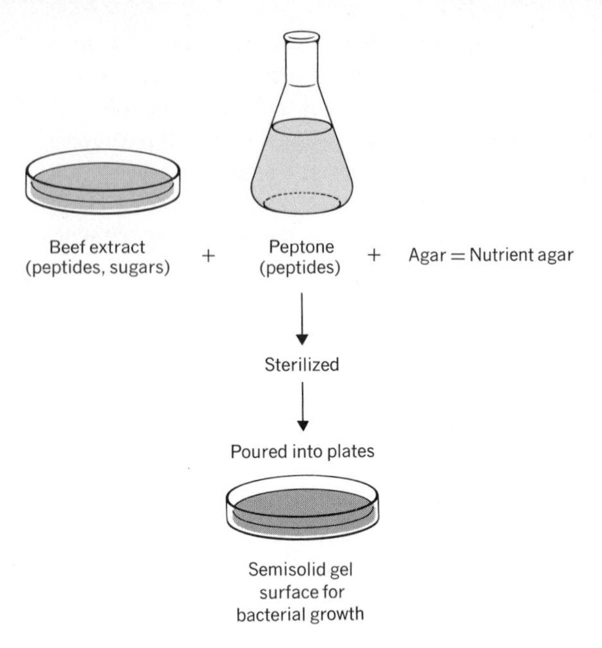

Fig. 1-16 Preparation of agar medium for bacterial growth.

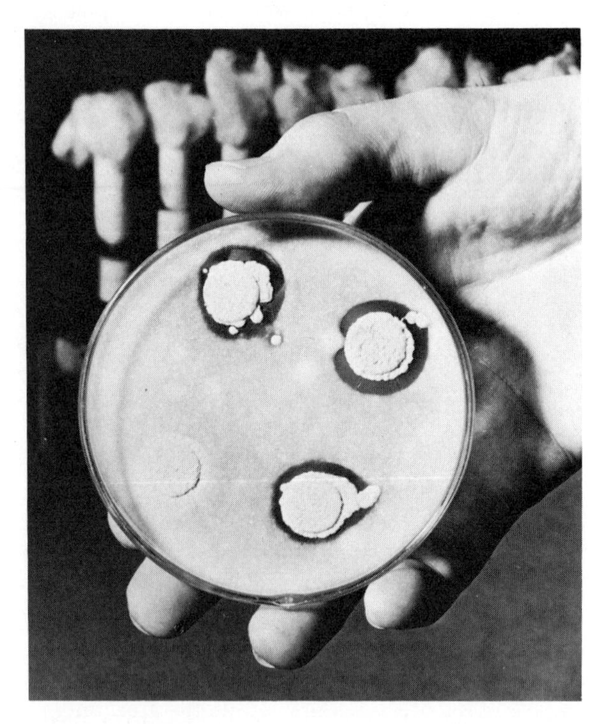

Fig. 1-17 Inhibition of bacterial growth by penicillium mold on an agar plate. Opaque areas on the plate are areas of bacterial growth. Clear areas indicate the absence of bacteria. [Courtesy of Chas. Pfizer & Co., Inc.]

methods of handling microorganisms were being developed. One means of growing bacteria was to spread them on a surface of nutrient agar. Agar is a material that is liquid at 100° C but solid at both room (25° C) and body (37.5° C) temperatures. Nutrient materials can be mixed with the agar, so that there is a source of "food" for the bacteria (see Fig. 1-16). One of the great problems encountered by the early microbiologists was contamination. The nutrient material could also support a variety of living forms, and very often, mold would contaminate the bacterial growth. In 1876, Tyndall described his observation that bacterial growth was inhibited by the presence of a particular species of the mold *Penicillium* (see Fig. 1-17). In 1896, Gosio isolated a bacterial inhibitor from the mold. In the early part of the 20th century, antibiotics generated a short-lived flurry of interest.

In 1929 Sir Arthur Fleming observed that the mold *Penicillium* inhibited the growth of his bacteria. As an observation, it was not new. Tyndall had described it long before. But by 1929 microbiologists had become acutely aware of the chemical basis of biological phenomena. Therefore, the tools to examine this inhibitory effect *and* the attitude of acceptance for the conclusions that could be drawn were both present.

Fleming extracted the **antibiotic** (anti = against, biotic = life) substance from the *Penicillium.* He found that as a pure chemical, without the presence of the mold, it retained its ability to inhibit bacterial growth. He proceeded to submit it to a variety of chemical analyses to determine its properties. Because of the thoroughness of his work, the ripeness of the time for accepting his discovery, and his own preparedness of mind, Fleming is credited with the discovery of *penicillin.*

PERSPECTIVES OF BIOLOGY AS A SCIENCE

While the natural sciences—physics, chemistry, and biology—are characterized by quantified methods of observation and conceptualization, biology is less quantitative than the other two. In the realms of chemistry and physics, order is difficult to find, and scientists must search for it. Anyone viewing the night sky sees a smattering of lighted points of different intensity. Early physicists concentrated on only two of these lighted points at a time, and sought some constant relationship between them. Order was found only by artificially considering one aspect of a phenomenon. Likewise, chemists found properties of substances through analysis of purified material rather than of the highly heterogeneous forms in which they naturally occur. Order in chemistry was only found after preliminary sorting. In biology, however, order is immediately evident at every level of organization: the coordination of limbs in movement, the patterns of migrating flocks, and the similarity of parent and offspring.

The type of order encountered in biology is more difficult to quantify because it involves the relationship of many components. A physicist can quantify the relationship between two heavenly bodies and a chemist can specify the properties of a pure substance by its relationship to other pure substances, one at a time. But in biology, the operation of one part usually has meaning only with respect to many other parts. A root of a plant has to be understood in the perspective of the soil, the stem, and the embryo from which it came. Man has learned how to quantify distance, weight, volume, and other simple features, but not to quantify complex interrelationships.

Hence, biology is highly quantified in areas of study where molecules or a number of individuals are encountered, but highly descriptive where multifaceted relationships are involved.

QUESTIONS

1. What are the different ways in which rigorous and universal concepts can be expressed?
2. Give several concepts in the history of biological thought and discuss how they changed this thought and led to advances in the science of biology.
3. Why does a scientist manipulate data, translating it to tabular and graphic forms, and submitting it to statistical analysis?
4. Estimate the number of cells in a mouse.
5. What is the difference between inductive and deductive reasoning? Which forms the basis of scientific inquiry and why?
6. When are controlled experiments mandatory, and when are they unnecessary in scientific inquiry?
7. Priestley's experiment was distinctly nonquantitative. If you were to redo his experiment with modern instruments, how and what would you measure? How much more information could you get from his system?

CHAPTER 2
THE CHEMICAL BACKGROUND FOR BIOLOGY

Only since the turn of the century have biologists recognized the full dependence of biological form and function on the chemical composition and the physical interaction of chemicals of an organism. Subsequent research, which focused on the organism as a complex chemical system, has resulted in a comprehensive and unified view of the biological world in chemical terms. If we are to understand the modern concept of an organism, and its form and function, we must study the chemical and physical basis of life.

In order to provide an orientation toward the world of chemistry, it is necessary to review a few terms.

Matter: Matter is the term used for all substances in the universe that manifest *mass* (a density per unit volume and size). Matter exists in three distinct states—liquid, solid, and gaseous.

Molecule: Pure substances are divisible into units called molecules. A molecule is defined as the *smallest chemical unit that manifests the property of a pure chemical substance.*

Element: An element is a substance that cannot be destroyed or decomposed by available levels of heat or electrical energy.

Atom: An atom is the smallest individual unit of an element.

All matter is made up of chemical entities, separable into pure substances, which are made up of molecules. And substances may, through a variety of treatments, *be decomposed into elements, which are made up of atoms.*

THE ATOM

Since all organisms are made up of molecules, which are in turn, made up of atoms, it is appropriate to begin the discussion of the chemical and physical basis of life with the atom. The first feature of atoms that should be noted is that they are not passive "building blocks" of molecules, but rather, dynamic units that exhibit specific properties. The cumulative effect of all the atoms of an organism is, in great part, responsible for the properties exhibited by that organism, and an understanding of the dynamic aspects of atomic structure is essential for a comprehension of biological entities.

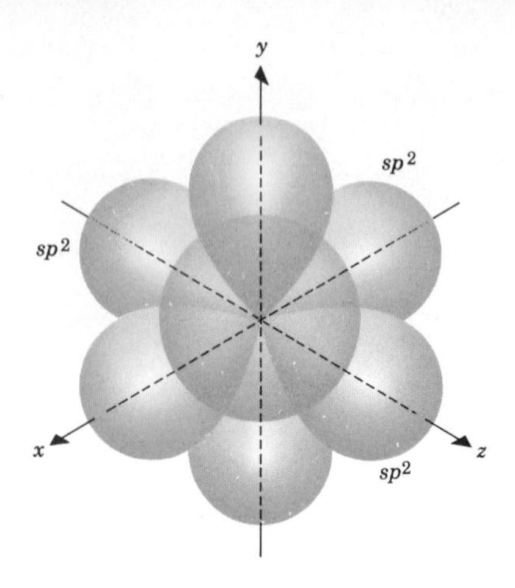

Fig. 2-1 A composite of the Bohr-Schroedinger atom.

No one has, of course, ever seen an atom. The atomic form envisioned today is a product of deductions made from experimental data, and serves as a model to explain many chemical phenomena. The somewhat static picture seen in a text may not be a true representation of the atom. It is merely a means for a scientist to say "if we visualize an atom in this way, we can account for the dynamics in which we know the atom is intimately involved."

The atom, as it is presently envisioned, is a composite of the theories of two men, Niels Bohr and E. Schroedinger. The greatest mass of the atom is located in the atomic nucleus. Orbiting the nucleus, and some distance from it, are electrons. Niels Bohr had postulated an atom of planetary design, in which a nucleus was somewhat analogous to the sun and the orbiting electrons were much like the planets. Although this was a valid picture of the atom in terms of atomic interactions, it did not account for other properties exhibited by electrons. Schroedinger modified the Bohr model to provide for a series of *electron clouds* surrounding the nucleus (see Fig. 2-1).

The Atomic Nucleus

The nucleus of an atom consists of two types of particles, **protons** and **neutrons.** These particles are essentially the same in mass (proton = 1.672×10^{-24} gm; neutron = 1.675×10^{-24} gm). Because the nucleus represents the largest area having mass in the atom, it accounts for *most* of the **atomic weight.** Each chemical element has a characteristic number of neutrons and protons in its nucleus, and consequently, there is a characteristic atomic weight for each element.

While the neutron and proton are similar in weight, they are different in electrical potential. The proton carries a single positive charge (+1), whereas the neutron does not carry any charge and contributes only to the mass of the atom. Since we know that particles with like charges repel one another, and that particles of opposite charges tend to attract one another, it is not easy to visualize the protons in the atomic nucleus in close spatial relationship. Recent discoveries have demonstrated that there are intranuclear forces that keep protons in place despite their natural tendency to move away from one another.

Electrons and Their Orbitals

The Bohr-Schroedinger model postulates the atomic nucleus and its surrounding electrons as a system somewhat comparable to the sun and planets. This analogy, however, must not be carried too far. The forces that hold the planets in their orbits are purely gravitational, whereas those that account for the orbiting of electrons are electrical. Planets follow predictable paths, but electrons do not.

Of great importance in the atomic scheme of things is the electrical charge of the electron. The electron is negatively charged, and although much smaller in mass than the proton, it carries the same *amount* of charge. Each atom represents a balanced electrical system, and there are the same number of electrons in orbitals around an atom as there are protons in the nucleus, so the atom is electrically neutral.

Fig. 2-2 Antiparallel spin of electrons in an orbital.

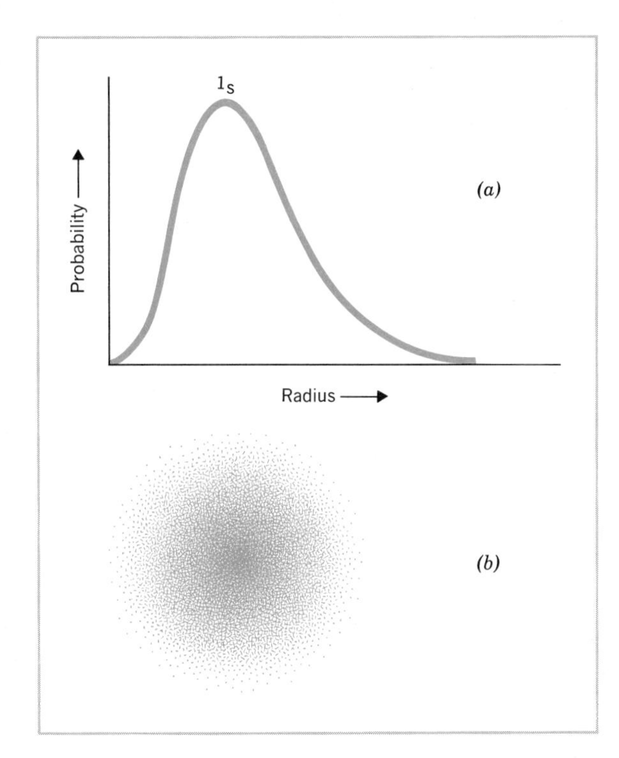

Fig. 2-3 The probability of finding an electron in a shell around the nucleus of an atom plotted against the radius of the shell. (*a*) Graph form of probability. (*b*) Electron patterns showing the relative of time an electron pair spends within the volume designated as the shell.

The atoms of each element have a characteristic number of protons balanced by the same characteristic number of electrons. The number of protons or electrons define the **atomic number** of an atom.

The electron is a rather elusive entity. Its size, as indicated by its behavioral pattern in different phenomena, seems to be dependent on where it is and what it is doing at a particular time. As a free electron it has a size of about 10^{-12} cm and acts as a particle. When it is an integral part of the atomic structure, however, it behaves as an entity 10,000 times this size, and manifests properties associated with wave mechanics. Rather than circumscribing the atomic nucleus in predictable paths, a pair of electrons, each spinning in opposite directions on its own axis (see Fig. 2-2), move about in a volume of space called an **orbital.** The volume assigned to an orbital is probabilistic, and decided on the basis of an electron being found in that region 90% of the time (see Fig. 2-3). It is this behavior of electrons that produces wavelike properties.

The position of the orbital with respect to the atomic nucleus is dependent on the atomic number of the element. As one proceeds up the elemental chart (see Table 2-1) toward atoms with higher atomic numbers, orbitals are found in different **subshells.** One or more subshells make up an **energy level.** The first energy level contains only one subshell (see Fig. 2-4). This is called 1*s* and is made up of only one orbital. Therefore, the first energy level is saturated by two electrons.

The atoms that contain more than two protons and two electrons (i.e., have an atomic number greater than two) have additional electrons found at higher energy levels. The next energy level contains two subshells, which are called 2*s* and 2*p* respectively. The 2*s* subshell contains only one orbital, and, like the 1*s*, is saturated when two electrons inhabit it. The 2*p*, on the other hand, contains three orbitals, and can hold six more electrons before being saturated. The atomic number of the element in which the first two energy levels are filled is ten (two electrons in 1*s*, two electrons in 2*s*, and six electrons in 2*p*). Those

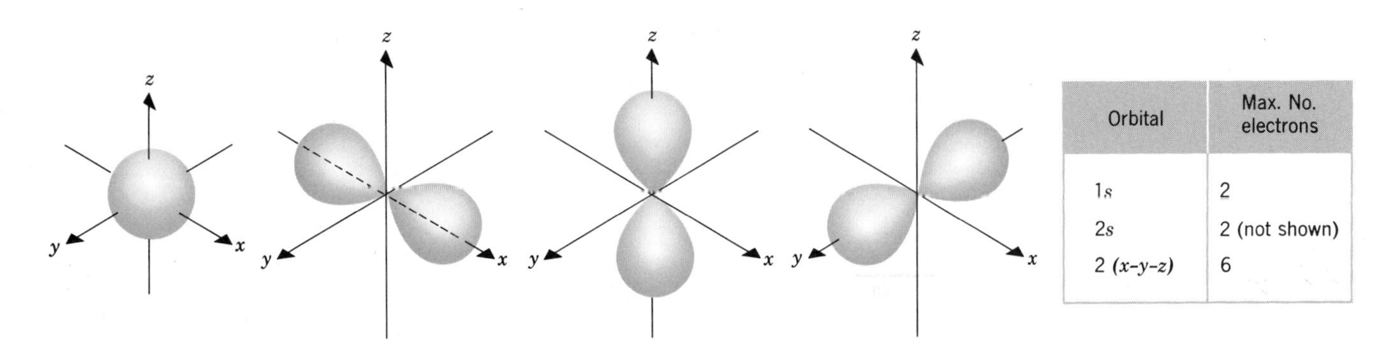

Orbital	Max. No. electrons
1*s*	2
2*s*	2 (not shown)
2 (*x*–*y*–*z*)	6

Fig. 2-4 Representation of the 1*s* and 2*p* orbitals. Notice that the 1*s* orbital is saturated by the presence of two electrons, while the 2*p* orbitals are saturated by the presence of six electrons, two in each suborbital.

TABLE 2-1 PERIODIC TABLE AND ELECTRONIC STRUCTURES[a]

KEY

f				×			
d			×	×	×		
p		×	×	×	×	×	
s	×	×	×	×	×	×	×
No. →	1	2	3	4	5	6	7

Group I	Group II	Group III B	Group IV B	Group V B	Group VI B	Group VII B	Group VIII	
H 1 1								
Li 3 2 1	Be 4 2 2							
Na 11 6 2 2 1	Mg 12 6 2 2 2							
K 19 6 6 2 2 2 1	Ca 20 6 6 2 2 2 2	Sc 21 1 6 6 2 2 2 2	Ti 22 2 6 6 2 2 2 2	V 23 3 6 6 2 2 2 2	Cr 24 5 6 6 2 2 2 1	Mn 25 5 6 6 2 2 2 2	Fe 26 6 6 6 2 2 2 2	Co 27 7 6 6 2 2 2 2
Rb 37 10 6 6 6 2 2 2 2 1	Sr 38 10 6 6 6 2 2 2 2 2	Y 39 10 1 6 6 6 2 2 2 2 2	Zr 40 10 2 6 6 6 2 2 2 2 2	Nb 41 10 4 6 6 6 2 2 2 2 1	Mo 42 10 5 6 6 6 2 2 2 2 1	Tc 43 10 6 6 6 6 2 2 2 2 1	Ru 44 10 7 6 6 6 2 2 2 2 1	Rh 45 10 8 6 6 6 2 2 2 2 1
Cs 55 10 10 6 6 6 6 2 2 2 2 2 1	Ba 56 10 10 6 6 6 6 2 2 2 2 2 2	La 57* 10 10 1 6 6 6 6 2 2 2 2 2 2	Hf 72 14 10 10 2 6 6 6 6 2 2 2 2 2 2	Ta 73 14 10 10 3 6 6 6 6 2 2 2 2 2 2	W 74 14 10 10 4 6 6 6 6 2 2 2 2 2 2	Re 75 14 10 10 5 6 6 6 6 2 2 2 2 2 2	Os 76 14 10 10 6 6 6 6 6 2 2 2 2 2 2	Ir 77 14 10 10 9 6 6 6 6 2 2 2 2 2
Pr 87 14 10 10 10 6 6 6 6 6 2 2 2 2 2 2 1	Ra 88 14 10 10 10 6 6 6 6 6 2 2 2 2 2 2 2							

***LANTHANIDES**

Ce 58 2 10 10 6 6 6 6 2 2 2 2 2 2	Pr 59 3 10 10 6 6 6 6 2 2 2 2 2 2	Nd 60 4 10 10 6 6 6 6 2 2 2 2 2 2	Pm 61 5 10 10 6 6 6 6 2 2 2 2 2 2	Sm 62 6 10 10 6 6 6 6 2 2 2 2 2 2	Eu 63 7 10 10 6 6 6 6 2 2 2 2 2 2	Gd 64 7 10 10 1 6 6 6 6 2 2 2 2 2 2
Tb 65 9 10 10 6 6 6 6 2 2 2 2 2 2	Dy 66 10 10 10 6 6 6 6 2 2 2 2 2 2	Ho 67 11 10 10 6 6 6 6 2 2 2 2 2 2	Er 68 12 10 10 6 6 6 6 2 2 2 2 2 2	Tm 69 13 10 10 6 6 6 6 2 2 2 2 2 2	Yb 70 14 10 10 6 6 6 6 2 2 2 2 2 2	Lu 71 14 10 10 1 6 6 6 6 2 2 2 2 2 2

[a] From D. H. Andrews and R. J. Kokes, *Fundamental Chemistry*, John Wiley & Sons, 1965.

			Group III	Group IV	Group V	Group VI	Group VII	Group 8
								Be 7 2
			B 5 1 2 2	C 6 2 2 2	N 7 3 2 2	O 8 4 2 2	F 9 5 2 2	Ne 10 6 2 2
	Group I B	Group II B	Al 13 6 1 2 2 2	Si 14 6 2 2 2 2	P 15 6 3 2 2 2	S 16 6 4 2 2 2	Cl 17 6 5 2 2 2	Ar 18 6 6 2 2 2
Ni 28 8 6 6 2 2 2 2	Cu 29 10 6 6 2 2 2 1	Zn 30 10 6 6 2 2 2 2	Ga 31 10 6 6 1 2 2 2 2	Ge 32 10 6 6 2 2 2 2 2	As 33 10 6 6 3 2 2 2 2	Se 34 10 6 6 4 2 2 2 2	Br 35 10 6 6 5 2 2 2 2	Kr 36 10 6 6 6 2 2 2 2
Pd 46 10 10 6 6 6 2 2 2 2	Ag 47 10 10 6 6 6 2 2 2 2 1	Cd 48 10 10 6 6 6 2 2 2 2 2	In 49 10 10 6 6 6 1 2 2 2 2 2	Sn 50 10 10 6 6 6 2 2 2 2 2 2	Sb 51 10 10 6 6 6 3 2 2 2 2 2	Te 52 10 10 6 6 6 4 2 2 2 2 2	I 53 10 10 6 6 6 5 2 2 2 2 2	Xe 54 10 10 6 6 6 6 2 2 2 2 2
Pt 78 14 10 10 9 6 6 6 6 2 2 2 2 2 1	Au 79 14 10 10 10 6 6 6 6 2 2 2 2 2 1	Hg 80 14 10 10 10 6 6 6 6 2 2 2 2 2 2	Tl 81 14 10 10 10 6 6 6 6 1 2 2 2 2 2 2	Pb 82 14 10 10 10 6 6 6 6 2 2 2 2 2 2 2	Bi 83 14 10 10 10 6 6 6 6 3 2 2 2 2 2 2	Po 84 14 10 10 10 6 6 6 6 4 2 2 2 2 2 2	At 85 14 10 10 10 6 6 6 6 5 2 2 2 2 2 2	Rn 86 14 10 10 10 6 6 6 6 6 2 2 2 2 2 2

ACTINIDES

		Ac 89 14 10 10 10 1 6 6 6 6 6 2 2 2 2 2 2 2	Th 90 14 10 10 10 2 6 6 6 6 6 2 2 2 2 2 2 2	Pa 91 14 2 10 10 10 1 6 6 6 6 6 2 2 2 2 2 2 2
U 92 14 3 10 10 10 1 6 6 6 6 6 2 2 2 2 2 2 2	Np 93 14 4 10 10 10 1 6 6 6 6 6 2 2 2 2 2 2 2	Pu 94 14 5 10 10 10 1 6 6 6 6 6 2 2 2 2 2 2 2	Am 95 14 6 10 10 10 1 6 6 6 6 6 2 2 2 2 2 2 2	Cm 96 14 7 10 10 10 1 6 6 6 6 6 2 2 2 2 2 2 2
Bk 97 14 8 10 10 10 1 6 6 6 6 6 2 2 2 2 2 2 2	Cf 98 14 9 10 10 10 1 6 6 6 6 6 2 2 2 2 2 2 2	Ea 99 14 10 10 10 10 1 6 6 6 6 6 2 2 2 2 2 2 2	Fm 100 14 11 10 10 10 1 6 6 6 6 6 2 2 2 2 2 2 2	Md 101 14 12 10 10 10 1 6 6 6 6 6 2 2 2 2 2 2 2

TABLE 2-2 BIOLOGICALLY IMPORTANT ELEMENTS

Element	Atomic Number	Distribution of Electrons				
		1*s*	2*s*	2*p*	3*s*	3*p*
Hydrogen	1	1				
Carbon	6	2	2	2		
Nitrogen	7	2	2	3		
Oxygen	8	2	2	4		
Phosphorus	15	2	2	6	2	3
Sulfur	16	2	2	6	2	4

elements with atomic numbers greater than ten have electrons in the third energy level, and so forth.

The patterns in which these subshells and their orbitals are filled is also important. It appears that the filling of a subshell in terms of active orbitals takes precedence over filling individual orbitals. Therefore, the atom with atomic number 7 has two electrons in 1*s*, two electrons in 2*s*, and one electron in each of the orbitals of 2*p* rather than two electrons in one of the 2*p* orbitals and one in another of the orbitals.

The distribution of electrons determines much of the dynamics of the chemical element. The combining power, or **valence** of the atom, in forming molecules is dependent on the state of saturation of the subshells and the highest energy level in the atom.

THE STRUCTURE OF CHEMICAL ELEMENTS

It is not the purpose here to present a thorough study of the field of chemistry, but rather to give sufficient background for handling the discussion of molecules of biological importance. Of the known chemical elements, which now number 103 (and have atomic numbers—protons—from 1 to 103 associated with each), only six warrant our consideration as elements found in appreciable quantities in biological material. To be sure, many other elements are present in trace amounts in material of biological origin, and their presence is essential to the functioning of the particular biological system. However, these are not representative of biologically active molecules in general. So for purposes of the present discussion, attention will be limited to these six (see Table 2-2).

Hydrogen

Hydrogen is the simplest atom known. It is characterized by a nucleus containing one proton and a single electron in the 1*s* shell. Because of this structure, both its atomic number and atomic weight are one (see Fig. 2-5). Notice that the 1*s* shell lacks one electron to be

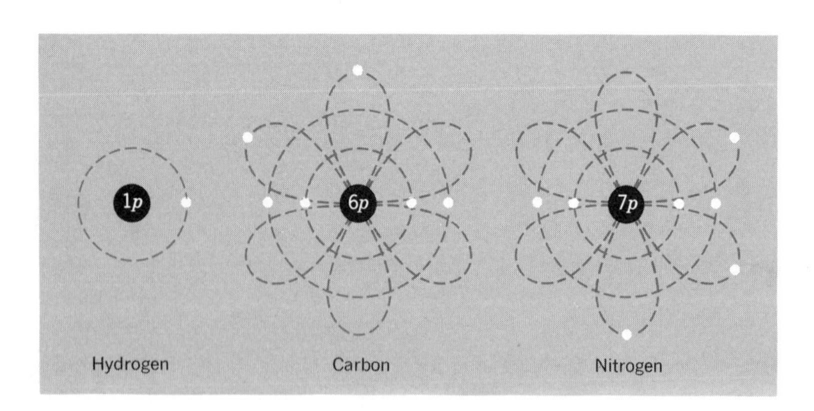

Fig. 2-5 The orbital configurations for hydrogen, carbon, and nitrogen.

full. Because atoms tend to combine until the occupied shells are saturated, hydrogen is said to have a combining power or valence of one.

One of the most important features of the hydrogen atom is its tendency to lose its electron. This feature is common to all elements that have only one electron at a particular energy level. This loss of an electron upsets the electrical neutrality of the atom, causing the positive charge conferred by its protons to be expressed. In hydrogen, the loss of the electron leaves only a single proton, which is then designated as H^+ and called the **hydrogen ion.** It is important to realize that *H^+, hydrogen ion, and proton are all equivalent terms for the same thing, and are used interchangeably.*

Carbon

Carbon has an atomic number of six, indicating that there are six protons in the nucleus balanced by six electrons surrounding the nucleus. These electrons occupy the energy levels indicated in Fig. 2-5. Notice that the carbon atom would need four more electrons in the $2p$ subshell for that subshell and the second energy level to be saturated. Therefore, carbon is considered to have a valence of four.

Nitrogen

Nitrogen has an atomic number of seven and contains seven protons in the nucleus and seven orbiting electrons. The distribution of these electrons in the energy levels may be seen in Fig. 2-5. Once again, notice that three electrons would be needed to saturate the $2p$ subshell and the second energy level. Nitrogen is assigned a valence of three.

Oxygen

The atomic number of oxygen is eight, indicating that the atom is composed of eight protons and eight electrons. These electrons are again distributed according to the established pattern seen in Fig. 2-5. The $2p$ subshell and the second energy level of the oxygen atom are two electrons from completion, indicating oxygen has a valence of two.

Phosphorus

Phosphorus has an atomic number of 15, since there are 15 protons and 15 electrons in its atomic structure. Once again, the electrons are arranged in the established pattern. The $3p$ subshell is three electrons short of completion, giving phosphorus a valence of three.

Sulfur

The sulfur atom has an atomic number of 16, and sulfur has a valence of two.

Although this discussion may resemble a game of putting the electron in its place, it is essential to remember that atomic configuration determines the properties of a particular element. Oxygen has the properties of its valence, its temperature of liquification, and its ability to support combustion, because it has eight protons, eight neutrons, and eight electrons. If it were possible to remove a proton, a neutron, and an electron from every atom in a particular volume of oxygen gas, one would not be left with some bizarre form of oxygen, but with nitrogen gas, which would manifest the properties of nitrogen.

Isotopes

The general properties of the different elements seem to depend on the number of protons and electrons of each element rather than on its number of neutrons. Therefore, the addition of neutrons in an atomic form does not change the atom to another type of element, but merely represents an **isotope,** another species of the same element. The isotope of nitrogen ${}_7N^{15}$ (having seven protons, as indicated by the subscript, and a total number of 15 nuclear particles, as indicated by the superscript) will have eight neutrons rather than the more common condition of seven. Although this isotope has the number of neutrons usually found only in oxygen, it still has the number of protons and electrons that characterize nitrogen. Therefore, it has the same valence and will take part in the same chemical reactions as nitrogen.

Certain isotopes occur naturally, although not in appreciable quantity, and are relatively stable. Some isotopes, however, represent unstable nuclear forms. These are the **radioactive isotopes,** which emit rays from the nucleus, and their ejection leads to a more stable nuclear state. The ejected rays are of three types: α (**alpha**) rays, which are a stream of particles identified as the close association of two protons and two neutrons (a helium nucleus) and which carry a positive charge; β (**beta**) rays, which are composed of electrons and carry a negative charge; and γ (**gamma**) rays, which are shortwave X-rays. The type or types of rays emitted from a particular radioactive element is characteristic of that element. For instance, radium is primarily an α emitter. As would be expected, the emission of the α particle from the radium nucleus (involving the loss of two protons and two neutrons) converts the radium to a totally different atomic form, in this case, lead. Those radioactive isotopes that emit β or γ rays usually undergo transitions to the non-isotopic form of the same element.

The emission of these rays from a radioactive isotope occurs over a characteristic period of time, and is the basis for a probability statement referred to as the **half-life** of the particular radioactive material. For example, P^{32} has a half-life of 14 days. At the end of this time, half of the original P^{32} atoms will be in the form of S^{32}, the stable form. At the end of another 14 days, half of the remaining material will be in the form of stable S^{32}, etc. Carbon^{14}, on the other hand, has a half-life of about 5000 years.

INTERACTION BETWEEN MOLECULES

Organisms are made up of atoms combined into molecules. The interaction of atoms to form molecules is, then, a prime feature of biological form and function.

Each atom is electrically stable. There are always the same number of protons (positive charges) and electrons (negative charges) associated with a particular atom. This does not imply, however, that it is chemically stable. The chemical stability of an atom is dependent on the degree of saturation of the outermost subshell. Atoms tend to combine with other atoms to form molecules until a state is reached in which there is both electrical and chemical stability.

In the course of combining physically, atoms form electrical associations with one another. These associations are called **chemical bonds,** and result in the formation of a molecule. These interactions between atoms by which the chemical bonds are formed involve *energy. The key to biological form and function is the energy phenomena of atomic interactions.* Before discussing the nature of these interactions, it is necessary to gain some understanding of the nature of energy.

Energy

The noted physicist H. A. Kramer once said, "My own pet notion is that in the world of human thought generally, and in physical science particularly, the most important and fruitful concepts are those to which it is impossible to attach a well defined meaning." This is true of such concepts as "life," "science," and, particularly, "energy."

The common definition given for energy is "the capacity to do work." The word "work" in physics refers to a force applied over a distance. Since force applied over a distance implies motion, then motion indicates that work has been done and energy has been expended.

The conceptual difference between work and energy arose as information on natural phenomena accumulated. Examination of molecules, atoms, and subatomic particles revealed that they exhibit patterns of motion. Studies on heat revealed that heat was directly related to the motion of the molecules. The greater the heat content of a system, the greater the velocity of the molecules therein; the greater the velocity of the molecules, the greater the heat content the system possesses. But this motion did not always produce work, at least not in the mechanical sense. Hence there arose a conceptual distinction between work and energy. The performance of work implies energy expenditure, but energy expenditure does *not* imply the performance of work.

Late in the nineteenth century, energy was thought of in two ways: "energy in use" (motion) or **kinetic** energy, and "energy available, but not in use" (position) or **potential** energy. A ball suspended some distance in the air has potential energy. Energy is there, but is not being expressed. If the string supporting the ball is cut, the ball will fall (a movement of a mass) and the energy will be expressed as kinetic energy—hence the definition of energy as the "*capacity* to do work."

Intensive studies of physical phenomena led to the acknowledgement of other forms of energy, such as *light, electrical, nuclear,* and *chemical* energy. Any of these forms of energy can be converted to any other form, indicating a basic interrelationship among the different forms of energy.

Chemical Energy

Most of the energy of a biological system is found in the chemical bond. An understanding of the energy relationships in biological form and function, then, depends on an awareness of the features of chemical energy.

The energy within chemical bonds is potential chemical energy. When a bond is broken, its potential energy is expressed as kinetic energy, usually in the form of heat. The amount of energy within a given bond is related to the affinity of the atoms for one another. For instance, if **A** represents one kind of atom, and **B** another, the amount of energy in a bond between two **A**s (represented **A**:**A**) is different from that between an **A** and a **B** (**A**:**B**).

It is possible to determine the amount of energy within a given bond. Potential energy cannot be directly measured so measurements must be done indirectly. Suppose one wanted to determine the strength of a piece of string. The strength is really the amount of energy that holds the molecules of the string together. One could add weights to one end of the string until it is broken, and then say that the string had a certain strength in terms of pounds. The strength attributed to the string would be the weight necessary to break it.

This same technique can be applied to the determination of bond energy. If a specific amount of energy holds two atoms together, an equivalent amount of energy is required to break the bond. This energy can be supplied in the form of heat, and the amount necessary to break the bond between two atoms can be measured in calories (see Chapter 1).

Chemical Reactions

A chemical reaction changes atomic combinations. A reaction always involves one or more chemical entities that undergo change (the **reactants**) and one or more chemical entities that result from that change (the **products**).

All chemical reactions proceed until a balance is established between reactants and products. This balance represents the **equilibrium** of the reaction. It is determined by the affinity of the chemical entities for one another and the temperature and pressure of the system.

The equilibrium is a dynamic one. The reactants are constantly forming products and the products are constantly forming reactants. The number of chemical forms in either state remains relatively constant, in accordance with a probability statement.

As with any equilibrium, a chemical equilibrium is upset if any of the initial conditions of the reaction are altered. For example, the addition of more reactants will drive the reaction toward the formation of more products, until the equilibrium is reestablished. It is also possible to shift the equilibrium point for a different ratio of reactants and products by changing the temperature or pressure.

Although chemical reactions are represented as taking place between molecules, the actual units undergoing change occupy the subatomic level of a single atom of the molecule. For example,

$$A{:}B + C{:}D \rightleftharpoons A{:}B{:}C{:}D$$

The chemical reaction indicates that a molecule of **A:B** has combined with a molecule of **C:D** to form a larger molecule **A:B:C:D.** In reality, however, **A** and **D** did not take part in the actual formation of the molecule. They only tagged along with the **B** and **C** atoms. The association of the **B** and **C** atoms is based on their electronic configurations. This phenomenon will be discussed in the next section.

The double arrows represent the dynamic equilibrium for all chemical reactions. Sometimes arrows of different lengths are used to indicate the preferential direction of a reaction and to reflect the relative rates toward the formation of product and reactant present when the equilibrium is established.

Chemical Reactions as Energy Conversions

An organism grows, moves, reproduces, and so forth. These activities are forms of work. The performance of work implies an expenditure of energy, that is, the conversion of chemical potential energy into kinetic energy and various forms of potential energy. Since biological systems are chemical systems, these energy conversions involve chemical reactions. Thus energy changes in biological systems follow the same physical laws as do all chemical reactions.

Once a chemical reaction begins and products start to form, the equilibrial state will inevitably be reached. If the system is left to itself, it will never again return to a state in which there are only reactants, and no products. Thus, a chemical reaction may be viewed as a system that is undergoing an irreversible change in state toward an equilibrial condition. The change that is occurring during such a reaction involves an alteration of the energy relationships of the system as well as the alteration of the particle relationships.

All energy conversions are determined by the laws of thermodynamics. The first law of thermodynamics states that energy cannot be lost or created, but is only converted from one form to another. Examine the energy conversions involved in the burning of a piece of wood, a type of chemical reaction. Within the wood, the energy is in a chemical form, inherent in the molecules of which the wood is composed. Heat and light energy, in the form of fire, are added to the wood to initiate the burning process. As the wood burns, one can feel heat and see light. If the system is scrutinized more closely, it can be shown that oxygen from the air is combining with the burning wood, and that carbon dioxide and water are given off. As chemical entities, both carbon dioxide and water contain chemical energy. So the burning of wood is essentially a conversion of chemical energy, heat, and light into heat, light, and alternative forms of chemical energy.

The conservation of energy permits a quantitative calculation of the energy relationships in a system. The energy in a system can be accounted for in the kinetic energy of that system, plus the bond energy of the product. In the case of the burning wood,

energy of the system (original chemical bond energy in wood plus heat and light used to initiate burning)	+	kinetic energy after burning (cumulative heat and light generated during change)	=	chemical bond energy in products

While the first law states that energy is conserved, it says nothing about the state of the energy resulting from an energy conversion in regard to its ability to do work. The second law of thermodynamics permits one to determine something about the work capacity of the energy resulting from a chemical reaction. According to the second law, the *energy content of a system undergoing change increases in entropy.* **Entropy** is an elusive factor that defies precise definitions because it cannot be measured directly. Therefore, some of the features of entropy, rather than entropy itself, will be discussed.

As entropy increases, *the degree of order of a system decreases.* Entropic changes are characterized as shifts from an ordered state of a system to a state of increased disorder. Chemical reactions, such as the burning of wood, can be considered to be a system going from a state of relatively high order to another state of increased disorder. The atomic forms in the wood are highly ordered in relationship to one another. The conversion through the chemical reaction to free carbon dioxide and water results from the chemical dissociation of large molecules. The relationship of atomic particles has, therefore, gone to a state of relative disorder, and the bond energy holding the atoms together is now in the form of heat, the random movement of CO_2 and H_2O.

Another aspect of entropy is that its value tends toward a maximum, and *the maximum entropy value that can be attained is dependent on the stability of the final state of the system.* In a chemical reaction, the equilibrium is reached when stable products are formed. This stability factor is the reason that the ordinary burning of wood produces carbon dioxide and water, rather than carbon, oxygen, and hydrogen. Carbon dioxide and water are stable products at combustion temperatures, and greater energy than is present in the burning system is necessary to disrupt the chemical bonds that hold the atoms together.

Another aspect of entropy is that as entropy increases, *the amount of energy available for work decreases.* In chemical reactions, much of the energy of the system is translated into heat. Heat is an energy form that is not efficiently translated into mechanical or chemical work.

The notation for a chemical reaction according to the second law of thermodynamics is

change in energy of the system	=	change in energy not available for work	+	change in energy available for work

The difference between the equation developed from the first law and that developed from the second is in terms of energy states. In the first equation, the term "kinetic energy" included both energy available and energy not available for work. The second equation permits the segregation of these energies into an entropy value (for energy not available for work) and a **free-energy** value (for energy available for work).

From the above word equation, it is possible to develop the standard thermodynamic equation for the energetics of a chemical reaction. Because one is studying the *change* in energy of a system, the symbol Δ (delta) is used. The change in energy of a system is referred to as the **heat of reaction** and is denoted as ΔH. In the burning of wood,

$$\Delta H = \left[\begin{array}{c}\text{initial chemical}\\ \text{bond energy}\\ \text{of wood}\end{array} + \begin{array}{c}\text{initial energy}\\ \text{used to start}\\ \text{burning}\end{array}\right] - \left[\begin{array}{c}\text{final chemical}\\ \text{energy after}\\ \text{reaction}\end{array} + \begin{array}{c}\text{kinetic}\\ \text{final energy}\\ \text{state of system}\end{array}\right]$$

The change in energy available for work is called the change in free energy, and is denoted as ΔG for Gibbs free energy. The ΔG of the burning wood would be the work capacity available in energy units. But the amount of energy available for work depends on the characteristic state of the system, part of which is the device used to measure work that is coupled to the system. For instance, burning the wood in a chamber of X volume with a piston will produce heat (in the random motion of CO_2 and H_2O molecules). These molecules can push the piston up a certain distance. If the chamber were $2X$ in volume, the piston would be pushed upwards a shorter distance. Therefore, the energy available for work (the ΔG in the smaller chamber) is greater, even though both chambers have the same energy content.

One cannot point to one portion of the energy and say "this is available for work, and the other portion is not." Both the ΔG and the ΔS comprise the total change in the kinetic energy of the system. The ΔG is detected by coupling a standardized device to the system. Universal agreement on the device then makes it possible to determine relative work capacities of different systems. A standardized device used for determinations of the ΔG of chemical reactions is the calorimeter, which measures caloric energy units of a system by measuring the work performed.

The entropy value is denoted as $T\,\Delta S$. The standard thermodynamic equation for a chemical reaction is

$$\Delta H = T\,\Delta S + \Delta G$$

or, by simple rearrangement to obtain the free-energy value,

$$\Delta G = \Delta H - T\,\Delta S$$

It is now possible to examine the equation in terms of the second law. In an energy conversion, the entropy value always tends toward a maximum. Since the change in the energy of the system is fixed, the change in free energy will tend toward a minimum. Thus, a chemical reaction will proceed in the direction that will produce the greatest possible entropy and/or the least free energy, and toward the formation of a product with an inherent stability that precludes the possibility of any further loss of order.

Examine the necessary conditions under which ΔG will be negative. This condition is met only when ΔH is negative, and its value is greater than that for $T\,\Delta S$. For example:

$$\begin{aligned} \Delta H &= -5 \text{ kcal (kilocalories)} \\ T\,\Delta S &= 3 \text{ kcal} \\ \Delta G &= -5 \text{ kcal} - (+)\,3 \text{ kcal} \\ \Delta G &= -8 \text{ kcal} \end{aligned}$$

What does this imply about the reaction? The ΔH is negative, indicating that there is a decrease of the energy in the reactant. Some of this loss constitutes entropic loss ($T\,\Delta S$), adding to the heat of the surroundings. Some of the energy is converted to an alternate chemical form in the bonds of the product. Clearly, such a reaction is behaving in accordance with the second law. Therefore, a reaction that produces these values *will occur spontaneously.* Reactions that have a negative ΔG value always give energy to the surroundings, and are said to be **exergonic** or **exothermic** reactions. Another aspect of such reactions is that the more negative the ΔG value, the more the reaction will be driven toward the formation of the products before equilibrium is reached.

Examine the necessary condition under which ΔG would be positive. This will occur only when ΔH is positive, and its value is greater than that for $T\,\Delta S$. For example:

$$\begin{aligned} \Delta H &= 5 \text{ kcal} \\ T\,\Delta S &= 3 \text{ kcal} \\ \Delta G &= 5 \text{ kcal} - (+)\,3 \text{ kcal} \\ \Delta G &= 2 \text{ kcal} \end{aligned}$$

In this case, the ΔH is positive, indicating that the energy content of the reactant *increases* at an expense to the environment. On the other hand, the positive ΔG value indicates that a product has been formed that manifests a higher potential energy than the original reactant. So the chemical system has *increased in order.* The only way in which this could happen is if energy is added to the chemical reaction. Therefore, the presence of a positive ΔG value indicates that the product will not be formed by the spontaneous action of the reactants, and the energy level of the reaction must be raised before such a reaction will proceed. Because such reactions draw energy from an external source, they are termed **endergonic** or **endothermic.**

Although endergonic reactions appear to be contradictions of the second law, they actually behave in accordance with it. The second law states that entropy increases when a system undergoes change. Some of the energy put into the chemical reaction is lost as heat (entropy), so that only part of it becomes chemical potential energy in the product. It is because of this necessary entropic loss that the ΔG values of reverse reactions are not simply reciprocal. For example:

$$\text{A:B:C:D} \xrightarrow{\Delta G = -8 \text{ kcal}} \text{A:B} + \text{C:D} - 2 \text{ kcal entropic loss}$$

Total energy originally in B:C bond = 10 kcal

$$\text{A:B} + \text{C:D} \xrightarrow{\Delta G = 12 \text{ kcal}} \text{A:B:C:D} - 2 \text{ kcal entropic loss}$$

Total energy in B:C bond is still 10 kcal, but it takes an input of 12 kcal to form the bond.

The ΔG value will be zero when the ΔH value is equal to that of $T\,\Delta S$, and opposite in sign. Under these conditions, the energy lost by the reactant is converted only to heat. The reaction is then in equilibrium, and no more product will be formed. This leads directly to the thermodynamic definition of equilibrium as the state at *which energy would have to be added in order to move the system in either direction.*

Energy of Activation

Each type of atom has an inherent stability. Its degree of stability is directly related to the energy content of the system (reaction vessel or molecule) in which the atom is found. A given atom may be quite stable at 10° C, at which point the energy inherent in the system is insufficient to bring about molecular speeds and a frequency of collisions high enough to disturb the stability. At 25° C (normal room temperature), however, the atom may be quite unstable. Other atoms may be stable for the energy levels associated with temperatures up to 500° or 1000° C.

If an atom is to take part in a chemical reaction, it must become unstable. The amount of energy required by a given atom for this loss of stability is called the **energy of activation.** Once the energy of activation is realized, the reaction will proceed in accordance with the second law of thermodynamics. If the reaction is exothermic, the reaction itself will generate sufficient energy to activate other molecules in the solution, and the reaction becomes self-feeding. If the reaction is endothermic, the energy of activation constitutes the energy that must be put into the system for the reaction to take place.

In exothermic reactions, the energy of activation is graphically expressed as an "energy hill" that must be overcome before the reaction will take place (see Fig. 2-6).

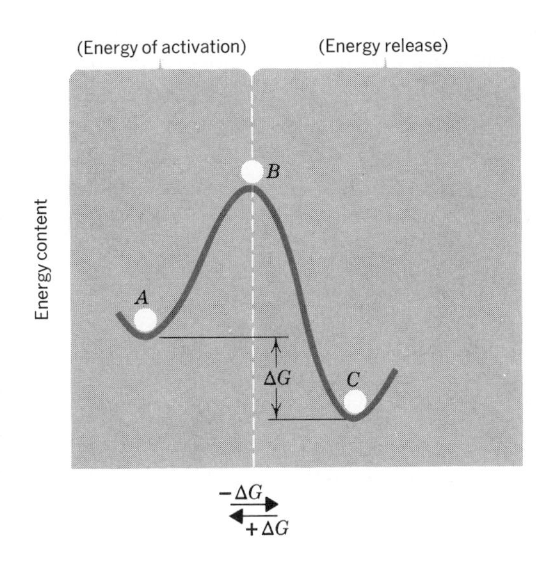

Fig. 2-6 Energy of activation. The energy added to a reaction is represented by the curve from *A* to *B*. The spontaneous culmination of the reaction is represented by the curve from *B* to *C*. Note that greater energy is derived from the reaction than was put in.

Limiting Factor

Another factor that determines whether atoms will interact is the presence of the necessary components of the reaction. The absence of a particular reactant serves as a **limiting factor** for the reaction. This may seem obvious, but often the obvious is the first to be overlooked.

THE FORMING OF MOLECULES—CHEMICAL BONDS

Ionic Bonds

An atom such as sodium (Na) with its atomic number 11 has a single electron in the 3*s* subshell (it takes 10 electrons to fill the first two energy levels). The stability of the sodium atom is of an order that the energy of activation necessary for the removal of the 3*s* electron is readily available in a reaction vessel of room temperature.

When elemental sodium is placed in water at room temperature, the energy inherent in the system brings about the removal of the 3*s* electron. The electrical stability of the sodium atom is disturbed by the loss of the negatively charged electron, and the remaining part of the atom expresses a positive charge, becoming the sodium **ion** (Na^+).

Water contains both H^+ and OH^- (**hydroxyl ion**), although most of the hydrogen and oxygen atoms are bonded as H_2O. In organic chemistry, the general name given to the OH^- group is the hydroxyl or alcohol radical. The hydroxyl ion, which is negatively charged, will move to a position near the Na^+. This movement expresses energy in the form of heat. The heat generated by the "coming together" of the Na^+ and OH^- is greater than the energy needed to remove the electron from the sodium atom and dissociate water into its ionic forms. Therefore, the reaction is exothermic and will proceed spontaneously in accordance with the second law of thermodynamics.

The reaction that has taken place is the formation of an electrical association between two ions. This electrical relationship is called an **ionic bond.**

$$Na^+ + HOH \longrightarrow Na^+\,OH^- + H^+ + e^-$$

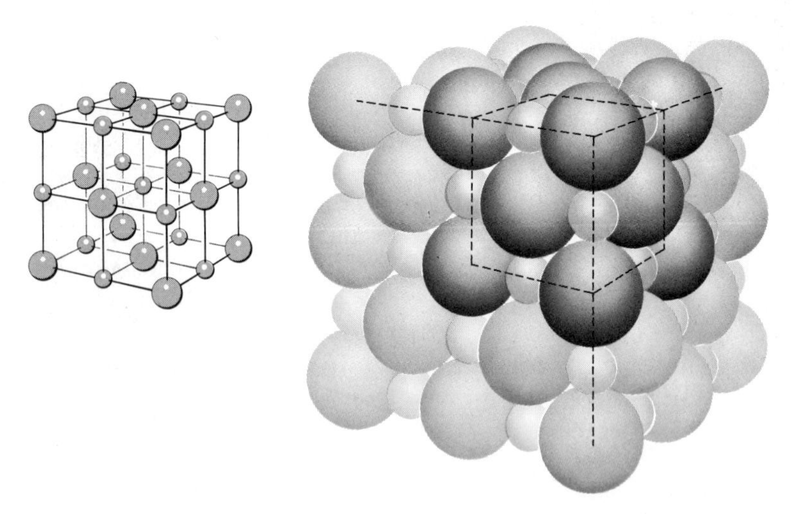

Fig. 2-7 Crystalline lattice of NaOH in which each Na^+ is associated with six OH^-, and each OH^- is associated with six Na^+.

But have they truly formed a molecule? In a water solution of sodium hydroxide, the individual ions are in constant movement, always associated with ions of opposite charge. However, they do not remain "faithful" to a single ion bearing that opposite charge, but keep changing their associations. When water evaporates from such a system, a solid crystal forms. The pattern of the sodium and hydroxyl ions within the crystal is very specific (see Fig. 2-7). Each sodium ion is surrounded by six hydroxyl ions and each hydroxyl ion by six sodium ions. There is no identifiable, discrete sodium hydroxide molecule in the solution or the crystal. In such cases ionic bonding does not lead to the formation of a true molecule, but rather to what are termed **ionic pairs.**

Some ionic bonding does lead to the formation of a true molecule in the sense that a discrete molecule is present in which the component parts exhibit great fidelity toward one another. These molecules undergo *partial* dissociation in water solution, and reach an equilibrium between associated and dissociated forms. Although the dissociated forms change their allegiances in much the same way as do Na^+ and OH^-, the associated form retains a close relationship much like that found in the solid crystal, but existing only between pairs of ions.

Acids, Bases, and Buffers

When ionic pairs are in water solution, they tend to undergo dissociation into their ionic components. If the ionic pair is made up of a hydrogen ion and a negatively charged ion or radical (a chemical form involving two or more elements that are bonded together in a manner to be discussed in the next section), this dissociation will bring hydrogen ions into solution. The hydrogen ion immediately combines with a water molecule to form H_3O^+, the **hydronium ion.** It is the presence of the hydronium ion that gives the solution acid properties, such as a sour taste and the ability to turn dyes to another color. The strength of the acid is directly related to the amounts of H_3O^+ in solution.

Certain molecules that contain hydrogen atoms undergo complete dissociation when placed in a water solution, thereby forming as many H_3O^+ ions as there were molecules put into the solution (HCl—hydrochloric acid, H_2SO_4—sulfuric acid; these are called **strong acids**).

$$n\text{HCl} \longrightarrow n\text{H} + n\text{Cl}$$

Other molecular forms that involve associations in ionic pairing with hydrogen exhibit a range of dissociation tendencies, each property of the individual form. These dissociations involve an equilibrium between the associated and dissociated form of the molecule, so that only a fraction of the potential H^+ is found in solution in the form of H_3O^+. The remainder is found in the form of H^+ in ionic association with the negative ion or radical. For example, H_2CO_3 (carbonic acid) may dissociate into two states:

$$H_2CO_3 \rightleftharpoons H^+ + HCO_3^- \rightleftharpoons H^+ + H^+ + CO_3^{=}$$

In each case there is an equilibrium state between the amount of

H^+ that will form an H_3O^+ and the associated states of the molecule, which do not contribute H_3O^+ to the solution and thus do not add to the acidity. Because of the relatively small amount of H^+ that is put into solution, carbonic acid is considered a **weak acid.**

The dynamic considerations that are characteristic of equilibrium states in general are also exhibited in dissociation equilibria. If the H_3O^+ were removed from the system, the dissociation reaction would shift to the right, until enough H_3O^+ had been formed to balance the system once again. On the other hand, if H_2CO_3 were removed from the system, the shift would be to the left, and the hydronium ion would break down into H_2O and H^+, which would then reassociate into the ionic molecular form with the available $CO_3^=$ or HCO_3^-. This is a reasonable expectation if one considers that when there is less acid in a solution, the solution exhibits the acid characteristics to a lesser degree. Note that it takes more H_2CO_3 than HCl, which undergoes complete dissociation, to give the same level of acidity to a solution.

Another important type of ionic pair is the one in which a hydroxyl ion (OH^-) is paired with a positive ion. The actual bonding between the oxygen and hydrogen atoms of the hydroxyl ion will be discussed later. Upon dissociation in a water solution, these ionic pairs show the characteristics of a **base,** the brackish taste, a characteristic changing of dye colors, etc. This **alkaline** (basic) condition is directly related to the presence of the OH^- in solution.

The strength of the base is determined by the concentration of the OH^-. Molecules that undergo almost complete dissociation are termed **strong bases** (NaOH—sodium hydroxide; KOH—potassium hydroxide), whereas those that undergo partial dissociation are considered **weak bases** (NH_4OH—ammonium hydroxide).

When equal numbers of molecules of a strong base and a strong acid are put into a solution,

$$NaOH + HCl \longrightarrow Na^+ \; Cl^- + H_2O$$

a neutral solution forms. If one looks at this reaction without the associated ionic forms,

$$OH^- + H_3O^+ \longrightarrow 2H_2O$$

it may be seen that in the course of the reaction, the H^+ is accepted by the OH^-, thereby forming a molecule of water. Because the H^+ is nothing more than a proton, the reaction may be viewed as a proton transfer.

Since all chemical reactions have states of equilibrium, even though these states may be highly preferential in one direction, it is reasonable to assume that water might undergo spontaneous dissociation to form OH^- and H_3O^+.

$$2H_2O \longrightarrow H_3O^+ + OH^-$$

Careful measurement indicates that this does indeed happen. Approximately one molecule of water in every 300,000,000 at 25° C is in an ionized state. A dynamic equilibrium is established between the H_2O and the ionized products, in which the rate of the reaction

$$H_2O \longrightarrow H_3O^+ + OH^-$$

is equal to the rate of the reaction

$$H_3O^+ + OH^- \longrightarrow H_2O$$

The addition of an acidic substance upsets the equilibrial condition. The dissociation reaction is driven toward the formation of water. In the course of this readjustment, OH^- will combine with H_3O^+ to form water molecules. The number of OH^- ions drawn out of solution is directly related to the number of H_3O^+ ions in the solution. Eventually, an equilibrium between intact water molecules and its ionic products is reestablished.

Under equilibrial conditions, the product of the concentrations of ions is constant, regardless of the relative concentrations of the individual types of ion.

$$[H_3O^+]^* \times [OH^-] = K$$

The constant K is equal to 10^{-14} at 25° C.

When the concentrations of H_3O^+ and OH^- ions are equal, the solution is neutral. According to Table 2-3, a solution is neutral when, at 25° C, there is a 10^{-7} concentration of both ions. At hydronium-ion concentrations above this level (10^0–$10^{-6.9}$), the solution is acid; and at hydronium-ion concentrations below this level ($10^{-7.1}$–10^{-14}), the solution is alkaline.

TABLE 2-3

H_3O^+ AND OH^- CONCENTRATIONS

When the H_3O^+ Concentration Is:	The OH^- Concentration Is:
$1\ (1 \times 10^0)$	1×10^{-14}[a]
1×10^{-1}	1×10^{-13}
1×10^{-2}	1×10^{-12}
1×10^{-3}	1×10^{-11}
1×10^{-4}	1×10^{-10}
1×10^{-5}	1×10^{-9}
1×10^{-6}	1×10^{-8}
1×10^{-7}	1×10^{-7}
1×10^{-8}	1×10^{-6}
1×10^{-9}	1×10^{-5}
1×10^{-10}	1×10^{-4}
1×10^{-11}	1×10^{-3}
1×10^{-12}	1×10^{-2}
1×10^{-13}	1×10^{-1}
1×10^{-14}	$1\ (1 \times 10^0)$

[a]Exponential values are multiplied by adding the exponents.

The standard expression for the acidity or alkalinity of a solution is the **pH.** The pH is defined as the *negative logarithm of the hydronium ion concentration.* Thus a solution containing 10^{-3} H_3O^+ per unit volume would be considered to have a pH of 3.

$$\log_{10} \text{ value of } 10^{-3} = -3$$
$$\text{negative } \log_{10} \text{ value of } 10^{-3} = -(-3) = 3$$

For pH values between zero and 6.9, a solution exhibits the characteristics of an acid, because the hydronium ion dominates in concentration and therefore dictates the properties of the system. Owing to the equal concentrations of the hydronium and the hydroxyl ion, the system is neutral at pH 7.0. From pH 7.1 to 14, it is alkaline, because of the predominance of the hydroxyl ion in the concentration. This nomenclature provides a relatively simple means of expressing the level of acidity of a system, but it is well to keep in mind that the pH value is reflective of the hydronium and hydroxyl ion concentrations.

The integrity of a chemical system is, in a great part, dependent on the appropriate amount of hydronium and hydroxyl ions present. Certain reactions will only take place in very narrow pH ranges. The majority of chemical reactions in biological systems have this narrow pH dependence. *It is essential for the maintenance of biological organization that the pH remain constant in biological systems.* The mechanism for the maintenance of pH levels is the chemical action of **buffers.**

*Brackets are standardly used to denote molar concentration.

Equilibrium states of a buffer

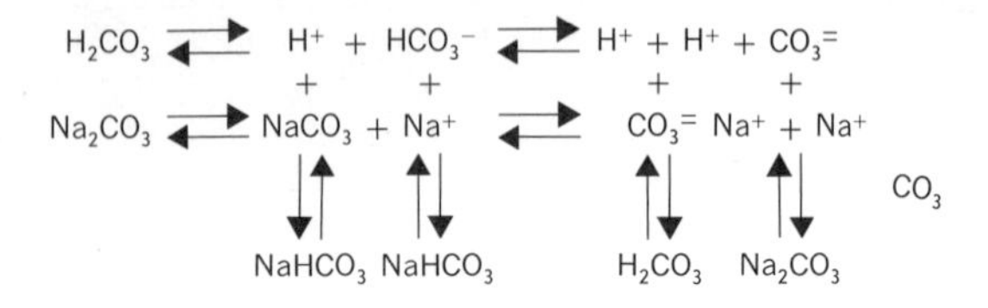

H3O+ added

H2CO3 ⇄ H+ + HCO3− ⇄ H+ + H+ + CO3=
Na2CO3 ⇄ NaCO3− + Na+ ⇄ CO3= Na+ + Na+
NaHCO3 NaHCO3 H2CO3 Na2CO3

colored arrows indicate preferred direction of respective reactions

OH− added

All reactions go in direction of H_3O^+ production

$H_3O^+ + OH \longrightarrow 2H_2O$

Fig. 2-8 Equilibrium states of a buffer. When acid is added to the system, the equilibrium shifts to the left so that more H_2CO_3 and $H^+ + HCO_3^-$ are produced.

These are mixtures of a weak acid and its salt. As we have seen, carbonic acid has several dissociation states, between which shifts occur in order to maintain equilibrium in accordance with the relative concentrations of the forms associated with each state. A salt of carbonic acid also exhibits equilibrium states. (A substance is considered a salt when it can be thought of as the product of a neutralization reaction between an acid and a base. In other words, it is considered to be the association of a **cation**—a positively charged ion—and a **anion**—a negatively charged ion—by an ionic bond.)

$$Na_2CO_3 \rightleftharpoons Na^+ + NaCO_3^- \rightleftharpoons 2Na^+ + CO_3^=$$
$$H_2CO_3 \rightleftharpoons H^+ + HCO_3^- \rightleftharpoons 2H^+ + CO_3^=$$

A water solution made up of the appropriate proportions of carbonic acid and sodium carbonate will tend to maintain its pH level (see Fig. 2-8).

The maintenance of pH levels is solely dependent on the amount of the weak acid and salt in solution. If sufficient amounts of a strong acid or a strong base were added to a buffered system, the dissociation reactions would be driven in one direction as far as possible. The remaining H_3O^+ *or* OH^- would then be *expressed*, and the system would undergo a change in pH. Buffered systems are effective means of maintaining pH levels in biological systems. That buffered systems may be overcome by the introduction of large concentrations of H_3O^+ or OH^- provides an important experimental tool for the study of biological organization on a chemical level.

For purposes of the present discussion, reference has consistently been made to the hydronium ion (H_3O^+) as being responsible for the acid properties of solutions. The organic chemist and biochemist find that reference to the hydrogen ion (H^+) provides a simpler nomenclature, and for the remainder of the book, references will be made solely to that ion, with the understanding that in aqueous solutions the presence of the H^+ ion leads directly to the formation of H_3O^+.

Covalent Bonds

Those atoms with two to five electrons in the $2p$ or $3p$ subshells (such as those given previous emphasis—carbon, nitrogen, oxygen, phosphorus, and sulfur) do not tend to undergo ionic pairing but, rather, form **covalent bonds.** The covalent bond involves the sharing of electrons between atoms in such a way that both atoms have completed subshells and maintain their electrical stability as well. The molecular form of many gases are formed this way. Hydrogen gas, for example, is made up of two atoms of hydrogen sharing their two electrons in a common orbital around both nuclei, thereby satisfying the 1*s* subshell that is now common to both. This formation constitutes the hydrogen molecule; it is a true molecule in the sense that there is a stable physical association between the two hydrogen atoms (see Fig. 2-9). This sharing of an orbital holds the two atoms together, and is the covalent bond itself. To indicate this physical association, the structural formula of the hydrogen molecule is given as H:H, and its chemical notation as H_2.

Hydrogen molecule

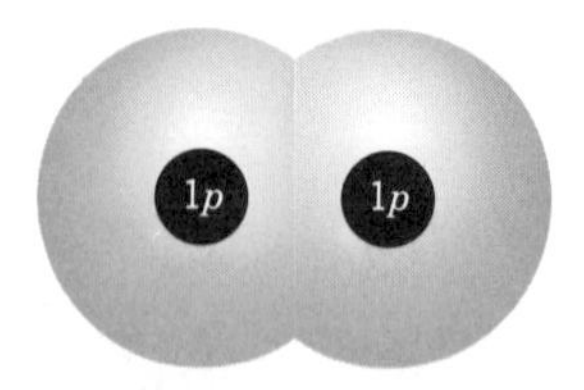

Fig. 2-9 The hydrogen molecule. The 1*s* orbital of shared electrons circumscribes the two atomic nuclei.

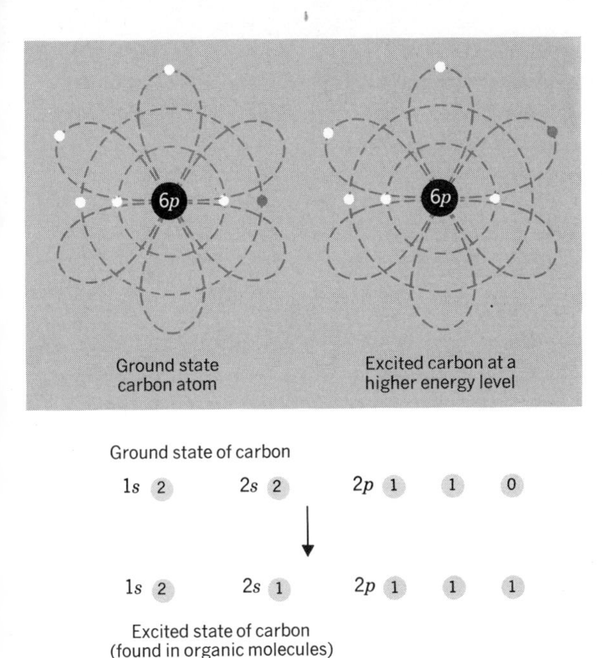

Fig. 2-10 The ground state and the excited states of carbon. The ground state is characterized by two electrons in the 2*s* shell, and one 2*p* shell lacking electrons. The excited state, which is found in organic molecules, is characterized by one electron in the 2*s* shell and one electron in each of the three 2*p* subshells.

Covalent bondings are not limited to atoms of the same type. Different atoms may also form covalent bonds. Each bond has a characteristic bond energy, related to the amount of energy that must be added to the system in order for the bond to be formed or broken. The covalent bond energy is considered to be in the range of 50 to 110 kilocalories, although some bonds are substantially lower.

The Bonding Potentials of Carbon

The atom of primary interest in dealing with biologically active molecules is the carbon atom. Therefore, it is of great interest to examine this atom's bonding possibilities. As noted, the carbon atom has two electrons in the 2*p* subshell and is said to have a valence of four. However, one orbital in the 2*p* subshell does not contain any electrons. Thus, in order to form a covalent bond, the carbon atom would need to borrow a complete orbital from another atom. This simply does not happen, because of the tremendous energy involved in such a transfer.

The carbon present in biologically active molecules has a slightly different electronic configuration from that of the elemental carbon atom. The elemental carbon atom with its noted electronic configuration is at its lowest energy level, or **ground state.** When energy is added to the atom itself, it goes to a higher energy level, with a consequent rearrangement of the electrons. This higher energy level of carbon involves the moving of one of the electrons in the 2*s* subshell up to the third and previously unoccupied 2*p* orbital (see Fig. 2-10). All the orbitals in the first two energy levels are now occupied—not filled, but occupied. The 2*s* and the three 2*p* orbitals each have only one electron, and four more are still required for completion of the energy level. The valence has not changed.

The Carbon-Hydrogen Bond. Each of these four orbitals may share electrons and thus form covalent bonds. The simplest sharing occurs

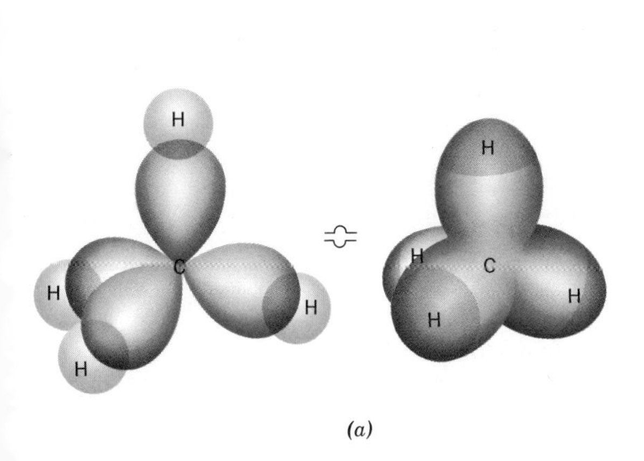

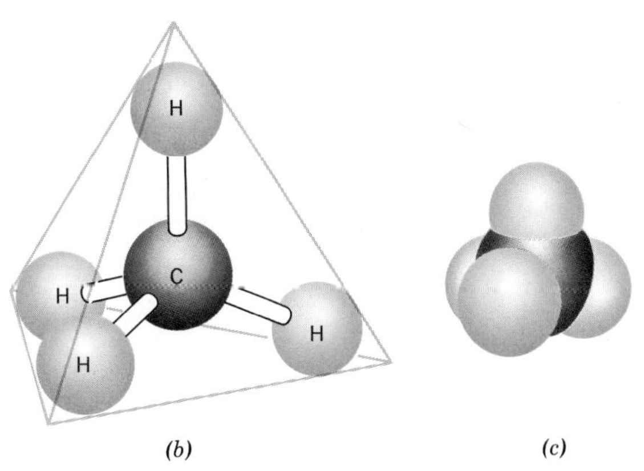

Fig. 2-11 Various representations of methane: (*a*) shows the carbon-hydrogen relationship of shared orbitals, (*b*) includes the bond angles, and (*c*) is the visualization of the molecular form.

when each orbital shares an electron with a hydrogen atom. Hydrogen nuclei move into a position whereby the electrons of the orbital simultaneously satisfy the energy levels of both the carbon and hydrogen atoms (see Fig. 2-11). The molecule **methane** has this configuration, and its chemical structure is given as

$$\begin{array}{c} \text{H} \\ \text{H}:\underset{\cdot\cdot}{\overset{\cdot\cdot}{\text{C}}}:\text{H} \\ \text{H} \end{array}$$

The Carbon-Carbon Bond. One of the most important types of covalent bonds for carbon is the bonding between two different carbon atoms. When two carbon atoms share an orbital, they are coupled by a single covalent bond. There are still three bonding sites on each of the carbon atoms:

$$\cdot\underset{\cdot}{\overset{\cdot}{\text{C}}}:\underset{\cdot}{\overset{\cdot}{\text{C}}}\cdot$$

If these bonding sites are each satisfied by hydrogen atoms, the resulting molecule is **ethane.**

$$\begin{array}{c} \text{H}\ \ \text{H} \\ \text{H}:\underset{\cdot\cdot}{\overset{\cdot\cdot}{\text{C}}}:\underset{\cdot\cdot}{\overset{\cdot\cdot}{\text{C}}}:\text{H} \\ \text{H}\ \ \text{H} \end{array}$$

A molecule may have three carbon atoms bonded together. In this case, each terminal carbon has three free bonding sites, whereas the middle carbon atom has only two such sites.

$$\cdot\underset{\cdot}{\overset{\cdot}{\text{C}}}:\underset{\cdot}{\overset{\cdot}{\text{C}}}:\underset{\cdot}{\overset{\cdot}{\text{C}}}\cdot$$

If these are all satisfied by hydrogen atoms, the resulting molecule is **propane.**

$$\begin{array}{c} \text{H}\ \ \text{H}\ \ \text{H} \\ \text{H}:\underset{\cdot\cdot}{\overset{\cdot\cdot}{\text{C}}}:\underset{\cdot\cdot}{\overset{\cdot\cdot}{\text{C}}}:\underset{\cdot\cdot}{\overset{\cdot\cdot}{\text{C}}}:\text{H} \\ \text{H}\ \ \text{H}\ \ \text{H} \end{array}$$

These carbon-carbon bondings permit the formation of long chain molecules that serve as the very basis for biological structure.

Two carbon atoms may also undergo a covalent association in which they share two orbitals, thus satisfying two bonding sites on each carbon atom. This is called the carbon-carbon double bond, the structural formula of which is

$$:\text{C}{=}\text{C}:$$

Notice that each of the carbon atoms still has two bonding sites that require associations. Molecular forms that manifest this type of carbon-carbon bonding are given an **ene** suffix, and

$$\begin{array}{c} \text{H}:\underset{\cdot\cdot}{\text{C}}{=}\underset{\cdot\cdot}{\text{C}}:\text{H} \\ \text{H}\quad\text{H} \end{array}$$

are called **ethene.**

The Carbon-Oxygen Bond. The carbon atom also undergoes covalent bonding with an oxygen atom. Oxygen has four electrons in the 2*p* subshell, and all its orbitals are occupied. Therefore, the elemental oxygen atom will readily combine in its ground state. Having a valence of two, oxygen requires two bonding sites, one for each of the incomplete orbitals. When one of these bonding sites is shared with a carbon atom and the other with a hydrogen atom, the molecule has the form

```
 ·
·C:O:H
 ·
```

If the three other bonding sites of the carbon atom are involved in covalent bondings with hydrogen atoms, the resulting molecule is **methyl alcohol:**

```
  H
  ··
H:C:O:H
  ··
  H
```

You will recognize this as the hydroxyl ion, which in water solutions confers alkaline properties. When this group is covalently bonded to a carbon atom, there can be no ionic form. These solutions *do not manifest the properties of a base.*

Molecules exist in which an oxygen atom shares both of its bonding sites with a carbon atom to form what is called a **double bond.** This double-bonded carbon-oxygen combination still leaves the carbon atom with two unsaturated orbitals. If these are also combined with another oxygen in the same fashion, the resulting molecule is **carbon dioxide:**

$$O{=}C{=}O$$

When oxygen is found double-bonded to the terminal carbon atom in a carbon chain, the molecule is considered an **aldehyde:**

```
  H H H
  ·· ·· ··
R:C:C:C:C=O
  ·· ·· ·· ··
  H H H H
```

(*The symbol* **R** *is used to represent a continuing molecule. It may indicate the presence of further carbon structure or simply another hydrogen. It is standard to generalize the structure of a group of molecules by the use of this symbol.*)

When oxygen is found double-bonded to a carbon atom along the chain, the molecule is considered a **ketone:**

```
  H H   H
  ·· ··   ··
R:C:C:C:C:H
  ·· ·· ·· ··
  H H O H
```

A third (and one of the most common) atomic arrangement linking carbon, hydrogen, and oxygen is the **carboxyl** or **acid group.** In the carboxyl group, two bonding sites of the carbon atom are covalently bonded to oxygen in the double bond. The third bonding site is bonded to an oxygen atom by a single covalent bond. This oxygen atom is in turn bonded to a hydrogen atom (hydroxyl group). This leaves the fourth bonding site of the carbon atom free, and when this is satisfied by hydrogen, the molecule is **formic acid:**

$$\begin{array}{ccc} \mathrm{H{:}} & \mathrm{C} & \mathrm{{:}O{:}H} \\ & \| & \\ & \mathrm{O} & \end{array}$$

Molecules that have the carboxyl group exhibit those properties associated with acids because of the tendency of the hydrogen from the hydroxyl group to form an H^+ ion. Because of the energy relationships within the molecule, the hydrogen part of the hydroxyl group has a bond that is not quite covalent, but not quite ionic either. Although it dissociates in the same fashion as an ionic bond, this dissociation does not occur as readily, indicating greater bond strength than is usually associated with ionic bonds.

There is one more significant way in which carbon and oxygen atoms are linked in covalent bonds. An oxygen atom may share one orbital with one carbon atom and another orbital with a different carbon atom:

$$\begin{array}{ccccc} & \mathrm{H} & & \mathrm{H} & \\ & \cdot\cdot & & \cdot\cdot & \\ \mathrm{R{:}} & \mathrm{C} & \mathrm{{:}O{:}} & \mathrm{C} & \mathrm{{:}R} \\ & \cdot\cdot & & \cdot\cdot & \\ & \mathrm{H} & & \mathrm{H} & \end{array}$$

In a molecule bearing this relationship between the carbon and oxygen atoms, both bonding sites of oxygen are filled, and the three remaining bonding sites on each of the carbon atoms can be filled by other atomic forms. When this relationship exists *within* a molecule, the molecule is called an **ether.** When this relationship is found between two discrete molecules, it is termed an **ester linkage.** Such a linkage contains about 1.8 kcal of energy.

The Carbon-Nitrogen Bond. Carbon atoms are also found in combination with nitrogen atoms. In these cases the four bonding sites of the carbon atom and the three bonding sites of the nitrogen atom are each completed in the formation of a molecule. When carbon and nitrogen atoms share two electrons (one orbital) and the other two orbitals of the nitrogen and three orbitals of the carbon are shared with hydrogen, the molecule is called **methyl amine:**

$$\begin{array}{cccc} & \mathrm{H} & & \\ & \cdot\cdot & & \\ \mathrm{H{:}} & \mathrm{C} & \mathrm{{:}N} & \mathrm{{:}H} \\ & \cdot\cdot & \cdot\cdot & \\ & \mathrm{H} & \mathrm{H} & \end{array}$$

The general name given to the NH_2 group is the **amino group** or **amine,** while the bond is called an **amide** bond. The amide bond contains 3.4 kcal/mol of energy. It is also found that the nitrogen atom shares two orbitals with the carbon atom to form a double bond. We shall see the importance of this type of bond in later discussions.

When a nitrogen atom shares all three of its unfilled orbitals with a carbon atom, and the final bonding site of the carbon atom is filled by a hydrogen atom, the molecule is **hydrogen cyanide,** which, as you are undoubtedly aware, is not found in organisms.

$$\mathrm{H{:}C{\equiv}N}$$

The remaining way in which nitrogen and carbon atoms are found in molecular forms is with nitrogen sharing one orbital with one carbon atom and a second orbital with a different carbon atom, while the third orbital is shared with a hydrogen atom:

```
  H H   H H
R:C:C:N:C:C:R
  H H H H H
```

The NH group is called an **imino** group.

Once again it should be noted that although putting these molecules together on paper may resemble a game of blocks in which one must follow definite rules in the building of structures (carbon has four sites, nitrogen has three, etc.), these atoms are not passive blocks, but dynamic particles. Part of the energy is inherent in the bonds themselves in the form of potential energy. Because of the energy relationships between the atoms involved in a molecule, a substance made up of such molecules (a **compound**) manifests certain properties. For instance, the presence of a hydroxyl group (OH^-) on a carbon chain (a series of carbon atoms) confers the property of an alcohol. The length of the carbon chain determines the properties of the particular alcohol.

MONOMERS AND POLYMERS

When material of biological origin is ashed, the products are found to be in elemental (carbon), ionic (sulfate ion – $SO_4^=$, phosphate ion – $PO_4^=$), and small molecular (carbon dioxide, water) forms. The heat and light generated in the ashing process results from the conversion of the bond energy of covalent bonds of larger molecules. This is consistent with what is known to be true. First of all, there is potential energy within covalent bonds; secondly, this energy may be released in the form of heat and light as the system goes to a state of greater disorder. The reduction of large molecules to smaller ones represents the transition from a state of high order (where individual atoms maintain an oriented position in relation to one another) to a more random one. Also, the elemental carbon residue found in ashing is carbon in a ground state, so that even this loses the energy of the "excited" state found in carbon associated with other atomic forms in an organic molecule. But this collection of elements, ions, and small molecules does not give the scientist any indication of the organized form in which these components were found in the original biological material.

If, rather than being ashed, the material is handled in a less severe way, larger, intact molecules can be found. Various methods have been developed to extract intact molecules from biological material. The first molecular forms to be extracted were a series of small units, made up of two to ten carbons bonded together in linear or branched chains. These small molecular forms are called **monomers.**

Within the monomer group are several classes. One class is made up of the sugars; another, of the amino acids; and so forth. The molecules classed as sugars all have basic structure and properties in common, and these are quite different from the basic properties and structure of molecules in the amino-acid class.

As extraction techniques became more refined, it was realized that the monomers were "building blocks" for larger molecules. The harsher treatments had disrupted the bonds holding the monomers together, so that earlier extractions had resulted in monomeric forms rather than the intact molecule present in the biological systems. The large molecules that are made up of monomer units are called **polymers.**

The polymers can also be divided into several subgroups. The sugar subgroup is composed of polymers in which many monomer sugars are bound together. The protein subgroup is composed of polymers in which many amino-acid monomers are bound together. As more information was accumulated about the molecular makeup of biological material, it was realized that polymers were the backbone of biological structure and function. Although there are some monomers naturally present in biological material, they do not exist as such for long periods of time.

Another group of molecular forms that were isolated were the **heteroconjugates.** Heteroconjugates are relatively small molecules made up of two or more classes of monomers. The sugar amino-acid complex that is found in bacterial cell walls is a heteroconjugate. The active portions of enzymes are often heteroconjugates. Most important in the biological scheme of things, the nucleic acids are polymerized heteroconjugates.

TABLE 2-4 CARBOHYDRATES AND THEIR EMPIRICAL FORMULAS

	Number of Carbons	Empirical Formula
Trioses	3	$C_3H_6O_3$ $C_3(H_2O)_3$
Tetroses	4	$C_4H_8O_4$ $C_4(H_2O)_4$
Pentoses	5	$C_5H_{10}O_5$ $C_5(H_2O)_5$
Hexoses	6	$C_6H_{12}O_6$ $C_6(H_2O)_6$
Heptoses	7	$C_7H_{14}O_7$ $C_7(H_2O)_7$

Carbohydrates

The name carbohydrate is given to the class of chemical forms in which carbon appears only in a hydrated form, CH_2O.

The carbohydrate monomer is called a **monosaccharide,** and all monosaccharides are **sugars.** They are generally categorized according to their **empirical formula,** which gives the number of each type of atom present in the structure. See Table 2-4. There are three chemical forms with three carbon atoms, six hydrogen atoms, and three oxygen atoms in their structure. Only two carbohydrates are found in biological material. The arrangement of the atoms determines the properties of the individual monosaccharide, and although on paper, the structures may seem to vary only slightly, the difference in atomic orientation is sufficient to give each type of molecule distinct properties (melting point, boiling point, chemical reactivity, etc.)

```
   CHO              CHO              CH2OH
    |                |                |
H:C:O:H         H:O:C:H              C=O
    |                |                |
   CH2OH            CH2OH            CH2OH
   (1)              (2)              (3)
```

Notice that the first two trioses are in the same form. Both have an aldehyde group on the first carbon atom, which characterizes them as **aldoses.** The second carbon, while having the same types of atoms, has the hydroxyl group oriented in different directions. The first triose has the hydroxyl group oriented toward the right and is called **D-glycer-**

aldehyde (D = **dextrorotatory,** rotating polarized light to the right), while the second triose has the hydroxyl group oriented toward the left and is called **L-glyceraldehyde** (L = **levorotatory,** rotating polarized light to the left). Because of the nature of the bonding, the middle carbon atom is held in a fixed position and the D- and L- forms tend to remain stable. The first two trioses are **stereoisomers** of glyceraldehyde, and have different reactive properties from one another. Only D-glyceraldehyde is extractable from biological material. The third triose is a **ketose** because of the presence of the ketone group on its second carbon atom. Notice that this is the only form the three-carbon ketose can take. Ketose and aldose forms having the same empirical formulas are called **epimers.**

There are four aldose forms (two distinct types with their associated stereoisomers) and two ketose forms (stereoisomers of the same type)

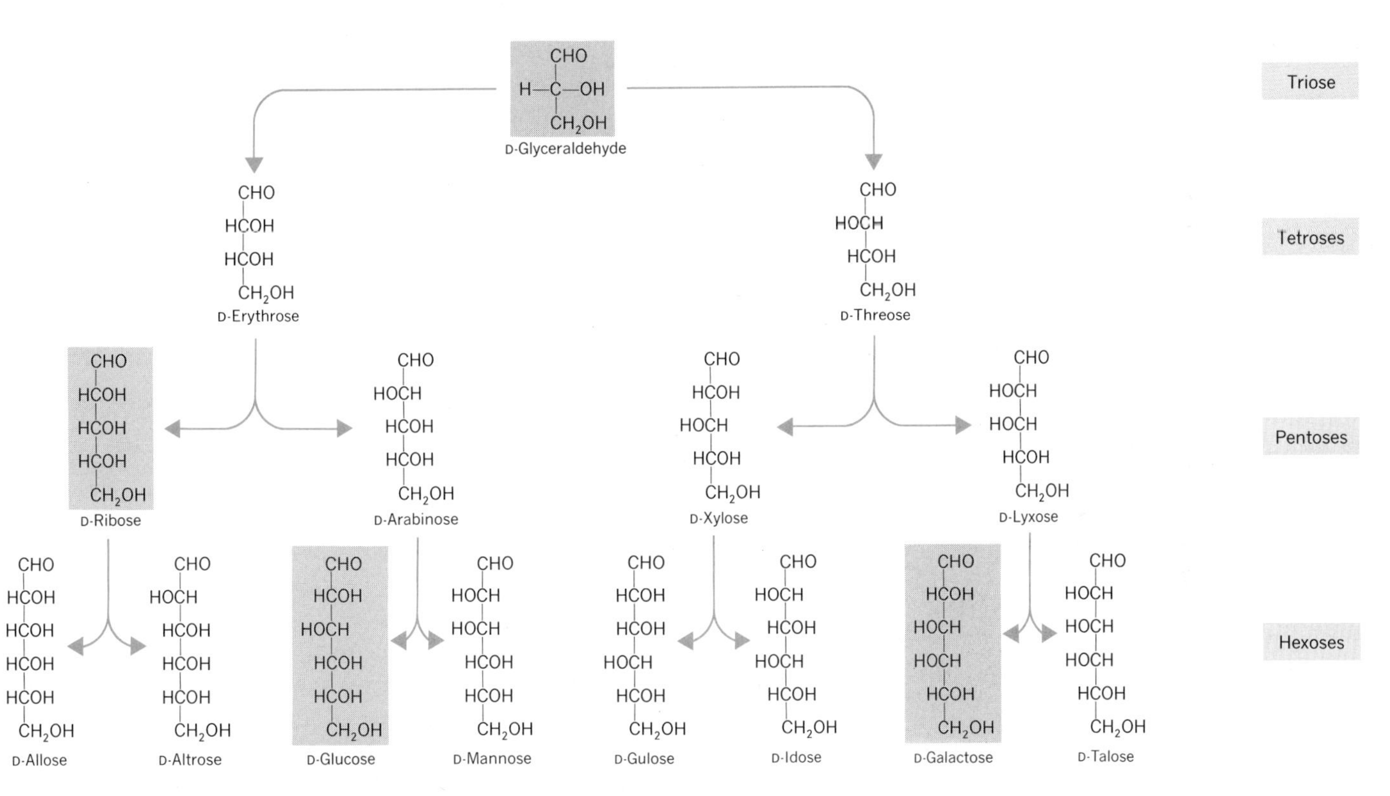

Fig. 2-12 Configurational relationships of the aldose D-sugars. Notice that the orientation of the CH_2OH groups determines the character of the particular sugar.

that have this empirical formula (see Fig. 2-12). Only one of these forms is found in appreciable quantities in biological material.

Of the pentoses (five-carbon monosaccharides), only two are of major importance in biological systems: **D-ribose** and its related form, **D-deoxyribose,** in which the second carbon atom lacks an oxygen atom. Although D-deoxyribose varies from the standard with its empirical formula of $C_5H_{10}O_4$, its chemical structure and properties are closely related to the carbohydrates and it is considered a carbohydrate form.

D-ribose	2-D-deoxyribose
CHO	CHO
H:C:O:H	H:C:H
H:C:O:H	H:C:O:H
H:C:O:H	H:C:O:H
CH_2OH	CH_2OH

The hexoses are the most prevalent form of monosaccharides found in biological systems. There are eight possible aldose epimers (and eight more stereoisomers of those epimers), of which only **glucose** and **galactose** are of prime importance.

(D-galactose)	(D-glucose)	(D-fructose)
CHO	CHO	CH_2OH
H:C:O:H	H:C:O:H	C=O
H:O:C:H	H:O:C:H	H:O:C:H
H:O:C:H	H:C:O:H	H:C:O:H
H:C:O:H	H:C:O:H	H:C:O:H
CH_2OH	CH_2OH	CH_2OH

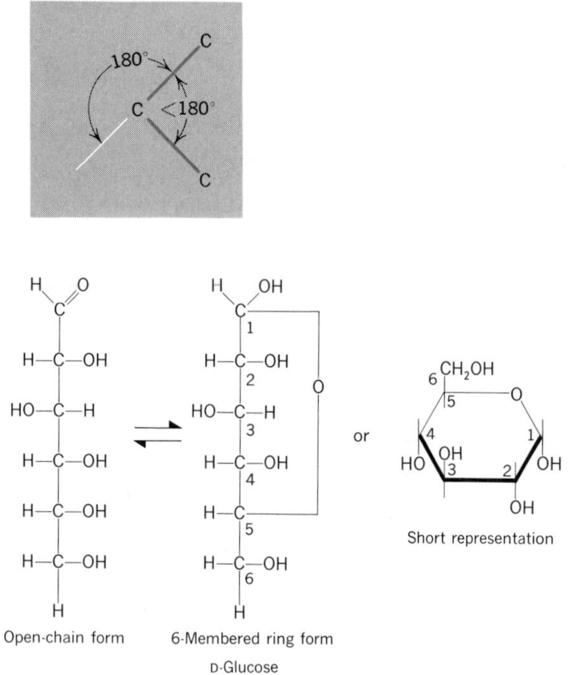

Fig. 2-13 Angles of carbon-carbon-carbon bonds. Carbon chains do not lie on a single plane because the C-C-C bond forms an angle of less than 180°. As a result, the stick-figure representation of sugars is inaccurate. Most references to sugars show the ring structure.

There are also four ketose epimers, of which **fructose** is the only one of major biological importance.

Although the heptoses (seven-carbon monosaccharides) exist, they are of relatively little importance to this discussion.

The structure of these monosaccharides has been represented in what is called the **open-chain form.** This is because it is easier to see the relationship of the bonded atoms to one another when the molecule is represented in this way. However, pentoses and hexoses occur naturally in a ring form. The molecule does *not* consist of carbon atoms bonded at sites 180° from one another. Indeed, each carbon atom forms an angle of *less* than 180° with those carbon atoms on either side of it (see Fig. 2-13). This causes the molecule to turn on itself, and the side groups—the hydroxyls and hydrogens—become physically close to one another. By a minor rearrangement requiring little input of energy (an amount that is readily available in terms of the system's heat content), a ring is formed. There are two possible types of ring formation in a hexose sugar. One is the five-membered ring in which the terminal carbon atom remains outside the ring as a side group; it is called the **pyranose** form of a monosaccharide. The second is the four-membered ring, in which

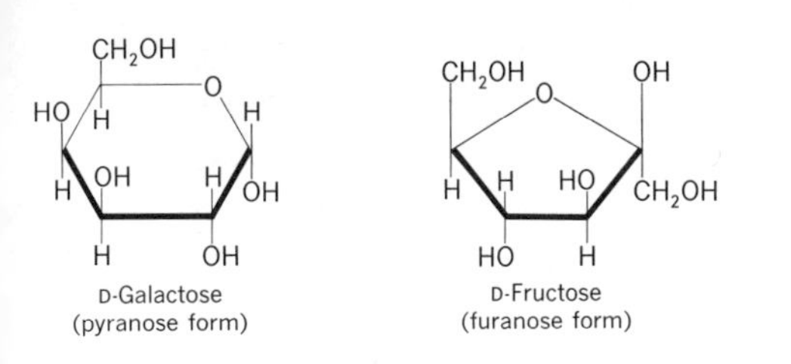

Fig. 2-14 Pyranose, the five-membered, and furanose, the four-membered ring.

the first and sixth carbon atoms remain outside the ring; it is the **furanose** form of a hexose. Fructose regularly forms a furanose ring, whereas glucose and galactose usually form pyranose rings (see Fig. 2-14).

These monosaccharides are also found bonded together in pairs to form **disaccharides.** Both the particular monosaccharides involved in the pairing and the type of bonding they undergo in the formation of this dimer determine the properties of the resulting disaccharide. For example, **cellubiose** and **maltose** are both disaccharides made up of two molecules of glucose. Both involve the bonding of the first carbon atom of a glucose molecule to the fourth carbon atom of the succeeding glucose molecule. The molecules, however, are oriented in different ways in this bonding process (see Fig. 2-15), and this difference in orientation is sufficient to make the properties of maltose and cellubiose very different. The disaccharide **sucrose** (table sugar) is composed of one molecule of glucose and one of fructose (see Fig. 2-16). The bond between the two (a 1-5 linkage—the first carbon atom of the glucose bonded to the fifth carbon atom of the fructose) is a relatively strong one, because much energy must be expended in order to bring about their dissociation into monomeric form. The bond between the two glucose monomers of maltose is relatively weak, and dissociation into the monosaccharides will occur with an input of only a moderate amount of energy, such as prolonged boiling. Disaccharides are unusual forms in the sense that they are not common to all biological material. Certain biological forms express their special characteristics in the formation of these disaccharides. For example, sugar cane contains large quantities of sucrose but no maltose, whereas barley contains large quantities of maltose but no sucrose. Neither contain any **lactose** (a glucose-galactose disaccharide), which is a primary constituent of milk (see Fig. 2-17).

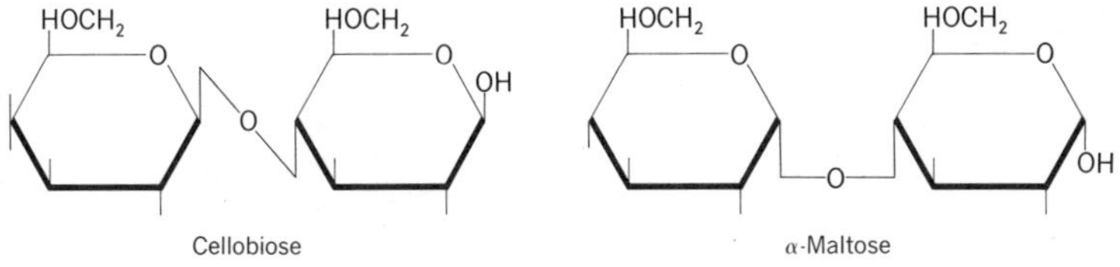

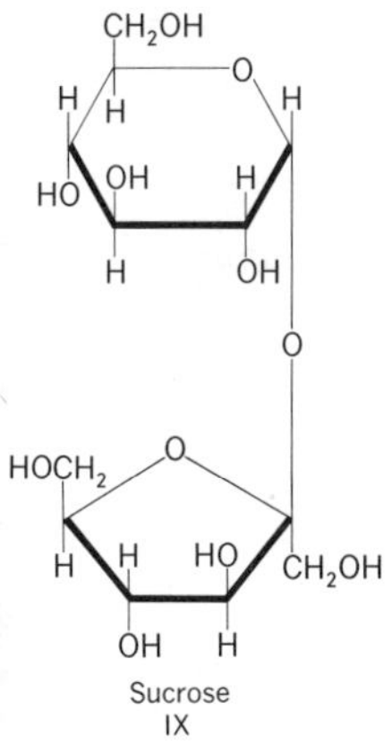

The more usual way in which carbohydrate material is found in biological systems is in the form of polymers composed of long chains of glucose molecules. In plant material, these polymers can take the form of long straight chains of **starch** molecules, in which the first carbon atom of one glucose is joined to the fourth carbon atom of the next (the 1-4 linkage or maltose linkage). Cellulose, another plant product, has the same linkage, but the individual glucose molecules are oriented in a different way (the cellubiose linkage). **Glycogen,** the animal counterpart of starch, is made up of glucose molecules joined together in 1-2, 1-4, and 1-6 linkages, forming a large molecule with many branches (see

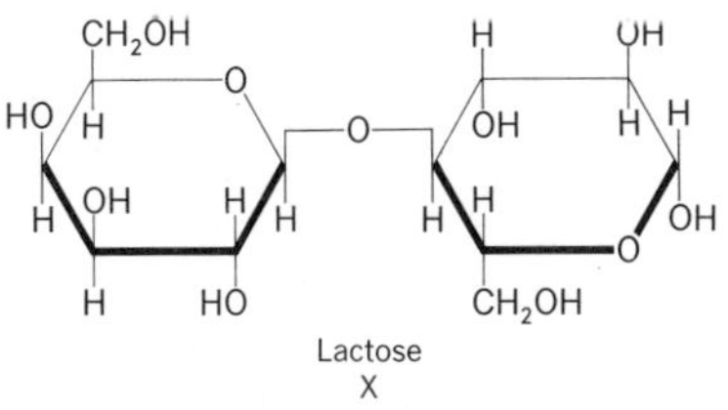

Cellobiose α-Maltose

Fig. 2-15 Cellobiose and maltose differ only in the carbons involved in bonding two glucose molecules together.

Sucrose
IX

Fig. 2-16 Sucrose is a disaccharide composed of glucose and fructose.

Lactose
X

Fig. 2-17 Lactose is composed of one molecule of glucose and one molecule of galactose in bond association.

Fig. 2-18 Different storage products of carbohydrates. Carbohydrate storage in the liver and muscle of animals is usually in the form of glycogen. Glycogen, as represented, is a highly branched molecule of glucose units.

$R{-}C(NH_2)(H){-}C{-}OH$

Fig. 2-19 The generalized amino acid. Amino acids are characterized by an amine group and a carboxyl or acid group.

A. Monoamino-monocarboxylic
- Glycine (Gly)
- Alanine (Ala)
- Valine (Val)
- Leucine (Leu)
- Isoleucine (Ileu)

B. Monoamino-dicarboxylic
- Glutamic (Glu)
- Aspartic (Asp)

C. Diamino-monocarboxylic
- Arginine (Arg)
- Lysine (Lys)
- Hydroxylysine (Hlys)

D. Hydroxyl-containing
- Threonine (Thr)
- Serine (Ser)

E. Sulfur-containing
- Cystine (Cys or Cy-S)
- Methionine (Met)

F. Aromatic
- Phenylalanine (Phe)
- Tyrosine (Tyr)

G. Heterocyclic
- Tryptophan (Tryp)
- Proline (Pro)
- Hydroxyproline (Hpro)
- Histidine (His)

H. Hydroxyl-containing
- Threonine (Thr)
- Serine (Ser)

A. Monoamino-monocarboxylic:

Glycine

$CH_2{-}NH_3^+$, COO^-

B. Monoamino-dicarboxylic:

D-Glutamic

$COOH{-}CH_2{-}CH_2{-}CH(NH_3^+){-}COO^-$

C. Diamino-monocarboxylic:

D-Arginine

$HN{=}C(NH_2){-}NH{-}CH_2{-}CH_2{-}CH_2{-}CH(NH_3^+){-}COO^-$

D. Sulfur-containing:

Cysteine

$HS{-}CH_2{-}CH(NH_3^+){-}COO^-$

Methionine

$CH_3{-}S{-}CH_2{-}CH_2{-}CH(NH_3^+){-}COO^-$

E. Aromatic:

Phenylalanine

$C_6H_5{-}CH_2{-}CH(NH_3^+){-}COO^-$ (ring: HC, CH, HC, CH, CH, C)

F. Heterocyclic:

Tryptophan

$C_8H_6N{-}CH_2{-}CH(NH_3^+){-}COO^-$ (indole ring: CH, HC, HC, CH, C, C, C, CH, NH)

Fig. 2-20 Classification of amino acids. Amino acids are classified by the chemical nature of their side groups.

Fig. 2-18). The bonds that form the glycogen will readily break down at room temperature, leaving a solution of glucose rather than of glycogen. Because of the small amount of energy that breaks the bonds, the glycogen is considered to be held together by weaker bonds than is the starch. Starch is fairly stable at room temperature. Both can be broken down by prolonged heating; this was the form in which monosaccharides were first found in biological material.

Proteins, Polypeptides, and Amino Acids

Another type of monomer found in biological material is the **amino acid.** All amino acids have the same structure on their first two carbon atoms. The remainder of the molecule, the **side chain,** differs considerably among the various amino acids. The first two carbon atoms are characterized by the first one bearing a carboxyl group and the second bearing an amino group (see Fig. 2-19).

Some twenty amino acids are found in biological material (see Fig. 2-20). The majority of them are associated with one another in linear chains as a **protein,** although some turn up in a transitional state (attached to another type of molecule) and in free monomeric form. Because of the preponderance of polymerized amino acids, those that are brought into solution as the result of treatment in the laboratory are called amino-acid **residues,** indicating that they are the breakdown products of the polymer.

Amino acids do not exhibit the regularity of form progression common to carbohydrates. They are categorized according to the structure and resulting chemical nature of their side chains. The association of amino acids with one another is through the bonding of the carboxyl group of one acid with the amino group of another acid. Unlike the polymer formed by monosaccharides, in which only one type of monosaccharide (glucose) undergoes bonding (thereby making a long chain molecule comprised only of glucose molecules), any one type of amino acid may join with any other type in the formation of the polymer. Such a polymer may take the form

alanine-glycine-serine-phenylalanine-glycine-methionine-R

A polymer containing up to 100 amino-acid units in the linear chain is considered a polypeptide, while 2 or more polypeptides are considered a protein.

The joining of two amino acids proceeds in the following way: Each amino acid is bonded by a weak covalent bond from its side chain to a large complex molecule that is commonly represented by the letters ***t*****RNA,** and which, for the present, we will term a "combining molecule." When two amino acids are in close proximity to each other, the reaction that leads to their bonding occurs. The bond between two amino acids is called the **peptide bond,** and the association is by a **peptide linkage** (see Fig. 2-21). Two amino acids bonded together in this way are called a **dipeptide.** Notice that the reaction resulting in the formation of a dipeptide also produces OH^- and H^+. In a buffered biological system, these atoms are immediately neutralized to form water. What would

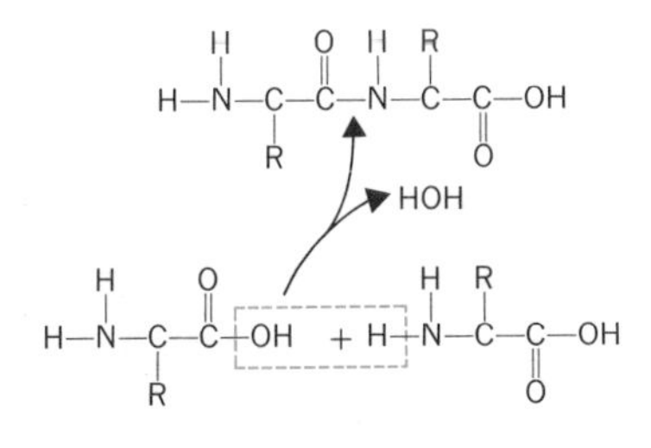

Fig. 2-21 The formation of the peptide bond.

happen, though, if the buffering system were overcome by the addition of large quantities of acid in a laboratory situation? This circumstance would upset the equilibrium and rather than going into complex with *t*RNA, the amino acids would directly accept the H^+ and OH^- groups, thereby becoming amino-acid residues. Thus, polypeptides and proteins can be broken down in the laboratory by treatment with acid (acid hydrolysis) to amino-acid residues.

Heteroconjugates

In addition to the formation of polymers, single monomers of different types may join together to form **heteroconjugates.** The presence of these heteroconjugates was discovered after the elucidation of the monomers and polymers of biological systems, because the techniques involved in their extraction from biological material as intact molecules is very complex. Unlike methods involved in the study of monomers, methods for extracting intact polymers or heteroconjugates require great chemical care to isolate the material without disrupting the bonds that give it a particular character. Therefore, the isolation of heteroconjugates had to await the development of sophisticated physical, rather than chemical methods, before a thorough study of these entities could be undertaken.

The most important heteroconjugates involve the bonding of a **purine** or **pyrimidine** to other monomeric forms. Purines and pyrimidines are called **nitrogen bases** because their rings contain nitrogen atoms

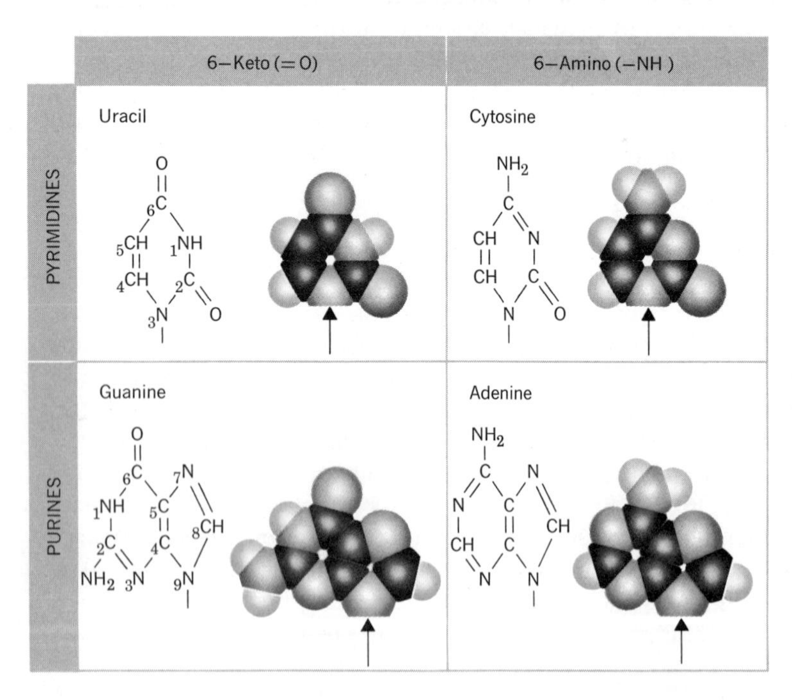

Fig. 2-22 Structure of purines and pyrimidines.

which, by virtue of their position on the molecule, act as proton acceptors. Thus purines and pyrimidines manifest properties associated with a **base.**

Three types of pyrimidines are commonly found in nature: **uracil** (which is generally abbreviated U), **thymine** (T), and **cytosine** (C). These differ from one another solely on the basis of the side groups attached to their ring structures (see Fig. 2-22). Two types of purines are biologically important: **adenine** (A) and **guanine** (G). These also differ from one another by their side groups.

Also present in biological material are the nucleotide forms with two phosphates (**nucleotide diphosphates**) and three phosphates (**nucleotide triphosphates**), the most common of which is **adenosine triphosphate** (ATP). The presence of three phosphate groups on a nucleotide leads to some interesting and important properties of the molecule.

An inorganic phosphate group (one not attached to an organic molecule) does not lend itself to a static description of its electronic configurations. There are, in fact, several states in which a phosphate group may be found, none of which is stable.

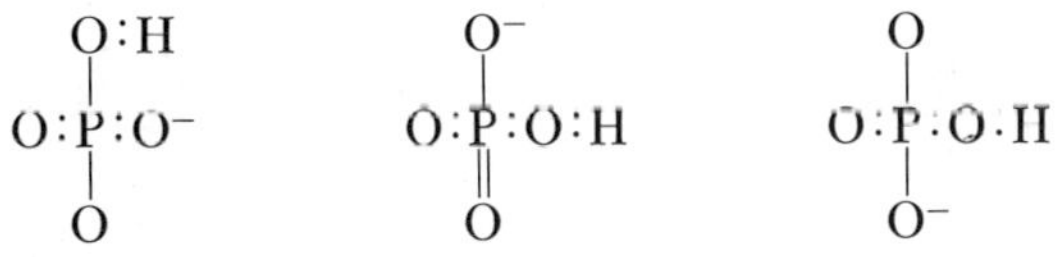

```
  O:H           O⁻            O
  |             |             |
O:P:O⁻        O:P:O:H       O:P:O:H
  |             ‖             |
  O             O             O⁻
```

Therefore, the electronic configuration of a phosphate group is in a state of constant flux, motion, and energy expression. Such a structure is termed a **resonating** form. The potential energy of the resonating form is lower than the energy of a stable form, because of the resonance itself. The terminal phosphate group of an ATP molecule exhibits less resonance than does inorganic phosphate. This seems to be due to the **opposing resonance** of the neighboring phosphate group to which it is attached. This reduction in kinetic energy and subsequent increase in potential energy requires a great energy expense on the part of the ATP molecule. Because of this reduction in resonance, and the energy contained within the ATP molecule that makes this reduction possible, ATP is considered a **high-energy molecule.** As such, it is essential in the functioning of biological systems.

The polymers formed by heteroconjugates are the **nucleic acids.** The structure of the nucleic acids is intimately related to their function, and any discussion of them at this time would be premature.

Two other important heteroconjugates that are never polymerized are those formed by the attachment of adenosine to a flavin to form **flavin adenine dinucleotide** (**FAD**), and the attachment of adenosine to a molecule of nicotinamide to form **nicotinamide adenine dinucleotide** (**NAD**) (see Fig. 2-23). The importance of these will become apparent in later discussions; now it is of interest to know that they are heteroconjugates.

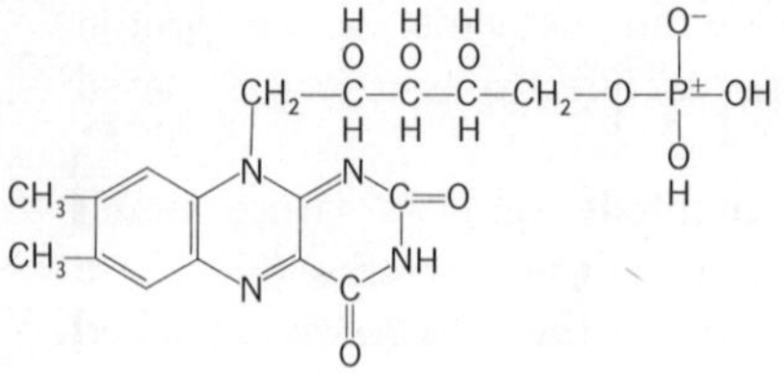

Flavin mononucleotide (FMN)
Riboflavin monophosphate

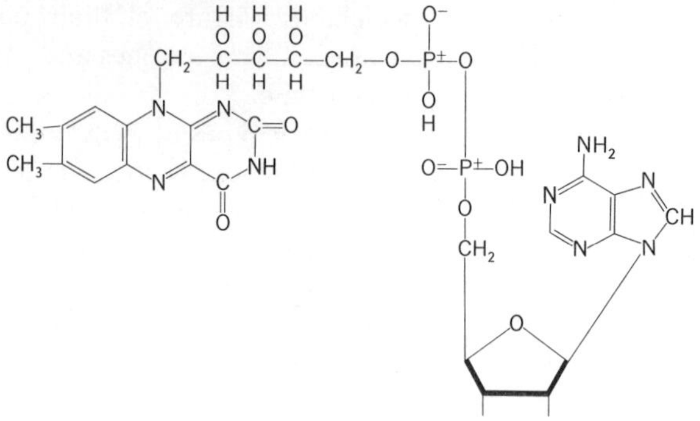

Flavin adenine dinucleotide (FAD)

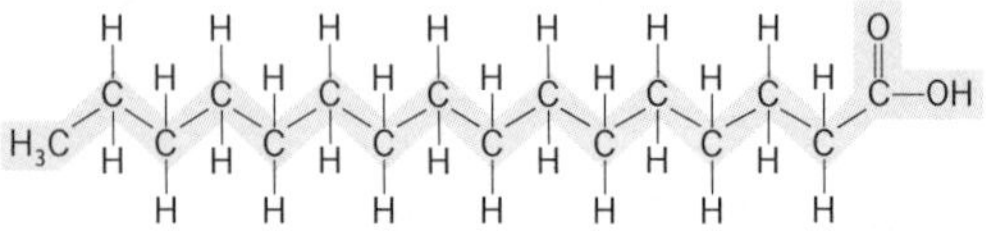

Fig. 2-23 **Structure of FMN and FAD.**

Palmitic acid
II

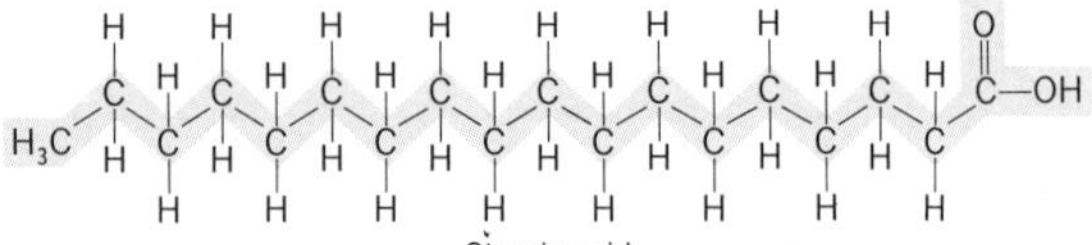

Stearic acid
III

Fig. 2-24 **Saturated fatty acids.**

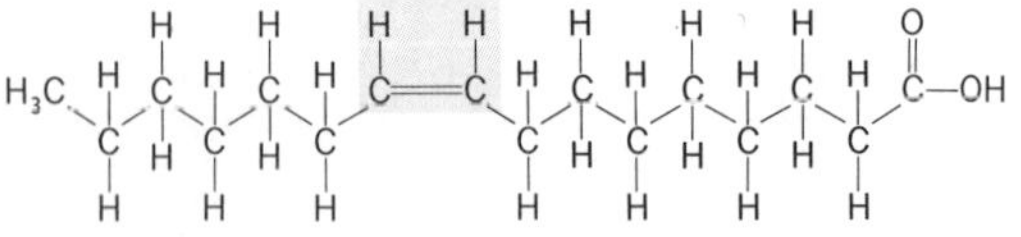

Palmitoleic acid

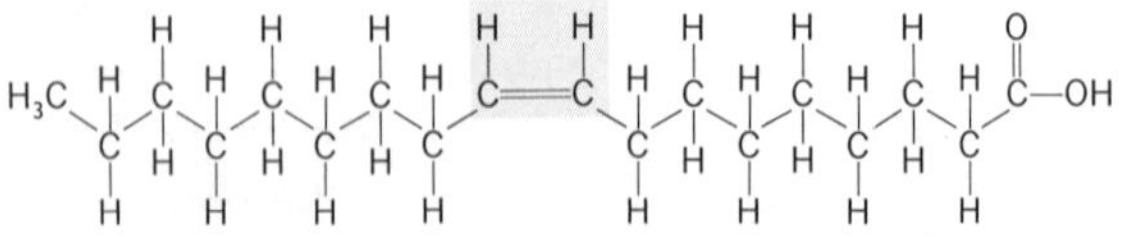

Oleic acid (*cis*)

Fig. 2-25 **Unsaturated fatty acids.**

NONMONOMERS AND NONPOLYMERS

The system of classifying biologically active molecules into monomers, polymers, and heteroconjugates makes it possible to view biological chemistry in a meaningful perspective. This will be seen more clearly in later discussions. There are, however, several molecular forms that defy categorization in this manner. In some cases, these molecular forms are much larger than monomers are considered to be. In other cases, these molecules are within the size range of monomers, but are never found polymerized.

Lipids

When the solution of fat is heated, the products are found to be **glycerol** and **fatty acids,** always in a ratio of 1:3. These are the component parts of the fat molecule. The fatty acids have the general form $CH_3(CH_2)_n$:COOH, and are long chains of hydrogenated carbon atoms with a methyl group (CH_3) at one end and a carboxyl group at the other. Two types of fatty acids occur naturally in biological material. The **saturated** fatty acids are those in which all the carbon-carbon bonds are single and covalent; as a consequence, there are two hydrogen atoms associated with each carbon atom. The most abundant of these saturated fatty acids found in animals are palmitic acid and stearic acid. Palmitic acid is made up of sixteen carbons, while stearic acid has eighteen (see Fig. 2-24). The **unsaturated fatty acids** are those in which some of the carbon-carbon bondings are formed by double covalent bonds. The unsaturated fatty acids are generally found in plant material, although there are two monounsaturated fatty acids (those with only one double bond in the structure) that are common to animal material; these are **oleic acid** and **palmitoleic acid** (see Fig. 2-25).

By means of their attachment to glycerol, fatty-acid monomers are joined to form the fat. Glycerol, a three-carbon alcohol, has the chemical form

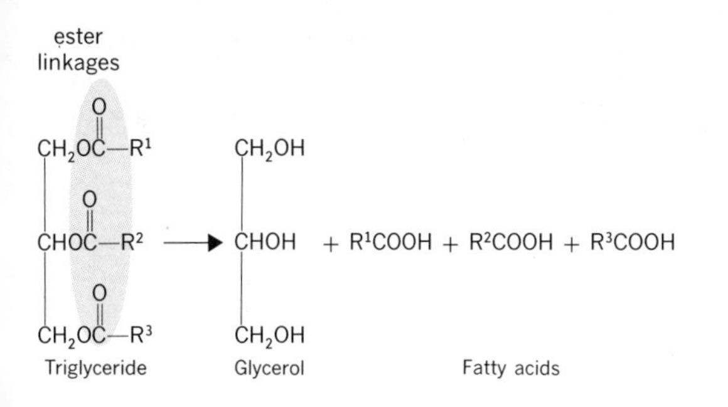

Fig. 2-26 Formation of a fat from three fatty acids and a molecule of glycerol.

CH_2OH
$CHOH$
CH_2OH

Each fatty acid attaches to a carbon atom of glycerol by means of an ester linkage. Thus, three fatty acids and a molecule of glycerol form a molecule of fat. The ester linkage provides a rather weak bond that is broken easily by heating (see Fig. 2-26).

The side chains of the fat, which are the fatty acids, are nonreactive, and therefore, unlike proteins, do not undergo reactions into secondary structures. Thus, the presence of fat molecules in biological material does not determine specific structure, and material that contains a great deal of fat tends to take on the form of an amorphous blob rather than to exhibit defined characteristics.

One important group of fats that warrants mention is the phospholipids. In a phospholipid, two of the three carbons of the glycerol are bonded to long-chain fatty acids. The third glycerol carbon is bonded to a phosphorylated organic molecule. The particular phosphorylated organic molecule determines many of the properties, and is used for the name of the phospholipid.

Fig. 2-27 Structure of a porphyrin.

Porphyrins

The **porphyrins** (see Fig. 2-27) are a group of large, complex carbon-nitrogen molecules that usually surround a metal. The porphyrin chlorophyll surrounds a magnesium ion; the porphyrin in hemoglobin surrounds an iron ion.

Fig. 2-28 Structure of a sterol (cholesterol).

Sterols

Another group of molecules that defies categorization as monomers and polymers is the sterols. Sterols are ring compounds (see Fig. 2-28) that are fat-soluble, and are never found in polymerized form. Many of the important biological regulators are sterols.

THE PROPERTIES OF WATER

The properties exhibited by water are of such importance that scientists believe that life could not exist on this or any other planet without a solvent displaying like properties.

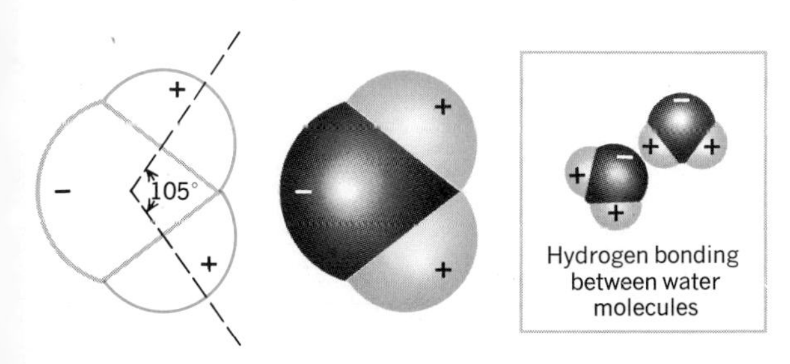

Fig. 2-29 Structure of water. Notice the dipolarity of the water molecule.

The water molecule is in **dipole** form (see Fig. 2-29). The bond orientations of the orbitals of the oxygen atoms lead to definite positive and negative sides of the molecule. Although the bonding of the hydrogen atoms to the oxygen atom is generally considered covalent in nature, in terms of bond energies, it is actually between covalent and ionic. Because of the ionic character of the bond, water will undergo dissociation into H^+ and OH^- (see p. 44).

The dipole character of the water molecule accounts for the **hydrophilic** (hydro = water, philic = loving) property of salts, and permits their dissolution in water. When a salt such as NaCl is placed in water, it will undergo almost complete dissociation into Na^+ and Cl^- ions. These ions will orient themselves according to the poles of the water molecule:

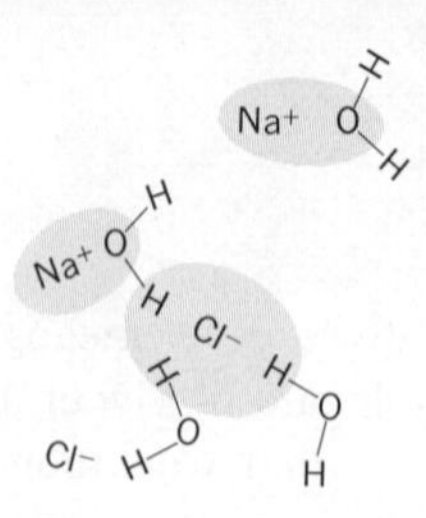

Fig. 2-30 **Salts in solution. Negative chloride ions are positioned at the positive pole of the water molecule, while positive sodium ions are positioned at the negative pole of the water molecule.**

$$CH_2{-}CHO{\cdots}H{\cdots}O\begin{matrix}H\\ H\end{matrix}$$

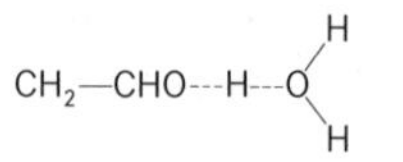

Fig. 2-31 **Alcohol in solution. The oxygens of the alcohol and water molecules share hydrogens through hydrogen bonding.**

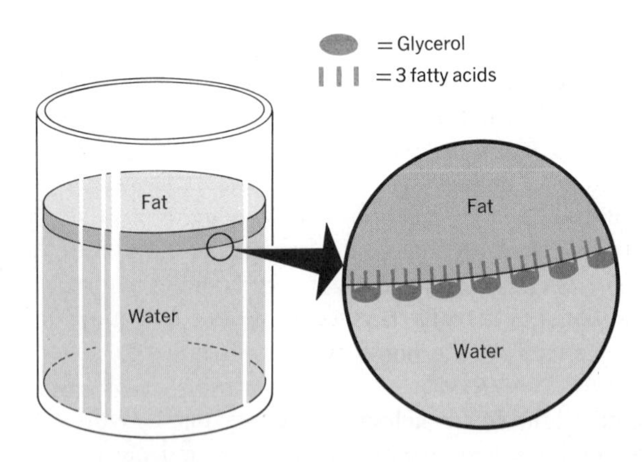

Fig. 2-32 **Position of fat molecules in a water suspension. Water-soluble glycerol is dissolved in the water, while the hydrophobic fatty acids are positioned above the water surface.**

the Na^+ toward the negative pole, the Cl^- toward the positive pole. The solubility of salts in water is directly related to their dissociation abilities. The greater the dissociation, the greater will be the solubility (see Fig. 2-30).

The solubility of organic compounds in water is related to the tenuous bonding of the hydrogen to the oxygen atoms of water. When alcohols are placed in water solutions, they tend to share hydrogen atoms with the oxygen atoms of the water molecule, thereby going into solution. This sharing of hydrogen is called **hydrogen bonding** (see Fig. 2-31).

Lipids are not soluble in water, and so are termed **hydrophobic** (phobic = hate). Their hydrophobic tendency is thought to occur because the strength of the C:H (hydrocarbon) bond does not permit hydrogen bonding, and thus they tend to stay away from water. The glycerol part of a fat, however, is soluble in water, and generally mixtures of lipids and water will have the fatty-acid portion of the molecule at the surface, and the glycerol portion of the molecule dissolved in the water (see Fig. 2-32).

Thus it is the dipole character and the bonding tendencies of the water molecule that determine the solubility of a substance, for the dissolving of a material in water is dependent on the ability of the material to undergo physical or chemical reactions, or both, with the water molecule itself.

PERSPECTIVES OF CHEMICAL BACKGROUND FOR BIOLOGY

The adoption and use of values and information of chemistry and physics has enabled biologists to view both organisms and the biological world in a new and more meaningful way. The chemical behavior of atoms, radicals, and molecules through their interactions can account for the structure and physiology of a biological system. The mammoth task of discerning the particular reactions or sets of reactions responsible for specific structure and physiology, and thus relating the chemistry to a biological perspective of organisms, has been underway for a century. This task has progressed far enough to present a comprehensive, though incomplete, picture of the generation of life in accordance with chemical and physical laws.

QUESTIONS

1. Explain the distribution of weight and charge in the atom.
2. What is the orbital distribution of electrons in an atom that has an atomic number of 5? Of 12?

   What is the valence of each of these atoms? Which type of bonds would they tend to form?
3. Distinguish between an element, an isotope, and a radioactive isotope.
4. What effect did the concepts of kinetic and potential energy have on the understanding of relationship between energy and work?
5. Why is a chemical reaction actually between subatomic units rather than whole molecules?
6. How does the first law of thermodynamics permit a quantitative calculation of the energy relationships of a system?
7. What does the second law account for that the first law does not?
8. How can a reaction be exergonic and still have a high energy of activation?
9. Compare and contrast ionic and covalent bonding on the basis of type of association of atoms and the features of atoms that undergo each type of bonding.
10. Outline the way in which H_3PO_4 and its salt, Na_2HPO_4 would act as a buffer.
11. What would be the pH of the following: hydronium ion concentration of 10^{-4}; hydronium ion concentration of 10^{-8}; hydroxyl ion concentration of 10^{-4}; hydroxyl ion concentration of 10^{-8}.
12. Following the ground rules for covalent bonding, give at least two different molecules that would have the empirical formula $C_4H_{11}ON$.
13. Name two types of monomers and the polymers they form. What type of linkage is found between the monomers in each polymer?
14. From the ground rules for monosaccharide structure, deduce the eight aldose epimers of hexose sugars.
15. How does the resonating capacity of the phosphate radical contribute to the high energy of ATP?
16. How does the dipole character of water account for the solubility of salts in water?

CHAPTER 3

THE STRUCTURE AND FUNCTION OF PROTEINS

In the late 1800s, Emil Fischer demonstrated that a protein contained a series of peptide linkages and could be broken down by acid hydrolysis into a number of amino acids. Different proteins yielded different ratios of amino acids (see Table 3-1). It seemed, therefore, as if each type of organism had a characteristic proportion of the amino acids. It became apparent that different combinations of amino acids in protein could be the basis for the difference between organisms and, therefore, the basis for the structure and function of organisms. So the search for the source of responsibility for the translation of chemical properties into biological organization focused on a study of protein structure.

By the end of the nineteenth century, many proteins were known to biochemists. They were found to be extremely large molecules, with molecular weights (the weight of all the atomic weights in the molecule) of 4000 to 4,000,000. Although other properties of proteins were recognized, they made little sense to the scientists of that time because so little was known regarding the actual structure of these molecules. For convenience they were classified with respect to the tissue in which they were found or the activity in which they were involved.

One of the major classes of proteins known at that time was the **enzymes,** proteins capable of mediating chemical reactions. This mediation involved an increase in the rate of the reaction (the amount of product produced per unit of time) when the enzyme was present. These enzymes were known to be highly specific. A single enzyme could control the rate of only a single reaction. For example, a **proteinase** such as trypsin could break down protein molecules into large polypeptides, but had no effect whatsoever on a molecule such as lactic acid.

Another major class of proteins were the so-called structural proteins, which did not serve any known enzymatic function. These seemed to be **fibrous** (long and narrow) rather than having the **globular** (spherical) shape of enzymes.

After Emil Fischer's work, research on proteins split into three approaches. All were based on the primary assumption that since each protein was highly characteristic for the individual type of organism, the organism must synthesize its own protein structure. This implies that an organism must "know" the particular arrangement of amino acids that constitute its protein. One approach was based on the additional assumption that there would be specific reactions within the organism for the incorporation of the amino acids into protein. This assumption agreed with all the known biochemical reactions at this time. A chemical structure must result from chemical reactions, and these reactions could be repeated *in vitro* (outside the living form). This approach involved an attempt to synthesize specific proteins under laboratory conditions by putting together proteins of desired preordained amino-acid combinations. Emil Fischer had synthesized polypeptides of lengths up to eighteen amino acids, but was unable to predict the sequence in which the amino acids would be incorporated into the polymer. Attempts in this direction were not successful, and biologists gained little insight into

TABLE 3-1 AMINO ACID COMPOSITION OF SEVERAL PROTEINS[a,b]

	Type of Protein			
Amino Acid	Gelatin	Rabbit Myosin	Calf Thymus Histone	Human Serum Albumin
Alanine	9.3	6.5	6.94	—
Amide N	0.07	1.20	0.87	0.88
Arginine	8.55	7.36	17.4	6.20
Aspartic acid	6.7	8.9	5.71	8.95
Cysteine	0.0	—	0.0	0.70
Cystine/2	0.0	1.40	0.0	5.60
Glutamic acid	11.2	22.1	4.30	17.0
Glycine	26.9	1.9	5.07	1.60
Histidine	0.73	2.41	2.69	3.50
Isoleucine	1.8	15.6 (Isoleucine + Leucine)	20.5	1.70
Leucine	3.4		5.21	11.0
Lysine	4.60	11.92	10.23	12.30
Methionine	0.9	3.4	0.0	1.30
Phenylalanine	2.55	4.3	4.08	7.80
Proline	14.8	1.9	4.04	5.10
Serine	3.18	4.33	4.71	3.34
Threonine	2.2	5.1	4.80	4.60
Tryptophan	0.0	0.8	0.0	0.2
Tyrosine	1.0	3.4	3.30	4.70
Valine	3.3	2.6	3.22	7.70

[a] From Cantarow and Schepartz, *Biochemistry*, 3rd Ed., 1962.
[b] The numerical values indicate the percentage of each type of amino acid present in the protein. The data illustrate the variability of amino acid constitution of different kinds of proteins.

the way in which the biological system managed to make its own protein. Later research indicated that the second assumption, regarding specific reactions for each incorporation, involves a very complex system that is not readily repeatable under laboratory conditions.

The second approach was also an attempt to uncover the biological mechanism for protein synthesis. If the organism did synthesize its own protein structure, and if this protein was put together in a stepwise fashion (one amino acid after the other), then careful analysis should indicate the presence of partially complete proteins. The attempts to find unfinished proteins were also futile, and it was later discovered that biological systems do not synthesize proteins in this stepwise fashion. It should be noted that these two lines of experimentation and the secondary assumptions on which they were based were quite reasonable in the early part of the twentieth century. That these workers did not outguess biological systems is not to their discredit. The very fact that they could not find an answer indicated that there had to be a reevaluation of the assumptions on which the experiments were based, and it was through this reevaluation and its correlation to the data of later workers in other areas that the mystery of protein synthesis started on its way toward a solution. This will be discussed more fully in a later chapter.

The third line of experimentation involved the study of ready made proteins. In order to study the proteins that organisms conveniently made available to biochemists, it was first necessary to obtain *pure* protein solutions. In a biological system there are a number of proteins, each with a unique structure. If one is to make any sense of protein structure, it must be in regard to a single protein type rather than a conglomeration of many proteins. When a pure chemical substance is in solid form, it takes on a characteristic and unique crystalline structure.

The basic assumption that underlay this line of inquiry was that the molecules of a particular protein were of the same structure. The significance of this assumption is overwhelming when the synthesis of proteins in a biological system is considered. Whereas biochemists had been able to synthesize peptides of only two or three predicted amino-acid residues in length, the biological system commonly synthesized proteins of 300 amino-acid residues. If the regularity of synthesis is such that it produces a great quantity of a single type of protein, all with the same structure, then biological systems are phenomenal factories indeed!

In 1926, James Sumner showed that this assumption was valid by isolating and crystallizing the enzyme **urease,** which had, of course, a protein structure. Subsequently, other proteins were also purified. Later workers developed very sophisticated techniques that took advantage of some of the properties of proteins to bring about their isolation and purification. Once it became possible to work on purified preparations of proteins, the elucidation of protein structure came within the grasp of the biochemist.

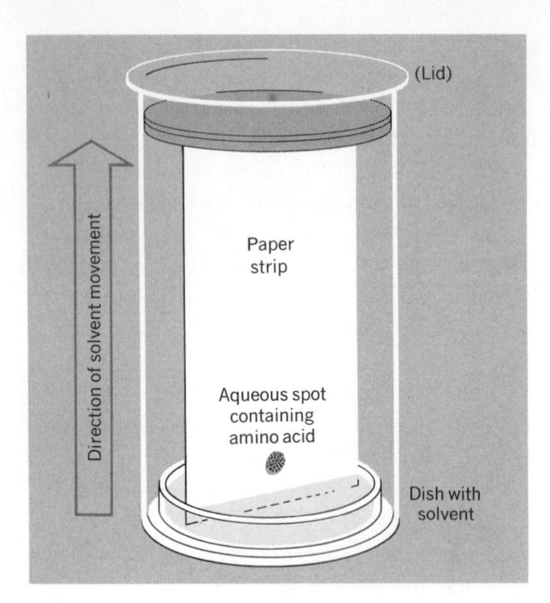

Fig. 3-1 Paper chromatography setup. Container is closed so that the paper is in an atmosphere of the solvent.

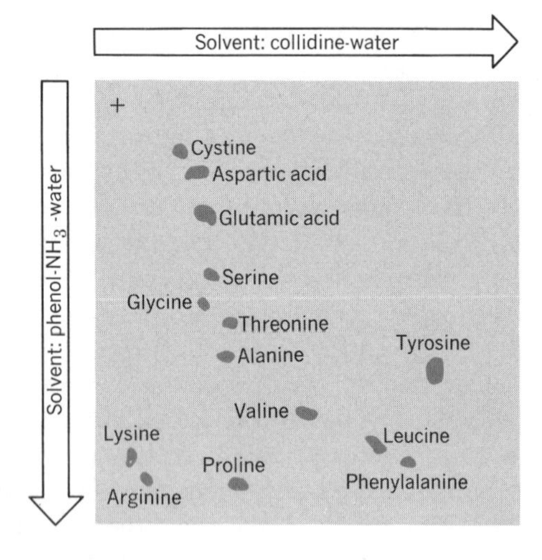

Fig. 3-2 Paper chromatograph of a mixture of amino acids. The rates of migration of the individual amino acids as well as their positional relationship to one another is highly characteristic, making it relatively easy to identify a specific amino acid by comparing an experimental result with the standard.

THE STRUCTURE OF PROTEINS

Primary Structure

The specific sequence of amino acids within the molecule provides the very basis for protein structure, and is considered the **primary structure** of proteins. In the 1940s, Sanger elucidated the primary structure of the protein **insulin.** His methods were somewhat complex, and the determination of the amino-acid sequence of this relatively simple protein, which is only 104 amino-acid residues long, took ten years. The insulin molecule, like all proteins, is characterized by a free COO^- group on one end and a free NH_2 group on the other (see p. 57). The first step in the procedure was to add the reagent 2,4-dinitrofluorobenzene (**DNFB**), which is yellow and combines with a free NH_2 radical. Then, by applying enzymes that break the peptide linkage between two specific amino acids, he would fragment the protein. He then separated the fragments, colored and noncolored, by the method of paper chromatography. Chromatography utilizes the property of molecular substances to combine physically (by electrical charge, etc.) with inert material. Minute amounts of an amino-acid mixture are placed at a single point near the edge of a strip of paper. After this has dried, the edge of the paper is placed in a compatible solvent, which will migrate up the paper (see Fig. 3-1). When it has reached the dried amino-acid mixture, the amino acids will dissolve in the solvent according to their individual affinity for the solvent. The amino acids will be drawn up the paper at different rates, which depend on the affinity of the individual amino acid for the solvent and for the paper, and on its molecular weight. This method of separating amino-acid residues has become so standardized that the position of a particular spot is universally recognized as an indication of the presence of a particular amino acid (see Fig. 3-2).

Sanger first separated the mixture of protein fragments into the colored and noncolored segments. The colored fragments were easily seen on the chromatograph, and once that part of the paper was cut out, the remainder of the material could be washed from the paper and brought back into solution (**re-eluted**). The procedure was then repeated on the fragmented material. The tagged amino acids (those attached to DNFB) were once again subjected to paper chromatography. It was a long and tedious procedure, because all the molecules did not fragment in the same way. Sanger was confronted with a situation rather like having hundreds of jigsaw puzzles, each with 104 pieces, all in the same box and in various states of completion. After many experiments of matching up the tagged amino acids with various fragments and subfragments, a consistent scheme emerged, which could only be interpreted as a highly ordered sequence of residues, beginning from the NH_2 terminus at one end and proceeding to the COO^- terminus at the other. Thus Sanger showed that individual molecules of insulin are linear (one-dimensional) sequences of amino acids, all the same length, rather than a branching structure that would account more readily for the three-dimensional configuration of proteins.

The major breakthrough of Sanger's research was received with enthusiasm. A number of other proteins were analyzed, with a consequent refinement of techniques. It is now possible to formulate certain general ideas about proteins with respect to their linear sequence of amino acids.

1. A protein is principally a one-dimensional structure made up of amino acids sequentially linked together by peptide bonds.
2. The order of the amino acids within the molecule is of prime importance in characterizing the particular protein. A new criterion for the purity of a protein is that all molecules in solution must have an identical sequence of amino acids. The properties regarding the specificity of protein are considered to be consequences of this primary structure.

It is interesting to note that the proteins with different activities have widely different primary structures, whereas those having similar roles, such as hemoglobin from different species of animals, have similar sequences, usually varying by only a few residues in a particular region (see Fig. 3-3).

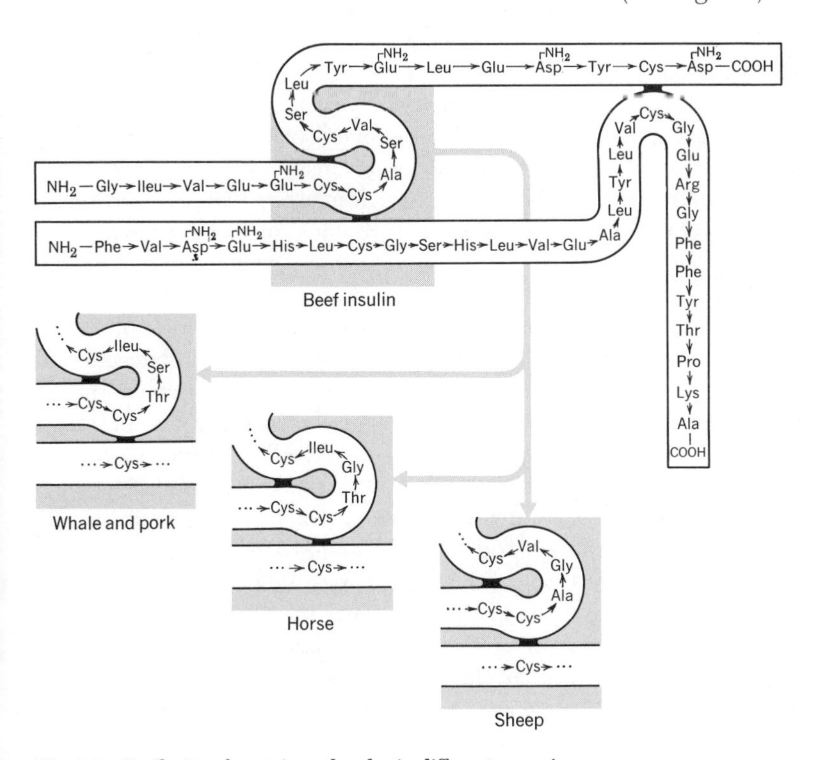

Fig. 3-3 Similarity of protein molecules in different organisms.

Secondary Structure

Since many proteins are globular (spherical) in conformation, there seems to be a discrepancy between the physical form and the one-dimensional linear structure of polypeptides. It seems most likely, and much evidence supports the notion, that the protein is somehow folded into a three-dimensional configuration. This problem was analyzed by

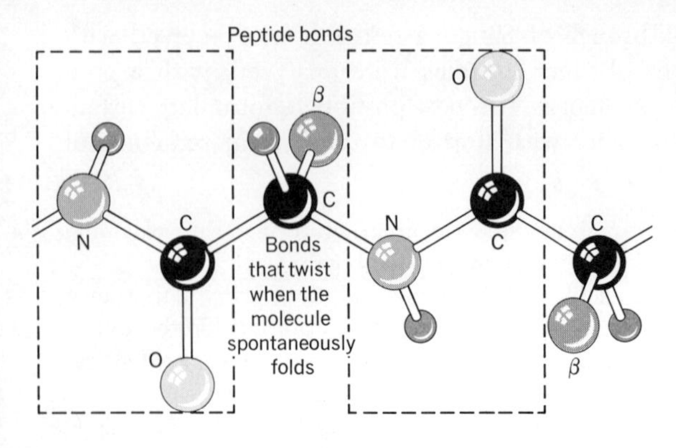

Fig. 3-4 Molecular model of a peptide bond. The peptide bonds twist when the molecule spontaneously folds.

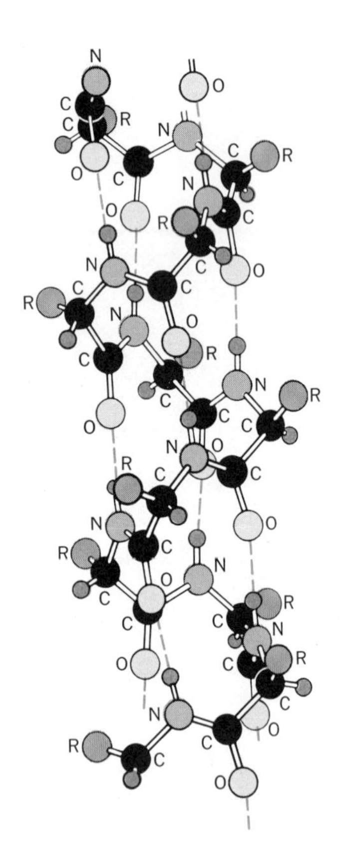

Fig. 3-5 The twisting of the peptide bonds create the helical structure of the protein.

Linus Pauling and his co-workers (1951) on both experimental and theoretical grounds. Purified proteins, or actually, small fragments of proteins, were crystallized and subjected to X-ray crystallography. This method involves bombarding the crystals with X-rays from different angles, and absorbing the diffracted elementary particles on a photographic plate. The deflected X-rays generate various geometric patterns. Each type of pattern indicates a particular molecular configuration, and since some translation of the abstract geometry into molecular shape is necessary, a great deal of interpretation is required.

The crystallography pictures indicated that the carbon and nitrogen atoms involved in the peptide linkage were on the same plane, whereas the carbon atoms adjacent to these were on a plane opposite to and diagonal from the plane of the linkage (see Fig. 3-4). This alteration in plane was due to the angle of the covalent bond, and caused a "twist" to the molecule. This twist generated a full circle (360°) every four amino acids, thereby creating a helical shape for the polypeptide. The twist of the polypeptide brings the hydrogen atom associated with the nitrogen of a peptide linkage into close proximity with the oxygen atom associated with the carbon of the peptide linkage four amino acids away. This proximity brings about a sharing of the hydrogen (hydrogen bonding—see p. 62) between the nitrogen and the oxygen, thereby stabilizing the helical shape (see Fig. 3-5). The discovery of the structure of the **α-helix** of proteins is considered a major advance in protein chemistry, and serves to explain the fibrous structure of many proteins. It does not, however, account for the spherical or globular protein structure.

Tertiary Structure

Pauling did not utilize the **R** groups, those various side radicals that account for the differences between amino acids. If the uniqueness of the sequence of amino acids distinguishes one polypeptide from another, then it is these side groups that account for the specificity of protein activity. Pauling's model would be the same for all proteins, the same for insulin or pure polyglycine (a polypeptide made up of only glycine monomers). So how is the three-dimensional structure of proteins generated? In solution, a polypeptide would manifest kinetic energy, wriggling about and assuming many different shapes. When an amide group and an oxygen become proximal, the helical shape is generated and stabilized by the formation of the hydrogen bond. As the secondary structure is generated, some of the side groups come in close proximity to one another, and, if these are potentially reactive with one another, weak associations will be formed, leading to the **tertiary structure.** It is the tertiary structure of proteins that accounts for the three-dimensional aspect of protein structure.

Weak associations are bonds extending over a range of 1 to 10 kcal/mol. They are sufficiently above the level of the kinetic energy of molecules at room temperature (0.6 kcal/mol) to remain relatively stable, but are much weaker than a typical covalent bond (50 to 100 kcal/mol). Many weak associations are only temporary. Several different types can be formed.

1. *Hydrogen Bonding:* If the two proximal side groups can potentially undergo hydrogen bonding, they will form this type of weak association. Hydrogen bonding can occur between two oxygens, between an oxygen and a nitrogen, or between two nitrogens. For instance:

   amino acid A *amino acid B*

   R:O·····H·······O:R
   R:O·····H·······N:R
   R:N·····H·······N:R

2. *Van der Waals forces:* Another form of secondary association is the van der Waals forces. A van der Waals force is the tendency of *any* two atoms to attract one another over a short range. The atoms will move toward one another until a point is reached where the similarity of atomic charges will cause a repulsion. This point of repulsion is considered to be where the radii of their electron orbitals overlap (the sum of their atomic radii). There is a point, however, where the attractive and repulsive forces are equal (the sum of their atomic radii), and the atoms will tend to maintain this physical relationship with one another. When several atoms of a molecule exert a van der Waals force with another radical, the force is greater. These forces are relatively small, about 1-2 kcal/mol, but cumulatively they may be as great as 10. If van der Waals forces are to generate any stability, it is necessary that the attracting groups have a complementary structure that can operate in a lock-and-key fashion (see Fig. 3-6). Hence, any radicals, regardless of chemical nature, can manifest stable associations with one another.

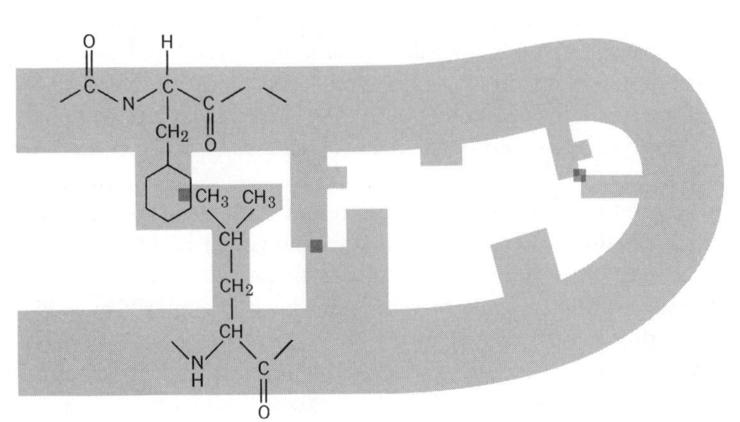

Fig. 3-6 Any two side chains, regardless of their chemical and physical properties, can be held in place by the attraction of an atomic nucleus of one molecule for the electrons of another atom of another molecule, a phenomenon known as van der Waals force. Because of the repulsive forces of the electrons of each atom for one another, the atoms position themselves at the sum of their atomic radii.

3. *Hydrophobic Bonding:* Another form of weak bond is the hydrophobic bond. This is actually not a bond but a repulsive force between a molecule that can form hydrogen bonds with water (serine) and one that cannot (phenylalanine). These **nonpolar** groups (those that cannot form hydrogen bondings) form an association with one another through van der Waals forces (see Fig. 3-6).
4. *Ionic Bonds:* Some of the amino-acid side chains manifest acid properties (aspartic acid, glutamic acid), while others manifest the properties of a base (lysine, arginine). When these groups are proximal to one another, they undergo a neutralization reaction, leading to the formation of an ionic bond between them. These bonds have greater energy than the previously discussed associations, and therefore tend to be more stable.

5. *Disulfide Bridge:* When two cysteines are closely associated as the result of the bending of a polypeptide, the terminal sulfhydryl groups (:SH) of each molecule can undergo a reaction with one another, forming a disulfide bridge (R:S:S:R). This generates a stable *covalent* bond (see Fig. 3-7).

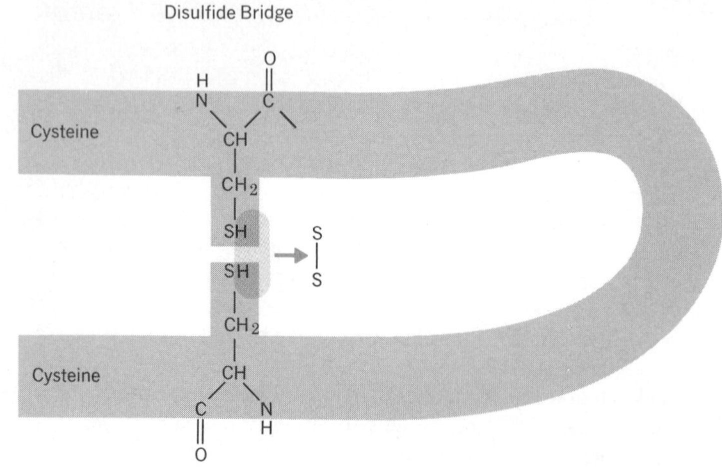

Fig. 3-7 When a peptide chain is folded in such a way that two cysteine molecules are proximal, a bond is formed between the terminal sulfur atoms. This bond is called a disulfide bridge.

It is presently visualized that the polypeptide will move through a number of changes, tending toward (as a statistical expression) the most likely (most stable) state, because it is thermodynamically that of the lowest energy content. It is also possible to envisage that the very sequence of amino acids affords some directed pathway to this final state.

The stability of the protein molecule is the result of the cumulative weak associations. One or a few weak associations, especially of only the van der Waals type, would not permit the specificity and stability required of the protein molecule. A number of different associations cumulatively, however, could account for a reasonably static form.

Experimental evidence tends to support the theories of secondary and tertiary structure. X-ray diffraction analysis of some proteins, namely **myoglobin** (found in muscle) and hemoglobin, reveal a highly twisted structure, having some regions with sausage shapes 20 Å in diameter (as predicted by Pauling), while other regions have irregularly twisted shapes, suggesting a number of tertiary bondings.

It had once been thought that this folding required enzymatic control, but two lines of reasoning tend to discredit this idea. First, some proteins can be **denatured** (unfolded) by high temperatures (60–100° C) or by treatment with 8*M* urea, but they can spontaneously fold back into their normal configuration when the urea is removed and resume their activity without introducing any enzyme. Second, if every protein requires at least one enzyme for its folding, then those enzymes, being proteins, would require other enzymes to fold them. The argument can extend backward indefinitely.

Quaternary Structure

An entire protein is not always one sequence of amino acids, but may be an aggregate of several peptide chains held together in covalent or ionic association. Insulin consists of two different peptide chains;

ribonuclease consists of one; and hemoglobin consists of four peptides that occur in pairs and are denoted as the α chain and β chain. It is, in fact, the association of similar or different polypeptide chains having a maximum molecular weight of about 60,000 (or about 300 amino acid residues each) that seems to account for the large size of many proteins. The associations of these polypeptides in the formation of the final protein are considered the **quaternary structure** of the protein. As with the folding process, the quaternary structure is thought to be a spontaneous formation, requiring no special controls. Studies on the reassociation of the inactive component polypeptides of hemoglobin *in vitro* indicate that the return to optimal activity is a fairly simple process.

The specific folding of a one-dimensional sequence into a three-dimensional form, and the self-assembly of the quaternary architecture, have been two factors in revealing the secrets of biological order. Given a proper sequence of amino acids, a uniquely contoured structure emerges, with surface sites of weak bondings that could form associations with other molecular forms in the cell. Proteins could associate together or with lipids to form visible membranes with specific properties; they could form supporting materials, such as hair, nails, tendons, etc. And so the chemical properties of these molecules (primarily their bonding tendencies) can generate form in biological systems.

There is one factor in the chemical basis for biological organization that cannot be overlooked. The sequence of amino acids determines many of the properties of the eventual protein that is formed, and a specific sequential arrangement of these amino acids is unique to biological systems. Just as the chemist, while realizing that chemical entities are forms of matter, and that their interactions follow basic physical laws, nevertheless, regards his system on a chemical level, the biologist realizes that although the formation of biologically active molecules conforms to the basic chemical and physical laws operative throughout the universe, such ordered sequences in macro-molecules can only be formed by a complex system of biological origin. Therefore, one of the primary features of the chemical basis of life is a purely biological one.

ENZYMES AS PROTEINS

The chemical basis of biological systems provides not only for the form of that system, but also for its function. The functions that can be considered the dynamic aspects of biological systems are thus founded on the kinetics—the chemical reactions—of the underlying chemical systems. All biological functions—movement, ingestion of food materials, excretion of wastes, etc.—are traceable to sets of chemical reactions.

These chemical reactions within biological systems are governed by the same factors as so-called "pure" chemical reactions, those that take place in a test tube on a chemist's bench. The two primary factors, as previously discussed (see pp. 37–41), are the energy of activation and the free energy (ΔG). It must be realized that biological systems are semiisolated, in the sense that they do not have an endless source of energy and are therefore dependent on conserving the energy within

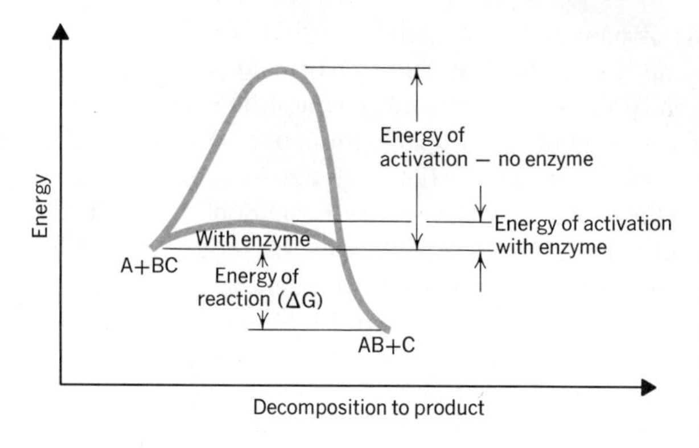

Fig. 3-8 Reduction of energy of activation by enzymes. Energy available in the system, usually in the form of heat, is often sufficient to provide the energy of activation for enzyme-mediated reactions.

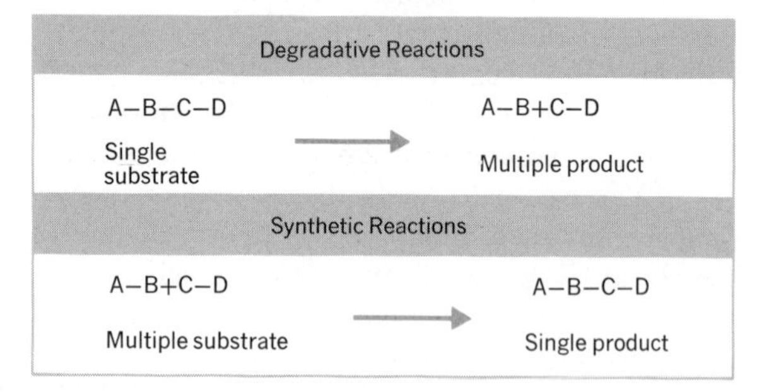

Fig. 3-9 Singularity of substrate or product. Degradation reactions usually involve the breakdown of a single substrate into two or more products, while synthetic reactions involve the joining of two or more substrates to form a single product.

TABLE 3-2 GENERAL CLASSES OF ENZYMES

Oxidoreductases	(oxidation-reduction reactions)
Transferases	(group transfer reactions)
Hydrolases	(hydrolytic reactions)
Lyases	(additions or losses of groups to double bonds)
Isomerases	(isomeration reactions)
Ligases	(syntheses by condensation of two groups requiring ATP)

the system for efficient function. One of the primary means developed to cope with this problem within the scope of physical laws is a reduction of the energy of activation of a reaction by means of enzymes. This reduction then permits the reaction to occur with less energy input on the part of the biological system (see Fig. 3-8).

One of the first properties recognized in enzymes was their extreme specificity. A single enzyme mediates a single reaction involving a specific reactant or reactants (called a **substrate**) and aiding in the conversion to a specific product or products (either the substrate or the product is singular; see Fig. 3-9). Degradation reactions usually have only one substrate that breaks down into two products. Synthetic reactions, on the other hand, involve the putting together of two or more substrates to form a single product.

Groups of enzymes are categorized and named according to the *type* of reaction in which they take part. **Oxidoreductases** catalyze oxidation-reduction reactions. **Transferases** catalyze group transfer reactions. **Hydrolases** catalyze hydrolytic reactions. **Lyases** catalyze the addition of groups to double bonds or the removal of groups from molecules so that double bonds are formed. **Isomerases** catalyze the conversion of D- to L- forms or vice versa. **Lignases** catalyze the condensation of two molecules.

Within each class are specific enzymes, such as urea amidohydrolase, which mediates the breakdown of urea to carbon dioxide and ammonia.

$$H_2N\text{—}\underset{\underset{O}{\|}}{C}\text{—}NH_2 + H_2O \xrightarrow{\text{urea amidohydrolase}} CO_2 + 2NH_3$$

The nomenclature of enzymes can be confusing. The original names given to enzymes as they were discovered are considered their trivial names. For instance, the name originally given to urea amidohydrolase was urease, a name that is still in common use.

Enzyme Structure

Enzymes are made up of two parts: a large protein portion known as the apoenzyme, and a nonprotein portion called the *coenzyme* or *cofactor*. The coenzymes of particular enzymes may be small organic molecules, inorganic ions, or heteroconjugates.

Enzyme activity is dependent on the presence of the appropriate apoenzyme and the appropriate coenzyme. Many enzymes have the same heteroconjugate unit. For example, many of the oxidoreductases have as their coenzyme a molecule of nicotinamide-adenine dinucleotide (NAD). Clearly, the cofactors of the enzymes could not account for their specificity in taking part in only one reaction and acting on only one substrate. It was reasonable to assume, then, especially in view of the structural characteristics of proteins, that the protein portion of the enzyme was responsible for the specificity of the enzyme reaction and the coenzyme was responsible for the activity.

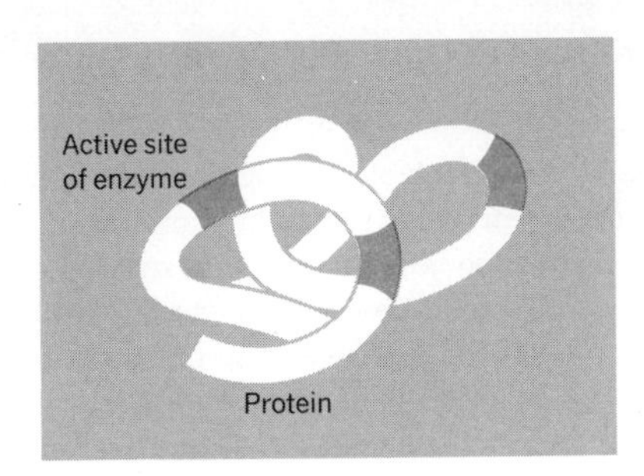

Fig. 3-10 Active sites of an enzyme. An enzyme may have one or more active sites, depending on the factors involved in the reaction it mediates. Each active site is composed of a specific group of amino acids that act together to attract and bond a factor of the reaction. One active site holds the substrate. Others may hold inorganic ions or energy donors necessary for that reaction.

The Enzyme-Mediated Reaction

Although the basic features of the theory of enzyme action were proposed before the elucidation of their protein structure, the theory was refined to its present state by taking the structural aspects of proteins into consideration. An enzyme, as a protein, has both a physical and a potentially reactive structure. The physical form is due to its secondary, tertiary, and quaternary bondings; its potential activity is due to the bonding tendencies of the side radicals of the amino acids on the surface of the structure. Each enzyme contains a single specific **active site** for each substrate and each molecular factor involved in the particular reaction. The active site of the enzyme is defined as those amino acids within the protein structure of the enzyme that come into direct contact with the participants of the reaction (see Fig. 3-10). The substrate combines with the enzyme at an active site, forming a temporary weak association called the **enzyme-substrate complex.** Kosland has discovered that upon contact with the substrate, the enzyme undergoes tertiary bond shifts so that the substrate "fits" into the enzyme structure. Complementarity between the enzyme and substrate is, therefore, induced by the substrate itself. In the formation of the enzyme-substrate complex, certain bonds within the substrate are under strain, and the energy content within the bonds undergoes a shift. This shift in bond energy results in the weakening of one bond, to the point where the thermal energy of other molecules is sufficient to break it. Upon breakage of the bond, the enzyme-substrate complex dissociates, yielding the enzyme in its original state, and the products (see Fig. 3-11). The reaction is usually represented:

$$\text{Enzyme} + \text{substrate} \longrightarrow \text{enzyme-substrate complex} \longrightarrow \text{enzyme} + \text{products}$$

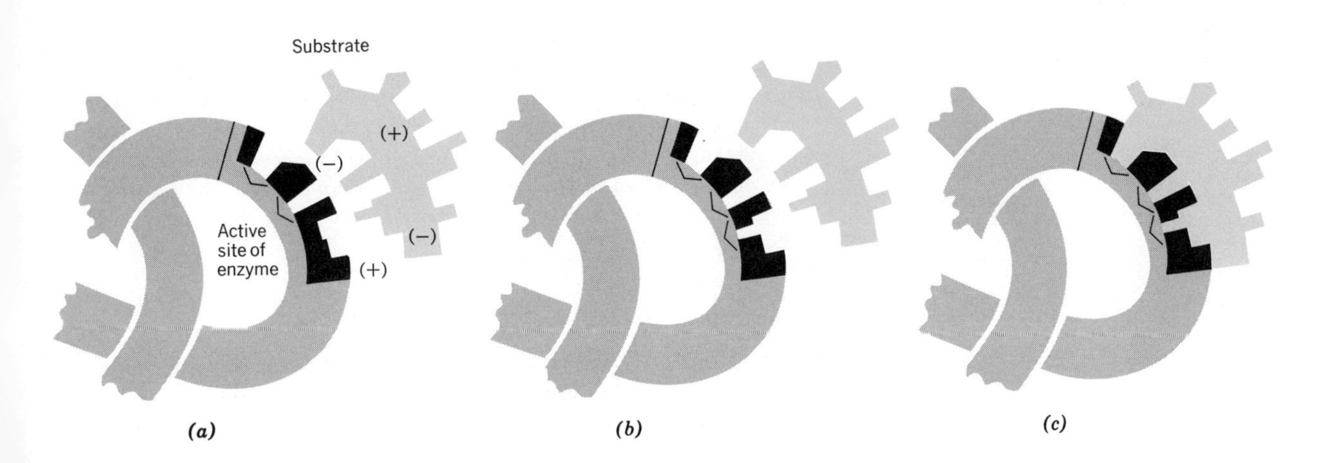

Fig. 3-11 Association of enzyme and substrate. The substrate "fits" into the active site of the enzyme through complementarity of shape and charge.

Several properties of enzymes emerge from this reaction and warrant consideration.

1. Notice that the enzyme as a product of the reaction is unaltered by the reaction process. This enzyme may now enter into another reaction with another substrate of the same species. In inorganic reactions, a **catalyst** is generally regarded as a factor that mediates a reaction without actively taking part in the reaction itself. In biological systems, the direct involvement of the enzyme in the reaction in no way diminishes its ability to mediate the reaction, for it is released in active form, and is therefore considered a "biocatalyst." In the 1950s it became obvious that some enzymes are not released from a reaction in an unaltered fashion. For example, in the course of dehydrogenation, the coenzyme NAD is converted to $NADH_2$ and is inactive (unable to dehydrogenate molecules) in this form. Through another reaction, $NADH_2$ itself undergoes dehydrogenation and returns to the NAD form, ready to act once again as an active enzyme. Because of the eventual regeneration of the active form of the enzyme, even though this regeneration involves another reaction, the enzyme is still considered a catalyst. Therefore, one of the primary characteristics of enzymes is *their emergence from a reaction or set of reactions in an unaltered state.*
2. The preceding considerations lead directly to another general characteristic of enzymes, that of *being active in small quantities.* Since the enzyme is reused, only a small quantity of an individual enzyme is required by biological systems. It has been estimated that a single enzyme can act upon 500,000 substrate molecules per minute. This value is refered to as the **turnover number.**
3. The third general property of enzymes is their *protein character.* The protein structure provides for weak associations by the very nature of the side groups of the amino acids within it. The enzyme-substrate complex involves the association of the enzyme and substrate by weak associations rather than covalent bondings. The turnover of the reaction proceeds at too low an energy level to involve the formation of covalent bonds between the enzyme and the substrate.
4. The weakening of the bond by alteration of the bonding energies within the substrate constitutes the decrease in the energy of activation required for the reaction. It is from this activity that another property attributed to enzymes emerges. An *enzyme changes only the rate of the reaction,* not the direction or the equilibrium. A reaction will proceed at a faster rate in the presence of an enzyme, but the equilibrium between the substrate and the product will remain constant with or without the intervention of the enzyme.

The Temperature Dependence of Enzyme Reactions

Another outstanding feature of enzyme reactions is their temperature dependence. The activity of an enzyme is determined by its ability to convert substrate into product. By mixing substrate and enzyme together, incubating them at different temperatures, and measuring the amount of product formed after a specified time period has elapsed, one can readily see this temperature dependence. Examine Fig. 3-12, which is a plotted curve of just such data. Notice that there is very little enzyme

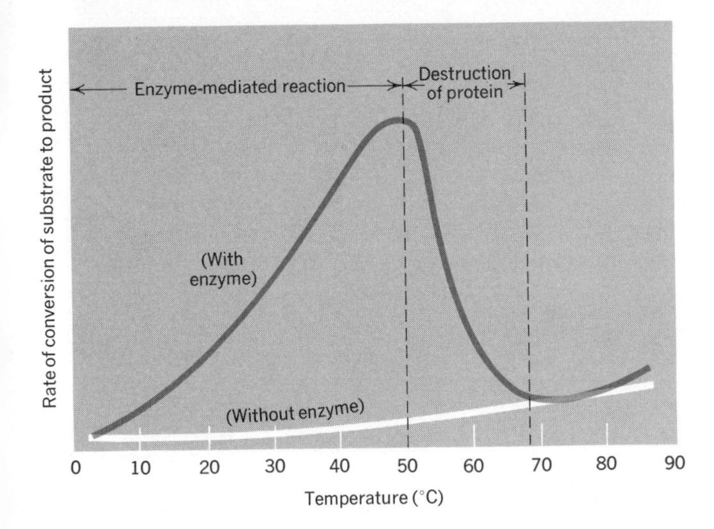

Fig. 3-12 Temperature dependence of enzyme activity. The rate of enzyme-mediated reactions increase until the energy content of the system is sufficient to destroy the enzyme. After enzyme destruction, the reaction proceeds at a rate that is solely temperature-dependent until the energy content of the system destroys the substrate or product.

activity near O° C, but as the temperature rises, the conversion of substrate to product increases substantially. The rate of reaction reaches a peak at 50° C, after which the rate declines. The curve reflects the occurrence of two sequential events. First, the increase of the reaction rate as kinetic energy in the form of heat is added to the system; and second, when the point is reached where the kinetic energy in the system is sufficient to disrupt the weak associations of the enzyme (the secondary, tertiary, and quaternary bondings), the enzyme becomes inactive and incapable of mediating the reaction (see Fig. 3-13).

The increase in rate of reaction is expressed by the Q_{10}. The Q_{10} of a reaction is the increase in rate per 10° C increase in temperature. In Fig. 3-13, the increase in reaction is two- to threefold over a rise of 10° C, indicating that this reaction has a Q_{10} of 2 to 3. The Q_{10} of the enzymatic reaction indicates the velocity of the molecules in the reaction vessel and the number of collisions leading to formation of the enzyme-substrate complex.

If one looks at the actual increase in the kinetic energy of a system by raising it 10 degrees, from 20 to 30° C, it is only about 4%. How can a 4% increase in kinetic energy account for a 200 to 300% increase in the rate of the reaction, as indicated by the Q_{10} value of 2 to 3? The explanation for this phenomenon lies in the probability distribution of the kinetic energy content of the molecules at the different temperatures. When a reaction vessel is at a particular temperature, the velocity (kinetic energy) of the molecules within the vessel is in accordance with a probability distribution. Some molecules are moving quite slowly, and others quite fast. The majority of the molecules will be moving at rates close to the kinetic energy within the system itself, but, as with any probability distribution, the rare events, the presence of fast- and slow-

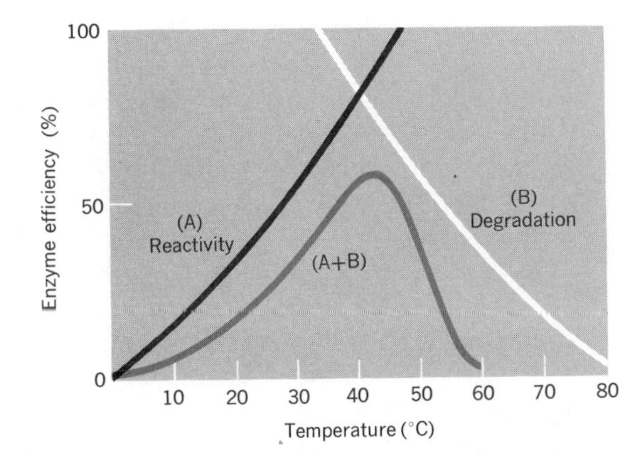

Fig. 3-13 Enzyme efficiency. Enzyme efficiency increases until the energy content of the system is sufficient to disrupt the quaternary, tertiary, or secondary bondings of the protein.

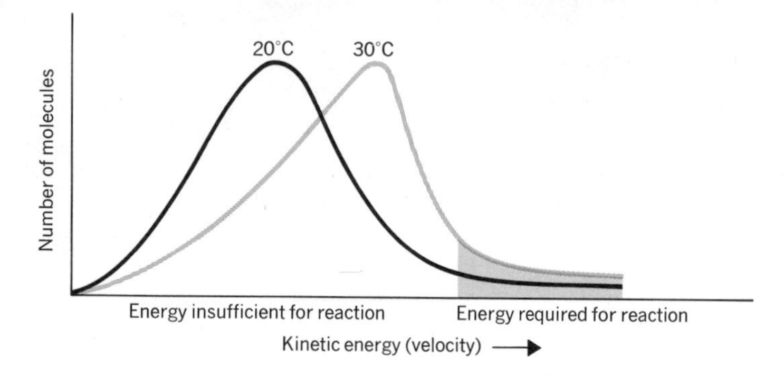

Fig. 3-14 The Maxwell-Boltzmann distribution. At 30° C, three times the area is under the curve than is under it at 20° C, and therefore three times the number of molecules have sufficient kinetic energy to undergo reaction than at 20° C. Thus this reaction has a Q_{10} of 3. The curves are essentially the same if nonenzyme- and enzyme-mediated reactions are considered. As the increased heat (as reflected by temperature) affects relatively small numbers of molecules, but increases the reaction threefold, enzymes combine with small numbers of molecules, bringing them to a reactive state, and similarly increase the rate of reaction.

moving molecules, will also occur. This type of distribution is given in Fig. 3-14, and is called the **Maxwell-Boltzmann distribution.** The two curves represent the distribution that would be expected from molecules in reaction vessels kept at 20° C and 30° C, respectively. The activation energy for the reaction to occur might correspond, let us say, to a temperature level of 50° C. At a temperature of 20° C, some of the molecules in solution manifest the kinetic energy associated with a 50° C temperature, and undergo reaction. At 30° C, two to three times this number of molecules manifest this kinetic energy and undergo reaction. So the Q_{10} is reflective not of the whole system, but only of that part of the system that encompasses the rare event, the manifestation of sufficient kinetic energy to undergo conversion into the products of the reaction. Since enzymes serve to decrease the amount of energy required for the reaction to take place, their effect may be thought of as shifting the curve in the same way. It is for this reason that enzymes tend to change the rate of reaction, permitting reactions to occur faster than in non-enzyme-mediated situations.

Enzyme Kinetics

The theories of enzyme action did not arise from purely descriptive forms of experimentation. Rather, they were the result of a mathematical analysis of data on **enzyme kinetics,** a quantitative study of enzyme reactions. Therefore, giving only a descriptive picture of enzyme activity leads to an inaccurate concept of the type of work actually being done in this very active field of biochemistry, and it becomes necessary to examine the kinds of experimental approaches that led to some of the more sophisticated ideas of enzyme activity in order to correct this concept.

In the early 1900s, experiments indicated that enzyme reactions showed a definite relationship to time. A given concentration of substrate (amount per unit volume) was mixed with a given concentration of enzyme (both enzyme and substrate concentrations were kept constant) and incubated at a constant temperature. At various time intervals, a portion (**aliquot**) of the incubating mixture was removed and analyzed

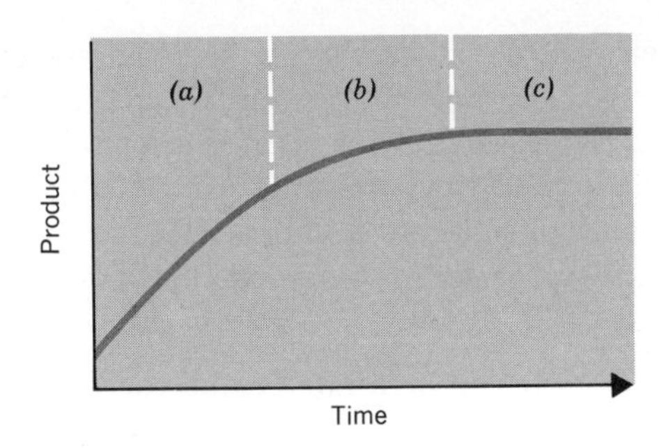

Fig. 3-15 Curves and equations. (*a*) The reaction is occurring according to first-order kinetics. The equation of the line is $y = ax^1$. (*b*) The reaction is occuring according to first- and zero-order kinetics. The equation for the line is $y = ax^1 + bx^0$, or $ax + b$. (*c*) The reaction is occurring according to zero-order kinetics. The equation for the line is $y = 0x^1 + bx^0$ or $0 + b$ or $y = b$.

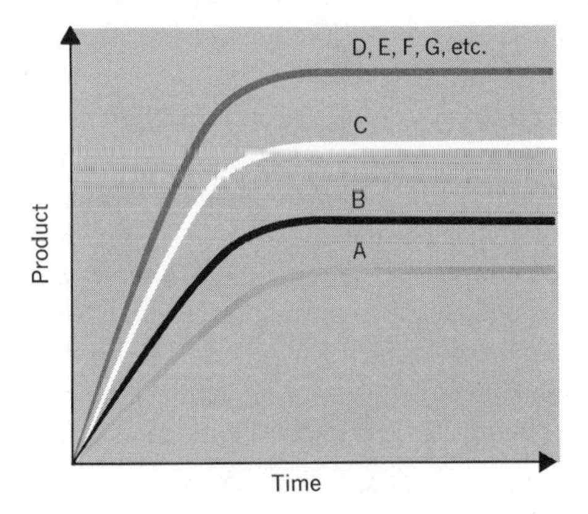

Fig. 3-16 Curves of different substrate concentrations. Curve A represents low substrate concentration, Curve B represents two times the substrate concentration, etc. Rate of reaction increases with an increase in substrate concentration up to a maximum concentration, after which no increase in substrate concentration will increase the rate of reaction.

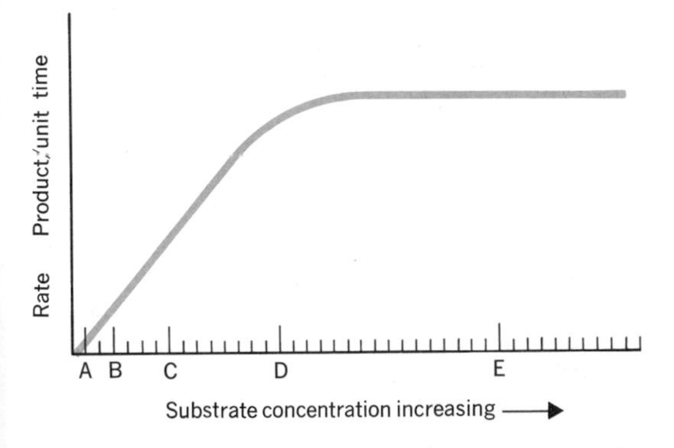

Fig. 3-17 Rate of reaction versus substrate concentration.

for the products of the reaction. When the concentration of the product produced was plotted on a graph against the time for a given substrate concentration, a curve was formed, resembling that in Fig. 3-15. Mathematical interpretation of the curve led to the concept of **first-order** and **zero-order** biochemical reactions. For a short period of time, the curve is essentially a straight line, the general equation of which would be $y = ax$, in which x appears in the first order form ($x^1 = x$). For this short time, then, the reaction may be viewed as being dependent on a single factor that is in abundant enough supply to cause the reaction to proceed—namely, the substrate. This is considered to be **first-order kinetics,** and reflects the reaction of substrate ⟶ product. After this time, however, the slope of the curve begins to decrease, indicating that the general equation now includes another factor. The general equation of this part of the curve is $y = ax - b$, where b is a constant. Another way of expressing this equation is $y = ax^1 - bx^0$ ($x^0 = 1$, so $bx^0 = b$). So the curve may be viewed as expressing a mixture of first-order and **zero-order kinetics.** (See Fig. 3-15) It was realized that the decrease in the slope of the curve indicated that there was a **limiting factor** involved in the reaction. Since no change was detected in the enzyme concentration, the limiting factor had to be an insufficient concentration of the substrate. As time progresses, the curve levels off and the equation is expressible simply as $y = b$ or $y = bx^0$, the zero-order reaction. The only part of the curve that directly reflected the reaction then, before the influence of the limiting factor became evident, was the first-order reaction. Determination of the equation of the first-order reaction then, would give the amount of product produced per unit time, or the rate of reaction for *that particular substrate concentration.* By repeating the experiment, using different levels of substrate concentrations in different reaction vessels, removing aliquots from them at specified times, analyzing for concentration of product produced, and plotting the data, a series of curves is generated (see Fig. 3-16). Determination of the first-order kinetic equation for each of these curves gives the reaction rate for each substrate concentration. (For those readers who are familiar with calculus, the determination of the first-order kinetic equation involves taking the first derivative of the equation of the curve. This gives the tangent line to the curve drawn at that point.) Once the values for the rates of reaction are determined, they may be plotted against the substrate concentration. When this was done, a curve resembling that in Fig. 3-17 was produced.

There are several interesting features about this curve. When the concentration of substrate was increased, the rate of reaction was increased only to a given level. After this level was reached, no additional substrate could make the reaction proceed at a faster rate. This level, then, represents the maximum velocity of the reaction, or the V_{max}. In catalyzed inorganic reactions, increase in the reactants leads to a constant increase in reaction rate. Obviously, this was not true of enzyme-mediated reactions. Something in the reaction was acting as a limiting factor, generating a curve that had both first-order and zero-order properties. In order for this to be true, the enzyme must act as a reactant

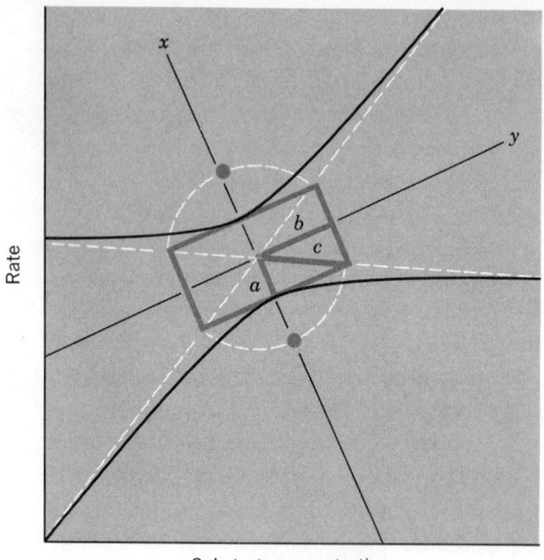

Fig. 3-18 Rate of reaction versus substrate concentration as a rectagular hyperbola. A rectangle of the dimensions 2a and 2b, with the asymptotes as diagonals, can be inscribed between the two parts of the hyperbola. A circle of radius c passes through the foci of the hyperbola.

and enter into the reaction, unlike catalysts in inorganic reactions. It was suggested that the enzyme and substrate formed an actual physical complex, and the $\mathbf{V_{max}}$ represented the point at which all the available enzyme was in complex with as much substrate as it could hold, so that an increase in the amount of substrate could not increase the reaction rate any further. This assumption led to the formulation of the theoretical equation

$$\text{enzyme} + \text{substrate} \longrightarrow \text{enzyme-substrate complex} \longrightarrow \text{enzyme} + \text{product}$$

Michaelis and Menten, upon graphing enzyme reactions in this way, realized that the generated curve was that of a rectangular hyperbola (see Fig. 3-18). One of the notable features of a rectangular hyperbola is the formation of a straight line when the reciprocals are plotted. The advantage of straight-line graphs is that they are easier to interpret than are curves. In plotting the reciprocals, the $\mathbf{V_{max}}$ may be read from the graph directly as the intercept of the **V** axis (the ordinate) (see Fig. 3-19). Could enzyme-mediated reactions be reduced to kinetic terms amenable to straight-line graphs?

Since all geometric curves are expressible in algebraic terms (see p. 15), Michaelis and Menten attempted to find an equation relating the rate of the reaction to the substrate concentration that would produce an equation of the form leading to a rectangular hyperbola, the general equation of which is $\mathbf{K = xy}$, where **K** is a constant. The equation probably was not evident immediately, and Michaelis and Menten undoubtedly did quite a bit of algebraic manipulation before arriving at the equation that now bears their name. The derivation of the equation and its mathematical form are included as a footnote* for interested students.

*Their starting point was the proposed enzyme equation: enzyme + substrate $\rightleftharpoons$ enzyme-substrate complex $\longrightarrow$ enzyme + product.

According to the **law of mass action,** which governs the rate of chemical reactions, the rate of reaction (k) is **proportional** to the concentration of reactants. In considering the rates of enzyme mediated reactions, they may be identified from three separate reactions:

$$\text{enzyme-substrate} \xrightarrow{k_1} \text{enzyme-substrate complex}$$
$$\text{enzyme-substrate complex} \xrightarrow{k_2} \text{enzyme-substrate}$$
$$\text{enzyme-substrate complex} \xrightarrow{k_3} \text{enzyme-products}$$

Notice that one does not have to consider a k_4 value for the rate of reaction where k = rate of reaction of the enzyme + product back to the enzyme-substrate complex, because this reaction does not occur at any detectable rate.

If one examines the rates of reactions and the concentrations of reactants at an equilibrium state, the nature of this proportionality becomes more evident. If, for instance, at equilibrium, the ratio between the concentration of reactants (enzyme and substrate) and the concentration of the enzyme-substrate complex is 100:1, then the ratio between the rates of the reactions, k_1 and k_2, is 1:100.

$$100 \text{ enzyme} + \text{substrate} \underset{k_2}{\overset{k_1}{\rightleftharpoons}} 1 \text{ enzyme-substrate complex}$$

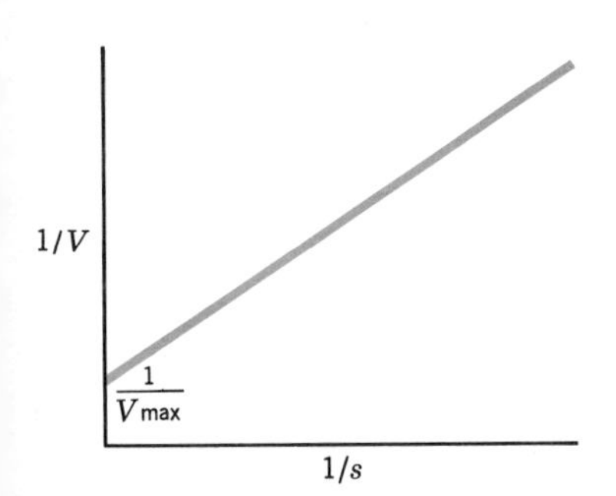

Fig. 3-19 Reciprocals of rate of reaction versus substrate concentration. By plotting the reciprocals of rate (velocity) and the substrate concentration, a straight-line graph is generated. $1/V_{max}$ may be read directly as the y intercept.

The Michaelis-Menten equation provided a very important tool for the study of enzyme kinetics. Experimental data indicated the validity not only of the equation but also of the assumption on which it was based, the formation of the enzyme-substrate complex. The first direct evidence of the existence of an enzyme-substrate complex did not come until 1936, when several investigators simultaneously reported that in a particular enzyme-mediated system, visual changes in the color of the solution could be noted as the reaction proceeded, and these color changes were attributed to the formation of the complex. The kinetic evidence for such a complex preceded this by twenty-three years, enabling workers in the field to utilize this important concept in the furthering of knowledge on enzyme action. The Michaelis-Menten equation represents the ideal enzyme reaction. When later workers found that their systems did not yield the perfect rectangular hyperbola, it permitted them to look for secondary reactions or factors that might be affecting their systems. The equation and its graphic interpretation also led to a mathematical and eventually descriptive explanation of factors that inhibit enzyme activity.

Competitive and Noncompetitive Inhibition

When certain substances were introduced into reaction vessels, the substances seemed to inhibit the activity of the enzymes in some way. It was shown that in some cases the inhibition could be overcome by the addition of more substrate, but in other cases the inhibition seemed to be of a more permanent nature. For example, it was shown that **malonic acid** acted as an inhibitor in the reaction converting **succinic**

If only one in every hundred molecules of enzyme and substrate are formed in an enzyme-substrate complex, the rate associated with this is $1/100\ (k_1/k_2)$. So the proportional of reactant to rate of reaction is an *inverse* one.

$$k_1 \alpha \frac{1}{[E][S]} \qquad k_2 \alpha \frac{1}{[ES]} \qquad k_3 \alpha \frac{1}{[ES]}$$

The brackets [ ] represent concentrations of the molecular forms within them; α is the symbol for "proportional to."

Upon examining the theoretical equation for enzyme-mediated reactions, Michaelis and Menten realized that the rate of the reaction was dependent on the stability of the hypothetical enzyme-substrate complex. The more stable the complex, the less readily it will separate into the enzyme and products, and the lower will be the rate of reaction. The stability of this complex is, in turn, dependent on the ratio of the reaction rates leading away from the complex (k_2 and k_3) to those leading toward its formation (k_1).

$$\frac{k_2 k_3}{k_1} - K_m$$

This ratio may be represented by a single constant, K_m, which is called the Michaelis constant. By substituting equivalent values,

$$\frac{1/[ES] + 1/[ES]}{1/[E][S]} = K_m$$

$$\frac{[E][S]}{[ES]} = K_m$$

(Continued)

acid to **fumaric acid,** which is mediated by the enzyme **succinic dehydrogenase.**

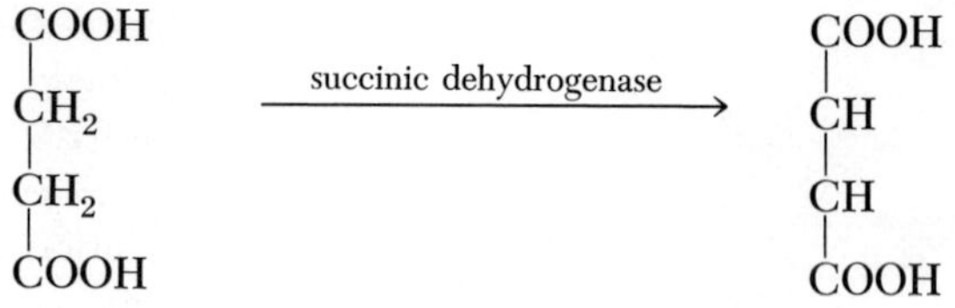

$$\begin{array}{c} COOH \\ | \\ CH_2 \\ | \\ CH_2 \\ | \\ COOH \\ \text{succinic acid} \end{array} \xrightarrow{\text{succinic dehydrogenase}} \begin{array}{c} COOH \\ | \\ CH \\ | \\ CH \\ | \\ COOH \\ \text{fumaric acid} \end{array}$$

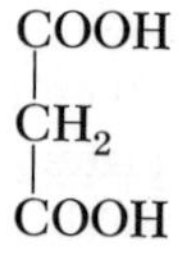

$$\begin{array}{c} COOH \\ | \\ CH_2 \\ | \\ COOH \\ \text{malonic acid} \end{array}$$

The $[E]$ represents the concentration of enzyme available for reaction with the substrate; therefore, this is really the total amount of enzyme minus the amount in complex with the substrate, $E - ES$. The same consideration does not apply to the substrate because of the preponderance of substrate compared to enzyme in a working system. The error incurred by omitting this consideration of the substrate is negligible. Therefore, the equation becomes

$$\frac{[E - ES][S]}{[ES]} = K_m$$

The equation may be rearranged to give a value for enzyme-substrate concentration (ES).

$$[E - ES] + [S] = K_m[ES]$$
$$[E][S] - [ES][S] = K_m[ES]$$
$$[E][S] = K_m[ES] + [ES][S]$$
$$[E][S] = [ES](K_m + [S])$$
$$[E][S] = [ES](K_m + [S])$$
$$\frac{[E][S]}{K_m[S]} = [ES]$$

The V_{max} is also a function of the enzyme-substrate complex, and represents a point at which all the enzyme is in complex with substrate.

$$V_{max} = [ES]$$

Therefore, the velocity of the reaction at any given time is equal to some value less than one times the total concentration of enzyme-substrate complex possible, or $k[ES]$, where $k = 1$.

$$V = k[ES]$$

By substituting the above equation for $[ES]$,

$$V = k\left(\frac{[E][S]}{K_m + [S]}\right)$$

But $k[E]$ reflects the V_{max}, since the V_{max} is dependent on the concentration of en-

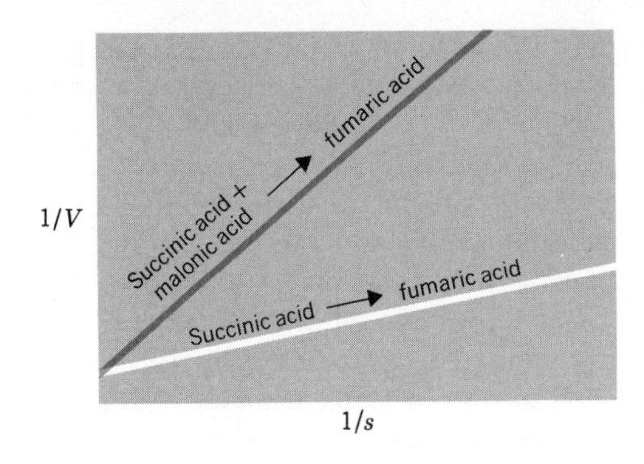

Fig. 3-20 Competitive inhibition. At high concentrations of fumaric acid, the V_{max} (y intercept) is the same for the reaction with or without the inhibitor. Thus the effect of malonic acid can be overcome, and it is considered to be a competitive inhibitor.

It is easy to look at the structures of these molecules and come to the conclusion that since malonic acid resembles succinic acid so closely, the enzyme might not be able to distinguish between them and might go into complex with the malonic acid. This is pure speculation. Analysis by means of the Michaelis-Menten equation and its graphic interpretation gives a more precise explanation. By following the standard procedures, the reciprocals of the reaction rates with and without the inhibitory factor (malonic acid) may be plotted against the reciprocals of the substrate concentration (see Fig. 3-20). The graph indicates that with high levels of substrate concentrations (succinic acid), the $\mathbf{V_{max}}$ (y-intercept) is the same for both reactions. This indicates that when there is a high concentration of substrate, the enzyme is in the form of enzyme-substrate complex in both cases. With a lower level of substrate, however, only a fraction of the enzyme is available for complex with the substrate when the inhibitor is present. An interpretation may be made directly from the mathematical and graphic considerations. The substrate and inhibitor are competing for the active site on the enzyme. The inhibitor forms an unstable inhibitor-enzyme complex that is in

---

zyme-substrate complex, which is, in turn, dependent on the total enzyme concentration. The equation may then be reduced to

$$V = \frac{V_{max}[S]}{K_m + [S]}$$

which is the **Michaelis-Menten equation.**

There are several factors that warrant consideration. First, the equation may be rearranged to give a value for K_m.

$$(K_m + [S])V = V_{max} + [S]$$

$$K_m + [S] = \frac{V_{max} + [S]}{V}$$

$$K_m = \frac{V_{max} + [S]}{V} - [S]$$

$$K_m = [S]\left(\frac{V_{max}}{V} - 1\right)$$

This is the form of the general equation of a rectangular hyperbola, $k = xy$, where $S = x$ and $(V_{max/V=1}) = y$.

So this is the equation of the actual curve derived from the experimental data. Second, the equation in its original form

$$V = \frac{V_{max}[S]}{K_m + [S]}$$

indicates the dependence of the reaction rate on the various factors involved in the reaction. When there is a low concentration of substrate available, very little enzyme-substrate complex will be formed, and the V_{max} (which is dependent on the enzyme-substrate complex concentration) will be low. Therefore, under these conditions, the rate of reaction (V) will be dependent on substrate concentration. When the substrate concentration is high, however, the enzyme-substrate complex concentration (as expressed by the V_{max}) becomes the dominant feature of the reaction, and the rate of reaction (as indicated by the experimental evidence) is independent of the substrate concentration.

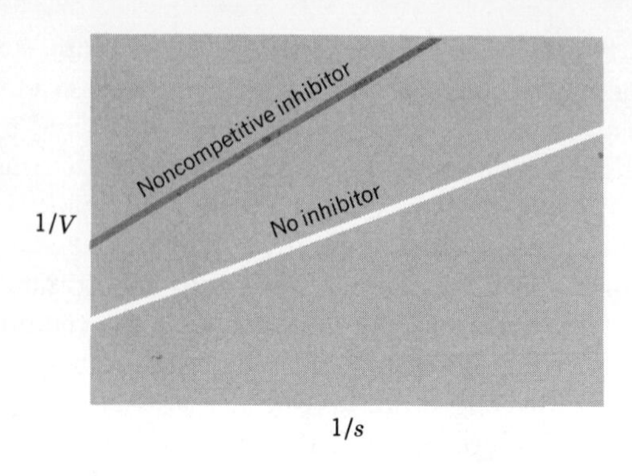

Fig. 3-21 **Noncompetitive inhibition. Regardless of the concentration of substrate, the V_{max} of the reaction with the inhibitor never achieves the V_{max} of the reaction without the inhibitor. The effects of noncompetitive inhibitors are consequently more devastating than those of competitive inhibitors.**

dynamic equilibrium with the inhibitor in solution (constantly forming and breaking apart). When more substrate is added to the system, the substrate gains the advantage in the competition, and with the addition of sufficient substrate, the reaction rate may be returned to its maximum. This type of inhibition is called **competitive inhibition.**

On the other hand, if the reaction in the presence of the inhibitory factor does not reach the V_{max} of the uninhibited reaction (see Fig. 3-21), it must be concluded that less enzyme-substrate complex is able to be formed under these conditions. The most probable way in which this could occur is if the inhibitor forms a stable complex with the enzyme, rendering it inactive. The enzyme then becomes unavailable for complexing with the substrate. This is called **noncompetitive** inhibition. Notice that in the case of competitive inhibition, addition of small quantities of inhibitor would have very little effect on the reaction, whereas in noncompetitive inhibition, even small quantities of inhibitor would have a profound effect on the reaction because the inhibition is directly related to the amount of enzyme present (which is relatively small) and is independent of the substrate concentration.

Enzyme inhibition is not merely a laboratory exercise. It is an important factor in the action of many poisons, drugs, and control mechanisms of living systems. Most poisons are noncompetitive inhibitors, which accounts for the lethal effect of small quantities of toxic material. Cyanide acts as a poison to living systems because it inhibits the enzyme **cytochrome oxidase** (the importance of which will become apparent in the next chapter) by binding with the metals associated with the enzyme, thereby rendering the enzyme inactive. In the enzyme **phosphotransacetylase,** arsenate also substitutes for the phosphate group, thereby affecting important phosphate transfers. Silver acts on **glutamic dehydrogenase** by forming a salt with the -SH groups of the cystine and methionine of the enzyme structure (indicating that these side groups must be exposed rather than convoluted within the protein structure). Carbon monoxide, on the other hand, acts as a competitive inhibitor. It competes with oxygen for the active site on the hemoglobin molecule. Therefore, carbon monoxide poisoning may be overcome by administration of large quantities of oxygen gas to displace the carbon monoxide. The role of inhibitors as drugs or for control mechanisms will be dealt with in later chapters.

PERSPECTIVE OF STRUCTURE AND FUNCTION OF PROTEINS

The elucidation of the primary, secondary, tertiary, and quarternary structures of protein was a major step in relating chemical structure to cellular physiology. One of the most significant features of protein structure is the spontaneous folding of the molecule into a physical configuration that represents its lowest, most stable thermodynamic level. Thus, biological specificity is generated as the protein molecule acts in accordance with chemical and physical laws.

Studies on enzyme kinetics led to an understanding of the energetics of the cell, specifically the way in which the energy of activation of cellular reactions is reduced. The perspective of protein chemistry, when combined with the concept of cell metabolism (the subject of the next chapter) is a vital contribution to a comprehensive picture of the cell as a functional unit.

QUESTIONS

1. Why did biochemists eliminate polysaccharides and fats from their consideration as to what was responsible for biological form and function?
2. What were the secondary assumptions of each of the three approaches to early protein chemistry research and how did they affect the outcome of each line of inquiry?
3. What methods did Sanger use to elucidate the structure of the insulin molecule?
4. How is the helical shape generated and stabilized in the secondary structure of proteins?
5. What types of bonds stabilize the tertiary configuration of a protein?
6. Why can the final form of a protein molecule be viewed as thermodynamically stable?
7. How do the component parts of an enzyme account for its specificity and its activity?
8. How do different temperature levels effect an enzyme-mediated reaction?
9. How do enzymes change the rate of the reaction, and how can this be related to the Maxwell-Boltzmann distribution?
10. What led biochemists to propose the formation of an enzyme-substrate complex?
11. Give a graphic and descriptive explanation of competitive and noncompetitive inhibition and provide examples of each.

CHAPTER 4

METABOLISM

Cells are primarily composed of polymers, such as proteins and polysaccharides, and large molecular complexes like fats and phospholipids. The maintenance and growth of cellular structures is dependent on the energy-requiring or **endergonic** reactions by which these polymers and large molecular complexes are synthesized. Energy is required for other activities too, and the processes by which they are driven also stem from endergonic chemical reactions. Therefore, a biological system must provide the energy necessary for its maintenance, growth, and various other capacities.

As chemical systems, organisms utilize chemical energy to build and maintain structure. This energy is derived from a series of reactions that take place within cells. Molecular forms brought into an organism by feeding are broken down, and some of the energy within these molecules is used for the biosynthesis of structural and enzymic parts of the cell. The sum total of all degradative and biosynthetic reactions in the cell is called **metabolism.**

Metabolic processes operate according to both chemical and biological laws. In chemical systems, one of the most common means of driving endergonic reactions is through **coupled reactions.** In a coupled reaction, the energy derived from an **exergonic,** energy-yielding reaction drives the endergonic reaction. For example, if the reaction

$$\mathrm{A} \longrightarrow \mathrm{B}$$

is endergonic (having a positive ΔG value) and the reaction

$$\mathrm{C} \longrightarrow \mathrm{D}$$

is exergonic (having a negative ΔG value), then

$$\mathrm{A} + \mathrm{C} \longrightarrow \mathrm{B} + \mathrm{D}$$

would represent the coupled reaction in which the energy derived from the exergonic reaction ($\mathrm{C} \longrightarrow \mathrm{D}$) drives the endergonic reaction ($\mathrm{A} \longrightarrow \mathrm{B}$).

Superimposed on the basic chemical phenomenon of the coupled reaction are the biological considerations of a system. For example, if a test tube reaction proceeds as

$$\text{A} \longrightarrow \text{B} \qquad \Delta G = +5000 \text{ kcal}$$
$$\text{C} \longrightarrow \text{D} \qquad \Delta G = -1000 \text{ kcal}$$

about one molecule of A will go to B for every five molecules of C converted to D, and the rate of the A $\longrightarrow$ B reaction will be about $\frac{1}{5}$ that of the C $\longrightarrow$ D reaction.

This does not occur in metabolic processes because of the low concentration of reactants and the small chance that five exergonic reactions would occur in close enough spatial proximity to the endergonic reaction to contribute sufficient energy. Therefore, *in biological systems, the exergonic reaction always has a greater ΔG value than the endergonic reaction that is being driven.*

Two other biological considerations that are superimposed on the chemical phenomenon of the coupled reaction in metabolic processes are the amount and kind of energy both used for, and derived from, degradative reactions. When a molecule such as glucose enters the cell, it must be broken down to carbon dioxide and water in order for the system to extract most of the energy from it. Glucose, however, has a high energy of activation, and in order to extract energy from it, considerable energy must be added to the molecule. Thus a cell must contribute some energy initially in order to obtain energy from degradation. The energy profit realized from glucose breakdown is dependent on the amount of energy derived from the breakdown minus that which the cell has directly contributed, and the efficiency of metabolism is judged on the basis of this net profit.

Also, when glucose is broken down in a one-step process, such as by combustion, there is a release (ΔG) of 686 kcal/mol (primarily in the form of heat). This is sufficient energy to destroy the integrity of adjacent biological structures. A cell breaks down a molecule such as glucose in a way (1) that there is *a minimum of energy expended,* and (2) that *the energy is released in a usable form.*

The three biological considerations of the coupled reactions used in metabolic processes are then:

1. The exergonic reaction always has a greater ΔG value than the endergonic reaction that it drives.
2. A minimum of energy is expended in the degradative process if a maximum energy profit to be realized.
3. Most of energy released from exergonic reactions is in a form that is usable to the cell.

Metabolism as a Series of Coupled Reactions

In coupled reactions there is always a molecular form that contributes energy (an **energy donor**) and a molecular form that receives the energy (an **energy acceptor**). While it is true that there are a variety of exergonic reactions taking place in a cell, the reactants of which could, theoretically, be used as energy donors, evolution has tended to favor the development of a specific type of molecule for the role of energy donor.*

The use of a single type of energy donor in biological systems imposes a strictly biological feature on the chemical phenomenon of the coupled reaction. Metabolic processes use a **double** coupled reaction. In the first step of the reaction, the energy is transferred from an exergonic reaction to the energy acceptor to form the primary energy donor in biological systems. This energy donor takes part in a number of endergonic reactions in the cell, or contributes its energy toward the fulfillment of the energy of activation of potential exergonic reactions.

The universal energy donors in biological systems all belong to the **nucleotide triphosphate** class of molecules. These molecules are heteroconjugates (see Fig. 4-1). Their most outstanding feature is their high energy content, which is thought to be the result of the decreased resonance of the phosphate radical due to its position on the molecule (see p. 59).

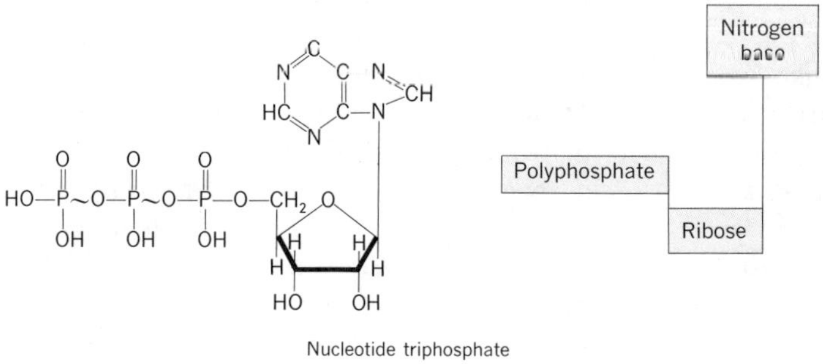

Nucleotide triphosphate

Fig. 4-1 Nucleotide triphosphate.

The most common of the nucleotide triphosphates used as energy donors, and the one that will be mentioned almost exclusively in this chapter, is **adenosine triphosphate** (**ATP**) (see Fig. 4-2). The phosphate ester linkage (P-O-P) between the two terminal phosphate groups of ATP is relatively weak, and is thought to break spontaneously from the kinetic energy of the molecule when ATP is in complex with an enzyme. The breakage of this bond causes a sudden and immediate shift in the bond energies within the resulting **adenosine diphosphate** (**ADP**) molecule, leading to an immediate release of 10 kcal/mol of energy. Because of

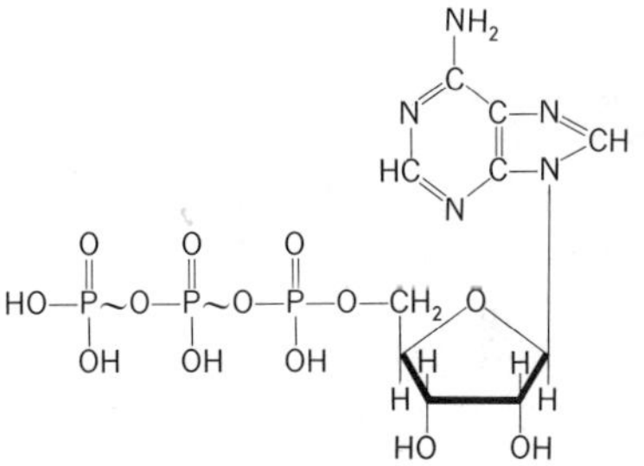

Fig. 4-2 Adenosine triphosphate, the nucleotide triphosphate most frequently used as an energy donor in biological systems.

*If endergonic and exergonic reactions were to take place randomly throughout a cell, essential endergonic reactions would be totally dependent on the chance ΔG of neighboring reactions. If, on the other hand, coupled reactions had certain sites within the cell, energy would be required to transport a wide variety of molecules first to one place in the cell for a specific reaction, and then to another, perhaps a distant site, for further coupled reactions. The use of only one type of molecule as an energy donor undoubtedly represents the operation of a cellular system at its lowest, and most thermodynamically stable (in this case, highly dispersed), energy level.

this high release of energy upon the breakage of the phosphate ester linkage, the bond is sometimes referred to as a **high-energy bond** or a **high-energy phosphate bond.** It should be remembered that the bond itself is relatively weak, for if the bond contained 10 kcal/mol, it would require at least that much energy to break it.

Upon breakdown, the terminal phosphate of ATP becomes an inorganic phosphate (usually denoted P*i*) manifesting its full resonance and thus becoming a lower-energy form. The reaction is reversible, and the formation of ATP requires adenosine diphosphate and inorganic phosphate.

$$\text{ATP} \underset{\Delta G = +12 \text{ kcal/mol}}{\overset{\Delta G = -10 \text{ kcal/mol}}{\rightleftarrows}} \text{ADP} + \text{P}i$$

The regeneration of ATP requires 12 kcal/mol, and occurs as the first set in the doubled coupled reaction in metabolic processes. Exergonic reactions yielding 12 kcal/mol of energy are used in biological systems to regenerate ATP, the universal energy donor.

In the second part of the double coupled reaction of metabolism, 10 kcal/mol of energy is released from the ATP molecule, indicating that there is an entropic loss of 2 kcal/mol in the regeneration. But of the 10 kcal/mol released in the bond breakage, only 8 kcal/mol can be converted to work; hence, there is another entropic loss of 2 kcal/mol. ATP can drive most metabolic reactions requiring 8 kcal/mol or less.

It is important to note that the release of the terminal phosphate group of ATP *does not* yield enough energy to break or form strong covalent bonds, such as the *typical* carbon-carbon or carbon-hydrogen bond (50 to 110 kcal/mol), and thus is limited in most cases to the breakage of formation of such bonds as ester linkages (R-O-R) (−1.8 kcal/mol) or amide bonds (O-N) (−3.4 kcal/mol).* The use of a molecule with such a limited energy release as the universal energy donor may be thought of as both an advantage and a disadvantage to biological systems. It is an advantage in the sense that ATP may be used in synthetic reactions in which monomers are bound by ester or amide linkages (as most are), without causing any overt damage to the molecular integrity of the individual monomers by releasing enough energy to disturb their carbon skeletons. It is a disadvantage in the sense that ATP could not act as an energy donor in the creation of the typical carbon skeleton *de novo* (from scratch). If one assumes that evolutionary processes have brought about the perpetuation of the most efficient biological systems possible, then it must follow that the advantages of using ATP as an energy donor outweigh the disadvantages. Indeed, there are other means of creating carbon skeletons in biological systems that use other sources of energy in conjunction with ATP (i.e., photosynthesis).

*There are some important metabolic reactions in which C-C bonds are formed and in which ATP acts as an energy donor. A class of enzymes, known as **lignases,** act in conjuction with ATP to join certain molecules, thus creating larger single molecules. For example,

$$\text{ATP} + \text{pyruvate} + CO_2 + H_2O \xrightarrow{\text{lignase}} \text{ADP} + \text{orthophosphate} + \text{oxaloacetate}$$

Minimum Energy Expenditure in Metabolic Processes

Minimizing the energy expenditure in metabolic processes involves not only the reduction of the amount of energy necessary to achieve the energy of activation of molecules that are to be degraded, but also the reduction of the amount of energy necessary for synthesis of polymers. In biological systems, minimizing the energy required for both aspects of the metabolic process is partially accomplished with the use of enzymes.

By subjecting a molecule of glucose to a series of stepwise, enzyme-mediated reactions, each with a low energy of activation, the molecule is broken down with a minimum expenditure of energy. Often, the energy necessary for the reaction is present in the kinetic energy of the reaction itself. Sometimes, however, the enzyme acts in conjunction with ATP, and the 8 kcal/mol of energy contributed by the ATP is sufficient to complete the reaction.

ATP can react with enzymes in both synthetic and degradative pathways. In synthetic reactions, the ATP molecule combines with the enzyme at one of its active sites, while the substrate(s) combine(s) at other active sites. The enzyme lowers the energy of activation of both the substrate bonds and the terminal phosphate linkage. The energy resulting from the removal of the terminal phosphate group of ATP is then used to drive the endergonic reaction joining the substrates.

$$\text{substrates} + \text{enzyme} + \text{ATP} \longrightarrow \text{ATP-enzyme-substrate} \longrightarrow \text{ADP} + \text{P}i + \text{enzyme} + \text{product (combined substrate)}$$

The typical way of indicating this as a coupled reaction is to ignore the enzyme in the notation (with the understanding, of course, that the reaction is enzyme-mediated).

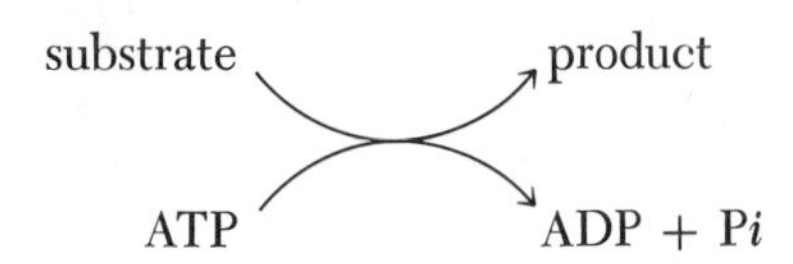

or

ATP ADP + P*i*

substrates ⟶ product

In **degradation** reactions, the terminal phosphate of ATP is often transferred to the substrate. The presence of the phosphate group on the molecule undergoing a degradative process leads to a rearrangement of the bond energy and brings the molecule to a more reactive state. In its new position in the organic molecule, the bond-energy relationships are such that the phosphate resumes its resonance. This type of transfer is sometimes referred to as the conversion of a *high-energy phosphate* to a *low-energy phosphate*.

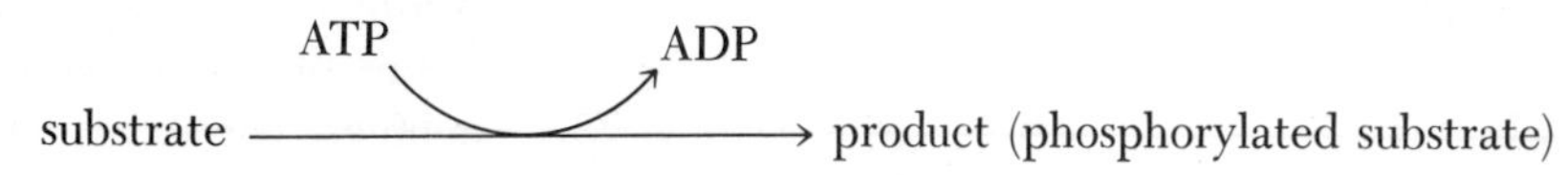

Energy in a Usable form for Metabolic Processes

The usable form of energy for metabolic processes is the nucleotide triphosphates, primarily ATP. ATP is used as the biological universal energy donor. It acts in conjunction with enzymes for both synthetic and degradative processes.

The regeneration of ATP is essential for the efficient function of biological systems. But what in the biological system could act as an energy donor in the first part of the double coupled reaction of metabolism? What exergonic reactions yield in excess of 12 kcal/mol of energy to drive the ADP + P*i* $\longrightarrow$ ATP reaction?

ATP may be regenerated in two ways, both of which are related to the breakdown of molecular forms in the degradative reactions of metabolism. The first way is by the transfer of a phosphate group from a high-energy molecule. In the course of the stepwise degradation of glucose, such high-energy molecules are formed, which act as *both energy and phosphate donors for the reaction.*

$$\text{high-energy phosphorylated molecule} + \text{ADP} \longrightarrow \text{ATP} + \text{low-energy molecule}$$

This type of reaction is referred to as **substrate-level phosphorylation.** The other way that ATP is regenerated is by the use of reactions that are auxiliary to the direct breakdown of glucose. These reactions involve the hydrogens that are removed from the glucose in the course of its degradation. The hydrogens are passed through a series of reactions, in the course of which ATP is regenerated. The auxiliary system responsible for the passage of the hydrogen electron is called the **electron transport system** or **ETS.**

Electron Transport System and the Regeneration of ATP

The reaction of ATP $\longrightarrow$ ADP and P*i*, and all other covalent degenerations, are known as **hydrolytic** reactions. This indicates that a dissociated molecule of water (H^+ and OH^-) is added to the degeneration products, giving them their completed molecular forms. For instance, consider the degradation of peptides:

$$\mathrm{H_2N{-}CHR{-}C({=}O){-}NH{-}CHR{-}COOH} \xrightarrow{\mathrm{H^+ + OH^-}} \mathrm{H_2N{-}CHR{-}C({=}O){-}OH} \quad \mathrm{H{-}NH{-}CHR{-}COOH}$$

or the degradation of ATP to ADP and P*i*:

$$\mathrm{R{-}O{-}P(OH)(O){-}O{-}P(OH)({=}O){-}O{-}P(OH)({=}O){-}OH} \xrightarrow{\mathrm{H^+ + OH}} \mathrm{R{-}O{-}P(OH)({=}O){-}O{-}P(OH)({=}O){-}OH} + \mathrm{HO{-}P(OH)({=}O){-}OH}$$

Another major class of reactions is the **oxidation-reduction** reactions (**redox reactions**), in which electrons are transferred from one atom to another. The ΔG values for some redox reactions are far greater than those of hydrolytic reactions and, therefore, these types of reactions can serve as a source of energy for ATP regeneration.

All redox reactions involve a **reducing agent** (an atomic form that tends to furnish electrons) and an **oxidizing agent** (an atomic form that tends to accept electrons). In the course of a redox reaction, the reducing agent becomes **oxidized** and the oxidizing agent becomes **reduced.** Because they are atomic forms, the reducing and/or oxidizing agents may be part of a molecular structure. Although redox reactions are sometimes noted as two separate reactions, the reducing agent that gives up electrons and the oxidizing agent that accepts them, it should be remembered that these reactions are always coupled, and involve the transfer of electrons from the reducing agent to the oxidizing agent. In some redox reactions, a proton tags along with the electron on its transfer, and the transfer then involves the transfer of a hydrogen atom.

The ΔG of a redox reaction is dependent on the tendency of the reducing agent to furnish electrons and the tendency of the oxidizing agent to accept them. The tendency for release and acceptance of electrons is called the **redox potential** and is the ratio of the reduced form of the atom (after having accepted the electrons) to the oxidized form (after having donated the electrons) or the ratio reduced/oxidized.

The redox potential is expressed mathematically as the value E_0. Because the transfer involves an electron, a particle bearing an electrical charge, the redox potentials are given in terms of volts. In order to provide a basis on which to judge the relative redox potentials of various chemical forms, hydrogen has been arbitrarily chosen as a standard. The participation of free hydrogen in any redox reaction may be represented

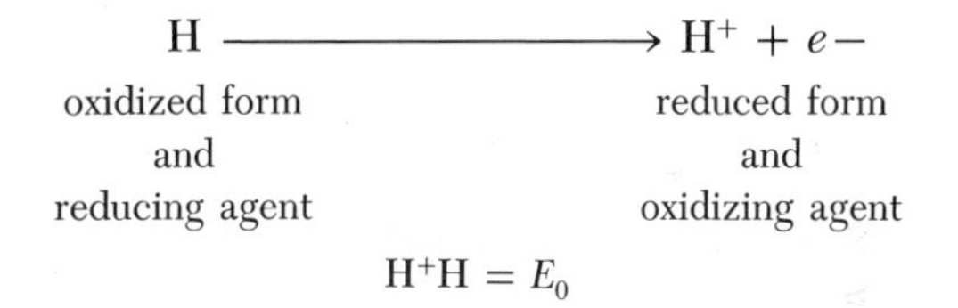

The redox potential arbitrarily given to hydrogen at pH 0 was 0.000. The E_0 of the hydrogen at pH 7.0 was then calculated to be -0.420 V. It was then possible to determine the potential of any other compound by comparison with the hydrogen standard. This comparison indicated that there was a range of redox potentials, and that the greater the negativity of the E_0 value, the greater the tendency of a particular compound to act as a reducing agent. Therefore, *any compound will act as a reducing agent for another compound with a more positive E_0 value.* In this way, a compound that might be a reducing agent in one reaction will act as an oxidizing agent in another reaction.

It should be noted that although the transfer of electrons occurs only on an atomic level, the structure of the molecule in which the atomic form resides determines the tendency of that atom to act as a reducing or oxidizing agent. For example, examine Table 4-1. The cytochromes that are listed are large, complex types of molecules of a general class known as porphyrins. The cytochromes, in addition to

TABLE 4-1 STANDARD REDUCTION POTENTIALS FOR SOME SYSTEMS OF BIOCHEMICAL IMPORTANCE

System	E'_0 (pH 7), volts
Oxygen/water	0.815
Ferric/ferrous	0.77
Cytochrome *a*; ferric/ferrous	0.29
Cytochrome *c*; ferric/ferrous	0.22
Cytochrome b_2; ferric/ferrous	0.12
Fumaric acid/succinic acid	0.03
Pyruvate + ammonium/alanine	−0.13
α-Ketoglutarate + ammonium/glutamic acid	−0.14
Pyruvate/lactate	−0.19
NAD^+/NADH	−0.32
H^+/H_2	−0.42
Acetate + carbon dioxide/pyruvate	−0.70

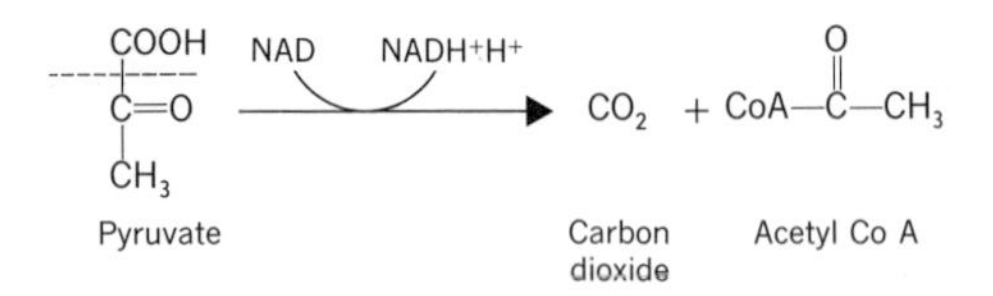

Fig. 4-3 The conversion of pyruvate. In the course of this reaction, the carboxyl group is lost by the pyruvate and the remaining two-carbon acetyl fragment is joined to Coenzyme A.

H^+ + Co A + COOH–C(=O)–CH_3 + NAD ⟶ CO_2 + Co A–C(=O)–CH_3 + NADH + H^+

Fig. 4-4 Source of hydrogen for NAD ⟶ NADH + H^+. NAD received one hydrogen from the carboxyl group of pyruvate and a hydrogen ion from the medium.

the basic porphyrin structure, have an atom of iron (Fe) bonded within the structure. Ionic iron may exist in two forms, Fe^{2+} and Fe^{3+}. These differ by one electron, and the conversion of one form to the other by means of a loss or gain of electron is a well-known redox reaction in inorganic chemistry.

$$Fe^{2+} \longrightarrow Fe^{3+} + e^-$$

The E_0 of this partial redox reaction, as you will note from Table 4-1, is +0.77. When a Fe atom is combined in an organic molecule such as a cytochrome, its redox potential changes. The Fe on the cytochrome molecule still acts as the electron donor or recipient, but its tendency to give up electrons is increased. The tendency to lose electrons is due to the particular molecular structure of the cytochrome. Cytochrome **a** has an E_0 value of +0.29; cytochrome **b** has an E_0 value of −0.04: cytochrome **c** has an E_0 value of +0.25. One of the most remarkable features of biological systems is, in fact, the presence of organic molecular structures with a variety of redox potentials that have been tailor-made for the efficient function of the system through the evolutionary process. There is nothing comparable to this variety of redox potentials in the inorganic world.

Undoubtedly, the most frequent reaction taking place in biological systems is the removal of a hydrogen atom from part of an organic molecule. In the course of this removal, NAD is the most common hydrogen acceptor, and thus acts as an oxidizing agent of the reaction. The feasibility of NAD as an oxidizing agent is indicated in Table 4-1. The reaction

$$\text{pyruvate} \longrightarrow CO_2 \text{ and acetate}$$

has a more negative E_0 value than NAD + H ⟶ NADH + H^+, and pyruvate acts as the reducing agent for a reaction between these two compounds (see Fig. 4-3). Therefore, the reaction

$$NAD^+ + \text{pyruvate} \longrightarrow NADH + H^+ + 2e^- + CO_2 + \text{acetate}$$

will proceed preferentially in the indicated direction. (When NAD^+ accepts a hydrogen, it also undergoes ionic bonding to a second hydrogen, often from the surrounding medium. Hence the product is NADH + H^+. See Fig. 4-4.)

Since the regeneration of ATP is dependent on hydrogen-removal reactions such as that described above, there should be some means of equating E_0 value with ΔG. It has been possible to derive an expression relating these values because both are directly related to the direction in which a reaction will preferentially proceed. In redox reactions, as in hydrolytic reactions, the sign (+ or −) of the ΔG does not indicate the energy of activation, and therefore does indicate whether a reaction will occur spontaneously. Instead, the negative sign indicates the release of energy from the molecule *when* the energy of activation is reached, while the positive sign indicates the gain of energy to the molecule *when* the reaction takes place. The mathematical expression relating the ΔG to the E_0 value is

$$\Delta G = -nF\,\Delta E_0$$

where **n** is the number of electrons transferred in an oxidation-reduction reaction, **F** is a constant that permits the conversion of voltage readings into caloric equivalents (23,063 cal/V) and ΔE_0 is the difference in the potential between the oxidizing and reducing agents. In the above example, there is one electron transferred in the course of the reaction, so $n = 1$. The oxidizing agent NAD^+ has an E_0 value of -0.320, while the reducing agent, pyruvate, has an E_0 value of -0.700, and the ΔE_0 (E_0 of the oxidizing agent minus the E_0 of the reducing agent) equals -0.380.

$$\begin{aligned} \Delta G &= -nF\ \Delta E_0 \\ \Delta G &= -(1)\ (23{,}063)\ (-0.380) \\ \Delta G &= 8740 \text{ cal or approximately } 9 \text{ kcal/mol} \end{aligned}$$

In biological systems, there are many compounds that have more positive redox potentials than $NAD^+/NADH$. When these compounds undergo reaction with NAD, the NAD serves as a reducing agent in its $NADH + H^+$ form. NADH, in fact, enters into a series of reactions, all involving redox potentials and the concomitant transfer of electrons, thus leading to the generation of a great deal of energy. This series of reactions is known as the **electron transport system** (**ETS**). In the course of these reactions, there are a series of molecular forms that first act as oxidizing agents by accepting electrons, and then as reducing agents by giving these electrons to another molecular form. The series of reactions usually terminates with the transfer of the electrons to oxygen, thus forming water (see Fig. 4-5). *It is because of the common use of*

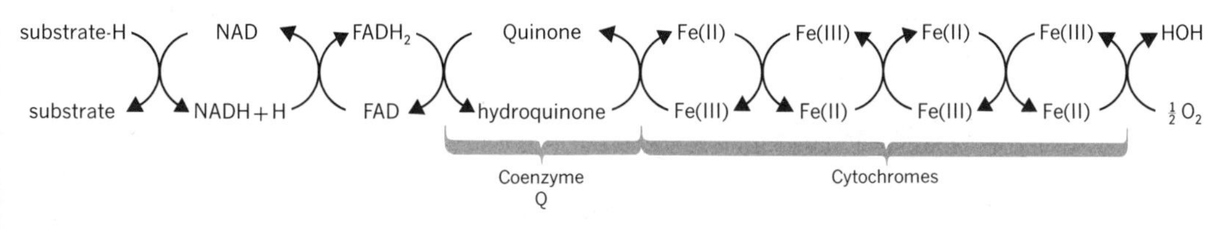

Fig. 4-5 General scheme of the electron transport system.

oxygen as a final hydrogen acceptor in the electron transport system that most organisms require oxygen to live. By temporarily disregarding the intermediary reactions, and considering only a reaction of $NADH + H^+$ with oxygen, one can obtain an idea of the amount of energy generated by the electron transport system.

$$NADH + H^+ + \tfrac{1}{2}O_2 \longrightarrow NAD^+ + II_2O$$

The ΔG of this reaction may be calculated as follows:

$$\begin{aligned} n &= 2 \\ \Delta E_0 &= +0.816 - (-0.320) \\ \Delta E_0 &= +1.136 \text{ V} \\ \Delta G &= -nF\ \Delta E_0 \\ \Delta G &= -(2)\ \ (23{,}063)\ \ (1.135) \\ \Delta G &= -52{,}350 \text{ cal/mol or approximately } -52 \text{ kcal/mol} \end{aligned}$$

In the course of the series of reactions, a total of 52 kcal/mol of energy is released. Some of this energy is used to drive the reaction ADP + Pi $\longrightarrow$ ATP, while some constitutes entropic loss to the system. *It has been calculated that for every NADH + H^+ entering the series of reactions of the electron transport system, there is a generation of 3 ATP molecules.* The generation of ATP molecules in conjunction with the electron transport system is called **oxidative phosphorylation.**

Notice that the generation of three molecules of ATP would take 36 kcal/mol (12 kcal/mol for each generation). Therefore, there is a loss of 16 kcal/mol of energy in the process. The efficiency of the system in producing energy in the usable form of ATP may be calculated as follows.

$$\frac{36 \text{ kcal/mol (energy put to work)}}{52 \text{ kcal/mol (total energy)}} \times 100 = 69.2\%$$

Since the amount of energy within the ATP that is actually available for useful work is only 8 kcal/mol, the efficiency is usually calculated in terms of this.

$$\frac{24 \text{ kcal/mol (energy in useful form)}}{52 \text{ kcal/mol (total energy)}} \times 100 = 48.0\%$$

This is still very high when one considers that the efficiency of an automobile is about 6%. *Therefore, the generation of the primary energy donor for synthetic reactions in biological systems results from a highly efficient transfer of energy derived from the most common reaction in biological systems, the removal of hydrogen atoms from organic molecules.*

Stepwise Reactions

One of the most important features in biological systems is the release of energy in a controlled manner. This implies that (1) the energy of a reaction must be converted into a usable form (ATP) and (2) the heat loss from the reaction must not be great enough to disrupt molecular structures important to the integrity of the system. A stepwise series of enzyme-mediated reactions is thermodynamically the most feasible way of dealing with these factors. For example, in the electron transport system, there is a loss of energy available for work of 16 kcal/mol. If this energy release were to occur at one time, it would be highly destructive. By having a series of stepwise reactions, in which the electron is transferred from one molecule to another, the loss of the 16 kcal/mol is spread out over a period of time, and represents the sum of many smaller energy losses as heat. In this way, there is never sufficient kinetic energy at any one time to act as a destructive factor.

On the other hand, however, at least three of the steps evidently have a sufficient energy release for the generation of ATP. ATP regeneration cannot occur as a summing of the energies from several reactions, but involves the release of energy from a single reaction in a 12 kcal/mol packet. The three reactions in the electron transport system that yield this energy can be noted in Fig. 4-6. So the enzymes and their substrates in the electron transport system have developed through the selective processes of evolution to fulfill both of these requirements.

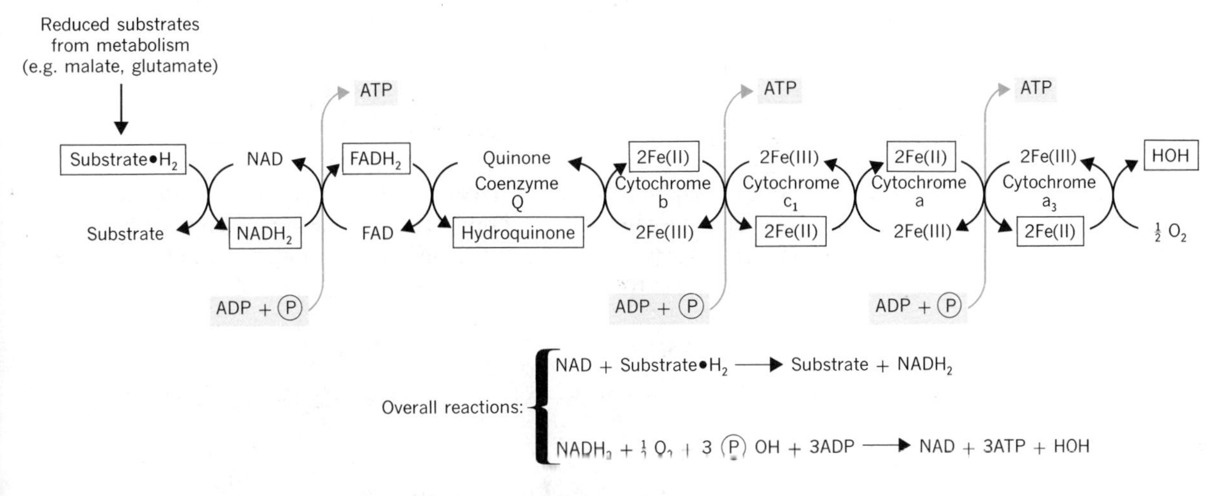

Fig. 4-6 ATP regeneration through the electron transport system.

Another important aspect of stepwise degradation is that such a system permits the linkage of degradation and synthetic processes. The source of many of the monomers used in synthetic processes in a biological system are the partial-degradation products of the nutrient material. The creation of these monomers is possible only through a stepwise degradation. For example, one of the twenty amino acids necessary in protein synthesis is **alanine,** the structure of which is

$$\begin{array}{c} COOH \\ | \\ HC\text{—}NH_2 \\ | \\ CH_3 \end{array}$$

alanine

The structure of a **glucose** molecule entering the system is

$$\begin{array}{c} HC{=}O \\ | \\ HCOH \\ | \\ HOCH \\ | \\ HCOH \\ | \\ HCOH \\ | \\ CH_2OH \end{array}$$

glucose

No single enzyme-mediated reaction could convert glucose to alanine; the energy requirement would be prohibitive. Yet in the course of the stepwise degradation of a glucose molecule in a biological system, the glucose is eventually broken down into two three-carbon molecules that take the form

$$\begin{array}{c} \mathrm{COOH} \\ | \\ \mathrm{CH_2} \\ | \\ \mathrm{CH_3} \end{array}$$

pyruvic acid

and the compound is called **pyruvic acid** or **pyruvate.** At this point, a single enzyme-mediated reaction leads to the **amination** (addition of NH_3) of pyruvate and its conversion into the amino acid **alanine.** The amine group is transferred from another organic molecule.

$$\begin{array}{l} \mathrm{COOH} \\ | \\ \mathrm{CH_2} \\ | \\ \mathrm{CH_3} \end{array} + \mathrm{R{-}NH_3} \longrightarrow \begin{array}{l} \quad\mathrm{COOH} \\ \quad | \\ \mathrm{HC{-}NH_2} \\ \quad | \\ \quad\mathrm{CH_3} \end{array} + \mathrm{H^+} + \mathrm{H^+}$$

pyruvate + amine ⟶ alanine

This linkage between degradative processes and synthetic processes is a primary factor in the conservation of the energy of the system. Not only is the direct conversion of glucose to alanine impossible from an energy standpoint, but the synthesis of alanine *de novo* would also require a great expenditure of energy. In the first place, the energy for the synthetic reactions involved would be prohibitive. Secondly, as these would undoubtedly be enzyme-mediated reactions, the system would have to expend energy for the production of the enzymes themselves. Thirdly, it would be a waste of energy to the system to take in an organized carbon structure such as glucose and break it down completely to carbon dioxide and water in order to extract most of the energy from it, only to use part of that energy for forming the same or highly similar carbon structures once again. In every chemical reaction there is a decrease of usable energy (increase in entropy) and therefore, the fewer the reactions in the system, the higher the efficiency of that system. The number of reactions within a system must, therefore, represent the lowest possible number in order for that particular system to obtain energy in a usable form by controlled energy release. A major facet in reducing the number of reactions is the linkage between degradative and synthetic processes, which can only arise from the stepwise degradative reactions.

THE CONCEPT OF A METABOLIC PATHWAY

A **metabolic pathway** is the series of stepwise reactions that lead to the formation of a final product. This product may be a biopolymer, in which case the pathway is **biosynthetic;** or it may be a waste product, such as carbon dioxide and water, in which case the pathway is **degrada-**

tive. (Some of the literature uses the term **anabolic** for synthetic processes and **catabolic** for degradative processes.)

In the course of a pathway, the product of one enzyme-mediated reaction becomes the substrate for another. This proceeds until the final product is formed. The sequence of reactions within the pathway is determined by the enzymes present in the system. Each enzyme is capable of recognizing only specific substrates and mediating specific changes (see Fig. 4-7).

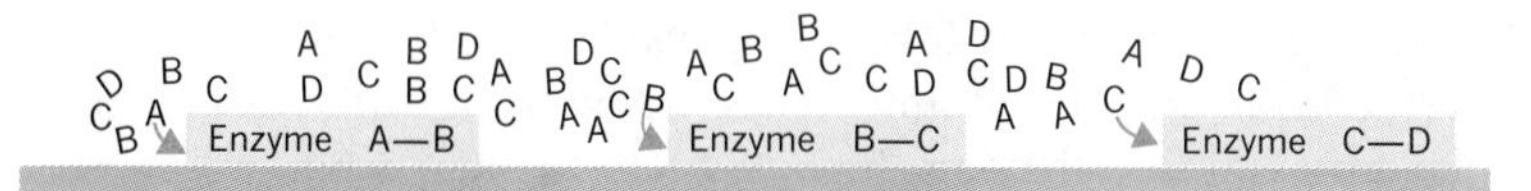

Fig. 4-7 Recognition of substrate by enzymes. Many different types of molecules collide with many different types of enzymes. Only those molecules for which the enzyme is specific will join with the enzyme to form the enzyme substrate complex necessary for the completion of the reaction.

The final product of a pathway is also determined by the enzymes present in the system. For instance, one organism may be able to synthesize the enzymes necessary for a sequence of reactions

$$A \longrightarrow B \longrightarrow C \longrightarrow D \longrightarrow E$$

and for that organism, E represents the final product. Another organism might only be able to synthesize the enzymes necessary for the conversion of

$$A \longrightarrow B \longrightarrow C$$

so for that organism, C represents the final product. If the pathway is a synthetic one, and the final product E represents a molecule necessary for further biosynthesis, the first organism is capable of synthesizing it if it is provided with A, B, C, or D in the course of another metabolic pathway or in its nutrient material. The second organism, however, cannot synthesize it and must take the completed form E in with its nutrient material. When E is unavailable to the second organism, it may die. Therefore, the ability of the organism to synthesize enzymes determines the nutritional requirements of that organism.

If, on the other hand, the pathway represents a degradative pathway, the first organism will eliminate E as a waste product, while the second will eliminate C as a waste product. E would represent a lower degradative product, and more energy would have been extracted from it for use in the system than from C. Therefore, the degradative pathway of the first organism is more efficient in terms of energy gain to its system than that of the second organism. And again, the efficiency of an organism is directly related to the enzymes it is capable of synthesizing.

One of the major features of metabolism is the interrelationship of the metabolic pathways. The great majority of organisms have one major pathway that starts with glucose, which is broken down in a stepwise fashion. Many steps of the breakdown have a low energy of activation that can be fulfilled by enzymes alone. Therefore, the breakdown is accomplished with a minimal energy expenditure on the part of the system. At various points in the breakdown, the molecular forms are such that they can be siphoned off and converted by minor rearrangements, addition, or subtraction of groups to forms needed by the system in biosynthetic activities. This major breakdown pathway for carbohydrates acts as a source for most of the monomers required by biological systems: amino acids, fatty acids, purines and pyrimidines, porphyrins, etc. By the same token, the auxiliary pathways provide a means for rearranging the atomic relationships in a wide variety of molecular forms so that they are funneled into the major pathway and undergo degradation.

Metabolism must be viewed as a large series of interrelated reactions. Each individual metabolic reaction has an associated equilibrium between the substrate and the product that determines the rate. Because the product of one reaction acts as a substrate for the next, each rate is dependent on the one before it. As long as the product of a given reaction is being siphoned off into another reaction, it will continue to produce the product. When one reaction slows down, all others will concomitantly slow down.

This interdependence is the reason why poisons have such a devastating effect on metabolic systems. For example, cyanide combines with cytochrome oxidase, one of the enzymes responsible for the transfer of electrons in the electron transport system. When this occurs, the electrons cannot be transferred. The reactions dependent on this step also stop. $NADH^+ + H^+$ cannot be converted into NAD^+. Therefore, it cannot act as an oxidizing agent in metabolic reactions; all reactions requiring NAD^+ will thus stop, and the organism will die. The interrelationship of metabolic systems is such that the entire metabolism can be affected by the alteration of the rate of one *critical* enzyme-mediated reaction.

It is the equilibria of the individual metabolic reactions that often determine the course to be taken by any given molecule. Many of the molecules that enter the system, or that are in various stages of degradation, have a variety of potential pathways. The one that is actually followed is the result of a series of equilibrial states within the system. For example, a molecule of A enters the system. A may follow either a synthetic pathway B or a degradative pathway C. The reaction $A \longrightarrow B$ is proceeding at a turnover rate of 1000 molecules per second, while the reaction $A \longrightarrow C$ is proceeding at the rate of 500 molecules per second. The chances are 2 to 1 that a single molecule of A will enter the faster reaction, $A \longrightarrow B$. If 300 molecules of A enter the system, about 200 will go into the $A \longrightarrow B$ reaction, and about 100 will go into the $A \longrightarrow C$ reaction. When the equilibria of these reactions shift, the chance of molecules entering particular reactions shifts accordingly.

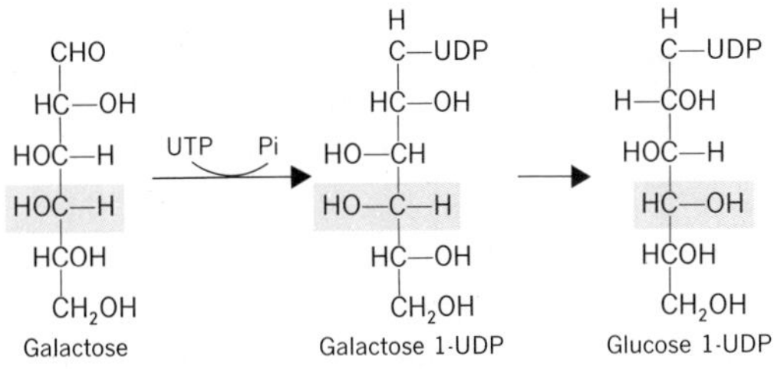

CHO / HC—OH / HOC—H / HOC—H / HCOH / CH_2OH — Galactose

UTP → Pi

H / C—UDP / HC—OH / HO—CH / HO—C—H / HC—OH / CH_2OH — Galactose 1-UDP

→

H / C—UDP / H—COH / HOC—H / HC—OH / HCOH / CH_2OH — Glucose 1-UDP

Fig. 4-8 Phosphorylation of galactose. Galactose is phosphorylated as uridine diphosphate attaches to the first carbon of the molecule and an inorganic phosphate is lost to the medium. The addition of UDP to the molecule evidently brings galactose to a more reactive state, so that an enzyme-mediated reaction can change the position of the hydroxyl group on the fourth carbon to generate glucose.

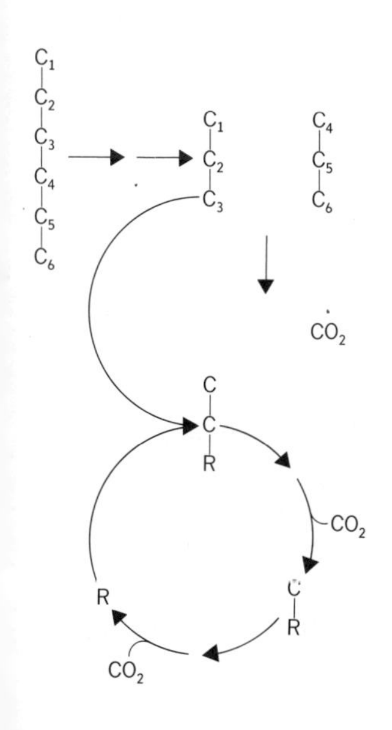

Fig. 4-9 General scheme for the degradation of glucose.

A Major Metabolic Pathway—General Considerations

A major metabolic pathway in most biological systems is that specific for glucose. Many organisms are capable of synthesizing the enzymes necessary for the conversion of other carbohydrates to glucose and, therefore, they can utilize these other carbohydrates as energy sources (see Fig. 4-8).

When glucose is present in the system, it may follow either a biosynthetic or a degradative pathway, as dictated by equilibrial conditions. The biosynthetic pathway for glucose is called **gluconeogenesis** and leads to the polymerization of glucose to glycogen or starch (depending on whether the organism is an animal or a plant). When the equilibrium of the system is such that glucose is required, glycogen or starch is broken down into glucose units, which then go into the degradative or **glycolytic** pathway. For this reason, glycogen and starch are considered energy-storage forms in biological systems. Their presence provides reserves of glucose until such time as it is needed by the system.

The metabolic pathway, starting with glucose and terminating with the formation of pyruvic acid, is called **glycolysis** (sometimes called the Emden-Meyerhoff scheme, honoring the scientists who worked out the sequence). Students generally find a detailed study of such a metabolic pathway quite overwhelming. Yet it is necessary if one is to achieve a coherent idea of the source of energy and monomers for biosynthetic processes. Therefore, let us first look at some of the general features of the glycolytic breakdown before starting a detailed study.

A glucose molecule is made up of six carbon atoms, twelve hydrogen atoms, and six oxygen atoms. First consider only the fate of the carbon skeleton in the degradative process (see Fig. 4-9). The six-carbon molecule undergoes a series of changes, until the arrangement of the groups associated with the carbon structure is such that there is a weakening of the bond between the third and fourth carbon. Then the six-carbon molecule is broken into *two* three-carbon molecules. The *two* three-carbon molecules then undergo a series of minor changes through stepwise reactions leading to the formation of pyruvic acid. At this point there are still as many carbons as there were on the original molecule ($2 \times 3 = 6$). The formation of pyruvic acid marks the end of glycolysis.

Pyruvic acid then undergoes a reaction in which the first carbon is removed. This first carbon is in a carboxyl group, whose removal causes an immediate rearrangement of the bonds.

$$-\overset{\displaystyle O}{\overset{\|}{C}}-OH \longrightarrow O{=}C{=}O + H^+$$

carbon dioxide

Carbon dioxide, therefore, is formed in metabolic reactions by the removal of the carboxyl group from organic acids.

There is now a two-carbon unit remaining from the original pyruvic acid (*two* two-carbon units from the original glucose molecule). These two-carbon units enter the next phase of metabolism, the **Krebs cycle.** It should be noted that the Krebs cycle is not specific for carbohydrate breakdown, but rather serves as the hub of metabolism; the sequence of reactions into which feed the breakdown products of amino acids, nitrogen bases, and fatty acids as well as carbohydrates. Each of the two-carbon units is joined to a four-carbon molecule, forming a six-carbon molecule. After a series of minor changes of the groups associated with the carbon structure, the energy relationships of the molecule are such that one of the carboxyl groups may be removed with little expenditure of energy. The removal of this carboxyl group leads to the formation of another molecule of **carbon dioxide** and a **five-carbon** molecule. The energy relationships on this five-carbon molecule permit the loss of another carboxyl group immediately. This brings about another molecule of **carbon dioxide** and a **four-carbon molecule.** The four-carbon molecule undergoes a series of changes that bring it to the molecular form capable of bonding with another two-carbon unit, thereby completing the cycle. *Two* two-carbon fragments from each original glucose molecule enter the Krebs cycle. In the course of the cycle, two carbon dioxide molecules are given off for every two-carbon fragment entering. Therefore, the original six carbon atoms of glucose are now in the form of carbon dioxide.

The removal of the six molecules of carbon dioxide from the system represents a removal of six carbon atoms and twelve oxygen atoms, six more oxygen atoms than entered with the glucose. These oxygen atoms are contributed by the system itself, primarily from phosphate groups and the dissociation products of water.

The hydrogen atoms originally associated with the glucose molecule have another fate. They are removed from the molecule by oxidizing agents related to the electron transport system, NAD^+ and FAD. In their reduced form, these now enter the electron transport system, generating energy through oxidative phosphorylation and leading to the formation of water. So the complete degradative products of glucose in a biological system are the same as those from direct combustion (carbon dioxide and water).

THE DEGRADATIVE PATHWAYS

Carbohydrate Breakdown—A Detailed Study

When glucose enters a biological system, it undergoes an immediate reaction in which it is phosphorylated on its sixth carbon to form a molecule of **glucose-6-phosphate** (see Fig. 4-10). In this reaction, ATP acts in its role as both an energy and a phosphate donor, and in doing so, lowers the energy of activation of subsequent reactions. If the glucose is to be polymerized to glycogen or starch, glucose-6-phosphate is dephosphorylated in the series of reactions that lead to polymerization. The phosphate group is released into the medium as glucose-1-P + UTP $\longrightarrow$ UDP-glucose + P-P_i, an inorganic phosphate,

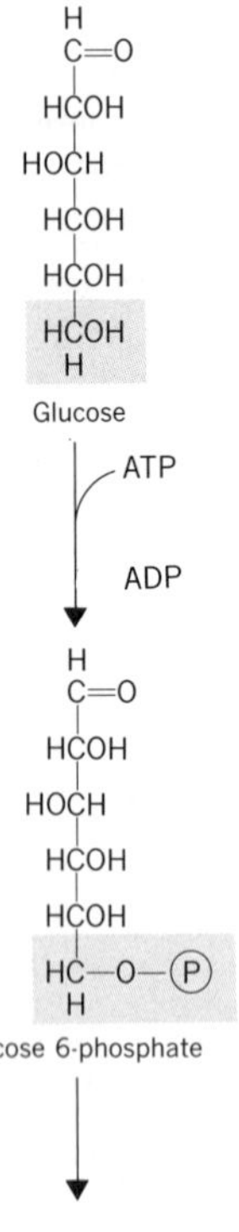

H
C=O
HCOH
HOCH
HCOH
HCOH
HCOH
H
Glucose
ATP
ADP
H
C=O
HCOH
HOCH
HCOH
HCOH
HC—O—Ⓟ
H
Glucose 6-phosphate

Fig. 4-10 Degradation pathway for glucose.

because there is insufficient energy to permit its reassociation to a molecule of ADP. The polymerization of glucose in a phosphorylated form may seem like a waste of precious energy. Actually, the energy required for a direct polymerization of glucose is much greater than that lost when a high-energy phosphate becomes an inorganic phosphate, so this method is energy-conserving.

If glucose-6-phosphate is to undergo degradation, the next reaction leads to its conversion to **fructose-6-phosphate.** This reaction involves a low-energy-requiring rearrangement of the hydrogens on the glucose molecule, and does not involve ATP, either directly or indirectly, but is solely under enzymatic control.

The fructose-6-phosphate is phosphorylated again, this time on the first carbon. This leads to the formation of a molecule of **fructose-1, 6-diphosphate.** Once again, ATP acts as both an energy and a phosphate donor, and in the transfer, the phosphate group moves to a lower energy level. In looking at the three changes that have occurred in the glucose molecule thus far in the pathway, it may seem to be nothing more than complicated molecular gymnastics. It should be firmly kept in mind that the determination of these, as of all the steps of metabolic pathways, was, in the course of evolution, on the basis of their efficiency for the biological system and its operation at the lowest possible and most thermodynamically feasible energy level for the *entire sequence.*

It is at this point in the pathway that the six-carbon molecule breaks into two three-carbon molecules. The structure of fructose-1, 6-diphosphate is such that the bond between the third and fourth carbon is weakened. Although carbon-carbon bonds represent the strongest covalent relationships in biological systems, certain arrangements of the molecular groupings associated with the carbon structure can cause the weakening of even these bonds. The next step in the breakdown is the enzymatic breakage of this bond and the formation of two three-carbon molecules, **3-phosphoglyceraldehyde** and **dihydroxyacetone phosphate.** Dihydroxyacetone phosphate may take one of two possible pathways. By a single-step conversion, it becomes **glycerol phosphate** which, as part of a fat molecule, is necessary for fat synthesis (see p. 61). If the equilibrial conditions of fat synthesis are such that more glycerol is needed by the system, dihydroxyacetone phosphate be converted to glycerol phosphate. If there is sufficient glycerol in the system, dihydroxyacetone phosphate will follow an alternative pathway and be converted into 3-phosphoglyceraldehyde. So the breakdown of fructose-1, 6-diphosphate can lead to the production of two molecules of 3-phosphoglyceraldehyde.

The energy arrangement of the bonds in 3-phosphoglyceraldehyde is such that in a simple enzyme-mediated reaction it can pick up an inorganic phosphate from the surrounding medium and reduce the resonance of this phosphate as it is bound to the first carbon. The resulting molecule, therefore, becomes a high-energy molecule, capable of great energy release upon the removal of this phosphate. In the course of this reaction, a hydrogen is removed from the aldehyde group of the

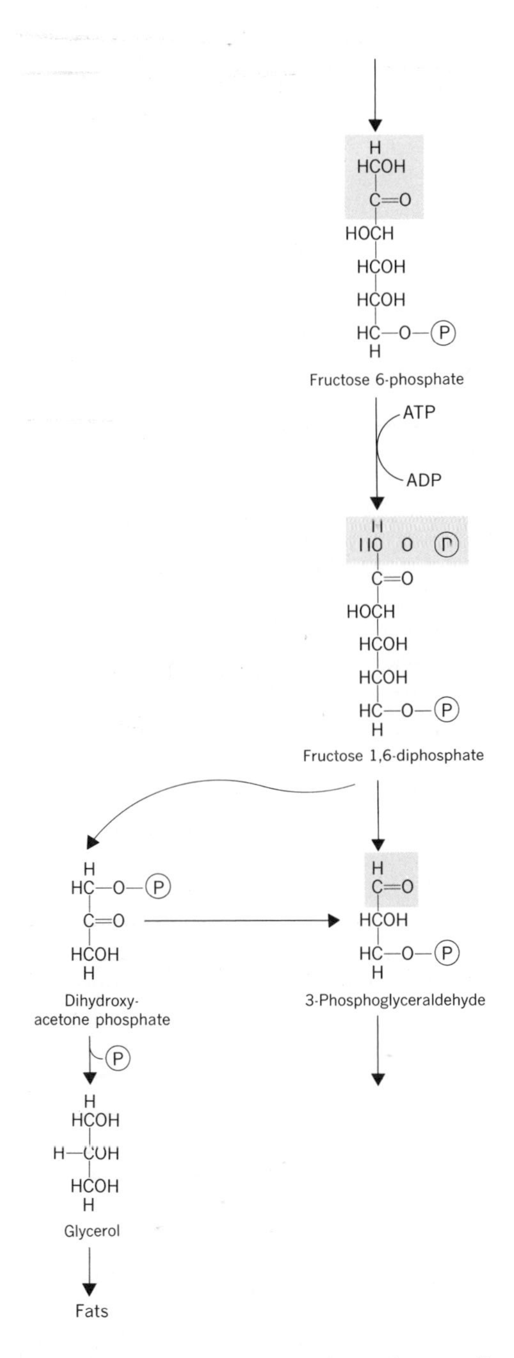

Fig. 4-10 Degradation pathway for glucose (*continued*).

first carbon by $NAD^+ \longrightarrow NADH + H^+$. Notice that the first carbon of this molecule is

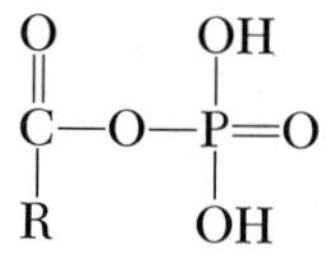

$$\mathrm{R{-}C({=}O){-}O{-}P({=}O)(OH)_2}$$

so that the aldehyde group has been changed to a carboxyl group by the addition of the phosphate radical. This molecule is called **1,3-diphosphoglyceric acid.**

The next reaction is the transfer of this high-energy phosphate to ADP, thereby converting it to ATP. Notice that when this reaction takes place, the phosphate group breaks off between the phosphorus atom and the oxygen atom, and the oxygen of the original phosphate group is left on the molecule.

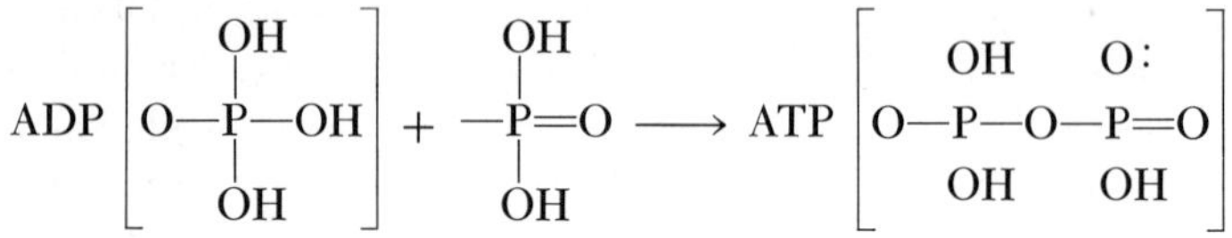

$$\mathrm{ADP}\left[\mathrm{O{-}P(OH)_2{-}OH}\right] + \mathrm{{-}P({=}O)(OH)_2} \longrightarrow \mathrm{ATP}\left[\mathrm{O{-}P(OH)_2{-}O{-}P({=}O)(OH)(O{:})}\right]$$

Therefore, the molecule resulting from this reaction retains its acid character and is now **3-phosphoglyceric acid.**

For every glucose entering the glycolytic pathway, 2 ATPs were converted to 2 ADPs in the original phosphorylations. Now, for every original glucose, there are two molecules of 1,3-diphosphoglyceric acid, *each* of which leads to the conversion of one ADP $\longrightarrow$ ATP. The biological system has recovered its lost energy, and at this point is "even." Any energy derived from the process after this point is profit.

The subsequent two rearrangements of the molecule bring about energy shifts that result in the formation of another high-energy phosphate molecule. First, the 3-phosphoglyceric acid is converted to **2-phosphoglyceric acid** by an enzyme-mediated reaction that transfers the phosphate to a position on the second carbon.

In the next reaction, a hydrogen is lost from the second carbon and a hydroxyl group is lost from the third carbon ($H^+ + OH^-$), leading to the formation of water. The resulting molecule is **phosphoenol pyruvate,** which is characterized by a high-energy phosphate group on the second carbon. This phosphate group is now transferred to an ADP molecule to form ATP.

There are two phosphoenol pyruvates formed for every original glucose molecule entering the pathway, so there are two ATPs formed per original glucose molecule. Since the system was "even" in terms of energy balance when it got back the two ATPs in the reaction converting 1,3-diphosphoglyceric acid to 3-phosphoglyceric acid, the ATP's formed from the loss of the high energy phosphate from phosphoenol pyruvate represent a profit of two high-energy phosphates to the system. The molecular form resulting from the loss of the high-energy phosphate is **pyruvic acid,** or **pyruvate.** This marks the end of the glycolytic pathway.

OH, (P), NAD → NADH + H+

C(=O)—O~(P) / HCOH / HC—O—(P) / H

1,3-Diphosphogyceric acid

ADP → ATP

C(=O)—OH / HCOH / HC—O—(P) / H

3-Phosphoglyceric acid

C(=O)—OH / HC—O—(P) / HCOH / H

2-Phosphoglyceric acid

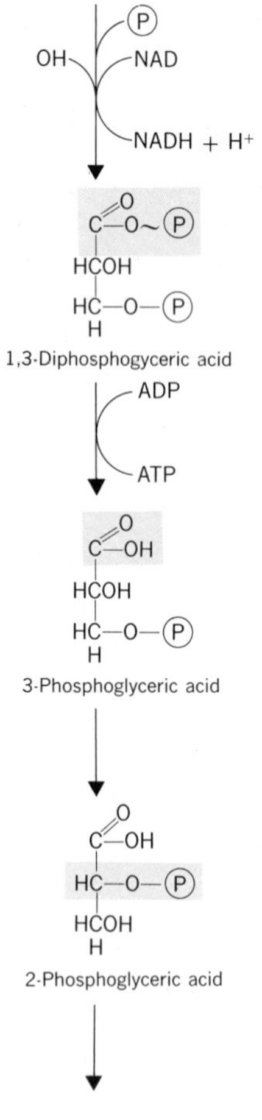

Fig. 4-10 Degradation pathway for glucose (*continued*).

There are many metabolic pathways that may be taken by pyruvic acid, depending on the enzymes and the balance of a variety of molecular forms within the system. At times, the Krebs cycle may be proceeding at its maximum rate, and pyruvate must be siphoned off in another direction. Some organisms, do not have the enzymes necessary in the electron transport system, and therefore cannot produce ATP by oxidative phosphorylation. In such cases, pyruvic acid may act as a hydrogen acceptor for $NADH^+ + H^+$, thereby removing excess hydrogens from the system and permitting the regeneration of the NAD^+ form for use by other metabolic reactions. In the course of this reaction, pyruvic acid is oxidized to **lactic acid.** Lactic acid is eliminated from the system as a waste product.

Other organisms, notably the yeasts, have enzymes capable of breaking the weak bond between the first and second carbon of pyruvic acid. This leads to the formation of carbon dioxide (by rearrangement of the bonds on the free carboxyl group) and a molecule of **acetaldehyde.** The acetaldehyde may act as a hydrogen acceptor for $NADH^+ + H^+$, thereby being oxidized to ethyl alcohol. In these organisms, ethyl alcohol is eliminated as a waste product.

For organisms that terminate the breakdown of glucose with minor rearrangements of the pyruvic acid molecule, the total energy profit to their system in terms of ATP is 2 ATP molecules generated per original glucose molecule. It should be noted that since these organisms do not have electron transport systems, they do not require oxygen, and are said to be **anaerobic** (without air); while organisms with electron transport systems that use oxygen as a final hydrogen acceptor are considered **aerobic** (with air). The glycolytic pathway is sometimes called anaerobic respiration or **fermentation** because at the point of its termination (the formation of waste products from pyruvic acid), oxygen is not needed

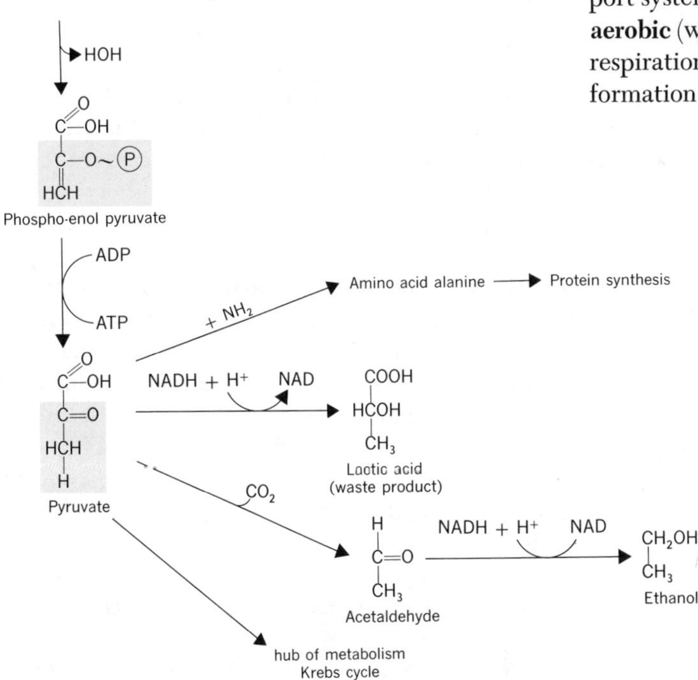

Fig. 4-10 Degradation pathway for glucose (*continued*).

by the system. Pyruvate may also be siphoned off from the breakdown pathway for use in biosynthetic reactions.

When the appropriate enzymes are in the system, pyruvate will be completely broken down. The first step in its degradation is the formation of a high-energy complex with a heteroconjugate, **Coenzyme A.** Coenzyme A is known to form this association through a sulfur linkage, and is usually represented as **CoA-SH.** In the course of the formation of this complex, the carboxyl group of the pyruvic acid is dissociated from the molecule. The immediate rearrangement of the bonds of the free carboxyl group leads to its conversion to carbon dioxide.

The complex of the two carbons from pyruvic acid and the CoA is called **acetyl CoA.** Acetyl CoA may follow biosynthetic pathways or enter into a series of reactions that form a cyclic process, ending where it began. This cycle is known as the **citric acid cycle,** (named after the first component of the cycle that was isolated), or the **Krebs cycle** (honoring the scientist who worked it out), or the **tricarboxylic acid cycle** (because of the presence of molecular forms within the cycle having three carboxyl groups). We shall use the term Krebs cycle for the remainder of the discussion, but as the others are found in much of the literature, it is just as well to be familiar with them.

Acetyl CoA undergoes reaction with a four-carbon molecule, oxaloacetic acid. In the course of this very complex reaction, which involves many additional organic and inorganic factors, the acetyl group is bonded to the second carbon of the oxaloacetic acid molecule, forming the branched molecule of **citric acid** (see Fig. 4-11).

The next reaction in the cycle is a minor change in which the hydroxyl group of the third carbon is transferred to the fourth carbon, thus forming a molecule of **isocitric acid.**

The next reaction involves the conversion of the hydroxyl group to a ketone group by the removal of the hydrogen and the formation of a double bond between the carbon and the oxygen. The removal of this hydrogen is accomplished by a molecule of NADP, a phosphorylated form of NAD^+, which acts in much the same way. The resulting molecule is **oxalosuccinic acid.**

The bond energies of the oxalosuccinic acid are such that the normally strong carbon-carbon bond between the third carbon and its branching carboxyl group is weak. The next reaction is the removal of that carboxyl group from the molecule, and its subsequent rearrangement to form carbon dioxide. The remaining molecular structure is that of **α-ketoglutaric acid.**

α-ketoglutaric acid may undergo an amination reaction to become the amino acid glutamine, or may undergo the reaction that constitutes the next step in the cycle.

When α-ketoglutaric acid continues in the cycle, the next step involves the removal of the terminal carboxyl group and formation of a high-energy complex of the remainder of the molecule with CoA. The free carboxyl group once again forms carbon dioxide. NAD^+ acts as an oxidizing agent, removing the hydrogen from the SH group of the CoA. The resulting high-energy complex is **succinyl CoA.**

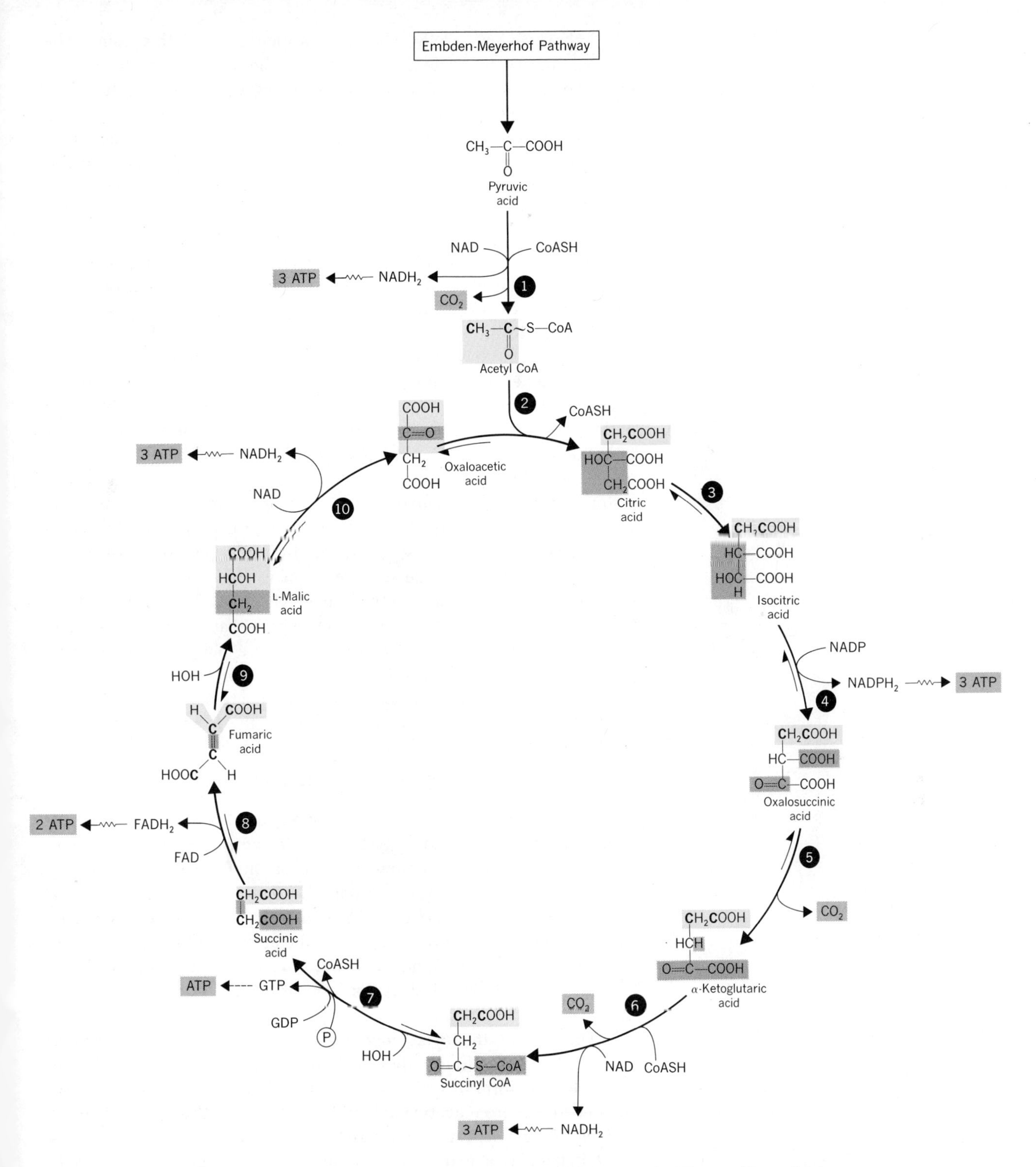

Fig. 4-11 The Krebs cycle. The overall reaction is:

$$C_3H_4O_3 + 1/2\ O_2 + 15\ ADP + 15\ P_i \longrightarrow 3\ CO_2 + 2\ H_2O + 15\ ATP$$

Let us review briefly what has happened up to this point. The original glucose molecule had six carbons. The six-carbon molecule was broken down into two three-carbon molecules. Upon entering the Krebs cycle, one of the carbons was lost as carbon dioxide. In the conversion of oxalosuccinic acid to α-ketoglutaric acid, another carbon was lost in the form of carbon dioxide. While it is true that the carbon lost in this case was not part of the original carbon structure of the glucose, it represents a trade of one carbon for another. In the reaction converting α-ketoglutaric acid to succinyl CoA, another carbon is lost as carbon dioxide. So for each three-carbon molecule of the original glucose (2), there have been three carbons lost as carbon dioxide, one prior to the entrance into the cycle, and two in the course of the cycle. Therefore, at this point in the breakdown process, all the carbons of the original glucose molecule have been converted to carbon dioxide.

From an energy standpoint, the biological system has gained useful energy in the form of two ATP molecules formed in the glycolytic pathway. But the primary means of producing ATP in biological systems is not through substrate-level phosphorylation, but by the oxidative phosphorylation that occurs in conjunction with the electron transport system. Therefore, the dehydrogenations that occur in the course of the glycolytic pathway and the Krebs cycle represent potential energy.

Succinyl CoA may also either follow a biosynthetic pathway or continue in the cycle. The next reaction in the Krebs cycle is both complicated and profitable to the biological system. The position of the CoA on the molecule of succinyl CoA is such that, upon its breakage, there is an energy release sufficient to drive a phosphorylation. This phosphorylation, however, is not to form ATP, but rather GTP (guanosine triphosphate), another nucleotide triphosphate that is active in some synthetic reactions. The reaction leading to the formation of GTP is similar to that leading to ATP

$$\mathrm{GDP} + \mathrm{P}i \longrightarrow \mathrm{GTP}$$

The inorganic phosphate for the reaction is taken from the medium in which the reaction is taking place. Also, in the course of the reaction, a molecule of water undergoes dissociation, and the H^+ is bonded to the third carbon while OH^- is bonded to the fourth. The molecule resulting from this is **succinic acid.** Notice that succinic acid is a four-carbon molecule. The cycle started by the attachment of a two-carbon molecule to a four-carbon molecule to form a six-carbon molecule. Now, two carbons have been lost in the form of carbon dioxide, and a four-carbon molecule remains.

The rest of the cycle involves a rearrangement of the groups associated with the four-carbon skeleton to the state where it takes on the molecular form of oxaloacetic acid, and another acetyl group can be bonded to it from acetyl CoA. In the first step of this rearrangement, succinic acid is converted to **fumaric acid.** In the course of this reaction, FAD is used as a hydrogen acceptor (oxidizing agent), and goes to $FADH_2$.

Fumaric acid then undergoes conversion into **malic acid** by bonding the dissociation products of water. The hydroxyl group is bonded to the second carbon and the hydrogen is bonded to the third carbon.

Malic acid now undergoes a change in which the two hydrogens on the second carbon are accepted by NAD^+. The second carbon becomes a ketone group and the NAD^+ is reduced to $NADH + H^+$. The resulting molecule is oxaloacetic acid, and is capable of either bonding another acetyl group from acetyl CoA to form citric acid once again, or following biosynthetic pathways. Thus, the cycle is complete, and the glucose is completely broken down.

Now look at a balance sheet in terms of the fate of the molecular components and the energy derived from the complete breakdown of a single glucose molecule (see Table 4-2). The original glucose molecule

TABLE 4-2 THE ENERGY-BALANCE SHEET

Reaction	Loss or Gain of High-Energy Phosphates to the System/ Original Glucose Molecule
Phosphorylation of glucose	−1
Phosphorylation of fructose-6-phosphate	−1
Transfer of high energy phosphate from 1,3-diphosphoglyceric acid	2
Transfer of high energy phosphate from phosphoenolypyruvate	2
	end of glycolytic pathway
$NADH + H^+$ from conversion of 3-phosphoglyceraldehyde to 1,3-diphosphoglyceric acid (ETS)	6
$NADH + H^+$ from conversion of pyruvic acid to acetyl CoA (ETS)	6
$NADPH + H^+$ from conversion of isocitric acid to oxalosuccinic acid (ETS)	6
$NADH + H^+$ from conversion of α ketoglutaric acid to succinyl CoA (ETS)	6
Formation of GTP from GDP and P*i* in the course of the conversion of succinyl CoA to succinic acid	2
$FADH_2$ from the conversion of succinic acid to fumaric acid (ETS)	4
$NADH + H^+$ from the conversion of malic acid to oxaloacetic acid (ETS)	6
	Total of 38 ATPs formed

had six carbons, six oxygens, and twelve hydrogens ($C_6H_{12}O_6$). Six carbon dioxide molecules ($6CO_2$) were eliminated from the system. In the course of molecular rearrangements, after the six-carbon structure of the original glucose was broken down into two three-carbon units, there were six dehydrogenations, four with NAD^+ as the oxidizing agent, one with $NADP^+$ as the oxidizing agent, and one with FAD^+ as the oxidizing agent. Therefore there were *twelve* dehydrogenations per original glu-

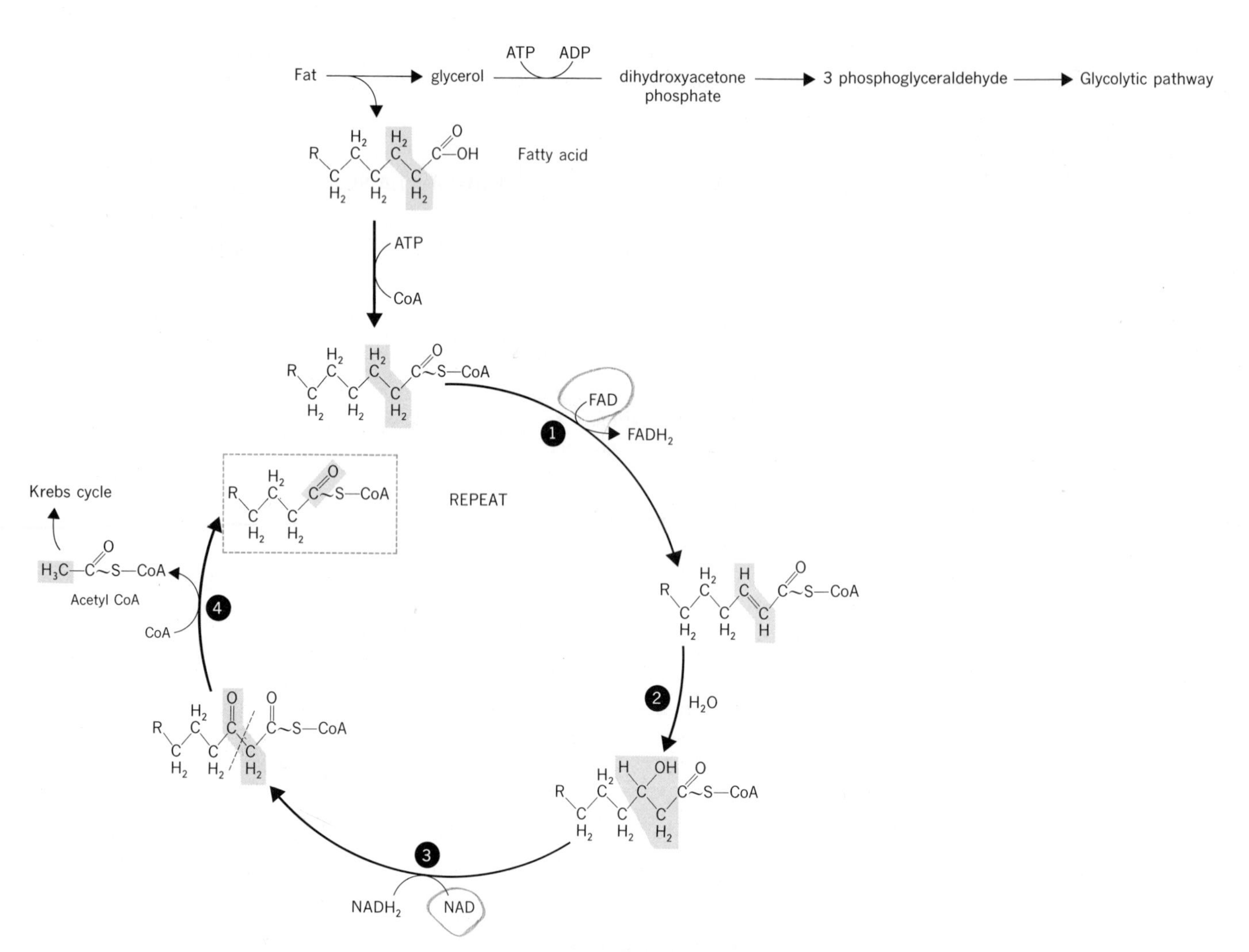

Fig. 4-12 Degradation of fats. Through a series of stepwise reactions, the bond between the second and third carbon of the fatty-acid chain is altered and the third carbon becomes double-bonded to an oxygen atom. The first two carbons of the fatty acid are removed as acetyl CoA, which may then proceed into the Krebs cycle. The remaining chain, now two carbons shorter, repeats the set of reactions. Thus fatty acids are broken down into two-carbon fragments.

cose molecule. These oxidizing agents enter the electron transport system in their reduced form ($NADH + H^+$, $FADH_2$ and $NADPH + H^+$) (see Fig. 4-6). Notice in Fig. 4-6 that there is an ATP formation generated in the reaction of $NADH + H^+$ and FAD^+. The same is true for $NADPH + H^+$ and its reaction with FAD^+. However, when FAD is used as an oxidizing agent in the metabolic reactions, the resulting $FADH_2$ can only undergo reaction with **coenzyme Q** (**quinone**). Therefore, the entrance of $FADH_2$ into the system generates *two* rather than *three* ATPs.

Other Degradative Pathways

Nutritional material also includes fats, proteins, and nucleic acids. The degradative pathways that handle these nutrient forms funnel, in many cases, into the major metabolic pathway for glucose.

For example, the glycerol portion of a fat molecule can be converted in a two-step reaction to dihydroxyacetone phosphate, and then to 3-phosphoglyceraldehyde. The latter will continue its breakdown through the glycolytic pathway. The fatty-acid chains are broken down into two carbon units, each of which is bonded to a molecule of CoA to form acetyl CoA. These enter the Krebs cycle (see Fig. 4-12).

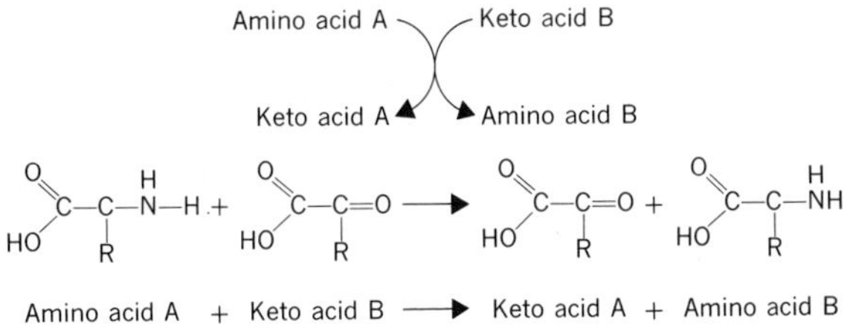

Fig. 4-13 Transamination of amino acids.

It is possible to compare the energy yield from the degradation of fatty acids and glucose. In order to afford a more direct comparison, only a six-carbon fragment of a fatty acid need be considered. The total number of ATPs generated by the breakdown of a six-carbon fatty-acid fragment is 50, twelve more than that generated by a carbohydrate with a comparable number of carbons. It should be noted that animals have localized fat depots that serve as energy reserves. Thus, energy-rich fat deposits lie both physically and metabolically at the outskirts of typical activities. The deposition of fat and the tendency of deposited fat not to take part in constant metabolic interchanges is a major factor in the difficulty obese people have in losing weight.

It is not possible to compare the energy yield of protein or nucleic-acid degradation with that of fats or carbohydrates. When one organism uses another type of organism for its food material, the former takes into its system proteins and nucleic acids that contain different ratios of monomers from its own. The protein and nucleic acids, therefore, must be reduced to their monomer form and rearranged in sequences that are native to the feeding organism.

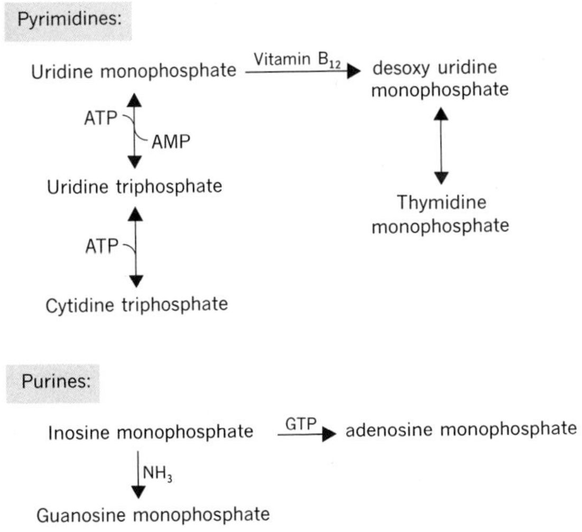

The depolymerization of proteins and nucleic acids yields monomeric forms that are in different ratios from those required by the organism. The equilibrium of the system dictates either the breakdown of the excess monomers or their conversion into usable forms. One of the more important means of converting amino acids is through transamination (see Fig. 4-13). The purines are interconvertible, as are the pyrimidines (see Fig. 4-14).

Fig. 4-14 Conversion of pyrimidines and purines. The pyrimidine uridine monophosphate may be converted either to thymidine or to uridine triphosphate and then to cytidine triphosphate. The purine inosine monophosphate may be converted either to adenosine monophosphate or to guanosine monophosphate. Each is then converted to the triphosphate form. All these reactions are reversible.

Excess monomers, and molecules resulting from conversion processes usually feed into either the glycolytic pathway or the Krebs cycle.

BIOSYNTHETIC PATHWAYS

Although the degradative pathways serve as important energy sources for the cell, they are also important as sources of monomers for the biosynthetic pathways. It has been estimated that only 2% of the glucose entering a system is totally degradated to carbon dioxide and water. The remaining 98% is siphoned off at various points in the degradative process and converted to useful monomeric forms in synthetic reactions (Fig. 4-15). Since this ratio represents that of a highly efficient system, the energy derived from the breakdown of only 2% of the glucose must be sufficient to drive the endergonic reactions in which the other 98% take part.

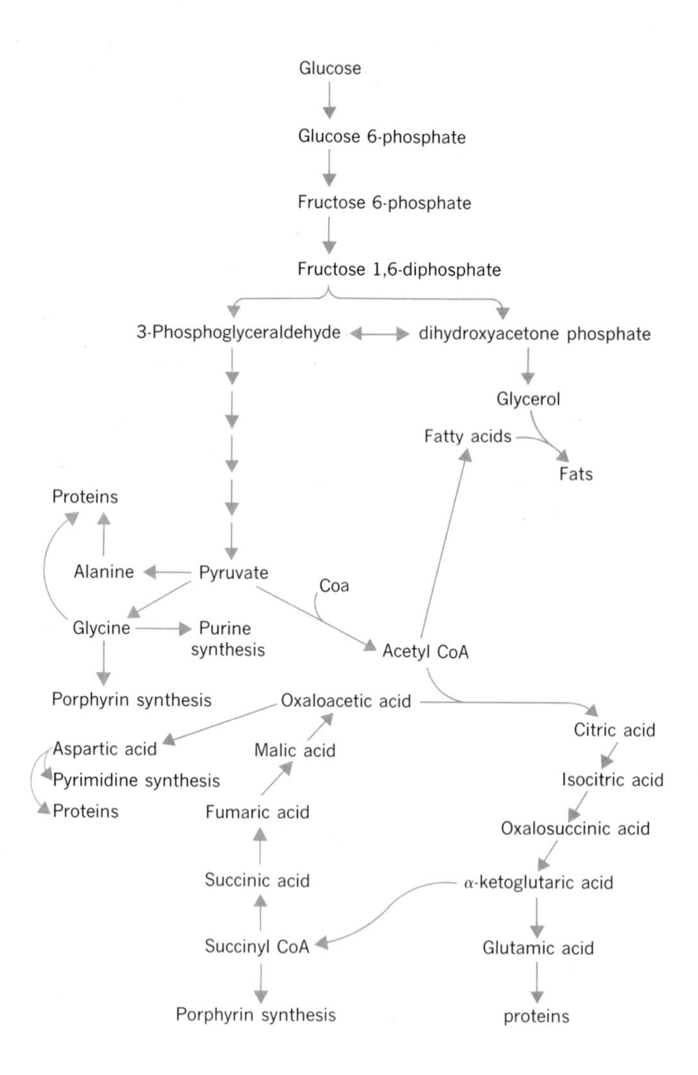

Fig. 4-15 General pathways with major siphoning and feed-in points.

The energy expenditure by the cell in the formation of its polymeric forms can arise from two sources. The first, and most obvious, is the direct use of ATP in a synthetic reaction. Its conversion, in the course of the reaction to ADP, represents a loss of one high-energy phosphate, while its conversion to AMP represents the loss of two high-energy phosphates. The second, and less obvious, means of expending energy is by the use of NADH + H^+ as a hydrogen donor in synthetic reactions. This means that NADH + H^+ will not enter the electron transport system and yield three ATP molecules through oxidative phosphorylation. Therefore, the use of this molecule as a hydrogen donor in synthetic reactions represents a loss of three ATP molecules to the cell.

Biosynthesis of Glucose Polymers

Most organisms store glucose in a polymerized form. The product of polymerization is dependent on the enzymes of the organism and the linkages between different glucose molecules that these enzymes foster.

Regardless of the product, the metabolic pathway is the same. Glucose-6-phosphate is converted to glucose-1-UDP, in which a molecule of uracil diphosphate contributes the phosphate in the 1 position of the glucose, and remains temporarily attached to the glucose molecule. Glucose-1-UDP is then incorporated on the terminus of a growing polymer chain (see Fig. 4-16).

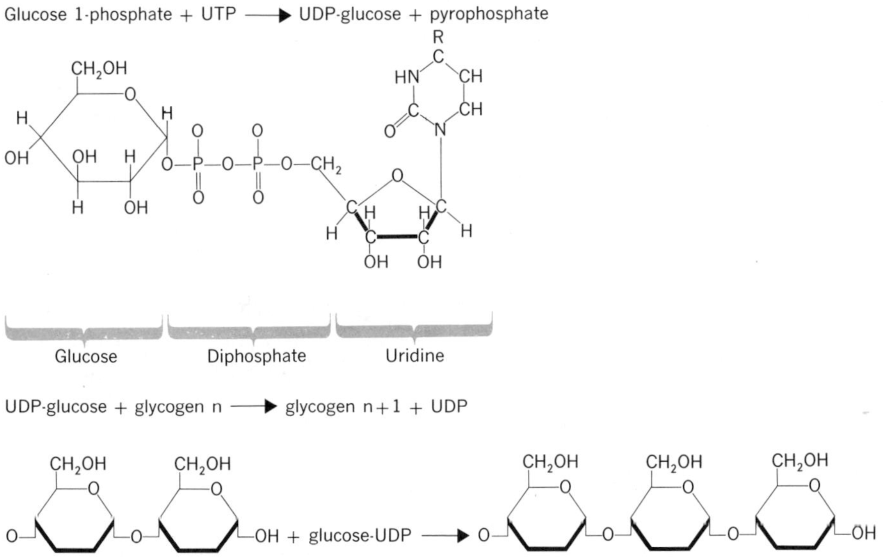

Glucose 1-phosphate + UTP ⟶ UDP-glucose + pyrophosphate

Glucose   Diphosphate   Uridine

UDP-glucose + glycogen n ⟶ glycogen n+1 + UDP

OH + glucose-UDP ⟶

Fig. 4-16 The synthesis of glycogen.

Unlike fat-storage deposits, glucose polymers remain very much in the mainstream of metabolism. Glyconeogenesis and glycolysis are constantly occurring, so that glycogen is constantly being formed and broken down. Glycogen, starch, or any of the other polymerized forms of glucose, act, therefore, as immediate energy-storage reserves.

Biosynthesis of Fats

The fat material taken into the animal is frequently broken down into glycerol and fatty-acid chains. These may recombine and proceed to the fat depots, where they are deposited and serve as remote energy reserves.

When there are excessive quantities of carbohydrates, these too can be converted into fat, and deposited in appropriate regions of the animal body. The integration of carbohydrate metabolism and fat synthesis is achieved through different parts of the glycolytic scheme and Krebs cycle. Dihydroxy-actetone-phosphate is readily converted to glycerol, a portion of the fat molecule. Fatty acids are formed by a complex pathway in which a molecule of malonyl-CoA, a three carbon unit, initiates a chain into which acetyl-CoAs, two-carbon units, are incorporated (see Fig. 4-17). Acetyl-CoA is the two-carbon unit formed from pyruvate at the start of the Krebs cycle.

$$\underset{\text{Acetyl CoA}}{CH_3\overset{\overset{O}{\|}}{C}\text{—S—CoA}} + n\ \text{Malonyl—S—CoA} + 2n\ \text{NADPH} + 2n\ H^+ \longrightarrow n\ \text{CoA—SH} + \underset{\text{Fatty acid—CoA}}{CH_3(CH_2)n\text{—}\overset{\overset{O}{\|}}{C}\text{—S—CoA}} + 2n\ \text{NADP} + nCO_2$$

Fig. 4-17 Biosynthesis of fatty acids. The synthesis of fatty acid originates with a three-carbon malonyl-Coenzyme A complex to which two-carbon units are bonded. These two-carbon units enter the reaction as acetyl Coenzyme A complexes.

Because of the complete integration of metabolic pathways, it is possible for excess nucleic acids, proteins, etc. to be converted into fat molecules.

Biosynthesis of Proteins

Protein synthesis, unlike that of carbohydrates and fats, requires the incorporation of specific monomers in specific sequences. Thus, the key to protein synthesis lies more in the realm of cellular control than in the chemical aspect of the process. Accordingly, the discussion of protein synthesis is delayed until Chapter 9.

One important aspect of protein synthesis, the source of amino acids, is, however, germane to a discussion of metabolism. In order to synthesize proteins, a cell must have a full complement of the twenty amino acids in the necessary ratios. These amino acids may be either brought into the cell with other nutritional material, produced within the cell through metabolic pathways, or both.

As previously discussed, the ratios of incoming amino acids may be altered through transaminations. Thus some amino acids can be generated by altering others.

The primary source for amino acids is through the metabolic pathways. In some cases, amino acids are produced quite directly. The amination of pyruvic acid yields the amino acid alanine. The amination

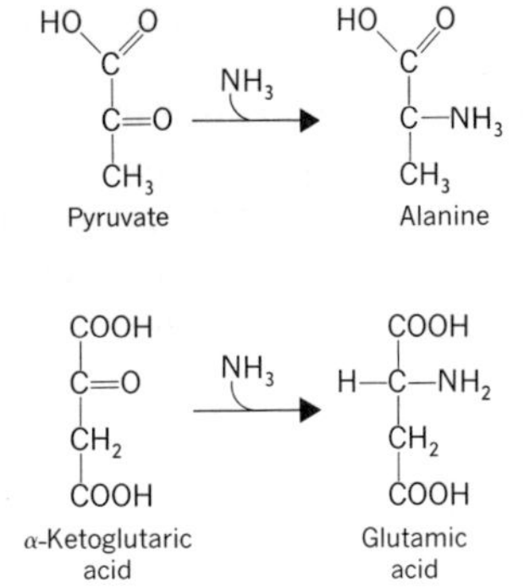

Pyruvate $\xrightarrow{NH_3}$ Alanine

α-Ketoglutaric acid $\xrightarrow{NH_3}$ Glutamic acid

Fig. 4-18 Amination of keto acids. Amination of pyruvic acid yields the amino acid alanine; amination of α-ketoglutaric acid yields the amino acid glutamic acid. The source of the amine group is other amino acids, and the reaction is usually by transamination.

of α-ketoglutaric acid yields the amino acid glutamine (see Fig. 4-18). In other cases, the production of an amino acid may require several steps from the major metabolic pathway.

Many organisms are incapable of producing all twenty amino acids. Those amino acids that are not within the metabolic capacity of the organism must be taken in with its nutritional material, and thus constitute part of the organism's nutritional requirement.

Biosynthesis of Nucleic Acids

A discussion of the incorporation of purines and pyrimidines into nucleic acids requires a knowledge of cellular control. Therefore, this discussion is delayed until Chapter 10.

Although some cells are incapable of synthesizing certain derivatives of purines or pyrimidines, all cells are capable of synthesizing the basic structure of purines or pyrimidines necessary for incorporation into nucleic acids. These syntheses build upon molecules in the Krebs cycle, and proceed to build appropriate ring structures (see Fig. 4-19).

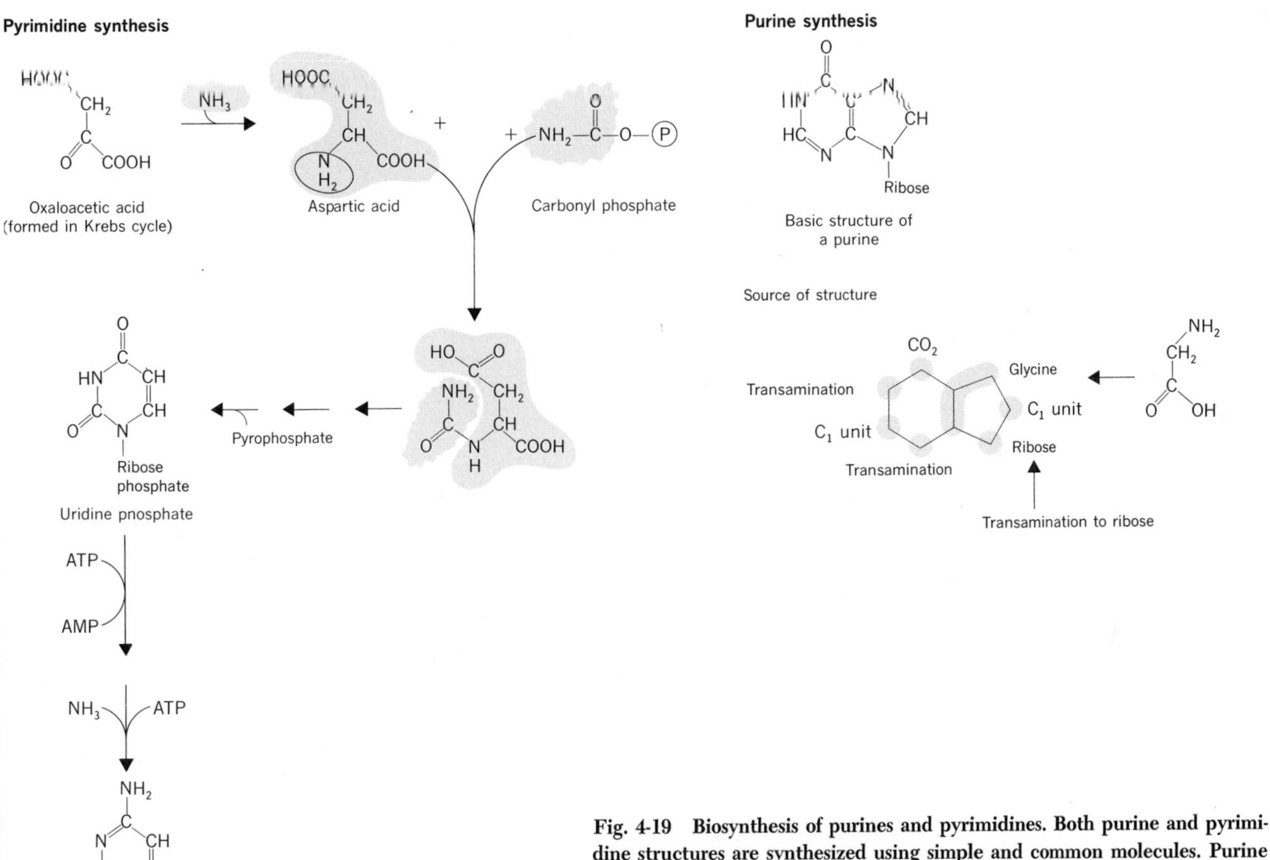

Fig. 4-19 Biosynthesis of purines and pyrimidines. Both purine and pyrimidine structures are synthesized using simple and common molecules. Purine synthesis begins with the amination of ribose phosphate and the bonding of the ribose phosphate amine to the simple amino acid glycine. Pyrimidine synthesis begins with the association of carbonyl phosphate and aspartic acid to generate the ring.

Biosynthesis of Porphyrins

Porphyrins are complex ring structures that usually surround a metal (see p. 61). The cytochromes in the electron transport system, chlorophyll, and hemoglobin, all have porphyrin rings as their active sites.

All cells that require a porphyrin structure for some aspect of their activity can synthesize the porphyrin ring. The synthesis of the porphyrin ring begins with two common, and therefore plentiful, molecules, succinyl CoA from the Krebs cycle, and the amino acid glycine (see Fig. 4-20).

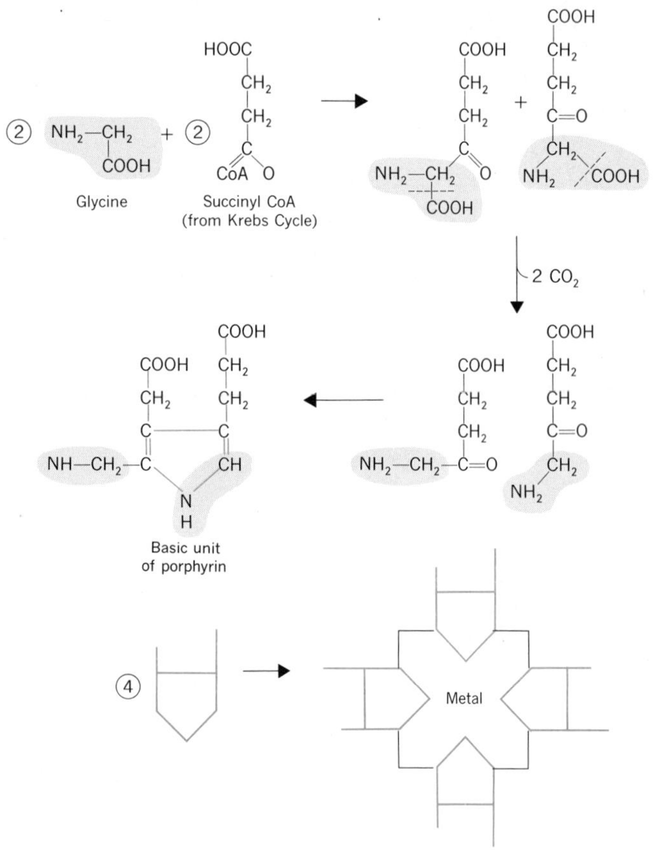

Fig. 4-20 Porphyrin synthesis. The basic unit of porphyrin, the porophobilinogen molecule, is formed by two molecules of the product formed by the combining of glycine and succinyl CoA. Four porphobilinogen molecules combine to form the porphyrin molecule. The specificity of the porphyrin is determined by the nature of the side groups and the proteins associated with it.

While most organisms are able to synthesize the basic structure of a porphyrin, many are not capable of synthesizing porphyrin derivatives. In such cases, these derivatives, which are usually the active units of enzymes, must be taken in, and are part of the nutritional requirement of the organism.

QUESTIONS

1. How does the use of a single class of molecules as energy donors in biological systems make one regard metabolic reactions as double coupled reactions?
2. How do enzymes work in conjunction with ATP to reduce the energy of activation of a reaction?
3. Distinguish between substrate-level phosphorylation and oxidative phosphorylation.
4. How does oxidative phosphorylation operate in conjuction with the electron transport system?
5. Justify that the following reaction will proceed in the direction indicated:

$$\text{succinic acid} + \text{FAD} \longrightarrow \text{fumaric acid} + \text{FADH}_2$$

6. How do step-wise reactions (1) inhibit large releases of energy in a single packet and (2) provide monomers for the biosynthesis of polymers?
7. What determines the number of reactions in a given metabolic pathway?
8. Account for the energy profit to a cell (a) after pyruvate is formed; (b) after one turn of the Krebs cycle.
9. What specific molecules are responsible for the links between the following metabolic pathways:
   (a) glycerol portion of fat molecule
   (b) fatty acid portion of fat molecule
   (c) amino acids for eventual protein synthesis (give two)
   (d) porphyrin synthesis
   (e) purine synthesis

CHAPTER 5
PHOTOSYNTHESIS

All organisms depend on the energy derived from the degradation of organic material for the polymer synthesis necessary for growth, maintenance, and function. One group of organisms has enzyme systems that enable the organisms to synthesize glucose from carbon dioxide and water. The glucose product may then follow a biosynthetic pathway, toward a storage product such as starch, or a structural product such as cellulose, or it may undergo degradative processes to produce both the energy and monomeric forms for other polymeric syntheses.

These organisms must have a source of energy for the synthesis of glucose. It would hardly be profitable for an organism to use the ATP derived from one glucose breakdown to synthesize another glucose molecule so that it too could be broken down. The second law of thermodynamics demands that every energy transfer involve an entropic loss (a loss of some useful energy). Using cellular energy to synthesize glucose *de novo* would involve many energy transfers, and hence the loss of a great deal of useful energy. The energy, then, must be obtained from the environment. Green plants use **light** energy for the synthesis of glucose through a process known as **photosynthesis** (photo = light).

The elucidation of the photosynthetic mechanisms has taken over two hundred years. During this time, thousands of scientific papers have been written on the subject. The history of inquiry has been marked by sequential major advances, and because of the linearity of development, it is an ideal subject for historical development.

A History of a Scientific Inquiry

The initial observations that led to the proposal of a photosynthetic mechanism were based on both experimental evidence and the "educated guesses" that were made by interpreting this data. In 1772, Joseph Priestley performed some of the earliest experiments on photosynthesis. He placed a sprig of mint in a glass of water and placed the glass in

Fig. 5-1 Setup for Priestley's experiment.

an inverted bell jar (see Fig. 5-1). After leaving it there for several weeks, he placed a burning candle in the jar. He found that, although a candle quickly burned out when it was placed alone in a bell jar, the candle in the bell jar with the mint burned brightly. He repeated the experiment, using a mouse instead of a candle (see p. 23). Priestley concluded from this series of experiments that vegetation revitalized used air. Priestley had also discovered oxygen and found that it was required for combustible processes, such as the burning of a candle or the respiration of animals. He therefore considered that this revitalization of bad air was an addition of oxygen to the system originating from the plant. It should be noted that Priestley considered oxygen to be a rare gas, and vehemently protested later workers' conclusion that it was an element.

In 1779, Jan Ingen-Housz repeated Priestley's experiment in further detail. He first found that oxygen production required only hours, rather than weeks. He also noted that the oxygen production by the plant occurred only in daylight. Shortly after sunset, the candle would burn out. He also found that the brightness of the flame of the candle was directly related to the light conditions of the day. The candle burned more brightly during a sunny day than when it was cloudy or overcast. From his work he proposed that the revitalization of used air (the production of oxygen) was dependent on light. He further extended his conclusions to say that the "greenness" of the plant was responsible for this production of oxygen, and the process was carried out only in the leaves and green part of the stem. Although this was speculation on his part, it was an "educated guess" that was later verified.

Jean Senebier in 1782 concluded that carbon dioxide was necessary for this production of oxygen by the green plant. He found that when the carbon dioxide was artificially removed from the environment, a green plant could not produce oxygen.

At this time, biologists endowed plants with a most altruistic virtue, supplying animals with the oxygen they required for life. Jan Ingen-Housz completely changed this point of reference in relating the work of Priestley, Senebier, and himself to some studies done several hundred years earlier. Von Helmont had, in the sixteenth century, performed a series of experiments to determine whether plants drew their nutritional needs from the soil. He weighed some earth, and then placed it in a bucket and planted a tree. The tree grew, increasing in both weight and size. After a growth period, he removed the soil and reweighed it. It did not vary by even one ounce from its original weight. Therefore, the soil was not the source of nutrient material for the tree. Although his conclusion was accepted as valid, biologists could not come up with an alternative means for the plant to obtain organic material for growth.

Jan Ingen-Housz proposed that carbon dioxide was the source of nutrition for the plant, and that the plant incorporated it into its living structure. His proposal led to the realization that oxygen was merely a waste product of the process, and the primary advantage was to the plant itself, in the formation of its organic structure.

In 1804, Nicholas de Saussure proposed that water was also required for this process. His conclusion arose from careful measurements of a plant in the limited environment of a sealed bell jar. He found that the gain in plant weight was greater than could be accounted for by the loss of carbon dioxide from the atmosphere alone. His choice of water to account for this additional increase in weight was a fortuitous one, based no doubt on the growing awareness of the necessity of water in chemical reactions.

In the late 1780s, Antoine Lavoisier amalgamated the available data on photosynthesis to form the first chemical interpretation. A plant absorbs carbon dioxide and water and produces organic material, liberating oxygen in the process. This could be represented by the equation

$$CO_2 + H_2O \xrightarrow[\text{light}]{\text{green plant}} \text{organic matter} + O_2$$

The responsibility for the determination of the component parts of the equation may be given as follows:

CO_2—Senebier
H_2O—de Saussure
O_2—Priestley
organic matter—Ingen-Housz
light—Ingen-Housz

It must be remembered that this equation for the photosynthetic process constituted a "hypothesis." Scientific inquiry after this time sought to verify, modify, and extend this basic equation to the point where all aspects of the process were thoroughly understood.

In the early 1800s, research on photosynthesis (a name not coined until 1898) split into three parallel approaches. The first was concerned with the effect of light, the second with the chemical reactions that led to the formation of organic material, and the third with the plant pigments themselves. The simultaneous work in these three areas eventually led to a comprehensive picture of the phenomenon.

THE STUDIES ON LIGHT AND CHLOROPHYLL

In the middle of the nineteenth century, the plant pigments were isolated and partially identified. The primary plant pigment was called **chlorophyll.** It was green and preferentially dissolved in organic solvents. (Later work showed that the chlorophyll fraction was actually a mixture of more that one molecular type, but the initial observations were made on a total chlorophyll fraction.) Also present in the plant were the **carotenoid** pigments. These included the red-orange pigment, **carotene,** and a variety of its oxygenated derivatives, called **xanthophylls** (see Fig. 5-2). The dominant pigment was chlorophyll, and it was assumed that its presence was responsible for the photosynthetic ability of the plant.

If, indeed, the chlorophyll was responsible for the utilization of light energy in the photosynthetic process, then it should, in some way, respond to light. To examine this response, it is wise to discuss some of the properties of light.

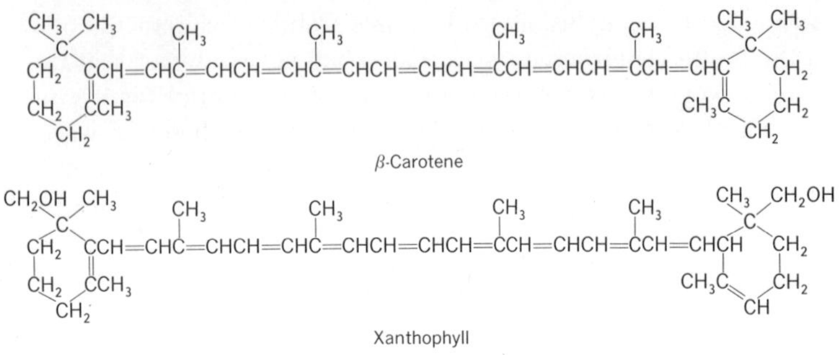

β-Carotene

Xanthophyll

Fig. 5-2 Molecular structure of xanthophylls and carotene.

Properties of Light

White light is separable into a spectrum of colors (see Fig. 5-3). As a physical phenomenon, light manifests properties of both waves and particles. When thought of as a wave, its wavelength can be defined as the distance between the peaks or, for that matter, any corresponding points of the waves.

A range of particular wavelengths produces an effect on the human eye that is interpreted as color. The eye sees wavelengths of 450 mμ as blue and those of 650 mμ as red. Therefore, the spectrum is really a dispersion of the wavelengths that make up white light in such a way that the shorter wavelengths appear physically separate from the longer wavelengths.

On the other hand, light may be thought of as the transmission of

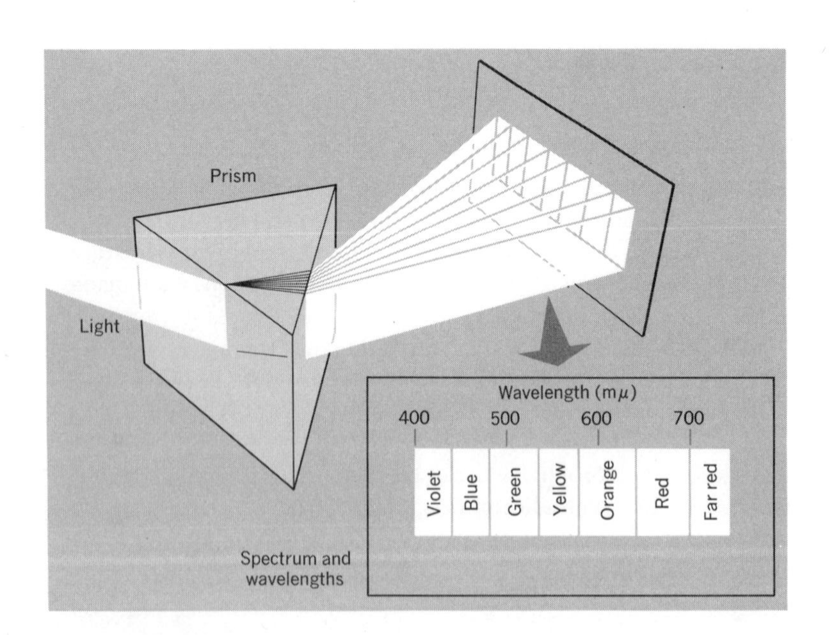

Fig. 5-3 The spectrum: the colors and their corresponding wavelengths.

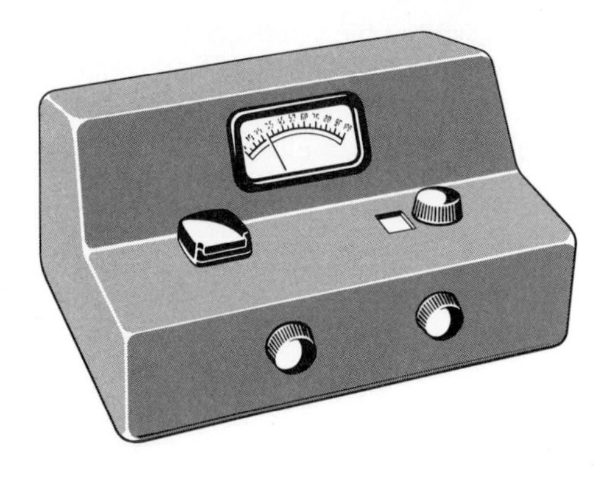

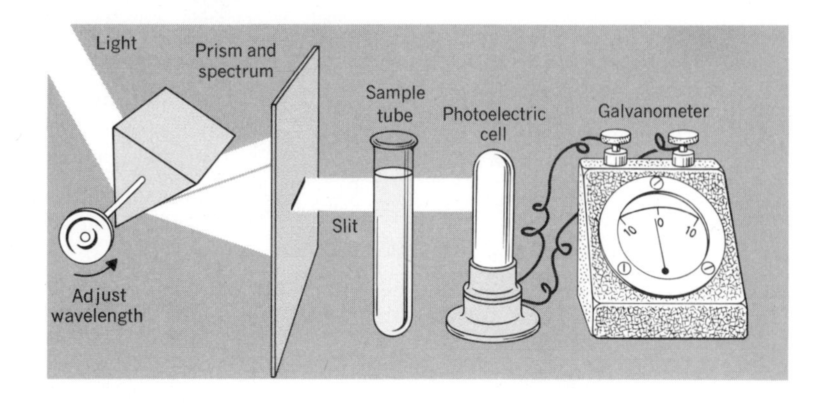

Fig. 5-4 A spectrometer. Light of a selected wavelength passes through a slit and is directed by mirrors through material in a quartz tube to a photoelectric cell. The amount of light passing into the photosynthetic cell registers as an electrical impulse and is translated into absorption and transmittance values on the meter of the spectrometer. Hence, the more light absorbed by the material in the tube, the less light passes through the tube, and the higher the absorbtion reading.

particles, known as **photons**. Here light is assumed to consist of discrete packets of energy, the energy in each packet being inversely proportional to the particular wavelength considered. Packets associated with blue light thus have more energy than those of red light. The reconciliation of the particulate and wave properties of light still constitutes a major problem in physics.

Chlorophyll and Light

Certain molecules have the capacity to strongly absorb particular wavelengths of light. When white light is passed through a solution of such molecules, portions of the light will be absorbed and the rest will pass through. The absorbing qualities of such molecules can be determined by the use of a **spectrometer** (see Fig. 5-4), which permits only specific wavelengths of light to be passed through a solution at one time. A detector placed on the other side of the solution registers the intensity (amount) of the light passing through it. When light of a given wavelength is absorbed by the molecules in the solution, the detector registers a decreased intensity. When a given wavelength of light is not absorbed, the detector registers an increased intensity. In this way, the preferential absorption tendencies of a particular molecular type may be determined.

When the data from the spectrometer reading is graphed, using the specific absorbance as the ordinate (the reciprocal of the intensity) and the wavelength of the light as the axis, an **absorption spectrum** for the particular substance is identified (see Fig. 5-5). When a solution of chlorophyll (the total chlorophyll fraction) was analyzed in this manner, it was found that there were two absorption peaks. These peaks corresponded to wavelengths of 450 mμ (blue light) and 660 mμ (red light).

The capacity of chlorophyll to absorb red and blue light was not a surprising discovery. It was known that the color of any object is

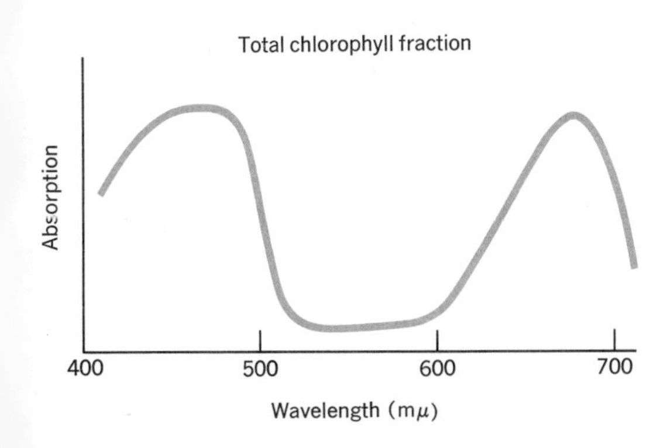

Fig. 5-5 Absorption spectrum for a total chlorophyll fraction indicates maximum absorption properties of the molecules in the blue and far red range.

determined by the light it *reflects*. A white object reflects all light, and since one sees only the reflected light, one judges the object's color to be white. Since chlorophyll is green, it was logical to assume that it reflects green light and absorbs light of other wavelengths. The absorption spectrum of chlorophyll indicated, however, that chlorophyll reflected orange and yellow light as well as green. These colors are indistinguishable from a visual standpoint. It remained to be shown that this absorbing capacity of chlorophyll was directly related to the photosynthetic process.

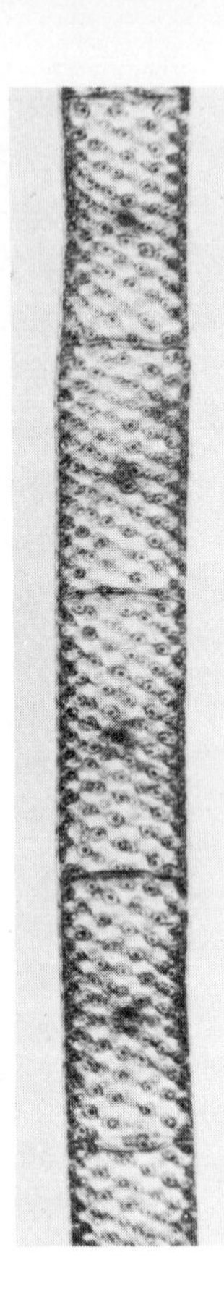

Fig. 5-6 *Spirogyra.* **The chloroplast of** ***Spirogyra*** **is oriented in a helical fashion. [Carolina Biological Supply House]**

Chlorophyll, Light, and the Photosynthetic Process

In 1882, an ingeniously devised experiment of von Englemann's provided a direct relationship between the absorption properties of chlorophyll and the photosynthetic process. He prepared a microscopic slide on which he placed several *Spirogyra* filaments. *Spirogyra* is a filamentous green alga made up of cells in a linear arrangement (see Fig. 5-6). He then spread the slide uniformly with some aerobic bacteria. In some ways this is analogous to Priestley's experiment with the mouse. The system contained a green plant form (*Spirogyra*) and an organism dependent on the presence of oxygen for survival (the bacteria). He then placed a prism in the path of the light entering the microscope in such

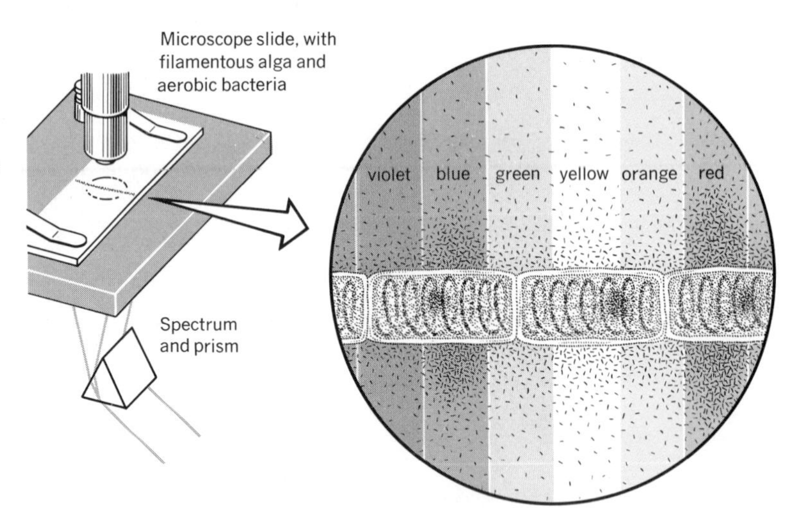

Fig. 5-7 Clustering of aerobic bacteria around a filament of ***Spirogyra*** **in von Englemann's experiment.**

a way that a spectrum formed in the microscopic field. After a few minutes he could see that the bacteria had moved, and were clustered in certain areas of the spectrum. The clustering could be interpreted as indicative of the availability of oxygen (see Fig. 5-7).

The bacterial clustering can be translated into a graph form in which the number of bacteria is the ordinate and the color region of

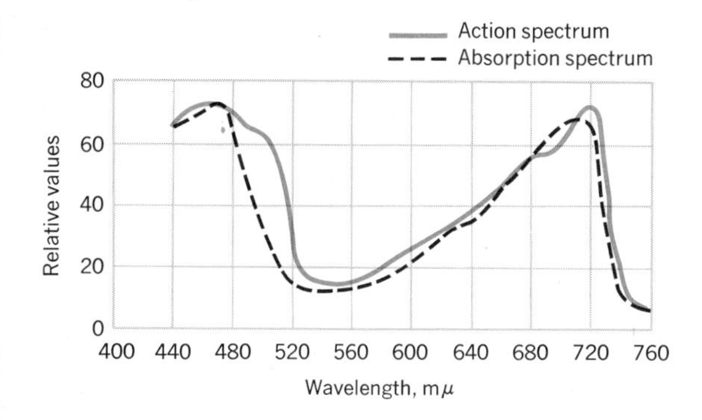

Fig. 5-8 Action spectrum for photosynthesis. The action spectrum shows the range of light in which an organism produces the products associated with the photosynthetic process.

the spectrum is the abscissa. This provides an **action spectrum** for photosynthesis (see Fig. 5-8). The action spectrum for photosynthesis and the absorption spectrum for chlorophyll were virtually identical. Both showed peaks in the red and blue light ranges. This information provided a definitive link between the ability of chlorophyll to absorb light and the efficiency of the photosynthetic process.

PHOTOSYNTHESIS AS A CHEMICAL REACTION

Even before von Englemann's work, the nutritional implications of the photosynthetic process became increasingly evident. In 1845, Julius Robert von Mayer, who had stated the first law of thermodynamics as a formal principle, was the first to see the physical function of photosynthesis as a conversion of light energy into chemical energy. He wrote:

"Nature set herself the task to catch in flight the light streaming toward the earth, and to store this, the most evasive of all forces, by converting it into an immobile form. To achieve this, she has covered the earth's crust with organisms, which, while living take up the sunlight and use its force to add continuously to a sum of chemical difference.

"These organisms are the plants: The plant world forms a reservoir in which the volatile sun's rays are fixed and ingeniously laid down for later use; a providential economic measure, to which the very physical existence of the human race is inexorably bound."

There was ample evidence that chlorophyll was indeed responsible for this translation of light energy into chemical energy. Through physical studies of light, it became possible to quantify light in terms of calories/mol so that a direct comparison could be made between light energy and chemical energy. The actual computations are not within the scope of this book; however, the results of these computations show that a mol of photons of wavelength 450 mμ has an energy content of 62 kcal/mol, and a mol of photons of wavelength 660 mμ has an energy content of 43 kcal/mol. The longer the wavelength, the smaller the energy content of the photons.

Chlorophyll is capable of absorbing only light with these two energy contents. One might logically argue that there should be a threshold of energy necessary for absorption by a molecule, after which all shorter wavelengths would be absorbed. This simply does not happen. The absorbing "preference" of molecules remains one of the great mysteries in physical chemistry.

Determination of the Photosynthetic Product. The other portions of the proposed photosynthetic equation were also placed under close scrutiny. The final product of the chemical reaction was the first aspect of the photosynthetic equation to be examined. In the middle of the nineteenth century, tests became available for the determination of the presence of many different types of organic molecules. One experiment was to keep one plant in the light and another in the dark for several hours. A leaf was then removed from both plants and boiled in an organic solvent to remove the plant pigments. Each leaf was then drenched in

Fig. 5-9 When iodine is added to a leaf that has been bleached, the starch present will turn black. Part of the leaf on the left was covered, while the remainder of the leaf was exposed to light. What is the relationship between photosynthetic activity and starch production? [Courtesy of Dr. Elliot Weier]

iodine solution, which gives a blue-black color when mixed with starch. The leaf from the plant kept in the dark showed almost no starch, while that from the plant kept in the light showed a high starch content (see Fig. 5-9). From this type of experimental evidence, it was concluded that starch is the organic product of photosynthesis. It was later realized that starch is a polymeric form of sugar. Therefore, biologists regarded the final product of photosynthesis to be glucose, rather than its polymeric form, starch.

Determination of Limiting Factors of Photosynthesis

In 1905, Blackmann performed a series of experiments on the reactants of the proposed photosynthetic equation. For these experiments he used a water plant then known as *Elodea* but presently known under the name of *Anacharis*. The use of this type of plant material had distinct advantages. In an atmospheric environment, neither of the products of photosynthesis could be detected. In an aqueous environment, however, the oxygen product is released and floats to the surface of the water in bubbles. Therefore, Blackmann could determine the rate of the reaction by counting the number of bubbles produced per unit of time.

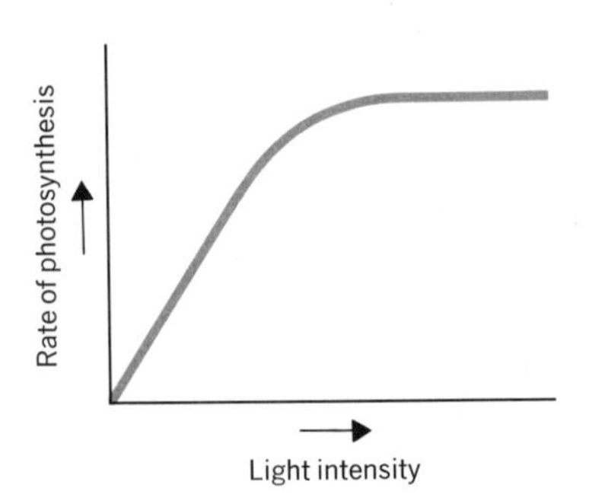

Fig. 5-10 Graph for reaction rate versus light intensity.

He adjusted his carbon dioxide source in such a way that a constant concentration was maintained despite its being drawn out of the system by the photosynthetic process. The system was then subjected to different light intensities. As the light intensity increased, the rate of reaction increased, and then leveled off (see Fig. 5-10). After this leveling had been reached, no increase in light intensity would increase the reaction rate. Since the concentration of carbon dioxide was constant, it could not be a limiting factor. Blackmann reasoned that if photosynthesis was the single reaction proposed by Lavoisier, with an availability of the reactants (carbon dioxide and water) and associated factors (green plant and light), there should be a constant increase in the rate of reaction. The fact that there obviously was not such a constant increase (see Fig. 5-10) led Blackmann to propose that there were at least two reactions operating in tandem in the photosynthetic process. One reaction or series of reactions was dependent on light, and was called the **light reaction.** The second reaction or series of reactions was not light-dependent, and was called the **dark reaction.** If the dark reaction were operating at its maximum velocity, the light reaction could occur only at a rate dictated by that velocity. The second reaction, therefore, constituted a **limiting reaction** for the process. This was the first time that the idea of limiting reactions was proposed for biological systems. The present concept of metabolic pathways as a series of reactions in which each reaction constitutes a limiting reaction for the one preceding it may give some indication of the powerful idea this concept presented to biologists.

Photosynthesis as a Redox Reaction

By 1929 it had been shown that a group of bacteria, known as the sulfur bacteria, could undergo a photosynthetic process in which carbon dioxide and hydrogen sulfide were the reactants. The products of this process were organic material, elemental sulfur, and water.

$$CO_2 + H_2S \xrightarrow[\text{light}]{\text{sulfur bacteria}} C_n(H_2O)_n + 2S + H_2O$$

Van Niel was struck by the similarity of this reaction to the photosynthetic reaction of green plants.

$$CO_2 + H_2O \longrightarrow C_n(H_2O)_n + O_2 + H_2O$$

(By this time water was recognized to be a product as well as a reactant of the reaction.)

Van Niel theorized that the reactants of the photosynthesis equation were, in reality, undergoing a redox reaction. He proposed that the reaction could be viewed as

$$\underset{\text{(acceptor)}}{CO_2} + \underset{\text{(donor)}}{H_2A} \longrightarrow \underset{\text{(oxidized product)}}{C_n(H_2O)_n + A}$$

In making this proposal, he focused attention not only on photosynthesis as a redox reaction, but also on the source of the oxygen product. Both of the reactants of the photosynthetic reaction for green plants contain oxygen. But which of them contributes the oxygen that is liberated as a product of the reaction?

The first indication of the oxidation-reduction character of the reaction, and the source of the oxygen as a product of the redox reaction resulted from a series of experiments by Robin Hill in 1939. He used the isolated chloroplast fraction of plant tissue. When these chloroplasts were placed in a solution of ferric oxalate (an organic oxidizing agent), a redox reaction occurred in which hydrogen was transferred. Concomitant with this transfer of hydrogen was a liberation of oxygen from the system. *There was no carbon dioxide in the system to be the source of the oxygen.*

He proposed that the chlorophyll molecule was causing a transfer of hydrogen from the water in the solution to the ferric oxalate, forming ferrous oxalate and oxygen. He further proposed that chlorophyll was responsible for the dissociation of water and acted in the same way *in vivo.* The dissociation of water due to the action of chlorophyll was called **photolysis** (breakage by means of light).

In 1941, radioactive tracers were starting to be used in biological experimentation. Ruben used radioactive oxygen (O^{18}) that was incorporated into carbon dioxide (CO_2^{18}). Plants were grown in an environment containing this carbon dioxide. The products of photosynthesis were then collected, separated, and prepared for the determination of the presence of radioactive material by the Geiger counter, an instrument used to detect the emission of high-energy particles from radioactive material, by translating this energy into sound waves that emit a series of clicking noises (the more frequent the clicks, the greater the number of high-energy particles being emitted). Products of photosynthesis containing radioactive material (O^{18}) would register on the counter, thereby giving an indication that these products, and only these products, contained the oxygen originally present in the carbon dioxide. Ruben found that the radioactive oxygen was limited to the organic material within the plant.

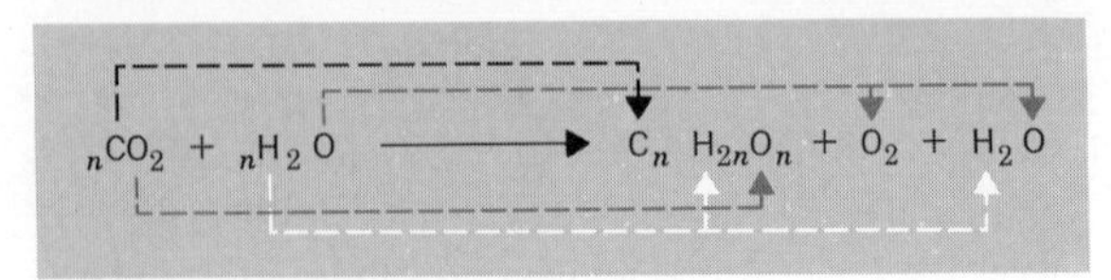

Fig. 5-11 The photosynthetic equation indicating the source of atoms.

$$CO_2^{18} + H_2O \longrightarrow C_n(H_2O^{18})_n + H_2O + O_2$$

He repeated the experiment, this time using O^{18} that was incorporated only into water (H_2O^{18}). This time the products that showed the emission of high-energy particles were the free oxygen and the water.

$$CO_2 + H_2O^{18} \longrightarrow C_n(H_2O)_n + H_2O^{18} + O_2^{18}$$

The equation could then be represented to show the source of each of the atomic forms in the products (see Fig. 5-11).

Insights into the Light and Dark Reactions

In 1952, Arnon succeeded in completely separating the light and dark photosynthetic reactions. He exposed a chloroplast fraction to light, providing it with ADP and NADP *but not with carbon dioxide.* He found that there was a formation of ATP and NADPH in conjunction with the light reaction. He removed these products and performed an enzyme extraction on the chloroplast fraction. The enzymes were extracted in a total fraction, since all the individual enzymes were not completely known at this time, and indeed, still are not. He discarded the remainder of the chloroplast fraction (including the chlorophyll). He then added radioactive carbon dioxide to the system, along with the ATP and NADPH that had been extracted, and found that there was an incorporation of the radioactive O^{18} into the organic structures. From this work, he concluded that the light reaction contributed energy in the form of ATP and a hydrogen donor, as NADPH, and that both of these were necessary in order for the dark reaction (the actual fixation of carbon dioxide) to take place (see Fig. 5-12).

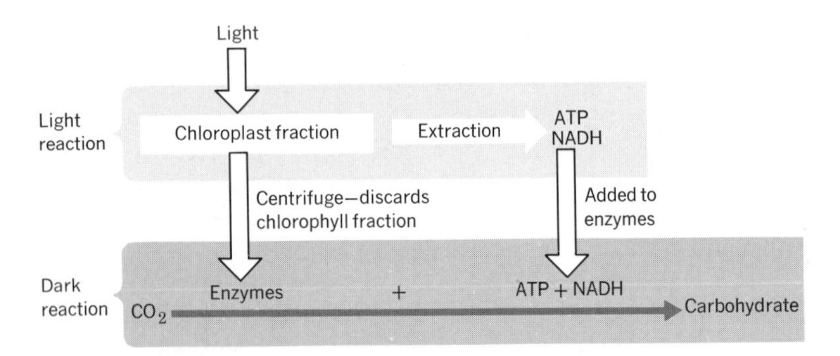

Fig. 5-12 Diagram of Arnon's experiment.

Thus, by the early 1950s, the picture of photosynthesis was beginning to become clear; the jigsaw pieces of the gigantic puzzle were falling into place. It was known that two major reaction systems were involved, a light one and a dark one. The light reaction in some way permitted the translation of light energy into chemical energy in the form of ATP and also provided for the generation of a necessary hydrogen donor in the form of NADPH. The dark reaction was the actual assimilation of carbon dioxide into the organic structures within the system. It was assumed that the light reaction involved a redox reaction in which chlorophyll acted as an electron donor. As the redox reaction proceeded, photolysis of water occurred. The photosynthetic pigments were known to be directly involved in the light reaction, but the true nature of this involvement was unknown.

From the determination of the metabolic pathways, an area that was simultaneously under investigation at this time, it became obvious that organisms were incapable of energy transfers of sufficiently high magnitude to permit the creation of a carbon skeleton *de novo*. Although the photosynthetic reaction was still represented as

$$6CO_2 + 12H_2O \longrightarrow C_6H_{12}O_6 + 3O_2 + 12H_2O$$

it was realized that the green plant did not simply put six molecules of carbon dioxide together by creating carbon-carbon bonds to form a molecule of glucose. A series of scientific inquiries were carried out, therefore, in which the intermediates of a stepwise series of reactions leading to the formation of glucose were sought.

The Elucidation of the Dark Reaction of Photosynthesis

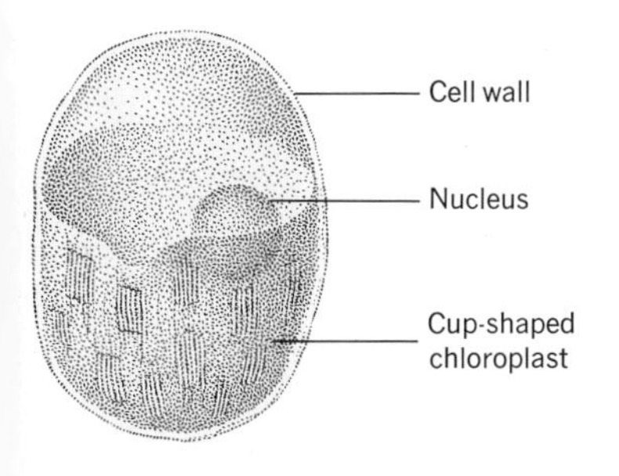

Fig. 5-13 ***Chlorella*, a simple unicellular green alga.**

In 1952, Calvin attempted to find the first product in the dark reaction of photosynthesis. For this series of experiments, he used a unicellular alga, *Chlorella* (see Fig. 5-13). Since he was interested in the organic products of photosynthesis, he used C^{14} incorporated into carbon dioxide. He regulated the flow of the radioactive carbon dioxide into the *Chlorella* culture, and then exposed the system to light for a short time. After five seconds, he removed a portion of the culture by permitting an aliquot to run into hot ethyl alcohol (EtOH). The hot EtOH served to kill the cells *and* stop all metabolic processes. He then subjected the material to chromatographic analysis, and looked for radioactive products (see p. 68). The radioactive areas of the chromatograph were re-eluted, and their chemical composition determined. He found that under these conditions, only 3-phosphoglyceric acid contained radioactive material.

When the process was permitted to go on for five minutes, he found that virtually every type of organic molecule in the system was radioactive.

Using this technique, a series of intermediates were isolated and eventually a consistent pattern emerged. The two most important features of this pattern are: (1) The pathway leading to glucose is indeed a stepwise series of reactions and (2) the pathway connects directly with other metabolic pathways of the system.

The incorporation of CO_2 into the system was shown to be a cyclic process, in which the molecular form to which the carbon dioxide was first bound was regenerated. This molecule was a five-carbon structure, **ribulose diphosphate.** The structure of ribulose diphosphate is such that carbon dioxide can be bound to the second carbon with a minimal expenditure of energy (see Fig. 5-14). The resulting molecule has a branched structure with a ketone group on the third carbon. Cleavage of this molecule occurs almost immediately, producing two molecules of 3-phosphoglyceric acid.

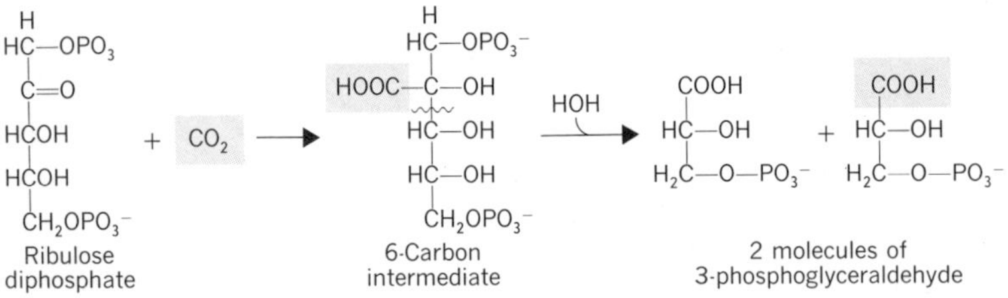

Ribulose diphosphate (H, $HC—OPO_3$, $C{=}O$, $HCOH$, $HCOH$, $CH_2OPO_3^-$) + CO_2 ⟶ 6-Carbon intermediate (H, $HC—OPO_3^-$, $HOOC—C—OH$, $HC—OH$, $HC—OH$, $CH_2OPO_3^-$) —HOH⟶ 2 molecules of 3-phosphoglyceraldehyde ($COOH$, $HC—OH$, $H_2C—O—PO_3^-$ + $COOH$, $HC—OH$, $H_2C—O—PO_3^-$)

Fig. 5-14 Ribulose diphosphate + CO_2 ⟶ 6-carbon branched molecule.

Fig. 5-15 Possible pathways from 3-phosphoglyceraldehyde.

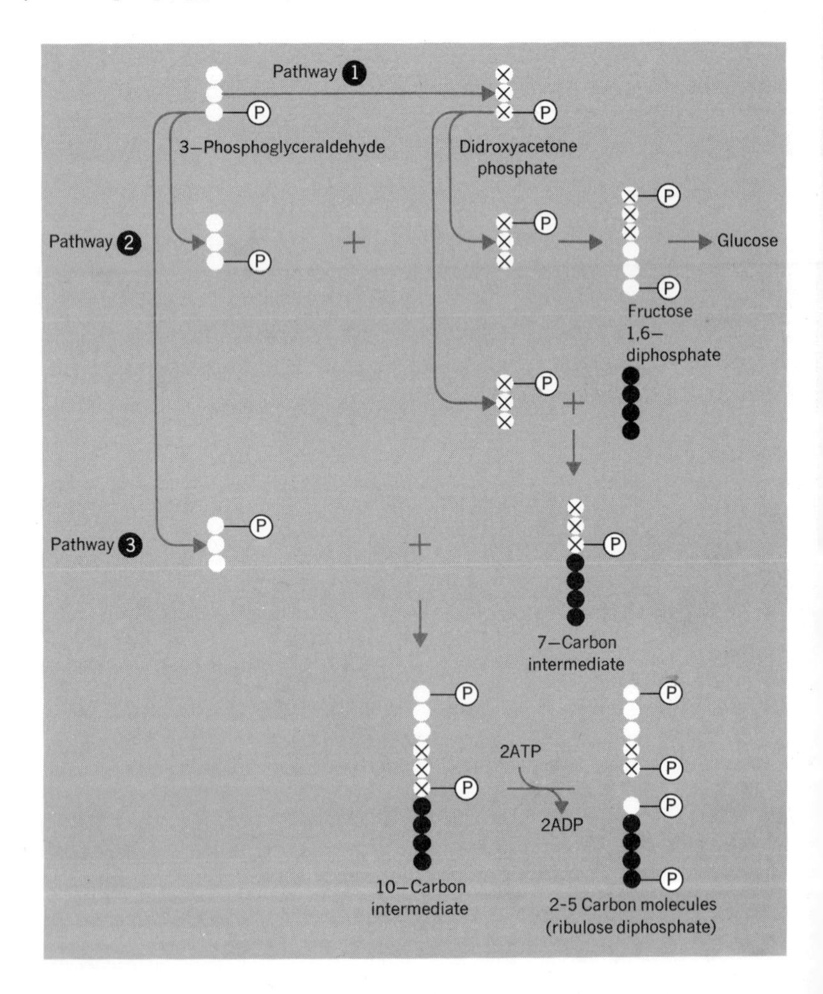

1 carbon (CO_2) + 5 carbons (ribulose diphosphate) ⟶
6-carbon intermediate product ⟶
(2) 3-carbon units (3-phosphoglyceric acid)

The resulting 3-phosphoglyceric acid molecules will follow synthetic pathways, leading either to the formation of glucose and ultimately starch, or to new ribulose diphosphate molecules. In either case, the first step is essentially a reversal of the glycolytic pathway.

In the conversion of 3-phosphoglyceric acid to 1,3-diphosphoglyceric acid, ATP is used as a phosphate donor. Notice that for every 3 phosphoglyceric acid molecules, one ATP ⟶ ADP occurs. Therefore, *two* ATP molecules are required for every CO_2 incorporation.

1, 3-diphosphoglyceric acid is then converted to 3-phosphoglyceraldehyde. In the course of this conversion, NADPH acts as a hydrogen donor. *It is these two steps in the dark reaction that require the ATP and the NADPH generated by the light reaction.*

3-phosphoglyceraldehyde may follow any one of three pathways, labeled A, B, and C on Fig. 5-15. It may be converted to dihydroxyacetone phosphate (A) and join with a molecule of 3-phosphoglyceraldehyde to form fructose 1,6-diphosphate (B). This would then continue the reversal of the glycolytic pathway, forming fructose 6-phosphate and then glucose 6-phosphate to ultimately be incorporated into starch. In plants such as sugar cane, fructose and glucose form a dissaccharide, so that the final product of photosynthesis is sucrose rather than starch. In barley, the glucoses are combined in a two-unit form, and maltose is the photosynthetic product. If the equilibrium of the metabolic pathways is such that glucose is to be broken down, it will leave the chloroplast and go to the cellular component in which the degradative processes are operative. In this way, the photosynthetic product can be converted to a myriad of molecular forms such as amino acids, nucleic acids, and fats.

Dihydroxyacetone phosphate (a three-carbon molecule) can combine with a four-carbon molecule to form a seven-carbon carbohydrate ($3 + 4 = 7$). This seven-carbon molecule can then combine with 3-phosphoglyceraldehyde (a three-carbon molecule) (C) to produce an unstable intermediate that breaks down into two five-carbon molecules ($7 + 3 = 5 + 5$). Although the products of this reaction have been determined, the proposed 10-carbon intermediate remains elusive. The two five-carbon molecules undergo a series of minor rearrangements, including a phosphorylation in which ATP acts as a phosphate donor. The result is two molecules of ribulose diphosphate. These may now pick up other carbon dioxide molecules from the environment.

Because of the alternative pathways in the system, it is impossible to make any sense of the process by attempting to follow a single CO_2 incorporation. Instead, let us review the process, using 600 molecules of carbon dioxide.

600 CO_2 + 600 ribulose diphosphate ⟶ 600 6-carbon molecules

600 6-carbon molecules ⟶
1200 3-carbon molecules (3-phosphoglyceric acid)

1200 3-phosphoglyceric acid molecules ⟶
1200 3-phosphoglyceraldehyde molecules

The phosphoglyceraldehyde may follow any one of three alternative pathways. One of these is its conversion to dihydroxyacetone phosphate. Let us consider that half of the available 3-phosphoglyceraldehyde is converted to dihydroxyacetone phosphate.

600 3-phosphoglyceraldehyde ⟶ 600 dihydroxyacetone phosphate

Now let us consider the alternative pathways themselves, using appropriate numbers of molecules. There are, at this point, a total of 600 molecules of 3-phosphoglyceraldehyde and 600 molecules of dihydroxyacetone phosphate.

300 phosphoglyceraldehyde + 300 dihydroxyacetone phosphates ⟶
300 fructose 1,6, diphosphate ⟶
storage product of photosynthesis

300 dihydroxyacetone phosphates + 300 four-carbon molecules ⟶
300 seven-carbon molecules

300 phosphoglyceraldehydes + 300 seven-carbon molecules ⟶
600 five-carbon molecules that are convertible to
ribulose diphosphate

So, starting with 600 ribulose diphosphate molecules, all of which incorporated carbon dioxide, the plant system ends up with 300 glucose molecules incorporated into starch (or whatever the end product for the particular plant system may be) and a regeneration of 600 ribulose diphosphate molecules.

This is the dark reaction of photosynthesis, the set of reactions that do not require light. Although it is obvious that the energy to drive this set of reactions arose from the light reaction in the form of ATP, the translation of light energy into chemical energy still constitutes a major problem. This is still an active area of research, but a series of logical proposals, derived from experimental data, permit more than just a speculation regarding the generation of energy by the light reaction.

The Elucidation of the Light Reaction of Photosynthesis

Upon careful analysis of chlorophyll solutions, it was found that the extracted chlorophyll fraction was a mixture of molecular forms. The major constituents were termed chlorophyll *a* and chlorophyll *b*, and were found to have slightly different chemical structures (see Fig. 5-16). Both were basically a porphyrin structure, with a 20-carbon grouping called a **phytol** extending from one part of the molecule.

It was found that the absorption spectra of chlorophyll *a* and *b* differed considerably (see Fig. 5-17). The absorption peak for chlorophyll *a* is 660 mμ, with only a minor peak at 450 mμ, while the major absorption peak for chlorophyll *b* is at 450 mμ.

Recent evidence indicates that the extraction of chlorophyll from plants alters the absorption properties. Hence the absorption spectra for chlorophyll *a* and *b* may not give a true indication of their properties. It now appears that, *in situ*, chlorophyll *a* primarily absorbs far red light 700 mμ) and chlorophyll *b* absorbs lower wavelengths of light ($<$700 mμ).

Because of the elaborate ring structure of both chlorophylls, they are capable of resonance. As resonating structures, they reside at their lowest energy level or ground state (see p. 59). One pair of electrons, the so-called **pi electrons,** circumscribe an orbit around the molecular structure in the course of this resonance. Therefore, the pi electrons do not "belong" to any given atomic form with the molecule, but are shared by the entire ring.

One of the well-established physical phenomena associated with light is its ability to move electrons. When photons strike a metal surface, electrons from that surface are released. It is thought that when photons strike a molecule that is capable of absorbing that wavelength of light (700 mμ for chlorophyll *a*, $<$700 mμ for chlorophyll *b*), the same phenomenon occurs. The light energy causes the displacement of the pi electrons to an orbit of a higher energy level. There is evidence that when this happens to one of the pi electrons, the electron changes the direction of its spin. This change in direction of spin is considered the **triplet state** of the electron. As the result of the triplet state, there is a total reorientation of the energy relationships within the molecule, and the molecule itself goes to a higher energy level or "excited" state. Concomitant with this excited state is the reduction of the resonance within the molecule.

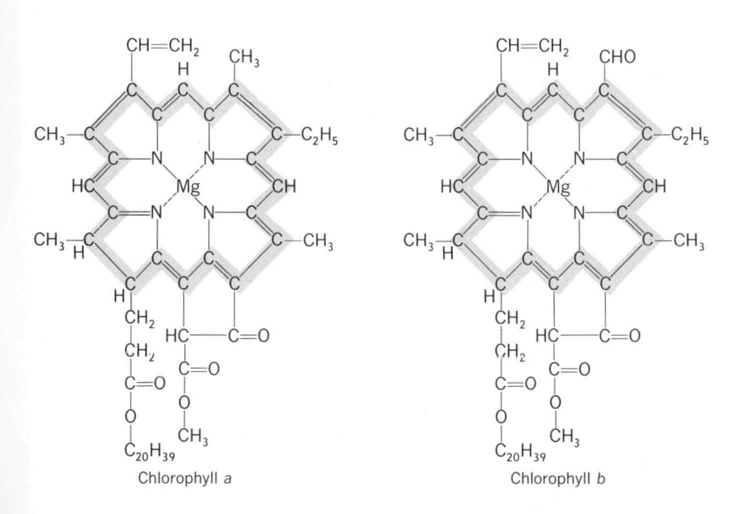

Fig. 5-16 Molecular structure of chlorophyll *a* and chlorophyll *b*. Rings of single and double bonds contribute to resonance for mobility of pi electrons.

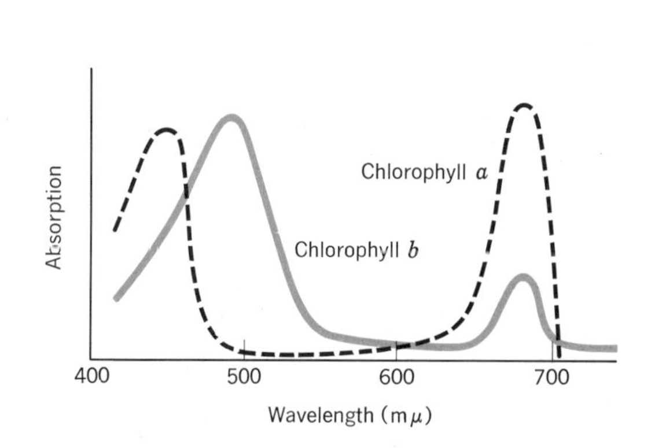

Fig. 5-17 Absorption spectra for chlorophyll *a* and chlorophyll *b*.

One way in which molecules may return to their ground state is by the spontaneous return of the pi electron to its original position and the dissipation of the energy in a random form. This dissipation is in the form of light, and is termed **phosphorescence.** In this case the energy transfer may be viewed as

light (photons) ⟶
chemical (triplet state and resulting molecular configuration) ⟶
light (as electron goes to ground state)

Another way in which molecules may return to their ground state is through a set of controlled reactions that involve the transfer of the triplet-state electron. Chlorophylls *a* and *b* return to their respective ground states by different sets of controlled reactions.

Chlorophyll *a* undergoes a redox reaction in which the triplet-state electron is transferred to the most powerful reducing agent yet found in biological systems, **ferridoxin.** Ferridoxin may transfer the electron into one of two possible systems. One possible pathway is the transfer of the electron to NADP to generate NADPH.

$$\text{ferridoxin "e"} + \text{NADP} \longrightarrow \text{ferridoxin} + \text{NADPH}$$

This reaction, or set of reactions, generates the NADPH necessary for the dark reaction.

The other pathway that can be followed by reduced ferridoxin is the transfer of the electron into a cytochrome system set of redox reactions very much like that involved in the electron transport system. In the course of the electron transfers through the cytochrome system, at least one molecule of ATP, and perhaps more, is generated. So the light energy that entered the system is converted to chemical energy in the form of the triplet-state electron, and then into a more stable form of chemical energy in the form of the high-energy phosphate bonded to ATP molecules. The phosphorylation of ADP through this process is called **photosynthetic phosphorylation** (see Fig. 5-18).

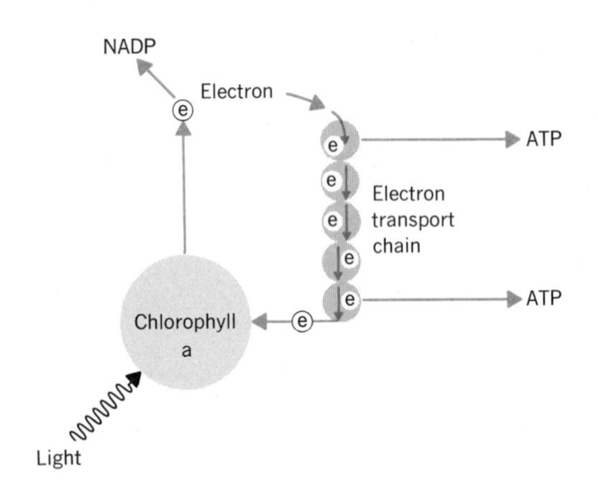

Fig. 5-18 Cyclic photosynthetic phosphorylation.

The electron emerging from the cytochrome system returns to chlorophyll *a*, thereby reestablishing its ground state. Hence, the triplet-state electron of chlorophyll *a* seems to have either one of two fates, the reduction of NADP to NADPH, or the return to the chlorophyll *a* molecule through a system in which ATP is generated.

Since the triplet-state electron can be siphoned off from the system to reduce NADP, chlorophyll *a* molecules must have another source of electrons for returning to their ground state. This other source involves the reactions of chlorophyll *b*.

The triplet-state electron from chlorophyll *b* is transferred to a molecule of **plastoquinone.** Plastoquinone can transfer this electron only into the aforementioned cytochrome system, in the course of which at least one ATP is generated. Upon emerging from the cytochrome system, the electron is transferred to chlorophyll *a*. Thus the electrons from chlorophyll *b* make up for those lost by chlorophyll *a* to NADP.

Because the electrons of chlorophyll *b* never return to reestablish the ground state of this molecule, the set of reactions involved in

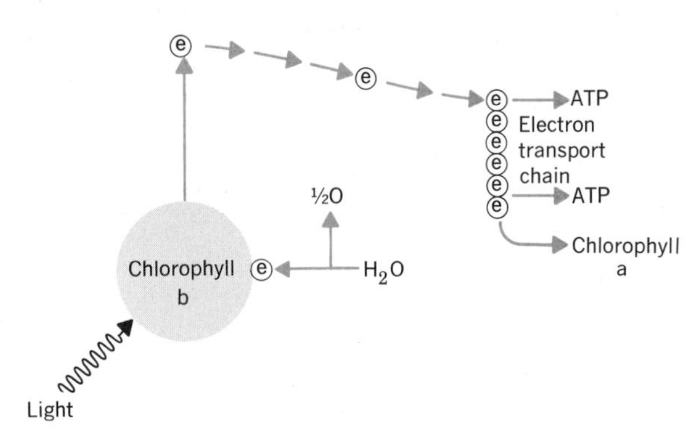

Fig. 5-19 Noncyclic phosphorylation.

transferring the chlorophyll *b* electron are called the **noncyclic phase** of the light reaction (see Fig. 5-19). Because chlorophyll *a* often receives electrons originally lost by chlorophyll *a* molecules, the reactions involving the transfer of the chlorophyll *a* electron are called the **cyclic phase** of the light reaction. It should be noted that several of these reactions overlap, and the terms cyclic or noncyclic provide a point of reference.

Of course, the contribution of an electron from chlorophyll *b* to chlorophyll *a* leaves chlorophyll *b* shy of electrons for returning to its ground state. The electrons transferred to chlorophyll *b* to reestablish its ground state are contributed by water.

Little is known about the water-splitting reaction because it is difficult to study. It is thought, however, that the OH^- radical is responsible for contributing the electron.

$$H_2O \longrightarrow H^+ + OH^-$$

$$OH^- \xrightarrow{\text{photolysis}} \tfrac{1}{2}O_2 + H^+ + 2e$$

The two electrons from the photolysis of water may be transferred to chlorophyll *b* to bring it to a ground-state condition.

So the requirements of the dark reaction, ATP and NADPH, are fulfilled by the light reaction. It has been estimated that the excitation of each of the chlorophylls twice results in one molecule of NADPH and at least two molecules of ATP. Since the dark reaction requires two NADPH for each carbon-fixation cycle, each chlorophyll must be transferred to its excited state four times.

Notice that photosynthesis depends on the raising of both chlorophyll *a* and chlorophyll *b* to higher energy levels. The advantage to the system of having two molecules, each brought to their higher energy levels by different wavelengths of light, is extraordinary. Plants are exposed to white light, which contains all the colors of the spectrum. Two separate wavelengths, occurring simultaneously, can bring about the rise in the chemical energy of the chlorophyll molecules *a* and *b*. The time factor for the photosynthetic process is increased substantially by the different function of the photosynthetic pigments.

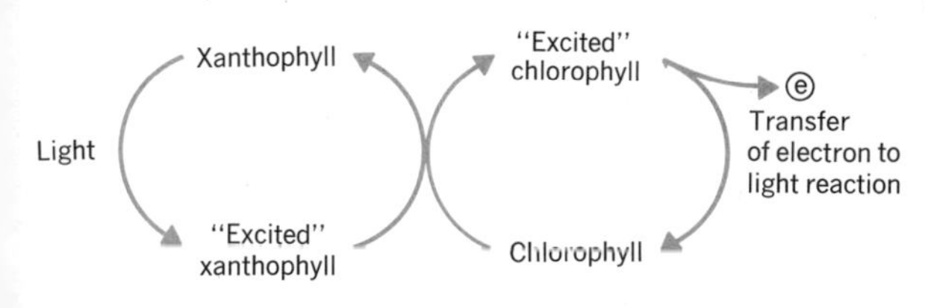

Fig. 5-20 Possible mode of transmission of excited states from auxiliary pigments to the chlorophylls.

It is thought that the auxiliary pigments add to the efficiency of the photosynthetic process by being raised to high energy levels by other wavelengths and transmitting their excited states to the chlorophyll *a* molecules directly, thereby increasing the wavelength range for which photosynthesis may occur (see Fig. 5-20). Although the absorption spectra for the carotenoids indicate that they are able to absorb different wavelengths of light than can the chlorophylls, any direct involvement has yet to be shown.

PERSPECTIVES OF PHOTOSYNTHESIS

The subject of photosynthesis is, in one sense, an extension of metabolism, and in another sense, an aspect of nutrition. Green plants, through their photosynthetic capacity, incorporate CO_2, a waste product of metabolism, into organic structure. As will be seen in the next chapter, this is important in the nutritional scheme of things.

The complex reactions of the photosynthetic process further exemplify the principles given in the preceding chapter on metabolism. Through stepwise enzyme-mediated reactions, light energy is converted into usable chemical energy, NADP is reduced to NADPH, and carbon dioxide fixation results in the generation of organic molecules.

QUESTIONS

1. What features did each of the following men contribute to the photosynthetic equation as it was originally stated by Lavoisier?
   (a) Priestley
   (b) Ingen-Housz
   (c) Senebier
   (d) De Saussure
2. Why was it necessary to correlate the absorption spectrum of chlorophyll and the action spectrum of photosynthesis before one could definitely state that chlorophyll played a role in the photosynthetic process?
3. (a) Why did Blackmann use a water plant rather than a land plant for his studies?
   (b) Why did the graphed data lead Blackmann to propose that there were two reactions to photosynthesis?
   (c) How did Blackmann's conclusion, arrived at through mathematical interpretation, affect biochemists who would perform future experiments on photosynthesis as a chemical reaction?
4. What led to the discovery that photosynthesis has a redox character?
5. Describe the fixation of carbon dioxide ending with the formation of fructose 1,6 diphosphate, and the regeneration of ribulose diphosphate molecules.
6. How are ATP and NADPH generated from the light reaction, and how are they utilized in the dark reaction?

CHAPTER 6
NUTRITION

Although biologists recognize that organisms require a variety of factors from the environment, such as sources of nitrogen, sulfur, phosphorus, and a host of trace elements, they have arbitrarily chosen only two factors as the basis of nutritional classification. These are the source of (1) energy and (2) electron donors for cellular redox reactions that generate ATP.

There are two kinds of energy sources: light and chemical. Those organisms that utilize light energy for the synthesis of glucose are nutritionally classified as **phototrophs** (*troph* = food, *phototroph* = food by means of light). Organisms that do not have the enzyme systems necessary for the photosynthetic process, and therefore must take in glucose and/or other organized carbon structures from the environment, are called **chemotrophs.** The organized chemical forms present in the environment are the energy sources for these organisms, and are often obtained by feeding on other organisms that contain them.

The other factor in nutritional classification is the type of electron donor, taken in from the environment, that is used in cellular redox reactions. Living organisms use redox reactions for the generation of ATP. Some use inorganic molecules as electron donors. The noncyclic phase of the light reaction of photosynthesis is dependent on the photolysis of water. Thus water, an inorganic molecular form, is an electron donor for redox reactions that result in the photosynthetic phosphorylation of ADP to ATP. Organisms that take in inorganic electron donors are referred to as **lithotrophs** (*litho* = stone).

Other organisms use organic molecules as electron donors for the generation of ATP. These electron donors are usually the organic degradative products of metabolism.

$$\underset{\text{(electron donor)}}{\text{pyruvate}} + \underset{\text{(electron acceptor)}}{\text{NAD}} + \text{CoA} \longrightarrow CO_2 + \text{acetyl CoA} + \text{NADH} + H^+$$

In such systems, the electron obtained from the donor is passed through a series of redox reactions (electron transport system) and results in the generation of ATP by oxidative phosphorylation. These organisms are called **organotrophs.**

Both lithotrophs and organotrophs take their electron donors from the environment. Lithotrophs generally take in a form that acts directly as an electron donor. Organotrophs generally take in a form that is readily convertible to the donor form (glucose $\longrightarrow$ pyruvate).

When the two criteria for nutritional classification are combined, four basic types of organisms emerge.

Photolithotroph: Requires light as an energy source and an inorganic electron donor (green plants).

Photoorganotroph: Requires light as an energy source and an oxidizable organic substrate as an electron donor (some purple bacteria).

Chemolithotroph: Requires organic energy sources and inorganic electron donors (several kinds of bacteria).

Chemoorganotroph: Requires organic compounds both as energy sources and electron donors (animals and most microbes).

Notice that the photoorganotroph group is limited to some of the purple bacteria. Their use of light as an energy source and organic molecules for electron donors represents an atypical kind of nutrition in the biological world at large, as do the chemolithotrophs. Therefore, the discussion of nutrition will be limited to the photolithotrophs and the chemoorganotrophs.

Within these categories, further subdivision is possible. Chemoorganotrophs may be classified according to the form in which they take in their nutritional material. **Holotrophs** are those chemoorganotrophs that can take in nutritional material in a solid form. Nutritional material in solid form is primarily polymeric structures such as starch, proteins, fats, etc. In order for nutritional material to enter a cell, it must pass through the membrane. Membranes are selective structures, and the passage of molecules through one is based on both the size and the properties of the molecules. The membranes that surround cells do not permit the entrance of large polymers. Therefore, in order for an organism to utilize solid nutritional material, it must have either (1) a means of bypassing the membrane to bring the solid material into the cell (such as the formation of food vacuoles in protozoans as depicted in Fig. 6-1), or (2) a means of secreting enzymes from the cell that operate in the environment (extracellular enzymes in some bacteria and the digestive tract of man). In either case, however, the use of solid nutritional material implies that the organism contains the enzyme systems necessary for reducing polymeric forms to their monomeric components.

ELEMENTAL CYCLES

Although biologists have arbitrarily chosen to use the criteria of energy source and electron donorship as the bases for nutritional classification of living forms, it must be remembered that these are not the only factors organisms must take in from the environment. A wide variety of other materials must be supplied from an external source.

TABLE 6-1 CHEMICAL ELEMENTS AND THEIR NECESSITY TO BIOLOGICAL MATERIAL

Element	Function
Hydrogen	Cell constituent, water, electron donor in bacteria
Oxygen	Cell constituent, water, electron acceptor in aerobes
Carbon	Cell constituent, electron donors in metabolism
Nitrogen	Proteins, nucleic acids, electron donor (NO_3, NO_2), electron acceptor (NO_3 NO_2) in some bacteria
Sulfur	Proteins, electron donor (H_2S, S, $SO_4^=$) and acceptor ($SO_4^=$)
Phosphorus	Nucleic acids, coenzymes, ATP
Potassium	Salt balance, cofactor
Magnesium	Cofactor in chlorophyll
Calcium	Cofactor, cell walls, bones, protein structure, salt balance
Iron	Cytochromes, cofactor, electron donor in some bacteria
Cobalt	Constituent of vitamin B_{12}
Copper	Cofactor
Zinc	Cofactor (alcohol dehydrogenase)
Molybdenum	Cofactor (nitrogen metabolism)
Sodium	Salt balance
Silicon	Wall structure of diatoms (single-celled plants)
Boron	? (Probably plant RNA metabolism)
Iodine	Thyroxin (hormone)

Table 6-1 lists the major and many of the minor chemical elements and their necessity to biological systems.

The enzyme systems present in an individual organismal type determines the form in which these elements can be handled. For example, most plants can utilize nitrogen in the form of nitrate ($NO_3^=$), and therefore require only inorganic nitrate sources. Animals, on the other hand, can only utilize nitrogen when it is incorporated into the ammonium radical (NH_3) in an organic structure, such as a protein. Thus an animal takes in both the necessary nitrogen and an auxiliary carbon structure at the same time. Many organisms, while requiring only simple carbon structures for their direct carbon requirement, also require a variety of elements in an organic form, and thus take in many auxiliary carbon skeletons.

Since matter can neither be created nor destroyed, the total amount of any element present on the earth is about all there is, and will be. Therefore, all systems, both biological and nonbiological, share the elements in such a way that there are many balances formed among them. These are the **elemental cycles.**

Elemental cycles exist for all the elements necessary for the efficient function of organisms. All the elemental cycles are totally related through the combinations of elements present in molecular forms. Carbon dioxide relates the carbon and oxygen cycles. Urea, $CO(NH_2)_2$, relates both the oxygen and nitrogen cycles to the carbon cycle. The cyclic nature of the passage of elements is accomplished by the fact that the waste product of one organism is often the source of nutrient for another organism. In a very real sense, the elemental cycles are the manifestation of the integration of metabolic processes of all living forms.

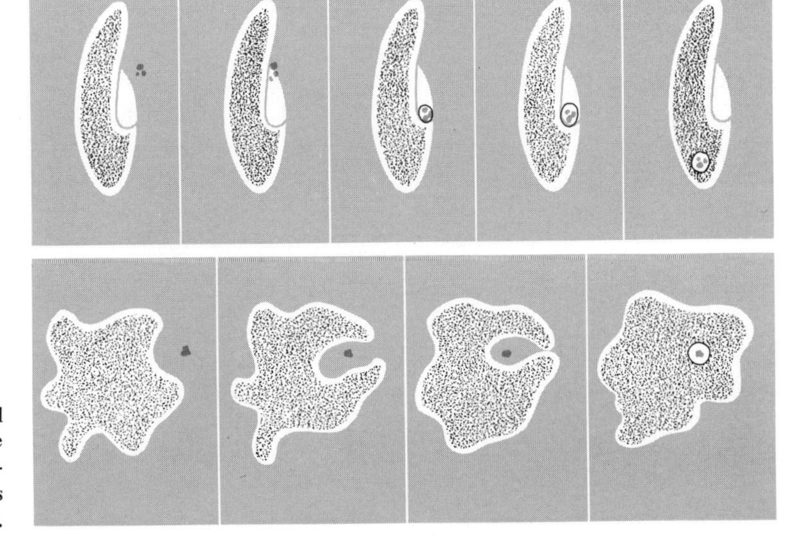

Fig. 6-1 Formation of food vacuoles in *Paramecium* and amoeba. Insoluble polymers in the form of a prey organism are introduced into the animal body within food vacuoles. Exoenzymes act upon the polymer, breaking down the molecules to monomer forms that pass through the vacuole membrane.

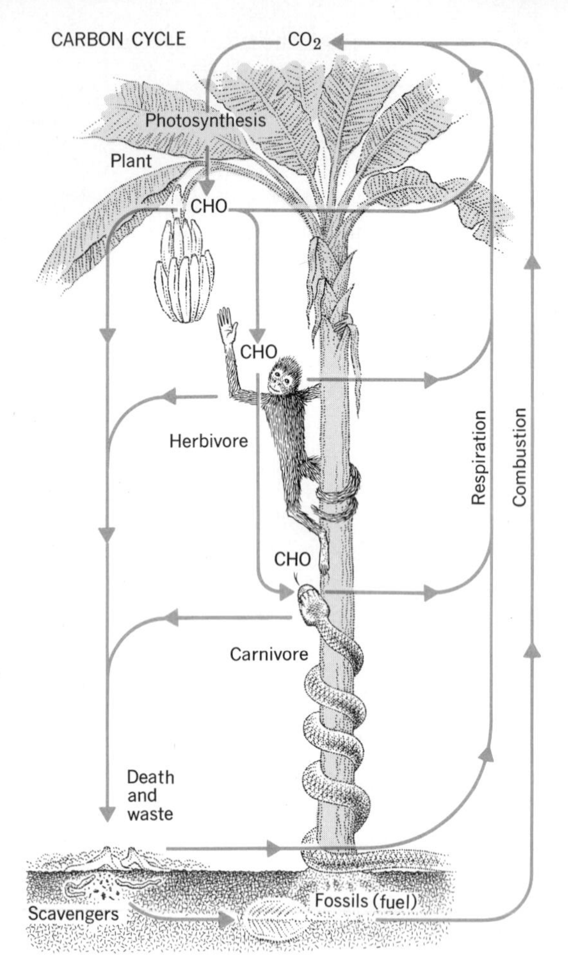

Fig. 6-2 The energy present in organic compounds is dissipated as these compounds are transferred through feeding and respiratory processes. The element carbon, however, is recycled through the photosynthetic process.

The Carbon Cycle

The classification system based on energy use and electron donors was initiated in the early 1960s. Before this, organisms were nutritionally classified according to their carbon source. Organisms capable of taking in carbon dioxide as their carbon source were called **autotrophs,** while those that require organized carbon structures were called **heterotrophs.** When discussing the carbon cycle, it is easier to use the older terms.

The carbon cycle is sometimes taken to be synonymous with the **food chain.** The term chain implies a linear sequence, with a beginning and an end. In terms of complexity of carbon skeletons, and the energy available in the bonds of these skeletons, a chain is formed.

In the food chain, the autotrophs, or **producers,** use light energy to produce organic compounds. The autotrophs are consumed by heterotrophs. Those heterotrophs that directly consume plant material are the **herbivores.** The herbivores are the prey of another group of heterotrophs, the **primary carnivores.** The primary carnivores are, in turn, the prey of the **secondary carnivores.** Upon the death of the autotrophs and heterotrophs, their organic carbon structures are reduced to carbon dioxide, an energy-poor compound. Thus, the sequential passage of carbon structures from organism to organism, and the energy inherent to these carbon structures, forms a chain.

In terms of the element carbon, however, the process is more correctly viewed as a cycle (see Fig. 6-2). Carbon dioxide arises from the respiratory processes of plants, animals, and bacteria. It is the waste product of the degradation of organic molecules, more specifically, from the cleavage of the terminal carboxyl group of an organic acid. In the photosynthetic process, carbon dioxide is incorporated into a molecule of ribulose diphosphate, and once again assumes its organic-acid character.

The phototrophs serve to replenish the supply of organic molecules. Indeed, if there were no autotrophs available, organic molecules would soon be converted to carbon dioxide and water, and life would cease to exist.

Notice also that some living forms do not degrade organic molecules completely. In yeast, ethyl alcohol is given off as a waste product. In lactic-acid bacteria, lactic acid is the waste product of metabolism. The balance within the carbon cycle, and indeed all the elemental cycles, is such that virtually every waste product of the metabolism from one type of organism serves as a carbon source for another type of organism.

Some of the waste products of the metabolic processes are molecular structures comprised of both carbon and nitrogen, such as urea.

$$NH_2{-}\overset{\overset{\displaystyle O}{\|}}{C}{-}NH_2$$

Such molecules link the carbon and nitrogen cycles.

The Nitrogen Cycle

The **nitrogen cycle** (see Fig. 6-3) is of a more complex nature than that of carbon, because it involves a change in the valence of the nitrogen. Nitrogen appears in the cycle in the following forms:

Valence of Nitrogen	*1s*	*2s*	*2p*	*3s*	*3p*	*Radical*
Saturated levels	2	2	6	2	6	
3	2	2	3			NH_2^-
5	2	1	3	1		NO_2
7	1	1	3	1	1	$NO_3^=$

Once again, the enzyme systems of the particular organismal type dictate which form of nitrogen can be handled by the system. For example, plants can utilize nitrogen in the form of nitrates ($NO_3^=$), while animal forms require it incorporated into organic molecules as NH_2^-.

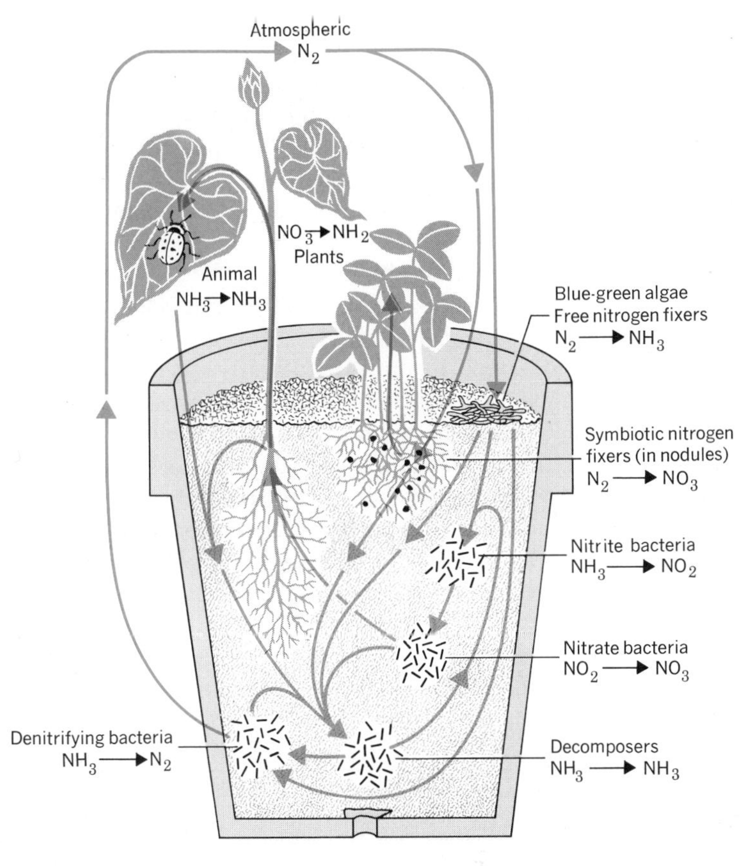

Fig. 6-3 Atmospheric nitrogen is converted to (1) NO_3 by bacteria residing in the root nodules of legumes, and (2) NH_3 for protein and waste in free nitrogen-fixers such as the blue-green algae. The nitrate form is utilized by green plants for their protein. The waste NH_3 from free nitrogen fixers is acted upon by nitrite and nitrate bacteria to bring about a regeneration of atmospheric N_2. All organisms are ultimately subjected to decomposition by decomposers that have an NH_3 waste product. The NH_3 produced by the decomposers follows the same pathway as that outlined for the NH_3 waste of free nitrogen fixers.

As noted before in the carbon cycle, a variety of microorganisms are active in many of the necessary transformations. The nitrogen-fixing bacteria and blue-green algae have metabolic processes capable of converting atmospheric nitrogen to nitrite form. Another group of bacteria are capable of utilizing nitrite, and excrete nitrate as a waste product. The process of converting atmospheric nitrogen to a final nitrate form by the tandem action of these two microbial groups is called **nitrification.** Plants and microorganisms utilize nitrogen in nitrate form, convert it to an atomic form with a valence of three, and incorporate it into protein material. These forms serve as an ultimate nitrogen source for animals and microbes requiring organic nitrogen sources.

The nitrogen-containing waste products of metabolism for one type of organism serve as the nitrogen source for the metabolism of other types of organisms, thereby contributing to the cyclic effect.

The cycle is completed by the **denitrifying bacteria** through a process called **denitrification.** Different groups of bacteria are active in the denitrification process. Some groups convert ammonium to nitrate; others convert nitrate to nitrite; others convert nitrite to atmospheric nitrogen. Still others convert ammonium, or nitrate, or nitrite directly into N_2. It should be emphasized that these conversions are inherent to the metabolic pathways of the individual groups of bacteria.

The Oxygen Cycle

The oxygen cycle is made up of two interrelated phases: one that involves oxygen as part of a water molecule, and one that involves oxygen as part of a carbon dioxide molecule.

As can be seen in Fig. 6-4, the oxygen generated by the photolysis of water in photosynthetic plants is used as a final hydrogen acceptor in the respiratory processes of all aerobic organisms. The photosynthetic and respiratory mechanisms both generate ATP in the course of this part of the cycle.

Oxygen bonded to carbon in the form of carbon dioxide is also recycled through the photosynthetic and respiratory processes. Photosynthesis provides for the fixation of carbon dioxide on a molecule of ribulose diphosphate. In the course of the dark reaction, glucose is formed. The degradation of glucose leads into many pathways, so that the CO_2 molecule, originally fixed, can ultimately be found as part of the carbon structure of an amino acid, a fat, etc. When these carbon structures are degradated, either by the metabolic processes of the individual organism or by the degradative process of bacteria and other scavengers feeding on the carcass, carbon dioxide is formed.

The links between the two parts of the cycle are the hydrolytic reactions in which the ionic products of water, OH^- and H^+, are incorporated into organic molecules or formed as a product of the degradation of polymers. For example, the breakdown of maltose, a disaccharide made up of two bound glucose units, involves the incorporation of the ionic products of water.

$$\text{maltose} + OH^- + H^+ \longrightarrow 2 \text{ glucose molecules}$$

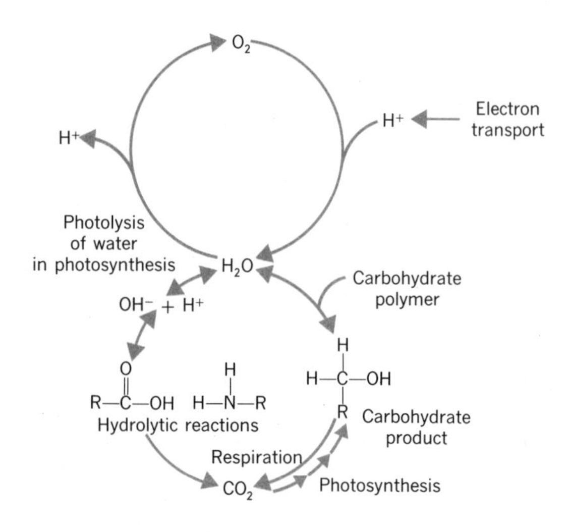

Fig. 6-4 Oxygen exists in three primary metabolic compounds: O_2, H_2O, and CO_2. Oxygen passes from a free state to a position in a water molecule and back again through the sequential occurrence of respiration and photosynthesis. As part of the water molecule, it takes part in ionization and hence in hydrolytic reactions. Because of its new-found position on an organic molecule, oxygen may leave the system as the combustion product CO_2. Subsequent photosynthesis brings the carbon and the oxygen once again into an organic molecule. Polymerization of this type of molecule may be a hydrolytic reaction in which water is removed. Once again, the oxygen atoms may be associated with hydrogen.

TABLE 6-2 THE NUTRITIONAL REQUIREMENTS FOR THE BACTERIUM *ESCHERICHIA COLI*

NH_4Cl	1 g
K_2HPO_4	1 g
$MgSO_4 \cdot 7H_2O$	200 mg
$FeSO_4 \cdot 7H_2O$	10 mg
$CaCl_2$	10 mg
Trace elements (Mn, Mo, Cu, Co, Zn as inorganic salts)	
Glucose	5 gm
Water	1 liter

The reaction is reversible, and the formation of maltose results in the production of the ionic products of water.

The element oxygen exists in a universal balance among the water, carbon dioxide, organized carbon structures, and finally, oxidized inorganic elements.

NUTRITIONAL REQUIREMENTS

The particular factors required from the environment by an organism are considered its nutritional requirements. The individual requirements are dependent on the enzyme systems present in the organism.

Microscopic forms exhibit a wide variety of metabolic patterns. For example, examine Table 6-2, which gives the nutritional requirements for the bacterium *E. coli.* In addition to a variety of inorganic materials, *E. coli* is able to use glucose alone as a carbon source. This means that it is capable of producing all the amino acids, all the coenzymes, all the porphyrin structures, all the nucleic acids, etc., from metabolic pathways that siphon molecular forms from the degradative products of glucose.

On the other hand, a protozoan like *Tetrahymena,* which is considered a higher organism on the evolutionary scale, has a wide variety of nutritional requirements (see Table 6-3). And man, supposedly the ultimate in animal forms, has about the same requirements (see Table 6-4).

TABLE 6-3 THE NUTRITIONAL REQUIREMENTS OF THE PROTOZOAN *TETRAHYMENA*

Inorganic salts		Carbon source	
K_2HPO_4	100.0 mg	Glucose	1 g
$MgSO_4 \cdot 7H_2O$	10.0 mg	Na acetate	1 g
$Zn(NO_3)_2 \cdot 6H_2O$	5.0 mg	Nitrogen bases	
$FeSO_4 \cdot 7H_2O$	0.5 mg	Guanylic acid	25 mg
$CuCl_2 \cdot 2H_2O$	0.5 mg	Adenylic acid	25 mg
Amino acids		Cytidylic acid	25 mg
L-arginine	150 mg	Uracil	25 mg
L-histidine	110 mg	Growth factors	
L-isoleucine	50 mg	Thiamine HCl	1000 μg
L-leucine	70 mg	Riboflavin	100 μg
L-lysine	35 mg	Calcium pantothenate	100 μg
L-methionine	35 mg	Niacin	100 μg
L-phenylalanine	50 mg	Pyridoxine HCl	2000 μg
L-serine	90 mg	Folic acid	10 μg
L-threonine	90 mg	Thioctic acid	1000 units
L-tryptophane	20 mg		
L-valine	30 mg	H_2O	1 liter

TABLE 6-4 ESSENTIAL FACTORS FOR MAN

Elements	*Transfer Groups*
Calcium (Ca)	Choline
Phosphorus (P)	Betaine
Magnesium (Mg)	Methionine
Sodium (Na)	
Potassium (K)	
Manganese (Mn)	
Iron (Fe)	
Copper (Cu)	
Cobalt (Co)	
Iodine (i)	
Sulfur (S)	
Zinc (Zn)	
Florine (Fl)	
Amino Acids	*Unsaturated Fatty Acids*[a]
Arginine	Linoleic
Histidine	Linolenic
Threonine	Arachidonic acid
Valine	
Leucine	
Isoleucine	
Lysine	
Methionine	*Carbohydrates*[b]
Phenylalanine	
Tryptophane	
Water (2.5 liters daily)	
Vitamins	
Vitamin A	
Vitamin D	
Vitamin E (tocopherols)	
Vitamin K (phylloquinone)	
Thiamine (B_1)	
Riboflavine (B_2)	
Nicotinic acid (niacin)	
Folic acid (pteroylglutamic acid)	
Pantothenic acid (B_3)	
Biotin (vitamin H)	
Vitamin B_6 (pyridoxine)	
Inositol	
Para-aminobenzoic acid	
Vitamin B_{12} (cyanocobalamin)	
Vitamin C (ascorbic acid)	

[a] At least one of three present.

[b] In sufficient amount so as not to impose on amino acids as caloric source.

Also because of the nature of their nutritional requirements, *E. coli*, *Tetrahymena*, and man are all classified as chemoorganotrophs. This serves to illustrate that the particular nutritional classification does not indicate the level of biological organization. Evolution has provided many solutions, at many different organismal levels, to the same nutritional needs.

VITAMINS

Although nutritional studies were started long before the determination of specific metabolic pathways, present inquiries into the nutritional requirements of organisms have been related to the absence of particular enzyme systems.

For example, it had been known for many years that some organisms required complex heteroconjugates from an external source in order to survive. The function of these heteroconjugates was a mystery for a long time, and they were given the trivial name **vitamins** (minerals necessary for life). Upon the elucidation of the structure of enzymes, it became obvious that some of these vitamins, notably those that are water soluble, served as sources of coenzymes that the organisms themselves were unable to synthesize. For example, **riboflavin** (vitamin B_2) can be converted into either FAD or FMN and **nicotinic acid** may be converted into NAD by reactions that are commonly found in biological systems (see Fig. 6-5). Organisms that manifest a vitamin requirement are termed **auxotrophs.**

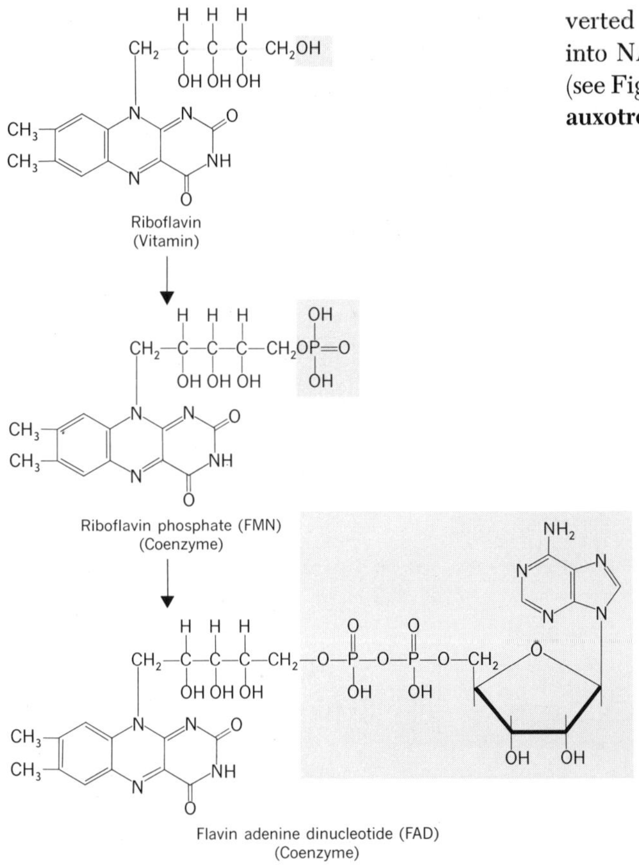

Riboflavin
(Vitamin)

Riboflavin phosphate (FMN)
(Coenzyme)

Flavin adenine dinucleotide (FAD)
(Coenzyme)

Fig. 6-5 When the vitamin riboflavin is phosphorylated, it becomes the coenzyme FMN, which is used in certain dehydrogenations. When adenine diphosphate (ADP) is added to riboflavin, it becomes FAD, a coenzyme active both as a dehydrogenator in respiratory reactions and as the first member of the electron transport system.

Vitamin requirements are known for many microorganisms (primarily those used in various aspects of research in which growth on a defined medium is desirable) and vertebrates (primarily those similar to man, on which research reflective of man's nutritional state could be performed). Of the water-soluble vitamins, only one or more of the B-complex vitamins may be needed by bacterial groups. Vertebrates require the full range of water-soluble vitamins.

The fat-soluble vitamins (A, D, E, and K) do not act as coenzymes. It appears that the aqueous environment of the cell demands the function of only water-soluble molecules in metabolic roles. Fat-soluble vitamins are active on a physiological level in vertebrates, in chemical reactions that are not present in microorganisms. Hence, the fat-soluble vitamins are neither present nor required by microorganisms.

Since fat-soluble vitamins have their effect on the organismal function of higher organisms, they will be discussed in later chapters, and the discussion here will be limited to water-soluble vitamins.

The elucidation of the chemical role of many vitamins was determined by the use of microorganisms, and subsequently related to higher animals. It should be noted that little is known of the translation of vitamin deficiencies on a chemical level to the physiological expression in either microorganisms or higher animals. Students, instructors, and authors, because of their taxonomic status, are usually interested primarily in the auxotrophy of the human. Therefore, this discussion includes the symptoms of vitamin deficiency in man, and is generally oriented toward the human.

B Complex

Vitamin B was originally isolated as a complex extract in 1911. A deficiency of this vitamin was associated with beriberi, a disease in which muscular coordination and balance are affected. Later work indicated that this complex extract was made up of nine compounds, each structurally different. Of the nine, only a lack of thiamine (B_1) was responsible for beriberi. Each of the other eight has a set of characteristic symptoms (syndrome) associated with a deficiency. Not all, however, are given formal disease names.

Thiamine (B_1)

Thiamine undergoes a reaction with ATP in which the two terminal phosphates (pyrophosphate) from ATP are transferred to the thiamine molecule to form thiamine pyrophosphate.

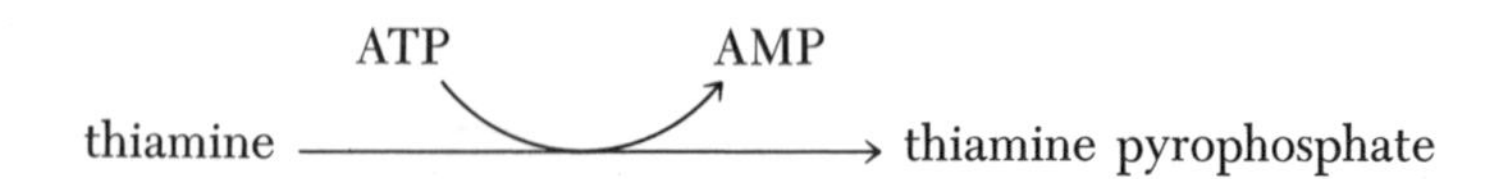

Thiamine pyrophosphate (TPP) acts as a coenzyme in both the glycolytic pathway and the Krebs cycle. It is involved in the decarboxylation of pyruvic acid and α-ketoglutaric acid. A reduction in the amount of thiamine available to the system will result in the decreased rate of the reactions of the major metabolic pathway and a concomitant decrease in the amount of energy available for the system.

On a human physiological level, thiamine deficiency results in weakness of the limbs and a decreased efficiency in the transmission of and response to impulses of the central nervous system (CNS). These are the symptoms of beriberi. These responses may be due to the loss of energy and/or the product accumulation due to the slowing of metabolism and the alteration of the equilibrium.

Riboflavin (B_2)

Riboflavin undergoes a reaction with ATP in which a single phosphate is transferred (see Fig. 6-5). Riboflavin phosphate (FMN) may remain as such in the system, or, under the dictates of equilibrial conditions, may undergo a further reaction with ATP to form FAD. The primary roles of both FMN and FAD are as hydrogen acceptors. FAD is a primary constituent of the electron transport system and is the only molecular form capable of accepting hydrogens from NADH + H^+. Therefore, a decrease in the amount of FAD in the system would severely hamper the efficiency of the electron transport system.

On the other hand, FMN is necessary in the biosynthesis of fats. One of the physiological effects of riboflavin deficiency is the breakdown of mucous membranes. Membrane structures have a high fat content. The inability of the system to synthesize fat-containing structures at a necessary maintenance rate may be responsible for the breakdown of membranes.

Nicotinic Acid (*Niacin*)

Nicotinic acid and its amine form, nicotinamide, both undergo reactions in which they attach to an adenine diphosphate.

nicotinic acid + ADP ⟶ nicotinamide adenine dinucleotide

NAD may undergo a further reaction in which the ribose structure is phosphorylated.

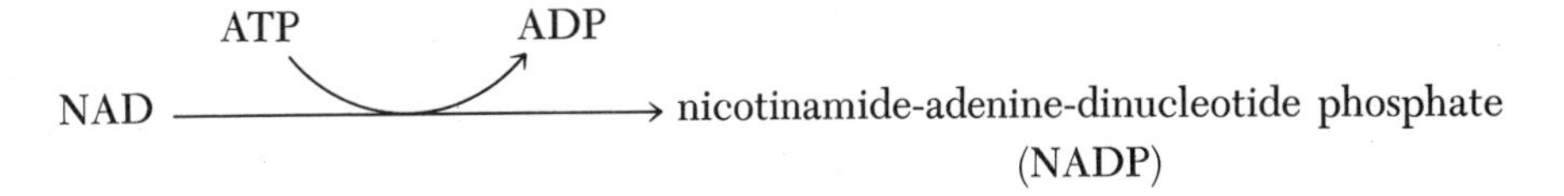

A niacin deficiency would result in a reduction of the amount of NAD and NADP available to the system. This reduction would greatly decrease the efficiency of respiratory metabolism.

Niacin deficiency is known to cause the disease pellagra in man, and black tongue in dogs. The physiological symptoms of the disease include effects on the skin, the digestive tract, and the CNS. The translation from the basic importance of NAD and NADP to metabolism and the physiological effects is unknown.

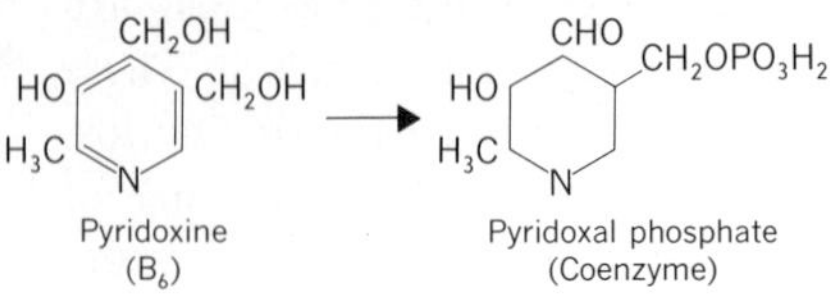

Pyridoxine (B_6)

Pyridoxine undergoes a reaction with ATP, in the course of which it is converted to pyridoxal phosphate (see Fig. 6-6).

$$\text{pyridoxine} \xrightarrow{\text{ATP} \;\rightarrow\; \text{ADP}} \text{pyridoxal phosphate}$$

Fig. 6-6 The phosphorylation of pyridoxine produces the coenzyme pyridoxal phosphate.

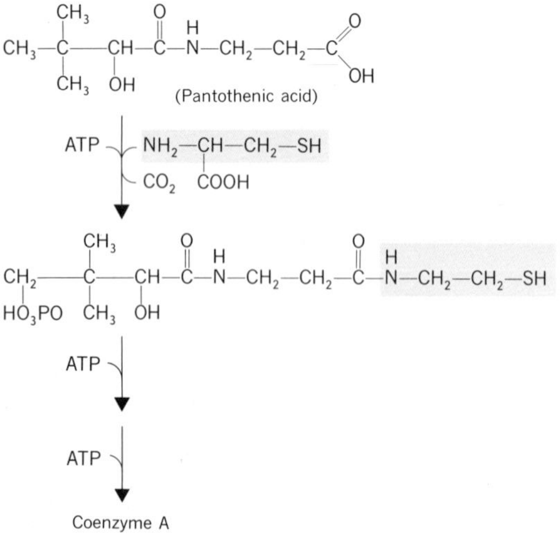

Pyridoxyal phosphate is a coenzyme active in the transaminations of amino acids, in decarboxylations, and in the biosyntheses of several amino acids. The known physiological effects of a deficiency are not given the formal name of a disease. The anemia caused by a pyrodoxine deficiency may be related to the involvement of the coenzyme in amino-acid metabolism. The synthesis of protein is dependent on the availability of the proper amounts of amino acids in the system. A reduction in this availability would decrease the capacity of the system to produce proteins. This would be especially noticeable in proteins with a high turnover rate, such as hemoglobin. A reduction in hemoglobin would lead to an anemic condition.

The other physiological symptoms of the deficiency, convulsive seizures and a retention of water by skin tissues (edema), have not been related to the metabolic functions of the coenzyme.

Pantothenic Acid

Pantothenic acid undergoes a reaction with a dinucleotide to form Coenzyme A (see Fig. 6-7).

$$\text{pantothenic acid} + \text{dinucleotide} \longrightarrow \text{Coenzyme A}$$

Fig. 6-7 Pantothenate becomes part of the Coenzyme A molecule. Coenzyme A is the link between the Embden-Meyerhoff scheme and the Krebs cycle, as well as a key factor in fat metabolism.

On a metabolic level, CoA is active in the synthesis and breakdown of fatty acids, in the entrance of pyruvic acid into the Krebs cycle, and within the Krebs cycle itself. Therefore a reduction in the amount of CoA available to the system would adversely affect both fat metabolism and the respiratory mechanism.

It is known that pantothenic acid is essential for the growth of microorganisms, including many pathogenic (disease-causing) strains. It has not been recognized as essential for the human diet. An important question arose as to whether the lack of a set of symptoms associated with this vitamin deficiency occurred because the human body lacks the enzyme systems for forming CoA from pantothenic acid and therefore has to take in another form or whether it was so universally present in food that no one had ever experienced a deficiency. When an antagonist to pantothenic acid (a molecular form that bonds pantothenic acid, making it unavailable for reaction in the system) was introduced into human subjects, it was found that a set of symptoms appeared. This indicated that the human body was indeed capable of utilizing pantothenic acid and that its omnipresence in food was probably responsible for the fact that no deficiency had ever been noted. The symptoms included fatigue, muscle cramps, and impaired coordination. The relationship between these physiological effects and the primary action of pantothenic acid is unknown.

Biotin

Biotin acts as a coenzyme in the form in which it enters the body. Its activities on a metabolic level are primarily concerned with carboxylation reactions in which CO_2 is fixed. (Fixation of carbon dioxide is not limited to photosynthesis, but is a necessary step in the biosynthesis of purines in nonphotosynthetic organisms.)

The source of biotin for the human is the action of the bacteria normally found in the intestine (intestinal flora). A deficiency of this vitamin was noted in the inmates of concentration camps and prisons in whom the intestinal flora had depleted because of the lack of nutrient material. The symptoms, however, were not separable from other factors, and a true picture could only be gained by testing experimental animals.

When the intestines of experimental animals were sterilized and the animals were placed on a controlled biotin-free diet, several physiological symptoms were noted. Among these were a cessation of growth, a loss of hair, and a loss of muscular control.

Choline

Choline can take part in several reactions upon entering the system. As a coenzyme it is active in fat metabolism and in **transmethylation** (transfer of a methyl group from one molecule to another).

It is also active on a physiological level, leading directly to organismal function. At this level it undergoes a reaction with acetyl-CoA to form acetylcholine.

$$\text{choline} + \text{acetyl-CoA} \xrightarrow{\text{ATP} \searrow\nearrow \text{ADP}} \text{acetylcholine} + \text{CoA}$$

This synthesis is endergonic, and ATP is used as the energy donor. Acetylcholine is responsible for the transmission of nerve impulses in the central nervous system.

Choline can be synthesized by the human body from the amino acid methionine. When methionine is a limiting factor in the diet of experimental animals, symptoms are manifested in which the fat content of the liver is greatly increased, leading to cirrhosis. No abnormalities have been found relating to the action of the CNS.

Inositol

Inositol has been assigned to the B complex because of its solubility in water and some evidence that it is of nutritional value. The physiological function and metabolic function are unknown.

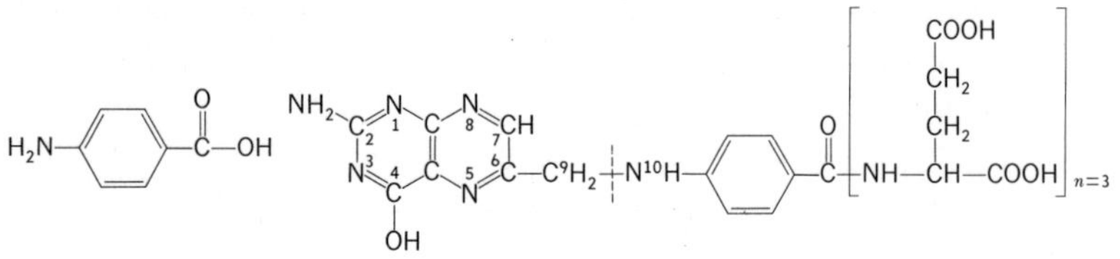

Para-amino-benzoic acid (PABA) (Pterin moiety) (PABA) Folic acid (Glutamic acid)

Fig. 6-8 Para-aminobenzoic acid (PABA) combines with a pterin moiety and glutamic acid to form folic acid, a cofactor active in single carbon unit transfers.

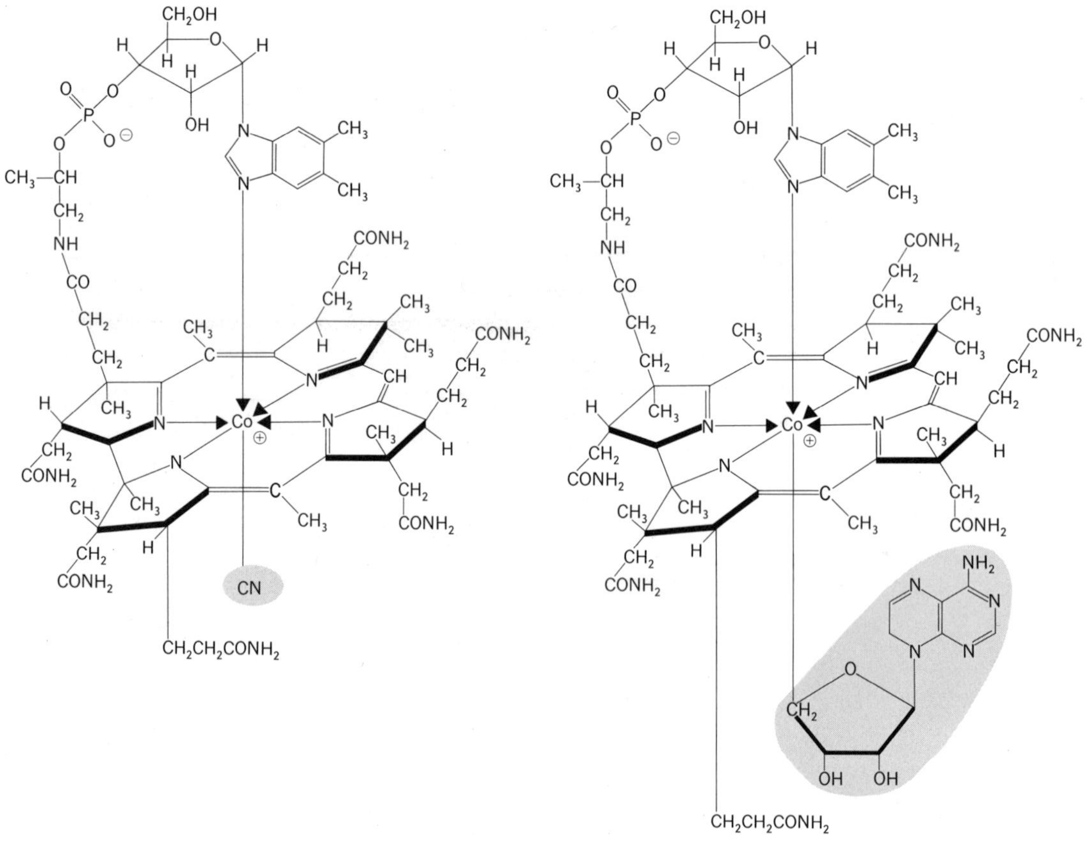

Vitamin B_{12} B_{12} Coenzymes

Fig. 6-9 Vitamin B_{12} is a porphyrin that combines with a ribonucleotide to become the B_{12} coenzyme.

Para-aminobenzoic Acid and Folic Acid

Para-aminobenzoic acid (PABA) is converted in bacterial systems to folic acid (see Fig. 6-8).

$$\text{PABA} + \text{glutamic acid} + \text{pterin moiety} \longrightarrow \text{folic acid}$$

The human body is incapable of producing folic acid from its component parts and must, therefore, take in the entire structure. Folic acid functions in metabolism in its hydrogenated form, tetrahydrofolic acid (FH_4). These hydrogens are transferred to folic acid by NADPH. Tetrahydrofolic acid is a C_1 or formate (CHOH) carrier. The formate unit is necessary in the biosynthesis of several important molecular structures, including purines.

No set of physiological symptoms has ever been related to a deficiency of PABA. It is believed that the intestinal flora provide the necessary folic acid for the human body.

Cyanocobalamin (B_{12})

Vitamin B_{12} has a porphyrin structure in which a molecule of cyanide is bonded to the cobalt ion in the center of the porphyrin ring (see Fig. 6-9). Its activity on a metabolic level seems to be related to rearrangements of branched molecular structures into linear ones. It also appears to be responsible for the conversion of a riboside to a desoxyriboside by removal of the oxygen on the second carbon of the ribose molecule.

It is interesting to note that cobalamine has only been found in animals and bacteria, never in plants. The physiological symptoms resulting from a deficiency of cyanocobalamine and termed pernicious anemia are found quite often in strict vegetarians.

The discovery of the metabolic activity of cobalamine has been too recent to permit speculation on its direct relationship to pernicious anemia.

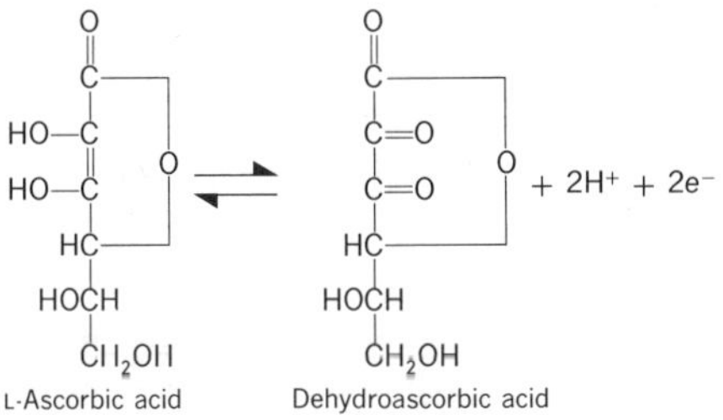

Vitamin C—Ascorbic Acid

Nobel Laureate Linus Pauling has advocated the administration of large quantities of vitamin C to guard against the common cold and complications that may result from the common cold, and there is some evidence to support his contention.

Ascorbic acid is a strong oxidizing agent (see Fig. 6-10).

$$\text{ascorbic acid} + 2e^- \longrightarrow \text{dihydroascorbic acid}$$

L-Ascorbic acid $\rightleftharpoons$ Dehydroascorbic acid $+ 2H^+ + 2e^-$

Fig. 6-10 Vitamin C readily converts to dehydroascorbic acid by giving up two protons and two electrons.

The actual site of its metabolic action is unknown, but it is thought to act in conjunction with the electron transport system. It may also serve to maintain sulfhydryl (-SH)-activated enzyme systems such as CoA by maintaining them in a reduced form. Vitamin C also maintains the intercellular ground substance. When there is a deficiency of ascorbic acid, the ground substance breaks down, becoming thin and watery. This breakdown may be responsible for the lesions that are found in all tissues of the body in a person with scurvy, the disease caused by the deficiency of vitamin C.

PERSPECTIVES OF NUTRITION

Although organisms are arbitrarily classified nutritionally according to their source of energy and the electron donor supplied by the environment for cellular redox reactions that generate ATP, within each class there may be a myriad of nutritional requirements. Venus flytrap (a plant capable of photosynthesis, and thus classified as a photolithotroph) requires the NH_3 form of nitrogen present in organic molecules. As fulfillment of this requirement, Venus flytrap captures and digests insects, and so is a carnivore.

Among the nutritional requirements of many organisms is the need for vitamins. The water-soluble vitamins serve as coenzymes for many apoenzymes, but their structure cannot be synthesized by the organism. Hence, the continued existence and function of the organism is dependent on taking in the coenzyme structure, or a precursor that can be converted to the coenzyme from the environment.

All organisms ever present on earth are linked by nutrition. The respiration of decomposing bacteria feeding on a carcass or plant link the living and the dead of the biological world. Organisms capable of fixing carbon dioxide provide a counterpart for their own respiratory processes, and those of organisms committed to taking in organic material. Organisms that require vitamins obtain them from feeding on other organisms that have either synthesized the coenzymes themselves, or have fed on other organisms who have. The pool of enzymes of all organisms present on the earth provides a means by which elements pass through different chemical states and into different types of molecules, generating the elemental cycles and the sharing of chemicals by all the biological world.

QUESTIONS

1. What are the classification criteria for the categories photolithotroph, chemolithotroph, photoorganotroph, and chemoorganotroph? For the categories autotroph and heterotroph?
2. Why is it feasible to regard the elemental cycles as an integration of all metabolic processes of all living forms?
3. Compare and contrast the food chain and the carbon cycle.
4. Justify the statement that nutritional classification does not indicate the level of biological organization.
5. Give evidence to support the contention that vitamins are coenzymes or are readily converted to coenzymes.
6. Only bacteria can convert PABA to folic acid. How then does the human body obtain folic acid?

SPECIAL INTEREST CHAPTER I

DRUGS

Although drugs have been used in various forms for thousands of years, it has only been since the late nineteenth century that the molecular basis of drug action was fully accepted. Prior to this, **pharmacologists** (scientists in the medical arts concerned with drug effects) were concerned solely with the physiological effects of drugs. These were listed in such terms as "increased heart rate," "vasodilator (dilates blood vessels)," "vasoconstrictor (constricts blood vessels)," "analgesic (pain killer)," etc. The realization that drugs acted on a chemical level within the body opened a new field within pharmacology, **pharmacodynamics,** the study of the primary reactions within a body that are altered by the presence of the drug.

The effective use of drugs would be greatly increased if the total spectrum of their activities were known. But pharmacodynamics is a relatively new science, and is confronted with many problems. The major problem is one of techniques. If one works with the whole animal, he is confronted with too many variables. It is very difficult to narrow experimentation down to a single variable in a highly integrated set of systems such as the mammalian body. If one works with parts of the animal, *in vitro,* he does not know whether he has disturbed the system to the extent to which it reacts differently than it does *in vivo. In vitro* and *in vivo* testing often give opposite effects. Also, *in vitro* experimentation requires heavier dosages of drugs to obtain similar effects to those obtained by *in vivo* experimentation. As a result, relatively little is known about drug action, and even less is known regarding the translation of drug action into drug effects.

Finding New Drugs

New drugs come from both the research units of drug companies and astute, unaffiliated chemists who then market their drugs to the drug companies. The discovery of these new drugs usually occurs in one of two ways.

The chemist synthesizes a new organic structure. He may suspect that the new compound will have an effect on the human body because of its structural similarity to a known drug. If he is part of a research unit in a drug company, he sends the compound to the pharmacologist for testing. If he is unaffiliated, he may do some preliminary tests or contact an associate to do such tests for him. If the testing proves that the compound is worthy of further investigation, it is tested clinically and patented. If all tests are passed satisfactorily, it is approved by the Food and Drug Administration and put on the market as a new drug.

But the patent and approval have been issued only for that particular drug having a specific chemical structure. It is well known in the drug industry that many molecular variations on a theme have similar effects. The action of a drug is thought to involve its participation in a chemical reaction within the system. In the course of the chemical reaction, the drug is bound to molecular forms within the system. Therefore, if the bonding site of an analogous drug is unaltered, it will have a similar effect (see Fig. I-1).

Another company then sets its chemists to work producing variations of the compound. A methyl group is added at one position in the molecule or an amino group is added at some other. Each of these "new" drugs is sent to the pharmacologist for testing, to determine whether the variant compound produces more marked effects than does the original product. If such a variation is found, it is patented, approved by the FDA, and marketed under a brand name. This is why the appearance of a new drug from one company is usually followed by a rash of so-called "me-too" drugs from other companies.

Although these two methods (the synthesizing of new organic structures and the variation of organic structures) account for the majority of drugs on the market, many of the most medically important drugs were found by chance. Fleming's discovery of penicillin (see p. 24) serves as a perfect example of this. The astute observations and testing done through basic research have done much to provide society with effective drugs.

Testing New Drugs

The pharmacologist tests new drugs in accordance with the manner in which they were discovered. When a newly synthesized chemical structure comes to him, he runs it through a generalized testing program. The pharmacologist is trained to see the structural relationships between the potential drug, other drug structures, and metabolic molecular structures. His choice of tests is based on the relationships he sees between molecular structures. Once he has found the potential drug to be active in a particular physiological area, he narrows his testing program and proceeds only as long as the drug remains promising.

In the case of a "me-too" drug, he is looking for specific effects. He submits the drug only to those tests that will indicate these effects. If the drug is being tested for its sleep-inducing activity (sedative), he may inject it into mice and keep a close record of how soundly and long they sleep. If, on the other hand, the drug is being tested as a local anesthetic, it will be injected into a restricted portion of an animal's body, such as a small area on the back of a guinea pig. Next to it will be injected the standard drug, either the one being copied or the most effective drug known as a local anesthetic. The animal may be tested by poking it with a sharp needle to determine when the area once again becomes sensitized.

If the "me-too" drug shows promise of having been more effective than the original compound at either a comparable or a lower dosage, it is tested further. Dosage is a very important factor in drugs. Because

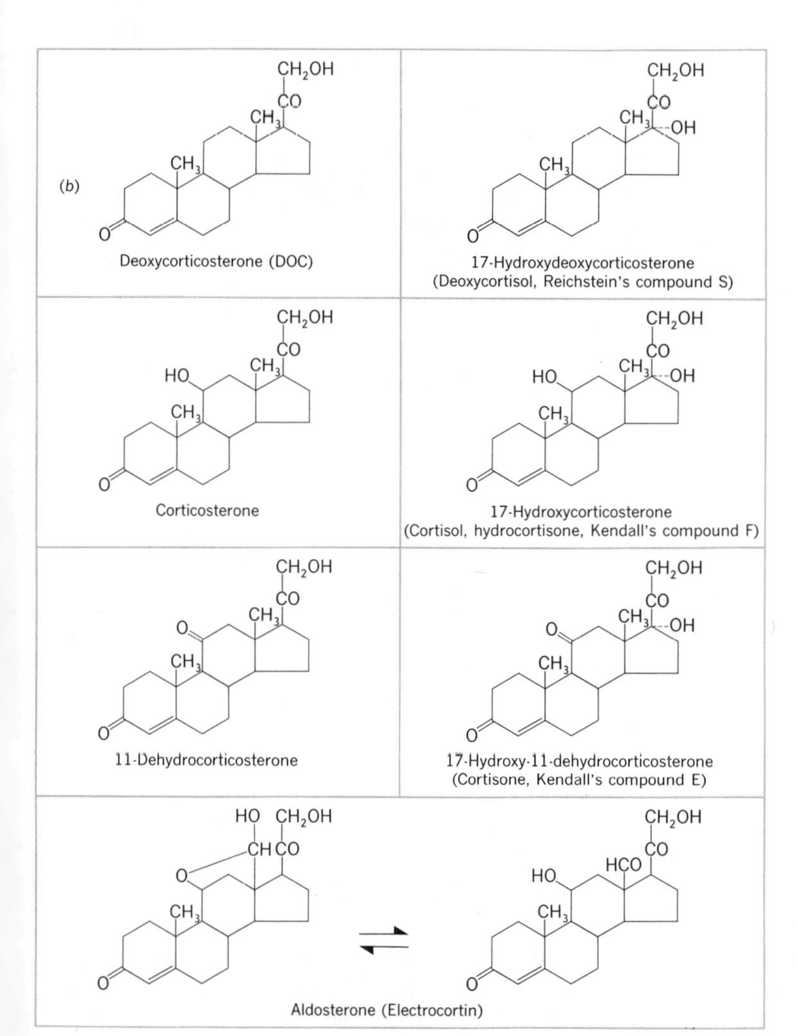

Fig. I-1 (*a*) The ring structure of the hormones secreted by the adrenal cortex (adrenocortical hormones) seems to be responsible for their general effect. (*b*) Different side groups enhance different features of the general effect.

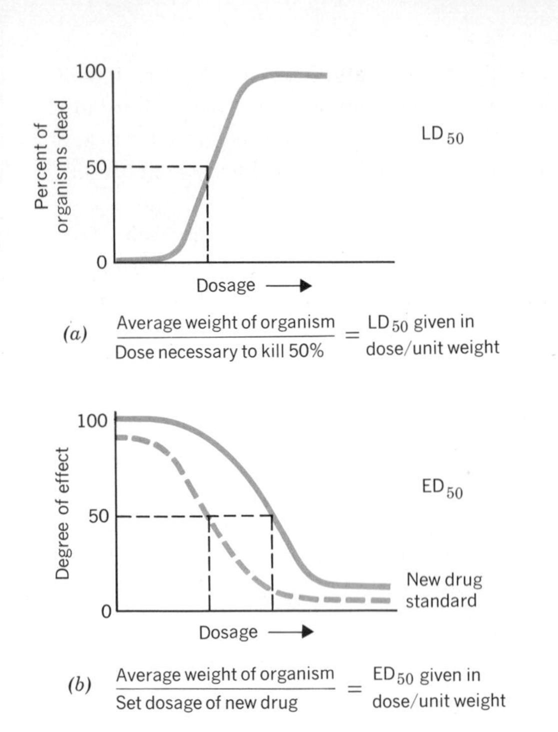

Fig. I-2 (*a*) **The LD_{50} represents a dosage at which 50% of the test organisms die.** (*b*) **The ED_{50} represents the dosage effective for 50% of the test organisms as compared to a standard drug used for that effect. In the illustration a lower dose of the standard than of the new drug is required to achieve the same effect. Hence the new drug is not as effective as hoped, and further research on it would not be anticipated.**

of the chemical nature of drug action, all the effects may not be desirable. The so-called "side effects" of a drug increase greatly with increased dosage. The idea of dosage is also important to the drug industry from an economic point of view. If the original drug is effective in dosages of the microgram range, but the "me-too" drug is effective only in the milligram range, then the manufacturer of the latter must put a thousand times as much drug material into the little bottle to get the same effect. This would mean that it is costing this manufacturer far more to market the drug. Therefore, the lower the effective dosage, the safer and more profitable the drug becomes.

Two dosage determinations are usually made for drugs, the LD_{50} and the ED_{50} (see Fig. I-2). The LD_{50} refers to the lethal dose and is the concentration of the drug per unit weight of the experimental animal that results in the death of half (50%) of the animals tested. The ED_{50}, which refers to the effective dose, uses a more arbitrary standard. If, for instance, the prescribed dosage for the original drug or a standard drug gives an effect that lasts for five hours in 50% of the laboratory animals, the ED_{50} of the new drug is the dosage of the "me-too" drug that gives the same effect. The ED_{50} is also given in terms of concentration of drug per unit of body weight of the animal tested. Generally, the LD_{50} and the ED_{50} must vary by several orders of magnitude if the drug is to be considered worthy of further testing.

Sometimes the drug industry takes advantage of a fortuitous happenstance. The original drug may be known to have a slight side effect. One of the variations of the chemical structure produced and tested by another company may show an exaggeration of this side effect to the extent that it becomes the dominant effect of the drug. This company may then focus attention on this side effect as being desirable in the treatment of other conditions, and the pattern of testing is altered. Now the effect of the drug paralleling that of the original drug is considered the side effect.

Clinical Testing

The ED_{50} and LD_{50} are first established for laboratory animals. The drug is, of course, meant for human consumption, and eventually it must be tested on the human animal. The LD_{50} provides a range of doses that must be eliminated from consideration. Since the LD_{50} is given in terms of dose per unit of body weight, it can be directly translated into the human application. The ED_{50} provides the starting point for dosages at the human level.

The initial clinical trials of a new drug are done very cautiously. These are usually carried out on volunteers as well as patients and are primarily to determine whether the drug is safe for further study rather than whether or not it is effective. Once this criterion is satisfied, the drug is run through a series of controlled clinical trials. In these, some patients are given the drug, while others are given *dummy* medication termed a **placebo.** The patients are not told which they were given, and sometimes, to avoid any psychological slips on the part of the technicians, those involved in dispensing the drugs are not told which are the placebos.

With certain drugs, the facilities for testing may be inadequate, and so testing may continue after the drug is in general clinical use. In some instances, the toxic effect of a drug may appear only after a long use or only in combination with other factors, and so an accurate picture of the drug may not be available until after the drug has been in use for several years.

Regulation of Drugs

Before a drug can be put on the market in the United States, it must be approved by the Food and Drug Administration. The FDA was formed in 1906 through the Food, Drug, and Cosmetic Act, at a time when there was a great deal of misuse and misbranding of both food and drugs. This act called for a complete listing of ingredients and experimental proof of the claims made by the manufacturer. In 1938 and 1962 the act was modified because the inadequate testing of drugs had resulted in numerous fatalities and maimings. These modifications (1) set testing standards for the marketable drug, (2) gave the FDA power to determine whether a drug could be sold only by prescription or "over the counter," (3) set standards in advertising drugs, and (4) called for a continuing evaluation of drugs after they were placed on the market. The FDA has the legal right to withdraw from the market drugs that are considered to be unsafe.

The Food, Drug, and Cosmetic Act of 1906 recognized two private organizations that set standards for drugs, the United States Pharmacopeia (U.S.P.) and the National Formulary (N.F.). The standards set by these organizations are primarily concerned with the purity and quality of drugs rather than with their effectiveness. The presence of the U.S.P. or N.F. signature on drugs is an assurance that they meet the standards of these organizations.

Some General Considerations

There are many factors that determine the dose of a drug to use. Children are often sensitive to drugs and require smaller dosages. Older people may respond in an abnormal fashion, usually because of their inability to eliminate excess drug from their bodies. Overweight and underweight individuals react differently to drugs. The manner by which a drug is administered may alter its effectiveness. The time of the day when it is given may also require consideration. Physiological factors or psychological states of the individual may modify the effects. All of these factors are considered by the doctor in prescribing medication, and should be considered by the untrained individual when he elects to buy and consume "over-the-counter" drugs.

ANTIBACTERIAL DRUGS

One of the major medical achievements of the late nineteenth century was the realization that many diseases are caused by bacterial invasion. The implications of this in terms of disease transmission were immediately obvious, and suitable preventive means greatly decreased the spread of disease. The treatment for diseases caused by pathogenic bacteria (disease-causing bacteria) was rest and hope.

In 1910, Paul Ehrlich discovered an arsenic compound that was lethal to the bacterium that caused syphilis, but relatively safe for the human. Ehrlich himself referred to his discovery as a "chemical knife," since it was capable of "cutting out" microbes from tissues, but was as dangerous to the individual as a surgical technique.

In 1935, Domagk discovered that a group of compounds, the sulfanilamides, were active against several types of pathogenic bacteria. About this time, extensive work was being done on penicillin. The "sulfur drugs" were marketed in the early 1940s, and the antibiotics in the late 1940s. It is well accepted that the action of these drugs is based on nutritional aspects of bacterial metabolism that differ from that of the human body.

The Sulfur Drugs (*Sulfanilamides*)

There are some 5500 sulfanilamide derivatives on the market. Sulfanilamide is the most often prescribed because of its high antibacterial activity and low toxicity to the human body.

Sulfanilamide closely resembles the structure of para-aminobenzoic acid (see Fig. I-3). It has been well substantiated that the enzyme responsible for the bonding of PABA to the pterin moiety in the formation of folic acid "mistakes" sulfanilamide for PABA and bonds to it instead, thereby producing a nonsense molecule, incapable of metabolic activity. This enzyme inhibition is competitive, and so is dependent on the concentration of the PABA and sulfanilamide present in the system. Because of this, relatively large doses of sulfanilamide must be used.

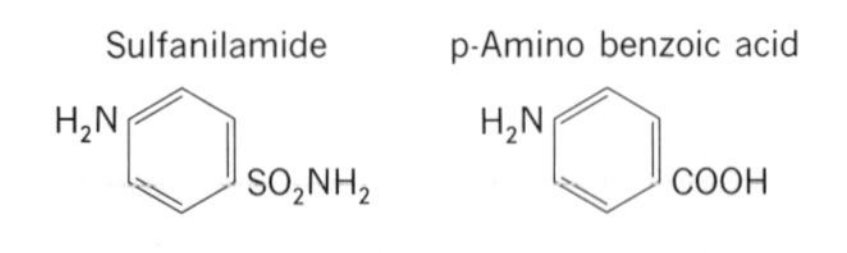

Fig. I-3 Similarity of structure of Sulfonilamide and p-amino benzoic acid.

As previously noted, the human body cannot synthesize folic acid from PABA, and must take in the entire molecular structure. But the source of folic acid for the human is in great part from the activity of the intestinal flora. As a PABA antagonist, sulfanilamide, when taken orally, affects the intestinal bacteria as well as the pathogenic bacteria. Therefore, prolonged usage of sulfanilamides can cause supplementary vitamin deficiencies in man.

Also, sulfur drugs tend to accumulate in the kidneys rather than being excreted. This accumulation severely hampers the function of the kidneys and results in bodily dehydration. One of the most dangerous aspects of dehydration is a rise in body temperature. When the temperature rises above 106° F., the possibility of protein and nucleic acid breakdown is markedly increased. This is particularly damaging to brain tissue. It is for this reason that physicians advise increased fluid intake when a sulfur drug is prescribed.

Antibiotics

Fleming's discovery of penicillin led to the search for other antibacterial microbial secretions. The most useful antibiotics found to date are produced by three types of microorganisms (see Table I-1).

TABLE I-1 MICROORGANISMS AND THEIR ANTIBIOTICS

Organism	Type	Antibiotic
Penicillium	mold	Penicillin
Streptomyces	bacteria	Streptomycin
		Tetracyclines
		Aureomycin[a]
		Terramycin[a]
		Tetracycline[a]
		Chloramphenicol (chloromycetin)[a]
Bacillus	bacteria	Tyrothricin
		Bacitracin
		Polymyxins
		Erythromycin
		Neomycin
		Viomycin

[a]Broad spectrum.

The tetracyclines and chloramphenicol are termed "broad spectrum" antibiotics because of their effect on a wide variety of bacterial types.

Penicillin

The action of penicillin is limited to those bacteria with cell walls made up of a mucoprotein. Part of the molecular structure of the mucoprotein is *n*-acetyl-muramic acid peptide. It appears that penicillin interferes with the incorporation of *n*-acetyl acid into the mucoprotein of the cell wall. The mechanism of this interference is unknown.

If the bacterial population is static (not growing), penicillin is not effective. From this it has been concluded that the turnover rate of cell-wall material is too low to manifest the exhibitory effect of the drug. But in a growing population, where new cell-wall material must be synthesized for the new bacteria being formed, penicillin has a devastating effect. The cell walls become weak and the material within the bacterial cell breaks through and is extruded (lysis) (see Fig. I-4).

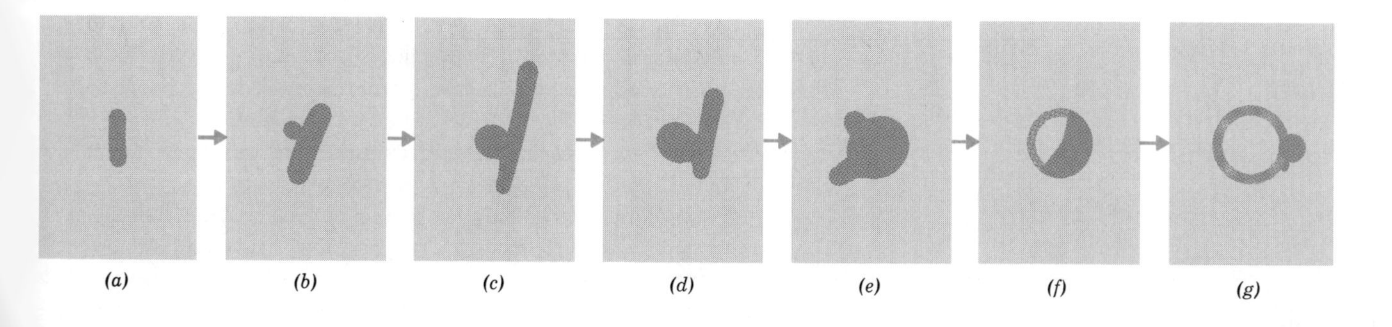

Fig. I-4 Effect of penicillin on *E. coli.* (*a*) Appearance of bacteria immediately after addition of penicillin. (*b-d*) Emergence of globular extrusions. (*e*) Formation of rabbit ear. (*f*) Cell partially vacuolized. (*g*) Ghost of membrane after cytoplasm has been extruded. [From "Penicillin-Induced Lysis of *Escherichia Coli*," by Hahn, F. E. and Crik, J. *Science*, Vol. 125, pp. 119–120, 18 January 1957.]

The inadvisability of an overuse of penicillin is twofold. First, bacteria have the ability to change (mutate) and become indifferent to the presence of penicillin. Some of these mutations take the form of the production of penicillinases, which convert the penicillin to a nonbactericidal form. Other bacteria have mutated to the extent that they became dependent on penicillin as a carbon source in nutrition. In these cases, the effectiveness of penicillin as a therapeutic measure is substantially decreased.

In addition to this, many individuals have manifested an allergic response to penicillin. Although this varies considerably with the individual, it has been associated with prolonged, and sometimes unnecessary, exposure to the antibiotic.

Chloramphenicol and Tetracyclines

These broad-spectrum antibiotics have their effect at a more basic metabolic site than does penicillin. The activity of chloramphenicol is in the inhibition of protein synthesis, and will be discussed more fully on page 250–254. Because of its basic activity, it affects not only growing bacterial populations (the invader), but also the host cells as well (the human). Chloramphenicol and the tetracyclines are cautiously prescribed by physicians and should be used with equal caution.

Polymyxin, Streptomycin, and Other Antibiotics

The remainder of the antibiotics seem to affect cell-membrane formation in bacterial populations. Although human sensitivity to these does not seem as prevalent as with penicillin, the ability of bacteria to mutate and become resistant to their effects limits their use.

FLUORIDATION

In 1943, the town of Bauxite, Arkansas changed its water supply. Within a few years it was noticed that the number of **dental caries** (cavities) in children increased substantially. Examination of the previously used water supply indicated the presence of low concentrations of fluoride. Subsequent research has definitely related the presence of fluoride in the water with reduction in the tendency of teeth to decay. Many communities have fluoridated their water; many others are involved in referenda and legislative battles over the fluoridation of water. The decision is ultimately a civic one.

Fluoride may enter the internal tissues of the body through several routes. It is absorbed by the blood stream either from the digestive tract or the lungs. The degree of absorption is directly related to the solubility of the molecular form in which the fluoride is incorporated. Sodium fluoride (NaF) is completely absorbed, whereas Na_3AlF_6 is poorly absorbed. Once in the body, fluoride appears to be absorbed by all the tissues. Most of these tissues rid themselves quite readily of excess fluoride; only the bones, teeth, and kidneys tend to concentrate it.

Large doses of fluoride in the tissues act as enzyme inhibitors, diminishing anaerobic glycolysis and respiration. As a result, there is a decreased oxygen consumption and carbon dioxide release in muscle. The lethal dose for the human body has been set at five grams. This dose is not reflective of cumulative intake and subsequent concentration, but rather of administration in a single dose.

The incorporation of fluoride into the bones and teeth seems to be directly related to intake and age. Growing bone tends to accumulate more fluoride than does bone in mature animals, so that incorporation is a function of the turnover. Recent research has indicated that fluoride replaces hydroxyl ions (OH^-) in the tooth structure, thereby strengthening the teeth. Therefore, the greater the turnover of hydroxyl ions, the greater the incorporation of fluoride.

There have been reports of fluoride poisoning from prolonged administration of relatively high dosages. This chronic effect takes the form of brittleness of bone and mottling of the enamel on the teeth due to an increased calcification. Continuous use of water containing 1.0 parts per million of fluoride has resulted in a mild mottling of the teeth of 10% of the children tested. When the fluoride in the water was increased to 1.7 parts per million, mild mottling was found in 40 to 50% of the children. At a fluoride level of 2.5 parts per million, the mottling incidence rose to 80% of the children. The standard fluoridation dose calls for 1.0 parts per million.

One of the major problems in the fluoridation of a water supply is the maintenance of the fluoride level. Tests on fluoridated reservoirs have indicated that the fluoride level varies from 0.4 to 6 parts per million from day to day and even within a given time at different parts of the reservoir.

Dental research has indicated that the primary benefit from a fluoride is a constant topical application rather than its internal consumption. The primary advantage to a fluoridated water supply is the constant passage of fluoridated water over the teeth rather than its transfer to tooth structure from internal mechanisms. Therefore, dentists consider water fluoridation preferable to administering fluoride tablets that are swallowed directly.

There have been several suggestions regarding the role of fluoride in the prevention of dental cavities: (1) a tooth is made more perfect in form and structure, (2) the chemical make up shifts to a more resistant composition, (3) the solution rate and probably the solubility of tooth minerals are reduced by the presence of fluoride, and (4) the acid production by bacteria is reduced. These suggestions form a series of hypotheses regarding the action of fluoride that have yet to be tested.

THE SALICYLATES AND ASPIRIN

Salicylic acid is so irritating that it cannot be taken internally. A variety of derivatives of this acid, however, are safe for consumption. The most common derivative used is one in which the hydroxyl group is replaced by acetate, forming acetylsalicylic acid, or aspirin.

The beneficial effect of the naturally occurring derivatives (extracted from plant material such as willow bark) in reducing the body temperature (antipyretic property) was known to the ancients. Aspirin, however, was not synthesized until 1899, and since that time has become the most frequently used drug in the world. It has been estimated that in the United States alone, some 200 million aspirin tablets are consumed daily. Although the clinical safety and effectiveness of aspirin has been very well established, the drug action and spectrum of effects remain elusive.

Metabolic changes from the administration of aspirin have been noted for doses far exceeding those that are normally recommended. In experimental animals, large doses result in an uncoupling of the oxidative phosphorylation mechanism from the electron transport system, thereby reducing the energy yield of respiration. It appears that cytochrome *c* is especially sensitive to the presence of the salicylates. In addition to this, there is an inhibition of ATP-dependent reactions. Whether this is an indirect result of the uncoupling of the oxidative phosphorylation mechanism or a direct action is unknown.

Nitrogen metabolism is also affected by large doses in experimental animals. There is a decrease in protein synthesis and an increase in protein degradation to amino acids. One of the confusing aspects of the effect of aspirin is that *in vitro,* low doses seem to stimulate protein synthesis, whereas *in vivo* no effect from low doses is detected.

The effect of aspirin on fat metabolism is to increase fatty-acid oxidation and decrease fat synthesis. In carbohydrate metabolism, there is a decrease in the formation of glycogen and an increase in glycogen breakdown and release of glucose into the system.

It has also been found that aspirin reduces the activity of a number of enzymes. This reduction in enzyme activity may, at times, be beneficial to the system. Many pathogenic bacteria produce an enzyme, **hyalouronidase,** which degrades hyalouronic acid. Hyalouronic acid is a mucopolysaccharide in a gel form that is found between cells and acts as a protective substance. The reduction of the activity of this bacterial enzyme is of obvious benefit to the human system.

On a physiological level, normal doses of the salicylates are effective in reducing pain and swelling of joints and reducing fevers without adverse effect on body function. Because of this, they are standardly used for arthritic patients and in diseases in which high body temperatures are common.

Overdoses of aspirin result in a salicylate intoxication that may be fatal. The symptoms of the intoxication are headache, dizziness, ringing in the ears, difficulty in hearing, dimness of vision, mental confusion, drowsiness, sweating, thirst, nausea, and vomiting. Certain physiological malfunctions can increase the susceptability to salicylate intoxication. Also, children with high fever and in a dehydrated state are more sensitive to aspirin.

The lethal dose varies with the individual. As little as 10 grams of aspirin has caused death in some adults. In children, the lethal dose appears to be much lower, and it has been estimated that 1000 children in the United States die from an overdose of aspirin each year.

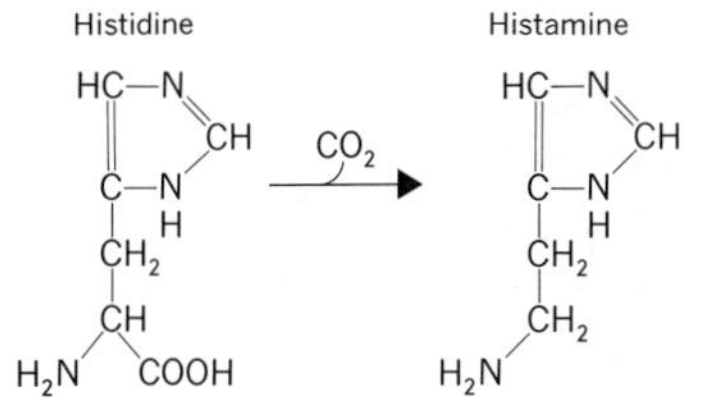

Fig. I-5 Decarboxylation of histidine to form histamine.

HISTAMINE AND ANTIHISTAMINES

Histamine is a "drug" produced by the body itself by a decarboxylation of the amino acid histidine (see Fig. I-5). As such, it is catagorized as an autocoid (autos = self; akos = medicinal agent). Although all mammalian tissues are capable of producing histamine, the skin, lining of the intestine, and lungs tend to release excess histamine into the system.

For many years pharmacologists were disturbed because the only pharmacological action manifested by histamine indicated adverse effects. An increase in the amount of histamine in the system was associated with headaches, increased flow of gastric juices to the extent that ulcers were formed, and pain and itching. The realization that histamine release was related to tissue injury and that it acted on a physiological level for activating systems and parts of systems to repair and restore the equilibrium of the body somewhat changed the point of reference. The adverse effects are now taken to be side effects of the drug.

A wide variety of drugs that are antagonistic to histamine have been developed to offset these side effects. These drugs are called the antihistamines. Although the chemical structure of the antihistamines varies considerably, they all have at one portion of their molecule a

$$R{-}CH_2{-}CH_2{-}N{-}R$$

group (see Fig. I-5). The antihistamines act as competitive inhibitors in physiological reactions that involve histamine, and so the effect is directly based on the dosage.

The antihistamines came into common usage in treatment of the common cold, many of whose symptoms can be traced to the action of histamine. Therefore, the inhibition of this action serves to relieve many of the symptoms.

But, antihistamines are not without side effects. The major side effect is a sedative or sleep-inducing property, making them quite dangerous to take when alertness is required. Other side effects include dizziness, incoordination, blurred vision, nausea, dryness of the mouth, tightness of the chest, tingling, and weakness of the hands. These side effects are so severe in some preparations that they are marketed for the major side effect rather than for the antihistamine capacity.

Some deaths have resulted from overdoses of antihistamine preparations. These have primarily been in children, where the lethal dose is self-administered by gaining access to the medicine cabinet (20 to 30 antihistamine tablets constitute a lethal dose for children).

HALLUCINOGENIC DRUGS—LSD

Lysergic acid diethylamide (LSD) was first synthesized by Hoffman in 1938. The original pharmocological studies on experimental animals did not reveal anything unusual. These studies were made in an attempt to relate LSD with the function of a similar molecular form normally present in the human system. In the course of these studies, Hoffman inadvertently ingested some LSD. He describes his experience as follows:

"In the afternoon of 16 April 1943, when I was working on this problem, I was seized by a peculiar sensation of vertigo and restlessness. Objects as well as the shapes of my associates in the laboratory appeared to undergo optical changes. I was unable to concentrate on my work. In a dream-like state, I left for home, where an irresistible urge to lie down overcame me. I drew the curtains and immediately fell into a peculiar state similar to drunkenness, characterized by an exaggerated imagination. With my eyes closed, fantastic pictures of extraordinary plasticity and intensive color seemed to surge toward me. After 2 hours this state gradually wore off."

Hoffman suspected a relationship between this state and the LSD compound on which he had been working. Although his initial ingestion had occurred accidently, he decided to administer the drug intentionally. He selected a dosage that was standard for a drug of similar molecular structure, one that is now known to be five to ten times the amount required for an effect. All that is said is that his reaction was quite spectacular.

LSD (lysergic acid diethylamide)

5-Hydroxytryptamine (serotonin)

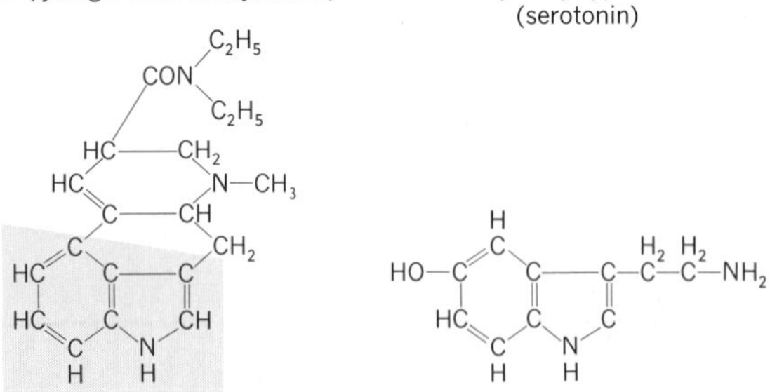

Fig. I-6 A portion of the LSD molecule is similar in structure to 5-hydroxytryptamine.

LSD appears to affect the metabolism of 5-hydroxytryptamine (a naturally occurring derivative of the amino acid tryptophane) in brain cells. The structural similarity between LSD and 5-HT in their indole moiety has led to speculation that LSD acts as a competitive inhibitor in an enzymatic reaction involving 5-HT (see Fig. I-6). This speculation is supported by the psychogenic effects of other indole-containing drugs.

Recent investigations into the chemical basis of mental disorders have led to the implication that aberrant 5-HT metabolism may underlie such disorders as schizophrenia and some types of cerebral palsy. The manifestation of symptoms common to schizophrenia after administration of LSD supports the theory that 5-HT is, indeed, the site of LSD action.

Upon administration, LSD produces anxiety-panic states and violent paranoid (persecution) reactions. There appears to be no adverse physiological effect from its administration. LSD is not readily eliminated from the system, and many effects are noted long after the administration. These include persistent mood changes and depressant states.

Although the FDA has placed restrictions on the use of LSD for clinical study of certain mental disorders, it is easily synthesized in a chemistry laboratory and has come into common usage. This intake has been of too short a duration and is under too little control for the determination of any chronic effects. It appears, however, that the body neither becomes conditioned to its presence, nor develops a dependence on the drug, so no withdrawal syndrome develops after a cessation of its administration.

In 1967 some rather startling evidence was obtained that LSD intake is related to chromosome breakage in cells. The number of cells containing broken chromosomes was three times higher in LSD users than in non-LSD users. Although this report was based on relatively few cases, extensive testing is being done to either confirm or refute its conclusions. The implications of chromosome damage will become obvious in Chapters 7 through 10.

Hallucinatory or psychogenic drugs were known long before the synthesis of LSD. Certain American and Mexican tribes used mescaline and ololiuqui extracted from plants to alter consciousness during religious rites. These have been shown to have a structure similar to that of LSD.

MARIHUANA

Cannabis, the scientifically acceptable name for the drug marihuana, is as old as recorded history, and was used for a variety of purposes in different societies. In China cannabis was employed as an anesthetic for surgery as long as two thousand years ago.

Marihuana is extracted from the resins secreted by the pluricellular hairs of the flower of the female hemp plant (genus *Cannabis*). The cultivation of hemp plants is presently under strict regulation by the federal government, and is confined to a few states once active in the production of hemp for the shipping industries.

A great furor has been raging for the past few years regarding marihuana and its inclusion in the narcotics laws. Proponents of the drug maintain that it is less harmful than cigarette smoking, while opponents claim that marihuana is addicting and leads individuals to eventual heroin addiction. Unfortunately, comparatively little research has been completed to validate either argument; but from the evidence available, it seems that there is some truth in both claims.

According to available information, there appears to be no correlation between violent crimes and the use of marihuana. Thus it would appear that marihuana use does not lead to the desire to do harm to others. On the other hand, the record abounds with accidental crimes, often involving harm to bystanders.

In terms of effect, the evidence indicates that marihuana is no more of an aphrodisiac than is alcohol, although far less of the former brings about the effect than of the latter. What is more, there is no evidence that marihuana usage leads to eventual morphine addiction, for while most individuals addicted to morphine started with marihuana, most marihuana users never go on to morphine.

Marihuana, unlike alcohol, morphine, and tobacco, is nonaddicting. In other words, the characteristic withdrawal symptoms experienced by habitués of these other drugs are not found when the individual stops smoking marihuana. In fact, habitués usually control the number of cigarettes they smoke, often having only 6 to 10 in order to maintain the desired psychic effects and avoid nullifying those effects by developing a tolerance for the drug.

Too little research has been done to state definitively the pharmocodynamic action of marihuana. Most investigations done to date report only the pharmocological results. The effects of marihuana appear to be restricted to the central nervous system, and since decerebrated cats

(cats in which the cerebral hemispheres have been destroyed) do not respond to the drug, it would appear that the cerebral hemispheres are the primary locus of action. Thus far, the experimental work on animals has failed to determine whether marihuana serves as a stimulant, a depressant, or both, depending on what drugs are used in combination with it.

Examinations of users tend to confirm the results on experimental animals in that there seems to be little physiological effect from use of the drug. Pulse rate increases and blood pressure is elevated, usually accompanied by redness of the eyes, but otherwise, the cardiovascular system is unaffected. The blood-sugar level rises, but not usually beyond the upper limits of the normal range. Kidney and liver function, as well as the blood-cell count and chemistry, are unaltered.

The effects on the central nervous system vary with several factors, including the individual and the variety and potency of the marihuana. Within a few minutes of smoking marihuana, the subject experiences a dreamy state in which ideas are uncontrollable, disconnected, and sometimes plentiful. Sometimes there is a feeling of exaltation, excitement, extreme well-being, and inner joyousness. Other times there may be a sinking into a state of depression or experiences of panic states. Time becomes disordered and seems to stretch out. Vivid hallucinations may result.

Many of the same signs of central nervous system disorders that are seen in laboratory animals are found in human subjects: tremors, vertigo, numbness of extremities, sensitivity to touch, pressure, and pain stimuli, and sluggish light reflexes. Toxic doses have resulted in disorientation and anxiety, and a true drug psychosis may develop from marihuana and persist for a time ranging from a few hours to several weeks. Under appropriate combinations of psychopathic personality and environmental factors, the use of the drug can cause a true and stable psychotic state in certain individuals.

Psychometric examinations of marihuana and nonmarihuana users with and without the drug indicate that simpler function tests, such as reaction time and speed of tapping, were only slightly affected by ever-larger doses of marihuana, while complex functions, such as static equilibrium, hand steadiness, and complex reaction time, were often adversely affected by even small doses of marihuana. The influence of marihuana on intellectual functions, as judged by intelligence tests and memory, form-design, and symbol tests, indicates that these functions were adversely affected in direct proportion to the size of the dose, higher intellectual capacities being more severely altered than simpler ones. Nonusers generally showed greater impairment than habitual users, suggesting that some degree of tolerance occurs.

Thus there is some truth in both the claim that marihuana use is less harmful than cigarette smoking or alcohol imbibition (in that, in the strictest sense of the word, marihuana is not addicting); and the claim that marihuana use can lead to hard drugs (although the path from marihuana to morphine is not as straight as, perhaps, some prefer to think). The connection nevertheless is there for the individual personality seeking more euphoric states than can be attained from marihuana. Far more research must be done before any definitive statement can be made on the effect of marihuana on the individual and on his society.

THE ADDICTING DRUGS: BARBITURATES AND NARCOTICS

There is evidence that opium was used to relieve pain (analgesic) before recorded history. Arabian traders introduced the drug to the Orient during the Middle Ages. By the sixteenth century, the effects of opiates were well known in Europe. So effective were they as analgesics that in 1680, Sydenham remarked "Among the remedies which it has pleased Almighty God to give to man to reduce his suffering, none is so universal and so efficacious as opium."

By the middle of the nineteenth century, morphine had been separated from the other chemical components of opium. It was found that only 18% of the total opium was active in analgesic ability. Ten percent of this fraction is morphine; the remaining 8% is a mixture of other compounds. Today morphine is considered the most effective analgesic known to man and is the standard by which all other analgesics are measured.

The primary physiological effect of morphine seems to be its inhibition of the hydrolysis of acetlycholine into acetyl and choline. Since this reaction is necessary for the transmission of nerve impulses in the central nervous system, its inhibition may be directly related to the analgesic effect of morphine. In addition to this, however, morphine is active on a metabolic level. It has been shown that morphine increases the incorporation of phosphate into phospholipids, thus making this inorganic ion unavailable for phosphate-dependent reactions. Another factor, which may be related to the alteration in the phosphate equilibrium, is an increase in glucose breakdown.

Morphine has many side effects, perhaps due to the basic metabolic effect on the system. After administration of morphine, an individual experiences a drowsiness and mental clouding. For individuals in pain, a euphoric state is induced. In pain-free individuals, however, a state of anxiety with accompanying nausea and vomiting is often induced.

These psychological effects of opiates may have been known to the ancients. The Sumerians (ca. 4000 B.C.) have as their ideograph for the poppy, *hul* (joy) plus *gil* (plant). Because of the sometimes pleasant psychological side effects, opiates came into common usage. They were consumed for years by imbibing and inhalation (smoking and snuff). The invention of the hypodermic needle and the improvident use of morphine for treatment of wounded soldiers in the American Civil War alerted many to the addictive properties of morphine and served as an impetus for finding nonaddictive analgesic drugs.

A drug is considered to be addicting when a consumer develops a *tolerance*, a *physical dependence*, a *behavioral pattern of compulsive use*, and a *high tendency for relapse after withdrawal.* Tolerance to a drug refers to the decreased effect after repeated administration and the increased dosages that must be given to obtain the same result as the original dosage. Physical dependence involves an alteration of the physical state of the system to the point where failure to administer the drug results in a withdrawal or abstinence syndrome. The behavioral pattern of compulsive use is characterized by an overwhelming involvement with the use and procurement of a supply of the drug. The high tendency to relapse is self-explanatory.

There are many reasons for the self-administration of addicting drugs. Clinical use of morphine may have brought about a dependence that requires satiation. The euphoric side effects may be sought by certain types of personalities. A summary of this latter class has been most effectively stated as follows:

"Narcotics do more than produce an indifference to pain. They also *suppress* those drives which motivate an individual to appease hunger, seek sexual gratification and respond to provocation with anger. In short, they seem to produce a state of total drive satiation. Nothing needs to be done because all things are as they should be. For certain types of personalities, but clearly not for all, such a state is extremely pleasant."

The attainment of such a state is not automatic, however. Many pain-free individuals have admitted to having trained themselves to emphasize the "euphoric" effects and ignore the unpleasant ones.

Tolerance to, and physical dependence on, narcotics appear to occur at both the systemic and the cellular level. Cells in tissue culture grown in morphine exhibit requirements for increasing levels of morphine for continued growth. When they are placed in a morphine-free medium, growth ceases completely. There have been many suggestions regarding the basis of tolerance to and physical dependence on narcotics, but thus far none can account for all the factors known.

A major factor in the dynamics of narcotic administration is cross-tolerance. An individual tolerant of one narcotic will be tolerant of another, even if the two are chemically dissimilar. Therefore, switching from one narcotic to another does not decrease the dosage required by the system.

There appears to be an upper limit of physical dependence, after which the same degree of withdrawal symptoms are noted. The length of time for withdrawal from narcotics is dependent on the cooperation of the individual. Withdrawal symptoms reach their peak in 48 to 72 hours after the last dose and subside completely after 10 to 14 days. The symptoms manifested during this time include crying, yawning, sneezing, sweating, tremor, restlessness, irritability, muscular weakness, fever, insomnia, hoarseness, vomiting, chills, and stomach cramps.

Overdoses of morphine have resulted in death. An overdose may result from an attempted suicide or from an inadvertent injection of a second dose by a clinician or addict because the first dose was not absorbed quickly enough. In the latter case, the gradual absorption into the system of the first dose may fatally increase the morphine level.

Barbiturates are a class of hypnotic drugs that are used as analgesics. They appear to uncouple the oxidative phosphorylation mechanism from the electron transport system, thereby decreasing the energy yield from respiration. The site of this uncoupling appears to be different from that of the salicylates because there is a decreased oxygen intake with barbiturates, indicating a slowing of the electron transport system; whereas there is an increased breakdown of glucose with the salicylates, indicating a speedup of the electron transport system.

In addition to this, barbiturates impair both energy storage and utilization by ATP. The precise mechanism and site of activity is not known.

Barbiturates exhibit the same type of tolerance as the narcotics. Therefore, ever-increasing doses must be administered to maintain the hypnotic effect. Barbiturates tend to boost the effect of narcotics, and many users are dependent on both. Barbiturate withdrawal has essentially the same symptoms as narcotic withdrawal.

PART TWO: THE CELL

In keeping with one of the objectives of the text, namely to provide experimental evidence whenever doing so can enhance the reader's comprehension, Chapter 7, "The Structure and Function of the Cell," is handled from a historical perspective. The orientation of this chapter is as much toward the development of the field of cell biology and a concept of research as toward a development of biological principles. Both objectives are accomplished through discussions of simple, compound, and finally electron-microscopic observations, and the necessity of correlating biophysical and biochemical techniques with these microscopic observations in order to obtain meaningful conceptualization of the cell.

Chapter 8, "Reproduction and Synthesis of Cellular Components," deals with the reproduction of mitochondria, chloroplasts, and membrane systems as well as nuclear reproduction. Chapter 9, "The Genetic Control of the Cell," introduces mendelian inheritance, and develops the concept of the gene as a functional unit. Detailed discussions of the concept of mutation and genetic analysis are considered more appropriate to an understanding of evolution, and are thus included in later chapters.

Chapter 10, "DNA and the Genetic Code," includes discussions of the DNA-RNA-protein sequence as an information system, as well as presenting evidence for, and a discussion of, the genetic code itself.

(Opposite) "Corona I," 1960, James Wines, Collection of Mr. & Mrs. Joseph Braun

CHAPTER 7

THE STRUCTURE AND FUNCTION OF THE CELL

In 1665, Robert Hooke saw, with the aid of a microscope, that the tissues of woody plants were made up of boxlike compartments that resembled a monk's chamber or cell. For almost two hundred years, scientists focused their microscopes on every conceivable type of biological tissue and observed this same structural unit. In 1839 the botanist Schleiden and the zoologist Schwann formalized the inevitable conclusion of these many years of observations in a statement of the cell theory; plants and animals are universally made up of cells.

It is doubtful that the early proponents of the cell theory were aware of its far-reaching implications. In the early nineteenth century, the recognition of a cell as a structural unit of life was in itself a great achievement. By the middle of the nineteenth century, however, the functional aspects of cells were well recognized, and came under close scientific scrutiny. In 1854, Rudolf Virchow stated that "Where a cell exists, there must have been a preexisting cell, just as the animal rises only from an animal and the plant only from a plant." The powerful idea that all cells come from preexisting cells imparted to cells the function of reproduction.

From the wealth of information that has accumulated since that time regarding the reproductive capacities of cells, the metabolic aspects of cellular function, the biochemical basis for cellular variation, the elucidation of cellular components, and a correlation among these various aspects, has emerged a picture of the cell as a structural and functional unit of life. The picture is, to be sure, incomplete. Nevertheless, the available information does generate a dynamic portrayal of this basic unit and serves to point toward the direction in which the other parts of the puzzle will be found.

THE LIGHT MICROSCOPE AND RELATED BIOCHEMICAL AND BIOPHYSICAL STUDIES

There are many types of cells, each exhibiting a characteristic structure. Reference to *the* cell is a biologists' attempt to universalize the concept of cells as basic units of structure and function for all organisms. *The* cell, is a *composite* cell, in which are all the cellular components collectively found in many kinds of cells. These cellular components may or may not be found in given cells, or they may be present in different ratios.

The most obvious differences in cellular structure are noted between the animal and the plant cell. This difference is of sufficient magnitude that they are usually handled separately. It is important to remember, however, that *the* animal cell and *the* plant cell are composite types in each class.

The Light Microscope

The determination of cellular structure is dependent on the quality of the microscope used for observation and the care and availability of specimen-preparation techniques. Two factors determine the quality of a microscope: **magnification** and **resolution.** Magnification is merely the capacity of the microscope to enlarge images. Resolution is the capacity of the microscope not only to present a clear image, free from distortion in form and color, but also to permit a distinction between objects. For example, microscopic studies have revealed that the chlorophyll in plants is confined to the chloroplasts within the cell. The limit of resolution of the human eye is too high to distinguish between the chloroplasts, and therefore we see the whole plant as green. Microscope manufacturers strive to produce microscopes with both *high* magnification and *low* resolution.

Human vision is based on a lens system. When light passes through a convex lens, such as that in the eye, it is bent (refracted) toward the center of the lens. The bending of the rays causes them to converge at a single point, which is called the **focal point.** The light rays continue in their paths and are projected on a surface of light-sensitive cells, the **retina.** The response of these light-sensitive cells produces a sensation of vision in the brain (see Fig. 7-1). As an object is brought closer to the eye, it is seen in greater detail. If the object is brought too close, however, the lens is unable to focus the image on the retina, and vision becomes blurred. The shortest distance from the eye from which objects can be clearly seen is about 10 inches. At this distance, the limit of resolution of the eye is about 0.2 millimeters, indicating that two objects having a distance between them of less than 0.2 millimeters will be seen as one.

The early microscopes were crude, single-lens instruments that permitted magnifications only up to 100 times ($100\times$). These were used by placing the lens very near the eye. When this was done, the refraction of light by the lens produced an image that appeared to be farther away from the eye, in the region of distinct vision (see Fig. 7-1).

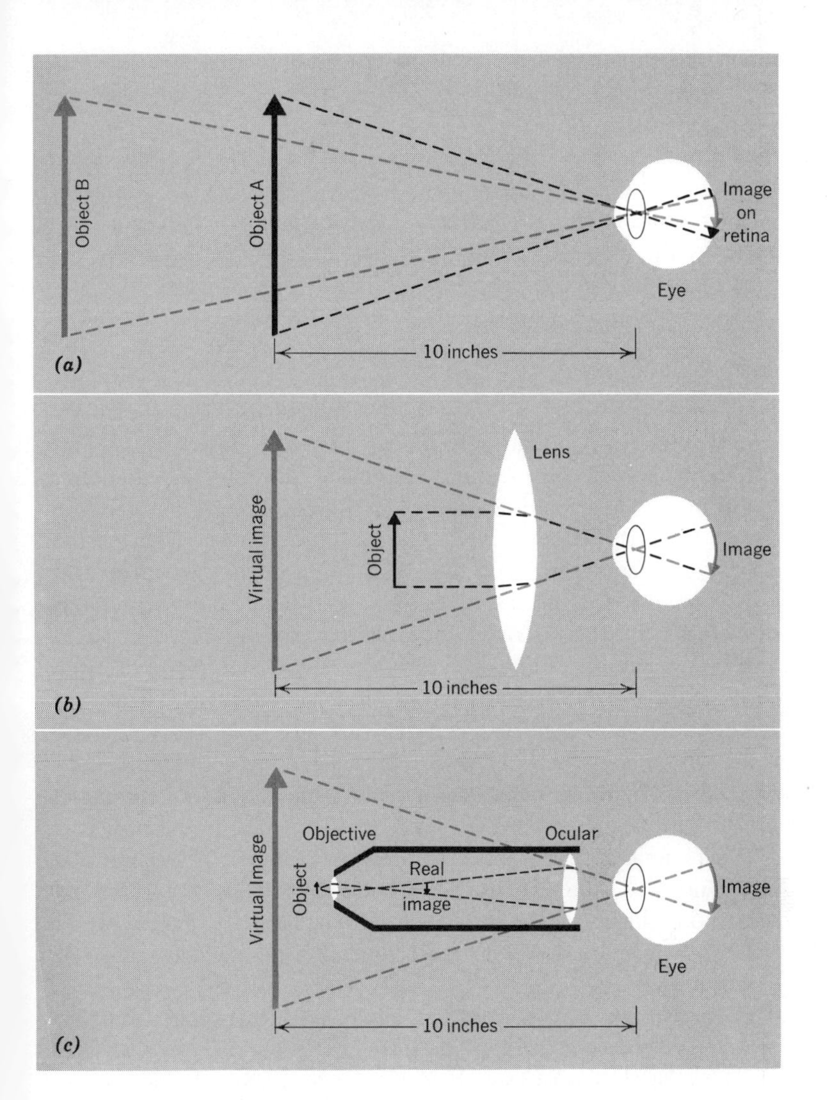

Fig. 7-1 (*a*) For a person with normal eyesight, an object ten inches away from the eye focuses most clearly on the retina. (*b*) A single lens magnifying an object permits it to be seen more clearly because it makes the object *seem* to be at the critical distance of ten inches. (*c*) The compound microscope not only magnifies the object itself but also magnifies the image of the object. The image of the object falls the same distance from the second lens as the object itself would be were it placed in a single-lens system. The second lens enlarges the image, creating the impression that it is ten inches away. Hence the real image lies some distance from the ocular within the body tube, while the virtual image lies below the stage of the microscope.

The limited magnification and restricted resolution of these early microscopes permitted little more than the determination of the general outline of cells. By the early part of the nineteenth century, a major advance was made with the invention of the **compound light microscope.** The compound microscope utilizes a double lens system: one lens is the **ocular,** through which the eye looks directly, and the other the **objective,** which is mounted at the bottom of the tube or **barrel.** The object to be examined is placed on a **slide,** and centered on the **stage** of the microscope. The amount of light passing through the slide is regulated by a **diaphragm** mounted below the stage. The light passing through the object forms an image at a point similar in position to the placement of an object in a simple single-lens system. This image is then magnified further by the ocular, and once again gives the impression of being in the region of distinct vision (below the stage of the microscope) (see Fig. 7-1). The highest magnification used for the compound light microscope is about 1500×.

The limit of resolution of the compound microscope has been greatly reduced by the improvement of lens systems and the alteration of the light source. The UV (ultraviolet) microscope achieves a finer resolution by the use of UV light. The phase microscope utilizes polarized light. In the best compound light microscope, the resolution has been reduced to near the theoretically obtainable value of 0.5 μ.

Specimen Preparation

Concomitant with the improvement of microscopes was the development of preparative materials for microscopic observation. This advance in specimen preparation was made possible because of a seemingly unrelated economic factor. In the nineteenth century, the yard-goods industry flourished in Germany. Government-supported chemists synthesized many new dyes that aided in the growth of this industry. The adoption of these dyes for staining microscopic preparations was based on the discovery that certain parts of the cell would take up the dye, and could be seen more clearly. It was found that the chemical nature of the dye and its known properties could lead to certain generalized statements regarding cell structure. For instance, an alkaline dye is known to combine only with material having a predominantly acid character. Therefore, when such a dye was taken up by an organelle, it could be said that the organelle was primarily acid in nature. (It is interesting to note that the major interest in France at this time was the wine industry, and the government-supported work by French chemists and biologists led to the concept of metabolic pathways while the German preoccupation with yard goods led to the elucidation of cell structure.)

The stains used for the preparation of specimens fall into two categories, **vital** (life-conserving) and **nonvital** (lethal). The use of vital stain permits the observer to see the cell components in a living system. Often, however, certain cellular components cannot be clearly discerned, and a **fixed** cell is necessary for observation. A fixed cell is one in which the life processes have been halted at a given time, and prepared in such a way that the structural integrity of the cell is undisturbed. There are a variety of ways in which cells and tissues may be fixed and stained with nonvital dyes. A discussion of this, however, is beyond the scope of this book.

The Animal Cell and The Plant Cell

In 1835, Robert Brown described a spherical body within cells that he termed the **nucleus.** In 1839, E. J. Purkinje described the viscous fluid within the cell, and termed all the living material of the cell **protoplasm.** The cell was envisioned as being composed of two types of protoplasm, **nucleoplasm,** which was resident in the nucleus, and **cytoplasm,** the medium surrounding the nucleus and bounded by the periphery of the cell. When it was realized that dead cells lacked protoplasm, the term became associated with the vitalistic theory of life. Protoplasm was the mysterious material that imparted life to its possessor. With the advent of the molecular interpretation of life, the connotation of the word protoplasm became inapplicable, and the word returned to its original meaning.

The compartmentalization of the cell into nucleus and cytoplasm is valid for both animal and plant cells. Also common to both are the structures found within the nucleus. Staining methods revealed that the nucleus contained at least one, and often more than one, **nucleolus** (pl. **nucleoli**), and a series of threadlike structures that take up alkaline stains and were called **chromatin.** The chromatin material was later identified as individual fibrous units that were called **chromosomes** (see Fig. 7-2).

Many cell structures presented cytologists with more of a problem. Although the presence of larger structures was universally agreed upon, the organelles that were at the limit of resolution (0.5μ) became a source of argument. Some claimed to have seen and accurately described these structures, while others claimed that what had been seen were "artifacts" due to the distortion of light, poor fixation and staining techniques, or the overactive imagination of the observer.

Among the generally agreed-upon cytoplasmic structures were the **vacuoles,** clear spherical bodies in the animal cell and large irregularly shaped structures around which the cytoplasm flows in plant cells. These vacuoles sometimes appeared to have granules and crystals within them.

Some animal and plant cells, notably those of unicellular organisms, had hairlike projections that originate in the cytoplasm and extend through the cell membrane out of the cell. When these hairlike projections were short, they were termed **cilia;** when they were long, they were termed **flagella.** Both cilia and flagella are directly related to the motility of the cell, and propel the organism through a liquid medium by beating or whiplike movements. Usually a motile cell contains many cilia, or one or two flagella. At the base of a cilium or flagellum, a particle was observed, called a **basal body.**

Other accepted structures found in animal or plant cells were found not to be universal for both. Animal cells were found to have a **central body** or **centriole** located near the nucleus. The presence of a centriole in plants seemed to be dependent on the stage of development and the type of plant tissue examined, and its universal presence was a topic of discussion for many years.

Plant cells, on the other hand, had pigmented bodies or **plastids.** When they contained chlorophyll, they were termed **chloroplasts.** When they contained another pigment or no pigment, they were called **chromoplasts** or **leucoplasts,** respectively. Certain plant tissues were found to contain bodies that stain black with iodine solution and are called **starch grains.** Also, the plant cells are bounded by a rigid **cell wall.**

Many other organelles were reported and described, only to be called "artifacts" by other observers. The most important of these were the proposed membrane systems of the cell. Proponents for membrane structures claimed to have seen a membrane system that bounded the cell (**cell membrane**), and membrane systems around the nucleus (**nuclear**

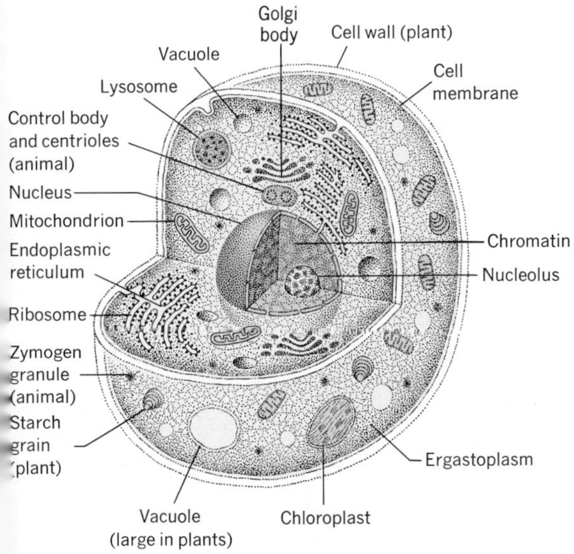

Fig. 7-2 Composite cell as seen by light microscopists.

membrane) and vacuoles (**vacuole membrane** or **tonoplast**). Other observers said that either the curvature of the cell gave the impression of a delineating region between the cell and the environment or the interphase surface of the cell spontaneously formed a selective barrier that was not a true structural unit.

Other organelles whose presence in the cell was a source of disagreement were the **chondriosomes** and the **golgi bodies.** In certain types of tissues, barlike structures of about 0.5 μ in length and called chondriosomes were described. The chondriosome is now called the **mitochondrion** (pl. **mitochondria**).

The golgi bodies were observed as amorphous structures found associated with the nucleus. They appeared to be more frequent in animal rather than plant cells. Although the size claimed for the golgi bodies was well within the limits of the light microscope, the indistinct nature of their structure produced differences in opinion regarding their existence.

Even in the early years of cytology (the study of cells), it became obvious that individual observation was insufficient. In an effort to eliminate the criticism of "artifact," cytologists welcomed the invention of photographic equipment that could be used in conjunction with their microscopes to produce photomicrographs of their specimens. Although this tended to confirm or refute some observations, many cytologists claimed that light dispersion or poor fixation gave the effect of a structure that was not really there.

The Genesis of Experimental Cytology

In the hands of the biochemists and biophysicists, the microscope became an experimental tool. Microscopists had identified the chloroplasts as localized units of chlorophyll. The presence and number of starch grains associated with the chloroplasts was directly correlated with the exposure of the plant tissue to light, thus supporting the plant physiologist's identification of chlorophyll as the photosynthetic pigment and implicating starch as the product of the photosynthetic process. If a chlorophyll fraction could be obtained, research could focus on these units. Their manifestation of a photosynthetic ability *in vitro* would conclusively prove that they were the site of the photosynthetic mechanism.

By the 1930s, methods of gently **lysing** (breaking open) cells were developed. If the cells were treated too severely, the organelles within the cell would be disrupted, and a general "soup" resulted. If, however, the treatment were sufficiently gentle, such as the lysing of cells with ultrasonic sound waves, the structural integrity of the organelles would be retained.

A cell homogenate (with intact organelles) could then be centrifuged and different cell fractions obtained. When a homogenate was spun for 10 minutes at 3000 *g* (3000 times the force of gravity), the cellular material that was of sufficient mass to respond to a force of this magnitude was brought to the periphery of the generated circle, into the base

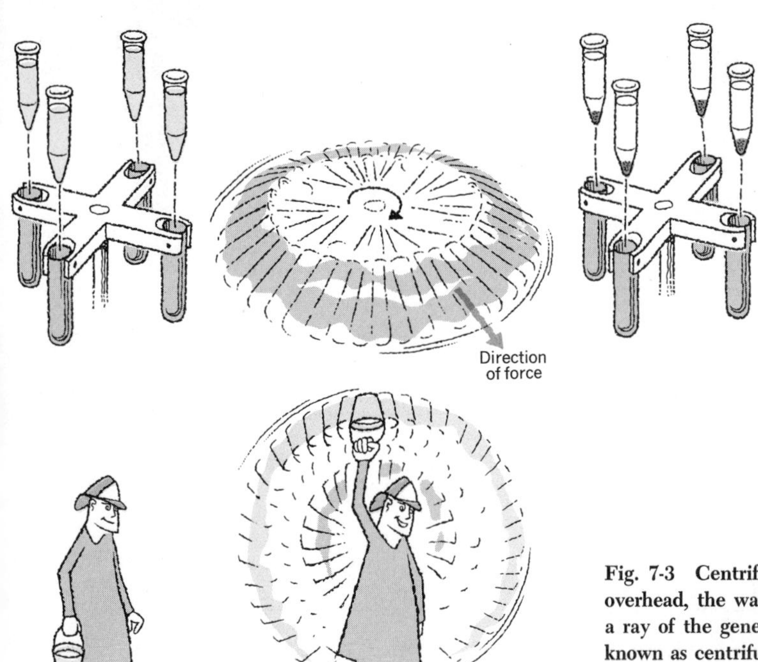

Fig. 7-3 Centrifugation. If a pail of water is swung rapidly overhead, the water will not spill. The force perpendicular to a ray of the generated circle, tangential to the circle itself, is known as centrifugal force. Centrifugal force creates a gravitational situation in which heavier materials are brought to the periphery of the circle. This principle is used in separating cellular components.

of the centrifuge tube, where it formed a **pellet.** Material of insufficient mass to respond in this manner remains in suspension, and is called the **supernatant** (see Fig. 7-3).

The supernatant is decanted off, and the pellet rediluted in an isotonic buffered salt solution (to maintain the structural integrity of the organelles in the pellet) and examined microscopically. When the original tissue was of plant origin, the pellet was observed to contain the nuclei and the chloroplasts.

The rediluted pellet could then be recentrifuged. Obviously, 3000 *g* was too great a force for the effective separation of the nuclei and chloroplasts, so a force of 2000 *g* was used. At this force, a pellet was formed which, when examined microscopically, revealed the presence of a nuclear fraction. The supernatant contained the chloroplast fraction. By repeating the 2000 *g* centrifugation several times and then submitting the supernatant to several 3000 *g* centrifugations in which the chloroplasts would come down as the pellet and be repeatedly rediluted in a fresh suspension medium (washing), a nearly pure chloroplast fraction could be obtained (see Fig. 7-4). It was on just such a chloroplast fraction that much of the more recent work on the photosynthetic mechanism has been done (see Chapter 5).

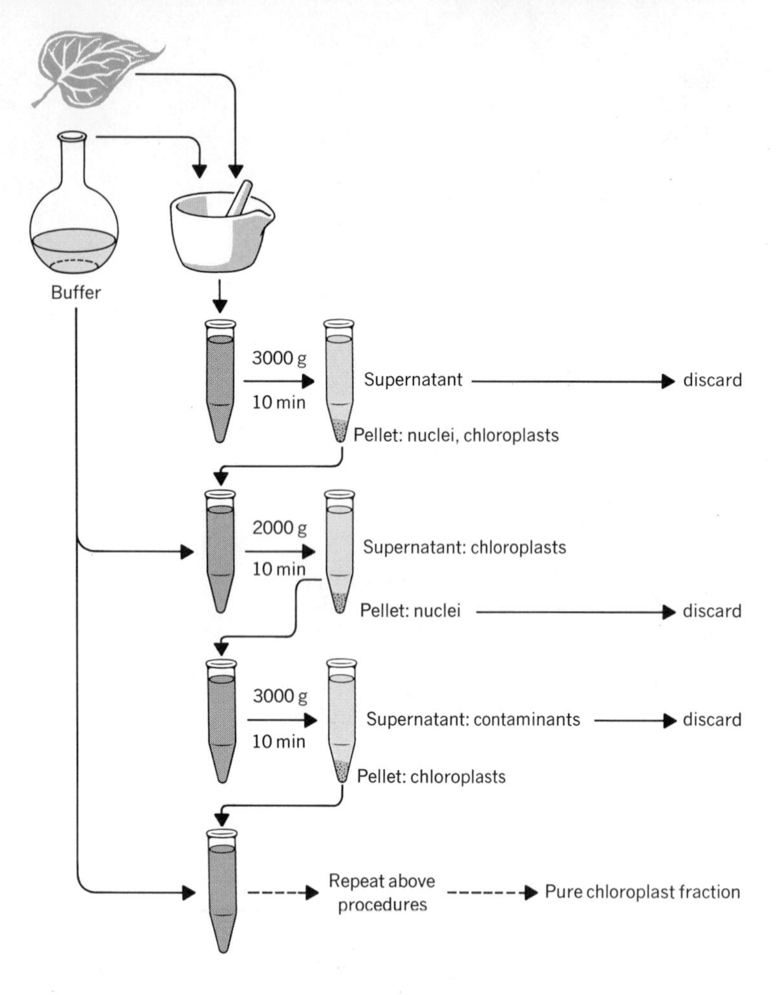

Fig. 7-4 Diagram for obtaining a chloroplast fraction by centrifugation. Centrifugation at 3000 g for ten minutes produces a pellet containing both nuclei and chloroplasts. After the supernatant is poured off, the pellet is resuspended and centrifuged at a lower speed. The result of this centrifugation is a supernatant containing the lighter chloroplasts, and a pellet containing the nuclei. This time the supernatant is saved. Further centrifugation serves to purify the fraction.

The labors of the early microscopists, physiologists, biochemists, and biophysicists laid the foundation for the type of scientific inquiry made after the invention of the electron microscope.

THE ULTRASTRUCTURE OF THE CELL

The necessity of coupling microscopic observations and experimental approaches became of prime importance when electron-microscopic study of the cell was begun.

The Electron Microscope

The invention of the electron microscope (EM) in 1935 propelled cytological studies into degrees of magnification and limits of resolutions never before possible. The electron microscope is capable of magnifying 500,000×, with a limit of 3 Å. This limit of resolution is almost at a molecular level (the bonding distance between two carbons in a single carbon-carbon bond is about 1.5 Å).

The sudden descent down to this level of microscopy created two problems for cytologists. First, living material cannot be observed in

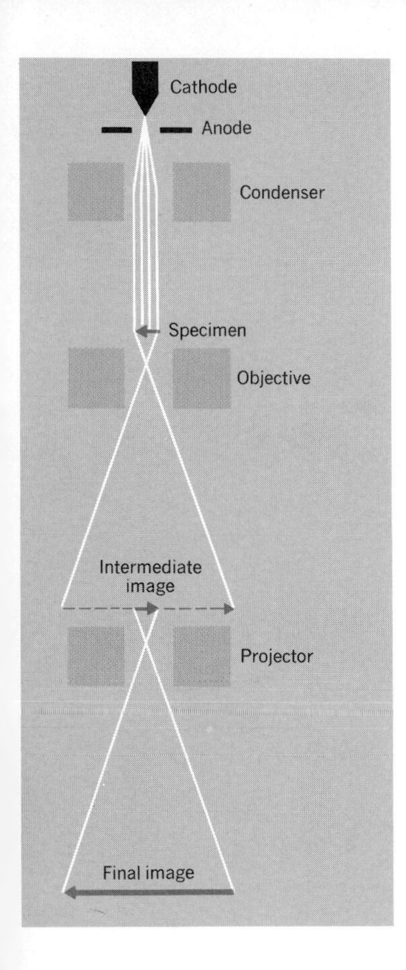

Fig. 7-5 Schematic representation of the electron microscope.

the electron microscope (EM). The light source is a beam of electrons that must pass through a vacuum (see Fig. 7-5). After material has been prepared and before it is inserted into the electron microscope, vacuum pumps evacuate the air from the system. No living tissue could survive under such a negative pressure. The inability to observe living material eliminates the possibility of determining function by direct observation.

Second, in going from a micron (10^{-3} mm) level to an angstrom (10^{-7} mm) level, a sense of continuity was lost. Although many structures could be directly related to those identified by light microscopy (i.e. mitochondria), many structures that had been only tentatively postulated or never observed were noted. One structure that had been postulated through light microscopy was the **endoplasmic reticulum,** which through electron microscopy was identified as a series of membranous shelves that extend throughout the cell. Palade saw the endoplasmic reticulum with the UV microscope in 1939 as a poorly defined structure. Associated with the endoplasmic reticulum, he noted small, dense particles. The nemesis of the light microscopists entered the realm of the EM observations: were the particles and endoplasmic reticulum "artifacts" from the preparation of the material, or did they in fact constitute an integral part of the cell? The solution to the problem of cellular function and integrity was found by the correlation of biochemical and biophysical techniques with the observations made with the electron microscope.

Experimental Cytology

The first centrifuges used in conjunction with the light-microscopic observations were capable of speeds that permitted the separation of cellular material into a nuclear, chloroplastic, and general cytoplasmic fractions. Since only nuclear and chloroplastic fractions could be confirmed by microscopic observations, there was little need for further physical separation.

When the electron microscope made it possible to examine structures in the cytoplasmic fraction (the supernatant of the 3000 *g* centrifugation) it became both propitious and necessary to adopt centrifuges that could effect a separation of this fraction. Ultracentrifuges developed by physicists early in this century, were capable of spinning material at the exceedingly high speeds, necessary for the physical separation of cellular components.

The cytoplasmic fraction (the 3000 *g* supernatant), when spun for 15 minutes at 10,000 *g*, forms a pellet (see Fig. 7-6). This pellet can, by light- and electron-microscopic studies, be shown to be a mitochondrial fraction.

When the supernatant from this centrifugation is spun at 15,000 *g* for four hours, an electron-microscopic examination of the pellet reveals the particles associated with the endoplasmic reticulum, and part of the membranous structure of the endoplasmic reticulum itself. This is called the **microsomal fraction.** Microscopic observations of the supernatant resulting from this centrifugation reveal that there is little material of particulate nature, and a chemical examination reveals the presence of enzymes, water, salts, etc.

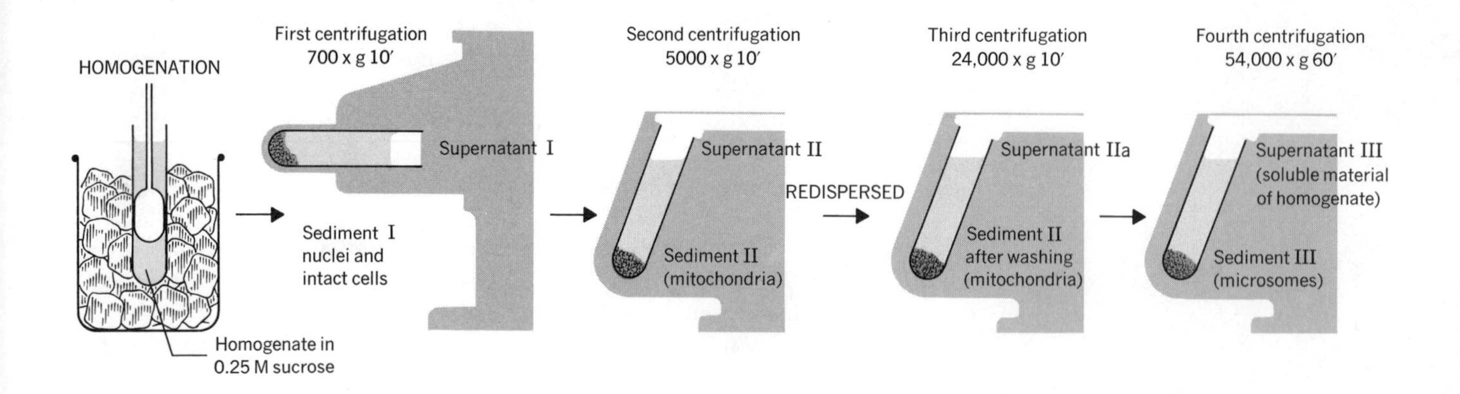

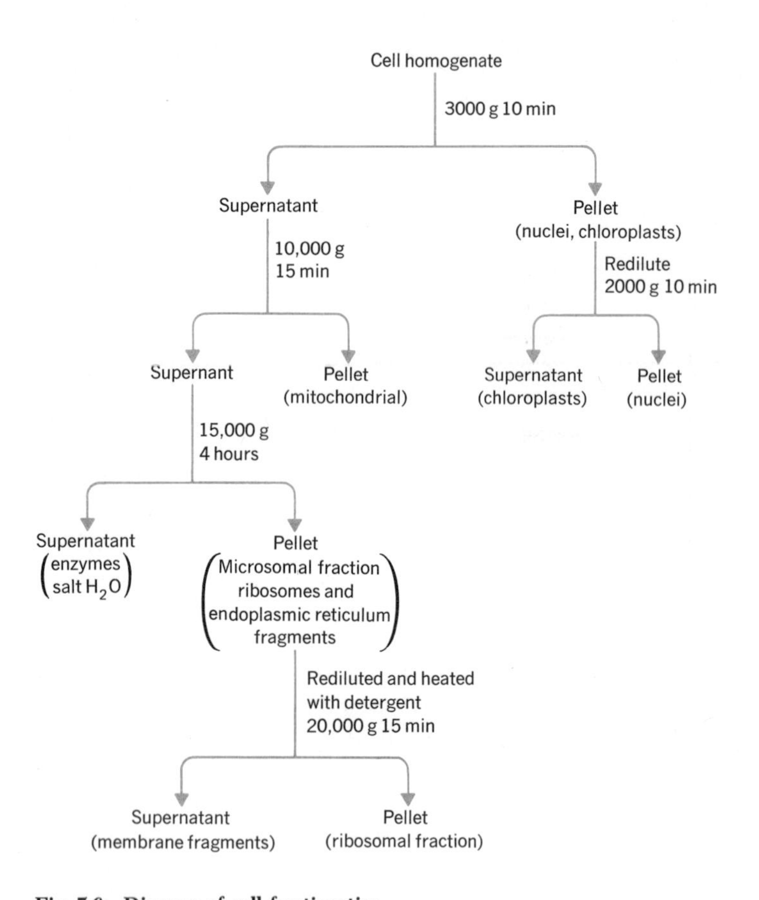

Fig. 7-6 Diagram of cell fractionation.

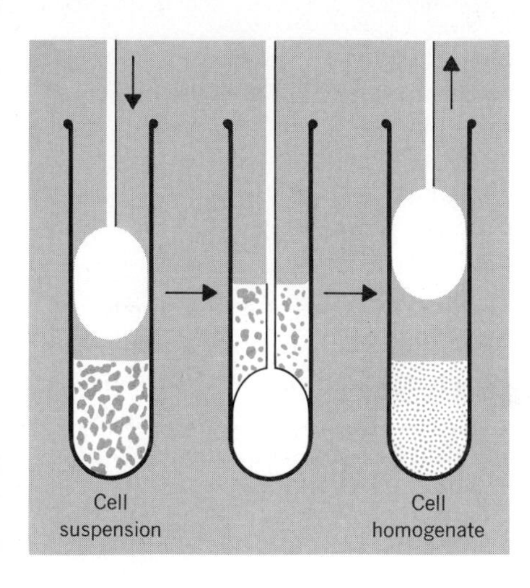

Fig. 7-7 The Dounce homogenizer typically used to make cell homogenates.

The microsomal fraction can be broken down further by the addition of a detergent, which releases the particles from the membranes. Subsequent centrifugation at 20,000 *g* for 15 minutes results in a pellet containing the so-called **ribosomal fraction** (microsomes without membranes) and a supernatant containing the membranes and enzymes with which these ribosomes had been associated.

The nuclear fraction obtained from the initial centrifugation can also be separated into different fractions. The nuclear fraction can be placed in a Dounce homogenizer, which is a glass tube with a bulb at the top, through which a rod is placed (see Fig. 7-7). Several thrusts of the homogenizer will disrupt the nuclear membrane, and the preparation can be centrifuged. At 5000 *g* a pellet is formed, which light- and electron-microscopic observation reveals to be the nucleolar fraction. The supernatant of this centrifugation contains the remaining nuclear material, primarily the chromosomes and nuclear envelope. A summary of the identification of cell structures is found in Table 7-1.

TABLE 7-1 CHRONOLOGICAL ORDER OF THE IDENTIFICATION OF CELLULAR STRUCTURES

Cell Structure	Identified by	Year	Method of Discovery
Cell wall	Robert Hooke	1665	Light microscope
Living cell	Anton van Leeuwenhoek	1674	Light microscope
Chloroplast	Anton van Leeuwenhoek	1674	Light microscope
Flagellum	Hamm	1677	Light microscope
Nucleus	Robert Brown	1831	Light microscope
Nucleolus	Wagner	1832	Light microscope
Chromosomes	Hofmeister	1848	Light microscope
Centriole	van Beneden	1875	Light microscope
Mitochondrion	Altmann, Benda	1897	Light microscope
Golgi body	Golgi	1898	Light microscope
Ribosome	Claude	1941	Ultracentrifugation
Endoplasmic reticulum	Porter	1945	EM
Mitochondrial cristae	Palade	1952	EM
Lysosome	De Duve	1952	EM
Chloroplast grana	Cohen and Bowler	1953	EM
Microtubule	Roth	1958	EM
Chloroplast quantosome	Park and Pon	1960	EM
Polysome	Warner, Rich, and Hall	1962	EM and ultracentrifugation
Mitochondrial particles	Chance and Parsons, Smith	1963	EM

The ability of biochemists and biophysicists to obtain purified fractions of individual cellular components provided conclusive proof that such components were an integral part of the cell, and not merely "artifacts." No longer could faulty fixation account for the appearance of certain structures, since biochemical and biophysical techniques were exceedingly gentle compared to the harsh fixation processes.

In addition to confirming the observations of the electron microscopists, the techniques of isolating purified fractions of cellular material provided the biochemists with material that could be directly analyzed, and lead to the integration of the structural and functional aspects of the cell.

The biochemical and biophysical studies of the cell, in conjunction

TABLE 7-2 SUMMARY OF CELLULAR COMPONENTS AND THEIR KNOWN CAPACITIES

Organelle	Macromolecular Complex	Known Capacity
Nucleus	Chromosome	Production of DNA and RNA
	Nucleolus	Changes chemical composition of some RNA molecules
	Nuclear envelope	Selective permeability
Mitochrondion	Electron transport particle	Krebs cycle, oxidative phosphorylation
	Membrane	Selective permeability
Chloroplast	Quantosome	Part of photosynthesis
	Membrane	Selective permeability
Lysosome	Membrane	Contains destructive enzymes
Cytoplasm constituents	Polysome	Protein synthesis
	Golgi body	Protein secretion
	Cell membrane	Selective permeability, excitation
	Endoplasmic reticulum	Supports polysomes, excitation (?), transport (?)
	Microtubules	In cell division, transport (?)
	Centriole	In cell division, in some cases produces basal bodies
	Basal bodies	Formation of cilia and flagella
	Cilia, flagella	Movement
	Vacuoles, vesicles	Storage of specific materials
	Cell wall	Conformation of cell

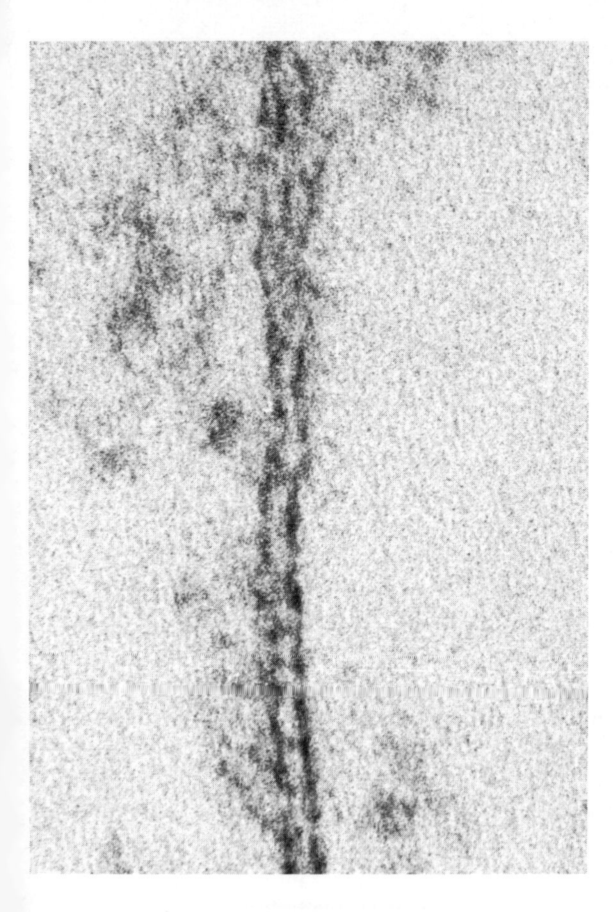

Fig. 7-8 The cell or unit membrane: rat-tongue blood capillary (760,000×). Electron microscopy revealed the cell membrane to be a double-layered structure. [G. E. Palade]

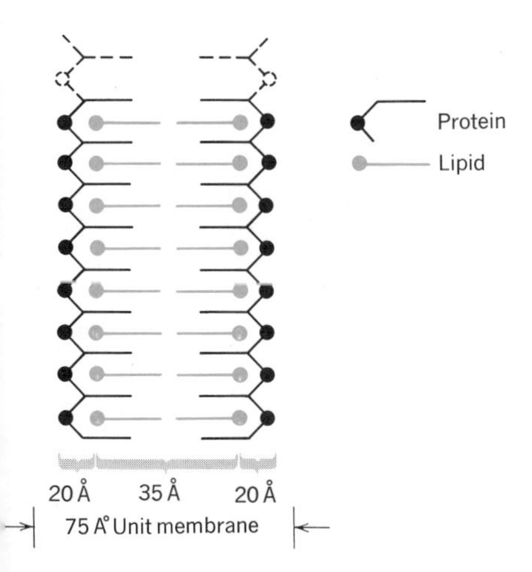

Fig. 7-9 Model of the cell or unit membrane.

with electron-microscopic observations, have brought cytologists to a new perspective of cellular structure and function. The organization of the cell is viewed as being composed of two levels of structural components, **macromolecular complexes** and **organelles.** Macromolecular complexes are physically associated macromolecules that possess a structural integrity. Some macromolecular complexes, such as the cell membrane, exist in a free state in the cytoplasm (see Table 7-2). Others maintain a spacial relationship with other complexes to form organelles. An organelle, then, is a unit made up of two or more macromolecular complexes.

A cell contains several types of organelles, the **nucleus, mitochondrion, plastid,** and possibly the **lysosome.** All of these organelles have one macromolecular complex in common: they are all surrounded by membranes. Rather than discussing membrane structure and function in the later section on macromolecular complexes, it is advantageous to discuss membranes now.

Membrane Structure and Function

The membrane systems of the cell include the extensive layers surrounding the cell (cell membrane), the chloroplast (chloroplast membrane), the mitochondrion (mitochondrial membrane), and the lysosome (lysosomal membrane), as well as membranous structures such as the endoplasmic reticulum and the golgi body (see Fig. 7-8). For many years, the question of whether a membrane existed remained unanswered. Some biologists argued that what appeared to be a membrane was in reality only an interphase between the cytoplasm and the environment that was formed in much the same way as the interphase between oil and water, which has properties of tension and permeability different from either phase. Others argued that the membrane was a real structure and supported this view by the fact that red blood cells can be broken open by placing them in pure water. The shell of the cell (ghosts) can be centrifuged and thus separated from the cell contents. The pellet is found to have a different chemical composition from that of the cytoplasm found in the supernatant.

Most studies on membranes assumed that they existed as real, physical entities, and attempted to determine various physiological properties. In 1935, Danielli proposed that the accumulated data suggested that the membrane is composed of three layers—an outer layer of protein, a middle layer of lipid, and an inner layer of protein. This sandwichlike structure was between 60 and 200 Å thick (see Fig. 7-9).

In the early 1950s, Robertson examined the membrane from many types of cells and found a recurrent pattern. He forwarded his **unit membrane** model in 1953, stating that a membrane is made up of three units: an outer dense unit 20 Å thick, an inner dense layer of the same thickness, and a light middle layer 35 Å thick. The unit membrane is 75 Å thick, and is much smaller than can be observed with the light microscope. Recent EM observations of the membrane from the top view seems to suggest another organizational feature. The membrane seems to be composed of hexagonal units arranged in a cobblestone fashion. Each hexagon is about 85 Å in diameter, and contains a central particle, thought to be a carbohydrate tetramer. This type of pattern is thought to be common in the membrane structures of chloroplasts and mitochondria, where specific chemical reactions units fit together.

The passage of material into the cell

Light microscopists had seen a cell boundary because the unit membrane is convoluted to form invaginations called **microvilli.** The convolutions produce a structure that is about 0.5 to 1.0 micron thick, within the limits of resolution of the light microscope. The convolutions enormously increase the surface area of the cell for the passage of materials.

Some electron micrographs have indicated that portions of the microvilli have pinched off in a process called **pinocytosis** to form vacuoles of **pinocytic vesicles** that appear to be on their way to the interior of the cell. No particulate material has been seen in the pinocytic vesicles, and it is assumed that dissolved materials or fluids are taken into the cell in this way (see Fig. 7-10). The entrance of dissolved material into a pinocytic vesicle does not bring the material into the cell. The only way material can gain access to the internal parts of a cell is by passage through the cell membrane directly, or through that portion of the cell membrane that surrounds the pinocytic vesicle.

If, indeed, the hexagonal construction is a major feature of membranes, then structure and function are inseparable. For instance, the passage of glucose through the cell membrane is now thought to involve biochemical activity. As glucose passes through the membrane, it is broken down by the glycolytic enzymes now thought to be localized there. Reduced molecules (such as NADH) act as electron carriers, and pass through the mitochondrial membrane that houses the enzymes for electron transport. Thus ATP and water are generated through membrane-associated enzymes that may be the very proteins that contribute to the structure of the membrane.

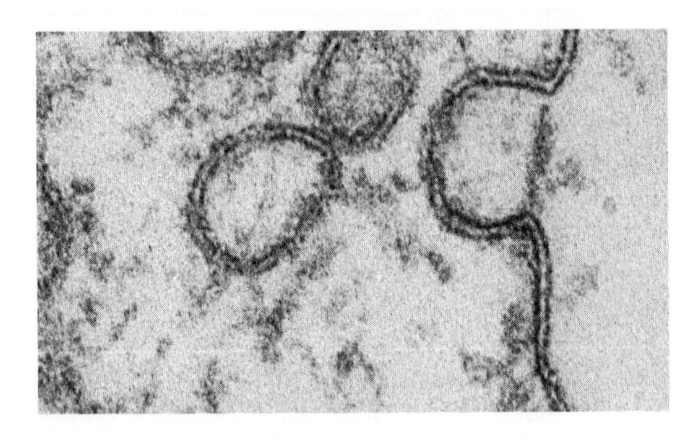

Fig. 7-10 Pinocytic vesicles of the cell: rat-tongue blood capillary (380,000×). Cells active in the transport of material have invaginations that either are continuous with the interior of the cell or appear to break off from the membrane and form vesicles that enter the interior of the cell. [G. E. Palade]

The passage of water through the membrane was described in early experiments, and early workers thought that the capacity of the cell membrane to establish unique steady-state conditions between the cell and the environment was primarily due to the difference in concentrations of solutes and water between these two factors. When a cell is placed in a solution containing a significantly greater concentration of water (**hypotonic condition**) than the cell, the cell tends to take in more water than it loses. As a result, the cell swells. If the concentration difference is too great, the cell may lyse. When a cell is placed in a solution in which the concentration of water is equal to that within the cell (**isotonic condition**), equal numbers of water molecules will pass through the membrane in both directions. When a cell is placed in a solution in which the water concentration is less than that of the cytoplasm (**hypertonic condition**), more water will pass out of the cell than into it, and the cell will shrink (see Fig. 7-11).

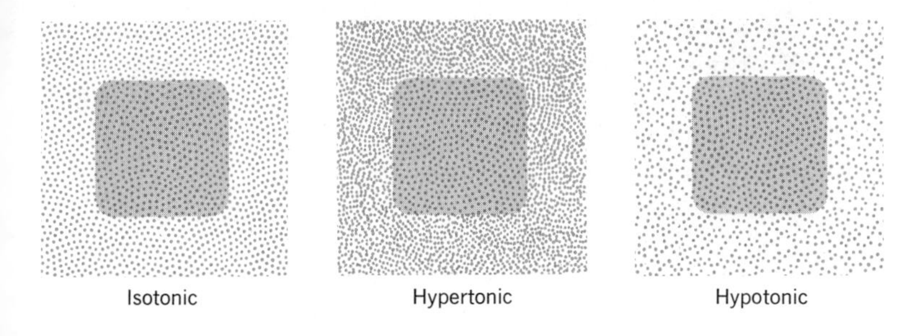

Fig. 7-11 Relative concentrations of materials inside and outside the cell under isotonic, hypertonic, and hypotonic conditions.

It should be noted that the concentration of solute molecules is inversely related to the concentration of water molecules in a solution. A hypotonic condition then represents a low concentration of solutes, and a hypertonic condition represents a high concentration of solutes.

Experiments that coupled physical treatment of membranes with EM observations indicated some interesting features. When cells were placed in hypotonic solutions before fixation, the cell membrane as well as the cell itself was swollen. The swelling of the cell that had been noted with light-microscopic observations was actually a response to a secondary hypotonic condition between the cell and its membrane. This characteristic swelling of membranes in response to hypotonic solutions became a major criterion for determining the presence of a membrane component in mitochondria, chloroplasts, and lysosomes.

While water, small water-soluble molecules, and highly fat-soluble molecules pass through the membrane passively, in accordance with the concentration gradient of the environment, it was found that large, water-soluble molecules and molecules with limited solubility in fats required **active transport,** utilizing energy-consuming mechanisms. Also, the cell membrane often works against a concentration gradient to pull in water, water-soluble molecules, and highly fat-soluble molecules through energy-consuming mechanisms. Thus the cell spends considerable energy in both attaining and maintaining precise steady-state conditions in regard to the molecular composition of the cell and the environment. Some measurements suggest that as much as 15% of the cell's energy is expended in active transport by pulling in certain materials against a concentration gradient, and pumping others out.

The energy required to maintain a given steady-state condition can be calculated as the amount of work necessary to maintain the respective concentrations of material inside and outside the cell. This energy, or work potential, of a membrane is given in electrical units. In most plant and animal cells, potassium (K^+) accumulates within the cell, and sodium (Na^+) is pumped out of the cell. The movement of small amounts of potassium out of the cell leaves negatively charged organic ions, and creates an electromotive force that can be measured by placing electrodes inside and outside the cell. The selective permeability of the membrane is dependent on the maintenance of opposite charges on either side of the three-layered structure.

The membrane components of cell organelles have many of the properties exhibited by the cell membrane. Perhaps the most significant of these properties is selective permeability, and hence, a work potential. The presence of a macromolecular complex possessing the property of selective permeability endows the organelle with the potential of having its own internal environment, qualitatively different from that of the cytoplasm.

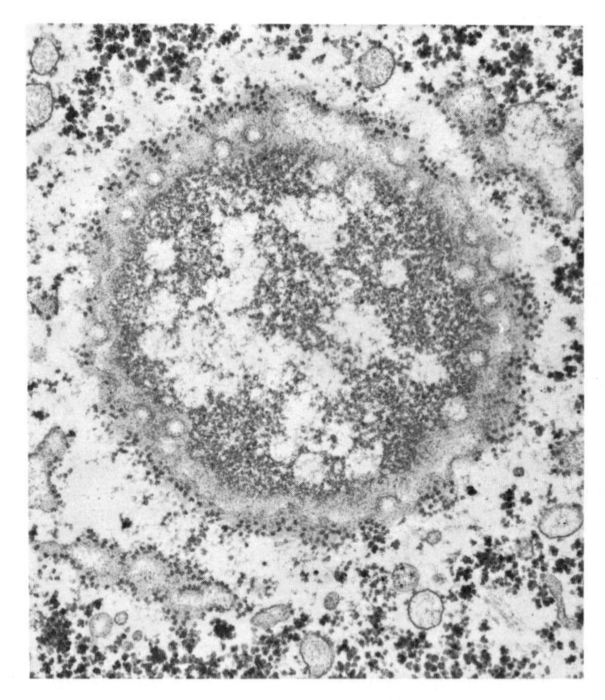

Fig. 7-12 The nuclear membrane: rat liver (28,000×). The nuclear membrane, also a double-layered structure, contains pores about 50 Å in diameter. This micrograph shows whole pores in the curvature of the nucleus. [G. E. Palade]

The Nucleus

The macromolecular complexes of the nucleus include the **chromosomes,** the **nucleolus,** and the **nuclear envelope.** The nucleolus appears in electron micrographs as an area filled with dense material, some of which appears as a knotted thread, and bounded by the nucleolar membrane. Chemical determination of the nucleolar matrix indicates that it is composed entirely of RNA and protein. In other words, RNA and protein are the molecules of the macromolecular complex, the nucleolus.

Few satisfactory electron micrographs have been made of chromosomes. Chemical determination of the chromosome fraction indicates that this complex is made up of DNA, RNA, and protein.

Electron microscopy has revealed that the nuclear envelope is a double membrane structure perforated with pores about 50 Å in diameter (see Fig. 7-12). The size of these pores would seem to indicate that the nucleus and cytoplasm are potentially chemically continuous.

The Mitochondrion

Electron micrographs of cells reveal the mitochondrion to be more than just the barlike structure described by light microscopists. The mitochondrion is surrounded by an external membrane, which gives it the barlike appearance. Cross sections reveal an internal membrane, from which the **cristae** are projected into the **central compartment.** Associated with the internal membrane are many minute "lollypop"-shaped structures. The central compartment of the mitochondrion is filled with a fluid medium (see Fig. 7-13).

Because of the ease with which a purified and intact mitochondrial fraction is obtained, these structures were among the first to be submitted to chemical analysis. The chemical composition of the mitochondrial membrane is very similar to that of the cell membrane, differing only in the presence of minute quantities of RNA and DNA.

Although mitochondria are found in most types of cells, the number of mitochondria seemed to be directly related to the activity of the cell. This led to the idea that mitochondria were in some way involved in the respiratory process. This suspicion was confirmed when several

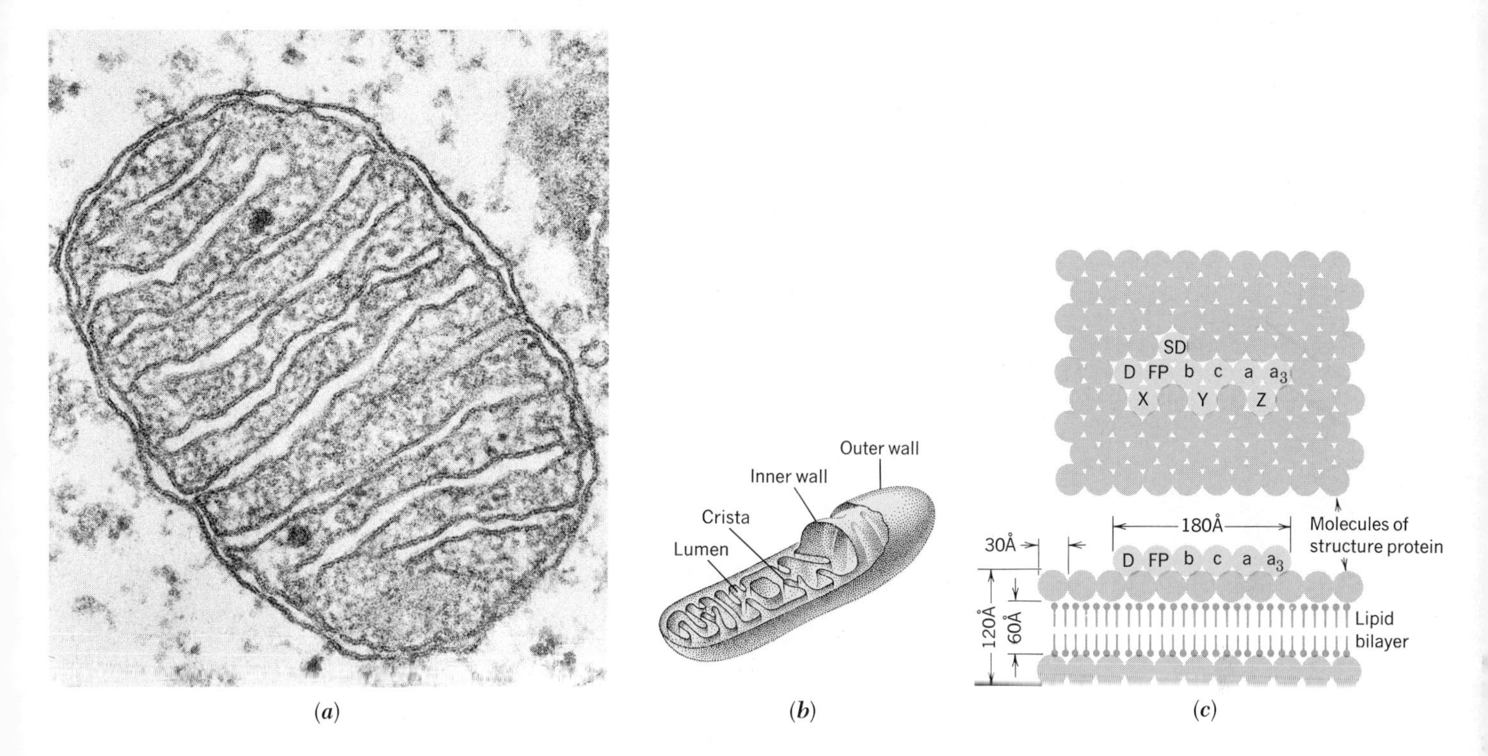

Fig. 7-13 Mitochondria. The double-layered structures visible in an electron micrograph (*a*) of mitochondria of a guinea pig pancreas (230,000×) are visualized (*b*) as part of a series of "shelves" within this organelle. These "shelves", or cristae, appear to be associated with dark-staining particles. The respiratory mechanism (*c*) is thought to be on the surface of the membranes. [(*a*), G. E. Palade. (*c*), From *Bioenergetics*, by Lehnninger, © 1968, W. A. Benjamin, Inc., New York]

workers in the 1950s found that $NADH^+ + H^+$ and ADP could be converted to NAD and ATP when mixed with a mitochondrial fraction. By the late 1950s, the entire electron transport system and oxidative phosphorylation system had been shown to reside in the mitochondria.

Advanced methods for handling the mitochondrial fraction permitted the separation of the two macromolecular complexes that comprise the mitochondrion. Biochemical assays on the separate fractions indicate that the enzymes for the Krebs cycle are present in the medium within the central compartment. The enzymes for the electron transport system are present on the membrane, probably associated with the numerous "lollypop"-shaped structures that have been tentatively termed **electron transport particles.**

There is also some indication that fat synthesis occurs in the mitochondria. The presence of these two systems operating side by side raises

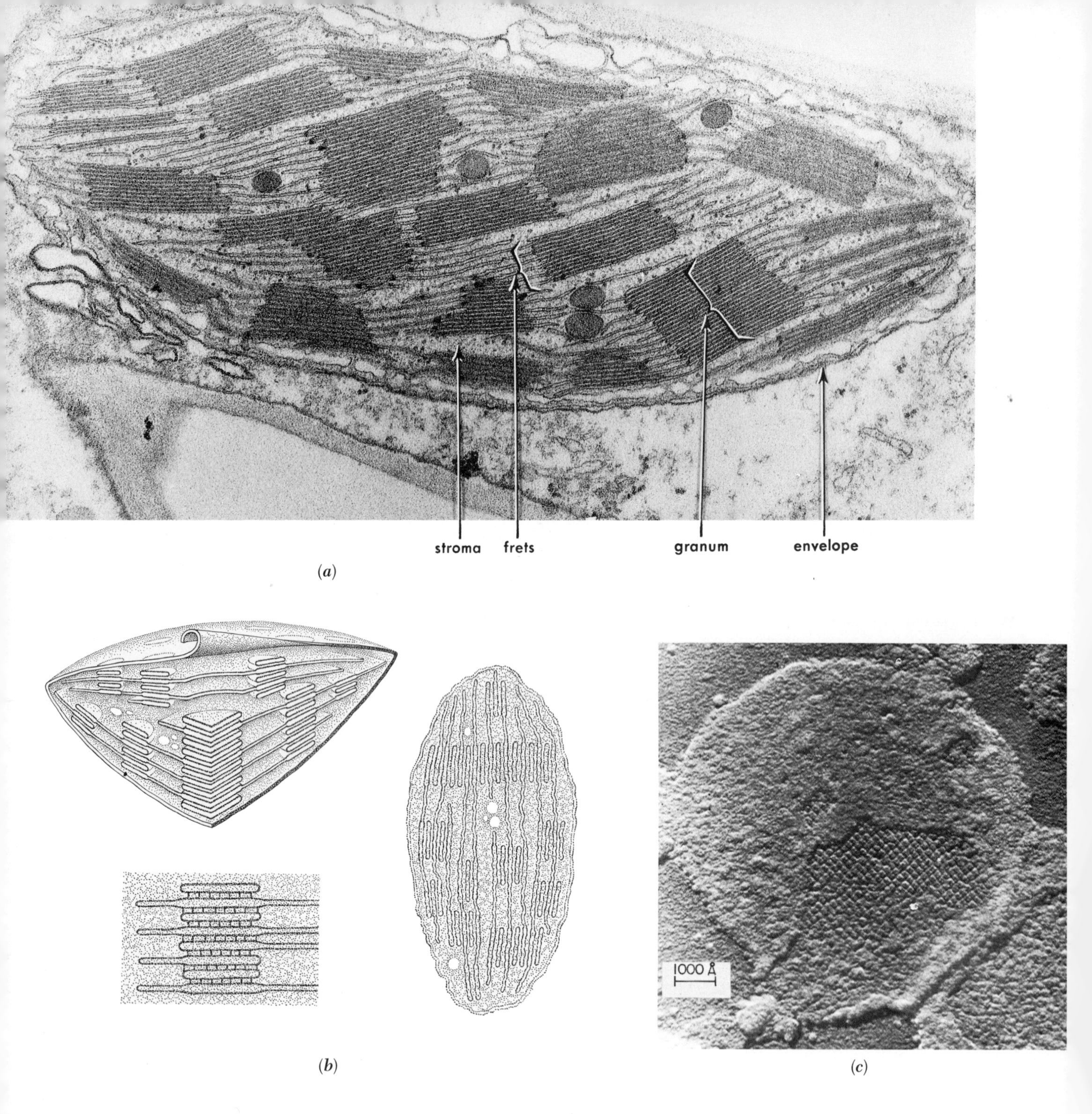

Fig. 7-14 Chloroplast. (*a*) An electron micrograph of a chloroplast of *Zea Mays* (21,000×) reveals layered packets which are organized in a specific pattern (*b*). The lamellae, or membranous packets, are associated with quantosomes (*c*). [(*a*) courtesy of Dr. Elliot Weier, (*c*) courtesy of Roderick Park-Lawrence Radiation Laboratory]

some rather interesting questions. Both systems are dependent on the presence of CoA. Are the mitochondria the only cell components in which CoA is operative? This and many other questions related to the complete correlation of the metabolic processes and cell structure are not yet answered.

The Chloroplast

Electron microscopy reveals that the chloroplast is a very elaborate structure (see Fig. 7-14). The chloroplast is surrounded by a membrane. Within the membrane are a series of membranous disclike layers, called **grana,** and a ground region termed the **stroma.** Embedded in the membranes of the disclike granum are numerous particles called **quantosomes,** which are the other macromolecular complex of the chloroplast.

Biochemical analysis of the membrane and quantosome fractions indicates that chlorophyll, cytochromes, and ferridoxin are localized in the quantosomes. It has been postulated that these particles are responsible for the light phase of photosynthesis, the conversion of light into chemical energy.

The stroma, on the other hand, is thought to contain the enzymes necessary for the dark phase of photosynthesis.

The Lysosomes

The membrane of the lysosome is the only known macromolecular complex in this organelle. Electron micrographs of lysosomes indicate that they are similar in size to mitochondria. Because of this similarity, lysosome preparations are difficult to obtain. Lysosomes are unique in having a single-layered membrane instead of a unit membrane. Chemical analysis of impure fractions reveals that lysosomes are little more than sacs of destructive enzymes, such as **lysozyme** (which causes a lysis of the cell) and **ribonuclease** (which breaks down RNA), among many others.

Because of the nature of the enzymes packed into the lysosomes, rupture of the lysosome membrane results in cell death. The advantage of having self-destructive enzymes within a cell is a subject for speculation. In higher plants, 90% of the cells are dead. Development of bird feathers occurs by differential cell death. Cells that engulf bacterial parasites may die in order to kill the intruder. Many examples of controlled cell death exist in the biological world, but their relationship to the lysosome organelle is unknown.

There is good evidence that lysosomes fuse with food vacuoles, and that the membrane of the lysosome breaks down, releasing enzymes into the vacuolar suspension. Thus food can be broken down without damage to the components of the cytoplasm. Undigested materials are released by the food vacuoles into small vesicles that are eliminated from the cell.

Macromolecular Complexes in the Cytoplasm

Many of the macromolecular complexes in the cell do not always exhibit a spatial relationship to other complexes, but exist as solitary units.

The Endoplasmic Reticulum

The endoplasmic reticulum is the extended shelving of membranes that extend throughout the entire cell (see Fig. 7-15). Since only very thin (0.1–0.2μ) sections can be used for electron-microscopic study, the endoplasmic reticulum has never been seen as a continuous system, and its three-dimensional configuration is envisioned through reconstruction of photomicrographs.

Because of the extensive nature of the endoplasmic reticulum (ER) within the cell, it has thus far been impossible to obtain an intact and purified ER fraction. There is good evidence, however, that the ER has a secretory function, and a close functional relationship to the golgi body (see Fig. 7-16).

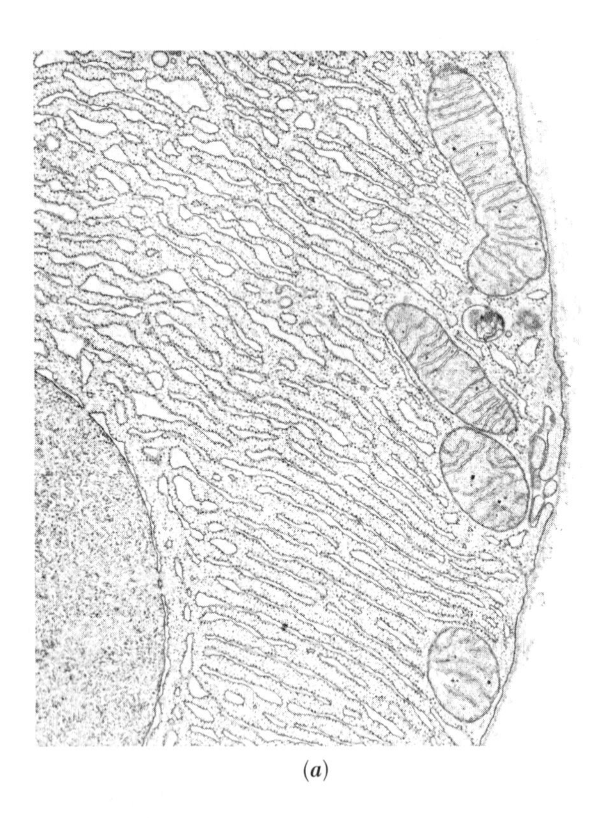

(*a*)

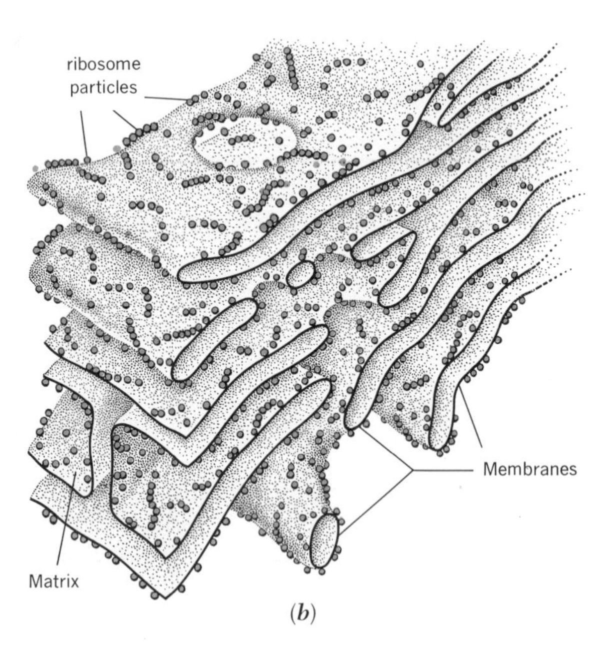

(*b*)

Fig. 7-15 Endoplasmic reticulum. (*a*) The series of layers in the electron micrograph of a guinea-pig pancreas (10,000$\times$) is visualized as a series of membranous shelves that run throughout the cell. (*b*) In cells that are actively engaged in protein synthesis, the endoplasmic reticulum has a rough appearance, due to the number of ribosomes associated with it. [(*a*), G. E. Palade, (*b*) From Keith R. Porter and Mary A. Bonneville, *An Introduction to the Fine Structure of Cells and Tissues*, © 1964, Lea and Febiger, Philadelphia]

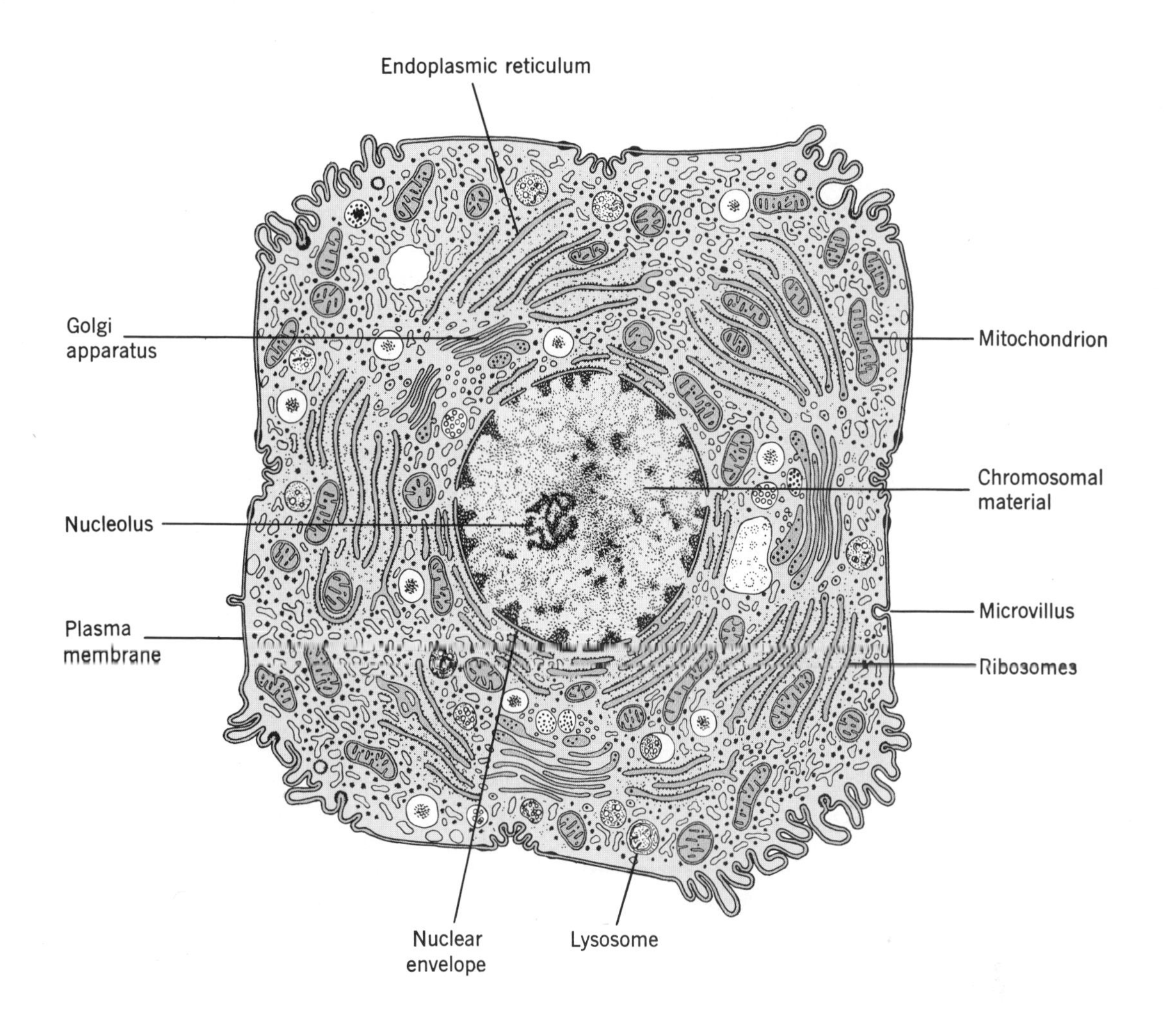

Fig. 7-16 Diagrammatic representation of the fine structure of a liver cell. The organelles include the nucleus and mitochondria. Macromolecular complexes include the endoplasmic reticulum and the Golgi apparatus. Small particles, or ribosomes, are abundant on the surfaces of the endoplasmic reticulum in some areas. The *cisternae*, or individual shelves of the endoplasmic reticulum, are arranged in stacks of 6 to 12 units. One margin of such an assemblage frequently lies adjacent to a Golgi apparatus. The expanded ends of the cisternae and the vesicles belonging to the Golgi proper, seem to constitute the mechanism of protein transport from endoplasmic reticulum to Golgi. The other margin of the stack of cisternae borders on glycogen particles. Typically this region of the endoplasmic reticulum is smooth (SER). There is evidence to suggest that the smooth ER is involved in the transport of glucose from the liver cell. Other components of the cell include lysosomes (LY) and microbodies (MB) of uncertain significance though probably rich in enzymes. [From Keith R. Porter and Mary A. Bonneville, *An Introduction to the Fine Structure of Cells and Tissues*, © 1964, Lea and Febiger, Philadelphia]

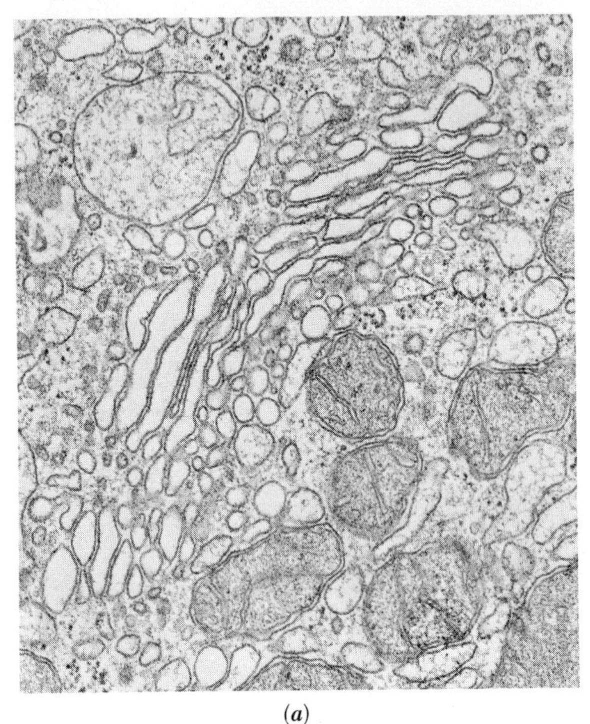

(*a*)

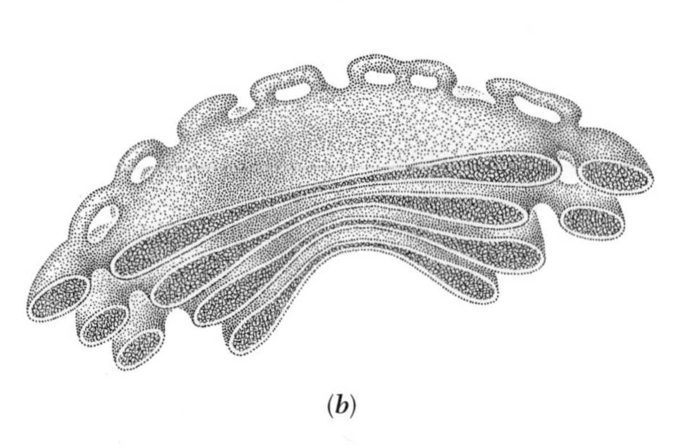

(*b*)

Fig. 7-17 Golgi apparatus. (*a*) The golgi apparatus of rat epididymis (46,000×) appears as a packet of membranous layers associated with several particles. (*b*) Schematic representation. [Courtesy of Dr. D. Friend]

The Golgi Body

Near the nucleus in many plant and animal cells, the endoplasmic reticulum takes on a lamellar structure and is surrounded by small vesicles and protein droplets (see Fig. 7-17). The size and position of this arrangement of the endoplasmic reticulum is similar to that proposed for the golgi body (now preferably called **golgi apparatus**).

Since the golgi apparatus appears to be little more than an organized portion of the endoplasmic reticulum, it has been difficult to obtain an intact and purified golgi fraction. The close proximity of protein granules and lysosomes to the golgi membranes indicates that its function is probably the secretion of protein formed on the neighboring ER.

Polysomes or Ribosomes

Electron micrographs revealed that the polysomes are often in close association with the endoplasmic reticulum (see Fig. 7-18). Although they have been found in most cells, they are so abundant in cells that are active in protein synthesis that the endoplasmic reticulum of these cells has a rough-surfaced appearance.

Closer examination of the individual polysome indicates that they are made up of many small, linearly arranged, snowman-shaped particles called **monosomes.**

Differential centrifugation techniques as outlined on p. 00 provided biochemists with pure microsomal fractions (polysomes attached to membranes of the endoplasmic reticulum) from which a pure ribosomal or polysomal fraction could be obtained. Chemical analysis of the polysomal fraction indicated that it was composed of about 60% RNA and 40% protein.

In 1960, Zaminik provided strong evidence that the polysomes were indeed the site of protein synthesis. He grew cells in a medium that contained C^{14}-labeled amino acids (amino acids in which radioactive carbon was incorporated into their structure). After the amino acids were permitted to enter the cells, the cells were transferred to a nonradioactive ("cold") medium for a short time, to give the "hot" amino acids a chance to become localized. Zaminik then homogenized and fractionated the cells.

When the cells had been placed in a cold medium for a short period of time, Zaminik found that the radioactive material was localized in the polysomal fraction. When the cells were permitted to remain in the cold medium for longer periods, he found that the radioactive amino acids had been incorporated into the protein structures of the cell (membranes, etc.). It appeared, therefore, that the amino acids originally went to the ribosomes, where they were bonded into proteins, and then proceeded in a polymer form to other parts of the cell (see Fig. 7-19).

A further indication that ribosomes provided the site for protein synthesis was given when, after the determination of the chemical events leading to the synthesis of proteins, it was found that the process did not occur unless ribosomes were introduced into the test system. Ribosomes are thought to act as a "table" on which the proteins are as-

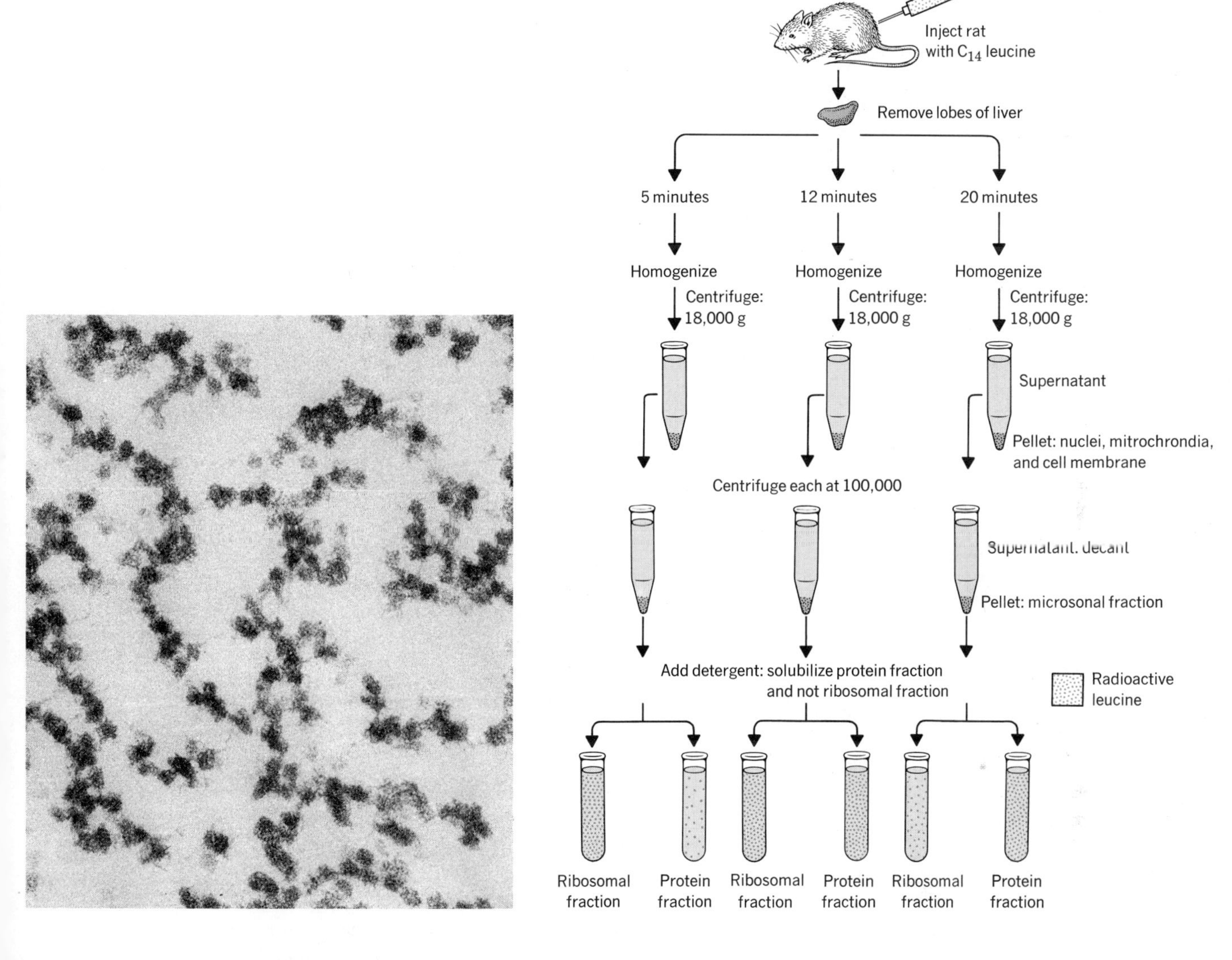

Fig. 7-18 Polysomes of guinea pig pancreas (136,000×). Ribosomes are frequently noted in linear arrangements. [G. E. Palade]

Fig. 7-19 Zamecnik's experiment demonstrated the essential role of ribosomes in the incorporation of amino acids for protein synthesis.

sembled. The structural and functional significance of the larger and smaller particles that make up the individual monosome is under active investigation.

Biophysical studies have indicated that a ribosomal fraction can be obtained from chloroplast and mitochondrial centrifugation. It appears to be this ribosomal fraction that is responsible for the minute amounts of RNA that have been found in both of these structures.

The ribosomes are brought down by centrifugation at 20,000 *g*. Since this speed is only obtained by the ultracentrifuge, and the particulate size is so small, **Svedberg units** (which take into consideration the natural tendency of minute particules to drift) are used rather than force (*g*) values. A force value of 20,000 *g* for these particles in animal and plant cytoplasm corresponds to 80S units, and these ribosomes are referred to by biochemists and biophysicists as 80S particles. The ribosomes of chloroplast and bacteria cells are found to be 70S. It is interesting to note that streptomycin impairs ribosomal function, thus interfering with protein synthesis in bacteria, and causing a bleaching of chloroplasts, but has no effect on animal cells. The relation of this activity of streptomycin with the 70S particle but not with the 80S particle is purely speculative at this time.

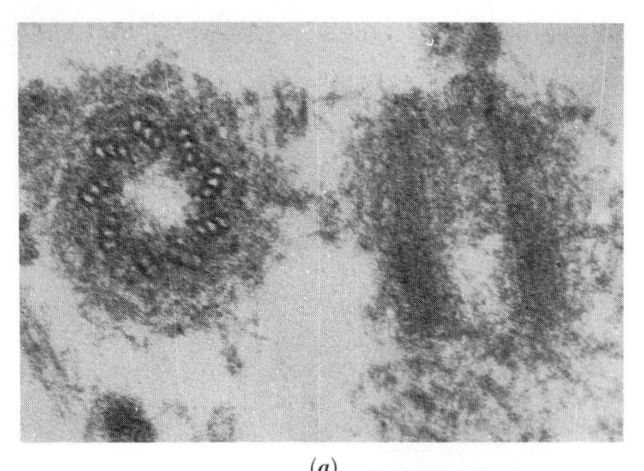

(*a*)

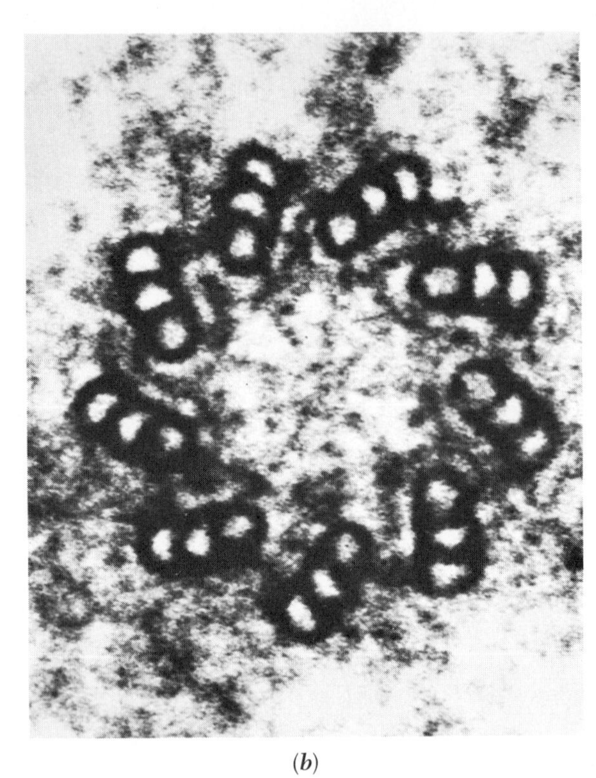

(*b*)

Fig. 7-20 The centriole. (*a*) A section through the entire centriole (87,700×). The two bars are at right angles to one another, thus in section you see a cross-section of one and a longitudinal section of the other. (*b*) Cross-section of one unit of the centriole showing the nine packets of fibers (278,000×).

[(*a*) Eva Nagy-Public Health Research Institute of the City of New York, (*b*) Photograph courtesy Dr. Etienne de Harven. From A. J. Dalton and F. Haguenau, *The Nucleus*. Copyright Academic Press, New York]

The Centriole

The centriole is a particulate macromolecular complex always located near or within the nucleus. It is made up of two separate barlike units located at right angles to one another (see Fig. 7-20). A cross section of these units reveals that they are made up of a series of nine triplet tubules forming a circle.

Analysis of impure fractions indicates that the tubules are composed primarily of nucleic acid and protein. Since a pure centriole fraction has never been obtained, no definitive chemical analysis has been done.

The function of the centriole seems to be directly related to the process of cell division. The chemical events leading to its role in this process are unknown.

*Basal Bodies, Cilia, and Flagella**

Electron micrographs revealed that cilia arise from **basal bodies.** These basal bodies are found near the cell membrane, and are thought to be the origin of growth of a cilium. The ultrastructure of the basal body is essentially the same as that of the centriole. Each basal body is made up of two barlike units located at right angles to one another. The cross section of these units indicates a pattern of eleven groups of tubules, nine forming a circle, and two centrally located.

Cross sections through a cilium from plants and animals indicates that these structures also manifest the 9:2 arrangement of fibrils (see Fig. 7-21).

No direct chemical analysis has been done on the basal bodies, since it has not been possible to obtain a pure fraction. Experimental evidence indicates that the basal bodies may contain nucleic acid as well as protein. Cilia and flagella appear to be protein structures.

Microtubules

Electron micrographs revealed that dispersed throughout the cytoplasm are a number of minute hollow cylinders called **microtubules.** In cross section, these microtubules appear to be composed of thirteen fibers running longitudinally and arranged in a circle (see Fig. 7-22).

Experimental evidence indicates that the microtubules orient themselves longitudinally to form the **spindle,** a cellular structure that is present only during cell division. The significance of this structure will be discussed more thoroughly in the next chapter.

Isolated spindles have been chemically analyzed, and have been shown to be composed of proteins that contain large amounts of sulfur-containing amino acids (methionine, cysteine). It is assumed that the analysis of spindle content reflects the chemical composition of the microtubules.

Vacuoles and Vesicles

The vacuoles of cells are primarily the storage areas for water-soluble material, while vesicles appear to accumulate protein material.

Electron micrographs of these cytoplasmic structures reveal a uniform internal composition, and results of chemical analysis (for the type of material stored) depend on the type of cell.

Fig. 7-22 Microtubules of the mitotic spindle in a mouse fibroblast cultured cell undergoing mitosis (metaphase stage) (42,500×). [Samuel Dales]

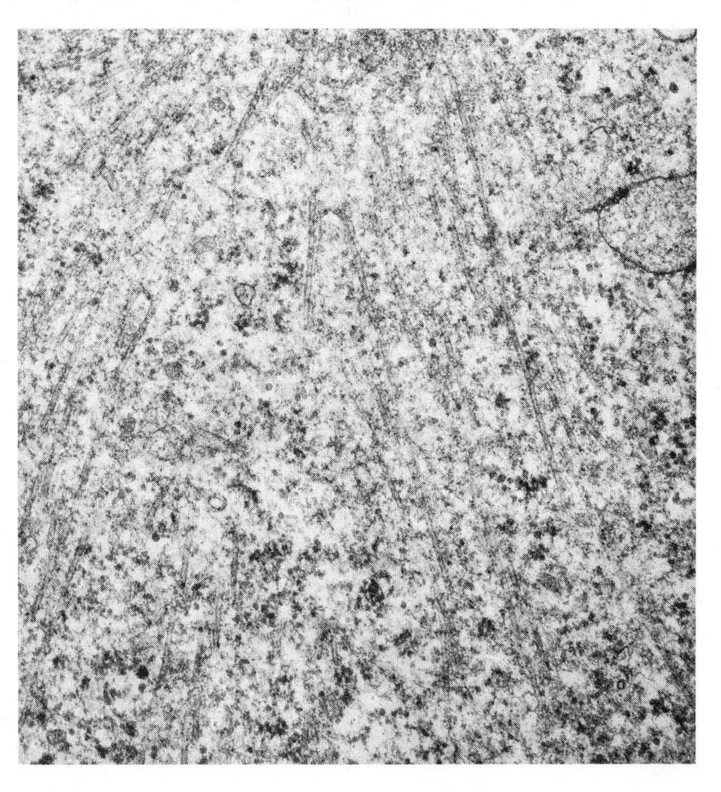

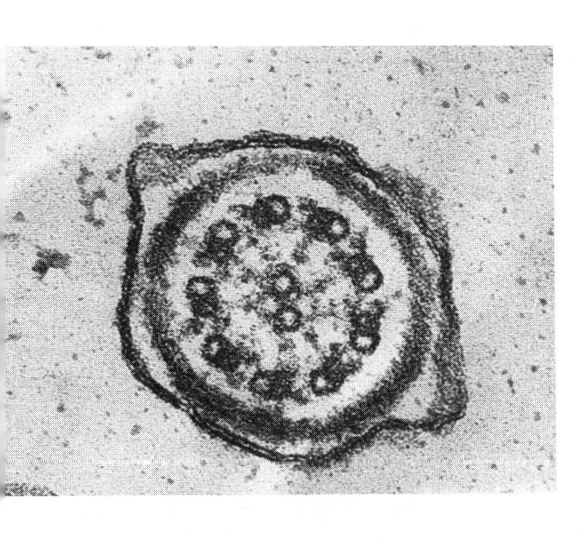

**ig. 7-21 Cross-section of a cilium showing the 9:2 arrange-
ient of fibers (127,200×). Electron microscopists regard the
:2 arrangement found in the cross-section of hair-like projec-
ons of higher organisms as the definition of a cilium. The term
agellum is reserved for the single-fiber structure found in
acteria. [Eva Nagy-Public Health Research Institute of the
ity of New York]**

Cell Wall

Electron micrographs indicate that the cell wall is a uniform structure. Chemical analysis reveals that the cell wall is a macromolecular complex of cellulose (glucose molecules bonded together in a cellubiose linkage), grouped into fibrils that in turn are specifically oriented in definite planes.

Cellular Dynamics

The cell is fundamentally a chemical factory in which various chemical reactions are highly localized. Most materials are either dissolved or suspended in the cellular medium. These conditions allow maximum exposure of molecules, and therefore, maximum reactivity. The speed of chemical changes that occur in the cell is beyond imagination. An enzyme molecule can react with over a hundred thousand substrate molecules in one minute. A protein molecule is put together at a rate of 300 amino acids per minute.

To add to this bewildering business of cellular activity, there are hundreds of different reactions occurring simultaneously, that involve millions of molecules. Yet, within this chemical cauldron, biochemists have identified distinct and specifically coupled pathways, and the cytologist has described highly organized and stable cell organelles.

The stability of cell organelles is actually the result of the dynamic considerations on the molecular level. Molecular forms in the cell are unstable, and remain in a specific location for short periods of time. (The exception to this is concentrated material such as glycogen granules and fat droplets that act as storage depots, and are temporally quite stable.) Most material enters the cell in monomeric or radical form, and tends to pass through a molecular hierarchy when it enters the cell. These simple molecules first become parts of heteroconjugates and/or macromolecules. Eventually, the macromolecules form macromolecular complexes, and may finally join into the structure of cellular organelles. This pattern can be diagrammed as shown in Fig. 7-23.

The structural components of the cell visible under the light and electron microscopes are complex structures of several levels of chemical

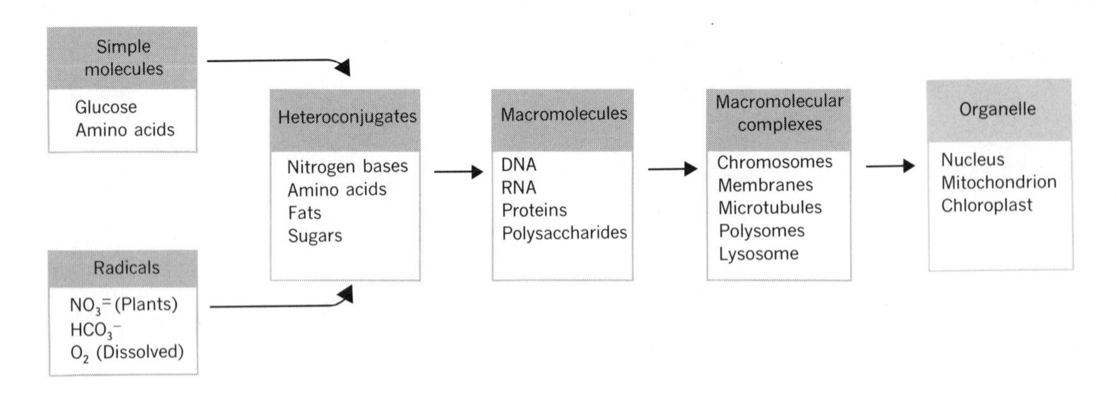

Fig. 7-23 Pattern of molecular stabilities.

organization. Even enzymes are known to exist as a team of molecules, held together as one functional unit, rather than as individual molecules.

The pathway up the cellular hierarchy from simple molecules to organelles is not unidirectional. Spontaneous destruction occurs at every level of organization. Macromolecules have a limited lifetime, and are replaced by newly synthesized ones. Enzymes may last for an average of a few days. Some RNA molecules last for but a few minutes. Only through the constant regeneration of cell structure on a chemical level can the integrity of the cell on a microscopic level be maintained.

As well as maintaining an internal steady-state condition, the cell maintains a steady-state condition with its environment. Materials are constantly entering and leaving the cell. When the environment changes, the cell must respond by changing to maintain the steady state with the new environment, or die. The achievement of a new equilibrial state is possible because of the dynamics of cell chemistry. In the presence of a new environment, the highly labile molecules, especially the proteins, will either shift in their structure to exhibit more appropriate properties, as they are programmed to do, or will degrade and be replaced by new and different proteins. In the long run, the spontaneous degradation of complex molecules, a major feature of the dynamics of cellular chemistry, is the most efficient means by which cell adaptation to new environmental conditions can come about.

The cell expends a great deal of energy in maintaining the steady state, both among its own component parts, and between itself and the environment. The energy expenditure has been estimated for the common colon bacillus, *Escherichia coli* (see Table 7-3).

TABLE 7-3 PERCENTAGE OF ATP REQUIRED TO SYNTHESIZE DIFFERENT CLASSES OF MOLECULES

Class of Molecules Synthesized	Percent of ATP Required for Synthesis
DNA	2.5
RNA	3.1
Protein	88.0
Lipids	3.7
Polysaccharides	2.7

The cell appears to be primarily a protein factory. But protein for what? Proteins can be enzymes, membrane units, etc. But what general cellular function requires the synthesis of enzymes, membranes, etc.?

It has been calculated that 15 to 20% of the ATP pool of a cell is directed toward active transport by membranes, and that the cell utilizes about 10% of its energy toward growth. The remaining 70 to 75% of the energy is directed toward maintenance of the cell. Therefore, the bulk of protein synthetic activity toward the production of enzymes, fibrils, membranes, etc. is for maintaining the structural and functional integrity of the cell within the confines of the dynamic chemical equilibria.

VARIATIONS ON A CELLULAR THEME

Through the course of evolution, cells have tended to develop specialized structures and functions. This specialization is nothing more than a variation on the basic cellular theme. Some specialized cells manifest different ratios of the various cell organelles, while others have additional structures or lack others. The range of variation extends from cell types that are quite similar to the "composite" cell to those that are difficult even to recognize as cells.

The **embryonic** cells found in all animal tissues and the **meristematic** cells found in all plant tissues are representative of the more generalized cell type. These cells function primarily as reproductive units, forming more embryonic or meristematic cells as well as cells that are fated to undergo a process called **differentiation,** leading to a specialized cell type.

The potential of the embryonic cell to become specialized by physical or quantitative alterations of its cell organelles and macromolecular complexes is an important aspect of the dynamic equilibria in which the cell is involved. The rapid turnover of molecular forms permits the degradation of labile molecules, and their replacement by newly synthesized molecules that are more harmonious with the functions of the specialized, differentiated cell.

The transition from an embryonic cell type to a differentiated cell, such as a liver cell, is a gradual process. Liver cells have unique combinations of enzymes, a cell membrane with a characteristic permeability, and extraordinarily large lysosomes (6μ). Through gradual changes in cell composition made possible by the temporally unstable features of cell chemistry, an embryonic cell can undergo a change to a liver cell with a minimum expenditure of energy and under maximum control.

In order to group the degree of variation that is possible in cellular structure, it is necessary to examine several types of differentiated cells.

The Liver Cell

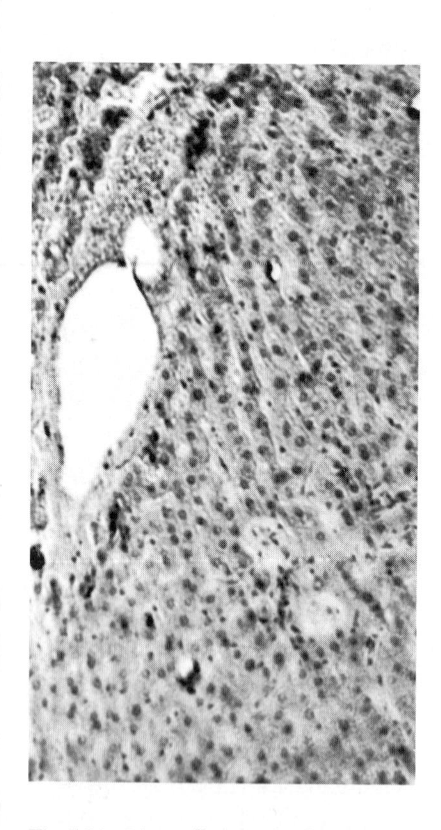

Fig. 7-24 Liver cells (500×). These cells have thin membranes; large, conspicuously stained spherical nuclei; and dense, granular cytoplasm characteristic of highly active cells. [Russ Kinne-Photo Researchers]

The liver in higher animals functions as a storage depot for glucose. It is the primary site in the animal body for glycogen deposits. In response to the controlling mechanisms of the body, it either stores glucose (by increasing the rate of glycogen formation—glyconeogenesis) or releases glucose into the blood stream for circulation to different parts of the body (glycogen breakdown).

The liver is the primary site in the body for fat metabolism, vitamin storage, destruction of red blood cells, production of bile salts, and detoxification of harmful substances.

The structure of the liver cell is well suited to its functions. First of all, the energy requirements for such an active system are very high. Both the formation and breakdown of glycogen are energy-requiring processes. The energy for these metabolic processes must come from the degradative breakdown of glucose and subsequent oxidative phosphorylation, processes operative on the cell membrane and the mitochondria, respectively. Therefore, it is not surprising to find that liver cells have a large number of mitochondria (see Fig. 7-24).

In order to perform its function efficiently, a liver cell must constantly replenish the enzymes necessary for its metabolic reactions. Since this replenishment involves the synthesis of the protein portion of the enzyme, the many ribosomes found in liver cells may be considered a structural adaptation.

The cell membrane also manifests a structural adaptation to its function. All of the functions of the liver involve a constant movement of materials across the membrane. Excess glucose, vitamins, and amino acids are brought into the liver cell. There they are stored or, in the

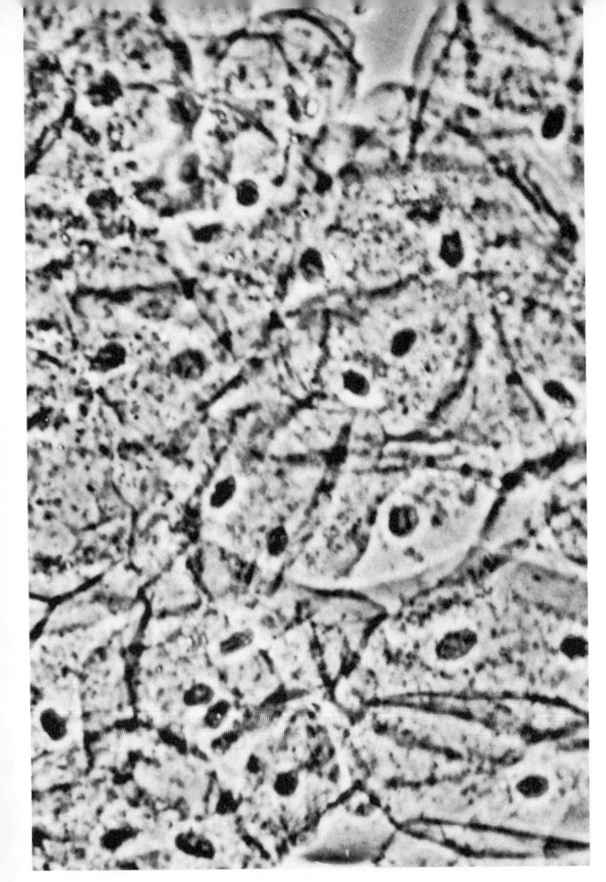

Fig. 7-25 Human cheek cells (200×). These cells eventually die and are sloughed off. Notice the distinct central nucleus and the somewhat dense cytoplasm. [Walter Dawn]

case of amino acids, deaminated. When the equilibrium of the body is such that these substances are needed, they are released from the liver cell into the blood stream. For this constant passage across the membrane to be as efficient as possible, a thin and highly permeable structure is necessary. Experimental studies have indicated that the cell membrane of the liver cell is indeed more permeable than most in terms of quantity of material transported per unit time. Electron micrographs have revealed that the membrane of the liver cell is thinner than those noted in other specialized cell types.

The Epithelial Cell

Epithelical cells are surface cells that seem to be primarily protective. Their shapes are characteristic for the surface on which they are found. **Squamous** epithelial tissue is found on the inside of the cheek, **cuboidal** epithelial tissue is found lining the kidney tubules, and **columnar** epithelial tissue is found in the lining of the stomach and the intestine (see Fig. 7-25). Because of the passive nature of their function, it is not surprising that one finds a scarcity of the cell organelles necessary for dynamic function. Although some mitochondria are evident, the number per unit volume is far less than that found in muscle or nerve cells. Electron micrographs indicate the presence of a smooth endoplasmic reticulum, almost free from ribosomal particles similar to the condition found in inactive liver cells (see Fig. 7-26).

An interesting variation of the epithelial cell type is found in the lining of the nose. Here the surface is lined with columnar epithelial cells whose exposed surface is covered with minute hairlike projections called **cilia.** These cilia beat in unison, and serve to remove dust particles from the air as it enters the upper part of the nose. Electron micrographs indicate that at the base of each of the cilia is a **basal granule,** which is made up of two barlike bodies at right angles to one another, manifesting the 9:2 relationship of tubules that is reminiscent of the centriole structure.

The Sperm

Perhaps the most obvious correlation between form and function is in the case of the male sperm. In mammalian reproduction, the sperm

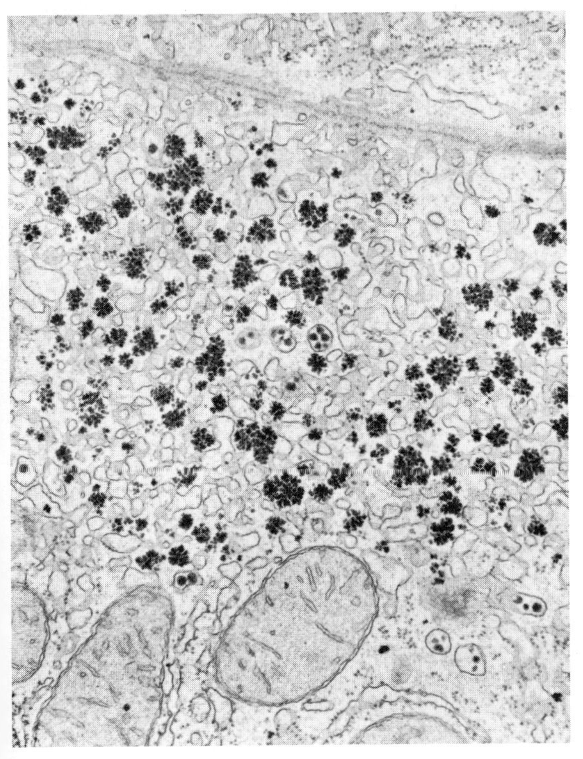

Fig. 7-26 Endoplasmic reticulum of rat liver cells not actively engaged in protein synthesis (20,000×). Notice the dark-staining glycogen granules. Compare the endoplasmic reticulum of these cells with those shown in Fig. 7-15. What cellular structures would then appear to be involved in the process of protein synthesis? [G. E. Palade]

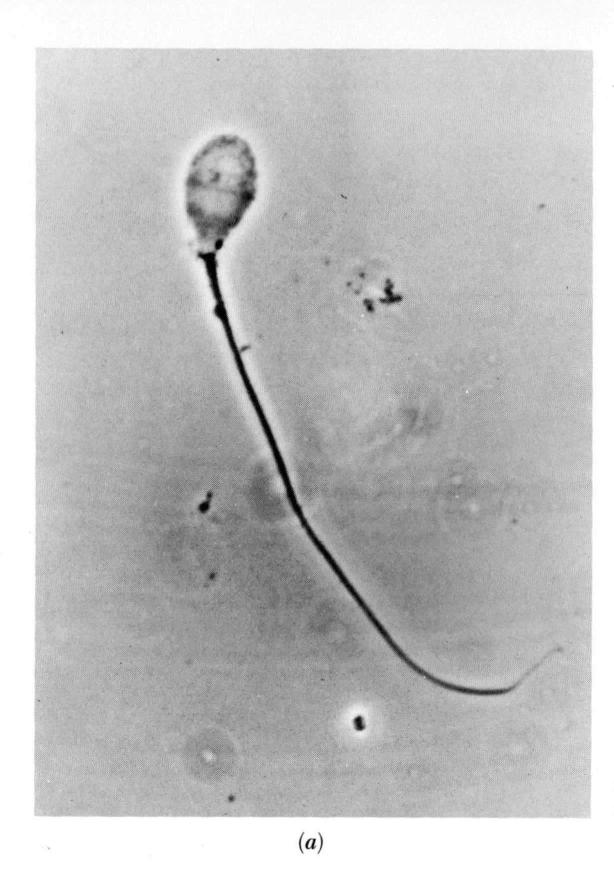

(*a*)

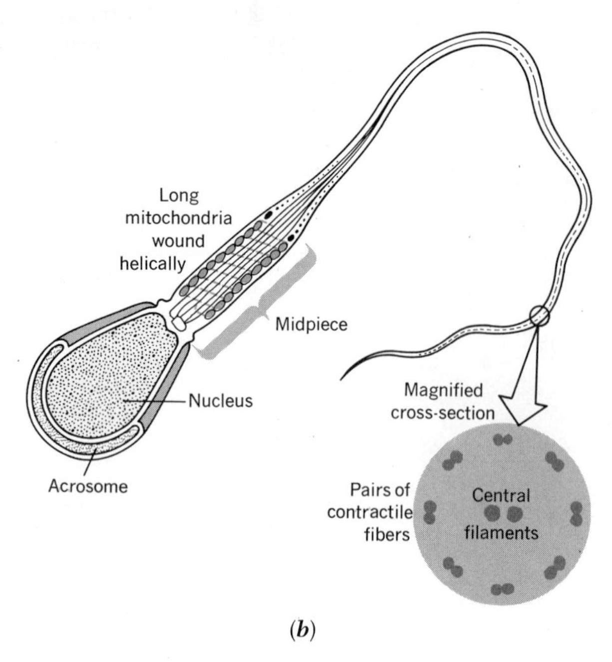

(*b*)

Fig. 7-27 Sperm cell of man (2,000×). The head contains the nucleus and acrosome (Golgi apparatus), the neck contains many mitochondria, and the long tail is essentially a set of fibers. [(*a*), Lester V. Bergman]

swims to meet the egg, with which it joins (fertilization), initiating the beginning of a new individual. For this role, it is structurally specialized in many ways.

First, the very shape of the sperm is specialized for its swimming function. The head is shaped like a flattened spoon and permits efficient penetration in a fluid medium. The sperm is propelled through the liquid medium by a single cilium (see Fig. 7-27).

Electron-microscopic studies of sperm revealed interesting specialization features. The first part of the head is made up of the **acrosome,** a vesicle from the golgi apparatus, which is followed by the nucleus. Directly behind the head is a narrow portion called the **middle piece,** from which the flagellum arises. Within the middle piece are stacked mitochondria, and at the base of the middle piece is a basal granule, from which the flagellum arises (see Fig. 7-27).

The sperm requires energy for its physical movement. The close spatial relationship between the mitochondria and the basal granule may provide an efficient transfer of energy in the compact system.

Because the sperm is made up of a minimum of cytoplasm and cytoplasmic structures surrounding a nucleus, it serves as an excellent carrier of nuclear material. DNA was first extracted in 1871 by Meischner from salmon sperm, and he chemically determined some of its properties.

Fig. 7-28 Xylem cells of the buttercup root (70×). These highly differentiated cells have thick secondary walls and total destruction of the nucleus and the cytoplasm. [Carolina Biological Supply House]

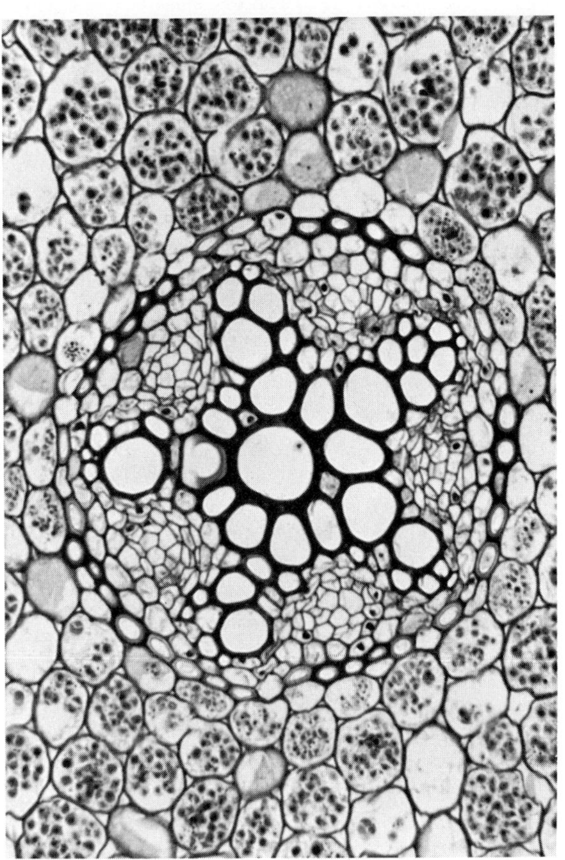

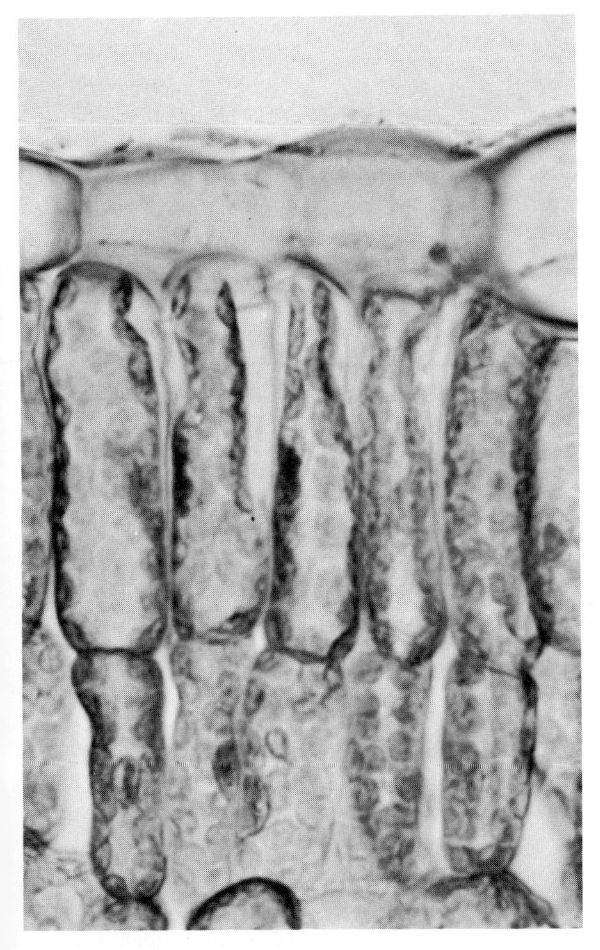

Fig. 7-29 Palisade cells (800×). The palisade cells, directly under the leaf epidermis, are the primary photosynthetic elements of the plant. [Lester V. Bergman]

The Xylem Cell

The xylem cells serve to transport water and minerals in the stems of higher plants. They are long, thin, tubular cells that are reminiscent of a straw. The most interesting feature of these cells is that they are not living structures. Because of the purely passive nature of their function, the dead cell, which has no energy requirements, is thought to be better suited to the purpose than a living cell. Therefore, after the potential xylem cell has been formed by the meristematic tissue of the plant, it undergoes a total degeneration in the course of its differentiation, leaving only its cell wall to act in the transport capacity (see Fig. 7-28).

It is possible that the lysosomes present in the presumptive xylem cell are triggered to break open, thus destroying the cell and leading to its differentiated state. The true mechanism for the initiation of this state is not known.

The Palisade Cell

Directly beneath the epithelial layer of cells on a leaf is a layer of long, narrow cells, called the **palisade cells.** (see Fig. 7-29). These cells are the primary sites in the plant for photosynthesis. Their very shape permits a greater number of chloroplasts to be exposed directly to the light. Also correlated with their function is the comparatively large number of chloroplasts present.

Electron micrographs reveal that the palisade cells have a large number of mitochondria and ribosomes, as would be expected of a cell with a chemically oriented function. In addition, the cell wall of the palisade cell is much thinner than that of transport cells (xylem). This is thought to be advantageous for the passage of light, and for the transfer of photosynthetic products out of the cell en route to other parts of the plant body as well as for exchange of gases and water vapor.

PERSPECTIVES OF CELL STRUCTURE AND FUNCTION

The biologist's concept of a cell has transcended the observations of light and electron microscopists. The day is long past when a drawing can truly represent a cell. A pictoral representation shows what the eye can see; it is a static unit, held in the moment of its existence. But the present biological concept of a cell is in terms of its continued existence, its changing structure and continuing function. The cell is to the biologists what the atom is to the physicist: if a person can draw it and be satisfied, then he doesn't really understand it. By viewing the cell as a unit capable of continual modification, its role as the building block in development and organismal activities becomes more meaningful.

QUESTIONS

1. Name five parts of the cell discerned by light microscopy and give the name of the investigator credited with the discovery and the approximate date of discovery.
2. Why did the coupling of experimental cytology with light- and electron-microscopic observations clarify the existence of certain cellular components?
3. In which direction across a membrane will water move when the concentration of material that cannot pass through the membrane is greater inside than outside the cell? What will happen to the size and/or shape of the cell?
4. Give two means by which materials pass into the cell.
5. What metabolic role do mitochondria play in the cell?
6. What portion of the chloroplast is responsible for the light reaction? For the dark reaction?
7. Give the experimental evidence that implicated the role of ribosomes in protein synthesis.
8. How is cellular adaption to a new environment related to the spontaneous degradation of complex molecules within the cell?
9. Give two examples of cellular function reflected in cellular form.

CHAPTER 8

REPRODUCTION AND SYNTHESIS OF CELLULAR COMPONENTS

The formal statement by Virchow that all cells come from pre-existing cells implied that within a cell exists the mechanism or mechanisms necessary to bring about the existence of new cells, similar in internal composition to the original one. At the time Virchow made this statement (1854), a cell was considered to be little more than a nucleus surrounded by cytoplasm. Cellular reproduction was thought to involve the growth of a cell to a given size, followed by a division of the nucleus and subsequently the cytoplasm in such a manner that each of the resulting cells received approximately half the material of the original cell (see Fig. 8-1).

Mitotic Division

Later in the nineteenth century, microscopic observations of embryonic and meristematic tissues demonstrated that nuclear division was a series of highly ordered events that resulted in specific parceling of the chromosomes into each of the resulting nuclei. This type of division was called **mitosis.** The manner by which mitotic division was accomplished assured both daughter cells of having the same chromosomal complement as that of the original cell.

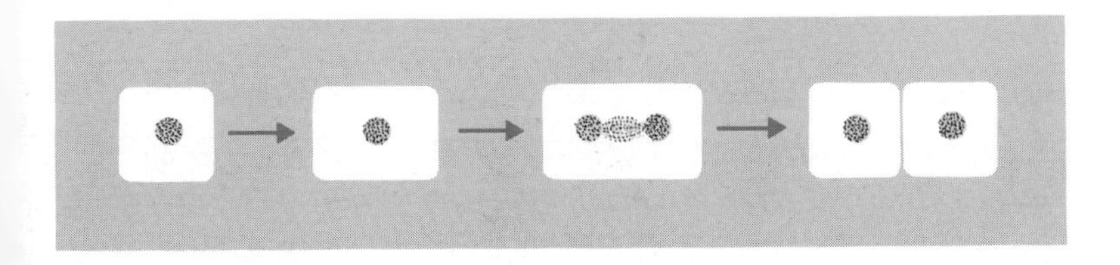

Fig. 8-1 Nineteenth-century concept of cell division.

Careful observations on organisms of different species revealed that there was a characteristic chromosomal complement for each species. This specificity existed both for the number of chromosomes and for their individual appearance. The cells of higher animals and plants (with the exception of their gametes, which will be discussed later) have chromosomes present in multiple of two (see Fig. 8-2). The human body has 46 chromosomes, or 23 pairs. The favorite research tool of geneticists for many years, the fruit fly (*Drosophila melanogaster*), has 8 chromosomes, or 4 pairs. Because of the duplicity of each chromosomal type, this condition is called **diploidy** (two sets), and organisms manifesting this type of chromosomal condition are referred to as **diploid organisms.**

Some lower organisms (bacteria, some protozoa and algae, etc.) have only one chromosome of each type. This condition is termed **haploidy** (one set) and such organisms are **haploid organisms.** Some plants (mosses, ferns, and seed plants) have both haploid and diploid phases. Regardless of the ploidy of the individual organism, nuclear division associated with simple growth processes always results in two daughter nuclei that are identical both to each other and to the mother nucleus in the number of chromosomal sets or genomes.

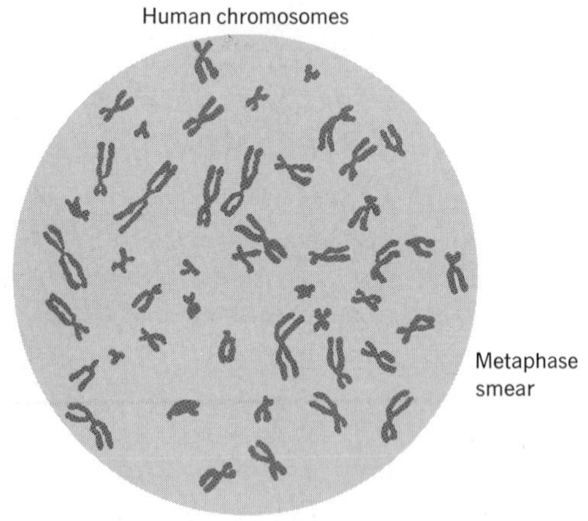

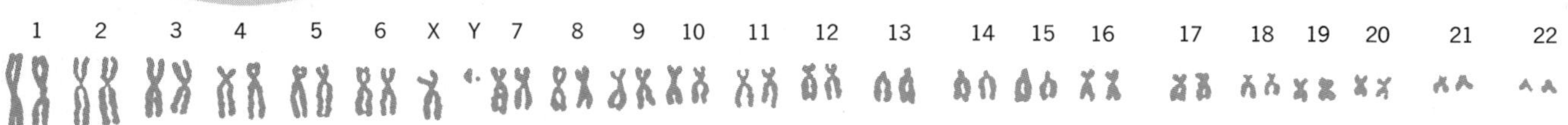

Fig. 8-2 Human chromosomes. The twenty-three pairs have been identified and numbered and are typically represented in this form, called an ideogram. Chromosome figures are cut out from a photograph, and are arranged and matched according to size and shape.

Meiotic Division

From breeding studies and general social observations came the obvious fact that two parents are required to produce an offspring. If nuclear divisions were always equational, then a joining of male and female cells (the **gametes**) during fertilization would double the number of chromosomes in the offspring. The fact that the number of chromosomes is constant for a given species, generation after generation, indicates that this predictable increase is incorrect. Examination of the **gametogenic** cells (those that upon division produce the gametes) of male and female diploid organisms brought about the elucidation of a second type of nuclear division, called **meiosis.** During the meiotic process one of each type of chromosome is parceled into the daughter cells, so the resulting daughter nuclei are haploid. When the haploid male gamete fertilizes the female gamete of the same species, the diploid condition is restored in the resulting **zygote.** By a series of equalizational or mitotic divisions and differentiations, the zygote is transformed into an individual bearing the characteristics of its species.

Chromosome Complements

During the early period of the twentieth century, hereditary studies revealed that structural characteristics of organisms were controlled by specific sites on the chromosomes. These specific sites were given the name **genes.** Biochemical studies implicated proteins as ultimately responsible for structural and functional characteristics of living cells (see Chapters 3 and 7). In 1941, Beadle and Tatum coupled genetic and biochemical studies to find that the production of several vitamins was regulated by specific chromosomal sites, thus providing a conclusive demonstration of the genic control of the cell on a molecular level. A coherent doctrine of cellular control was founded: *genes on the chromosomes in the cell nucleus are responsible for the regulation of protein synthesis.*

The specificity of organismal type as determined by the amino-acid sequences in their respective proteins is directly related to the specific number and individual character of the chromosomes. Since the genes on the chromosomes ultimately determine the amino-acid sequence, it is reasonable that their arrangement should also exhibit organismal specificity. If two cells resulting from a nuclear division are to have the same functional capacities as the parent, they must have the same potentials for protein synthesis. Therefore, *they must have the same chromosome complement, in both quality and quantity, as the original nucleus.* The mitotic division appears to provide quite well for this phenomenon.

Parceling of the Cytoplasm

Because of the preoccupation with the nuclear events of cell division, little attention was paid to the fate of the cytoplasmic structures. Since the nuclear control of the cell is by control of protein synthesis, it was assumed that the cytoplasm simply divided, and that the nucleus controlled the restoration of cytoplasmic structures and the preparation of the cell for the next division.

The first indications that this perspective was incorrect arose from microscopic observations of the reproduction of the cell organelles themselves. Like most of the phenomena based solely on observation, claims for the reproductive capacities of the individual organelles were doubted. In the 1950s, the chemical analysis of mitochondrial and

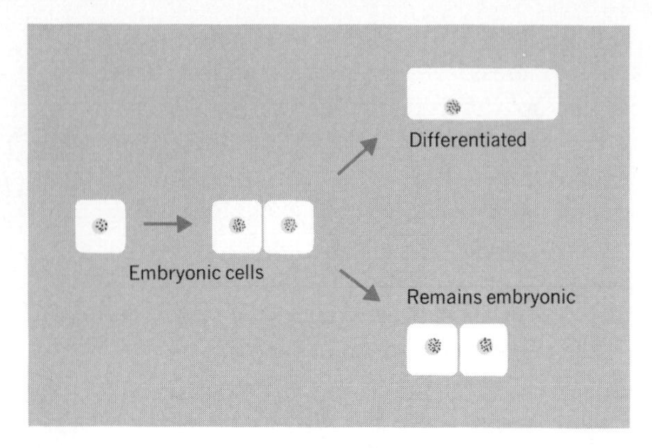

Fig. 8-3 The product of a cell division may continue as embryonic cells or may undergo differentiation. Differentiated cells rarely divide.

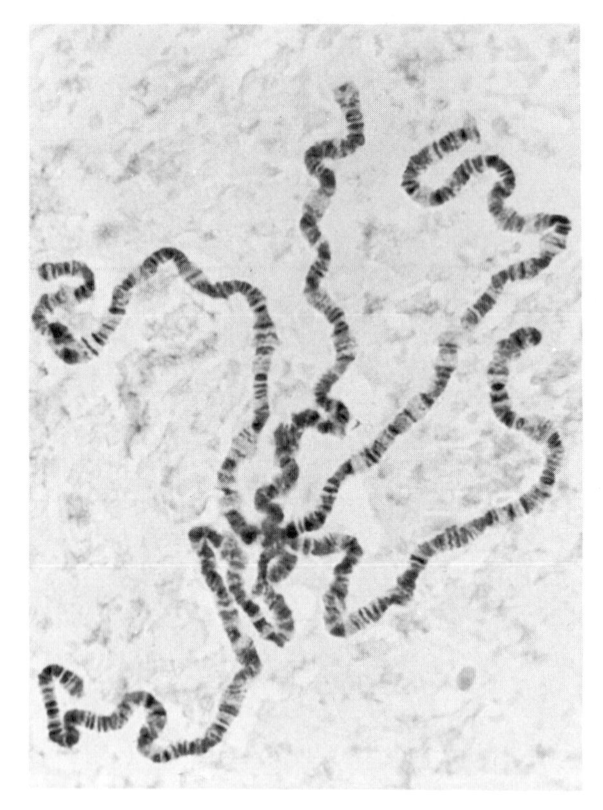

Fig. 8-4 Polytene chromosomes found in cells of the salivary gland of the fruit fly (300×). Each chromosome is actually 1000 to 4000 identical strands; hence any discontinuities along the strand assume the appearance of bands and interband regions. [Dr. Berwind P. Kaufmann]

chloroplast fractions of the cell indicated the presence of minute quantities of DNA, the only chemical form known to impart a reproductive capacity to its possessor. Subsequent studies have indicated that these organelles are indeed capable of reproducing their own type; however, their mechanisms of reproduction remain unknown.

The means of parceling cellular components into each of the daughter cells resulting from a nuclear and cytoplasmic division differ considerably. The nuclear components (the chromosomes) undergo an *active* separation, whereas the cytoplasmic components undergo a *passive* separation and their inclusion in a newly formed cell is often on the basis of chance alone.

NUCLEAR DIVISION

Growth processes of organisms are dependent on the mitotic cycle of the embryonic cells of animals and the meristematic cells of plants. These cells undergo repeated divisions, producing daughter cells that are fated either to become differentiated and thereby undergo a rearrangement of their structural components, or to remain as an embryonic or meristematic cell (see Fig. 8-3). Because of the constant repetition of the events leading to division, the process is regarded as a cycle. The nuclear events of this process are cumulatively called the **mitotic cycle.**

The Mitotic Cycle

In the course of the mitotic cycle, distinct nuclear changes are visible. The heavily granulated nucleus characteristic of both differentiated and nondividing embryonic cells begins to exhibit structural alteration when the latter starts the division process. The chromosomes contract and become visible as distinct structural entities. It is, in fact, only at these stages during the mitotic cycle that the chromosomes ever assume a visible structural character. (An exception to this is the chromosomes found in the salivary-gland tissue of insect larva. These chromosomes are formed by duplication without separation and are called **polytene** chromosomes. See Fig. 8-4.) The chromosomes perform a series of movements, eventually resulting in the inclusion of a total chromosome complement in each of the two nuclei that result from the division. The chromosomes then lose their visible structural form, and the granular state of the nucleus reappears. The cells then undergo a growth phase, during which time they increase in mass. After sufficient time has elapsed, the nucleus will once again exhibit the structural alterations that indicate a predivision period.

Because of the cyclic nature of the mitotic process, one could technically start a discussion at any point (see Fig. 8-5). The most logical point, however, is with the granular **interphase** or **metabolic** nucleus. For many years this phase of nuclear division was termed the "resting stage," which conveyed the idea of an inactive nucleus. The presence of this granular nucleus in differentiated cells that do not divide but are actively involved in metabolic processes led to the term "metabolic nucleus." Biochemical studies of nuclei in this granular state led to the realization that it was neither "resting" nor *only* involved in the regula-

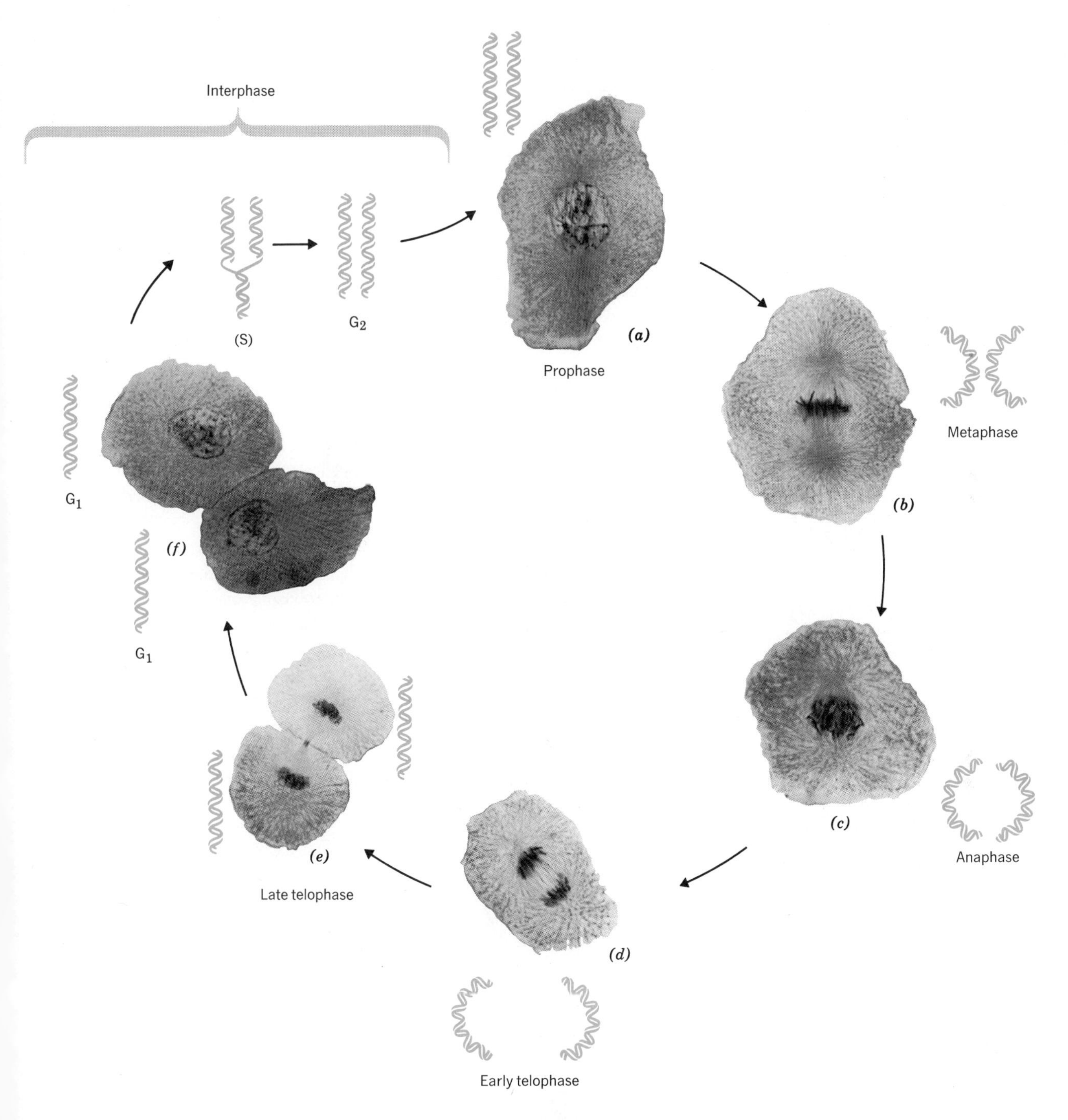

Fig. 8-5 The mitotic cycle. Drawings show the state of DNA during the cycle. [Courtesy of The Upjohn Company]

tion of metabolic processes. It was undergoing those divisional processes at a molecular level that later become the visually discernible events.

On the basis of these biochemical studies, interphase has been subdivided into three distinct stages. The first and often the longest is the $\mathbf{G_1}$ stage (growth or gap). During this period the nucleus is in a metabolic state and appears to be regulating the growth processes of the cell. This is followed by the **S** or synthetic stage. During the **S** stage, the primary chemical constituent of the chromosomes, DNA, is replicating. This replication results in doubling the number of DNA strands in each chromosome that are identical in both the sequence and the number of nucleotide pairs (adenine-thymine and guanine-cytosine). If each of the nuclei resulting from a division is to have the same chromosome complement as the original nucleus, *there must be at least two of each of the chromosomes in the original nucleus to be parceled into the new nuclei.* The replication of the DNA during the **S** stage of interphase is thought to be the first step in the formation of duplicated chromosomes.

The **S** stage is followed by another growth phase, the $\mathbf{G_2}$ period. It is thought that during this time the protein material and energy pools associated with the chromosome structure and movement are established. After the completion of chromosome duplication, the visually discernible events commence.

The chromosomes become shortened and appear as thick, deeply staining bodies. The stage at which the chromosomes become visible structures is called **prophase.** The prophase chromosome has several distinct regions (see Fig. 8-6). It has two snakelike **arms** (which are arbitrarily called the left and the right) bound together by a constriction body called the **centromere.** The position of the centromere is characteristic for each chromosome. Some chromosomes have centromeres that are almost terminal, and therefore have one long arm and one short one, while others have more centrally located centromeres. In diploid organisms, the pairs of chromosomes can be distinguished by matching those

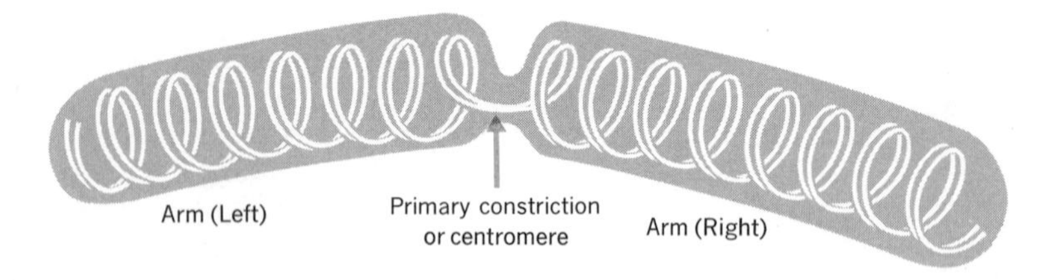

Fig. 8-6 Chromosomes are visible at prophase because the chromatin material coils, giving it a thickness discernible with microscopic techniques.

with similar lengths and the same centromere position, even though the individual chromosomes are spatially independent of one another. Closer microscopic observation reveals that each arm of the chromosome at prophase is duplicated into **sister chromatids.** The duplication of the centromere has not happened at this time.

Another feature of prophase is the disappearance of the nucleolus. The significance of this event will be discussed in Chapter 10.

While these nuclear events are occurring during prophase, one of the products of centriolar division (that has taken place in late interphase) migrates around the nucleus and takes a position on the same plane, 180° from its sister centriole. The sister centrioles form the poles toward which the chromosomes will move.

The disappearance of the nuclear membrane occurs during late prophase. The duplicated chromosomes move toward the central plane of the cell, defining the onset of **metaphase.** As they move, a series of fibers appear. These fibers converge at each pole, and lie parallel between the poles. This fibrillar structure is the **spindle.**

Isolated spindles have been chemically analyzed and appear to be protein structures containing large amounts of sulfur-containing amino acids. Electron micrographs reveal that the spindle is made up of bundles of microtubules arranged in longitudinal bands. Very little is known about the chemical events leading to this orientation of the microtubules.

In addition to the spindle, animal cells have a series of fibers, called **asters,** that radiate outward from the poles. The origin, function, and structure of the asters is unknown, although it is assumed that they anchor the spindle poles in some manner.

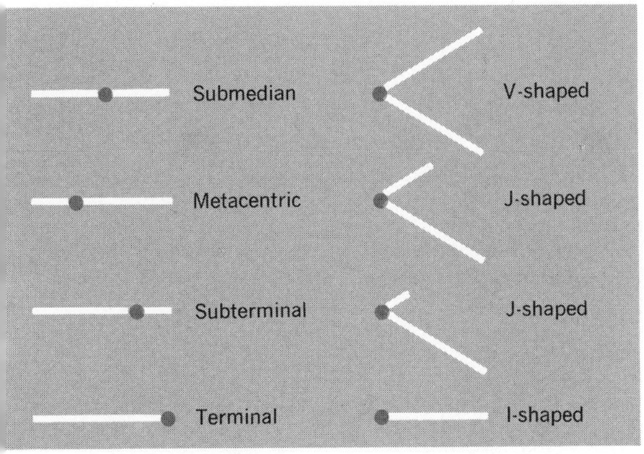

Submedian V-shaped
Metacentric J-shaped
Subterminal J-shaped
Terminal I-shaped

Fig. 8-7 The major identifiable characteristic of an individual chromosome is the position of its centromere, which is most obvious at anaphase, when the centromere precedes the chromatids to the pole. A chromosome with a centrally located centromere will appear V-shaped as it goes to the pole, while one that has a centromere located off to one side will appear J-shaped as it goes to the pole.

Once the spindle is laid down, chromosomal movement becomes more directional and the chromosomes assume a position on the equatorial plane or **metaphase plate** of the spindle (see Fig. 8-5). The forces responsible for the movement of the chromosomes into this position are not known, although it appears that the centromere becomes physically associated with the microtubules.

Anaphase begins when the sister chromatids of the duplicated chromosomes separate from one another and move along the longitudinal axis of the spindle to the opposite **poles of the spindle.** The new chromosomes (formed as the duplication products of the original chromosomes) appear to be pulled to the poles by their centromeres, while their arms flay passively, pointing in the direction from which they came (see Fig. 8-5). Because of the manner in which they go to the poles, the position of the centromere is most clearly seen at this time. A chromosome with a centrally located centromere will have a V-shaped configuration en route to the pole, while one with an almost terminal centromere will have a J-shaped appearance (see Fig. 8-7).

Once the new chromosomes reach the poles, **telophase** begins. The chromosomes begin to lose their visible structural characteristics, the nucleus acquires a new membrane, and the chromosomes once again become granular in appearance. Nucleoli are formed at this time, too. Biochemical and genetic studies have suggested that this structure is formed at a specific locus on the chromosome, which is called the

nucleolar organizer region (see Fig. 8-8). Nucleoli are not reproductive entities, but are instead constructed at prescribed periods by the chromosome. The nucleus then enters the G_1 stage of interphase.

In most tissues, **cytokinesis** (cytoplasmic division) accompanies nuclear division (**karyokinesis**). In animal cells, the formation of two complete cells is accomplished by a "pinching in" of the cell membrane at the periphery, forming a furrow that serves to establish the two new cells. In plant cells, a **cell plate** is laid down on a plane corresponding to the equatorial plane of the spindle. The cell plate gradually grows outward from the center toward the cell membrane, eventually separating all material into two newly formed cells (see Fig. 8-9).

Some tissues, however, are characterized by a temporary or permanent multinucleate condition. This condition is thought to arise by repeated nuclear divisions that are not followed by cytokinesis. In the endosperm cell of plant seeds, the multinucleate condition is temporary, and terminates when a series of membranes are formed around the nuclei and the cytoplasm that immediately surrounds each (see Fig. 8-10). In skeletal muscle tissue, the multinucleate condition is a permanent characteristic of the cell, and appears to be directly correlated with its function.

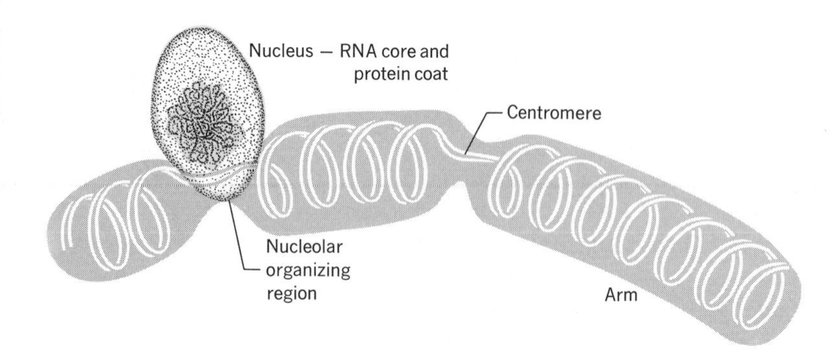

Fig. 8-8 The nucleolar organizer region is located at a specific site on one pair of chromosomes. This site is activated in late telophase.

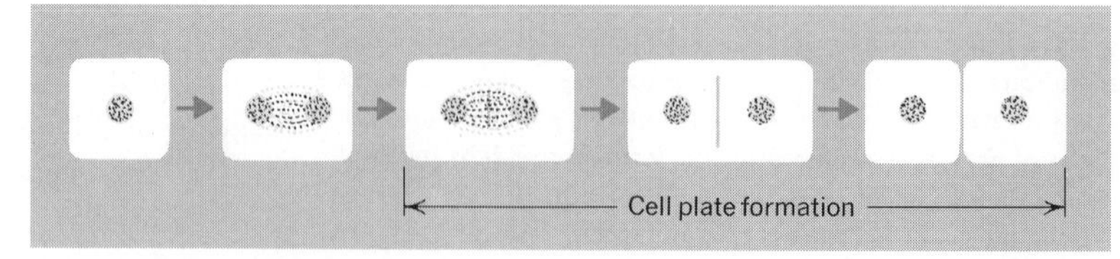

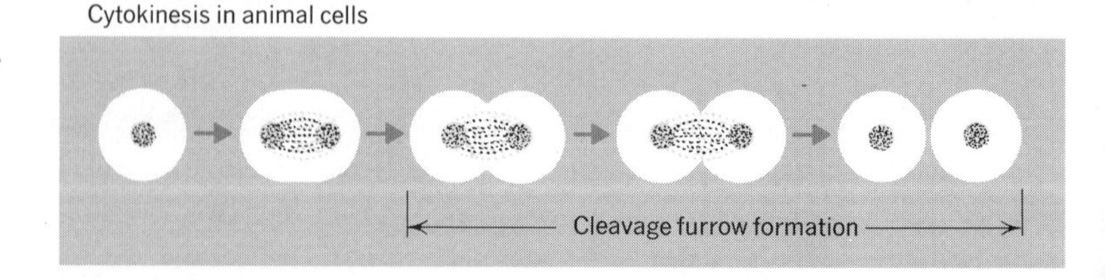

Fig. 8-9 Cytokinesis in plant and animal cells.

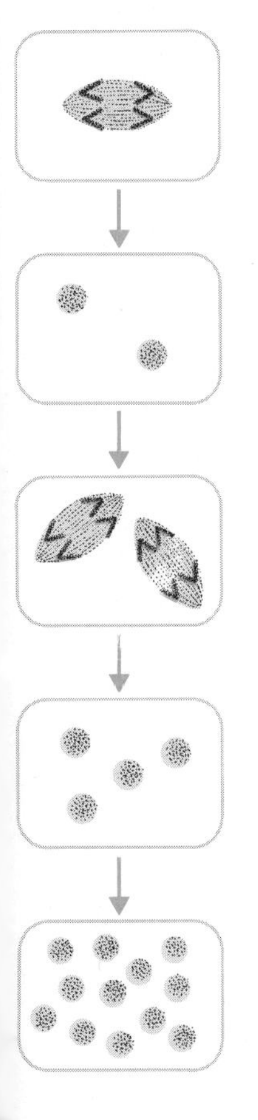

Fig. 8-10 Origin of the multinucleate condition.

Meiosis

The gametogenic cells of diploid organisms constitute a special type of embryonic tissue. The daughter cells resulting from the **mitotic** (equational) division and subsequent cytokinesis of these cells are fated either to remain as gametogenic cells, to form more of their own kind, or to undergo differentiation by a subsequent division process that reduces the number of chromosomes to the haploid number. The division process, in which the chromosome complement is reduced to a haploid condition, is called **meiosis.** After the meiotic process, the resulting cells will undergo a structural differentiation to form the male sperm (in animals), pollen grain (in higher plants) antherozoid (in some lower plants), or the female egg or **ovum** (general term used for the female gamete of plants and animals). When the male and female gametes fuse at fertilization, the diploid character is restored in the zygote. Subsequent mitotic divisions and differentiations result in the formation of a new individual.

Some haploid organisms are capable of undergoing fertilization directly. In some species the organisms themselves act as the gametes; in other species, the individual organisms produce, as the result of mitotic division, specialized gametic cells. In the latter case, the larger of the two gametic cells is the female gamete and the smaller is the male gamete.

When fertilization occurs between the gametes of *two haploid* organisms, a diploid condition *not* characteristic of their species is produced. In these instances, the diploid zygote undergoes **meiotic** divisions directly, thus restoring the haploid condition characteristic of the species in the resulting daughter cells (see Fig. 8-11).

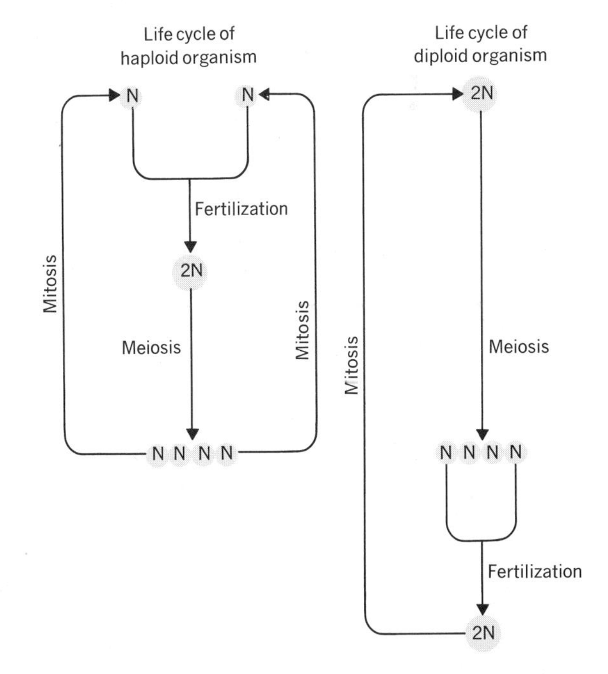

Fig. 8-11 For haploid organisms, the diploid condition created by fertilization is atypical for the species. Haploidy is restored when the diploid zygote undergoes meiotic division, and retained when the haploid forms resulting from meiosis undergo mitotic processes. In diploid organisms, the haploid condition created by meiotic division produces the gametes. Fertilization then restores the typical diploid condition, and subsequent mitotic divisions create the population of cells, or groups of tissue characteristic of the organism.

Meiotic Division

Meiosis is characterized by *two* nuclear divisions that result in the haploid state of the resulting daughter cells (see Fig. 8-12). Because of the two nuclear divisions, *four haploid cells always result from meiosis.*

The first interphase of the meiotic process is the same as that for the mitotic cycle. There is a growth phase, $\mathbf{G_1}$, followed by an **S** phase during which the DNA of the chromosomes replicates. The subsequent growth phase, $\mathbf{G_2}$, is also typical.

At the beginning of **prophase I,** the chromosomes take on a thickened and dense appearance. At the point where they are seen as threadlike structures, the nucleus is considered to be in the early stages of prophase. The prophase of meiotic division is arbitrarily divided into five stages, during which the chromosomes become more definitive and their arrangement in the nucleus more orderly. The most important features of the prophase of meitotic division are (1) that homologous (identical in structure) chromosomes pair and move as a single unit, indicating that there is a physical association between them and (2) that each of the paired chromosomes is duplicated. At late prophase, sister chromatids are "joined" by one parental centromere. *The centromeres have not yet duplicated.* Because the duplicated chromosome pairs are difficult to discuss, the *centromeres of the chromosomes are standardly used as reference points for the reductional and equational phases of meiosis.*

The disappearance of the nuclear membrane and the appearance of the spindle signal the beginning of **metaphase I.** The events leading to the formation of the spindle for the meiotic process are thought to be analogous to those of the mitotic process. The chromosome pairs move toward the equatorial plane of the spindle.

When the separation of the undivided centromeres of the chromosome pairs occurs, the nucleus is considered to be in **anaphase I.** The sister chromatids remain attached to the centromere and pass to a pole; those chromosomes that paired earlier separate and now pass to opposite poles.

As cytokinesis separates the two products of the first division, each cell sets up a new spindle, oriented at 90° to the plane of the original spindle. Therefore, the cells do not return to an extended interphase. In the few instances where the first and second meiotic divisions are interrupted by an interphase, the stage is brief, during which time DNA synthesis does *not* occur. One of the major characteristics of a meiotic division is that *two divisions occur with only one DNA replication.*

When there is essentially no lapse between the first and second meitotic divisions, the chromosomes immediately orient themselves on the equatorial plane of the new spindle. This is **metaphase II.** Each of the centromeres now duplicates.

The separation of the duplicated centromeres marks the beginning of **anaphase II.** When the sister chromosomes have reached the poles, the second division of the meiotic process is in **telophase.** New nuclear membranes reappear, the chromosomes lose their visible characteristics, nucleoli are formed, and the interphase nucleus is thus restored (see Fig. 8-12).

Each of the four cells resulting from the meiotic division undergoes differentiation into cell types characteristic of the organism.

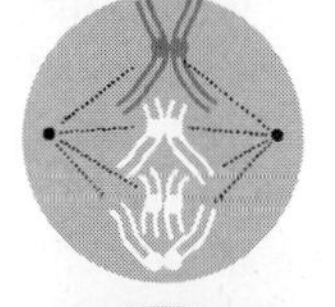

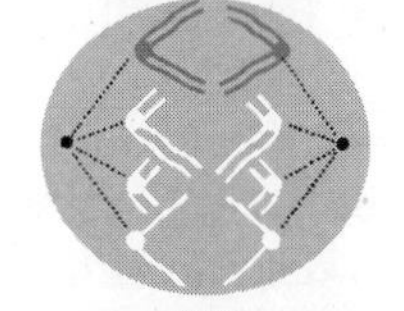

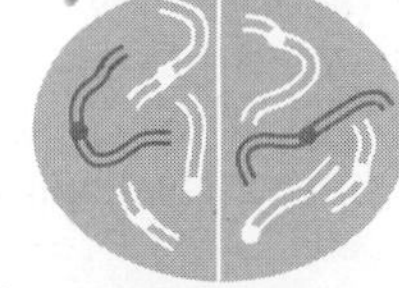

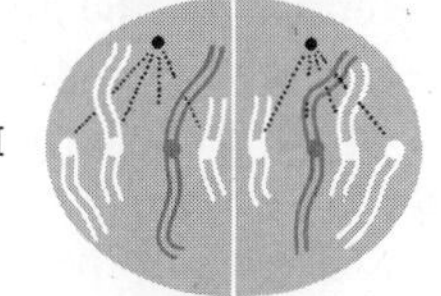

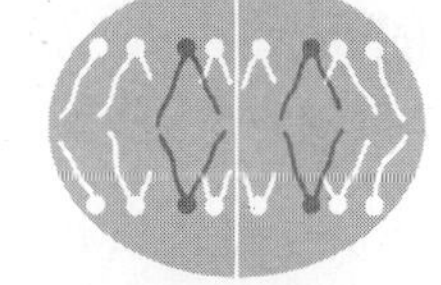

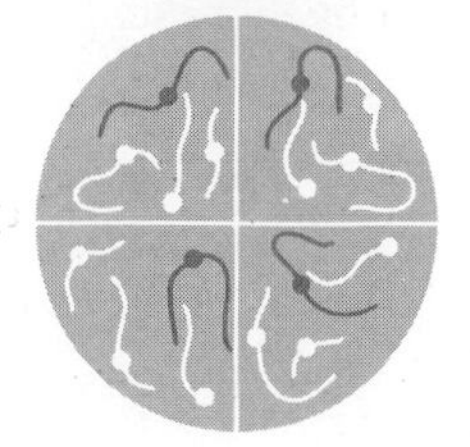

Fig. 8-12 Meiosis.

CYTOPLASMIC DIVISION

Cytokinesis distributes the cytoplasm of the parental cell approximately equally between the two resulting daughter cells. (Exceptions to this occur in the cell division of yeast and the meiotic division leading to the formation of the eggs in the females of higher animals, where the nuclear divisions take place very near the cell membrane, and only a small portion of the cytoplasmic material is pinched off). By the pattern through which the cytoplasm divides, the inclusion of individual organelles in the resulting daughter cells is dependent on their position and number at the time of cytokinesis.

Cellular Roulette

The position of the organelles within a cell appears to be determined by somewhat random events, and follows the distribution similar to that of the binomial expansion. An imaginary line may be drawn across the middle of a cell and the halves labeled **a** and **b** respectively ($\mathbf{a} + \mathbf{b} = 1$). If there are only two chloroplasts per cell in a large number of cells, their distribution would be expected to follow a binominal pattern of $\mathbf{a}^2 + 2\mathbf{ab} + \mathbf{b}^2$. Therefore, one would expect to find that $\frac{1}{4}$ of the cells would have both chloroplasts in the **a** half and *none* in the **b** half; $\frac{1}{2}$ of the cells would have one chloroplast in the **a** half and one in the **b** half; and $\frac{1}{4}$ would have both chloroplasts in the **b** half and *none* in the **a** half of the cell (see Fig. 8-13).

But what are the results of such a situation? If cytokinesis occurred according to the arbitrarily drawn line, half the resulting daughter cells would have no chloroplasts ($\frac{1}{4} + \frac{1}{4} = \frac{1}{2}$). This would be of no important consequence if the nucleus were capable of bringing about chloroplast formation *de novo,* for then new chloroplastic structures could be synthesized and the cell would regain its full functional capacity. But cytological and genetic evidence reveals that once a chloroplast is lost, it cannot be replaced by the cell.

The solution to the problem of the retention of the maximum capacity of cellular function in resulting daughter cells lies in the number of individual organelles present. If, instead of there being only two chloroplasts in each dividing cell of a tissue, there were 10, then the binomial distribution would follow the expansion of $(\mathbf{a} + \mathbf{b})^{10}$ and the chance that a chloroplast would be present on both sides of the arbitrarily drawn line (classes that have both **a** and **b**) would be enormously increased. Therefore, the greater the number of each type of organelle within the cell, the greater the chance that at least one will be in a new daughter cell. The distribution pattern expected from the random positioning of cell organelles has been born out by microscopic observation.

THE REPRODUCTIONS OF CYTOPLASMIC ORGANELLES

One of the most important additions to Virchow's statement regarding the reproductive capacities of cells is the corollary formed by the overwhelming evidence in the 1950s and '60s that "nuclei, mitochondria, chloroplasts, centrioles and basal granules come only from preexisting nuclei, mitochondria, chloroplasts, centrioles and basal granules, respectively." The origin of a given organelle can be traced back to another organelle of the same or similar structure and function. The fate of a given organelle is to divide, thus giving rise to two structures having the same function and structure as the original one.

Mitochondria

Because of their minute size, it is impossible to obtain evidence for mitochondrial multiplication with the light microscope. If one sees one 0.5 μ particle and later two, it can be argued that one was directly under the other, and simply moved away. Electron micrographs have

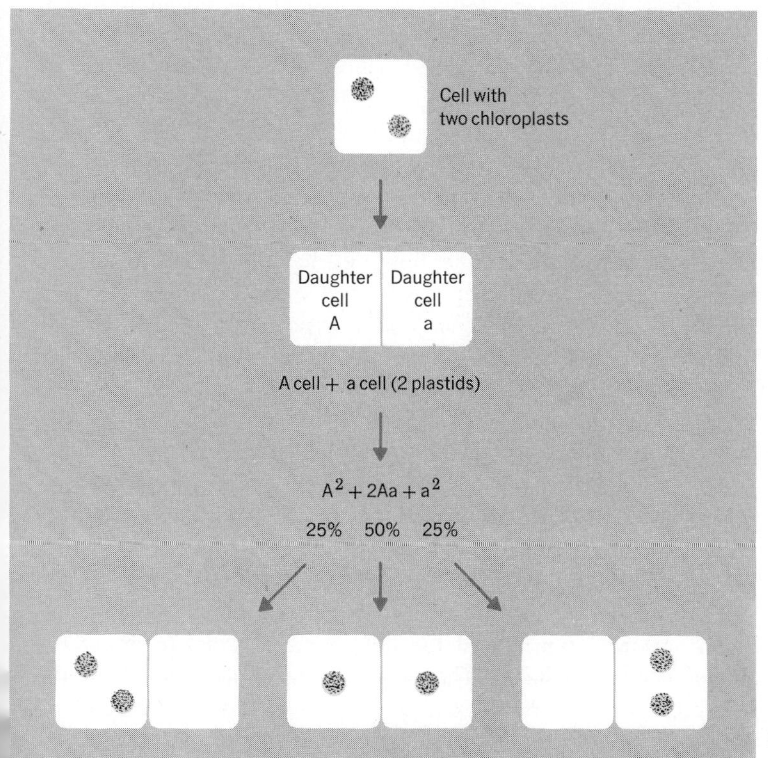

Fig. 8-13 Cellular roulette.

not offered any conclusive solution to this problem, either. The inability to observe living processes under the electron microscope eliminates the possibility of watching a mitochondrion increase in size and actually divide. Electron micrographs of mitochondria indicate that they sometimes have an irregular structure, but who can say that these irregular structures are in the process of elongating and dividing? There is no unequivocal cytological evidence for the reproductive capacity of mitochondria.

There is, however, substantial biochemical and genetic evidence for such a phenomenon. One of the first indications that the mitochondrion might have reproductive capacities was the detection of a small amount of DNA, similar in structure to that found in the chromosomes, in mitochondrial fractions.

D. Luck obtained a particular strain of the mold *Neurospora* that required choline for growth in addition to the other nutritional requirements of the typical strains found in nature. He grew the choline-requiring strain on a medium containing the full complement of nutritional requirements and C^{14}-labeled choline ("hot" choline). After a period of time, the mold was transferred to a medium containing non-radioactive ("cold") choline. The free radioactive choline was diluted out of the cells and only that already incorporated into intact structures remained.

He then homogenized the mold tissue and fractionated it by differential centrifugation (see p. 182), and examined the various fractions for the presence of radioactive material. He found, as he anticipated, that much of the radioactive choline was localized in the mitochondrial fraction. This had been anticipated, because it was well known that choline was a major constituent of a phospholipid found in large quantities in the mitochondrial membrane. The localization suggested to Luck a means of determining whether or not mitochondria divide.

He repeated the procedure of growing some of the choline-requiring strain in "hot" choline. This time, however, he transferred portions of the mold into three flasks, each containing "cold" choline. He fractionated the mold material in one flask after permitting sufficient time for the fixation of "hot" choline. The molds in the other two flasks were permitted to grow. After ten hours of growth, he homogenized and fractionated the mold from a second flask. After twenty hours of growth, he followed the same procedure with the final flask.

After obtaining the three mitochondrial fractions, he submitted each to a **density-gradient centrifugation.** This is a very sensitive biophysical technique for separating materials that have slightly different molecular weights. A centrifuge tube is prepared by carefully pouring a 5% and a 30% sucrose solution together at different rates in such a way that a gradient varying from 30% at the bottom of the tube to 5% at the top of the tube is produced. The most typical solutions used for the gradient are those where sucrose or cesium is the solute. Random movements of the sucrose or cesium will cause only slight disturbances in the continuity of the gradient. Because of the low molecular weight of the sucrose or cesium chloride molecules, the gradient is undisturbed by centrifugation.

Mitochondria (or other cellular material) are placed on the surface of the density gradient. It is then drawn through the gradient by the force of centrifugation. As it is pulled down the tube, toward the periphery of the generated circle, the material will reach a level beyond which it cannot move. At this point the concentration of the gradient may be thought of as acting as the base of the centrifuge tube for the particular material.

After centrifugation, the bottom of the tube is punctured, and succeeding layers of the tube removed and tested for the presence of the cellular components.

"Cold" mitochondria (those not containing radioactive material) form a layer in the upper region of a density gradient. Because of the increased molecular weight contributed by the C^{14}, which replaced the C^{12}, "hot" mitochondria form a layer at a lower level in the density gradient. Luck reasoned that if the mitochondria had no reproductive capacities, and new mitochondria were formed solely under nuclear control, those mitochondria formed during the growth period in "cold" choline (after exposure to the "hot" choline) would not contain the heavier C^{14}. Therefore, the density-gradient centrifugation of the mitochondrial fraction of the mold tissue that was permitted to grow for ten or twenty hours would show two distinct mitochondrial layers, a "hot" and a "cold." If, on the other hand, the mitochondria did have a reproductive capacity, and new ones were formed from a division of existing ones, then the limited amount of "hot" choline would be constantly parceled between the daughter mitochondria resulting from a division. In this case, the radioactivity would be divided among all the mitochondria. Therefore, there would be a single mitochondrial layer in the density gradients positioned somewhere between that expected for the totally "hot" and "cold" layers.

When the mitochondrial fractions of the mold tissue that had been grown in "cold" choline for 0, 10, and 20 hours, respectively, were submitted to density-gradient centrifugation, Luck found different mitochondrial layers in each case. The tube containing the mitochondrial fraction of the mold grown for 0 hours showed a mitochondrial layer at the level expected for "hot" mitochondria. The tube containing the mitochondrial fraction of the mold grown for 10 hours showed a single layer at the level above that expected for "hot" mitochondria; and the tube containing the mitochondrial fraction of the mold grown for 20 hours showed a single layer above that found for the 10-hour mitochondrial fraction, but below that expected for "cold" mitochondria (see Fig. 8-14). The inescapable conclusion was that the original mitochondria had contributed structural material for the next generation of mitochondria. They were undergoing a reproductive process.

Although the biochemical work by Luck indicated that mitochondria were capable of reproduction, genetic studies have indicated that the control of the reproductive process may be partially nuclear. It is well established that DNA can, when exposed to certain adverse conditions (ultraviolet light, mustard gas, X rays) undergo structural changes in which some of the nucleotide pairs are altered or deleted. This change is manifested in an altered protein regulated by the affected region of DNA, and may be detected as a change in a structural characteristic

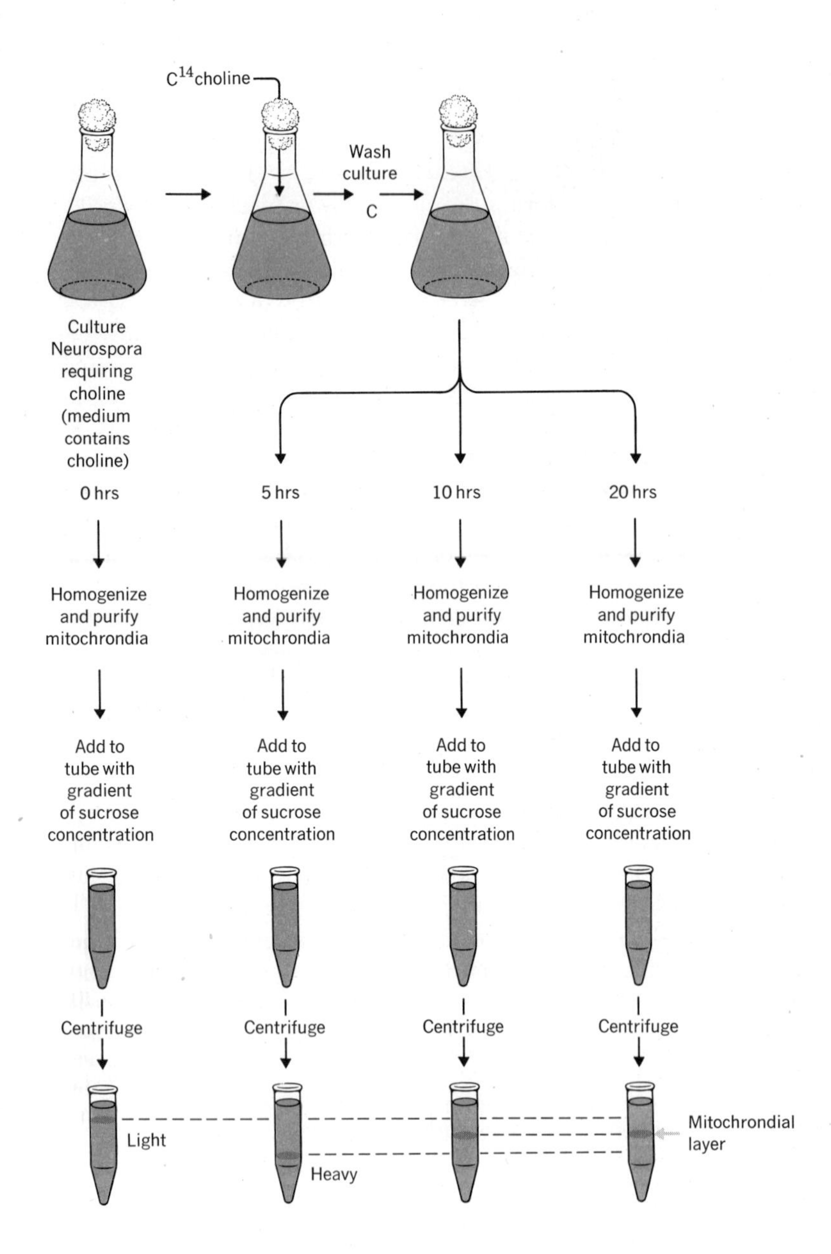

Fig. 8-14 Luck's experimental setup to determine whether mitochondria have reproductive capacity.

or some enzymatic capacity of the individual. Such a change is called a **mutation,** and the new characteristic or altered individual is called a **mutant.** Because of the way in which DNA is passed from one cell generation to the next (mitotic cycle and meiotic process), the mutant condition is inherited.

Boris Ephrussi produced several mutations in different yeast cells, each of which manifested the same characteristics. These mutant cells produced cells that were smaller than those typically found in the yeast population, and because of this feature, were called "petites." Electron-microscopic examination of the independently produced "petite" mutants revealed no structural alteration. Chemical analysis, however, indicated that several enzymes associated with the mitochondria were absent. Clearly, the portion of DNA that controlled the synthesis of the enzymes had been altered. But was it an alteration of the nuclear or the mitochondrial DNA that produced this condition?

Yeast cells are haploid organisms. Under certain environmental conditions, two yeast cells will undergo fertilization and form a diploid zygote. Although the two cells that combine to form the zygote appear structurally similar, it has been found that not just any two cells will mate. There must be a functional difference between the two cells, which is associated with their **mating types.** A yeast cell of mating type plus (+) will combine only with a yeast cell of mating type minus (−). When the cells of either mating type undergo *mitotic* division (asexual reproduction), they produce only cells of their own mating type. When fertilization occurs, the zygote formed between the two mating types will undergo a meiotic division to produce four haploid yeast cells, two of which will be mating type (+), and two of which will be mating type (−). This illustrates that the mating types are not only inherited, but are also localized on chromosomes.

When fertilization occurs between (+) and (−) cells, the two cells fuse into a single large zygote. There is some evidence that the minus strain does not contribute cytoplasm in the formation of the zygote, so that the cytoplasm of the plus strain is primarily active in the formation of new cells.

If the *mitochondrial* (cytoplasmic) DNA is responsible for the "petite" condition, then when a "petite" (+) cell fuses with a normal (−) cell, all yeast cells resulting from the division of the zygote would be "petite." When a normal (+) cell fuses with a "petite" (−) cell, the zygotic products would all be normal. On the other hand, if the *nuclear* (chromosomal) DNA were responsible for the "petite" condition, two of the yeast cells formed from the meiotic division of the zygote would be "petite" regardless of which of the two mating types contributed the trait (see Fig. 8-15).

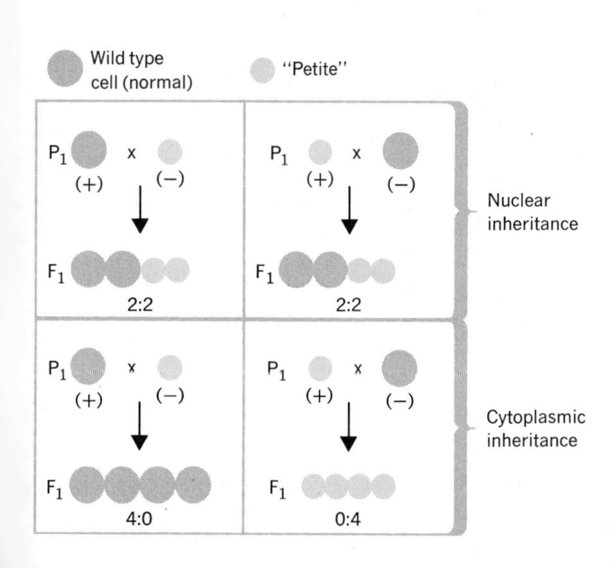

Fig. 8-15 **Ephrussi's genetic experiment to determine the reproductive capacity of mitochondria.**

When the appropriate matings were done, Ephrussi found that in some of the mutant strains the "petite" trait was inherited through either parent, because the division of the zygote resulted in two cells that were "petite" and two that were normal. In these strains, the "petite" trait, like the mating-type determination, was of chromosomal origin. Other "petite" strains he crossed gave different results. "Petite" (+) crossed with normal (−) produced only "petite" offspring, and normal (+) crossed with "petite" (−) gave no petite offspring. Only cytoplasmic transmission could account for these results. Although more information regarding the mechanism of inheritance in yeast is needed before any clear conclusion can be offered, it seems that the synthesis of mitochondrial enzymes is under both nuclear and mitochrondrial control.

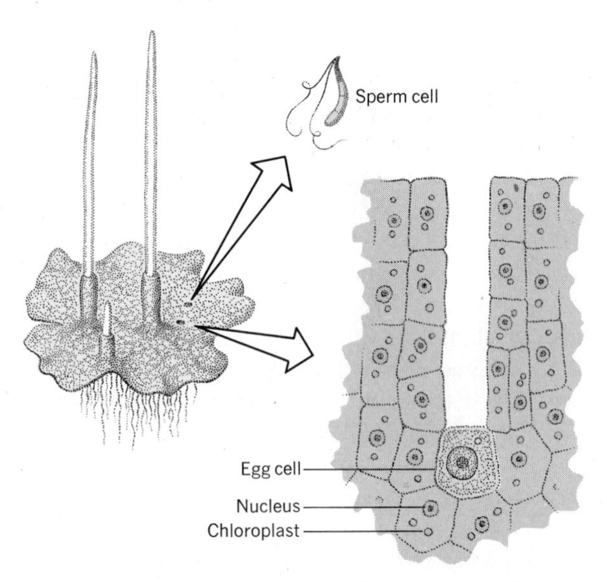

Fig. 8-16 ***Antheroceros.*** **The egg of *Antheroceros* remains within the plant body and is nurtured by it. The sperm, produced elsewhere on the plant body, migrate through the moisture covering this low-growing plant to fertilize the egg. The egg contains a chloroplast; the sperm does not.**

Chloroplasts

Direct evidence for the reproductive capacity of chloroplasts was obtained by the observation that the egg cell of the lower plant *Antheroceros* (a liverwort) always contained two chloroplasts, while the antherozoid (sperm) of the same species had none (see Fig. 8-16). It was reasoned that if the nucleus is capable of producing chloroplasts *de novo*, then the presence of a chloroplast in either gamete would be unnecessary. This pattern would be of no value and hence not perpetuated through evolution.

Other cytological evidences were provided by many observers who saw chloroplasts actually dividing. Repeated observations in many plants by many different cytologists reaffirmed the phenomenon of chloroplast division.

Biochemical evidence supports this view. Analysis of the chloroplast indicates that about 1% of its chemical composition is DNA. Although this DNA does not appear to be completely *homologous* (exactly the same in structure) to nuclear DNA, its presence in the chloroplast is thought to be indicative of a reproductive mechanism.

The primary source of evidence for the reproductive capacity of chloroplasts is genetic. When the unicellular alga *Euglena* is treated with streptomycin, the chloroplasts become bleached. (This effect is now thought to be due to the action of streptomycin on the 70S RNA ribosome present in the chloroplast; see pages 250–254.) When cells of *Euglena* are transferred to a medium *not* containing streptomycin, the bleached condition persists in all members of the growing population because the chloroplasts cannot divide and only nonfunctional chloroplasts are passed into daughter cells. No new chloroplasts are formed. Therefore, it seems that once a chloroplast is lost or inactivated, the nucleus does not have the capacity to replace it.

In some plants, mutant strains have been produced that are characterized by a yellow coloration of the leaves and stem. Microscopic observation reveals that the yellow coloration is localized in the chloroplasts, and biochemical analysis indicates the presence of the carotenes, but no chlorophyll. The fertilization mechanism in higher plants is analogous to that of most organisms. The male gamete contributes only its nucleus to the fertilization process; the cytoplasm for the resulting individual is mainly from the female gamete.

When appropriate plant crosses are carried out, it is found that the trait for yellow chloroplasts is inherited only when the yellow trait is contributed by the female parent. Apparently, chloroplastic DNA had been affected in the production of this mutant trait.

Evidence accumulated thus far indicates that the chloroplasts are, indeed, capable of reproducing, mutating, and perpetuating themselves in a mutant state. These patterns are exactly parallel to those of the nucleus.

Centrioles and Basal Bodies (*Kinetosomes*)

The centrioles and basal bodies will be discussed together because of their similarity of structure. The minute size of these structures does not permit direct observation of their division, although some claims have been made for the observation of this phenomenon for the centrioles of animal cells. The products of a centriolar division are thought to form the poles at metaphase.

A group of unicellular parasites, called **trypanosomes** (one member of which is, incidentally, responsible for African sleeping sickness) are characterized by a flagellum that arises from a basal body or **kinetosome.** L. Wolff found that by treating these animals with the dye **acridine,** one of the daughter cells produced upon division will have the old kinetosome, and the other daughter cell will have none. Apparently, the dye inhibits the division of the kinetosome. The trypanosome in which there is no kinetosome will not develop a cilium, and upon subsequent divisions, will produce an entire population of immobile trypanosomes. It may be concluded that the cilium arises from this type of basal body, and once the body is lost from the cell, it can never be replaced.

Biochemical analysis indicates that there may be a small amount of DNA in both the centriole and basal bodies. Geneticists, however, have failed in their attempts to obtain aberrant centriole or basal-granule structures that could be used for genetic study.

SELF-ASSEMBLY OF MEMBRANE SYSTEMS AND PARTICLES

The division of the nucleus and cytoplasmic organelles still has many poorly understood features. The division of a nucleus to form two daughter nuclei involves far more than the duplication of chromosomes. It involves precise timing for the appearance and disappearance of membranes, spindle fibers, furrows or cell plates, and nucleoli. Although the replication of the chromosomal DNA and the deposition of protein material around the chromosomes have been supported by an overwhelming amount of experimental evidence, mechanisms for the formation of the auxiliary structures in the mitotic and meiotic cells remains unknown.

This same consideration is true of the cytoplasmic structures. The presence of mitochondrial or chloroplastic DNA indicates that there is a sort of cytoplasmic chromosome within each of these structures. But these are membrane-based structures, and their growth and duplication are dependent on the formation of new membranes.

And what of the cellular components whose chemical analysis has not revealed the presence of DNA? How do cell membranes and endoplasmic reticula increase in size? How do the numbers of ribosomes increase?

Biologists have devised a scheme based on the indirect evidence of *de novo* appearance of membrane systems and ribosomes that they believe to be consistent with the chemical and physical properties of the molecular forms of which these organelles are composed. The scheme of **self-assembly** first suggested itself to biochemists for the formation of the secondary, tertiary, and quarternary structure of proteins, where there is evidence that enzymatic control is unnecessary for protein folding (see p. 72). If an individual polypeptide could fold and take on a three-dimensional character by chance associations of charges and nonpolar groups, then why could not a structure made up of many proteins be formed by the chance association of the exposed charges and nonpolar groups of the individual proteins (see Fig. 8-17)?

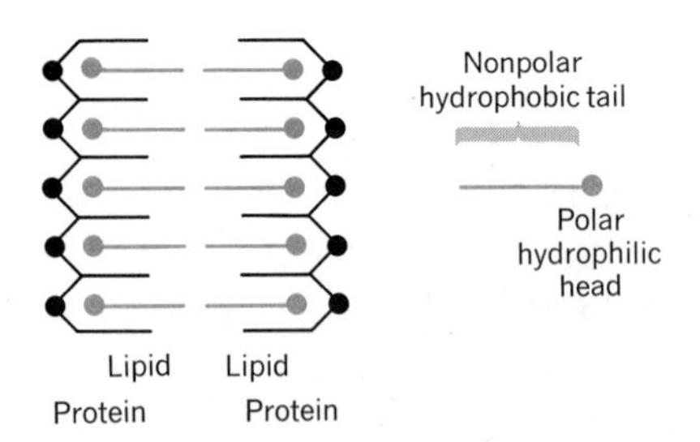

Fig. 8-17 The structure of the unit membrane.

A Proposed Model for the Self-Assembly of Membranes

The membrane systems in the cell include the cell membrane, the endoplasmic reticulum, the nuclear envelope, the golgi apparatus, and the mitochondrial and chloroplast membranes. The formation of the protein components of these membranes is traceable to DNA. In the formation of the cell membrane, endoplasmic reticulum, nuclear envelope, and golgi apparatus, it is thought that nuclear DNA is responsible for the protein components. In the case of mitochondrial and chloroplastic membranes, the DNA responsible for the protein components may be either nuclear or cytoplasmic or both (see p. 222).

In addition to regulating the formation of the structural protein directly, the DNA is thought to bear the responsibility for the formation of the lipid and phospholipid portion of the membrane structure by regulating the synthesis of the protein structure of the enzymes necessary for the production of the lipid material. Again, the position of the DNA in regulating the production of the enzymes may be either nuclear and/or cytoplasmic for chloroplastic and mitochondrial membranes.

It is believed that once the components of the membranes are synthesized, their properties spontaneously generate a series of events that terminates in the assembly of a structural unit. Lipid material is characterized by its hydrophobic property. The tendency of lipids to accumulate on the surface of an aqueous solution is thought to be due to the polar nature of the glycerol and its ability to form hydrogen bonds with water, and the nonpolar character of the hydrocarbons of the fatty-acid chains that do not form hydrogen bonds with water (see Fig. 8-18). Because of this inherent property, the nonpolar portions of lipid molecules form van der Waals forces among themselves, and the polar groups are found directly associated with the aqueous environment.

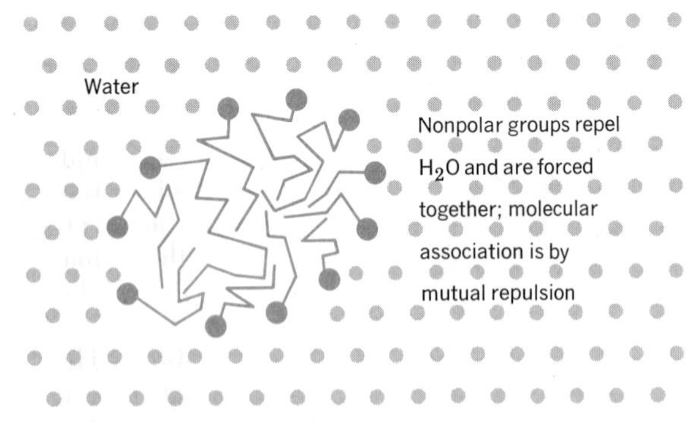

Fig. 8-18 Orientation of lipids in water.

Phospholipids, on the other hand, are **polar lipids.** Molecules of phospholipid contain long nonpolar and therefore, hydrophobic carbon chains in addition to groups that are polar, and therefore, **hydrophilic** (water-loving) (see Fig. 8-19). The phospholipids orient themselves in accordance with their tendency to form polar associations with the aqueous environment of the cellular material. In the course of this orientation, the nonpolar groups form van der Waals forces, and the polar groups are aligned along a single plane.

The association of the protein material with the lipid and phospholipid portion of the membrane is thought to be by a spontaneous association determined by the charge fields set up by the polar groups of the phospholipids reacting with the exposed charges of the protein molecules.

This type of mechanism is thought to account for the *de novo* formation of membranes and the growth of membrane structures that are already established in the cell.

Proposal for the Self-Assembly of Ribosomes

Experimental evidence seems to indicate that at least part of the ribosomal structure is capable of self-assembly. The 80S monosome associated with the endoplasmic reticulum of both plant and animal cells can be fractionated by treatment with a high magnesium concentration. Subsequent centrifugation yields 55S and 35S particles. When these are resuspended in a low concentration of magnesium and mixed, a reassembled product is found. Since these preparations were devoid of any enzyme material, the reassociation of the 55S and 35S particle to form the snowman-shaped 80S particle (monosome) is thought to occur

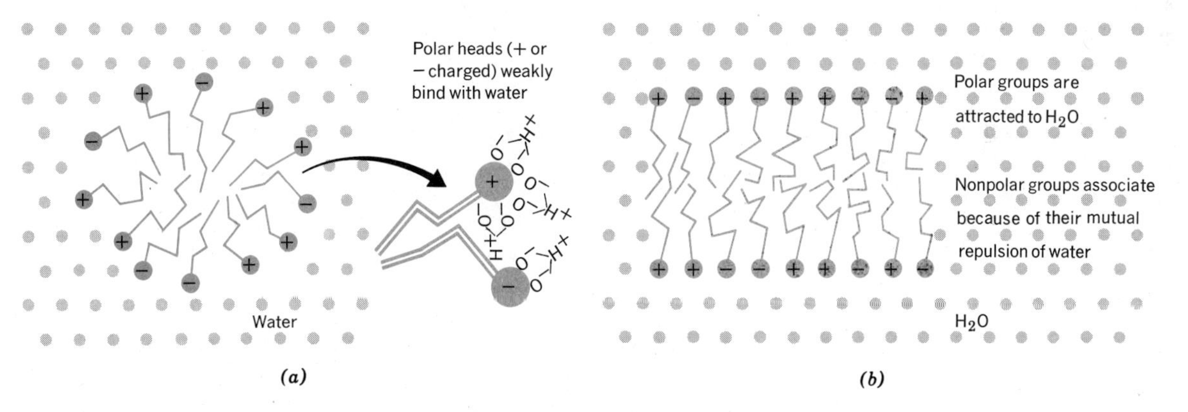

Fig. 8-19 Orientations of polar lipids in an aqueous solution. (*a*) Polar groups are attracted to water molecules. (*b*) Nonpolar groups combine through van der Waal's forces.

purely by a self-assembly process (see Fig. 8-20). Further degradation and reformation by suitable Mg concentration is also possible.

The mechanism proposed for the complete assembly of ribosomes involves the transport of RNA from the nucleus to the cytoplasm, where, by the affinity of the nucleotides and protein for one another, the ribosome is formed.

Cell Walls of Plants and Bacteria

The cell walls of plants and bacteria also appear to involve self-assembly. In plant cells, new cellulose material appears outside the membrane throughout the old wall.

In bacterial cells, the growth of the cell wall appears to take place around the center of the cell. After the new material is deposited, the bacterium divides and the cell wall is half old, half new, similar to a doubly colored medicine capsule (see Fig. 8-21).

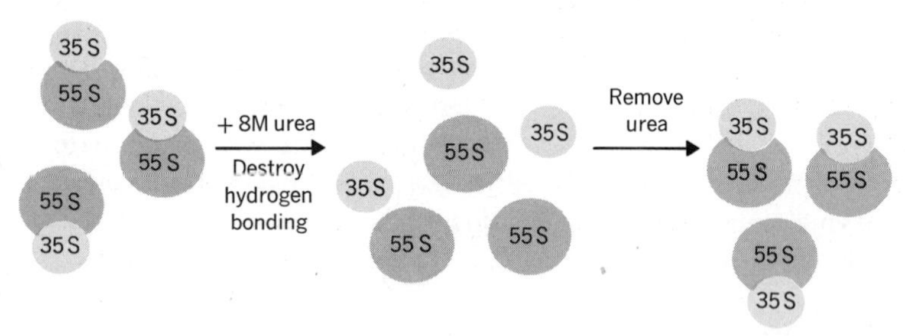

Fig. 8-20 Self-assembly of 80*S* particles.

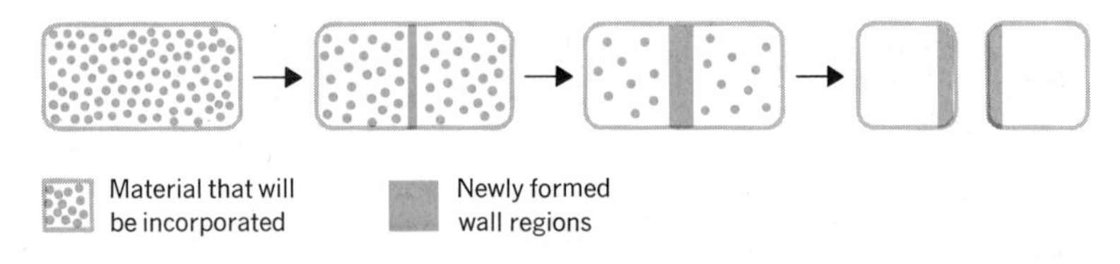

Fig. 8-21 Formation of bacterial cell wall after a division of the genetic material.

Some General Considerations

The formation of two daughter cells by cell division involves the reproductive capacity of the chromosomes and cell organelles, and the self-assembly of associated structures. Cells utilize control mechanisms for the synthesis of structural and functional components with specific properties, which permit them to associate into structural entities by self-assembly processes. The final result is the formation of two cells, identical in structural and functional capacity to that from which they arose.

QUESTIONS

1. Define haploidy and diploidy.
2. How is the metabolic capacity of a cell related to its genetic constitution?
3. How does the mitotic cycle provide for the production of two daughter cells that have the same metabolic capacity as the parental cell?
4. How does meiotic division and subsequent union of gametes in diploid organisms provide a means by which species integrity is maintained?
5. What feature of prophase in meiosis is distinctly dissimilar from prophase in the mitotic cycle?
6. Justify the statement "The higher the multiplicity of organelles the greater the chance that two functional cells will be produced upon the division of the cytoplasm."
7. Give the experimental evidence that mitochondria, chloroplasts, and kinetosomes are at least partially self-reproducing units.
8. What chemical properties make possible the self-assembly of a membrane structure?

CHAPTER 9

THE GENETIC CONTROL OF THE CELL

The concept of the gene as the unit of heredity and its action in the control of protein synthesis constitutes one of the major breakthroughs in biology since 1900. Initially, a knowledge of the patterns of inheritance and their relationship to the behavior of chromosomes during meiosis and fertilization provided biologists with an explanation of how certain traits were passed from generation to generation. Early geneticists realized that these inheritance patterns were intimately related to the theory of Darwinian evolution, and their work was primarily concerned with delimiting mechanisms by which evolution had occurred.

When the nature of genetic control of cellular function was revealed, basic genetic studies branched into two lines of inquiry: one concerned itself with the evolutionary mechanisms and the other related control of cellular form and function to gene action. Because these two inquiries deal with different areas of biology, they will be handled separately. This chapter will relate the background from which both inquiries originated, and then will develop only the cellular aspects of gene function. The genetic studies that provide mechanisms for evolution will be delayed until Chapter 13.

Gregor Mendel

The first step toward the formulation of the gene theory was taken by an obscure Austrian monk, Gregor Mendel, when he published a paper on breeding experiments with the garden pea. At the time of publication the papers aroused little, if any, interest. It was a case of the time not being ripe for the man and his work.

In 1900, DeVries of Holland, Correns of Germany, and Von Tschermak of Austria performed similar experiments. In their perusal of the literature in preparation for their publications, they independently rediscovered Mendel's work. The strength of the data and the clarity of his conclusions left little doubt that Mendel had determined the patterns of inheritance and deserved full credit for the discovery. In 1902 and 1904, Sutton and Montgomery independently related these hereditary patterns to the behavior of chromosomes during meiosis and fertilization; thus contributing to the emergence of a magnificently coherent concept of heredity.

Breeding experiments had been performed for thousands of years before Mendel. Yet none of these workers had postulated any precise pattern of inheritance. The genius of Mendel was that he alone saw the value of the single variable in inheritance studies. Previous workers had attempted to study the inheritance of whole organisms. Mendel chose to focus his attention on a single trait and follow the path of inheritance for that trait from generation to generation, ignoring all other features of the plant. It was because his experimental design provided for the single variable that he succeeded in detecting the inheritance patterns that had eluded his predecessors.

It is unlikely that Mendel began his study with a preconceived notion of what he would find. In his initial studies, he crossed (mated) two pea plants (the parental or $\mathbf{P_1}$ generation) that differed from one another by a single trait and examined their offspring (the first filial or $\mathbf{F_1}$ generation) for the expression of the trait. He then crossed two individuals of the $\mathbf{F_1}$ generation (symbolically denoted as $\mathbf{F_1} \times \mathbf{F_1}$) and examined their offspring (the second filial or $\mathbf{F_2}$ generation) for the expression of the trait. From his extensive data, he was able to draw certain conclusions that led him to the formulation of a group of principles that now bear his name: the **mendelian laws of heredity.** It should be noted that once he formulated his principles, he no longer maintained his scientific objectivity and actually discarded data that did not coincide with his conclusions. In retrospect, it is fortuitous that he did so, for the inclusion of all data would have confused the picture, and it is unlikely that any coherent scheme would have emerged. The aberrant data that Mendel ignored can now be explained by mechanisms that could not have been postulated from breeding experiments alone.

Mendel worked with seven sets of characteristics of pea plants (see Table 9-1), performing breeding experiments on the sets individually. When he crossed a plant that produced axial flowers (borne along the stem) with one that produced terminal flowers (occurring at the tip of the stem), he found that the offspring ($\mathbf{F_1}$) produced only axial flowers (see Fig. 9-1). The capacity to produce terminal flowers was completely absent in the $\mathbf{F_1}$ generation. There could be several possible explanations for this phenomenon.

1. The hereditary mechanism is such that only one trait is passed on to the offspring.
2. The trait-producing factors meld together, and only one is expressed.
3. The factor that produces terminal flowers is diluted out in the $\mathbf{F_1}$ generation.
4. Both factors are present, but that for axial flowers masks the one for terminal flowers.

The first three ideas involve the total loss of the factor for the production of terminal flowers. The fourth involves the retention of the factor.

TABLE 9-1 MENDEL'S RESULTS IN THE F_2 GENERATION OF GARDEN PEAS

Trait	Dominant	Recessive	Ratio of Dominant to Recessive
Seed color	6022 yellow	2001 green	3.01:1
Stem length	787 long	277 short	2.84:1
Seed shape	5474 round	1850 wrinkled	2.96:1
Flower color	705 red	224 white	3.15:1
Pod shape	882 inflated	299 wrinkled	2.95:1
Pod color	428 green	152 yellow	2.82:1
Flower location	651 axial[a]	207 terminal[a]	3.14:1

[a] *Axial flowers* are distributed along length of stem, whereas *terminal flowers* are located at the end of stems.

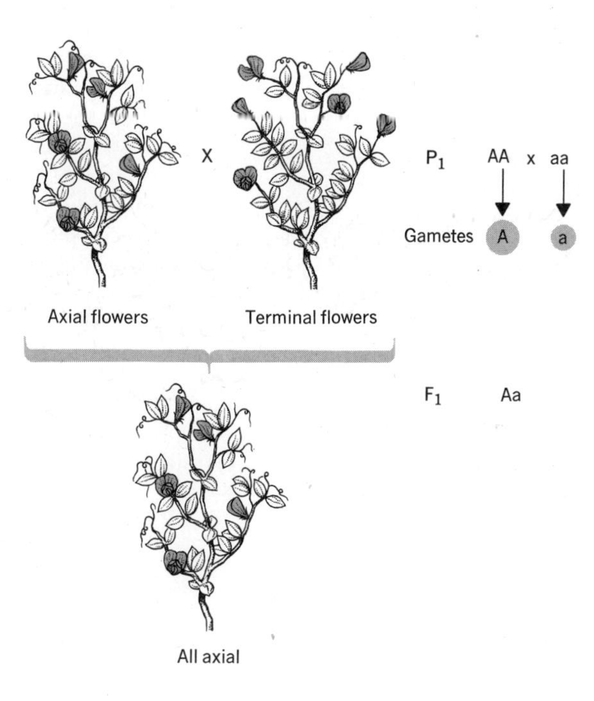

Fig. 9-1 Cross between a plant with axial flowers, which are borne along the stem, and a plant with terminal flowers.

Mendel tested the permanence of the loss of this trait by an $\mathbf{F_1} \times \mathbf{F_1}$ cross. Examination of the $\mathbf{F_2}$ generation revealed that there were both axial and terminal flowering plants. Hence, the trait was present in the $\mathbf{F_1}$ generation, and had somehow retained its integrity, even though it had not been expressed. Mendel chose to call the capacity for the trait that was expressed in the $\mathbf{F_1}$ generation the **dominant factor** and that for the trait that was masked and not expressed, the **recessive factor.** Although the term gene (Johannsen, 1908) was not coined by Mendel, it is now the acceptable term for the hereditary factor, and will be used in the remainder of the discussion.

Upon counting the number of plants of each type in the $\mathbf{F_2}$ generation, Mendel discovered an interesting pattern. Approximately three-fourths of the $\mathbf{F_2}$ plants produced axial flowers, and one-fourth of them produced terminal flowers (his actual ratio was 3.14:1). When he performed this series of crosses ($\mathbf{P_1} \times \mathbf{P_1}$, $\mathbf{F_1} \times \mathbf{F_1}$) for each of the other six pairs of trait differences, he found that one of the trait pairs was dominant, and expressed in every member of the $\mathbf{F_1}$ generation, and that the $\mathbf{F_2}$ groups always manifested the dominant characteristic in approximately a 3:1 ratio with that for the recessive character (see Fig. 9-2).

Mendel's data suggested that the factors operating in heredity had the following characteristics.

1. Each plant contains two genes for a given trait, because only a duplicity could explain the reappearance of the recessive trait in the $\mathbf{F_2}$ generation.
2. Each gene maintains its integrity, whether or not it is expressed. The dominant and recessive traits were the same in the $\mathbf{P_1}$ and $\mathbf{F_2}$ plants. The gene had not undergone a dilution or alteration by being associated with the gene for axial flowers (now referred to as its **allele**) in the $\mathbf{F_1}$ generation. Therefore, the genes were **particulate,** and were manifested as particulate units from generation to generation.
3. The characteristic expressed by the plant (now called its **phenotype**) was determined by the genes it contained (now called its **genotype**). A plant that phenotypically had axial flowers could have two genes for axial-flower production (**homozygous dominant**) or one gene for axial-flower production and one for terminal-flower production (**heterozygous**). A plant that phenotypically manifests terminal-flower production has two genes, both of which must be for terminal flowers (**homozygous recessive**).

This explanation is simplified by the use of genetic symbols. Let **A** represent the gene for axial flowers, and **a** represent the gene for terminal flowers.

$\mathbf{P_1}$ generation	**AA** (axial)	×	**aa** (terminal)
$\mathbf{F_1}$ generation		**Aa** (axial)	

The $\mathbf{P_1}$ cross then is between a homozygous dominant and a homozygous recessive plant. The $\mathbf{F_1}$ individuals resulting from the cross are all heterozygous, and express only the dominant trait.

The ratios of the traits in the $\mathbf{F_2}$ generation led Mendel to further conclusions. Such patterns could only be generated if there were:

1. An orderly separation of the two genes in an individual during the formation of the gametes.
2. A random process of fertilization (that is, the hereditary factors present in different female gametes have an equal chance of combining with hereditary factors in male gametes—the egg and sperm that form the zygote do so independently of the genes that they bear).

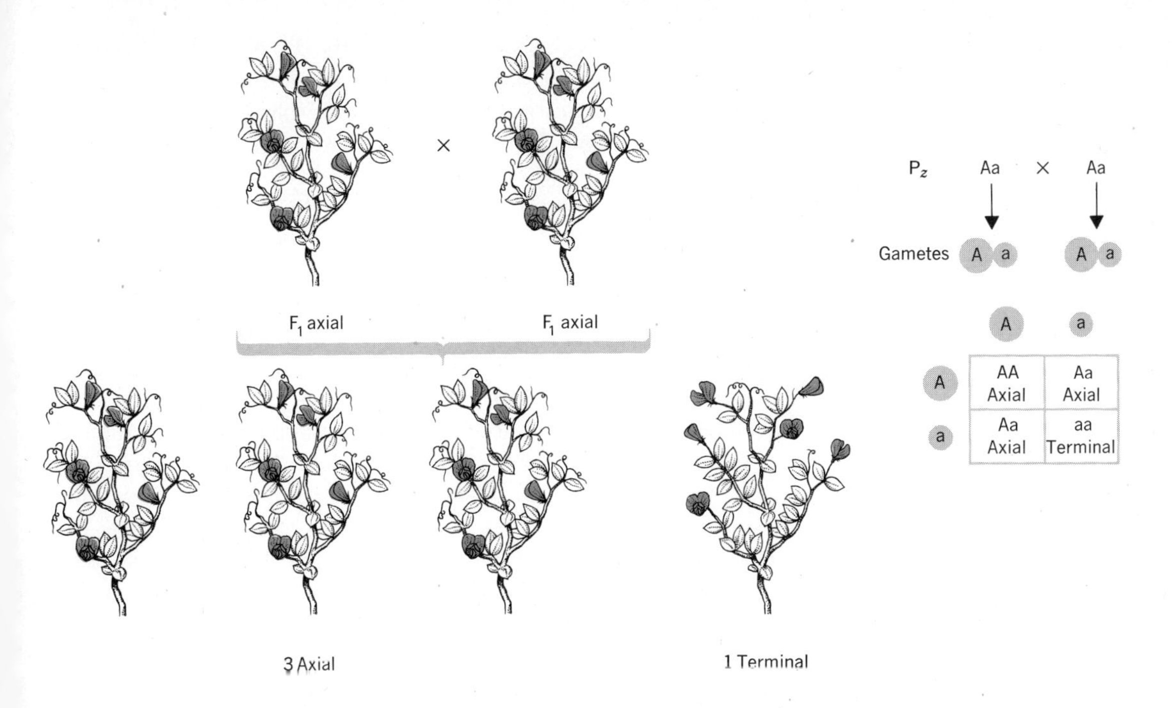

Fig. 9-2 Phenotypically axial-flowered plants resulting from the cross shown in Fig. 9-1 are crossed with one another.

The orderly separation of genes in the formation of the gametes had to occur in such a manner that each of the gametes contained one gene for the determination of flower position. Therefore, when the $\mathbf{F_1}$ generation produced gametes (either male or female), some of the resulting gametes would contain the **A** gene and the others would contain the **a** gene. This orderly separation of alleles is the first law of mendelian inheritance, the **law of segregation.**

Associated with the segregation of alleles is a random combination of the gametes. If this is true, then by tabulating all possible gametic combinations, an expected value could be obtained. The usual way in which this is diagrammed is with the Punnet square. All possible female gametes are noted across the top, and all possible male gametes are noted along the side. The possible offspring from the combination of the different gametes is then noted in boxes corresponding to the row and column of the Punnet square (see Fig. 9-2).

The expected phenotypes would be three axial-flowering plants to one terminal-flowering plant, which coincided with the result of the actual cross.

The hereditary scheme Mendel proposed permitted the prediction of certain results. If the $\mathbf{F_1}$ plants were heterozygous, then by crossing

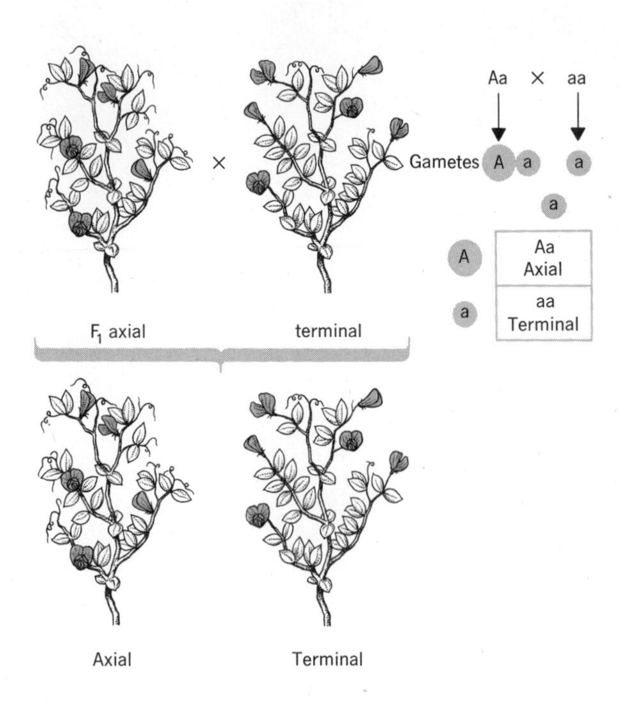

Fig. 9-3 A phenotypically axial-flowered plant resulting from the cross shown in Fig. 9-1 is crossed with a plant with terminal flowers.

them to homozygous recessive individuals (the $\mathbf{P_1}$ terminal-flowering plants), he should obtain a ratio for axial- : terminal-flowering plants of 1:1 (see Fig. 9-3). Because of the use of the plant like one of the members of the recessive $\mathbf{P_1}$ generation, this type of cross is called a **backcross.**

The results of the actual cross coincided with the expected values (1:1), thus confirming Mendel's hypothesis of an orderly segregation and a random combination of gametes in fertilization.

Once Mendel had established the pattern for the inheritance of a single trait, he turned his attention to the inheritance pattern found for the offspring of plants that differed by two pairs of traits. He crossed a plant that produced red axial flowers with one that produced white terminal flowers. The $\mathbf{F_1}$ generation was made up of plants that produced only red axial flowers. The $\mathbf{F_1} \times \mathbf{F_1}$ cross yielded an $\mathbf{F_2}$ generation in which the traits appeared in all combinations: red-axial, white-axial, red-terminal and white-terminal.

Once again the ratio in which the traits combined generated a pattern. There were approximately 9 red-axial : 3 red-terminal : 3 white-axial : 1 white-terminal. Mendel hypothecated that the only way in which this could have happened was if the two pairs of alleles, one for flower position and the other for flower color, were separate and hereditarily independent sets and that the segregation of one pair of alleles did not influence the segregation of the other pair. A short mathematical proof for this thesis is as follows: If the expected ratio for each pair of traits is 3:1, then the expected ratio for these traits combining independently would be

$$(3:1) \times (3:1) = 9:3:3:1$$

This genetic relationship can be represented as follows:

F_1 generation (heterozygous for both traits)		AaRr		
possible gametes	AR	Ar	aR	ar

Since gametes combine randomly during fertilization, any of four possible gametic types of eggs have an equal chance of uniting with any of the four gametic types of sperm at fertilization. The Punnet square that gives the values expected to result from such a cross is given in Fig. 9-4.

Once again, the expected values coincided with the results of the actual cross. On the basis of the independent inheritance of genes governing separate traits, Mendel predicted the outcome of a backcross (see Fig. 9-5). The results of the actual cross once again coincided with the expected ratio of 1:1:1:1, thus confirming Mendel's hypothesis. This hypothesis formed the second law of Mendelian inheritance, the **law of independent assortment.**

A summation of Mendel's contribution to the concept of the gene is as follows:

1. The gene is a particulate unit of inheritance and expression that maintains its integrity from generation to generation.
2. There are two genes present in each organism that govern the expression of a single trait.
3. The phenotype of an individual is based on the presence of dominant and/or recessive genes within that individual. The expression of a trait governed by a dominant gene may indicate a homozygous dominant or heterozygous genotype. The expression of a recessive trait indicates the presence of two recessive genes.
4. During the formation of gametes, alleles separate in such a way that each gamete receives one gene for the determination of each trait. This is the law of segregation.
5. Which male and female gametes unite during fertilization is totally random and is not determined by the genetic constitution of individual gametes.
6. Allelic genes governing different traits segregate independently from one another. This is the law of independent assortment.

P_1 AA RR × aa rr

F_1 Aa Rr (Axial-red flowers)

$F_1 \times F_1$ Aa Rr × Aa Rr

Gametes AR Ar aR ar

	AR	Ar	aR	ar
AR	AA RR Axial-red	AA Rr Axial-red	Aa RR Axial-red	Aa Rr Axial-red
Ar	AA Rr Axial-red	AA rr Axial-white	Aa Rr Axial-red	Aa rr Axial-white
aR	Aa RR Axial-red	Aa Rr Axial-red	aa RR Terminal-red	aa Rr Terminal-red
ar	Aa Rr Axial-red	Aa rr Axial-white	aa Rr Terminal-red	aa rr Terminal-white

Axial-red = 9
Axial-white = 3
Terminal-red = 3
Terminal-white = 1

Fig. 9-4 The dihybrid cross in which a plant with axial red flowers (AARR) is crossed with a plant with terminal white flowers (aarr). The offspring of this cross are then crossed with one another. The Punnett square shows the chance of recovering different offspring according to random fertilizations.

F_1 Aa Rr × aa rr

Gametes AR, Ar, aR, ar — ar

	ar	
AR	Aa Rr Axial-red	= 1
Ar	Aa rr Axial-white	= 1
aR	aa Rr Terminal-red	= 1
ar	aa rr Terminal-white	= 1

Ratio = 1:1:1:1

Fig. 9-5 Backcross of the F_1 from the cross diagrammed in Fig. 9-4 (AaRr) with a homozygous recessive (aarr).

A summary of genetic terms is found in Table 9-2.

TABLE 9-2 BASIC TERMS IN GENETICS

Dominance a gene or trait expressed in the heterozygous state (*Aa*); that which appears in the P_1, F_1, and F_2 generations.
Recessive a gene or trait expressed only in the homozygous state (*aa*); not expressed in the F_1 generation, but expressed in P_1 and F_2 generations.
Homozygous both genes or alleles alike (*AA*, *aa*); characteristic of a pure line.
Heterozygous both genes different (*Aa*); a pair of alleles.
Allele genes contrasting or different in expression that control the same trait (for blue and brown eyes); genes that segregate at meiosis; different genes on the same site of homologous chromosomes.
Homologous chromosomes (homologues) morphologically identical chromosomes; genetically similar chromosomes in a cell; chromosomes that pair and segregate at meiosis.
Phenotype the expression of the genetic elements through time (sex, hair color, skin color, pattern of balding, etc.).
Gene the unit factor controlling a trait; a site on a chromosome; a unit read to form an RNA polymer.
Genotype all the genes or genetic elements or hereditary factors, whether or not they contribute to the phenotype (*AaBBccDd* etc.).
Segregation the separation of alleles into different gametes; the "coming-apart" of different allelic genes into different cells during meiosis ($Aa \longrightarrow A + a$).
Independent assortment the way one pair of alleles segregates has no effect on how another pair of alleles segregates ($AaBb \longrightarrow AB + ab$ or $Ab + aB$).
Autosome a chromosome present in the cells of both sexes in equal numbers; they are not primarily involved in the determination of sex.

Mitosis, Meiosis, Mendel, and Morgan

In 1902, some 18 years after the death of Mendel, the elucidation of the cytological events of nuclear divisions brought about a resurgent interest in the work done by the Austrian monk. It was found that the mitotic cycle resulted in two cells that were identical in both number and types of chromosomes. Higher organisms, both plant and animal, were diploid. Could Mendel's hereditary factors, the genes, be on the chromosomes?

The meiotic process, which leads directly to the formation of the gametes of higher organisms, provided further implications that the pattern of Mendelian inheritance could be related to the passage of chromosomes from generation to generation. Chromosome pairs (homologous chromosomes) separated in an orderly fashion at the first division of meiosis. Each of the resulting cells received one chromosome of each type. If the genes were on the chromosomes, the meiotic division provided a means for allelic segregation of genes into different cells.

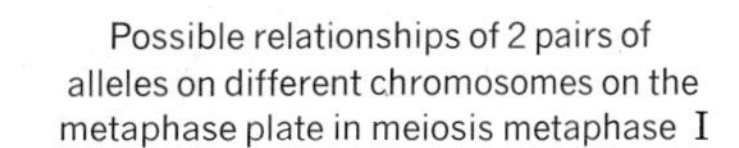

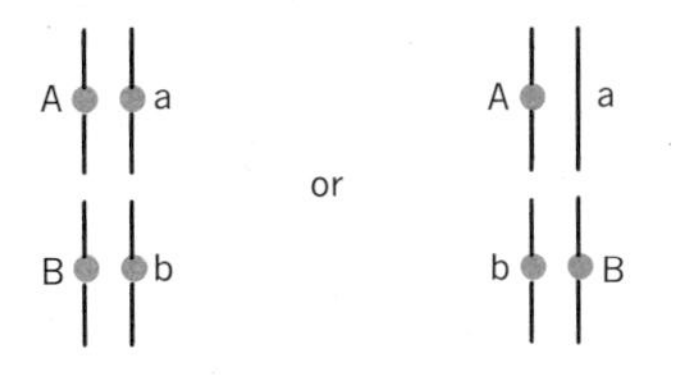

At anaphase I

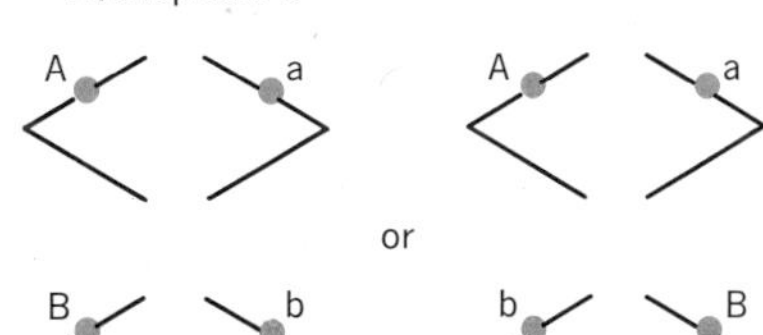

Fig. 9-6 There is an equal chance of *A* going to the same pole as *B* as there is of *A* going to the same pole as *b*. The separation of alleles in meiosis is genetic segregation. The equal chance of different alleles going to the same pole is termed independent assortment.

Also, the chromosomes behaved independently of one another in passing to the spindle poles. Therefore, if the genes for different characteristics (nonallelic genes) were on different chromosomes, it would seem that a given allele could go to the pole with either allele of the other pair of genes (see Fig. 9-6). This would provide a mechanism for Mendel's theory of independent assortment.

This association of Mendelian inheritance patterns and the events of the nuclear division processes was partly speculative on the part of Sutton and Montgomery. The first conclusive evidence that chromosomes contained genetic sites was given in the work of Thomas Hunt Morgan in 1912 (for which he was awarded the Nobel Prize in 1917).

Morgan was the first to use the common fruit fly (*Drosophila melanogaster*) for genetic research. The fruit fly can be handled easily in the laboratory. Virgin females can be identified and collected and then mated with the type of male selected by the investigator. The flies can be anesthetized with ether for the few minutes it takes to manipulate them into tubes or vials in appropriate mating pairs. The parents of a given mating can be transferred from the tube after a few days, and within two weeks, the fertile eggs will give rise to larvae that metamorphose into flies. Thus a large number of offspring from a single mating can be reared and evaluated in a relatively short time. (Plants, such as peas, take months for growth and maturation.)

Another advantage of *Drosophila* was that cytologists had determined their chromosome complement. They are diploid organisms, composed of four pairs (eight) of chromosomes. The female of the species has four distinct pairs. The male has three pairs of homologous chromosomes, and a fourth pair in which one of the chromosomes is like that found in the female, and the other is unusually small (see Fig. 9-7).

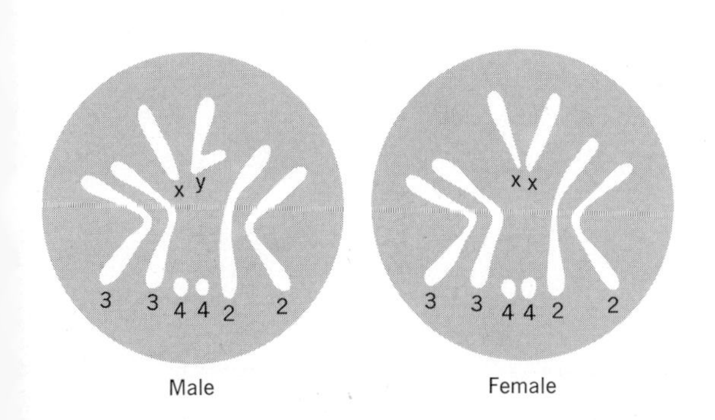

Fig. 9-7 *Drosophila* chromosomes. *Drosophila* has four pairs of chromosomes. The female contains two X chromosomes and the male contains an X and a Y.

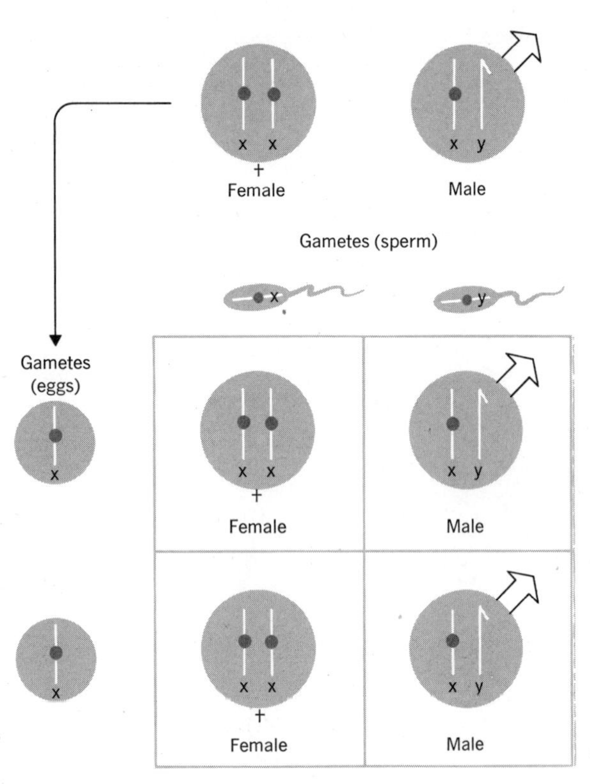

It had been proposed that the chromosomal differences between male and female insects, such as *Drosophila*, are responsible for the sex differences. This proposal led to the distinction between chromosomes as **sex chromosomes** *and* **autosomes.** The sex chromosomes of the female were called the X chromosomes, and those of the male were called the X and Y chromosomes. The sex of the individual fly was thought to be determined by the chromosome complement of the zygote from which it was formed (see Fig. 9-8).

Upon examining large numbers of *Drosophila*, Morgan came upon a male fly that had white, rather than the typical red, eyes. This was a product of a **spontaneous mutation** that had arisen in a natural population rather than one that had been induced (induced mutations were not produced until 1926). Morgan performed a series of standard Mendelian crosses with the white-eyed male. He found that when it was mated with a red-eyed female, all the offspring were red-eyed. This was an expected phenomenon if the white-eyed trait were recessive.

The $\mathbf{F_1} \times \mathbf{F_1}$ cross produced an $\mathbf{F_2}$ generation with a surprising feature. Although the ratio of red-eyed flies to white-eyed flies was in the expected 3:1 ratio, *the entire class of white-eyed flies was made up of males.*

Morgan hypothecated that this could only have come about if the gene that controlled the white-eyed characteristic was on the X chromosome, and that the Y chromosome in the male did not have a corresponding allele. The recessive trait for white eyes would then be expressed whenever the gene was present in the male (see Fig. 9-9).

Fig. 9-8 Sex determination in *Drosophila*. The female produces eggs that contain only X chromosomes. The male produces gametes, 50% of which contain an X and 50% of which contain a Y (the X and Y pair at meiosis). When an egg is fertilized by a sperm containing an X, the offspring will be female; when an egg is fertilized by a sperm containing a Y, the offspring will be male.

Fig. 9-9 The genotypes of flies containing the allele for white eyes. There are three possible genotypes for the female, only one of which would produce a white-eyed female, while there are only two possible genotypes for the male, one of which produces a white-eyed male.

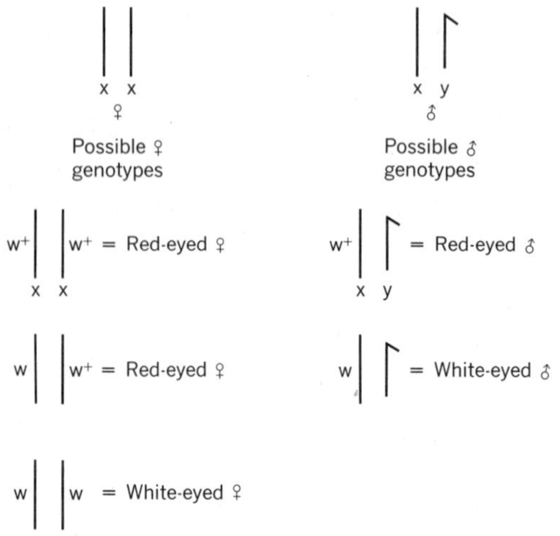

x x
♀
Possible ♀ genotypes

x y
♂
Possible ♂ genotypes

w^+ | w^+ = Red-eyed ♀ (x x)

w^+ | = Red-eyed ♂ (x y)

w | w^+ = Red-eyed ♀

w | = White-eyed ♂

w | w = White-eyed ♀

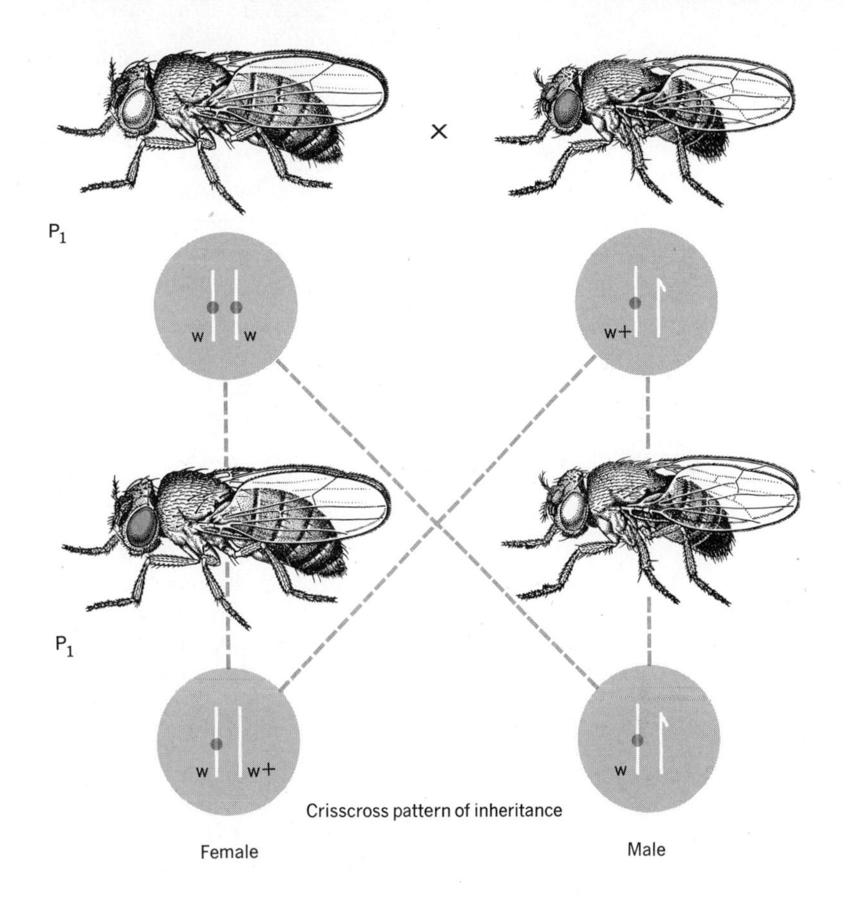

Fig. 9-10 A cross between a white-eyed female and a normal-eyed male.

If this were true, then if a white-eyed female (which would be homozygous recessive) were mated to a red-eyed male, resulting offspring would consist of red-eyed females and white-eyed males (see Fig. 9-10). But in order to perform this cross, Morgan had to first obtain a white-eyed female and initiate a white-eyed stock (in which both males and females had white eyes and which, upon mating, produced only white-eyed offspring). He obtained both white-eyed males and females from a series of random matings, established a stock, and proceeded to test his hypothesis.

In performing the actual cross between a white-eyed female and a red-eyed male, he found that the $\mathbf{F_1}$ flies approximated the expected results. This was the first time that a genetically controlled trait had actually been associated with the specific behavioral pattern of a chromosome, and provided conclusive evidence that the genes and the chromosomes were intimately related.

Subsequent investigations of genetic phenomena by Morgan and many of his students, as well as others who utilized other organisms for research, brought about the elucidation of both Mendelian and non-

Mendelian inheritance patterns. A brief categorization of these may be given as follows:

Mendelian: Genes on different chromosomes that manifest segregation and independent assortment.

Non-Mendelian: (1) Genes on the same chromosome that do not assort independently (linkage); (2) cytoplasmic inheritance that is characterized by the exhibition of an all-or-none expression (either a cell receives a mitochondrion or it does not).

As a result of much work, specific genic sites were localized on chromosomes and the genes of many organisms were mapped (see Fig. 9-11). The implication of this work, however, is more closely related to the mechanisms of evolution and will be discussed as such in Chapter 13.

Chromosomes, Genes, and DNA

What was this "intimate spatial relationship" of the genes on the chromosome? Did the genes reside on the chromosome as distinct particulate units or was the chromosome nothing more than a series of genes linked together? Regardless of which was the case, biologists fully anticipated that biochemical and biophysical analysis of the chromosomes would reveal particulate units that could be interpreted as genes.

When cells were submitted to biochemical and biophysical analysis, however, it was found that the deoxyribonucleic acid (DNA) and protein of which the chromosome was composed exhibited complete uniformity of structure. Although this finding led to the conclusion that the chromosomes were a series of linked genes and not a structure on which the genes resided, the problem arose of reconciling the uniformity of the genetic material with the particulate expression of genes. The resolution of this problem is only now becoming clear, and will be discussed later.

The identification of DNA and protein in chromosomes brought about a great controversy in biological circles. Was the DNA or the protein the genetic material? The first experimental evidence that implicated DNA as the genetic material was supplied by a series of experiments performed by Avery, McCleod, and McCarthy in 1941. Their experiments were based on earlier experiments by Griffith in 1928.

Griffith worked with virulent (disease-causing) and nonvirulent (non-disease-causing) strains of bacteria of the genus *Diplococcus*. The generic name diplococcus is derived from their structural characteristic of occurring in spheric pairs (diplo = two; coccus = sphere) (see Fig. 9-12). When Griffith killed the virulent cells with heat and injected them into live rats, the rats continued to live. When he injected the nonvirulent bacteria into live rats, the rats retained their healthy condition. When he combined the heat-killed virulent and the live nonvirulent cells and injected them simultaneously into the rats, the rats became sick and

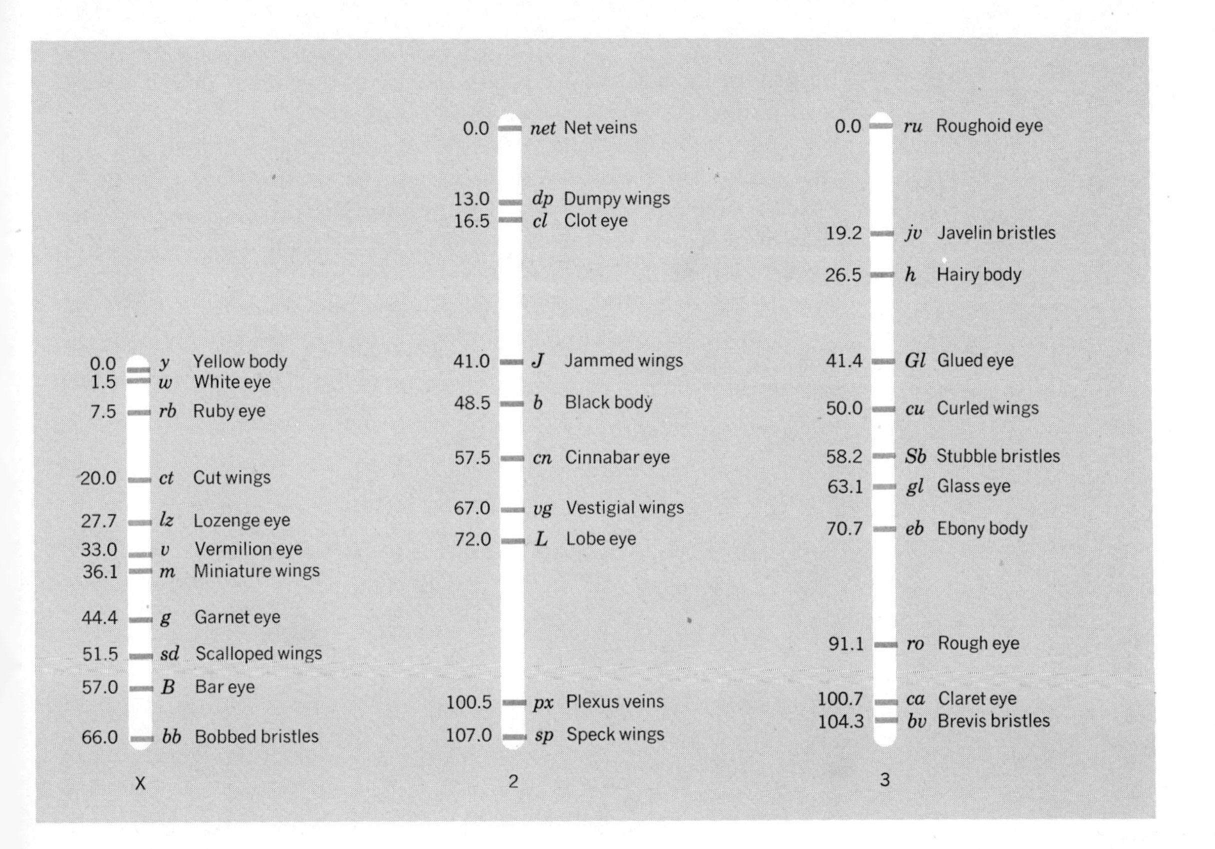

Fig. 9-11 The position of genes controlling a variety of traits have been mapped by crossing-over analysis, a subject that will be discussed in Chapter 14. Neither the Y nor the fourth chromosome is shown.

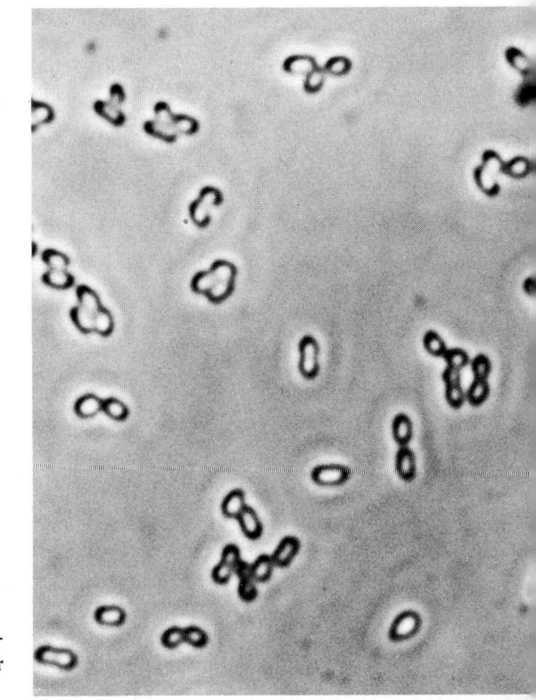

Fig. 9-12 *Diplococcus* (4500×), spherical bacteria most frequently found in pairs. In older cultures, groupings of four or more bacterial cells appear. [Walter Dawn]

died (see Fig. 9-13). Griffith concluded that some chemical entity in the virulent strain had **transformed** the nonvirulent strain to a virulent state, and termed the phenomenon **transformation.**

Avery, McCleod, and McCarthy recognized that the chemical entity responsible for the transformation must carry the genetic factor for virulence, so that the process of transformation was really the introduction of the gene or genes for virulence into the nonvirulent strain. This recognition permitted them to carry the experiment a step further, toward the determination of the genetic material itself.

Among their laboratory stocks of bacteria were virulent and nonvirulent pneumococcal strains. (The virulent strain of *Pneumococcus*

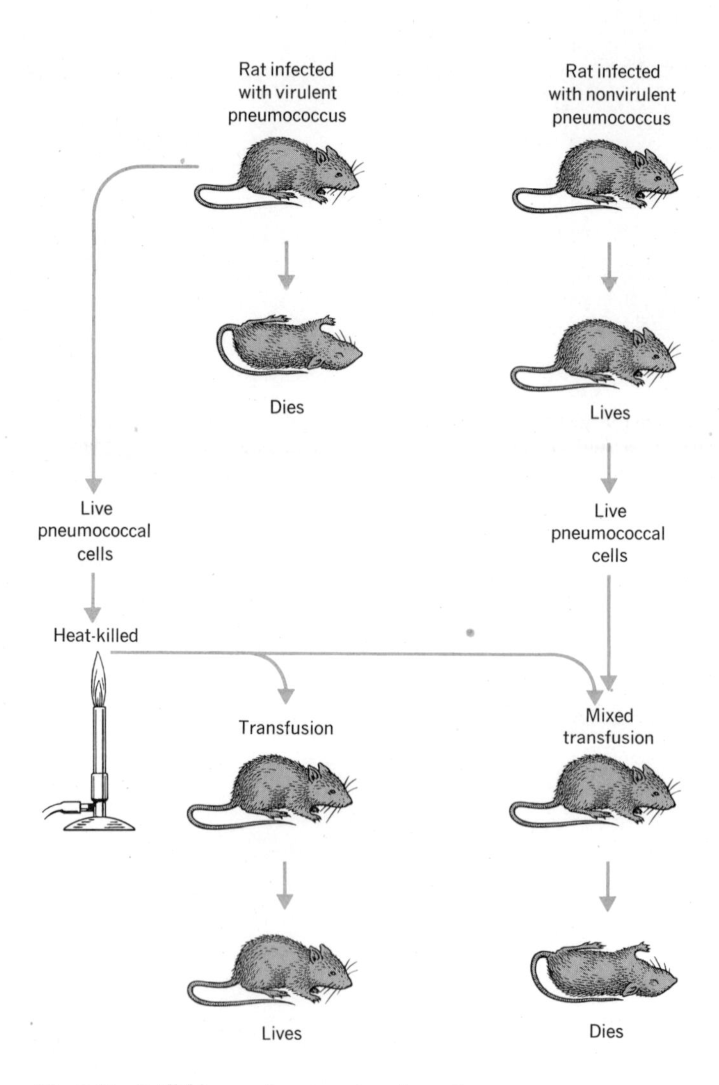

Fig. 9-13 Griffith's experiment on transformation.

causes bacterial pneumonia.) In addition to their differences in virulence (a characteristic manifested only when the strains are injected into rats), these two strains manifested colony differences *in vitro*. When a single bacterium is placed on an agar surface, it will undergo repeated divisions. Since most bacteria are sessile, the products of their divisions remain localized on the agar surface. After many divisions, the bacterial growth becomes visible to the naked eye. This is called a **colony,** and contains in the neighborhood of 10^7 bacteria, all of which are derived from the single bacterium originally placed on the surface. Different types of bacteria can be distinguished by the type of colony they produce (large or small, raised or flat). The virulent pneumococcal colony is characterized by a rough surface, while the nonvirulent strain produces a smooth colony. The colony characteristic is directly related to the strain's virulence, so that the two traits cannot be separated.

When the virulent (rough-colony) bacteria were killed by heat and injected into a rat simultaneously with the live, nonvirulent (smooth-colony) bacteria, the rats died. The bacteria were isolated from the dead rats and grown once again *in vitro*. The only strain that could be isolated was that which produced the rough colony.

The experimenters next made an extract of the virulent strain, thus destroying its structural integrity completely. The extract was separated into protein, DNA, and carbohydrate fractions. Each of the fractions was mixed with the live nonvirulent strain *in a flask*. The culture was left undisturbed for a short time, and then plated (the bacteria spread on an agar surface). The plates revealed only smooth colonies for the nonvirulent cultures when the virulent protein or carbohydrate extracts were used, but the plate with the nonvirulent culture and the virulent DNA contained both smooth and rough colonies! (see Fig. 9-14.)

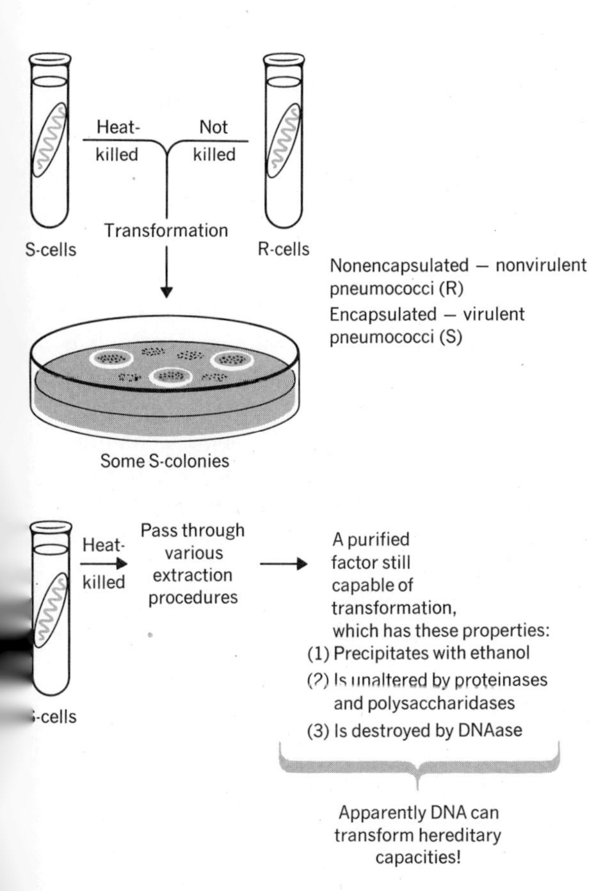

Fig. 9-14 Avery, McCleod, and McCarthy's experiment to demonstrate that the transforming principle is DNA.

Although the transformation experiments demonstrated that DNA is the genetic material, the mechanism of transformation was an atypical means of heredity, and the issue remained in doubt. Conclusive evidence that DNA and only DNA is the genetic material has since been obtained in a number of ways, but perhaps most convincingly by an experiment with viruses.

A virus is the smallest and simplest living form. It is made up of only nucleic acid (either DNA or RNA, depending on the individual type of virus) and protein. Because of the paucity of molecular types, viruses exhibit no metabolism and are totally parasitic. They depend on the host cell to provide the nucleotides for their nucleic-acid structure, and on the amino acids for their protein structure.

There are viruses that attack almost every type of cell. Those that use bacterial cells as hosts are called bacteriophages (phage = eat). Each type of bacterium appears to have a phage or set of phages that is capable of penetrating the cell and taking over the metabolic processes. For the favorite bacterium of researchers, *E. coli*, there is one group of seven phages known as the T series (T_1, T_2, etc.).

By the mid-1950s, the infection process by which the phage attached to and entered the cell was well known. Electron micrographs had revealed that the inner portion of the phage was "injected" into the bacterial cell, and the "shell" remained outside (see Fig. 9-15). Once the infection was complete, the infected bacteria could be placed in a Waring blender. The agitation of the blender was sufficient to detach the shells from the bacterial cell. Subsequent centrifugation effected a separation of the infected bacteria (which formed the pellet) and the viral shells (which remained in the supernatant). Chemical analysis of the shells revealed that they were made up solely of protein.

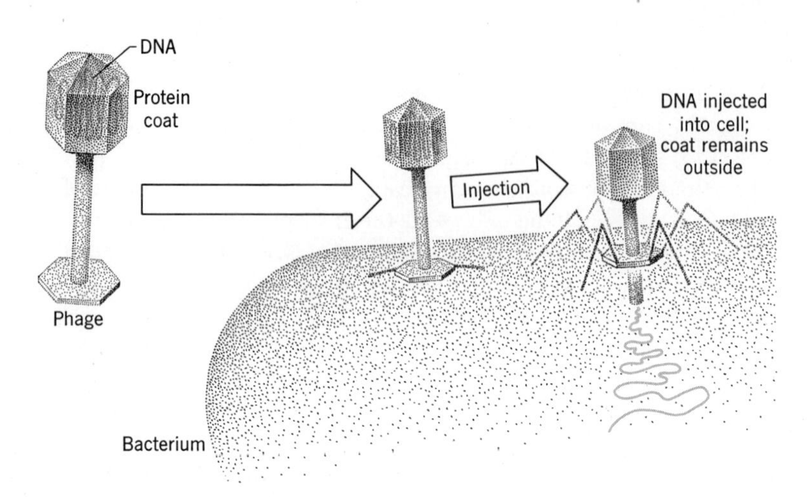

Fig. 9-15 Infection of a bacterium by T_2 bacteriophage.

The obvious question was, did some of the protein enter the bacterium in addition to the DNA? Hershey and Chase grew bacteria in a medium containing S^{35}, radioactive sulfur. The bacteria incorporated the "hot" sulfur into their protein structure in the sulfur-containing amino acids (methionine, cysteine). The "hot" bacteria were then infected with phages. Since the phages utilized the protein of the bacterium for the formation of phage protein, the phage protein became labeled with radioactive sulfur.

The "hot" phages were then used to infect "cold" bacteria. After the infection had taken place, the bacterial culture was placed in a Waring blender and the viral protein shells were detached from the bacterial cells by agitation. The pellet and the supernatant of the centrifugation that followed were examined for radioactive material. If protein material had entered the bacterial cell in the course of the infection process, Hershey and Chase expected to find radioactive material in both the supernatant and the pellet. If, however, only DNA entered the bacterial cell, only the supernatant would manifest radioactivity. The latter proved to be the case, thus indicating that no protein entered the bacterial cell in the course of the infection process (see Fig. 9-16).

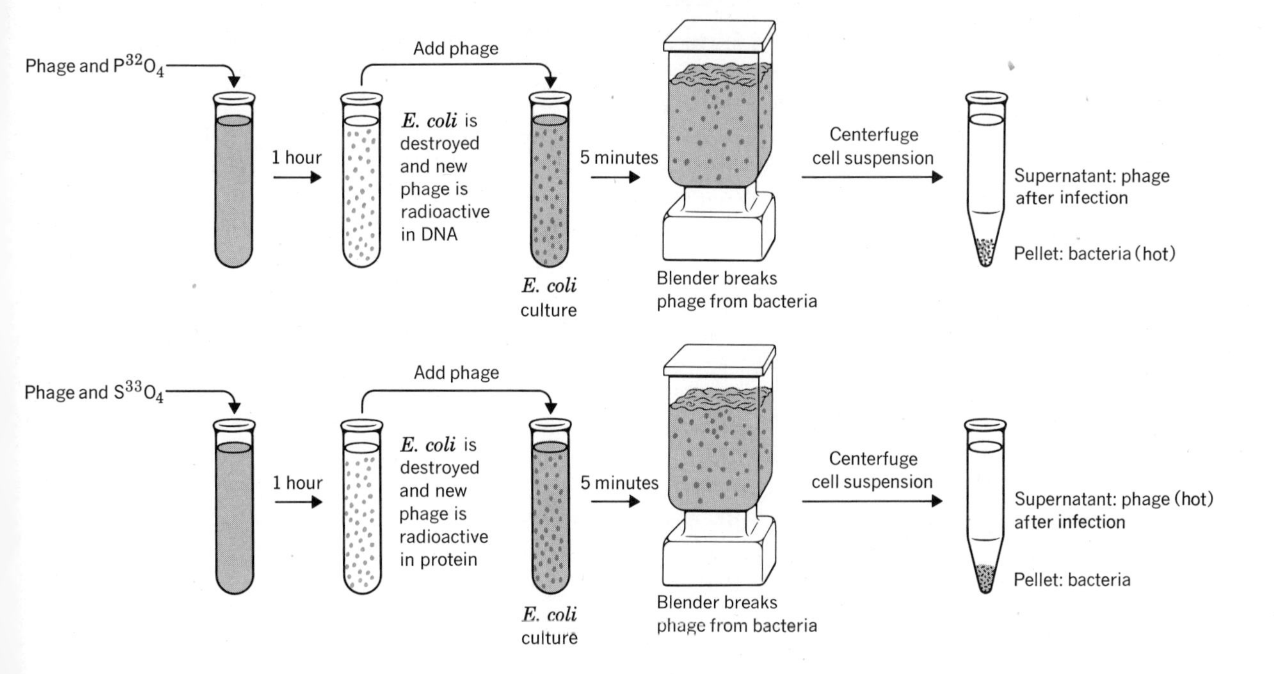

Fig. 9-16 The Hershey-Chase experiment.

The two scientists repeated the experiment, growing the bacteria on P^{32}, radioactive phosphorus, which is incorporated primarily into nucleotide structure, but not protein. Infecting these bacteria with phages resulted in "hot" phages in which only the DNA was labeled. Subsequent use of these phages for infection of "cold" bacteria, and the repetition of their procedures for shearing off the shells, centrifuging the culture, and testing the pellet and supernatant for radioactive content, revealed that only the infected bacteria in the pellet contained the radioactive phosphorus. Therefore, the experimenters concluded that all the DNA, and only the DNA, entered the bacterium and brought about the production of new phages. DNA was the genetic material.

GENE ACTION

The patterns by which the chromosomes are transmitted from generation to generation and the overwhelming evidence that the DNA within the chromosomes was the genetic material presented biologists with a mechanism by which the capacities of organisms are perpetuated. But what is the mechanism by which these traits are expressed in individuals? How does the DNA function within an individual organism to bring about the molecular structures that are peculiar to that organism alone?

The answer to this question is sought by the group of biologists, biochemists, biophysicists, and geneticists whose primary research interest is gene action. From the experimental evidence that the genetic material controls protein synthesis, a relatively complete picture of gene action has developed.

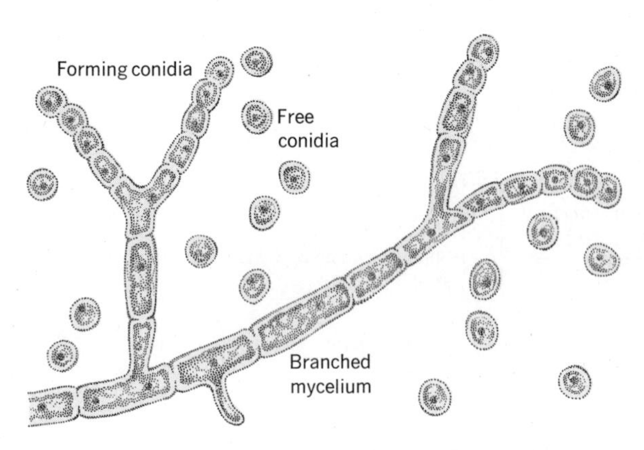

Fig. 9-17 ***Neurospora*** **forming conidia. Conidia are formed directly from the terminus of the hyphal branches, and are haploid spores from which a new mycelium will develop.**

Protein Synthesis and the Gene

In the late 1920s, George Beadle attempted to perform both genetic and biochemical analyses of the eye color in *Drosophila.* Although he found that the brick-red eye color was in reality a combination of a bright-red pigment and a brown pigment controlled by different genes, and although he managed to isolate these pigments, he became dissatisfied with *Drosophila* as a research tool for this type of work. The successful determination of the pigments responsible for the eye color in *Drosophila* had substantiated his feeling that the basis for gene action was on a chemical level.

Beadle and Tatum teamed up to study the problem of gene action by using the mold *Neurospora.* (*Neurospora* appeared to be a promising system because it can be grown on a defined medium containing inorganic salts, a carbon source such as sucrose, and the vitamin biotin.) If they were correct in believing that genetic control was mediated through enzyme synthesis, it should be possible to induce a mutation so that an enzyme would not be produced. The absence of such an enzyme would be expressed by the *Neurospora* requiring organic compounds that it was heretofore able to synthesize itself.

Beadle and Tatum irradiated the **conidia** (haploid spores) of *Neurospora* (see Fig. 9-17). Because each spore is haploid and grows into a haploid plant, any genetic change induced by high-energy radiation would be directly expressed. The spores were spread on an agar surface, where they were permitted to germinate and grow. The agar contained a *complete medium,* having many organic compounds unnecessary for the growth of the normal strain, in addition to those that were. When each plant grew into a small mass of hyphal threads, a portion was transferred to a defined medium, which contained only the nutritional requirements for the normal strain. Those plants that did not grow on the defined medium were suspected of having undergone a nutritional change to a condition in which more organic material was needed. These plants were then transferred to media containing the basic nutritional requirements *plus* one additional organic compound. One agar plate contained defined medium plus one vitamin; another contained defined medium plus one amino acid, etc. If the plant grew, the experimenters would know that they had succeeded in supplying it with the one organic compound that it had lost the ability to synthesize.

After testing thousands of plants reared from irradiated spores, Beadle and Tatum found one plant that grew only when the vitamin para-aminobenzoic acid (PABA) was present with the defined medium. When they crossed the PABA$^-$ (PABA minus) strain to the normal (PABA$^+$), both PABA$^-$ and normal plants were produced. This indicated that the trait followed the pattern for chromosomal inheritance. Two other plants were later found that also manifested vitamin requirements, and the controlling mechanism for these were also found to be chromosomally located.

The additional nutritional requirements of vitamins indicated that these strains had, indeed, lost the ability to synthesize the enzymes for the production of these vitamins. Beadle and Tatum concluded that genes

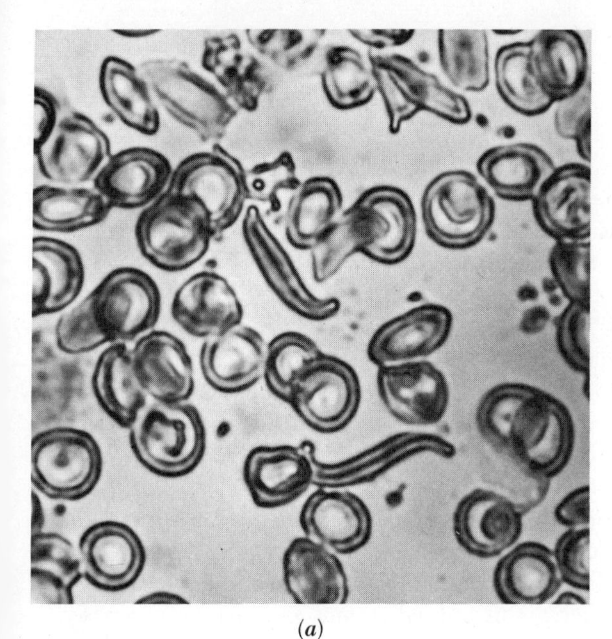

(*a*)

Position	Hemoglobin	
	Normal	Seckle-celled
1	Valine	Valine
2	Histidine	Histidine
3	Leucine	Leucine
4	Threonine	Threonine
5	Proline	Proline
6	Glutamic acid	Valine
⋮	⋮	⋮
150	Histidine	Histidine

(*b*)

Fig. 9-18 Sickle-cell anemia. (*a*) Red blood cells from a heterozygous individual showing normal red cells mixed with sickled red cells (1400×). The sickle-celled condition is due to a genetic alteration that results in a change in the primary structure of the protein. (*b*) Difference in amino acid sequences between normal hemoglobin (Hb-A) and sickle-celled hemoglobin (Hb-S). [(*a*), Walter Dawn-National Audubon Society]

on the chromosomes control the synthesis of enzymes, and further postulated that each of the genes governs the synthesis of one enzyme. This dictum has become known as the **one-gene-one enzyme hypothesis.**

Subsequent work indicated that the idea of one gene-one enzyme is somewhat of an oversimplification of the translation of genetic information into cellular function, and the relationship is more accurately one gene-one polypeptide. Yet Beadle and Tatum did provide the first real demonstration of gene action at a molecular level.

The ultimate proof that genes control the incorporation of specific amino acids into proteins was given by Ingram in 1958. There exists in Negro populations a type of anemia characterized by red blood cells shaped like the blade of a sickle. Because of the appearance of these red blood cells, the disease is called **sickle-cell anemia** (see Fig. 9-18a). The condition is inherited as a Mendelian blending of two traits. The homozygous recessive condition (**ss**) is fatal. The heterozygous (**Ss**) condition manifests the characteristic sickle-shaped red blood cells. The condition is both functional and structural, affecting the ability of the hemoglobin to carry oxygen as well as the shape of the red blood cell.

Ingram analyzed the hemoglobin protein in both sickle-celled and normal individuals. Normal hemoglobin is made up of four polypeptides, two α-chains and two β-chains. Ingram found that the α-chains for both the normal and the sickle-cell hemoglobins were identical. The β-chains however, differed by a single amino acid. At position 7 in one of the peptide fragments, the valine found in normal hemoglobin was replaced, in sickle-cell hemoglobin, by a glutamic acid (see Fig. 9-18b). This change in a single amino acid was of sufficient importance to produce the sickle-celled anemic condition.

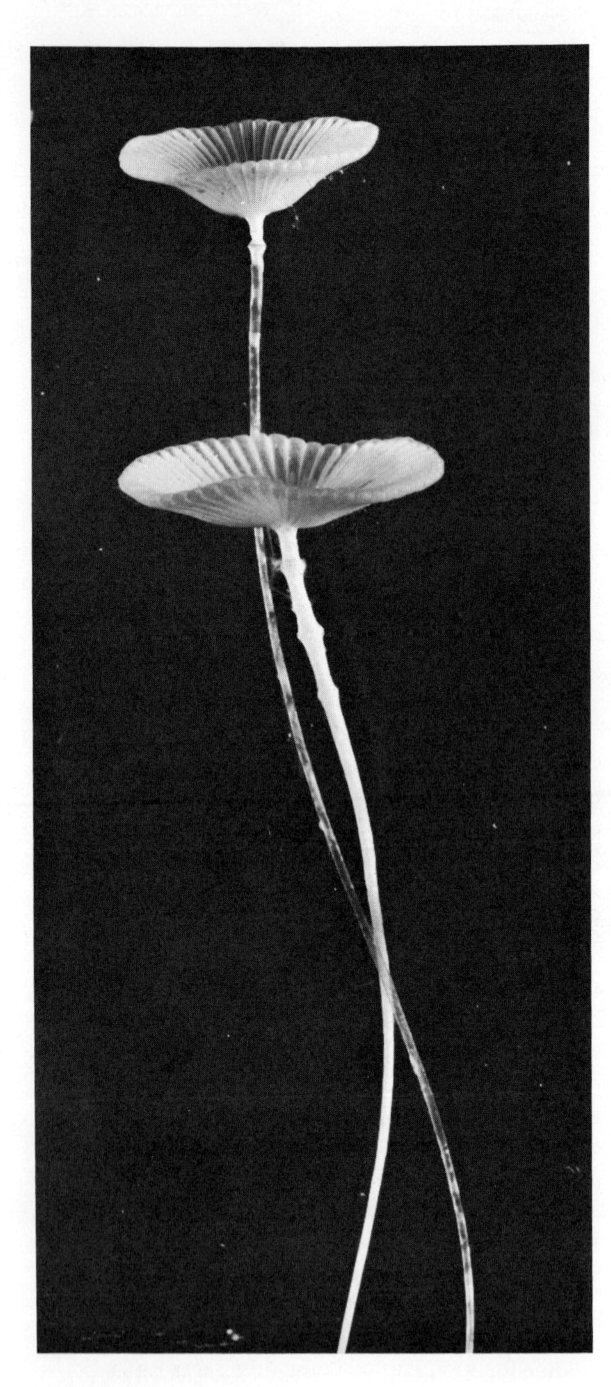

Fig. 9-19 *Acetabularia* (3×). [Lester V. Bergman]

Ingram's work was received with great enthusiasm in biological circles, for it demonstrated not only the genetic control of protein synthesis, but also the extreme specificity of proteins that was required for efficient function of the individual.

The Elusive Intermediate

The discovery that DNA provides the cell with the information for the sequence of amino acids in proteins was a major step toward the understanding of the genetic control of the cell. But the problem remained as to how the DNA information is translated into protein structure.

The first indication that there was an intermediate informational substance through which this translation was effected was supplied by a series of experiments by Hammerling on the development of the alga *Acetabularia*. *Acetabularia* is an unusual organism in that it is composed of only one cell, about an inch and a half long, with the appearance of an umbrella. At the base of the cell is a group of rhizoids (filaments) that serve to anchor the plant to the rocks at the bottom of shallow marine waters in the tropic and near-tropic regions. The remainder of the cell is composed of a stem topped by an umbrellalike cap (see Fig. 9-19). The rhizoids and stem of different species are very similar, but the species are (taxonomically) separated by the appearance of umbrellalike caps. The nucleus of the *Acetabularia* cell resides in the basal rhizoidal region.

Two of the species Hammerling worked with were *A. crenulata* and *A. mediterrana*. Members of the *A. crenulata* species have a lobular cap, while those of the *A. mediterrana* have a discoid cap (see Fig. 9-20). Hammerling found that when he cut off the caps of the plants, new caps regenerated that were characteristic of the species. *A. crenulata* individuals always regenerated lobular caps, and *A. mediterrana* individuals always regenerated discoid caps. Since the cap shape was consistent, and therefore genetically controlled, Hammerling realized that the nucleus in the rhizoidal region was influencing the cap formation some $1\frac{1}{2}$ inches away, up the stem. But what was the nature of this influence?

Hammerling next performed a series of grafts between two plants of different species. He cut off the top portion of the stem of both

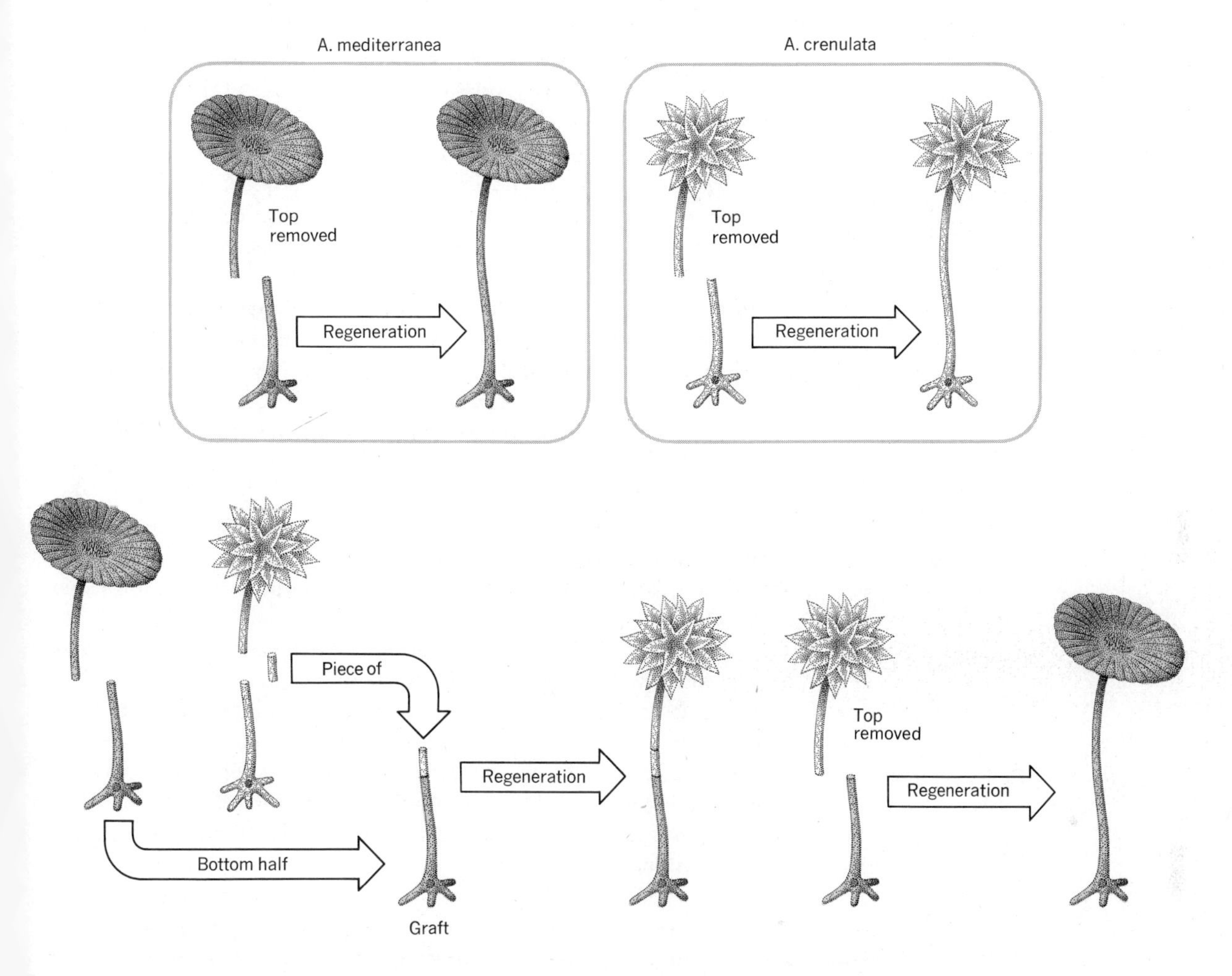

Fig. 9-20 ***Acetabularia mediterranea*** **and** ***Acetabularia crenulata,*** **and experiments of Hammerling on the control of cap shape.**

individuals when they were about to form their caps, and grafted portions of the stem. One plant had an *A. crenulata* base, and a stem segment from *A. mediterrana;* the other had an *A. mediterrana* base and a stem region of *A. crenulata.* When cap formation followed, *both plants formed caps characteristic of the terminal stem portion rather than the type associated with the base.* When he cut off these caps, and permitted the cells to form new ones, subsequent caps appeared *intermediate in type between the two species.* When he once again cut off the caps and permitted the cells to form new ones, these were of the *species type of the base and exhibited no influence from the grafted stem.* In the numerous other grafts he performed, he always found that regardless of the intermediate types, the cell eventually produced a cap that was characteristic of the species of the base.

Hammerling concluded from this series of experiments that there was material of a chemical nature, produced in the nucleus, that moved up the stalk and controlled the differentiation of the cap. When he had performed the graft just prior to cap formation, this material was concentrated in the upper portion of the stem. Even though this stem was grafted to a new base, the material already dictated that the cap would be like that of the previous base, and so a cap foreign to the present base was formed. When this cap was removed, the nucleus produced more differentiating material, which then moved up the stem. Now there was both the old differentiating material (from the previous base) and new differentiating material in the terminal portion of the stem, and an intermediate cap was formed. When this cap was removed (and the nucleus once again produced the differentiating material that moved up the stem), the old material was either diluted out or broken down by spontaneous action, so that only the effect of the new material was noted in the cap formation. It was, in fact, this dilution or spontaneous breakdown of the old differentiating material that led Hammerling to postulate that it was of chemical nature, and to term it a **morphogenetic substance.**

Neither Hammerling nor subsequent workers have been successful in their attempts to isolate and identify the morphogenetic substance. This work on *Acetabularia* did, however, provide an indication that there is an elusive intermediary substance involved in the translation of genetic material (in the nucleus) to expression in cellular form.

DNA, RNA and Protein—The Central Dogma

Biochemical and biophysical analysis of cellular material had revealed that DNA was to be found in the nucleus (except for the minute amounts found in mitochondria, chloroplasts, etc.), and protein and RNA occur throughout the cell. The role of proteins as structural and enzymatic molecules provided the logical reason why they should be found in both the nucleus and the cytoplasm. RNA, on the other hand, was difficult to study and to assign a function, and so its omnipresence was inexplicable.

Chemical analyses carried out on RNA did not clarify the function of this substance in the cellular mechanism. RNA was revealed to be made up of four nitrogen bases, adenine, guanine, cytosine and uracil. Hence it differed from the chemical composition of DNA by the presence of ribose instead of deoxyribose and one nitrogen base; DNA contains thymine instead of uracil. Biochemists capitalize on this nitrogen base difference in labeling experiments by using C^{14}-labeled uracil when they desire to label only the RNA, C^{14}-labeled thymine when they desire to label only the DNA, and P_{32} when they wish to label both.

During the 1950's, research on RNA served to complicate rather than clarify matters. A major forward step toward understanding RNA was made when it was realized that there were many different kinds. When RNA is isolated from cellular material and centrifuged in a density

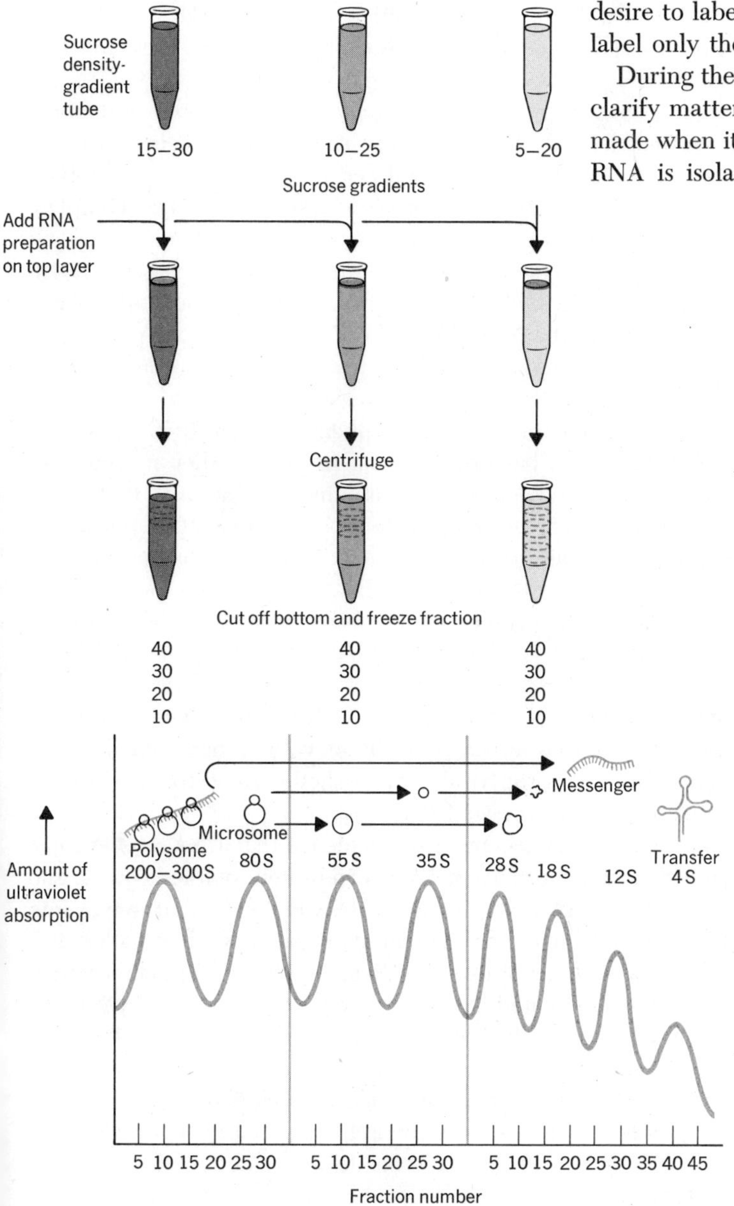

Fig. 9-21 Diagram showing method of obtaining different types of RNA and the typical graph produced by plotting fraction against UV-light absorption.

gradient tube, different RNAs can be ascertained at different layers according to the UV light they absorb (see Fig. 9-21). The different RNAs are identified by their Svedberg units, and electron microscopy and biochemical analyses led to the assignment of certain RNAs to certain structural entities. The 100-1000S RNAs are identified as polysomes; the 80S RNAs are identified as the snowman-like monosomes; and the 55S, 35S, and 5S are the fragments of the monosomes (see p. 196). The 28S and 18S RNAs are fragments of the 55S particle. But what are the 12S, 4S, and the so called polydisperse RNAs, and what are their function in the cell?

T-2 phage is a virus that infects only *E. coli.* Hershey and Chase had shown that the T-2 phage, which is made up of protein and DNA, injects only the DNA into the E. coli cell upon infection. After about twenty minutes, the bacterium bursts, releasing about 100 complete phages, with both a DNA core and a protein shell. Biochemical analysis of phage infection indicated that shortly after injection of the viral DNA, and prior to any detectable synthesis of the protein shells, there was a marked increase in the production of RNA. Was the RNA production a response of the bacteria to infection, or was it the production of a morphogenetic substance dictated by the virus?

Astrachan and Volkin analyzed this phenomenon in 1956 to discover whether the RNA was bacterial or viral in origin. They placed both *E. coli* and T-2 phage into a medium containing P_{32}, reasoning that newly synthesized RNA would draw from the pool of radioactive phosphate and the new RNA would be labeled. Phage infection was permitted to occur for ten minutes, after which the RNA was isolated, purified, and hydrolyzed into the nitrogen bases. The base ratios of the labeled RNA were compared to the known base ratios of bacterial and viral DNA (see Fig. 9-22). The base ratio comparison showed a close correlation between the labeled RNA and the viral DNA. The RNA synthesized before protein synthesis appeared to be of viral rather than bacterial origin, and synthesis of the viral protein shell seemed to be mediated through this rapidly synthesized RNA.

In the light of the accumulating evidence that RNA was in some way an intermediary substance necessary in protein synthesis, Jacob and Monod postulated that DNA makes an RNA molecule that moves into the cytoplasm and codes for protein synthesis, and they called this **messenger** RNA (*m*RNA). Subsequent work with the Astrachan-Volkin and other systems indicated that the rapidly synthesized RNA was the 12S RNA, and that 12S RNA is the messenger that provides the code for protein synthesis.

A somewhat related picture was being developed for the 4S RNA by several workers. This RNA was found to be of relatively low molecular weight (25,000 to 30,000), and made up of about 85 nucleotide residues. There appeared to be about sixty different kinds of 4S RNA, each of which had the same terminal bases: a guanine at one end and two cytosines and a terminal adenine at the other (see Fig. 9-23).

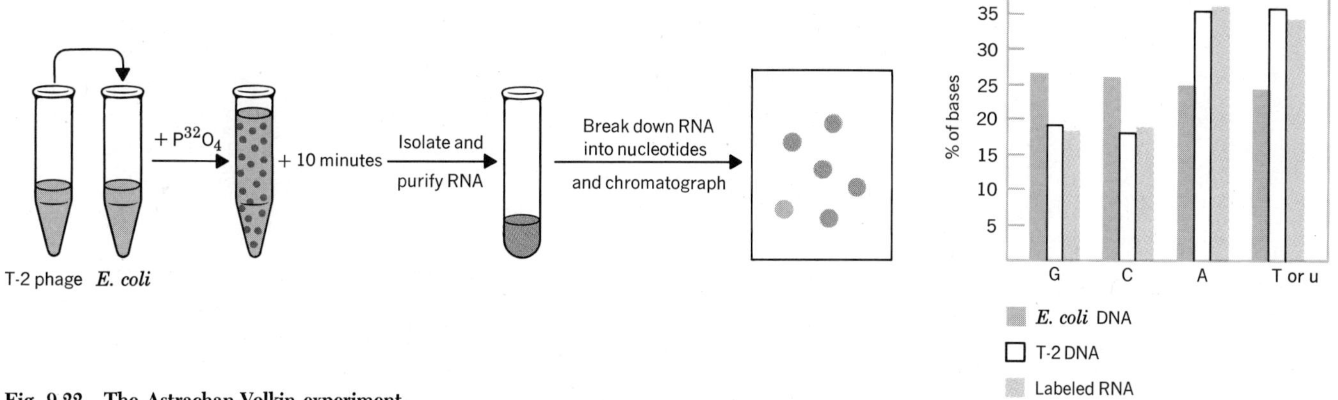

Fig. 9-22 The Astrachan-Volkin experiment.

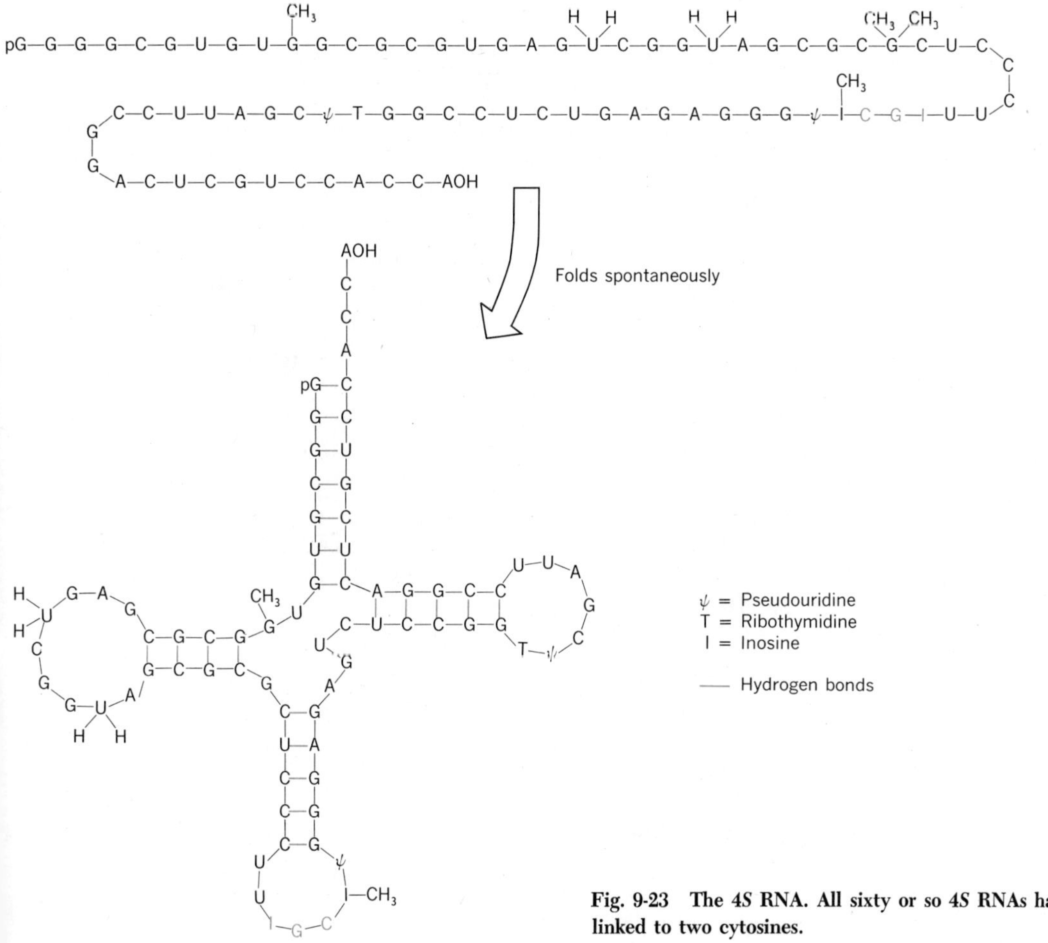

pG—G—G—G—C—G—U—G—U—G—G—C—G—C—G—U—G—A—G—U—C—G—G—U—A—G—C—G—C—G—C—U—C—C—C—U—U—I—G—C—I—ψ—G—G—G—A—G—A—G—U—C—U—C—C—G—G—T—ψ—C—G—A—U—U—C—C—G—G—A—C—U—C—G—U—C—C—A—C—C—AOH

Folds spontaneously

ψ = Pseudouridine
T = Ribothymidine
I = Inosine
— Hydrogen bonds

Fig. 9-23 The 4*S* RNA. All sixty or so 4*S* RNAs have the same terminus, an adenine linked to two cytosines.

Each type of 4S RNA is capable of binding to a specific amino acid. It is thought that the different base sequences in the middle portion of the 4S RNA molecules contain the information for which amino acid is bonded. Once bound, the amino acid-RNA complex proceeds (by random movement and short-range attractions) to specific sites on the messenger RNA, associated with the polysome. As the amino acid-RNA complex molecules take specific positions on the messenger, the peptide bonds are formed, and the complex then dissociates. Thus protein is synthesized.

Because of the role of the 4S RNA in picking up and transferring amino acids to the messenger, this RNA fraction is called **transfer** RNA (*t*RNA). Before an amino acid is bonded to the *t*RNA, it becomes activated. This activation is accomplished by a reaction between the amino acid, GTP, and the amino acid-specific activating enzyme.

GTP + amino acid + activating enzyme ⟶
amino acid-GMP-enzyme complex + P-P

When the bond between the amino acid and the specific *t*RNA is formed, the enzyme is not released, but the GMP goes into the surrounding medium.

amino acid-GMP-enzyme complex + specific *t*RNA + GTP + ATP
⟶ amino acid-*t*RNA-enzyme + AMP + P-P + GDP + Pi

Therefore, the actual bonding of the amino acids together to form a protein does not require an energy transfer. The activation and incorporation of each amino acid involves the loss of three phosphates, two from GTP and one from ATP. A protein that has some 100 amino-acid residues has cost the cell 300 high-energy phosphates to synthesize.

At one time it was thought that *t*RNA was formed in the cytoplasm. Geneticists however, have found several chromosomal sites for the production of specific *t*RNAs. Therefore, both the *m*RNA and the *t*RNA are formed by nuclear DNA.

The general picture of gene action that has emerged is more elaborate than that postulated by Monod. The DNA in the nucleus transcribes information to a messenger RNA (**transcription**). The messenger RNA passes into the cytoplasm and associates with many monosomes to produce one polysome. The transfer RNAs, each in complex with their specific amino acid, are attached to specific coding sites on the *m*RNA. Once in place, the bonds between the amino acids and the *t*RNA break, and the peptide bonds between the amino acids are formed (**translation**). The protein dissociates from the *m*RNA as it forms, and moves away to function in the cell (see Fig. 9-24). This total picture of the transcription of genetic information into RNA and the translation of RNA into protein synthesis is called the "**central dogma.**"

It should be noted, however, that no functional role has yet been attributed to the polydispersed fractions of RNA.

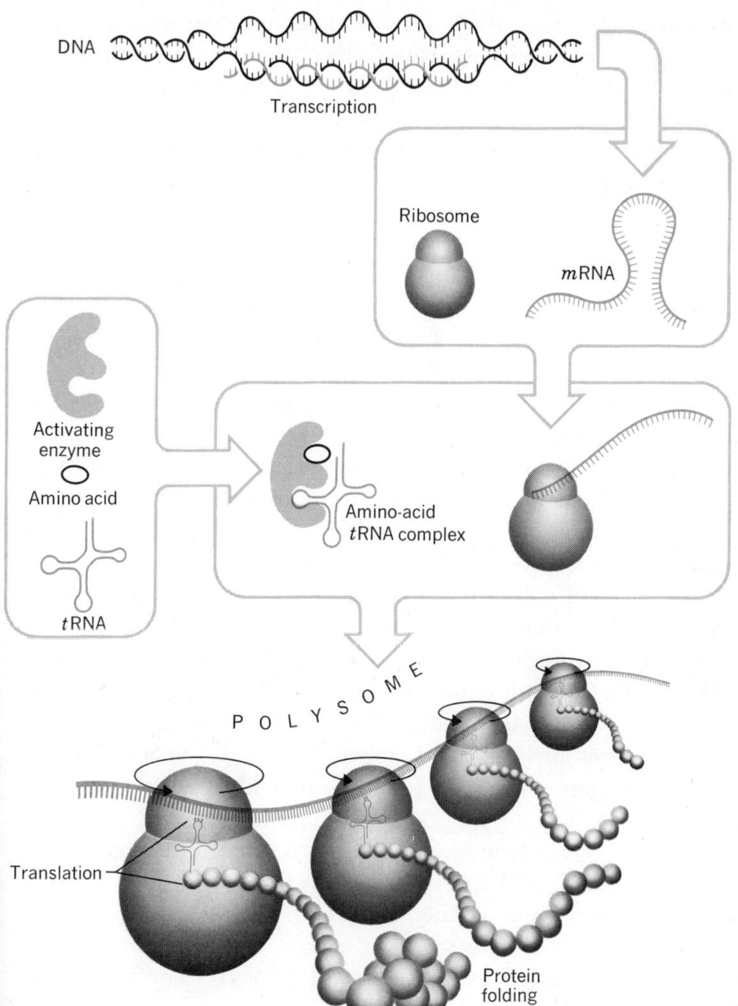

Fig. 9-24 Diagram of the central dogma.

PERSPECTIVES OF GENETIC CONTROL OF THE CELL

The dynamic unit of biological structure, the cell, is a product of biological specificity and chemical phenomena. Biological specificity provides a means by which identical sequences of DNA are passed from cell generation to cell generation, and a means by which DNA information is transcribed to an RNA messenger, and ultimately translated into protein. Chemical laws can account for the means by which proteins of specific amino-acid chains fold into their secondary and tertiary conformations, ultimately achieving a thermodynamically stable structure that represents a low energy level, and join together spontaneously to form the final quarternary structure. The chemical behavior of molecules also provides for the spontaneous joining of carbohydrates, proteins, and lipids for self-assembly of membranes. Hence, cellular form, function, and stability are the products of biological specificity superimposed on known chemical phenomena.

QUESTIONS

1. What role did elimination of variables play in Mendel's determination of the genetic basis of heredity?
2. Yellow-seeded peas are crossed with green-seeded peas. The offspring are all yellow-seeded. Members of the F_1 are crossed to one another, and the results are as follows: 32 yellow-seeded : 9 green-seeded. How are the yellow- and green-seeded traits inherited? Explain your reasons for coming to your conclusion.
3. Explain the rationale of the law of segregation of alleles on the basis of a monohybrid cross, and of the law of independent assortment on the basis of a dihybrid cross.
4. An individual with one pair of heterozygous alleles will form two types of gametes; with two pairs, four types; and with three pairs, eight types. What is the mathematical formula relating the number of heterozygous alleles to the number of types of gametes formed? How many types of gametes would an individual of the genotype AABbccDDEeFfgghhiiJjkkllMMnnooPpqqrrSS form?
5. According to genetic considerations, does the male or female parent determine the sex of the offspring? Explain.
6. How did experiments on the "transforming principle" implicate DNA as the genetic material?
7. Give the experimental evidence provided by Hershey and Chase that implicated DNA as the genetic material.
8. Give the experimental evidence that led Beatle and Tatum to the "one gene-one enzyme" hypothesis.
9. How did Ingram provide the ultimate proof of the linkage between the genetic complement and the protein structure?
10. How did Hammerling's work on *Acetabularia* indicate the presence of an intermediate between genetic information and protein synthesis?
11. How did Astrachan and Volkin's work indicate the presence of a messenger RNA?
12. Explain the "central dogma."

CHAPTER 10

DNA AND THE GENETIC CODE

Even before there was conclusive evidence that DNA is the genetic material, the growing awareness of its importance in living processes attracted the interest of many biochemists. Through their work, many of the properties, and eventually, the structure, of DNA were determined.

The Structure of DNA

Hydrolysis of DNA revealed that it was made up of four nitrogen bases: *adenine, guanine, cytosine,* and *thymine.* These appeared to be in nucleotide form, bonded by ribose phosphate (*AMP, GMP, CMP,* and *TMP*) (see Fig. 10-1).

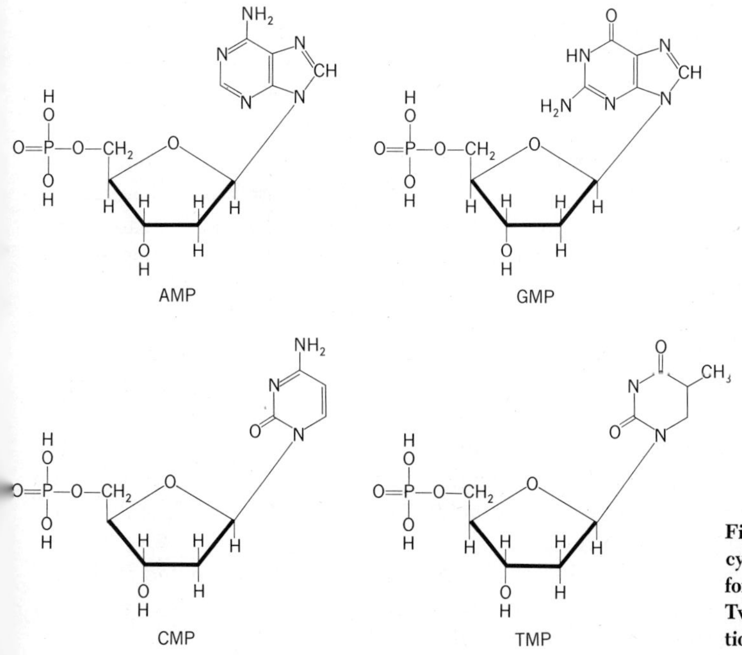

Fig. 10-1 Incorporation of adenine, guanine, thymine, and cytosine into DNA requires the presence of the triphosphate form. Hydrolysis of intact DNA yields the monophosphate form. Two high-energy phosphates are expended for each incorporation of a base into DNA.

In the late 1920s, Levine had found that the nucleosides (nitrogen base + ribose) of DNA were bonded to one another by means of phosphate groups. A phosphate was bonded to two adjacent nucleosides at a site on their riboses.

In 1951, Chargoff extracted the nitrogen bases of DNA from several types of tissues (see Table 10-1). Each type of tissue had a characteristic ratio of the number of bases. Regardless of the tissue, however, the molar amount of adenine always equaled the molar amount of thymine, and the molar amount of guanine always equaled that for cytosine. Because of this relationship, the concentration of purines in the tissues was always the same as the concentration of pyrimidines.

Purine		*Pyrimidine*
number of adenine molecules	=	number of thymine molecules
number of guanine molecules	=	number of cytosine molecules

adenine + guanine = cytosine + thymine

TABLE 10-1 THE CHARGOFF DATA[a]

Organism	Organ	A	G	C	T	A/T	G/C
Man	Thymus	30.9	19.9	19.8	29.4	1.05	1.00
Sheep	Thymus	29.3	21.4	21.0	28.3	1.03	1.02
Herring	Sperm	27.8	22.5	20.7	27.5		
Salmon	Sperm	29.7	20.8	20.4	29.1		
Wheat germ		27.3	22.7	22.8	27.1		
Yeast		33.5	14.8	16.1	25.7	1.04	1.01
E. coli		24.7	26.0	25.7	23.6		

[a] Table shows the proportions of adenine (A), guanine (G), cytosine (C), and thymine (T) extracted from different sources of DNA. Note that the adenine:thymine ratio (A/T) and the guanine:cytosine ratio (G/C) equal approximately 1.1.

X-ray crystallography data of DNA that had been obtained by Wilkins and Astbury indicated that DNA was a long molecule (20,000 Å), uniformly narrow (20 Å), symmetrical, antiparallel (running in opposite directions somewhat like the signs along the opposite sides of a road that face oncoming traffic), and with some pattern repeated every 28 Å.

In 1953, Watson and Crick utilized the data of Chargoff, Wilkins, Franklin and Astbury to postulate a model for DNA. Their model consisted of two long, narrow strands that were wrapped around one another in such a way as to form a helix. One complete twist of the helix was 28 Å long, thus accounting for the 28 Å repeats that had been detected by X-ray crystallography. The strands were antiparallel, running in opposite directions.

The uniform width of the DNA molecule was accounted for by the pairing of specific bases on one strand of the helix with specific bases on the other strand (see Fig. 10-2).

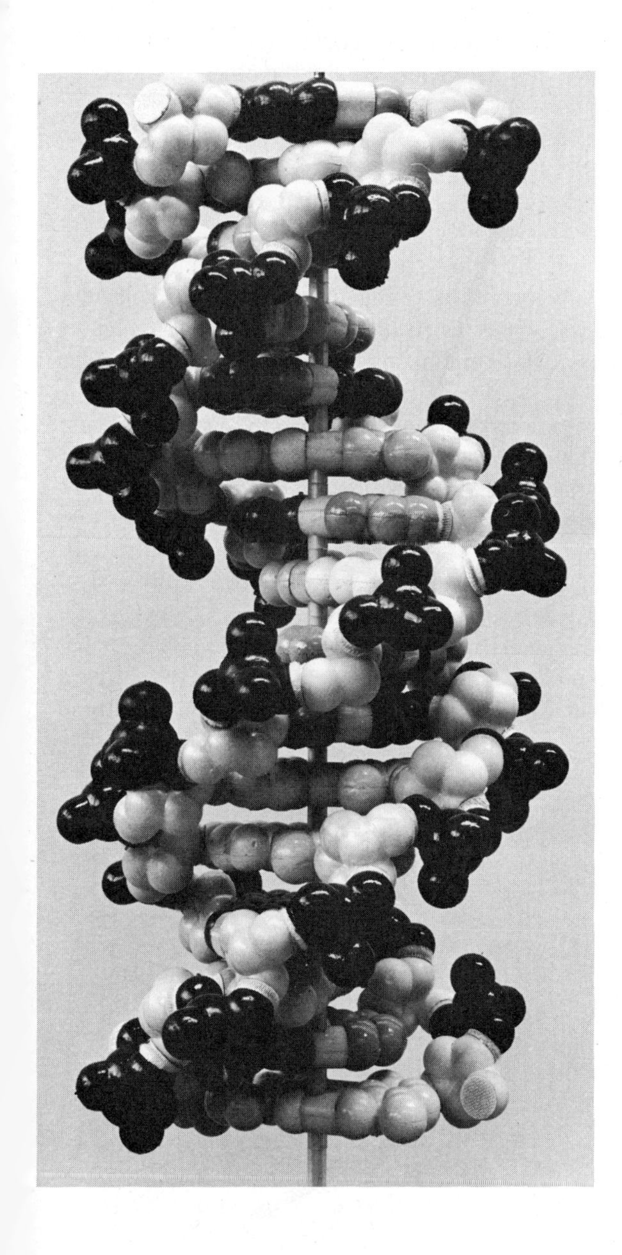

Fig. 10-2 A model of the DNA molecule. [John Oldenkamp-Psychology Today]

adenine pairs only with thymine; adenine = 12 Å, thymine = 8 Å
adenine + thymine = 12 Å + 8 Å = 20 Å diameter
cytosine pairs only with guanine; cytosine = 8 Å, guanine = 12 Å
cytosine + guanine = 8 Å + 12 Å = 20 Å diameter

The Watson-Crick model for DNA was partially supported by the work of Todd. He found that the strands of DNA have a 3′-5′phosphate-ribose bonding. Although his interest was not in confirming the Watson-Crick model, his data could be interpreted as accounting for the antiparallel strands of the proposed model in which one strand manifested a 3′-5′ ribose phosphate bonding, and the other a 5′-3′ bonding.

Inheritance Patterns and DNA Replication

It had become obvious that the chemical material responsible for the inheritance of characteristics had to exhibit specificity and be able to replicate in such a way that the resulting molecules were identical to the original. The subsequent parceling of each of the molecules resulting from replication into sister chromatids and then into daughter nuclei would then impart the same genetic potentials to the daughter cells.

Watson and Crick proposed a means by which DNA could replicate to produce two identical molecules. They suggested that the strands of DNA unwound from their helical condition. As this unwinding proceeded, nucleotides present in the nucleus in a free state (comprising a **nucleotide pool**) would pair with each of the nucleotides in the strands in a complementary fashion. Free adenine would pair only with a strand-bound thymine; free guanine with a strand-bound cytosine, etc. (see Fig. 10-3). The phosphate ester bonds would form, thus polymerizing DNA. The result would be two double helices, each of which was identical to the original, and each of which contained one old strand and one new strand.

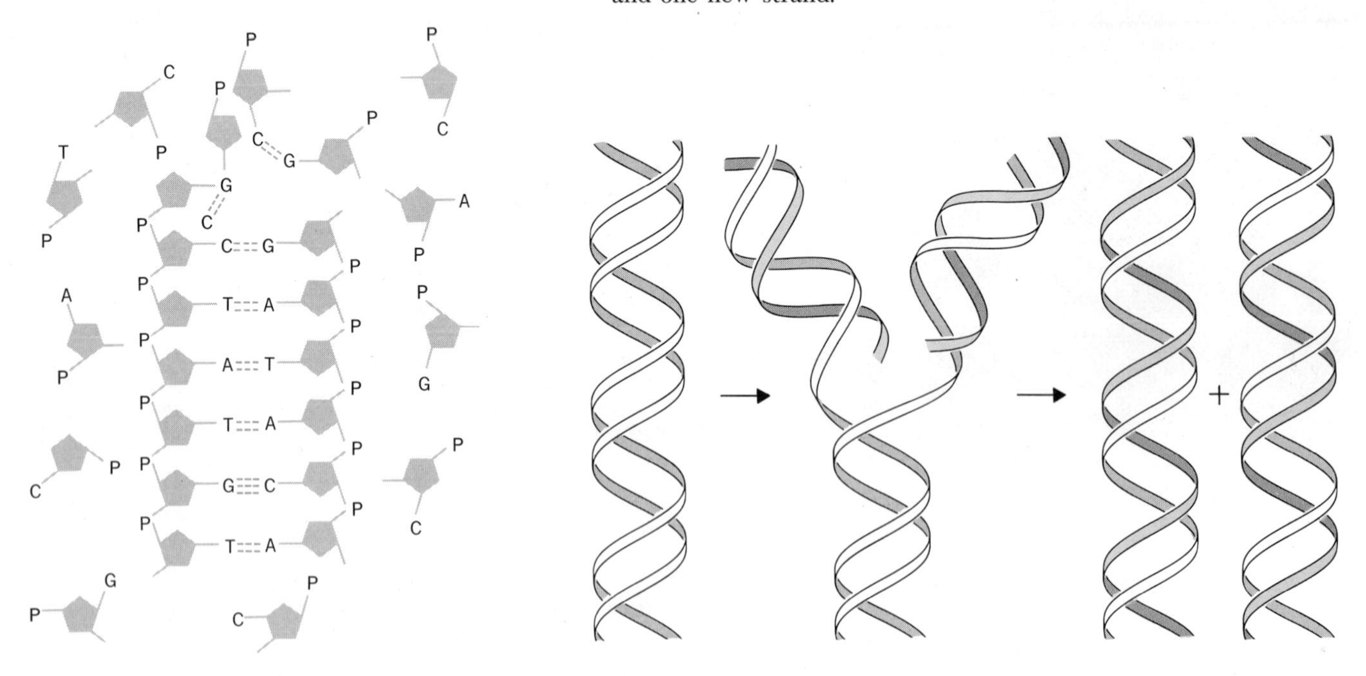

Fig. 10-3 Adenine tends to pair with thymine; guanine tends to pair with cytosine.

Max Delbruck suggested that there are three possible modes of DNA replication for the Watson-Crick model. The Watson-Crick proposal constituted a **semiconservative** pattern, but **conservative** and **dispersive** patterns must also be viewed as possibilities. The conservative pattern would involve the retention of the helical form, and the attraction of nucleotides from the pool in such a manner that they oriented into a totally new double helix. Such a means of replication would result in two double helices, one of which was of old material, and the other of new material. The dispersive pattern involves the breakage of the DNA at many points; each segment duplicates in either a conservative or a semiconservative manner, and then rejoins in such a way as to form two intact helices, each of which would be made up of both old and new material (see Fig. 10-4).

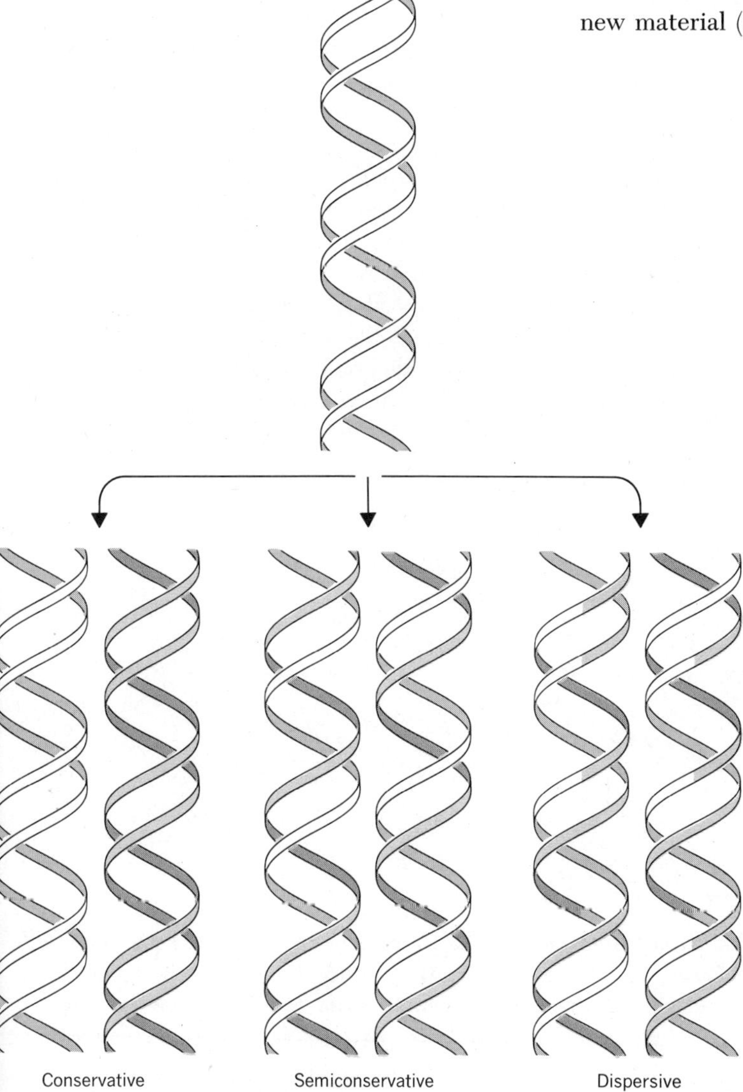

'ig. 10-4 The possible patterns by which DNA can replicate: conservative, semicon- ervative, and dispersive.

A test to determine how new material is incorporated into cellular structures can be made by the use of radioactive labeling. Taylor, in 1956, grew young bean plants in a medium containing **tritiated thymidine** (radioactive thymidine), which is incorporated only into DNA. He used the bean root because the mitotic figures of the meristematic tissue can easily be seen, and the sister chromatids (which are the duplication products of the chromosomes) can be sufficiently separated by smearing the root material on a slide.

After allowing time for the incorporation of the "hot" thymidine into the nucleotide pool and then into the DNA, he washed the roots and placed them in "cold" thymidine. He left the roots in the "cold" thymidine long enough for the cells to pass through the S stage of interphase. The labeled DNA would then constitute the "old" material, and the nonlabeled DNA would constitute the "new" material synthesized during the last S period. He then treated the roots with colchicine, a drug that allows cells to continue through prophase, but not to enter metaphase. Therefore, the mitotic process is stopped at a point where the chromosomes, or their sister chromatids, are most easily seen. This treatment leads to a buildup of a maximum number of cells at metaphase.

After smearing the material, Taylor placed a photographic emulsion over the slide, and left the slide in a dark room for several days. The emission of the radioactive particles from a cell structure exposed the photographic emulsion in localized regions. When the photographic emulsion was developed, an **autoradiograph** was produced in which the dark spots overlying the cellular structures indicated the presence of radioactive material within those structures (see Fig. 10-5).

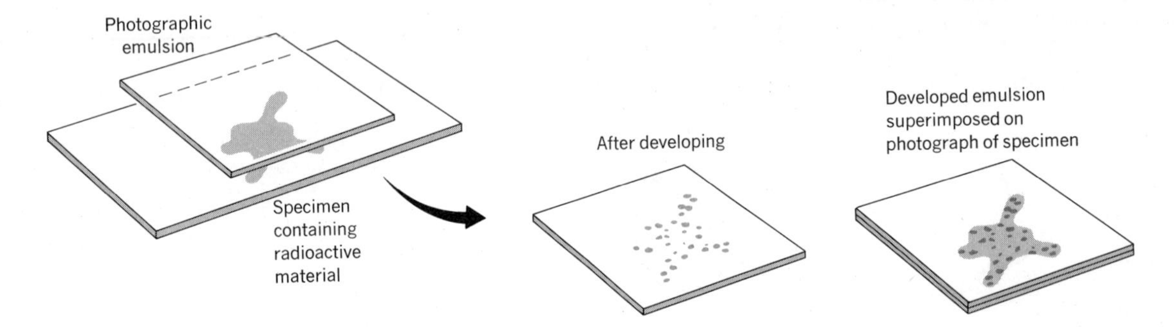

Fig. 10-5 Autoradiography techniques.

If DNA replicated in a conservative manner, then one of the helices would be made up of labeled DNA, and the other would not. Since each of the products of DNA replication forms the basic structure of each of the sister chromatids, one would expect to find that one of the chromatids would emit radioactive particles and the other would not. This would be detected on the autoradiograph.

If DNA replicates in a semiconservative or dispersive manner, then both products of replication would contain old and new material. Therefore, both sister chromatids would emit radioactive particles.

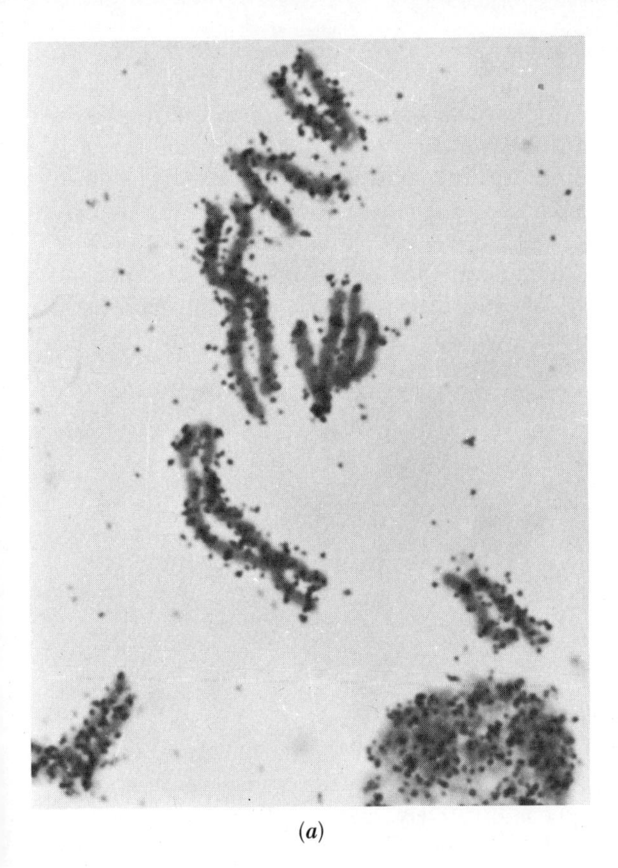

(*a*)

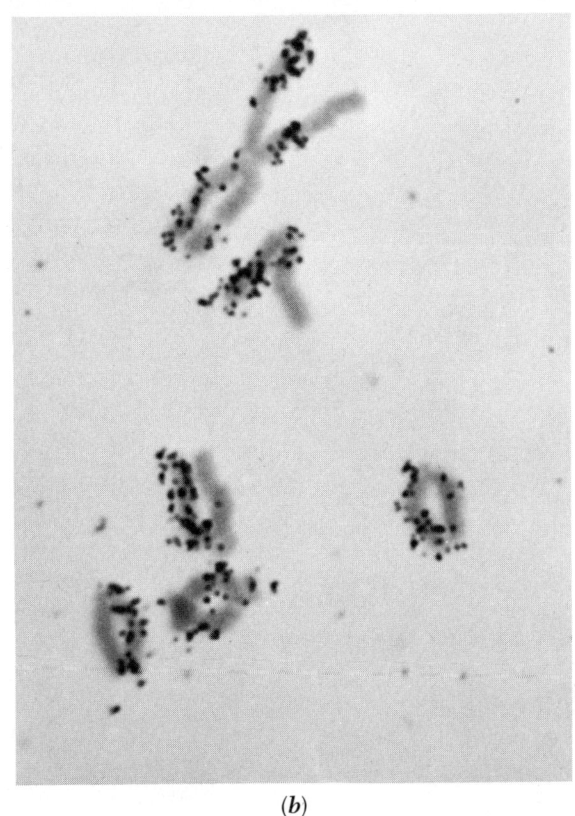

(*b*)

Fig. 10-6 Autoradiographs. (*a*) One division after removing bean root from tritiated thymidine (2000×). (*b*) Two divisions after removing it from tritiated thymidine (2000×). [Courtesy Dr. J. Herbert Taylor, The Florida State University]

The autoradiographs (see Fig. 10-6) indicated that both sister chromatids emitted radioactive particles, thus eliminating the possibility of replication in a conservative manner. In order to determine whether replication is in a semiconservative or a dispersive manner, Taylor repeated the experiment, this time permitting the root cells to pass through two division cycles after treatment with tritiated thymidine.

If DNA replication occurred in a semiconservative manner, then after one division cycle, one of each of the strands of the double helix would be labeled. The next DNA replication would involve an unwinding of the two strands, and the complementary pairing of "cold" nucleotides from the nucleotide pool. Replication would result in one chromatid having one labeled helix and one unlabeled helix, and the other chromatid having both helices containing nonlabeled material. The autoradiograph would reveal one of the sister chromatids as "hot" and the other as "cold." If DNA replication occurs in a dispersive manner, both products of replication would contain old and new material, and therefore would emit radioactive particles. The autoradiograph would show both sister chromatids as "hot."

The autoradiograms demonstrated quite clearly that one of the sister chromatids emitted radioactive particles, while the other did not. Therefore, DNA appears to replicate in a semiconservative manner, in accordance with the proposal by Watson and Crick.

The semiconservative pattern of replication was further confirmed by the work of Meselson and Stahl in 1964. They grew *E. coli* bacteria in a medium containing N^{15}, a nonradioactive isotope of nitrogen (see p. 34). They extracted the DNA from an aliquot of the culture, and divided the remaining culture into four flasks, each of which contained the lighter N^{14}. After permitting the bacteria to undergo a single division (one generation), they extracted the DNA from the bacteria in one of the flasks. After two generations, they extracted the DNA from the second flask of bacteria; after three generations, from the third flask; and after four generations, from the fourth flask (see Fig. 10-7).

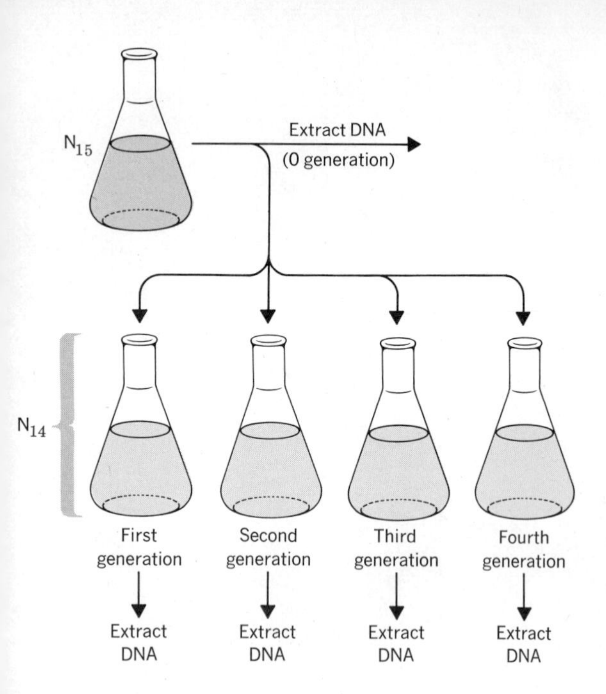

Fig. 10-7 Setup of Meselson-Stahl experiment.

The DNA extracts were separately submitted to cesium chloride density-gradient centrifugation (see p. 218 for the rationale of this procedure). If DNA replicated conservatively, they would expect to find two layers in the "one-generation" tube. The lower layer would contain N^{15}-labeled DNA, and the upper layer would contain the lighter DNA with only N^{14}. The tubes of succeeding generations would always show these two layers. If DNA replicated semiconservatively, they would expect to find one layer in the "one-generation" tube. This layer would be made up of DNA in which one strand was labeled with N^{15} and the other contained N^{14}. The "two-generation" tube would be expected to have two layers—a lower layer in which old N^{15}-containing material was associated with new N^{14}-containing material, and an upper layer that would have only N^{14}-containing DNA. The tubes for subsequent generations would all exhibit these two layers. If DNA replicated dispersively, then all the tubes would be expected to show a single layer, in which all the DNA contained both old and new material (see Fig. 10-8).

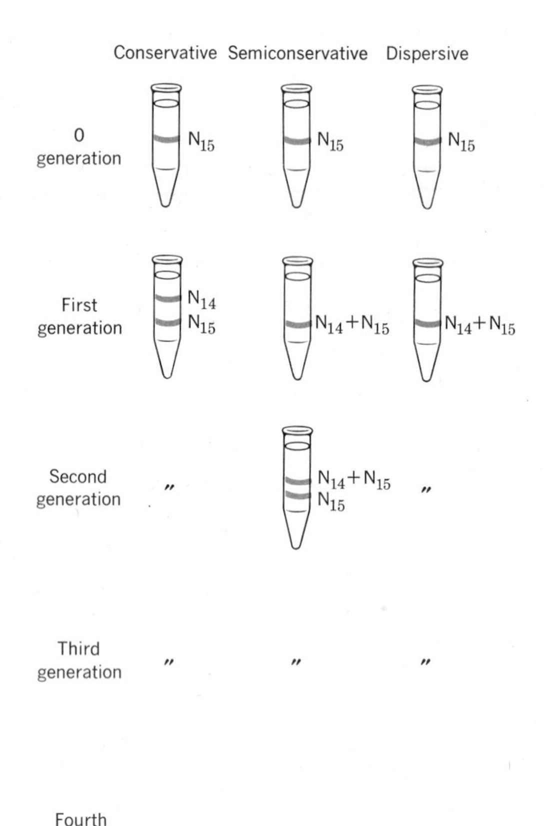

Fig. 10-8 Expected results of Meselson-Stahl experiment according to whether DNA replicates in a conservative, semiconservative, or dispersive manner.

The results were as follows: The zero-generation tube contained a single layer characteristic of the maximum N^{15} content, and serves as a reference point for the other tubes. The "one-generation" tube also contained a single layer, but the position of this layer was somewhat above the position of that for the zero-generation tube, indicating the presence of some N^{14} in the DNA structure. The "two-generation" tube had two layers, the lower one corresponding to the single layer of the "one-generation" tube, and the upper one being identical to the standard N^{14} DNA. After two generations, the N^{14} level showed a greater accumulation of material. The interpretation of this result is that the DNA of *E. coli* replicates in a semiconservative manner.

The determination of a semiconservative pattern of replication in such diverse organisms as bean plants and bacteria, and subsequent demonstrations of this pattern in algae and viruses, indicates that this mechanism of DNA replication is probably universal.

The full picture of mitosis can now be appreciated. The DNA in a mammalian cell is about two meters long. The DNA in human cells is distributed among forty-six chromosomes, so that each chromosome is made up of DNA that is some 43,000 microns long.

$$\frac{2 \text{ meters}}{46} = \frac{2 \times 10^6 \text{ microns}}{46} = 43{,}000 \text{ microns}$$

This is a rather astounding fact, since the cell is only 20 microns in diameter and the nucleus is only 5 microns in diameter. The semiconservative replication of the DNA in a mammalian cell would involve the unwinding of each of the 46 DNA double helices—surely a phenomenal feat, the mechanics of which is far from understood.

The chromosomes at metaphase are only five microns long, indicating that they have shorted some 8000 times. This is accomplished by the chromosomes coiling up into a helix, which then coils up once again into a larger helix (see Fig. 10-9). Imagine trying to separate 92 pieces of string, each four miles long, in a 16 × 16 foot room so that there are 46 pieces on opposite sides of the room. How much easier it would be if each piece of string were rolled up into a coil only eight feet long. The coiling of the chromosomes, then, is an efficient means of separating the DNA in the sister chromatids. The DNA is narrow (20 Å), and because of the coiling it is possible for the chromatids to be visible under the light microscope.

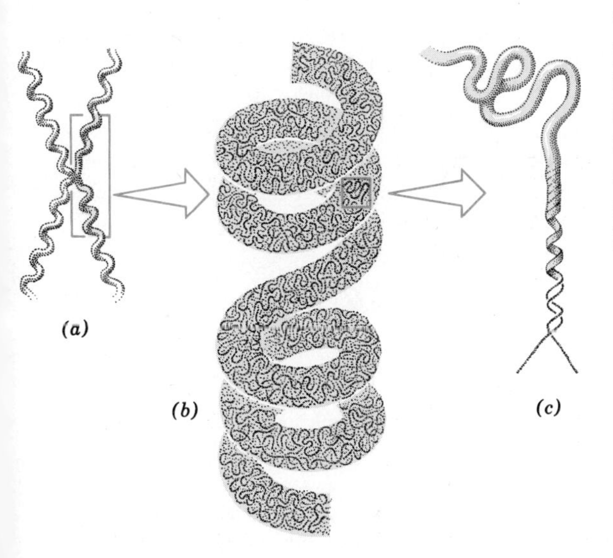

Fig. 10-9 Helix coiling into a visible chromosome.

THE GENETIC CODE

The paper in which Watson and Crick proposed their model for DNA also included an idea of how such a model would provide for cellular control. They postulated that the sequence in which the bases were incorporated into the DNA structure determined the sequence in which amino acids were incorporated into proteins. A sequence AGTC would have a different message than would the sequence GCTC, etc.

Their hypotheses held a host of implications. First, the DNA molecule could be viewed as a code in molecular form—a language composed of four letters (adenine, guanine, cytosine, and thymine). Just as different combinations of the twenty-six letters of the English alphabet transfer information from one individual to another, the genetic language transfers information to the cytoplasm, thereby controlling cellular function. The presence of a genetic language permits analysis by means of information theory and its methods, which were originally developed for decoding primitive languages, military secrets, machine communication, etc.

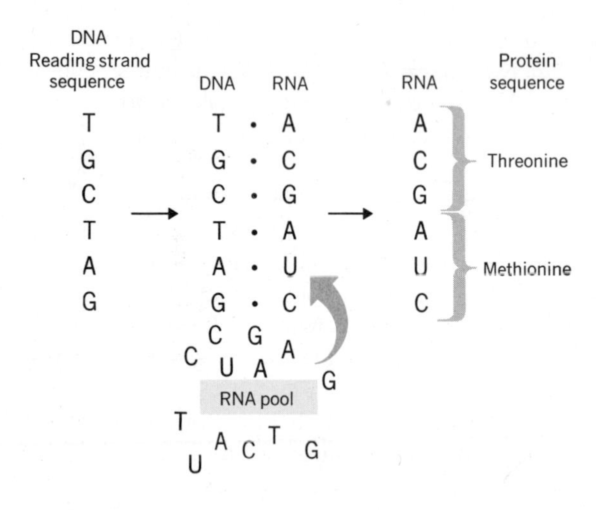

Fig. 10-10 Transcription of a message from the DNA to the RNA code.

Second, the Watson-Crick model presented biochemists and geneticists with the task of deciphering the genetic code. But it was not a singular code that required deciphering, but three variations: the DNA, *t*RNA, and *m*RNA variants. From the sequence of events in the central dogma, it could be inferred that the code is operationally a system of codes and anticodes, and that once one was determined, the others would fall into place. The paired structure of DNA suggested that there was a natural attraction of two bases for one another: adenine for thymine, and cytosine for guanine. In the transcription to RNA, the attractive forces would be between adenine and uracil, since RNA does not contain thymine. As will be discussed more fully later, only a region of one strand of the DNA double helix is read at a time. A sequence in a strand that is TGCTAC would be transcribed into an *m*RNA sequence of ACGAUG (see Fig. 10-10). The subsequent lining-up of *t*RNAs on the messenger would manifest the same code as the DNA, but in RNA form, and therefore be **UGCUAC.** Since the experimental breakthrough in the cracking of the genetic code came with the elucidation of that for *m*RNA, the *m*RNA has been termed *the* code, and the corresponding form for *t*RNA is termed the anticode. DNA is then also an anticode. Biologists are somewhat unperturbed that this terminology is the opposite of the operational sequence of function.

THE GENETIC CODE AND INFORMATION THEORY

One of the most interesting speculations in the 1950s was on the similarity, if not the parallelism, of gene organization to the structure of language. That messages analogous to the written word exist at a chemical level is a fascinating phenomenon to the twentieth-century biologist.

The basic factors involved in the communication of information are collectively called **information theory.** The study of these factors was initiated to find the most efficient means of sending messages by machine by analyzing symbolic language. The analysis of symbolic languages, such as the English alphabet, the 0-9 number system, and the Morse code, led to an awareness of a number of factors in both the symbols and the transmission system through which information is conveyed that affect the efficiency of communication.

Is DNA An Information System?

Information is truly symbolic in that it has no physical correspondence to reality unless it is translated by some appropriate device. The word *cat* means nothing in itself, but the mind has been programmed to translate this message into an image corresponding to something in reality.

If DNA represents a truly informational system within the cell, then it must satisfy this requisite for information systems. Is DNA truly symbolic, or does it really correspond physically to the protein it forms? When models of amino acids and their corresponding RNA are compared, no spatial or charge complementarity can be found between them. Also, mixtures of *m*RNA and a pool of twenty amino acids do not lead to protein synthesis. The activating enzymes for each of the twenty amino acids must be added. It is presently assumed that the correspondence between the amino acids and the *m*RNA exists in the structure of the activating enzymes, and that these act as the translating devices. Therefore, it would seem that *DNA and RNA are totally symbolic* in nature, and can be considered components of an information system.

Factors in Symbolic Language that Affect Efficiency of Communication. One of the prime requisites for an efficient information system is that it *uses the fewest number of symbols possible.* As the number of symbols in a message increases arithmetically (1, 2, 3, 4, etc.), the number of possible messages from that set of symbols increases exponentially (1^1, 2^2, 3^3, 4^4, etc.). For example, given the single letter *A*, there is but one message. Given the two letters *A* and *B*, there are four combinations, *AA*, *AB*, *BA*, and *BB*. Given three letters, *A*, *B*, and *C*, there are twenty-seven combinations that form messages (see Table 10-2).

TABLE 10-2 POSSIBLE COMBINATIONS OF THREE LETTERS IN A WORD

FIRST LETTER	SECOND LETTER: A	B	C	THIRD LETTER
A	AAA	ABA	ACA	A
	AAB	ABB	ACB	B
	AAC	ABC	ACC	C
B	BAA	BBA	BCA	A
	BAB	BBB	BCB	B
	BAC	BBC	BCC	C
C	CAA	CBA	CCA	A
	CAB	CBB	CCB	B
	CAC	CBC	CCC	C

The most efficient communication system is one with only two symbols, a *binary system.* Any teacher knows that the easiest type of examination to grade is a set of true-false questions. The efficiency of the binary system is further exemplified by the Morse code, in which two symbols, a dot (·) and a dash (–), are used.

A second requisite of an efficient information system is that *the symbols contain a minimum amount of information, or a minimum information content.* The average number of questions that must be asked in a given system in order to determine a given amount of information is termed a **bit** (**bi**nary digi**t**). The idea of a bit of information introduces a means for mathematically computing the complexity of a language system. For example, Morse represented two letters of the alphabet by a single symbol, **E** by a dot (·) and **T** by a dash (–). Given a single-symbol message, one need only ask "is it a (·)"? If the answer is yes, then the letter **E** is known; if the answer is no, then it is known that the symbol must be that for the letter **T.** For distinguishing between an **E** and a **T** in the Morse code, it takes but one question. Therefore, **E** and **T** have an information content of *one* or contain one bit of information.

Morse represented four letters of the alphabet by two symbol messages. In order to determine which of the four letters is to be conveyed, *two* questions must be asked. The first question determines whether the first symbol in the message is a (·) or a (–). The second question determines the character of the second symbol of the message. Therefore, a two-symbol message in a binary system contains two bits of information. Eight letters of the alphabet are represented in Morse code by three symbols. It would take three questions to determine a given letter from a three-symbol message. Therefore, a three-symbol message in a binary system contains three bits of information.

Notice that with *one* question in a binary system, it is possible to distinguish between *two* symbols. With *two* questions, it is possible to distinguish among *four* symbols, and with *three* questions, it is possible to distinguish among *eight* symbols. The equation is

$$x = 2^H \qquad \text{where } x = \text{total number of symbols in a code}$$
$$H = \text{number of bits}$$

This equation can be converted into a logarithmic form

$$H = \log_2 x$$

For the Morse code, the total number of symbols in the system is 2. Therefore, the number of bits in a single symbol message is

$$H = \log_2 2$$
$$H = 1 \text{ bit of information}$$

This same equation can be used to determine the information content of a letter of the alphabet, or a number in the 0–9 numerical system. The English alphabet contains 26 letters, so the number of symbols in the system is 26.

$$H = \log_2 26$$
$$H = 4.7 \text{ bits}$$

Notice that the number of letters in the alphabet (26) is between 2^4 and 2^5. Hence one would expect the number of bits to be between four and five.

There are 10 symbols in the numerical system (0–9), and 10 is between 2^3 and 2^4.

$$H = \log_2 10$$
$$H = 3.2 \text{ bits}$$

A comparison of the information content of the alphabet, the numerical system, and a binary system such as the Morse code indicates that the fewer the number of symbols in the code, the more quickly and accurately one can determine the information that is to be conveyed, and the more efficiently the system is decoded.

The third requisite for an information system is the *use of redundancy only as an advantage to the system.* The more often a message is repeated, the greater the chance that it will be received and translated correctly. Some types of redundancy, or too much redundancy, is only a waste in communication practice.

The fourth requisite for an information system is that *recognition of more than one symbol unit at a time be possible.* In a system where two symbols are read simultaneously, the two-unit symbol is called a **digram.** In English, *st, ch, is,* and *te* are examples of digrams. The **trigram,** or three-letter symbol unit read simultaneously, also increases the efficiency of a system by saving time. Examples of trigrams in English are *ing, per, ant,* and *ion.*

In summary, maximum efficiency of a communication system is dependent on the following:

1. The use of the fewest number of symbols.
2. The use of symbols that contain small bits of information.
3. The use of redundancy only as an advantage to the system.
4. The recognition of more than one symbol unit at a time.

DNA and RNA as Information Systems

DNA and RNA are each composed of four different kinds of bases, so the number of symbols in the genetic code is four. *Since the efficiency of a communication system is inversely proportional to the number of symbols used in the code, DNA and RNA represent a highly efficient means of communication.*

Although DNA is composed of adenine, guanine, cytosine, and *thymine,* and RNA is composed of adenine, guanine, cytosine, and *uracil,* the number of bases used in each coding system is the same. Accordingly, the information content in one base is

$$H = \log_2 4$$
$$H = 2 \text{ bits of information}$$

The DNA-RNA code, therefore, fulfills the requisite of efficient communication systems by using symbols that contain small bits of information.

It has been calculated that a single gene contains approximately 1000 bases. Therefore, the amount of information in a single gene would be

$$H = a \log_2 4$$
$$H = 1000 \times 2$$
$$H = 2000 \text{ bits of information}$$

The amount of information in various organisms may also be calculated. The virus T-4 was found to have 200,000 base pairs (only one base of a base pair is informational—see p. 277).

$$H = 200{,}000 \times 2$$
$$H = 400{,}000 \text{ bits of information in a T-4 phage}$$

The bacterium *E. coli* has 2.5 million base pairs.

$$H = 2.5 \text{ million} \times 2$$
$$H = 5 \text{ million bits of information in a cell of } E.\ coli$$

A mammalian cell contains 10^9 base pairs.

$$H = 10^9 \times 2$$
$$H = 2 \times 10^9 \text{ bits of information in a mammalian cell}$$

To appreciate these values more fully, one can compare them to the information content of a book. An average printed page has approximately 3000 symbols. Each letter has 4.7 bits, but by complicated manipulation to include letter frequency, the figure can be reduced to 4.129 bits of information.

$$H = 3{,}000 \times 4.129$$
$$H = 12{,}387 \text{ bits of information on a single written page}$$

A virus, therefore, would be similar in information content to a 16-page pamphlet; a bacterial cell would be similar in information content to a 416-page book; and a human cell would be similar in information content to 500 books, each with 332 pages!

That this much information can be stored, transmitted, and translated in so small a physical entity as a cell indicates the extreme efficiency of DNA as an information molecule, as well as the chemical complexity of a cell.

It is now possible to examine the breaking of the genetic code, and the features of DNA, according to the perspective of DNA as an efficient informational system.

BREAKING THE GENETIC CODE

The hypotheses offered in the 1950s for the genetic code were based on the growing awareness of information theory. There are twenty different amino acids. Therefore, there must be a code specific for each of them. If the code operates in a manner similar to the Morse code, there would be four single symbols (a, t, c, or g) and sixteen double symbols (at, ag, ac, aa, ta, tt, tc, etc.) This would add up to the necessary twenty. But how would the RNA discriminate between reading a single-symbol unit and a double-symbol unit? If the base sequence was ACGGTAT, was the first amino acid coded by A or AC? Of course, the correct coding could be enforced by spacing, so that the code would read A CG G T AT, but there seemed to be no evidence for a spacing system within the DNA molecule.

The coding for amino acids appears to be in units of a standard number of bases. George Gamov theorized that since combinations of two bases would yield only sixteen possible combinations (4^2), a combination of three bases would be the most probable. Although this would offer a greater variation than necessary ($4^3 = 64$), it is the lowest set capable of generating twenty combinations without requiring spacing.

Experimental Work on the Code

The first breakthrough on an experimental level was made by M. N. Nirenberg in 1961.

Using an RNA-polymerizing enzyme, Nirenberg made an RNA polymer composed only of uracil (**poly U**). He then placed the poly U, ribosomal fraction, *t*RNA, and cell-free enzymes in each of twenty tubes. To each of the tubes he added a full complement of amino acids, one of which was labeled. For instance, in the first tube, the alanine was labeled, in the second, the serine, etc. (see Fig. 10-11).

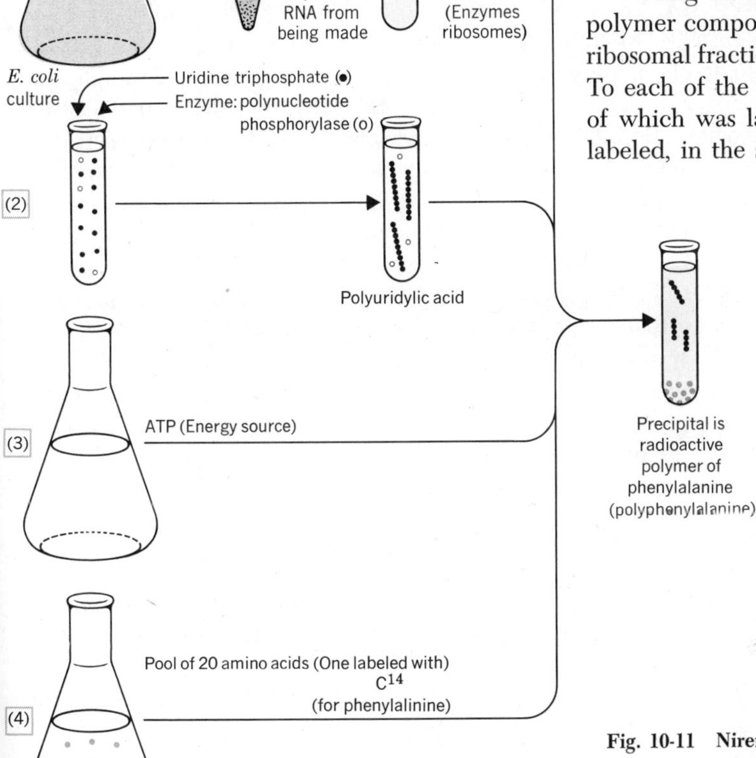

Fig. 10-11 Nirenberg experimental setup.

After a short time, he tested each of the tubes for the presence of labeled protein. The only tube in which he found this was the one containing "hot" phenylalanine. Analysis of the protein showed that it was nothing more than a polymer of phenylalanines linked by peptide bonds (polyphenylalanine). Nirenberg had succeeded in introducing an artificial *m*RNA, and producing an artificial protein!

When Nirenberg calculated the molar concentration of the poly U and polyphenylalanine, he found that the molar ratios were approximately 3:1 (they varied from 2.6:1 to 3.2:1). This supported Gamov's view that the code was in a triplet form. Nirenberg postulated that the code in the messenger for phenylalanine was UUU.

Through a series of brilliantly conceived experiments, Nirenberg proceeded to determine the code for the other amino acids. He was able to determine the numbers and types of bases that would code for a given amino acid, although he did not ascertain the sequence of the triplets. For instance, 2 uracils and 1 adenine code for isoleucine, leucine, and tyrosine. There are three different ways in which 2 uracils and an adenine may form a triplet (UUA, UAU, and AUU), but which of these three coded for leucine, isoleucine, or tyrosine could not be discerned.

Another line of evidence that supports the RNA triplet code involves the types of bases that code for a specific amino acid. Some triplets contain only one type of base, which is a triplet repeat (AAA, UUU, etc.); others contain two bases, and contain a double repeat of one of the bases (AAC, GGU, etc.); while still others contain three different bases (GCA, CUG, etc.). No amino acid requires four different bases.

In 1964, Khorana developed an experimental method by which he could synthesize messenger RNAs of known sequences. Through such messengers he determined the triplet-base sequence that coded for each amino acid. Both Nirenberg and Khorana received Nobel Prizes in 1968 for decoding RNA (along with Holley, who determined the nitrogen-base sequence for a transfer RNA). It is now known that UUA codes for leucine, UAU for tyrosine, and AUU for isoleucine. The **codons,** as the triplet nucleotide code is called, for each of the amino acids may be seen in Table 10-3. It should be noted that several of the amino acids have more than one codon, and thus redundancy is present.

The transfer of genetic information as determined by the base sequence of a three-base unit is directly comparable to the transfer of information in the English language. Given three letters, A, P, and T, there are six different ways in which they can be put together (using each letter once) to form an informational unit (a word). Some of these letter combinations have meaning; others are nonsense words. For instance:

P A T Name, or to remain unchanged, as in *stand pat,* or to bring ones hand down lightly on an animate or inanimate object, as in *pat* a dog's head.
T A P Sound made by striking a foot against something solid.
A P T Apropos of a situation.

TABLE 10-3 THE GENETIC DICTIONARY: KHORANA SCHEME FOR THE GENETIC CODE

First Base	Second Base U	C	A	G	Third Base
U	phenylalanine	serine	tyrosine	cysteine	U
	phenylalanine	serine	tyrosine	cysteine	C
	leucine	serine	terminate[a]	tryptophane	A
	leucine	serine	terminate[a]	tryptophane	G
C	leucine	proline	histidine	arginine	U
	leucine	proline	histidine	arginine	C
	leucine	proline	glutamine	arginine	A
	leucine	proline	glutamine	arginine	G
A	isoleucine	threonine	asparagine	cysteine	U
	isoleucine	threonine	asparagine	cysteine	C
	methionine	threonine	lysine	arginine	A
	methionine	threonine	lysine	arginine	G
G	valine	alanine	asparagine	glycine	U
	valine	alanine	asparagine	glycine	C
	valine	alanine	glutamic acid	glycine	A
	valine or met	alanine	glutamic acid	glycine	G

[a] Punctuations in the code that will be discussed in Chapter 10.

P T A Initials for Parent Teachers Association.
A T P Abbreviation for Adenosine Tri Phosphate.
T P A Abbreviation for Tri Phenyl Acetic acid.

Each of the combinations transfers information totally different from the others. By viewing genetic information in the same way, it is not surprising to find that UUA, UAU, and AUU lead to the incorporation of totally different amino acids.

Yet another point warrants consideration in the above example with three-letter words using the letters A, P, and T. The factor that determines whether the words make sense is the comprehension of the individual doing the reading, not the combinations of the letters themselves. To the nonchemist, TPA is a nonsense word; but for the chemist, it has meaning. The same is true of ATP. Before this semester began, the word may have been meaningless; but by this point in the semester, it should be fraught with meaning. Symbols are only dark scratches on a page, and one scratch is as good as another. Only a translation mechanism (such as the human mind) with suitable experiences, or, as in the case of DNA, the activating enzyme that combines an amino acid with a *t*RNA, can transform information into action.

The Reading of the DNA Molecule

In the final analysis, the efficiency of any communication system is determined by the success with which messages are transmitted. For DNA molecules, the transmission efficiency is, in turn, dependent on the accuracy of DNA transcription to RNA. How is the DNA molecule organized for reading, and how is the molecule read?

There is some evidence that DNA does not contain any symbols that act as spaces or commas within a message. Schlesinger determined the sequence of amino acids in a protein, and then proceeded to alter the appropriate gene in such a way that one base was deleted. He then determined the amino-acid sequence in the protein produced by the mutant. If the message contained commas or spaces, the shift of the reading frame would affect only the one altered codon, and he would expect to find essentially the same amino-acid complement (see Table 10-4). If, on the other hand, there were no spaces or commas, then the bases after the deletion would all be moved up one space, and different codons would be formed. This would cause the incorporation of totally different amino acids into the protein (see Table 10-5). Schlesinger found that the amino acids after the alteration were different from those found in the protein produced by the original DNA sequence. Therefore, it appears that the gene does not contain any internal punctuation, and the codons are read as three-symbol units or trigrams. Thus the RNA-DNA system fulfills the fourth requisite for efficient communication systems.

But there must be some form of punctuation between genic messages, for if there were not, the entire DNA molecule would be read out into one long message, and individual genes would not have a particulate character. Such a terminus of the reading sequence has been identified in mutant bacteriophages.

A number of mutations that produce a wide variety of phenotypic changes were found for bacteriophages. Upon examining the proteins produced by these so called **ochre** and **amber** mutant forms, it was found that they were a great deal shorter than the corresponding proteins of the normal phage. It would seem, therefore, that the mutation had caused the readout mechanism to be turned off before it had completed its message. This "turning off" was the mutation, and was traceable to a change in the base sequence of the DNA.

Experimental evidence indicates that the RNA triplets CUG and AAU cannot be translated. There are no amino acids for these two combinations in the code. When this combination of bases is present in *m*RNA, the readout stops. An RNA code of CUG would be the complement of GAC in the DNA code. It is thought that the mutation that brings about an unfinished protein is an alteration of the reading frame by substitution of a base in such a way that a DNA codon of GAC or TTA is formed. The DNA codons GAC or TTA constitute nonsense triplets and are assumed to be the genetic equivalent of a period for ending a sentence.

TABLE 10-4 RESULTS OF CHANGING A SINGLE BASE IN A CODE WITH SPACES OR COMMAS

Original DNA	AAA	GGG	GAC	ACG	ACC
RNA complement	UUU	CCC	CUG	UGC	UGG
amino acids	phen	thre	met	cyst	trypt
	"A" Omitted in First Codon				
Altered DNA	AA–	GGG	GAC	ACG	ACC
RNA complement	UU	CCC	CUG	UGC	UGG
amino acids	– – –	thre	met	cyst	trypt

TABLE 10-5 RESULTS OF CHANGING A SINGLE BASE IN A CODE WITH NO SPACE OR COMMAS

Original DNA	A A A	G G G	G A C	A C G	A C C
RNA complement	U U U	C C C	C U G	U G C	U G G
amino acids	phen	thre	met	cyst	trypt
	Omit "A" in first codon				
Altered DNA	A A G	G G G	A C A	C G A	C C
RNA complement	U U C	C C C	U G U	G C U	G G
amino acids	phen	thre	cyst	ser	ala

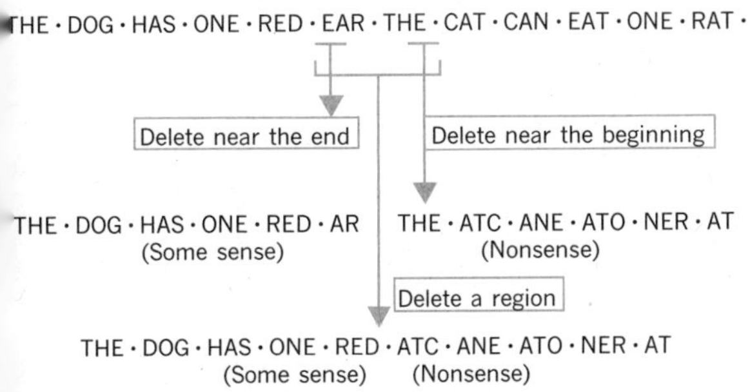

THE · DOG · HAS · ONE · RED · EAR · THE · CAT · CAN · EAT · ONE · RAT ·

Delete near the end

Delete near the beginning

THE · DOG · HAS · ONE · RED · AR (Some sense)

THE · ATC · ANE · ATO · NER · AT (Nonsense)

Delete a region

THE · DOG · HAS · ONE · RED · ATC · ANE · ATO · NER · AT
(Some sense) (Nonsense)

Fig. 10-12 If the reading frame is thrown off at the beginning of a message, the entire message is meaningless. If only the end of the message is thrown off, and if that segment does not contain an amino acid essential in the tertiary structure of the final protein, the protein may be functional.

The long, narrow shape of the DNA molecule suggests that the readout begins at a given point and proceeds in one direction toward the terminus, in much the same way as one reads a written sentence. Evidence for this pattern of readout was provided by Crick.

Crick found a single mutation in T-4 phage that appeared to affect two traits. The mutation appeared to be due to the deletion of a sequence of bases along the DNA strand. In addition, it was previously known that the two genes that controlled the affected traits were adjacent.

There was also a difference in the severity of the alteration. One of the genes appeared to be badly misread, and produced a phenotypic alteration that was quite different from the normal one. The other gene appeared to produce only a slight phenotypic modification. Crick reasoned that this would be true if the DNA strand were read in one direction, and that the deletion of bases had occured in such a way that one gene was missing a terminal portion, while the other was missing the initial portion. The gene missing the initial portion would have its reading frame thrown out of step, and would produce a totally aberrant protein. The gene missing only the terminal portion would produce a slightly abnormal protein, so that its phenotypic expression would not be severely altered (see Fig. 10-12).

The situation is somewhat similar to communication in the English language. If a message was "John went to the store to buy eggs and bacon," and the only part of the message that was printed was "John went to the store to buy eggs and ba," one could still discern the meaning. If, however, the three deleted letters were missing an earlier part of the message, and the same number of letters per word were used, the message would become "Nwen ttot he sto retob uy egg sand bac on" and the meaning would be essentially lost.

One of the questions that arose while the genetic code was being deciphered was whether the code was overlapping. An overlapping code would be one in which adjacent codons had certain bases in common. For example, in the sequence UCGACUA, the UCG would be the first triplet, the CGA the second, the GAC the third, etc. The overlapping triplets of RNA would then be represented

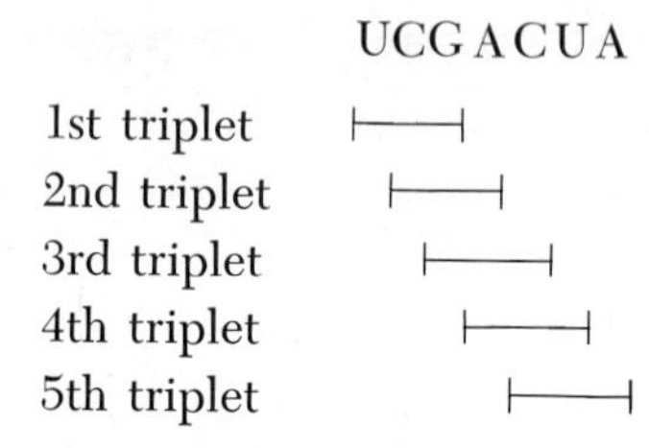

whereas the nonoverlapping triplets would be designated as

U C G A C U A

The overlapping-codon scheme was appealing because it provided for much conservation of space in the genetic material. It was finally discarded, however, because the mutations caused by a single base substitution resulted in the substitution of only one amino acid in a protein. If the code were overlapping, three amino-acid substitutions would be expected in most cases (see Fig. 10-13).

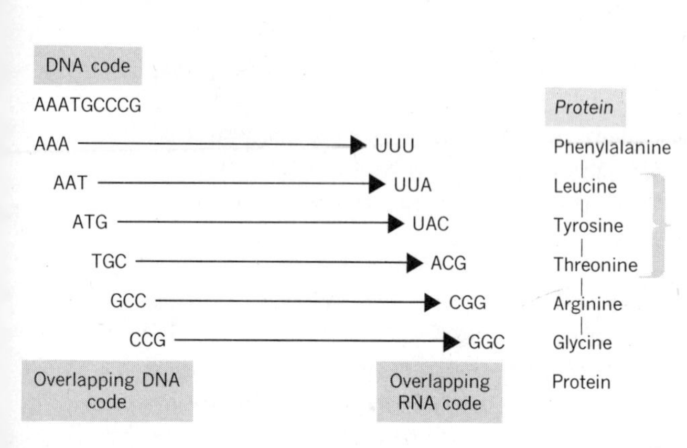

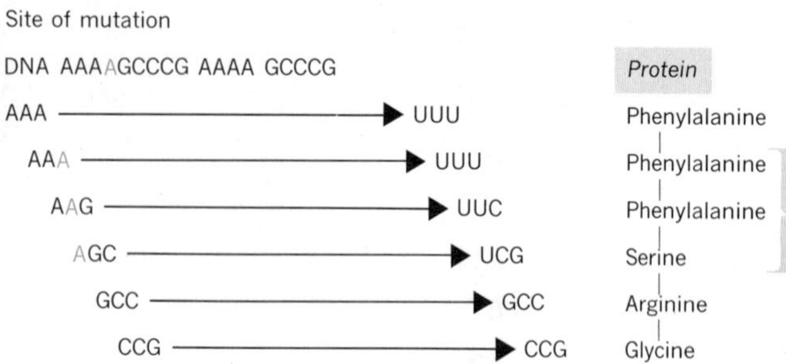

Fig. 10-13 Consequences of an overlapping code. With an overlapping code, a single base alteration would alter *three* amino acids in the final protein.

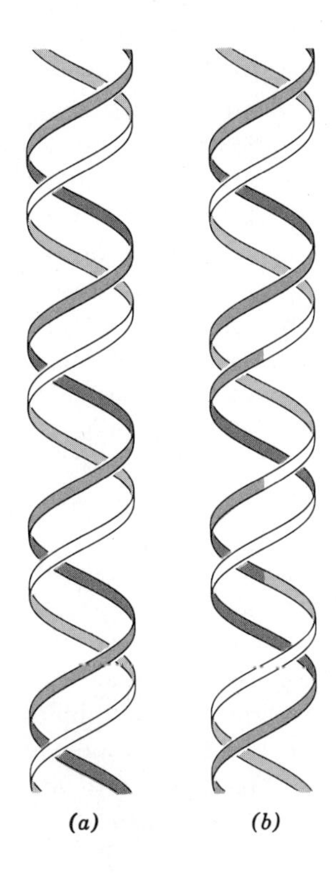

Fig. 10-14 Reading of DNA. It is not known whether a single strand is read or whether reading switches back and forth from one strand to another. It is known that only one strand is read ***at a time.***

General Considerations of DNA as Information

The DNA sequences in a cell contain some redundancy. This redundancy can be **lateral redundancy**, which involves the repetition of the same message in different parts of the cell. In diploidy, two homologous chromosomes carry information for the synthesis of the same proteins, and this is one type of lateral redundancy. Another example of lateral redundancy is the DNA molecule itself. It has been determined that only one strand of the DNA is read. It is now known that one full strand is read incompletely, but the surface of the DNA molecule is read so that the reading switches from one strand to another (see Fig. 10-14). Regardless of how the actual reading is accomplished, however, a region of a strand that is not read, but still carries a message, is a form of lateral redundancy. Of course, the non-read-out strand bears a reproductive function, rather than an informational one, and the redundancy is essential for the passage of the information from one cell generation to the next.

In the formation of the eggs of higher animals, the numerous DNA regions responsible for the formation of ribosomes replicate and dissociate from the nuclear DNA. These free genes pass into the nucleoplasm. The enormous increase in protein synthesis in the egg is thought to be supported by the presence of the abnormally large number of ribosomes. The free genes in the nucleoplasm that produce ribosomes are really repeats of messages that are already contained on the chromosomes. Therefore, their presence constitutes a form of lateral redundancy that is advantageous to the developing embryo.

Another form of lateral redundancy is the multiplicity of chloroplasts and mitochondria in a cell. Each individual chloroplast or mitochondrion contains the same information as another. The multiplicity is, of course, advantageous in assuring that the cells resulting from a cytokinesis will each have at least one unit containing the necessary information (see Chapter 8).

Another way in which redundancy can exist is in the form of **linear redundancy.** There appear to be from 10 to 100 identical sites of the gene that code for one subunit and an equal number of sites for another subunit of the ribosome (the number of sites varies with different organisms). This means that there are a number of sites that are repeated units. The advantage to the cell of such redundancy is, of course, that the greater the number of ribosomes produced per unit of time, the greater the efficiency of protein synthesis. *Since protein synthesis is the final product of the information communication, an increased efficiency of protein synthesis represents an increased efficiency of communication.*

Another example of linear redundancy, the advantage of which is not fully understood, is that in some codons, only two of the three bases seem to be important. For example, GGG, GGA, GGU, and GGC all code for the amino acid alanine. Clearly, only the presence of the two Gs is important; the third base merely tags along, and is redundant. The importance of this in the functional system may be related to the spacing for a triplet code, so that even if only two symbols are necessary, the code must have three symbols to permit uniformity.

An example of the possible disadvantage of a lack of linear redundancy occurs in haploid organisms. The lack of repeated units makes the genotype more susceptible to change. For example, if a haploid organism is subjected to mutagenic treatment (UV light, a mutagenic agent) and a base on the DNA is either altered, added, or deleted, the message will be read incorrectly and the mutation will be expressed. If, however, the message were repeated in some other part of the DNA, the chances that both would be affected by the same treatment would be remote. One message would then be read incorrectly, producing an aberrant protein, but the other homologous gene would be read correctly, producing the normal protein. The expression of the mutant trait would in this case be dependent on whether the amount of "normal" protein is sufficient to retain the original capacity of the organism. Such is the case with diploid organisms, where lateral redundancy serves to decrease the expression of mutations. (Geneticists term the gene responsible for the production of sufficient protein for the expression of a trait in a heterozygous condition, a dominant gene.)

Because haploid organisms have neither lateral (by definition) nor linear redundancy, they are more subject to mutagenic change. If, by chance, the mutation were beneficial to the organism, then the absence of redundancy in haploid organisms would be considered an advantage, but this is rare. The fact that evolution has tended to favor diploidy, and thus lateral redundancy, indicates that the inhibition of mutagenic expression is more valuable than not. *The use of redundancy to advantage in the DNA-RNA system greatly increases its efficiency.*

The DNA system as an information center, and the RNA as a communication system, approach the ideal by which computer scientists judge the efficiency of machines, an ideal that far surpasses any language used for communication in today's world. The DNA molecule has been tailor-made for its function as an information carrier during the early stages of evolution. This is not to imply that life sought out a DNA molecule that would endow living things with a sense of knowing what they need (called **teleology**), but rather that life could not thrive until an efficient information molecule had been formed.

THE DNA MOLECULE

The efficiency of DNA as an information molecule goes beyond the language of the code. Its efficiency is in great part related to the physical chemistry of the molecule itself: those chemical properties that contribute to the stability, rigidity, and efficiency of readout.

The stability of DNA is thought to be correlated with its double-stranded state. When DNA is heated, the hydrogen bonds that hold the base pairs together are broken, and the DNA strands separate. This phenomenon permits the examination of DNA in both single- and double-stranded states.

Double-stranded DNA is highly viscous in aqueous solution. Upon heating, however, the strands separate and the viscosity is greatly decreased. It is thought that the high viscosity exhibited by the double-stranded form is the result of interlacing the large, stiff molecules to form a three-dimensional "mesh" through which water cannot flow. The single-stranded state, on the other hand, is less viscous. This is thought to be because it tends to bend and fold, thereby forming less of a barrier for the water. It would seem that the double-stranded state would be less subject to the effect of deleterious chemicals in aqueous solution than a single-stranded state—a feature that would increase the stability of the molecule.

The idea that the double-stranded state serves to stabilize the DNA molecule by inhibiting exposure to deleterious conditions is also supported by UV studies. Double-stranded DNA is characterized by the absorption of certain amounts of UV light. This absorption is localized in the nitrogen bases of the DNA structure. The single-stranded DNA formed on heating, absorbs more UV light than does the double-stranded form. It is thought that the nitrogen bases are more exposed in the single-stranded form, and this exposure permits the increased absorption. Concomitant with the absorption of UV light is the possibility of molecular alteration. Therefore, the double-stranded state serves to "protect" the molecule from either light-induced or chemically induced alterations, and endows it with a greater stability.

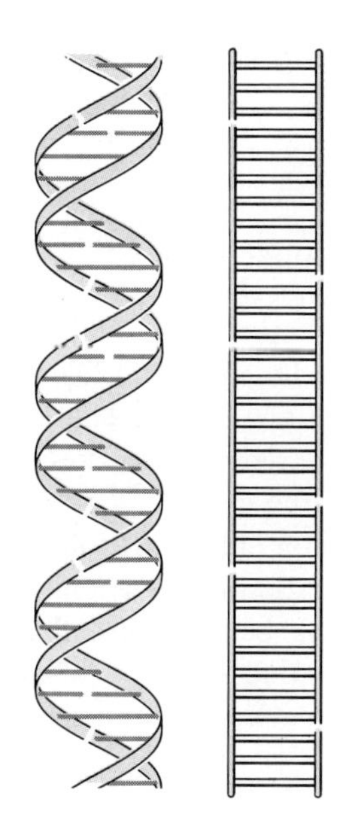

Fig. 10-15 The double chain lends strength to the DNA molecule in providing a structure that is not easily disturbed. If there is one chance in 100,000 (10^5) of a treatment breaking one strand of the DNA, then the chance of both strands being broken would be $(10^5)^2$ or 10^{10}.

The linear stability of DNA is also attributed to its double-stranded condition. When DNA is treated with dilute concentrations of the enzyme DNAse, it loses its high viscosity, but does not change in molecular weight. This is explained by the action of the enzyme on the phosphoester bonds at different points along the chains. It would be unlikely that two bonds at the same level on complementary strands would be broken (see Fig. 10-15). The enzyme attack would then be much like sawing through both legs of a ladder at different levels. The ladder would be weakened but not shortened.

The rigidity of the DNA molecule appears to be, in part, explained by the hydrogen bonds. There are *two* hydrogen bonds that form spontaneously between paired adenine and a thymine, and *three* that form between paired guanine and a cytosine (Fig. 10-16). It would be expected that the more guanine and cytosine present in the DNA, the more rigid the molecule would be, and the more resistant it would be to strand separation. Indeed, the data given in Table 10-6 support this notion.

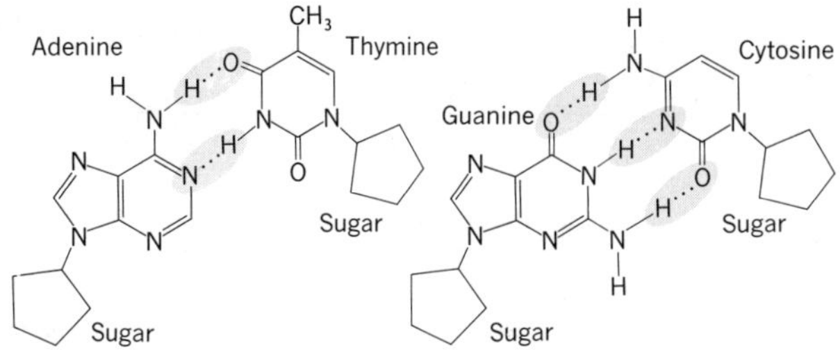

Fig. 10-16 Hydrogen bondings between adenine and thymine, and between cytosine and guanine.

TABLE 10-6 PERCENTAGE OF GUANINE AND CYTOSINE AND TEMPERATURE NECESSARY FOR STRAND SEPARATION

Organism	Temperature Necessary for Strand Separation (T_m)	Percentage of G + C
Pneumococcus	85° C	39%
Salmon sperm	89° C	41%
Calf thymus	89° C	42%
E. coli	90° C	50%
Serratia	93° C	60%
Micrococcus phlei	97° C	70%

The size of the DNA molecule affords ample opportunity for an enormous amount of information to be stored. The DNA of a virus appears, from electron micrographs, to be about 50 micra long, having a molecular weight of 100 million (see Fig. 10-17). The bacterial DNA is also a single molecule with a molecular weight of one billion. Therefore, the DNA molecule is much larger than the largest protein (~5 million), and by far the largest single molecule in the cell.

The linear feature of DNA is also advantageous for efficient reading. DNA is essentially one-dimensional, and the information is sequentially arranged. A molecule tending to remain linear can be read directly.

The physical and chemical properties of DNA that endow it with stability, rigidity, and linearity are essential in the creation of an efficient information molecule.

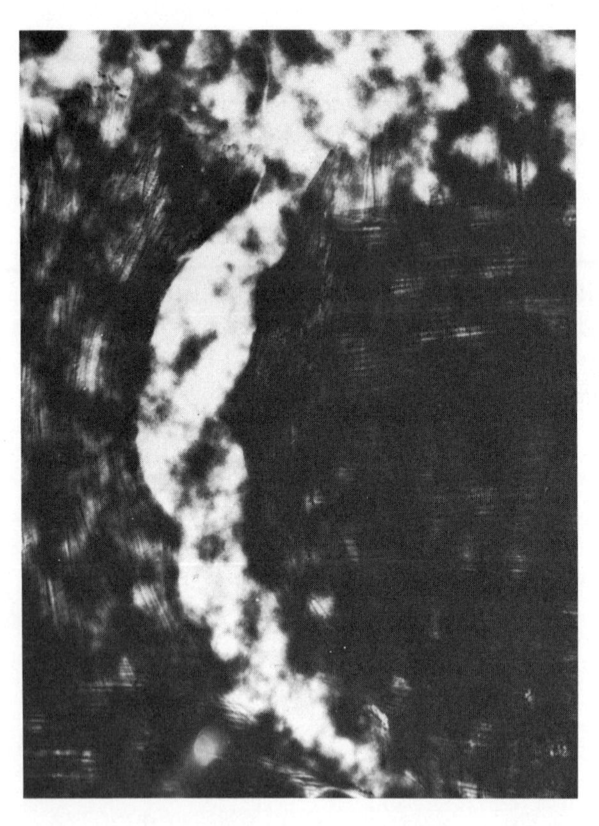

Fig. 10-17 Actual micrograph of DNA (5,600,000×). Pictorial confirmation of the structure of DNA as proposed by Watson and Crick was provided by this photograph obtained in 1969. [Wide World Photos]

PERSPECTIVES OF DNA AND THE GENETIC CODE

The DNA molecule has a number of significant features to contribute to the activity of cells. Its specificity in the sequence of nitrogen bases is symbolic, and therefore informational in nature. Its enormous size (a thousand times greater in advanced cells over that of proteins) and its linearity make it ideal as an information center. Additional features contribute to its stability, replication, and efficiency for readout.

The research emphasis on DNA has been primarily one of deciphering the genetic code, one of the truly great achievements of man. Unraveling this mystery, however, is but the first step, for it reveals the manner of spelling and punctuation. The job of molecular biologists for the next several decades is to discover the rules of grammar. How are the genes positioned to determine when they are to be read and when to be left untranscribed? How are the genes read to produce the optimal number of RNA products? The new perspective of molecular genetics is to view DNA not only as informational but also as instructional: what is the recipe for building cells?

QUESTIONS

1. How did Chargoff's data provide for the uniform diameter of the DNA molecule?
2. Describe the experimental evidence provided by Taylor that DNA replicates in a semiconservative manner.
3. Give the experimental evidence provided by Meselson and Stahl that DNA in *E. coli* replicates in a semiconservative manner.
4. What characteristics of DNA justify its being considered an information system?
5. What are the criteria for the maximum efficiency of a communication system? How does DNA fulfill these criteria?
6. On what basis did Gamov propose a three-symbol (trigram) code?
7. How did Nirenberg experimentally show that (a) a code exists, and (b) the code is a triplet?
8. What sequence of amino acids would appear in a protein, the RNA code of which is AAAACAGCCCCUUUUGGGGGAUCA?
9. What evidence is there that the code is not overlapping?
10. Give two examples of lateral redundancy in the DNA information system.
11. Give an example of linear redundancy in the DNA information system.
12. What aspects of the structure of DNA contribute to its stability in form and function?

SPECIAL INTEREST CHAPTER II

RADIATION

On August 6, 1945, a parachute bearing a 20-ton cargo quietly descended over a little-known Asian city. Within a few seconds an immense ball of flame erupted, followed by an enormous shock wave and heat blast. A towering mushroom unleashed a blanket of radioactive waste. About 100,000 people died and 10,000 buildings were destroyed in the holocaust. This was the birth of the nuclear age. The first atomic bomb rocked civilization; man had finally succeeded in harnessing enough energy to destroy himself and all of life.

The decision as to whether and when to use nuclear weaponry was, is, and probably always will be essentially political. But a government is not blind to public opinion. A citizenry unaware of the vast destructive potential and duration of radiation effects, and very tired of war, applauded the use of the bomb in 1945. This same citizenry, upon returning to peacetime moral concepts and realizing the overwhelming power of such weapons, has since expressed disapproval. The fact that nuclear weaponry has not been used in any conflict between two countries since 1945 indicates, to some degree, the power of public opinion. It is possible, however, that under circumstances of war similar to those in 1945, even an informed public might support the use of nuclear weapons.

In the final analysis, the validity of using destructive nuclear power involves one's personal philosophy as to whether the value of another individual's life is relative or absolute regardless of the circumstances (a philosophical discussion that is outside the realm of this text); it also should be a decision made with a knowledge of nuclear "kill" power and long-range biological effects on the world population. Nuclear "kill" power is mentioned almost daily in the newspapers. The less-discussed long-range biological effects are the topic of this special chapter.

Nuclear power is a two-sided coin, on one side destructive and all-powerful, and on the other side constructive and useful (radiation therapy, etc.). Both sides have a common factor, the emission of ionizing radiations, which are always destructive to biological systems. In some cases, (for example, cancer), the destructive aspect of radiation is highly desirable. In most instances, however, this effect is undesirable.

Man depends on his senses to perceive environmental factors that may be harmful. He sees fire, or feels heat, and instinctively moves away from its source. Unfortunately, ionizing radiation cannot be detected by the human senses. In the past, there was never sufficient radiation in concentrated areas to constitute a danger, so man's inability to detect radiation did not threaten his continued existence. The increased levels of radiation that permeate the atmosphere from nuclear testing and frequently unnecessary use of radiation in medical practices have made his inability to directly detect ionizing radiation a distinct disadvantage.

The intelligent use of nuclear energy depends on a common awareness of the different types of radiation emitted from explosive devices, controlled devices (medical instruments), and natural sources (the sun and cosmic rays), and their effect on biological systems.

Types of Radiation

Radiation always involves the movement of subatomic particles in such a way that they exhibit wave properties. As with any wave, there is an energy value, measured in **quanta.** In general, the shorter the wavelength, the greater the energy content of the radiation. Therefore, as one proceeds down a listing of radiation types, such as that found in Fig. II-1, the energy levels of the radiation decrease.

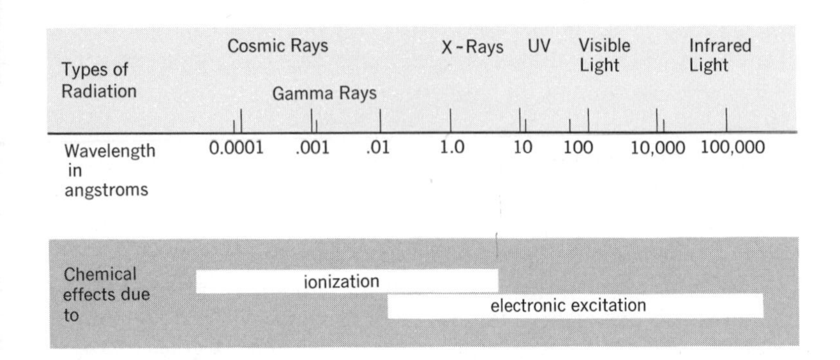

Fig. II-1 Types of radiation.

Several properties of radiation are dependent on the energy level. Those forms with higher energy levels are capable of traveling distances varying from a few feet (for lower high-energy radiations such as X rays) to a few miles (for higher high-energy radiations, such as those emitted by radioactive decay—α and β particles).

Those forms of radiation that have relatively high energy levels (neutrons to X rays) are capable of directly changing the atomic configurations with which the radiation comes in contact. These changes result in a loss of an electron to the atomic structure and the formation of an ion. In some instances, an electron pair from an inner orbital may be moved to an orbital in an outer shell, bringing the atom to an "excited" state in which it is highly reactive, and is called a **free radical.**

Because of the effect of this type of radiation, it is called **ionizing radiation.**

The other group, **nonionizing radiation,** is composed of photons with wavelengths in the ultraviolet and visual range (radio waves and sound waves). Although it is true that light in the red and blue range forms free radicals in the chlorophyll molecule, the chlorophyll molecule is specifically adapted for this effect. The energy content of red and blue light is insufficient to affect a nonspecific molecule.

Ultraviolet light is considered a form of nonionizing radiation, and as such it is of sufficient energy content to bring about indirect alterations of atomic structures. The effect of ultraviolet light is dependent on the specific molecular configurations that tend to absorb it. The most important biological molecules that absorb UV light are DNA, RNA, and (to a lesser extent) protein.

CHEMICAL EFFECT OF IONIZING RADIATION

Once ionizing radiations emanate from their source, they will pass undisturbed along a path until they come into contact with some matter. Since matter at the atomic level is mostly space, these radiations may move considerable distances before some interaction occurs.

The kind of interaction depends on the properties of both the radiation and the matter it comes into contact with. Nonionizing ultraviolet radiation has a relatively low energy level, and will tend to be absorbed only by susceptible molecules. The energy inherent in UV radiation causes a displacement of an electron pair in one of the atoms of such a molecule to a higher shell, thus forming a free radical. Often, one of the displaced electrons is lost. Since atomic stability is in part due to paired electrons with opposite spins (that nullify each other's effects) the loss of one electron creates a highly unstable atomic state, an ion (see Fig. II-2). Ultraviolet light does not break molecular bonds or "knock out" atomic particles directly, but, rather, causes an atomic instability by the addition of energy.

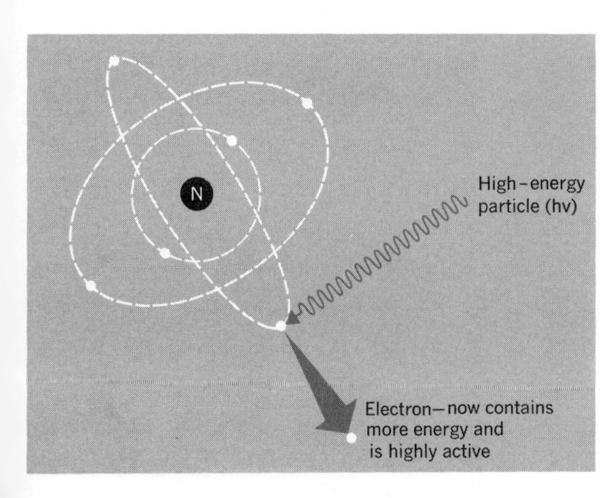

Fig. II-2 Loss of an electron brings the atom, and subsequently the molecule of which the atom is a part, to a higher energy level.

Ionizing radiations have sufficiently high energy levels to alter atomic structure directly. When this occurs, the form of radiation acts in accordance with its particular wave properties and the energy associated with them. When X-rays or gamma rays come into contact (undergo a collision) with an atomic nucleus, the stability of the nuclear structure is of sufficient magnitude to remain undisturbed by these particles of such small mass. However, when these same rays undergo collision with the electron of an atom, it may be physically thrown out of the atomic configuration. This leads directly to the formation of an ionized atom within a molecule.

The ionizing radiations of even higher energy levels with particles of greater mass (α and β particles) have a more devastating effect. Although their collision with an atomic nucleus does not usually give rise to an alteration, their collision with an electron causes it to move with sufficient speed to knock out other electrons of other atoms, so that a number of ionizations occur. (It should be noted that the effect of the collision of either an α or β particle with an atomic nucleus is a function of the stability of that nucleus. Generally, the more protons and neutrons in the atomic nucleus, the greater its instability. Therefore, upon the collision of an α or β particle with an unstable nucleus, the nucleus itself will degenerate by throwing off α and/or β particles that are capable of undergoing collision with other unstable nuclei and producing a similar effect. This is the basis of the so-called **chain reaction** on which the explosion of nuclear bombs is based.)

In short, the higher the energy level of the ionizing radiation, the more immediate and more profound is the damage to atomic structure.

Radiation Effects on Biologically Important Molecules

Relatively little study has been made of the effects of α and β emissions on biologically important molecules. Indeed, most of the work that has been carried out in this area has utilized UV or X-rays as radiation sources. It is assumed, however, that the damage produced by these lower-energy radiations is of the same nature and quantitatively less than the damage produced by the higher-energy radiations.

Proteins. When one site of a protein becomes ionized, as the result of X-radiation, electrons from adjacent parts of the molecule move to the incomplete orbitals. When they do so, their orbitals are left incomplete. Again, electrons migrate from adjacent sites, thereby leaving *their* orbitals incomplete. In this way, the ionized site moves along the protein structure (see Fig. II-3). The movement will continue at the rate of 1000 sites per 10^{-9} sec until some site important for folding is altered. If movement of the ion hole does not proceed beyond this point, the protein will unfold and its shape will be altered permanently. This theoretical consideration explains the inactivation of proteins by X-rays.

The effect of the alteration of a protein on cellular function is dependent on the abundance of the protein type in the cell and the

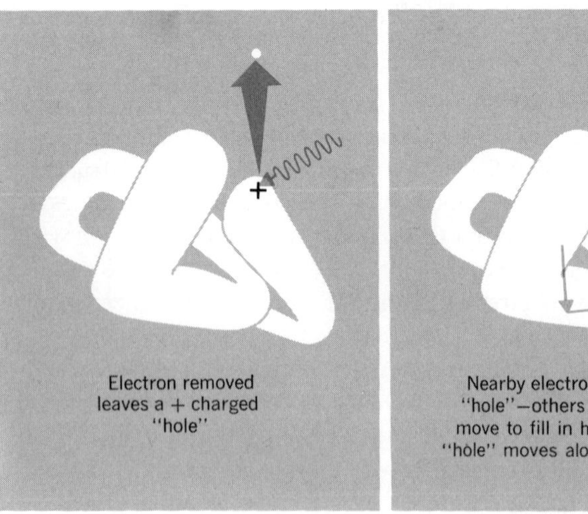

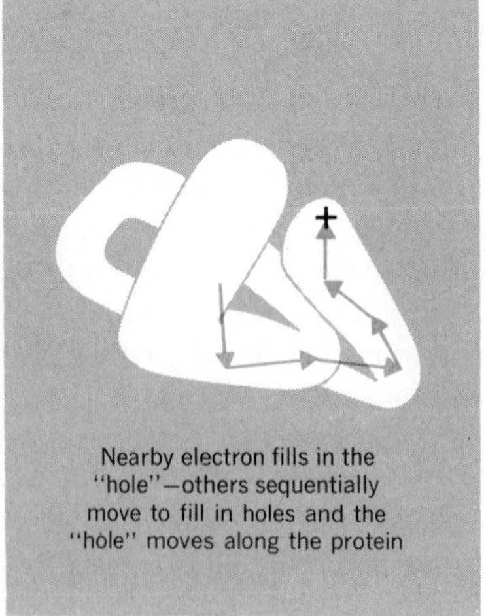

Fig. II-3 The movement of an ionized site along a protein after irradiation.

importance of its position (if it is a structural protein) or function (if it is an enzymatic protein). In general, each type of protein is present in sufficient multiplicity (thousands) so that cellular function is not severely disturbed by the destruction of some protein molecules.

It is conceivable, however, that a sufficient number of proteins can be damaged to significantly slow metabolic and growth processes. If the genetic sites for the production of these proteins remain undisturbed, more protein structures may be synthesized, and the metabolic and growth rates will return to normal.

Nucleic Acids. Because the function of the nucleic acids is such that it governs cellular integrity, any alteration of the atomic structures within the nucleic acid molecules can have a long-range effect. Therefore, the alterations of nucleic acid caused by radiations are of the greatest importance.

Because of the chemical nature of their nitrogen bases, both DNA and RNA preferentially absorb UV light. With the absorption of UV light, the nitrogen bases will become free radicals, and therefore, be highly reactive with other molecules in the cell. This instability will often lead to the conversion of one base into another, in such a way that a purine base is converted to another purine, or a pyrimidine base is converted to another pyrimidine (see Fig. II-4).

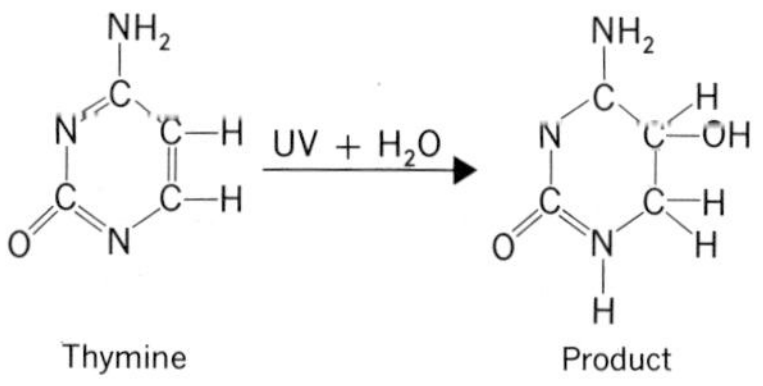

Fig. II-4 Conversion of pyrimidine (thymine) by irradiation in an aqueous environment.

If such an event occurs in an *m*RNA molecule, the triplet code for a given amino acid will be changed, and an aberrant protein will be formed. But a given *m*RNA molecule is active for the formation of only about ten proteins, after which it is depolymerized. Therefore, only a few aberrant proteins would be formed, and the effect on cellular function is trivial.

On the other hand, if the base substitution occurs in the DNA molecule, the effect can be lasting. When the readout strand is altered by radiation, the DNA will be consistently read incorrectly, and all the proteins formed from that genetic site will be aberrant. The severity of the consequences is dependent on the importance of the particular protein. If, for instance, the protein served an enzymatic function for the synthesis of an amino acid, the cell could continue to live as long as that amino acid were supplied by the environment. The cell would simply manifest the additional nutritional requirement of the amino acid. If, however, the protein were an essential enzyme in the electron transport system, the inability of the cell to produce the enzyme would constitute a lethal factor.

Another way in which the base substitution could act as a lethal mutation is if it affected a DNA site that produces either the *t*RNA or activating enzyme for a given amino acid. This would mean that the amino acid could not be incorporated into protein structure, and all protein synthesis would cease.

When the nonreadout strand (the replicative strand) of DNA is altered by ionizing radiation, the effect is not immediately detected. The replication of the DNA, in the course of nuclear division, will bring about

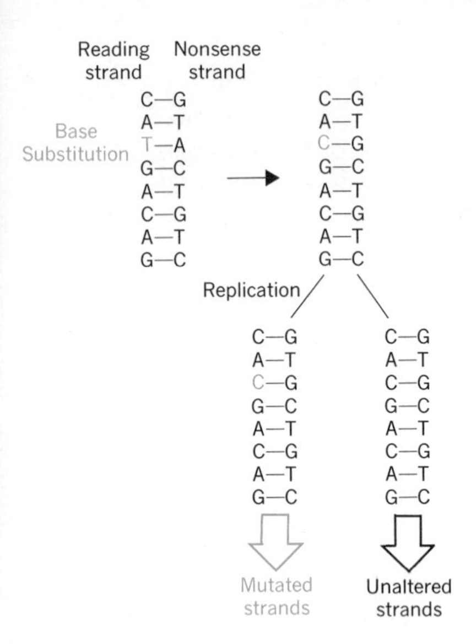

Fig. II-5 Effect of an altered single strand.

the formation of one correct DNA molecule and one mutant DNA molecule, in which the reading strand is altered (see Fig. II-5). In such a situation, half the descendant cells will be affected.

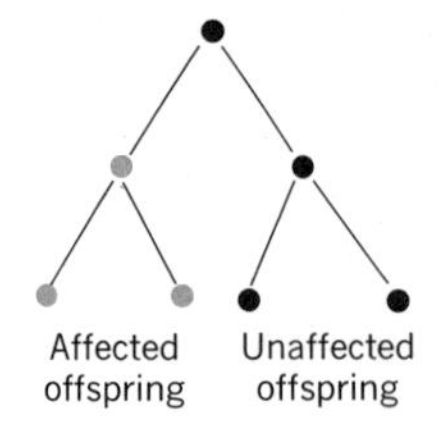

This alteration of a single base by UV light is called a **point mutation.** UV light can also cause a single- or multiple-base deletion. If such a deletion is not a multiple of three, it will throw the entire reading frame off. If the deletion extends over a region of more than one gene, the synthesis of more than one protein may be affected.

X-radiation appears to cause more severe structural damage to the nucleic acids. It is generally accepted that X-rays cause **deletions** and **rearrangements** rather than point mutations. Their effect is generated by the breakage of the phosphate ester linkages at one or more sites along the DNA strands (see Fig. II-6). This results in the loss of a large number of bases and subsequent loss of the cell's ability to synthesize the protein associated with those genetic sites. While the survival of the cell is dependent on the importance of those proteins in cellular function, the means by which X-radiation alters the nucleic acids increases the chances of lethality.

The chemical effect of other ionizing radiations with higher energy content is purely conjectural. It is assumed, however, that in dosages equivalent in potency (**roentgens**)* to X-rays, they would cause more deletions, and the chance of lethal effects would concomitantly increase.

Other Molecular Forms. Radiation produces ions and free-radical formations in many molecular forms other than proteins and nucleic acids. If the dosage is high, there may be sufficient alteration of organic buffers, carbohydrates, nucleotide triphosphates, porphyrins, free amino acids, etc., to cause a tremendous shift in the electrical and energetic equilibria of the cell. This shift can result in the dissociation of important macromolecules and the formation of aberrant molecular aggregates that cause gross malfunction and eventually the death of the cell.

This effect has been dramatically demonstrated by irradiating cell-free solutions of nucleic acids, nitrogen bases, etc., and then adding cells. Even though the cells themselves had not been irradiated, they died. Their death appears to have been caused by the entrance of ions and free radicals into the cell, causing the imbalance described above.

* A measured unit of high energy radiation; one roentgen equals the amount of energy absorbed by one gram of biological material that dissipates 83 ergs.

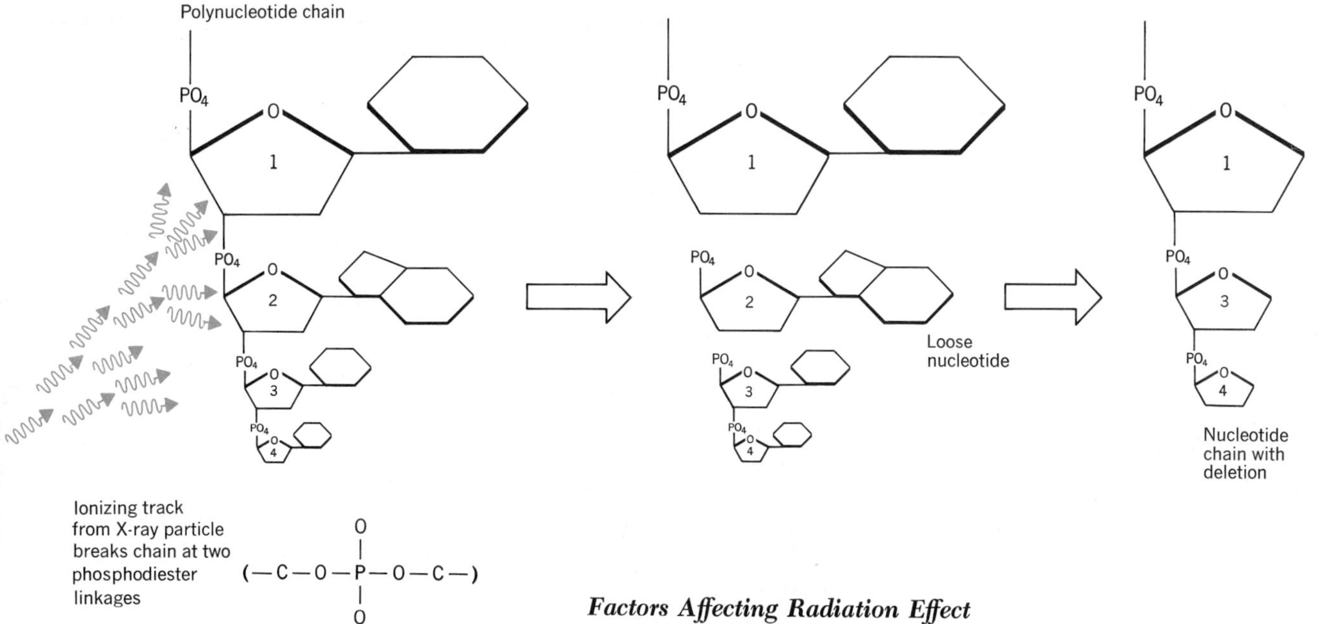

Fig. II-6 Effect of X-rays.

Factors Affecting Radiation Effect

Several factors appear to alter the effects of ionizing radiation. Certain of these have been studied exhaustively, and are fairly well understood.

It has been found that radiation effects increase directly with the oxygen content of the system. Oxygen tends to incorporate into molecular structures, forming aberrant molecules called **peroxides.** The additional oxygen present in an organic peroxide acts as an oxidizing agent. Peroxides are powerful reactants that react indiscriminately with molecules. Hence, the elusive energy in radiations is converted to bond energies of relatively stationary molecules. On the other hand, reducing agents, which give electrons to the cell, tend to decrease the effects of radiation, presumably by interacting with the few peroxides that are formed.

A factor that alters the radiation effect of UV light is the presence of blue light. When cells have been exposed to UV light, and are subsequently exposed to blue light, the effect of the UV irradiation is markedly decreased. This phenomenon is called **photoreactivation.**

One of the consequences of UV irradiation is the formation of free radicals of thymine. Often two thymine molecules will bond together to form a **thymine dimer,** which is then incorporated into the DNA structure as a single unit. This leads to an alteration of the sequence of bases in the DNA, the prologue to a mutation. It has been shown that blue light stimulates the production of a repair enzyme that is capable of recognizing the thymine dimers in the DNA molecule, and removing them. It is thought that it is by this effect that blue light decreases UV damage.

The major factors that affect the extent of radiation damage are the dosage and duration of exposure. All ionizing radiations have what is called a **cumulative** effect. This is primarily because of their effect on nucleic acids. If a given site on a DNA molecule has been altered by a single dose of radiation, and another site is altered by another dose of radiation, there would be two alterations of the molecule and at least two aberrant proteins formed. Effects of subsequent irradiations would add to the already-present effect.

On the other hand the alteration of proteins and other molecular forms has a **transitory effect** that lasts only until these molecules are replaced by others. It should be noted, however, that this replacement may take several days to occur. Additional exposure to radiation during this time may bring the number of altered molecular forms in the system to a critical level, causing the death of the cell. In this sense, transitory effects may also be cumulative.

The most common example of cumulative damage is seen in the epithelial cells of "outdoor" type individuals. After many years of exposing the cells of the skin to UV irradiation by sunbathing or beach activities, the molecular damage is so extensive that the DNA cannot replace the damaged proteins, and the cells undergo degeneration. Although light in the visible range does tend to decrease the UV effect by photoreactivation, constant exposure precludes significant repair. The overall effect is a premature dryness and wrinkling of the skin.

RADIATION EFFECTS ON CELLULAR COMPONENTS

The effects of ionizing radiation on cellular components are directly related to the chemical composition of the organelles themselves.

Cytoplasmic Organelles

The multiplicity of the different types of cytoplasmic organelles serves to protect the cell from the deleterious effects of irradiation. If the DNA of one mitochondrion is affected in such a way that it can no longer produce functional daughter mitochondria, it is of little consequence to the cell. Theoretically, as long as one functional mitochondrion survives irradiation, reproduction of that one will eventually build up the full complement of mitochondria, and the cell will return to its initial metabolic rate.

Nuclear Components

The deletion of bases by UV or X-radiation is dramatically reflected on a cellular level. Examination of cells that have been irradiated reveals the presence of chromosomal fragments. The deleted portions of the DNA will still manifest themselves in chromosomal form.

But what happens to these deletions during a nuclear division? When the chromosomes align themselves on the equatorial plane during metaphase, they are oriented by their centromeres. Accordingly, when the chromosomes go to the poles during anaphase, it is their centromeres that show active movement. Chromosome fragments often do not contain centromeres. As a result, they neither line up on the equatorial plane nor go to the pole, but merely drift in the general region where the separation takes place. Subsequent formation of the nuclear membranes at telophase usually excludes these fragments (by chance alone), and they

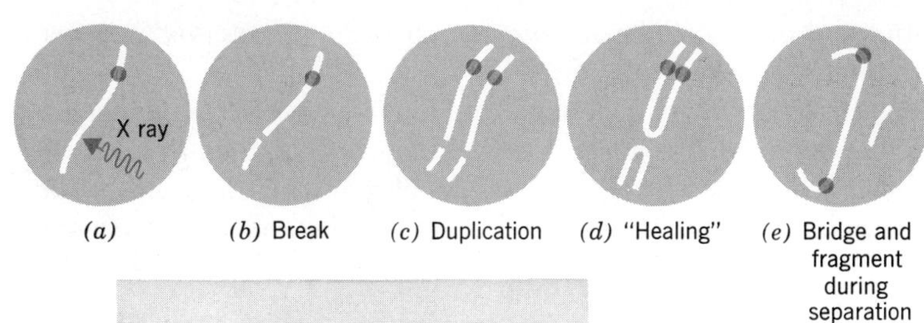

(*a*) (*b*) Break (*c*) Duplication (*d*) "Healing" (*e*) Bridge and fragment during separation

(*f*)

Fig. II-7 Chromosome breakage from X-radiation: (*a-e*), Previous events that lead to (*f*) at division. [(*f*) Brookhaven National Laboratory]

end up in the cytoplasm of the daughter cells (see Fig. II-7). The alien cytoplasmic environment does not permit either their readout or their duplication, so they are essentially lost to the cell.

It is important to note that as long as the cell does *not* undergo division, the DNA complement of the nucleus is complete and, although the points at which the breaks in the DNA molecule occurred may cause some misreading, there is no quantitative loss of the cell's protein-synthetic capacities. This is the reason why differentiated cells are relatively unaffected by chromosome breaks, while the same phenomenon is a death blow to embryonic cells.

EFFECT OF RADIATION ON CELLS

The survival of a cell that has been exposed to ionizing radiation is, of course, directly related to the extent and type of damage that has been done on a molecular level. When isolated cells are bombarded with high-energy radiations, the two most noticeable effects are death and delay of cell division. The delay of cell division is thought to represent the time required for the cell to overcome the radiation effects (establish the normal complement of mitochondria, replace aberrant proteins, wash itself free of ions and free radicals by exchange with the environment, etc.). Death, on the other hand, refers to a **genetic death.** This is defined as the *inability of a cell (or organism) to produce reproductively capable offspring.* An irradiated cell may never divide, or may divide once into two cells incapable of further division. In either case, the irradiated cell is considered dead.

Genetic Death

A variety of chemical alterations may result in genetic death. The alteration may have affected metabolism in such a way that a sufficient metabolic rate necessary for division cannot be attained. Or the mutation may have affected the synthesis of a protein necessary in the division process. Or the chromosome fragments lost during the division process may have contained information essential for the cell's continued existence.

It is important to realize that an individual cell may manifest a high rate of metabolism and other features that one usually associates with living systems. For instance, the protozoan *Paramecium,* when given a lethal dose of X-rays, will continue to swim and exhibit feeding activities and other cellular actions for up to two years; yet it never divides, and so is genetically dead.

The consequences of genetic death of individual cells in the human is very important. In the adult, relatively few cells in the human body undergo division. Therefore, genetic death of these cells does not impair their functionality, and little (if any) effect may be noted. The obvious exception to this, in the human, is the formation of the gametes. The gametogenic cells are undergoing repeated divisions, and the genetic death of a sufficient number of these can greatly reduce the reproductive capacity of the individual. The genetic death of the individual gametes (egg or sperm) would mean that a zygote formed from the union of a genetically dead gamete and a live gamete would be unable to divide and form a new individual.

Although the gametogenic cells are not functional in the human until puberty, and effects on these cells will not be noticed until then, many other types of cells are actively multiplying in the course of growth processes. The genetic death of a large number of these cells can have a serious effect (see Table II-1).

TABLE II-1 RELATIVE SENSITIVITY OF NORMAL TISSUES TO X RADIATION[a]

Cell Class	Cell Type	Radiation Sensitivity
White blood cells	Lymphocytes	2.5
	Polymorphs	2.4
Germinal cells	Ovarian	2.3
	Testicular	2.2
Blood-forming cells	Lymphatic	2.0
	Spleen	2.1
	Bone marrow	1.9
Endocrine glands	Thymus	1.8
	Thyroid	1.7
	Adrenal	1.6
	Endothelium of blood vessel	1.5
Dermal cells	Hair papillae	1.4
	Sweat gland	1.3
	Sebaceous gland	1.2
	Mucous gland	1.1
	Skin	1.0
	Serous membrane	0.9
Miscellaneous	Liver, pancreas	0.7
	Fibrous tissue	0.5
	Muscle	0.4
	Bone	0.3
	Nerve	0.2
	Fat tissue	0.1

[a] Skin = 1.0; all other values are relative ratios.

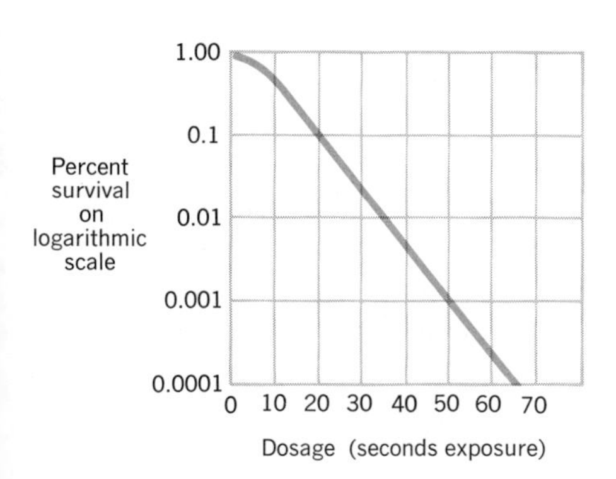

Fig. II-8 Survival curve.

Relationship of Radiation Dosage to Genetic Death

A million bacteria may be spread on a petri plate and given a dose of either 5000 r (roentgens) of X ray or its equivalent in UV light (20,000 ergs). After incubation, only 100 colonies would have been formed; only 0.01% of the bacteria would have survived the irradiation.

$$\frac{10^2}{10^6} = 10^{-4}$$

$$10^{-4} \times 100\% = 0.01\%$$

The remainder of the bacteria, although on the plate, would not produce enough offspring to be identified as a colony.

Original studies done in the 1930s indicated that the survival rate of cells varied inversely with the dosage. The data may be converted to a graph form called a survival curve, as in Fig. II-8.

The cumulative effect of radiation may also be demonstrated on a cellular level. If a given genetic death rate is noted when a dose is given for twenty consecutive seconds, that same death rate is noted when the twenty-second treatment is given in a series of four *closely* spaced five-second treatments. The cumulative effects of many small doses are *not* equal to one large dose if the time intervals between treatments are long enough to permit repair. The previously explained chemical effects of radiation provided the explanation for this phenomena (see pp. 285–290).

Radiation and Nonlethal Mutations

Although the majority of mutations induced by exposure to radiation are lethal, some genetic alterations permit the continued existence of the cell in a changed form. The number of nonlethal mutations increases with radiation dose.

Most of the mutant forms of bacteria, viruses, *Drosophila*, mice, etc., that geneticists work with have been induced by UV or X-radiation (UV for bacteria and viruses; X-radiation for *Drosophila* and *mice*). For scientific inquiry, radiation has produced an important and valuable tool.

The explosion of the atom bomb at Hiroshima made an innocent world all too aware of the possible effects of irradiation. Although the confusion after the war made actual statistical analysis difficult, a qualitative, if not a quantitative, study revealed important consequences of the release of high-energy radiation.

One of the first effects noticed was the decreased fertility of men and women within a hundred-mile radius of the explosion. Although this analysis is based on the low number of births following the war, which may have, in great part, been due to the lack of desire to have children in a war-torn country, and the legality of implementing this by abortion, it is thought that at least some of it was due to the effects of the bomb itself. The other factors make it difficult to quantitatively determine the actual reduction in reproductive capacity.

Another effect that was noted was the propensity of female births in areas within the radius of a few miles. Although this, once again, may have been due to the random abortions that by pure chance may have terminated the growth of more male than female fetuses, it is thought to have, in part, been a direct result of the emission of radioactive substances. Geneticists have speculated that sites on the X chromosome could have been affected in such a way as to cause a recessive lethal mutation. When the X chromosome was present in a single form in the male, the lethal effect was expressed, and the potential male offspring either never developed beyond a zygote, or died early in development. When the altered X chromosome was present in the female, however, and its homologous X did not contain the same recessive lethal (the chances of which would be remote), a normal fetus could develop. The proof of this hypothesis requires the examination of the offspring of these female children born from 1946 to 1950, and perhaps of their offspring.

There also appeared to be an increase in the number of cases of cancer, primarily leukemia. The effects of radiation in producing cancer will be discussed in Special Interest Chapter IV.

THE EFFECTS OF RADIATION ON THE ORGANISM

The effects of radiation on the organism can be divided into two categories, genetic and metabolic.

Genetic Effects

Because the genetic effects of radiation are cumulative, constant exposure can bring about a degradation of the genetic material to the point where it can no longer produce essential proteins. When the minimum (necessary) number of cells in a given individual have been incapacitated in this way, the individual may die.

Each type of organism appears to be able to cope with a characteristic amount of radiation, which constitutes a radiation LD_{50} for the species (for a definition of LD_{50}, see Special Interest Chapter I). The LD_{50} for the human is 400 r. Cumulative exposure above this level is extremely dangerous.

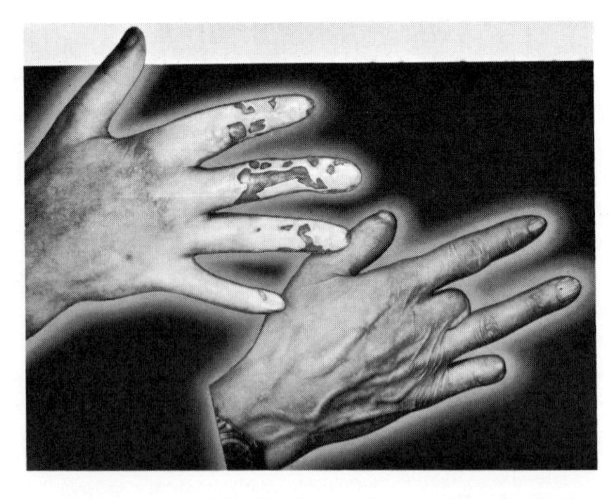

Fig. II-9 Early radiologists frequently experienced radiation damage to their hands. [American Cancer Society]

Metabolic Effects

Metabolic effects are usually noted only when an individual has been exposed to a large dose of radiation over a relatively short period of time. It is thought that a large dose of radiation produces an abundance of ions and free radicals that cause imbalance in the cells and the systems, which is ultimately detected by the individual as a series of symptoms that are collectively called **radiation sickness.** The various organs of the human body manifest a range of susceptibility to radiation, and therefore manifest symptoms at different radiation dosages. These symptoms become more severe with increased dosage (see Table II-2 and Fig. II-9).

Radiologists and X-ray technicians usually wear badges containing a material that absorbs radiation and changes color when the total absorption reaches a given level. When the change in color occurs, these people must remain away from radiation sources for a sufficient length of time for their bodies to recover from the transitory effects of radiation that might cause radiation sickness.

TABLE II-2 SYMPTOMS OF RADIATION SICKNESS IN MAN IN WEEKS AFTER EXPOSURE

Time After Exposure in Weeks	Probable Moderate (100–300 r)	Possible Median (400–700 r)	Probable Lethal (800 r)
1	Latent phase; no definite symptoms	Nausea, vomiting	Nausea, vomiting, diarrhea, inflamed mouth and throat, ulcers, fever, emaciation, and rapid death
2		Epilation, loss of appetite, sore throat, diarrhea, emaciation, death	
3	Epilation, loss of appetite, malaise, sore throat		
4	Diarrhea, emaciation (moderate)		

Some General Considerations of Radiation

In many cases, one cannot control his exposure to radiation. It is certainly beyond the control of an individual if the U. S., Russia, France, or China tests nuclear weapons and permeates the environment with radioactive wastes. Exposure to many other forms of radiation is not beyond the individual's control.

The average dentist, doctor, and X-ray technician have their X-ray equipment calibrated when it is installed, but rarely have it checked. The innocent-looking X-ray machine may be giving out many times the dosage necessary for an X-ray plate and may be scattering rays throughout the room.

Also, X-rays are often given unnecessarily, and even when they are important for diagnosis, the recipient is not given adequate protection for other parts of his body.

In addition to this, the value society places on the sun-tanned Adonis or Greek goddess causes many individuals to soak up the sun, exposing their skin to its ultraviolet rays. To make sure the skin receives the maximum exposure, some people use reflective devices.

From a biological standpoint, nothing would be more ludicrous than a group of people in bathing suits on a beach in the bright sunshine, carrying signs that say "stop nuclear testing."

PART THREE: GROWTH AND CELLULAR DIFFERENTIATION

The biological principles and implications of Chapters 11 and 12, "Growth" and "Cellular Differentiation," are so broad that convincing arguments could be made for their placement in any one of several places in this text. The rationale for placing them directly after the major discussion of the cell, and prior to the discussion of evolution, was (1) that growth and cellular differentiation are the immediate products of cellular processes, (2) that research either directly on growth or differentiation or on using growth or differentiation as research tools, is usually carried out by cell biologists, and (3) that these chapters serve to bridge the cell theory with the population concept used in the following section.

The implications of growth and cellular differentiation permeate several major areas, and appropriate references are made in future chapters to Chapters 11 and 12. The inclusion of a separate chapter on cellular differentiation represents a major departure from more classical approaches, in that much of the material of this chapter is usually found in chapters on development. A discussion of the expression of cellular potential through differentiation is crucial to discussions of evolution.

(Opposite) Brassai / Rapho Guillumette

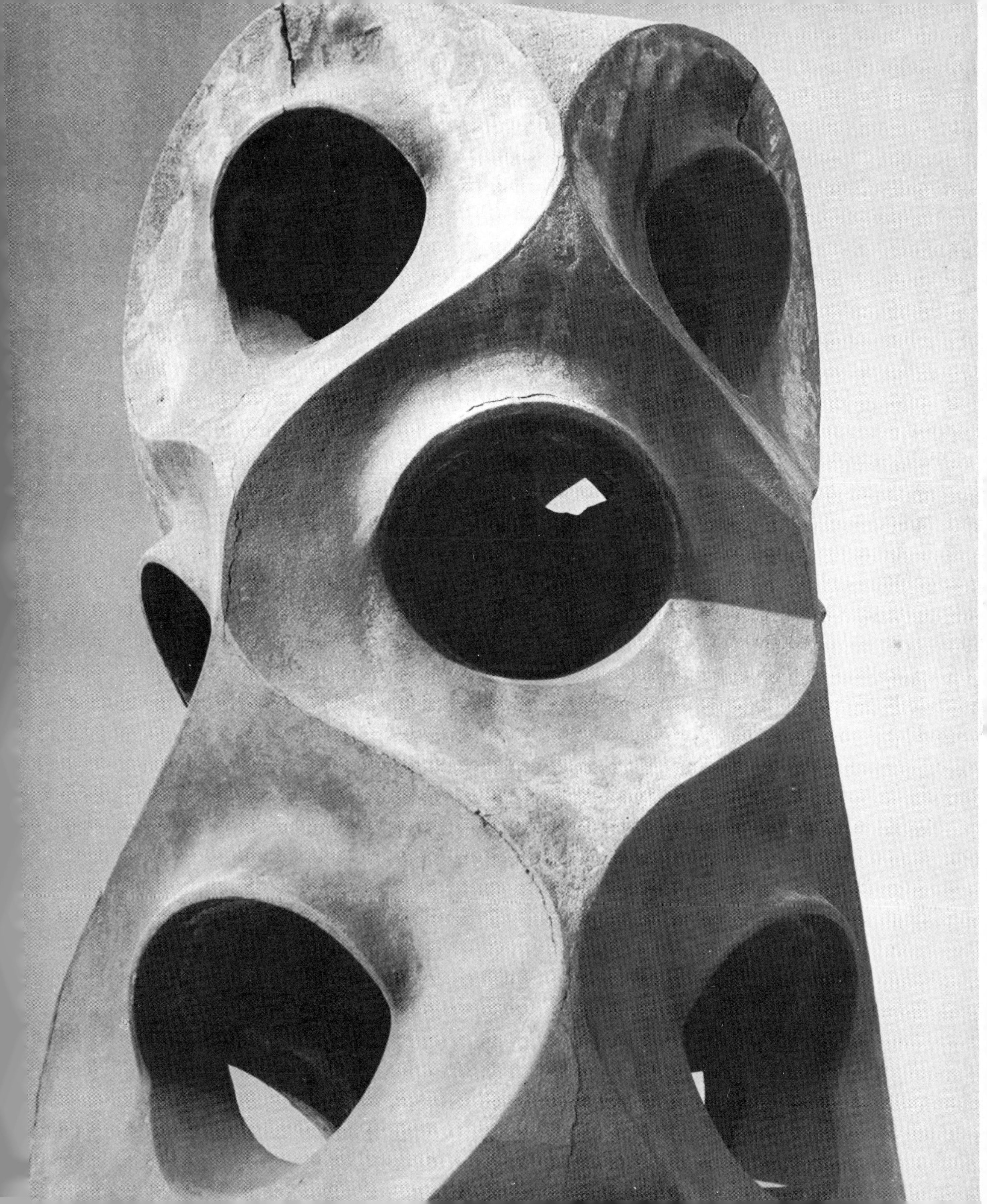

CHAPTER 11

GROWTH

As discussed in Chapter 7, the energy that a cell derives from metabolic processes is largely used in the synthesis of cellular material. As cellular material is synthesized, the protoplasmic mass of the cell increases. *This increase in cellular mass is called* **growth.** Growth is manifested either by an increase in size of the individual cell, or by the division of a cell into two daughter cells. Although there are some examples of "giant" cell growth in organisms, most growth processes at the cellular level involve a cyclic repetition of increases in cell mass followed by the division of this mass into two cells.

The biological definition of growth is somewhat more restricted than that in normal usage. The biological concept refers only to the increase in protoplasmic mass, not to the cytoplasmic rearrangements that characterize differentiation. Thus, when a cell divides and differentiates, growth considerations dictate that attention be focused on the cell number resulting from a division, rather than on the types of cells that result.

Growth Phenomena as a Research Tool

In order to undergo a successful division, a cell must have sufficient nutrient material from which to derive carbon skeletons, and energy in the form of ATP. It must form the *t*RNAs and the *m*RNAs necessary for protein synthesis, and must synthesize a whole host of enzymes necessary for efficient cellular function and the division process itself. DNA must be replicated. Finally, the cell must form a spindle, undergo cytokinesis, form new nuclear and cell-membrane material, etc. If any of these functions is not completed, the cell cannot complete the division process.

To the biologist, the completion of cell division indicates that the normal sequence of events has occurred, and an incomplete division indicates that one or more of the cell activities has been interrupted. This simple concept permits biologists to *use* growth as a means of measuring the effects of experimental procedures on a cellular level. In a typical experiment, an investigator may compare the optimal growth rate of an organism or population of organisms with the growth rate when some factor is added to the medium. If the growth rate is diminished by the addition of the factor, he knows that one or more of the features leading to a successful and repetitious division has been impeded. He can then go on to perform other types of experiments to determine the feature that has been altered. In another type of experiment, an investigator may be looking for a means to overcome some factor that diminishes growth. Here he would compare the optimal growth rate with the growth rate achieved when that factor and its inhibitor are added to the medium. Hence growth, which can be studied as a biological phenomenon itself, is more often used as a means of measurement.

It is often important that an investigator select a growth system that is not complicated by processes of differentiation. Most of the work, either on growth itself or on growth as a means of measurement, has been done on unicellular organisms, particularly the bacteria. Bacterial lineages are easier to follow than those of the embryological cells found in higher organisms because bacteria undergo repeated division instead of some of the progeny being siphoned off and undergoing differentiation.

Bacteria are also used extensively in various types of research because they can be manipulated in large numbers. The importance of large numbers rests on the nature of experimental results. In growth experiments, for instance, a variety of factors diminish the growth rate to varying degrees, but few factors eliminate it completely. If an investigator wants to find the precise percentage of survival of a population and knows that it is in the neighborhood of 0.1%, he cannot use 1 cell (0.1% $\times$ 1 = 0.001 cell), 100 cells (0.1% $\times$ 100 = 0.1 cell) or even 1000 cells (0.1% $\times$ 1000 = 1 cell). Only when the number of cells used is sufficiently large for him to obtain a number of dividing cells can he accurately determine the extent to which the treatment affected the population. Therefore, the suitability for manipulation of large numbers of bacteria, in the order of 10^8, makes bacteria an ideal system for studies both directly and indirectly related to growth.

Although bacteria represent an almost ideal system for some types of growth experiments, they are unsuitable for others. For instance, the bacterium is so small that detailed light-microscopic observation of the division process is impossible. Bacteria do not have mitotic figures, the tell-tale signs of division in higher cell types. Thus, an investigator could not recognize an effect on the state of cellular organization or on a particular organelle if he wanted to follow up initial growth experiments with cytological observations. For more definitive studies, more advanced forms of cells are used.

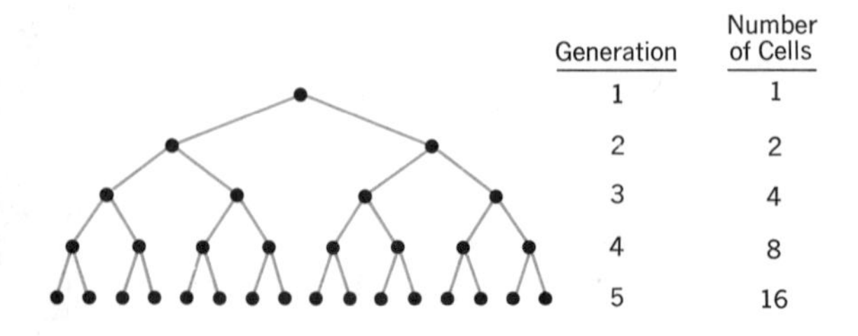

Fig. 11-1 A clone is a population of cells derived from the division of a single original cell.

It is possible to isolate plant and animal cells and grow them on the glass surface of a culture vessel. Once these cells are isolated from their native tissue, they remain in an embryonic state, so observations are not impeded by differentiation processes. Such cells will divide by mitosis to form genetically and physiologically homogeneous cell populations that are called **clones** (see Fig. 11-1). Since the growth rate of a clone can be interpreted as the activity of a single population rather than a mixture of cell types, investigators can analyze clones in the quantitative manner once reserved for bacteria. Cell cultures are slightly more difficult to manipulate than bacterial cultures, but large numbers (10^{6-8}) can be studied with relative ease.

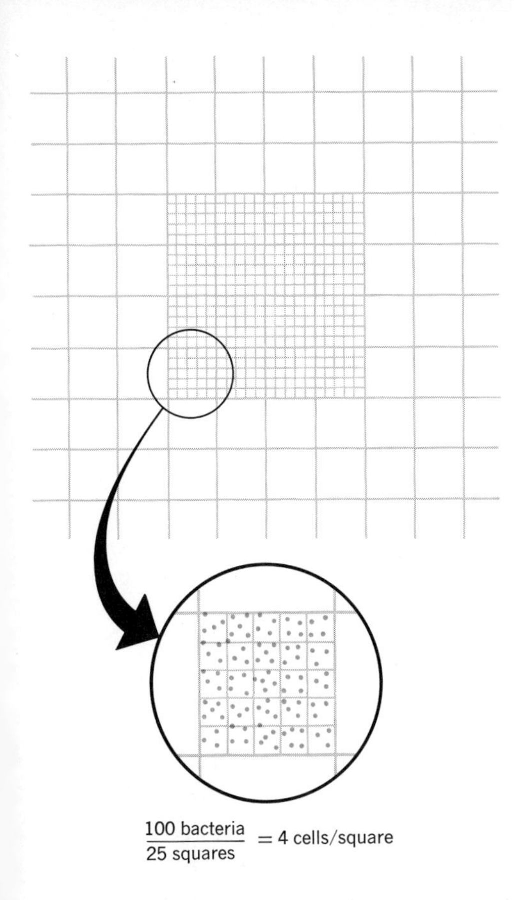

Fig. 11-2 Petroff-Hauser counter. This counter has a factor of 2×10^7. There is an average of four bacteria per square. Four times 2×10^7 equals 8×10^7, meaning that in the culture from which the sample was taken there are 8×10^7 bacterial cells/ml.

MEASUREMENT OF GROWTH

There are a number of techniques available for measuring the number of organisms present in a culture tube at a given time. The particular method used in measurement is dependent on the system an investigator studies and the accuracy he demands. Generally, growth-measurement techniques can be divided into direct and indirect approaches.

Direct Measurement of Growth

The most direct means of measuring the number of organisms present at a given time in a culture vessel is to count them by placing a known quantity of culture on a slide under the microscope. If, for instance, $\frac{1}{25}$ ml of culture fluid is placed on a slide, and all the organisms are counted, then the number of organisms per ml is 25 times the count.

As a population increases in number, counting can become almost impossible. This problem is overcome by the use of accurately calibrated cell counters and appropriate dilutions of suspensions of growing cells. One type frequently used is the Petroff-Hauser counter. When a drop of fluid is placed on the counter, the fluid seeps between the calibrated glass slide and the cover slip, forming a thin monocellular layer. With a microscope, the series of lines that mark off the calibrated squares can be seen (see Fig. 11-2). The counter is calibrated so that the number of bacteria per square multiplied by 2×10^7 gives the number of bacteria per ml in the original culture. In Fig. 11-2, there is an average of four bacteria per square. The number of bacteria per ml of the culture from which these bacteria were taken is then

$$4 \times (2 \times 10^7) = 8 \times 10^7 \text{ bacteria/ml}$$

When the investigator returns an hour later, he may find that there are too many bacteria to count, even by this method. He then takes a known amount from the culture, and dilutes it tenfold, placing a drop of the diluted culture on the counter. Again he finds an average of four bacteria per square; but in order to determine the number of bacteria in the original culture flask, he must include the dilution factor in his calculations.

$$\underset{\text{no./square}}{4} \times \underset{\text{counter factor}}{(2 \times 10^7)} \times \underset{\text{dilution factor}}{10} = \underset{\text{cells/ml}}{8 \times 10^8}$$

In this way the number of bacteria can be directly counted and over a period of time, the pattern of growth is determined. Comparable techniques are available for cultured animal cells.

The major disadvantage of this technique is that it gives a **total** cell count, including both organisms that are undergoing division, and those that will never again divide. Usually, only the viable cells are of interest to an investigator.

It is in obtaining **viable** counts that bacterial cultures have the advantage over cell cultures. Bacteria withstand manipulative techniques quite readily, whereas tissue-culture cells are more fragile, and damage easily. A viable count of a bacterial population is taken by removing an **aliquot** (a sample of known volume) from a culture flask, or sacrificing one of a series of identical flasks. From 0.1 to, at most, 0.5 ml of the culture is spread on the agar surface of a petri plate, and the plate is incubated at a temperature that favors rapid growth. Each bacterium on the plate that is capable of division (viable) will reproduce repeatedly. Since these cells are unable to move freely on the agar surface, their growth is highly localized, and a colony is formed (see Chap. 9). By counting the number of colonies, the number of viable bacteria placed on the plate (and hence the number of bacteria in the culture flask at the time the aliquot was taken) can be determined.

When a determination of a growth pattern is desired, aliquots are removed from the culture flask and plated at various time intervals. As the size of the population increases, a point is reached at which there are too many colonies on a plate to permit accurate counts. Bacteriologists agree that 5×10^2 is the maximum number that can be accurately counted, and most prefer a maximum of 2×10^2. In order to avoid the problem of too many colonies to count, the investigator makes a series of dilutions of the aliquot for plating. The dilution factors and the data from such a series is presented in Fig. 11-3

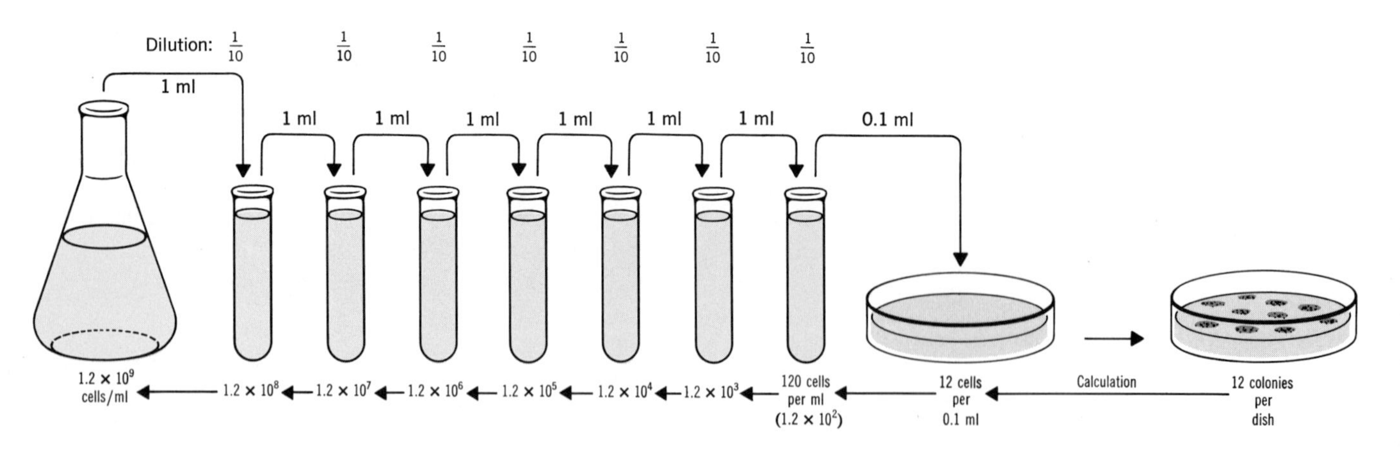

Fig. 11-3 The number of cells obtained after dilution is multiplied by the dilution factor(s) to arrive at the number of cells/ml in the original culture.

Another method of directly measuring growth is by weight. An aliquot of a growing culture can be centrifuged. The pellet of cells can be blotted to remove excess water and weighed for the **wet weight,** or for greater accuracy, it can be dried by heat and weighed to give the **dry weight.** Although there are many shortcomings to the measuring of growth by weight (total rather than viable count, inaccuracies stemming from physical manipulations, etc.), this method may represent the most accurate way a given cell system can be handled. For instance, because of the availability of plating techniques, bacterial growth is rarely measured by weight. On the other hand, cultured cells that are too fragile for plating (and do not exhibit colonial growth) are standardly measured by weight.

Laboratories in which growth experiments are standardly run (either for studies on growth or for using growth as a tool for studying related phenomena) use a Coulter counter for determining the number of organisms in a growing culture. Organisms are grown in a culturing unit of the counter, and pass through a narrow tube in the course of their circulation. A photoelectric cell is beamed at one point along the tube, and each cell passing that point interrupts the beam. Each interruption is automatically registered by the counter, and totals are tabulated on a recorder according to the time units chosen by the investigator. The Coulter counter can register thousands of cells per minute, giving, of course, total rather than viable counts.

Indirect Means of Growth Measurement

Indirect means for growth measurement utilize factors that are related to growth. One such measurement utilizes a quantitative determination of the oxygen consumption. The reasoning behind this is that the greater the protoplasmic mass, the higher the respiratory rate, and the higher the respiratory rate, the greater the volume of oxygen consumed.

Another more laborious means of indirectly measuring growth is by biochemical analysis of aliquots from a growing culture taken at different times. The analysis is usually limited to protein or organic nitrogen determinations, but sometimes DNA and RNA content is used.

By far the most typical indirect method is that which uses the turbidity of the growing culture as an indication of growth. A tube containing an aliquot of the growing culture is placed in a **spectrometer,** a device that passes a beam of light through the tube and into a photoelectric cell. The more cells there are, the less light will pass through the cell suspension to excite the photoelectric cell.

The amount of light absorption by the tube is proportional to the cell density. The readings of the spectrometer can be directly correlated with given numbers of cells in the culture (see Fig. 11-4). Once again, only total cell counts or mass values are possible, and turbidometric methods are usually used just to obtain some idea of the approximate level of growth.

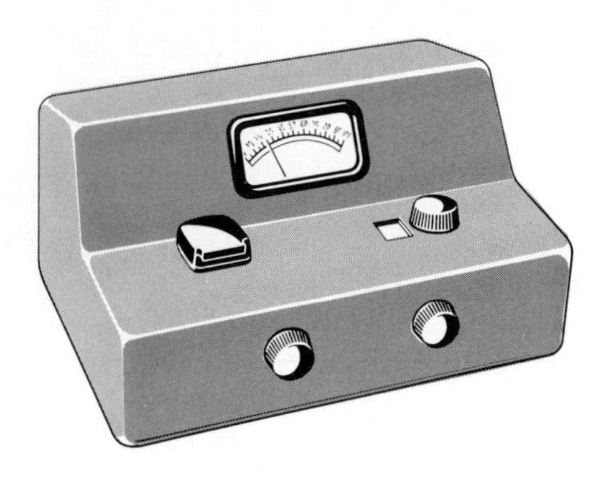

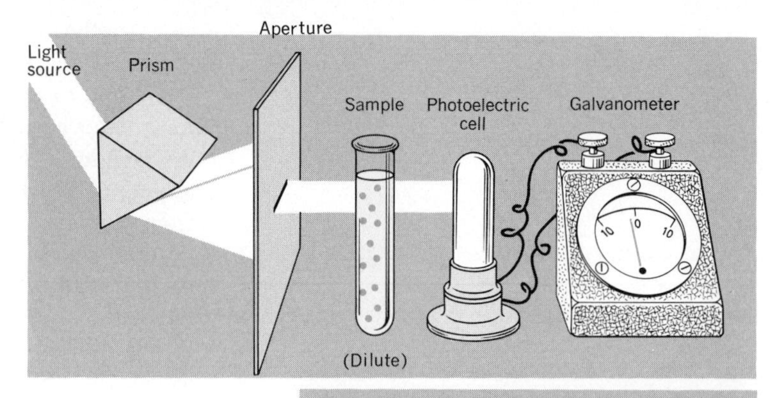

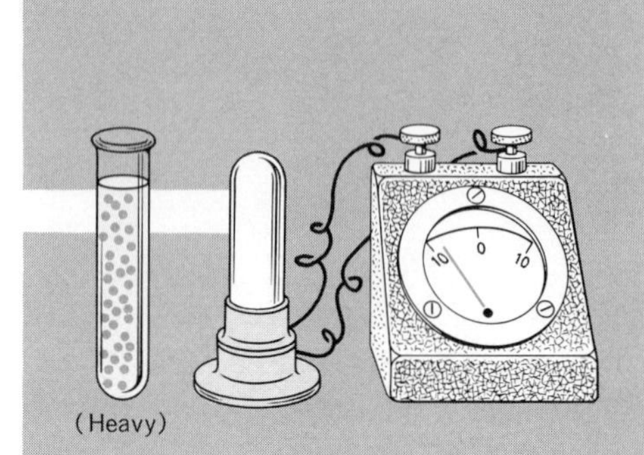

Fig. 11-4 A spectrometer. Light is reflected through a prism, and then through a slit to pass through the quartz tube containing the sample. The light passing through strikes a photoelectric cell, which converts the light to an electrical impulse. The less turbid the sample, the more light passes through, and the higher the transmission will be.

THE GROWTH CURVE

There are two ways of viewing growth: from the cellular level at which the processes of mitotic and meiotic divisions provide elegant schemes by which a single cell divides to form two daughter cells, and from a population level, where the overall effect of the divisions of many cells generates a pattern that can be translated into a graphic representation known as the **growth curve.**

The growth curve is formed by plotting data of growth measurements. The abscissa is used to represent time, and the ordinate to represent protoplasmic mass or some equivalent biological parameter correlated with mass. The particular notation for the ordinate depends on how growth is measured. If counts (either total or viable) are used, the ordinate represents numbers of organisms; if weight is used, the ordinate is given in grams, etc.

Regardless of the type of cell used, or the means of measuring growth, the growth curve is essentially the same. It is divided into four distinct regions, according to the angle of the slope (see Fig. 11-5). These regions are the **lag phase,** the **log phase,** the **stationary phase,** and the **death phase.** Each will be investigated briefly.

The Lag Phase

The **lag phase** is that period of time (after cells are introduced into a culture flask) during which little or no growth occurs. The length of the lag phase depends on the immediate preceding history of the cells. If they were in a dormant state, the lag phase is long; if they were actively dividing, the lag phase is relatively short or practically nonexistent.

Physiological studies of cells during the lag phase reveal that they are undergoing considerable RNA and protein synthesis, a fact that indicates that the cells are metabolically quite active. This finding has been interpreted to mean that the cells are in an unbalanced state, and once a particular steady state of synthetic processes is achieved, the cells divide. One might imagine (and there is data to support the idea) that the available energy is channeled into repair and production of general cellular constituents, and after this has been accomplished, only then is energy directed into processes specific for division.

The Exponential or Log Phase

After the lag phase, there is a gradual increase in the rate of growth until a sustained, maximum rate is achieved. The cells at this time can be viewed as passing through a cyclic series of steady-state conditions, during which new cellular material accumulates and is depleted by the division of a cell into two cells. The period during which the maximum rate is sustained is the **exponential** or **log phase** of growth, a name derived from the fact that measurement of growth during this time produces a curve like that expected from an equation with exponential factors. This pattern is directly related to the way in which the cell population increases (namely, by repetitive cell divisions). The division of two cells will produce four cells, the division of four will produce eight, etc. Therefore, the growth of the population occurs as $\mathbf{2^n}$, where $\mathbf{n}$ is the number of cell divisions. Thus, in the second generation there would be 2^2 or 4 cells, in the third generation there would be 2^3 or 8 cells, etc.

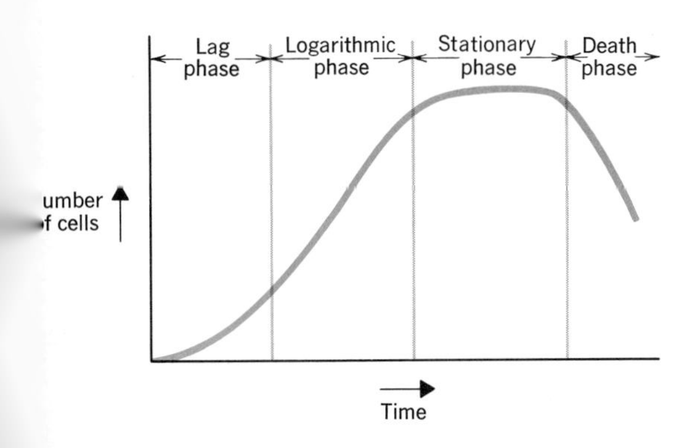

ig. 11-5 The growth curve.

This period of the growth curve is of extreme interest to experimenters for several reasons. First, when cells are grown in a complete or defined medium (a medium that contains all the nutritional requirements of the cell), the rate achieved during log phase represents the maximum rate that can be achieved by that population under the most ideal conditions. Diminished growth in cultures where some nutritional factor is eliminated or some foreign agent added is reflected in a change of the slope of the log phase of growth. Therefore, the quantitative relationships and mathematical expressions of the log phase provide important and sophisticated tools for research.

The first step in developing a meaningful mathematical expression for the log phase is to relate the number of generations (**n**) to a time factor (**t**). One way to do this is to find the average **generation time** for a given population. The generation time can be obtained by selecting a point along the curve, and finding another point that represents a

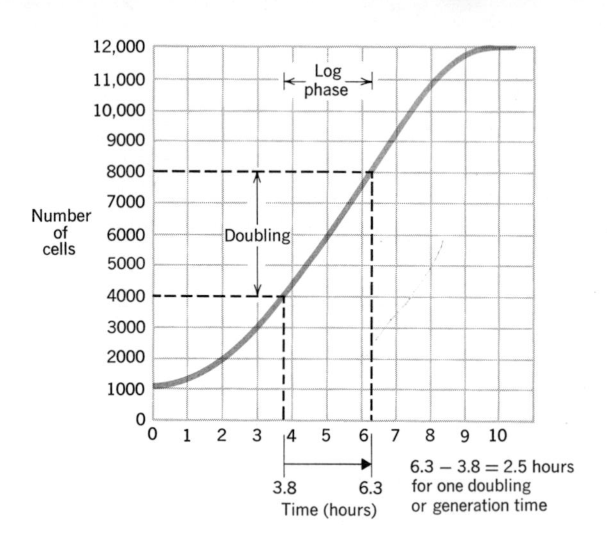

Fig. 11-6 The generation time is the time, during the log phase of growth, that it takes for a population to double.

doubling of the population. The difference between the corresponding times is the generation time (see Fig. 11-6). A more accurate and useful way of deriving generation times, however, is by manipulation of the growth expression 2^n.

For growth studies in microorganisms, time is usually expressed in hours. For organisms that complete a single mitotic cycle ($n = 1$) in one hour ($t = 1$), $n = t$. But the generation time for a given population of cells under a given set of conditions is seldom exactly one hour. Therefore, the equation relating the number of divisions (n) with time (t) must include a factor that will automatically adjust the time factor. This factor, k, gives the number of generations per hour, and the reciprocal of **k** represents the fraction of an hour necessary for the doubling of the population. Thus,

$$\text{number of generations} = (\text{no. generations in one hour})(\text{number of hours})$$

or

$$n = kt$$

The exponential expression for the log phase of growth is then

$$2^{kt}$$

To find the value of k when the initial number of cells (N_o), the final number of cells (N), and the time that has passed, are known, the following equation is used.

$$N = N_o \times 2^{kt} \quad \text{where} \quad N = \text{the final number of cells}$$
$$N_o = \text{the initial number of cells}$$

For example, suppose 1000 cells are placed in a culture flask, and five hours after the initiation of the log phase, the investigator finds that there are 32 million cells.

$$32{,}000{,}000 = 1000 \times 2^{k(5)}$$
$$32{,}000 = 2^{k(5)}$$
$$32{,}000 = \text{approximately } 2^{15}$$

Therefore,

$$15 = k(5)$$
$$k = 3$$

So this population doubles three times in an hour, and the generation time for the population is obtained by multiplying the reciprocal of k ($\frac{1}{3}$) and one hour (1), which is 20 minutes.

Exponents become too cumbersome when large numbers are encountered. In addition, the curved shape of the graph for the log phase of growth makes direct graphic interpretation difficult. Both these problems are solved by converting the equation and the graphic values to logarithmic form.

When the equation is converted to logarithmic form, and the data are plotted on a graph with a logarithmically scaled ordinate, a straight line is produced to represent the log phase of growth. Both **extrapolation** (extending the line past time periods considered in the experiment) and **interpolation** (determining points at time not actually counted) are better done from such a graph.

The logarithmic expression of the log phase equation is developed as follows. By algebraic rearrangement,

$$N = N_o \times 2^{kt}$$

becomes

$$\frac{N}{N_o} = 2^{kt}$$

The log of both sides of the equation is then taken

$$\log \frac{N}{N_o} = kt \log_{10} 2$$

or

$$\log \frac{N}{N_o} = kt\ (.3010)$$

With the values previously used

$$\log 32{,}000{,}000 = 7.5051$$

$$\log 1{,}000 = 3$$

$$\text{Hence, } \log \frac{7.5051}{3.000} = k(5)\ (.3010)$$

$$4.5051^{*} = 1.5\ k$$

$$k = 3$$

The growth equation provides a quantitative basis for evaluating growth. Once the value of **k** for a given population under certain conditions has been determined, the equation may be used to calculate the number of organisms expected after any given growth period.

The k value is also used as an indicator of a change in population growth. For example, when an investigator has determined the k value for a population of cells under ideal conditions, and then grows the same type of cells at a lower temperature, he may find that the k value is lower than in the control. This can either mean that the generation time is longer at lower temperatures or that fewer cells of the population divide at lower temperatures. The change in the k value reveals that by altering the conditions within the culture flask, the investigator has quantitatively altered the growth kinetics of the population. In order

*Numbers are divided by subtracting their log values.

to determine the cause of the change in the k value, he would have to use means other than growth experiments.

The Stationary Phase

The growth rate detected in all phases of growth is actually the difference between the death and division rates. In the exponential phase, the death rate is so low (compared to the division rate) that it passes unnoticed. In the **stationary phase,** however, the growth rate decreases to zero and there is no detectable increase in cell number. Cell division is still occurring, but it is counteracted by death processes.

Two physiological interpretations for the zero rate of growth have been suggested: that the nutrient supply has been exhausted, or that the by-products of metabolism have accumulated in toxic concentrations in the medium. However, the same growth pattern is found in multicellular organisms, and here these explanations are not justified. In multicellular organisms, it appears that growth has been turned off in accordance with the genetic program for development. In unicellular populations, a static state appears as a result of the release of organic acids, which lower the pH of the medium. It appears that the initiation and duration of the stationary phase is dependent on the kind of organism and the degree of environmental change brought about by growth itself.

The Death Phase

Eventually the death rate becomes greater than the division rate, and there are fewer and fewer viable cells. When this occurs, the population is in the **death phase.** It is thought that the death phase begins when destructive processes outweigh growth processes.

The death phase is revealed by viable counts rather than total counts. If the growth curve is generated from data gathered by the use of total counts, no death curve will be noted. It is impossible to distinguish between a living bacterium, capable of dividing, and a dead one on the basis of microscopy, weight, or culture turbidity. Therefore, by using any of these methods, the stationary phase will appear to continue.

The death phase is often expressed exponentially. The slope of the death curve is, however, different from that of the growth curve.* When the equations for the growth and death phases are combined, they generate an asymmetric bell-shaped curve. The combined equation is termed the **logistic law of growth.**

*The equation for the death rate is

$$\log \frac{N_d}{N_{do}} = mt \log 2$$

where N_d = final number of dead cells
N_{do} = initial number of dead cells
m = arbitrary constant that corrects t value for actual time of death

The Stochastic Analogue of the Growth Model

The logistic law of growth depends on the assumption that the generation time for every member of the population is the same. This is like saying that since the average age at which a man marries is 24 years, all men with the same birthday will marry on the same day. This is, of course, fallacious.

It is equally fallacious to suppose that all cells in a population would have the same generation time. All biological material shows fluctuations from the mathematical average, and generation time is nothing more than an average. For the past several centuries the attitude in science has been to interpret fluctuations in nature as deviations. This position was adopted because the easiest statistical value to obtain is the average or mean, and measured discrepancies could easily be viewed as deviations. Today scientists are not so naive as to think that even the simplest entities must all behave identically, and hence, differences from the average should be accepted as (1) natural, (2) explainable, and (3) mathematically tractable. Particles and cell populations, as well as paramours lose much of their interest when treated too abstractly.

The smooth curve generated during the log phase is due to the nonsynchronous growth of the population. If all the cells divided synchronously, a series of steps, each of which would be a near-doubling of the previous population number, would be obtained. A cell does not follow a rigidly timed schedule of division. It is instead like a woman preparing for an evening out. Sometimes the preparation takes fifteen minutes, sometimes an hour. The time depends on the state she was in when she started, her mood, etc. So it is with a cell undergoing division. Sometimes it will divide within twenty minutes of its formation, and

---

It is also possible to express the entire growth curve as the difference between the growth and death equations.

$$\log \frac{N}{N_o} - \log \frac{N_d}{N_{do}} = kt \log 2 - mt \log 2$$

$$\log \frac{N - N_d}{N_o - N_{do}} = t \log 2\ (k - m)$$

This equation may be simplified by using $\tilde{N}$ to denote the *net value*, which derived by subtracting the death rate from the growth rate:

$$\log \frac{\tilde{N}}{\tilde{N}_o} = t \log 2\ (k - m)$$

Since the death rate is not constant, but increases with the exhaustion of nutrients or the buildup of toxic substances, the death constant must be transformed into a variable by multiplying it by $\tilde{N}$. The equation is then written

$$\log \frac{\tilde{N}}{\tilde{N}_o} = t \log 2\ (k - \tilde{N}m)$$

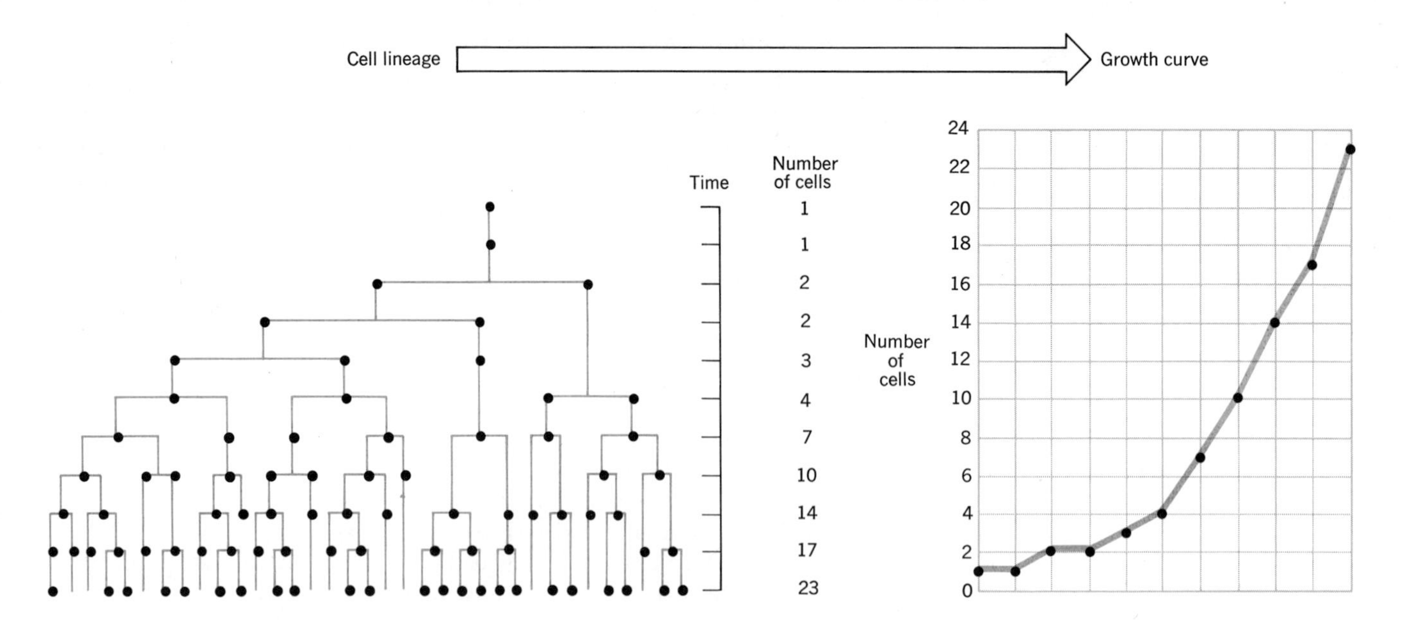

Fig. 11-7 If cell division of a population were synchronous, the growth curve would appear as a series of steps. The randomness of the divisions produces the effect of a smooth curve.

sometimes it will take an hour. It is the randomization of generation times that produces the smoothness of the curve (see Fig. 11-7).

It is impossible to see the progression of a cell through the sequence of events leading to a division. Therefore, one can only assign a probability value to the length of time necessary for it to complete the processes that must precede division (see Fig. 11-8). The probability that a cell will divide fifteen minutes after being formed is low. As time increases toward the average noted for the population (in Fig. 11-8, the average is twenty minutes), the probability increases. As time passes beyond the average, the probability again begins to decrease, and eventually becomes extremely small.

Superimposed on these probabilities are others that pertain to the state of individual cells. As an individual cell passes through the sequence of events that lead toward division, the chance that division will successfully occur increases. A cell that has completed the S_1 phase of interphase has a greater probability of completing division than one that has not yet reached that point.

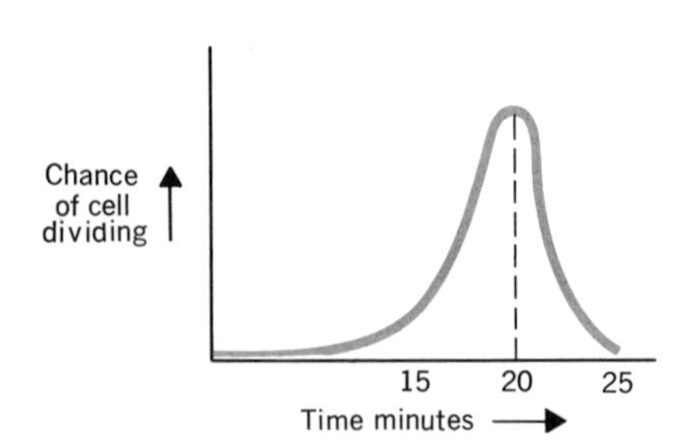

Fig. 11-8 As time passes, the chance increases that any cell within a population will complete a division. After the average generation time of the population is past, the chance lessens that a cell that has not yet undergone division will do so.

The logistic law of growth discussed in the last section does not reflect, mathematically, the probabilities involved in growth, but is concerned only with the average, which is really an expression of the most probable event, and because of that is considered a **deterministic** equation. On the other hand, equations that mathematically include probability considerations are called **indeterministic** equations. When change, as in this problem of growth, is a statistical phenomenon, the change is called **stochastic.**

The difference between the logistic and stochastic models for growth is extremely difficult to express in mathematical terms, and can best be illustrated in verbal and graphic terms. For example, how many cells will be present after three generations if a culture is started with one cell? According to a deterministic equation, such as $\mathbf{2^n}$, the answer would simply be eight.

$$2^n = \text{total number of cells} \qquad n = \text{number of divisions}$$
$$2^3 = 8$$

Any departure from this value would be submitted to statistical analysis to determine the significance of the deviation.

According to the stochastic model, anywhere from one to sixteen cells are expected; but there is a certain probability associated with each cell number (see Fig. 11-9), and eight (cells) is the number with the greatest probability. Stochastic interpretations of growth and other natural processes are far-reaching, and the reader may find it worthwhile to view previously discussed processes (protein folding, metabolic reactions, DNA transcription, etc.) from this perspective.

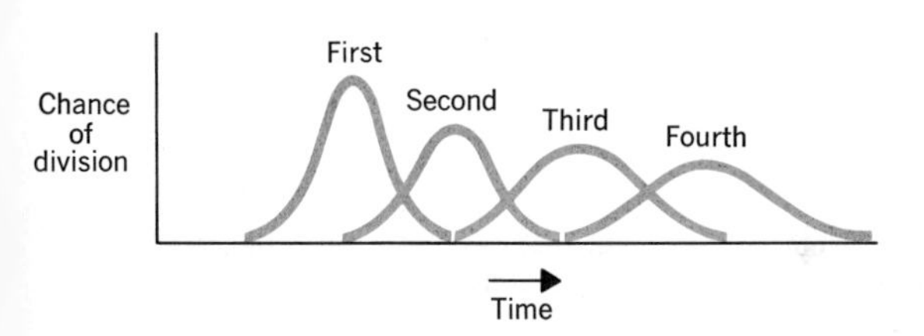

Fig. 11-9 Probability expression of a stochastic model for growth of a population.

GROWTH PATTERNS

The growth phenomenon, as expressed in either the logistic or stochastic models of growth, is universal for both uni- and multicellular forms. Certain developmental patterns of multicellular organisms can be interpreted without reference to differentiation mechanisms. Multicellular organisms can be viewed as a group of different cell populations, each manifesting their own growth rates. The differences between the growth rates of different cell populations can result in the conformation of the individual.

Conformation by Unequal Growth Rates

Most biological structures are not spherical, because the growth rate differs in various regions and directions. Imagine a circle enlarging at the same rate in all directions; the circle will simply increase in size. But imagine a circle enlarging so that points *A* and *B*, 180° apart, move outward at a faster rate than do points *C* and *D* (see Fig. 11-10). An ellipsoidal figure will be formed.

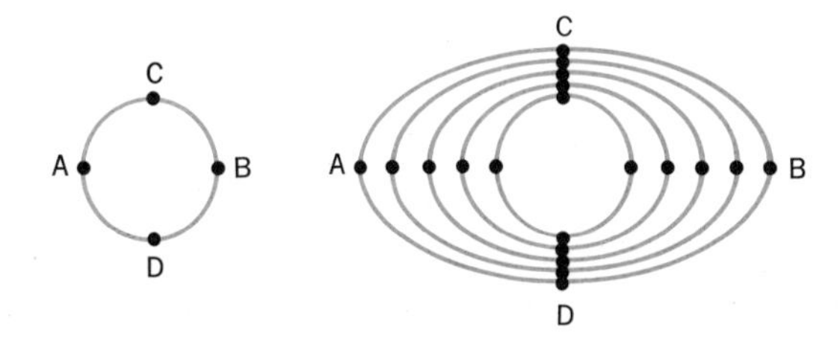

Fig. 11-10 Allometric growth. The zygote of most organisms is spherical. A differential growth rate on different planes of the sphere can produce a variety of shapes.

In multicellular organisms, various loci enlarge in different directions at different rates to generate the structural appearance associated with the individual. This phenomenon of conformation by unequal growth rates, called **allometric growth,** was postulated by D'arcy Thompson to explain the difference between similar species of fish (see Fig. 11-11). The same explanation has been used for the growth of stems along a vertical plane, or, for that matter, for the conformation of any biological form.

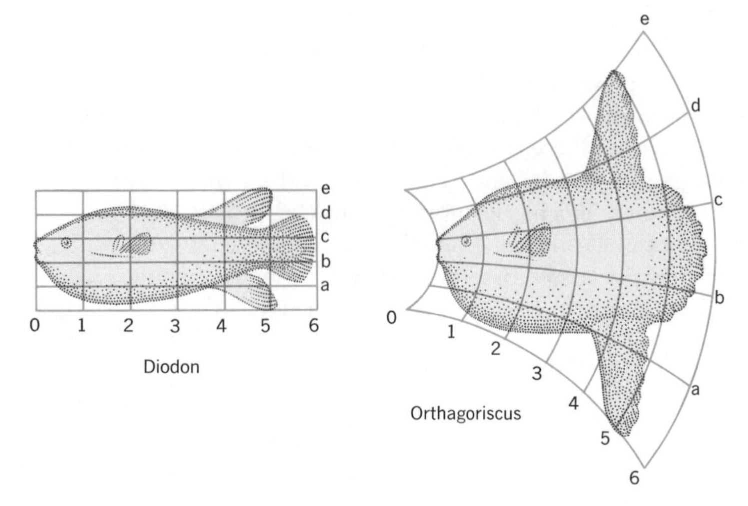

Fig. 11-11 Thompson proposed that differential growth rates were responsible for certain shapes in fish.

Allometric growth can be the result of an increase in either cell size or cell number. In a gourd, for example, the rate of cell division is the same in all parts of the gourd, but cells in certain areas accumulate more material and become distended. The shape of fish, on the other hand, is due to an increase in cell number along one plane. The cells in different parts of the fish are approximately the same size; there are simply more of them in given areas.

The size and number of cells is presumed to be genetically controlled. But how can two cells that have arisen from the same original zygote, and that carry identical genetic information, manifest different growth rates expressed as allometric growth? It is obvious that different cells express different degrees of quantitative rather than qualitative information in terms of the synthesis of cellular components, but little is known of the mechanisms underlying these activities. The next chapter examines some of the research studies on cellular differentiation, the mechanisms of which probably account for such phenomena as allometric growth.

Organismal Aberrations by Deviant Growth Patterns

Growth can be thought of as the result of particular balances between various synthetic processes of a population or group of populations: the accumulation of many materials, the dilution of these materials by reproduction, the use of cellular materials as energy sources, and the passage of materials into the environment to keep the cellular house clean. Thus, the maximum growth rate achieved by microorganismal populations under ideal laboratory conditions (complete complement of nutritional requirements, optimal temperature, optimal pH, unlimited supply of oxygen) and the conformation of multicellular organisms depends on the steady-state conditions among cellular components and the steady-state conditions between the cell and its environment. The maximum growth rate in microorganisms, and the proper conformation of cell populations in multicellular organisms, is considered **balanced growth.**

When one or more of the steady-state factors is disturbed, the growth patterns associated with that particular population change. A change in the typical growth pattern associated with a population is called **unbalanced growth.**

Unbalanced growth can, theoretically, result in three patterns: (1) an increase in the growth rate, (2) a decrease in the growth rate, or (3) a change in cell size. Because of the difference in the typical growth of different cell populations, unbalanced growth is detected in different ways. For example, unicellular organisms grown under ideal laboratory conditions are growing at their maximum rate during log phase. Any alteration of the steady-state conditions of these cells can result either in a decreased growth rate or an increase in the size of individual cells, but not in an increase in the growth rate. Thus, unbalanced growth in unicell populations refers only to decreased growth rate or aberrations of individual cells.

The conformation of multicellular organisms, on the other hand, depends on differential growth rates of different populations, which implies that few, if any, of the populations are proceeding at their maximum possible rates. Any alteration of the steady-state conditions of these populations would tend to increase the growth rate, adding clumps of cells known as **tumors.** A slight decrease in the growth rate of a tissue would pass unnoticed. The presence of aberrant cells in a particular tissue might not be detected. Therefore, unbalanced growth in multicellular organisms refers almost exclusively to tumor formation. It should be noted that some tumors, notably malignant ones, also contain large aberrant cells.

Each example of unbalanced growth reveals something about the steady-state conditions of cell populations and the mechanisms that stabilize the steady state. Tumors, for instance, can be induced by radiation or treatment with other mutagenic agents, a fact that indicates that the regulation of cell division is genetically controlled. Spontaneous or induced mutations of the controlling genes render them nonfunctional, and the cell becomes incapable of inhibiting the division process.

X-ray and ultraviolet-light treatments cause bacteria to elongate into filaments, and cause unicellular algae to enlarge and eventually lyse. The control mechanisms of these cells have been shut off by the mutation of certain genes. Hence, they uncontrollably continue synthetic processes, but cannot divide.

It appears that the altered control mechanism need not be directly associated with growth processes (such as production of a protein essential for division or the replication of DNA). The absence of a simple metabolic factor may disturb the cellular balance sufficiently to cause aberrant growth. Strains of *Neurospora* that have a deficiency of an enzyme in some metabolic pathway will accumulate the reactant for that enzymatic step (see Fig. 11-12). Even though the investigator provides the product of that reaction, distorted hyphae are produced. The abundance of the reactant, it seems, affects the rates of other cellular reactions.

Fig. 11-12 There is an accumulation of reactant C when the enzyme for converting C to another product is missing.

PERSPECTIVES OF GROWTH

The concept of balanced growth as the result of suitable steady-state conditions can be extended to natural populations. As long as the environment supplies the necessary nutritional requirements, a suitable temperature and pH range, etc., populations will grow. When the necessary nutrients are depleted, or the physical factors of the environments change, the growth of the population slows and may stop (unbalanced growth). Populations can, and have, increased to such vast numbers that the food supply becomes limiting—so limiting that extinction can occur.

Growth phenomena come into play in many aspects of biology. First, growth is a study in its own right. Second, growth kinetics are used extensively in nutritional and genetic studies as research tools. Third, the concepts of growth kinetics and balanced growth are important for the perspectives of evolution and ecology, the subjects of the next two sections of this book.

QUESTIONS

1. What are some of the advantages and disadvantages of using a bacterial culture for growth experiments?
2. Distinguish between direct and indirect measures of growth, and give examples of each.
3. Explain the phases of the growth curve on the basis of the condition of the cells in culture at each phase.
4. Define generation time.
5. At zero time, 10^6 cells are placed in a culture flask. Eight hours after the initiation of log phase there are 2.2×10^9 cells. What is the generation time of this culture?
6. Compare the logistic and stochastic approach to the growth phenomenon.
7. How can the conformation of certain organisms be explained by growth rates?
8. Give two examples of organismal abberations by deviant growth rates.

SPECIAL INTEREST CHAPTER III

VIRUSES

The field of virology, perhaps more than any other in the area of biology, exemplifies the advances that can be made by the mutual exploitation of the discoveries of both medically and nonmedically oriented scientists. The medical sciences have relied on related fields to provide techniques and hypotheses. Conversely, basic researchers have relied on the medical sciences for discoveries of phenomena that are not detectable in their systems.

It is impossible to present only the medical aspects of viral infection. A true and comprehensive picture must by necessity include the discoveries in medical and basic research that have led to both the present picture and the speculations for the further development of that picture.

The Discovery of Viruses

Pasteur and Koch had discovered that bacteria were the causative agents of many diseases. Koch's procedure for establishing a given bacterium as a pathogen was to (1) remove a sample of tissue fluid from the diseased animal, (2) culture the bacteria isolated from the fluid, (3) infect a healthy animal with the disease from the cultured bacteria, and (4) isolate and grow the bacteria from the infected animal. If the symptoms manifested by the animal infected from the laboratory-grown culture were the same as those of the original diseased animal, and, if upon isolation, the bacteria manifested the same morphological and physiological characteristics, it was judged to be the pathogen.

Subsequent workers illustrated that it was often the presence of the bacteria, and not simply bacterial secretions, that caused the disease. Bacterial filters were constructed with such minute pores that bacteria could not pass through. A suspected pathogenic culture was placed in the filter funnel and the liquid drawn through by a suction pump. When the filtrate (the culture fluid) was injected into a healthy animal, no symptoms indicative of the disease occurred. Thus, the disease-causing agent was a discrete particle of definite size, the bacterium.

In 1898, Bijerinck discovered that the agent that caused mottling and lesions (holes) in the leaves of tobacco plants *passed through* a bacterial filter. When the filtrate was rubbed onto healthy tobacco leaves, the symptoms of mottling and lesions appeared later. He called the infectious agent a **filterable virus,** which literally means a poison capable of passing through a filter. He did not realize the full significance of his discovery: a new kind of organism at a previously unsuspected level of biological organization.

It soon became apparent that many animal (including human) diseases are caused by filterable viruses. The first fifty years of medically oriented viral research consisted primarily of isolating viruses from diseased individuals and introducing the viral solution into healthy ones. The subsequent manifestation of disease symptoms by these individuals indicated that the causative agent was in the solution. Because it is illegal (and immoral) to use human subjects, experiments on human diseases were done with animals that are evolutionarily close to man.

Several interesting observations regarding viral infections were made during this time. First, viral infections cause alterations at the *tissue* rather than the *systemic* level. Second, viruses differed considerably in their effect on the tissue. Some caused tissue degeneration in which the cells of the tissue disappeared. Other viruses caused the tissue to undergo overt growth (tumor). Thirdly, a given virus was capable of infecting only a given group of tissues in a limited group of organisms (both tissue- and species-specific). Some viruses isolated from humans could be grown in other animals, but most could not.

Because of the minute size of viruses (see Fig. III-1) and the difficulties in establishing and keeping isolated tissues growing and free from bacterial contamination, it was impossible to study the infective process directly. The giant steps toward the determination of the dynamics of

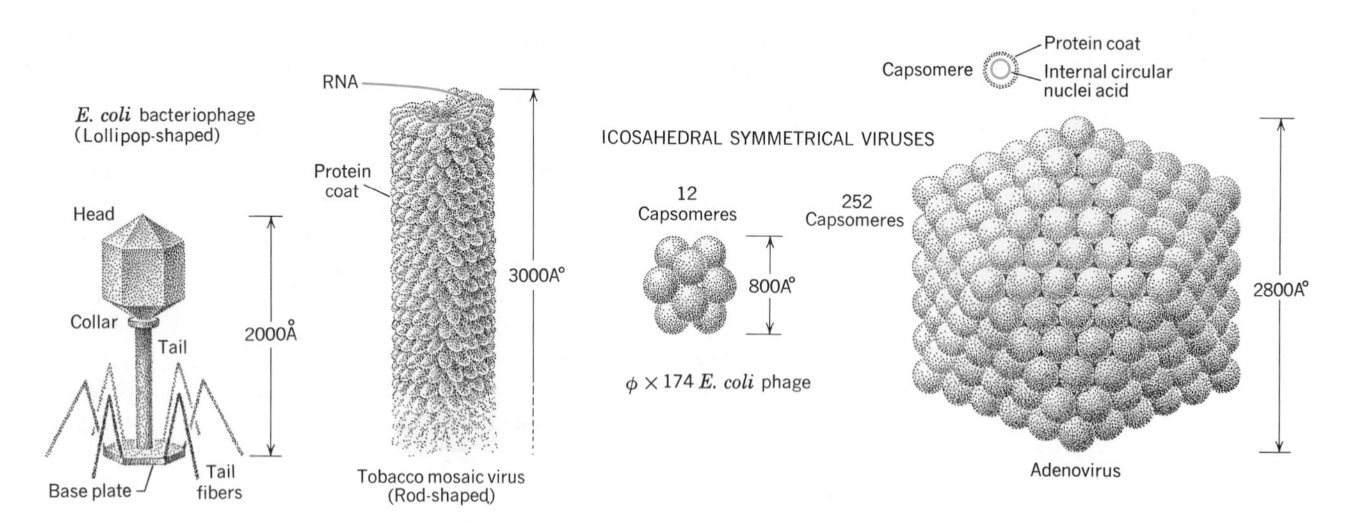

Fig. III-1 Sizes of various viruses.

viral infection came from areas of research far removed from the direct studies of the medical sciences.

TABLE III-1 RNA AND DNA VIRUSES

Viruses	Type of Nucleic Acid (Number of Strands)
Bacteriophage	DNA (2)
T series	
φX174	DNA (1)
MS2	RNA (1)
Tobacco mosaic	
(TMV)	RNA (1)
Herpes simplex	DNA (2)
Polyoma	DNA (2)
Poliomyelitis	DNA (1)
Influenza Type A	RNA (1)
Reovirus	RNA (2)
Adenovirus	DNA(2)

The Chemical Nature of Viruses

In the 1930s, Rawden and Pirie demonstrated that the tobacco mosaic virus (TMV), which causes lesions in tobacco leaves, is composed of 95% protein and 5% nucleic acid. The subsequent analysis of other viruses indicated that this composition was universal. However, the specific nucleic acid found in general classes of viruses varied. Most **bacteriophages** (viruses that infect bacterial cells) contain only DNA; plant viruses contain only RNA, and animal viruses contain either RNA or DNA (see Table III-1). The similarity of chemical type suggested a universal mode of action, and this in turn suggested the feasibility of utilizing information gained from experiments with one system on the others.

The elucidation of the chemical nature of viruses had far-reaching effects. It drew the attention of many scientists interested in basic research to the elements of life. Here was a system in which the transmission of genetic information and its translation into macromolecular structures could be studied. For medical scientists engrossed in the isolation and reinfection by filterable solutions, it pointed the way to more fruitful research through studies of viral infection at a biosynthetic level.

The Morphology of Viruses

The application of the electron microscope to biological problems was a boon to virologists. The high magnifications permitted them to examine this hitherto hypothetical world of creatures.

The determination of viral shapes and the use of the EM in conjunction with experimental procedures enabled scientists to elucidate the infection process in bacteriophages, and ultimately permitted sound speculation on the infection patterns of other viruses.

Experimental Techniques Used for Viruses

Single-celled organisms (such as bacteria) that can be grown in culture, have provided workers with an ideal host system. Bacteria can be transferred from one medium to another, changed from liquid to agar-surfaced cultures, and grown in such a way that a thin layer of bacteria is formed on an agar surface.

Because of the ease with which the host organism can be handled, the study of bacteriaphages advanced rapidly. Animal virologists were seriously hampered by their inability to remove cells from an animal and grow them successfully under bacteria-free conditions. Many attempts to initiate a successful and easy-to-maintain tissue culture were made. It was not until the 1950s that tissue-culture techniques became standard operating procedures in most laboratories.

The ultimate success of tissue culturing techniques was based on many things. First was the realization that a quantitative study of viral

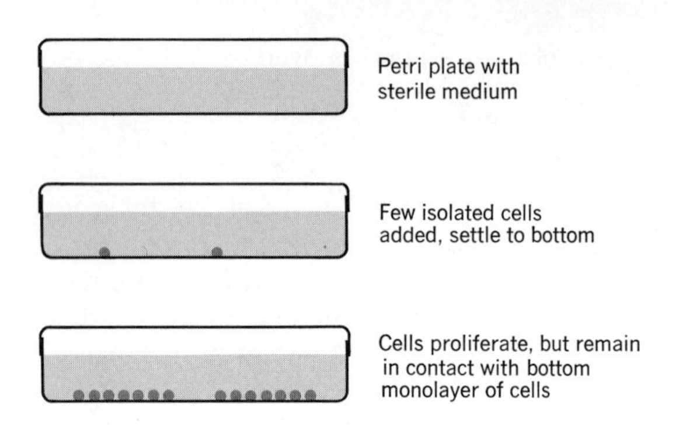

Fig. III-2 (*Above*) **Technique for producing monolayer culture.** (*Below*) **A monolayer culture of neurons (420×). [(*Below*) Eric V. Gravé]**

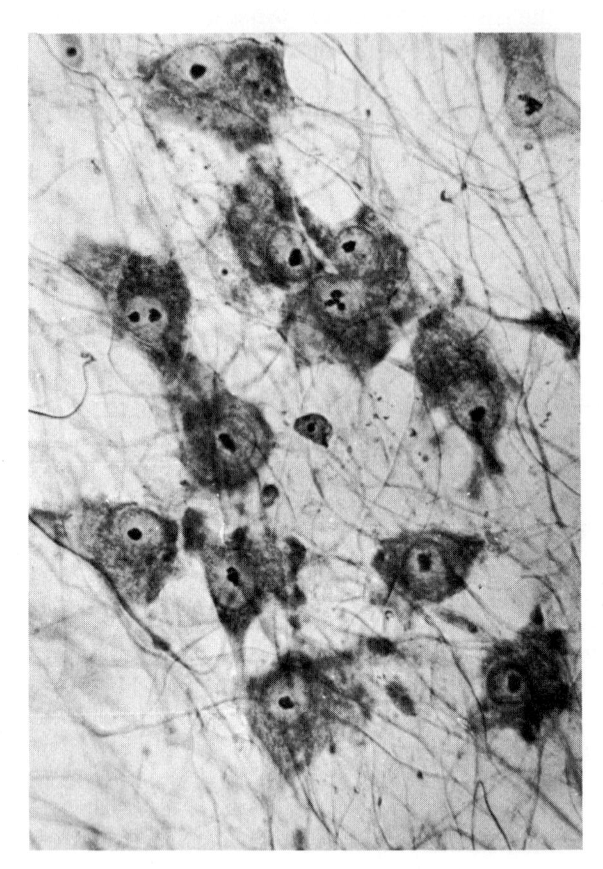

infection required a system that was analogous to that used for bacteriophages. This realization made many increase their efforts toward that end. Second was the awareness of the advantages of a tissue culture, in which individual cell changes resulting from infection could be determined. Thirdly, the common use of antibiotics made it possible to free isolated cells from bacterial contamination without damage to the cells themselves.

By using standard techniques, many different kind of cells can be cultured. The easiest, and therefore most commonly used, are from monkey kidney or human amniotic membrane (a membrane in the placenta or afterbirth).

The tissue is mixed with dilute **trypsin,** which depolymerizes the protein "glue" that cements the cells of a tissue together, but does not overtly damage the cells themselves. After several hours of agitation, so that the trypsin has a maximum effect, the tissue is centrifuged, and the free cells in the supernatant are collected and counted with appropriate devices.

The cells are then diluted so that there are about 2.5×10^5 per ml, and placed in suitable culture vessels. The cells will adhere to the glass surface of the culture vessel.

A medium (usually contains blood plasma, tissue extracts, salts, etc.) is added to the culture vessel. To provide maximum aeration and exposure to the medium, the vessel is usually agitated gently. Within five to ten days, a monolayer of cells forms (see Fig. III-2).

Once a tissue culture has been established, the cells can be handled in much the same way as the bacteriologist handles bacteria. New cultures can be started from older ones by placing a few cells in a fresh culture vessel. This permits the establishment of a **cell line** traceable back to an original tissue that may have been isolated years before.

In addition to this, certain types of culture vessels permit direct microscopic observations, thus allowing the investigator to directly determine cytological changes as they occur.

MODE OF VIRAL INFECTION

The most complete information on viral infection is from the phage-bacterial system.

Infection by Lytic Bacteriophages

Upon the introduction of bacteriophages into a susceptible bacterial culture, the phages and bacteria undergo chance collisions. When the "attachment" part of a phage (the **tail**) comes into contact with a bacterial cell wall, it adheres. The DNA of the phage is "injected" into the bacterial cell through pores in the bacterial cell wall; the protein coat remains outside (see Fig. III-3).

An individual phage appears to completely take over control of the bacterial cell, destroying its structural integrity (see Fig. III-4) and redirecting its cellular activities for making viral rather than bacterial material. Bacterial DNA and RNA are depolymerized, and the nitrogen bases enter the general nucleotide pool for the phage DNA replication. The protein of the bacterial cell is also depolymerized, and the amino acids are reassociated to form new viral protein shells.

Within twenty minutes to an hour (depending on the particular phage and bacterium involved), the bacterial cell lyses, releasing from one hundred to two hundred new phages. These new phages randomly collide with other uninfected bacteria, until their "attachment" points, by chance, come into contact with a bacterial cell-wall site, and the process of infection begins again.

There are many aspects of bacteriophage infection that are not known. How does the phage "inject" its DNA into the bacterial cell? Electron micrographs do not indicate that the phages squeeze the DNA into the cell; yet the entrance of the DNA does not seem to be by purely random processes. How does the DNA replicate a hundredfold in twenty minutes? A semiconservative mode of replication would involve an extremely rapid unwinding of the DNA strands over and over again. One would anticipate that a certain degree of internal turbulence would be detectable, yet none has been. These are only a few of the physical problems of phage infection. There remain a whole host of other physical as well as the chemical problems related to phage control of the bacterial cell that are, at this time, only partially answered.

Fig. III-3 Infection of *E. coli* by T_2 phage.

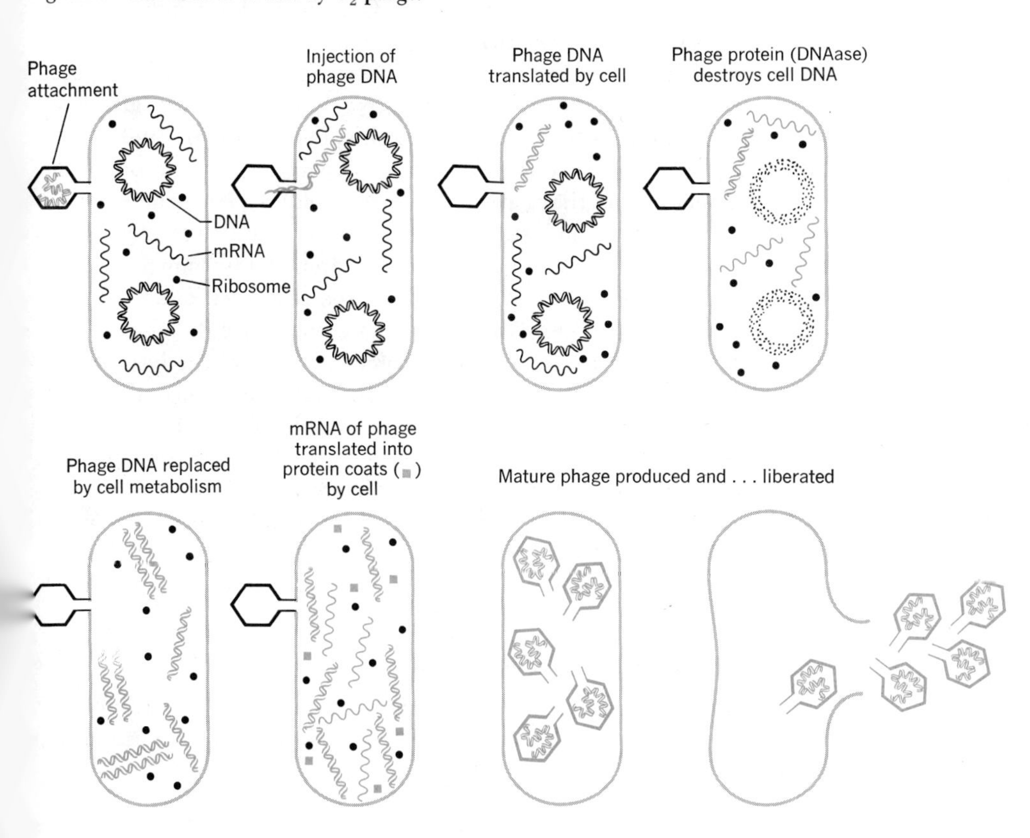

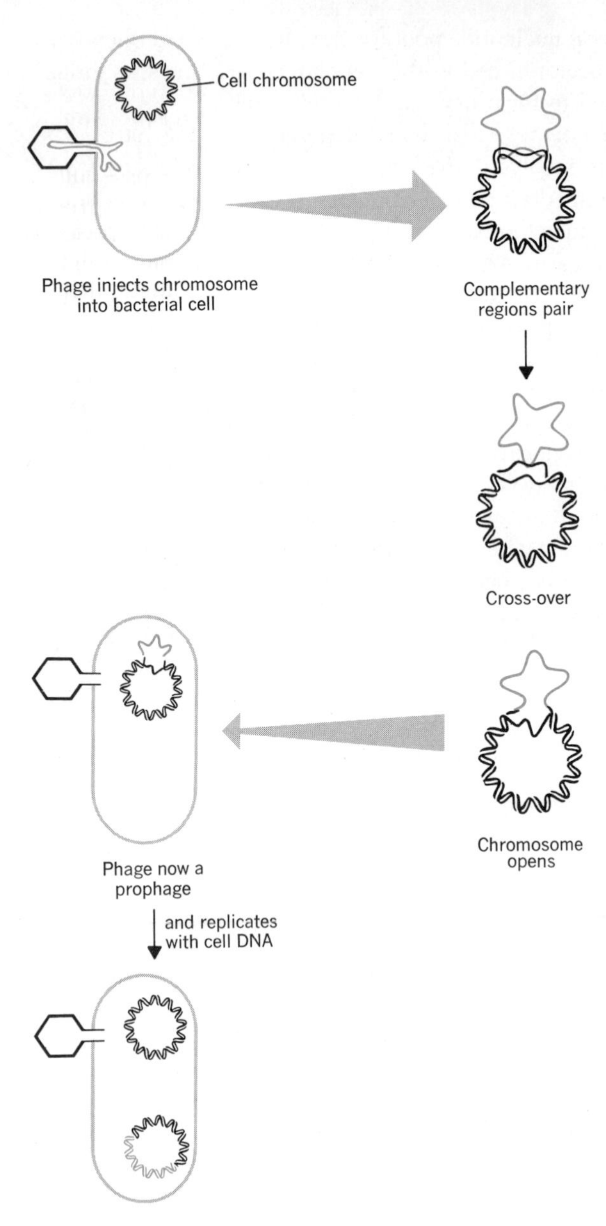

Fig. III-4 Bacterial infection by lysogenic bacteriophage.

Infection by Latent or Lysogenic Bacteriophages

One of the most interesting mode of infection and transmission of viruses was found in bacterial systems. It was noted that pure cultures of certain strains of bacteria would act as if they were infected with phages. That is, some of the bacteria would lyse spontaneously. When the bacteria were filtered, and the filtrate innoculated with fresh bacteria, some of these bacteria would lyse, while others would continue to grow with no viral interference. After some time, some of these others would lyse.

Exhaustive experiments indicated that when these viruses infected a bacterial cell, they either followed a pattern very much like that of the lytic bacteriophage (growing and lysing the cell) *or* the phage DNA attached to the bacterial DNA and essentially became part of the bacterial chromosome. In this case, the phage DNA replicated when the bacterial DNA replicated, and was accordingly parceled into the daughter cells in the course of the cell division. The phage DNA retained its integrity, and remained latent throughout many bacterial generations. Then, for some unknown reason (probably related to the chemical equilibria in the bacterial cell), the phage DNA would detach from the bacterial chromosome and act as a lytic phage, lysing the bacterial cell as it released 50 to 100 lysogenic phage into the environment.

The point of attachment for the lysogenic phage DNA to the bacterial chromosome is highly specific. Genetic evidence indicates that it attaches in such a way that only one point of the phage DNA is attached, and the rest forms two DNA arms on either side of the point of attachment (see Fig. III-4). When two lysogenic phage attach at proximal sites on the bacterial chromosome, there is the possibility that they will exchange genetic information. It was, in fact, the discovery of this phenomenon that gave birth to the field of viral genetics.

It is possible to induce the detachment of the phage DNA from the bacterial chromosome by a variety of treatments known to cause DNA breakage. This would indicate that the relationship between the bacterial and phage DNA is a very intimate one. Once the phage DNA detaches from the bacterial chromosome, it goes into a lytic phase.

There are, of course, many things that are not known about the lysogenic mode of transmission and infection. What determines whether, upon entering a bacterial cell, a phage will act in a lytic or lysogenic manner? How is the attachment to the bacterial chromosome accomplished? How does the bacterial DNA replicate semiconservatively with a piece of phage DNA attached? What chemical events determine the detachment of the phage DNA? What triggers the lytic mode of action? And so forth.

Infection of Plants by Plant Viruses

Morphologically, plant viruses are not adapted for the infective process in the same way as bacteriophages. Most forms of bacteriophage,

ig. III 5 A tobacco leaf infected with tobacco mosaic virus (TMV). Some of the lesions caused by the virus are circled. ach lesion indicates a single infective site. [USDA photo]

for example, have a tail structure. It is thought that the passage of the phage DNA into the bacterial cell is mediated by this structure. The lack of a comparable structure in plant and animal viruses had lead to the speculation that the protein as well as the nucleic-acid portion of the virus enters the cell upon infection.

This speculation is based on several lines of evidence. First, when the nucleic acid (in the case of plants, the RNA) is isolated from the protein and used to artificially infect cells, it is noted that the viral generation time (time for lysis of a cell) is much shorter. It is thought that the nucleic acid alone, therefore, is responsible for the multiplication of the virus and that the longer generation time noted with infection by whole viruses is due to the time required for the shedding of the protein coat and the exposure of the viral nucleic acid to the cell environment.

Plant viruses fall into two major categories: those that cause a mottling or spotting (sometimes with lesions) of leaves and are called **mosaic** viruses, and those that cause a curling or yellowing of leaves, and are called **leaf curl** and **yellow** viruses. Within each category, the viruses cause specific damage to specific plant types. Tobacco mosaic virus causes lesions only in tobacco leaves, never in a tomato plant.

One of the interesting features of plant viral infections is that the plant cell must be damaged before the infection will take place. In fact, in order to infect a plant with a virus, the experimenter must mix the virus with an abrasive agent such as carborundum, and rub it on the leaves, indicating that the virus is not capable of penetrating a healthy cell.

Before the refinement of bacteriophage techniques, most of the quantitative research on viruses was done with plants—more specifically, with the tobacco mosaic virus. For instance, the concept of "one virus, one infective site" was first developed in this research system. The presence of a single lesion represents the site where one virus originally infected the plant. Because of this idea, it is possible to count the number of viral infections by merely counting the number of lesions on a plant leaf (see Fig. III-5). The plant viruses, therefore, provided the initial system for handling (organic) tissues experimentally.

Another important contribution of plant virology to the total virus picture was the elucidation of the chemical composition of the tobacco mosaic virus (TMV). The reason that a chemical analysis was first carried out for TMV was because large enough amounts of TMV could be isolated by pressing infected tobacco leaves to extract the juice. The system itself provided a means by which manipulative techniques could be developed that led to the chemical analysis. Much of the credit for foreseeing the importance of the chemical composition of viruses as far back as 1935 must go to Wendell Stanley, who first crystallized the TMV virus, and the TMV infection system presented a means for him to translate this foresight into reality.

The discovery that TMV was made up of RNA and protein came at a time when scientists were first beginning to think in terms of the chemistry of genetic material. Subsequent analyses of viral solutions indicated that all plant viruses have this same composition, whereas bacteriophages are composed of DNA and protein. In some ways, the elucidation of the chemical composition of TMV served to confuse the issue, for it was presupposed that the genetic material would be universal to all organisms, and the presence of the different types of nucleic acids in different viruses tended to support the idea of protein rather than DNA as the genetic material. Although it is now accepted that DNA is the genetic material, the idea of an RNA-based organism is not totally reconcilable.

Some latent plant viruses have been isolated and studied. In some cases, the latency seems to be due to a slow but constant release of viruses from a plant cell. In other cases, however, seem to resemble the process active for the lysogenic bacteriophage. It is thought that the virus may attach to the plant chromosome, or to some organelle within the cytoplasm. Studies on corn seed indicate that the virus associates with a region of the chromosome, and in doing so, masks several seedling characteristics simultaneously. It is interesting to note that the lack of a nuclear envelope in bacteria permits a lysogenic phage direct access to the bacterial chromosome, whereas the presence of the nuclear envelope in higher cell types, such as a plant cell, demands that the virus first penetrate into the nucleus before it could attach to the chromosome. The additional membrane penetration leads to the speculation that the site of the latent phage may be in the cytoplasm.

Infection by Animal Viruses

Once techniques of tissue culture were developed, many aspects of viral infection on a cellular level could be determined with the use of animal viruses. These aspects, however, are primarily descriptive, and the dynamics of infection are thought to be somewhat analogous to the bacteriophage system.

Animal viruses take on many fantastic geometric shapes, each of which is specific for a given virus (see Fig. III-6). None of the animal viruses, however, has a structure analogous to the tail of the bacteriophage.

Experiments similar to those performed with plants have been made. Artificial infection with the nucleic-acid portion of the virus gives a shorter generation time than does infection with the whole virus particle. This leads scientists to believe that only the nucleic-acid portion of the virus is active in actual viral multiplication, and that the time difference is due to the shedding of the protein before the nucleic acid is exposed to the cell environment.

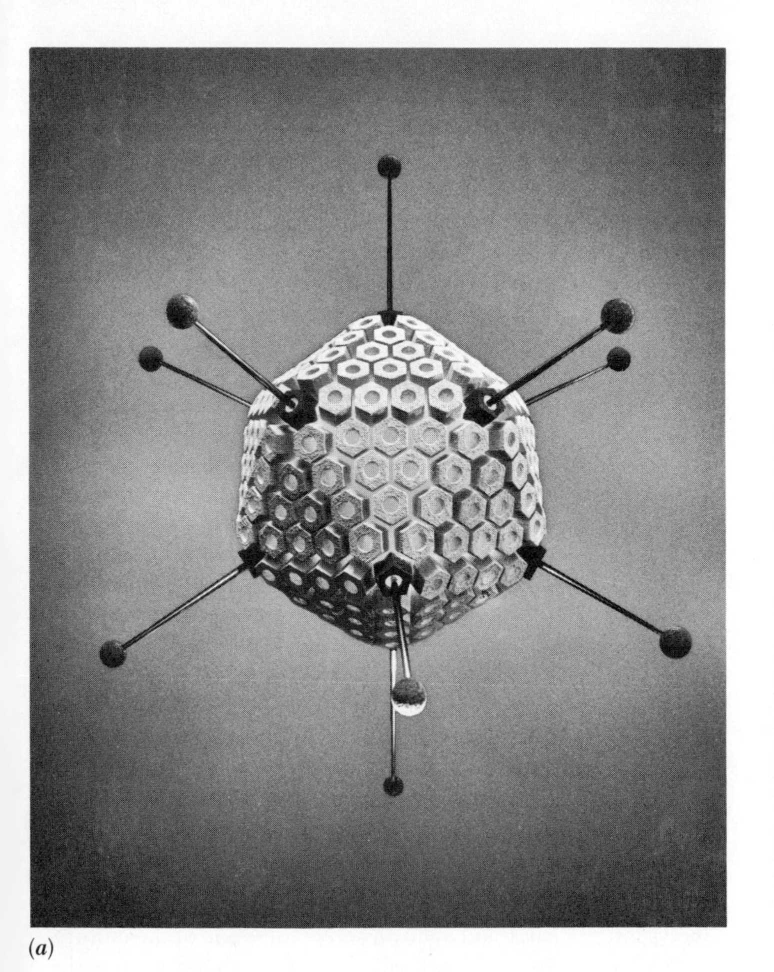

(a)

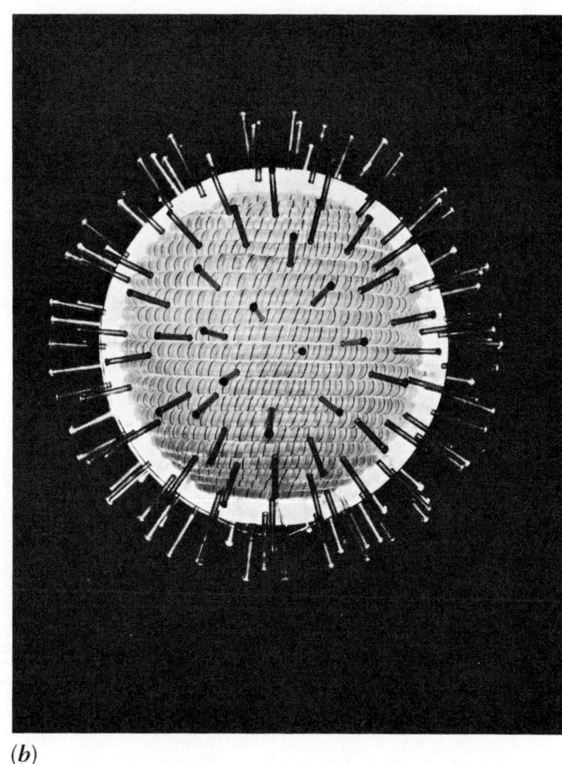

(b)

Fig. III-6 Models of the shapes of animal viruses. (*a*) **Adenovirus, which causes respiratory infections;** (*b*) **flu virus. [Yale Joel—Life Magazine, © Time Inc.]**

EFFECTS AND TRANSMISSION OF ANIMAL VIRUSES

The effect of viruses on the cells of a tissue is called the **cytopathic effect** or **CPE.** The CPE is thought to be due to the growth of the viruses in the nucleus and/or cytoplasm and the disruption of normal metabolism. The microscopically visible cytopathic effects fall into three categories.

1. **Crenation** or shrinkage of cells. The sequential cytoplasmic events have been extensively studied. The cells exhibit a "wrinkling" of the nucleus, which eventually forms a lobular body (see Fig. III-7*a*). A new cytoplasmic inclusion is formed that stains with **eosin,** indicating that there is a basic metabolic change in the cell. The cells shrink and fall off the glass of the

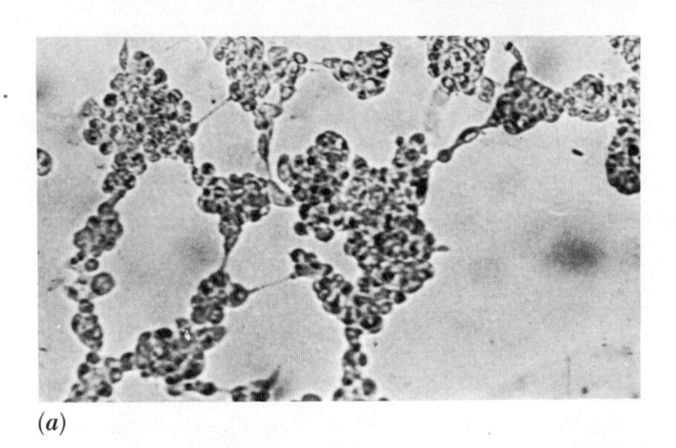

(*a*)

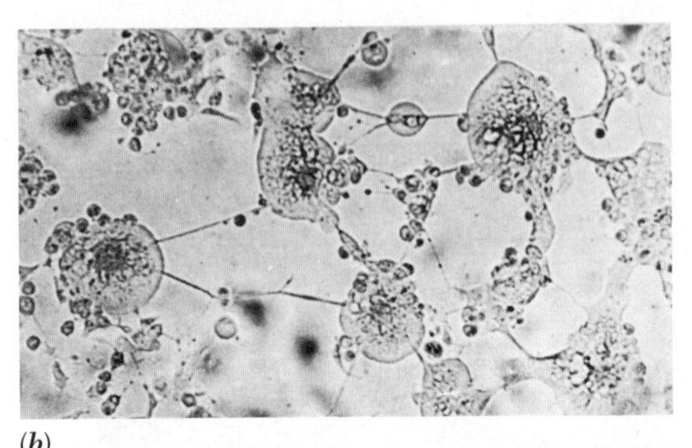

(*b*)

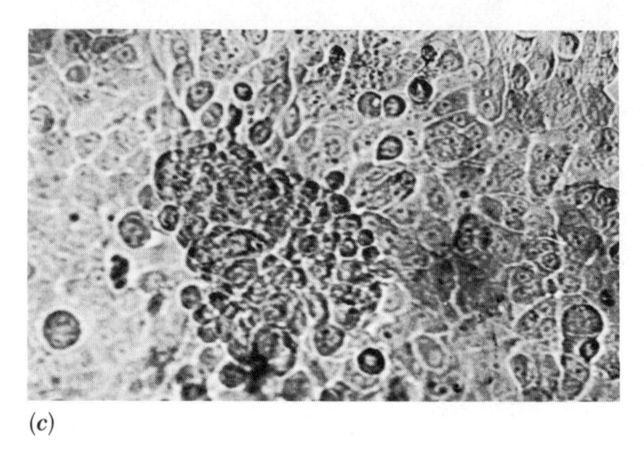

(*c*)

Fig. III-7 (*a*) Certain viruses cause a shrinkage or crenation of cells (60×). (*b*) Certain viruses cause a breakdown of cell membranes, creating a syncetial tissue (60×). (*c*) Proliferative growth and the formation of giant cells are caused by specific viruses (100×). [Courtesy of Dr. Alan Gray]

culture vessel, indicating that either the proteinaceous material that acts as a cell "glue" has been dissolved, or that production and therefore replenishment of it has ceased.

This type of CPE is noted with several viral infections in tissue culture. The most familiar virus that shows this effect is the **poliomyelitis virus.**

2. **Syncetial** CPE is the formation of giant multinucleate cells. These giant cells contain inclusions in their nuclei, that stain with eosin, once again indicating an alteration of their metabolism (see Fig. III-7*b*).

   The giant cells seem to be formed by one cell engulfing a neighboring cell in much the same way as an amoeba feeds, and bringing about a dissolution of the cell membrane. The nucleus of the engulfed cell remains intact, leading to the multinucleate condition.

   The syncetial CPE is noted with **measles virus, mumps virus, variola** (chicken pox) **virus,** and **herpes simplex** virus (which causes cold sores). It is the typical effect, therefore, of viruses that give systemic effects of localized spots or swellings. The site of these spots is thought to be the site of giant cell formations, and to be traceable back to a single virus infection.

3. **Proliferative growth** of cells. This involves the piling up of giant cells in regions of the monolayer culture. The cell clumps are attached to one another by long fibrils of cytoplasmic bridges. Some of the cells in the clump are small multinucleate forms (see Fig. III-7*c*). It is not known whether these multinucleate cells form by the engulfing of other cells or by repeated nuclear division unaccompanied by cytokinesis. Proliferative cell growth is noted with herpes simplex.

An interesting and surprising consequence of tissue culture was the isolation of latent animal viruses. Pure, experimentally uninfected tissue cultures would suddenly show signs of viral infection. Although the cells for the tissue culture had been taken from an apparently healthy animal, they contained a latent virus that began an active infection of the culture.

Among the latent viruses found was MINIA, the monkey measles virus. It has been shown that the protein of the MINIA virus was identical or almost identical to that of the human measles virus.

Systemic Effects of Viral Diseases

The systemic effect of a given viral disease is directly related to the type of tissue the virus (1) has access to, (2) is capable of attacking, and (3) actually attacks. For instance, measles is a fairly mild disease as long as only the epithelial cells are affected. If the virus has access to and actually attacks nerve cells in the brain, it can be fatal. Even in cases where measles encephalitis (brain involvement of a measles infection) is not fatal, the recovering patient often manifests the scars of mental deficiencies traceable to the viral infection. On the other hand, kidney cells in tissue culture readily become infected with measles, whereas in the body, the kidney is rarely affected. It appears that the virus either does not have access to or is incapable of attacking these cells in their natural state.

Certain viruses are transmitted through the placenta, from the mother to the developing fetus. Although these viruses may have only minor affects on the mother, the effect on the unborn child is often devastating. Undoubtedly the most familiar of these is the **rubella** or German measles virus. If the mother contracts German measles in the first three months of pregnancy, the probability is fairly high that the child will be affected. Such congenital abnormalities as deafness, heart lesions, and congenital cataracts (clouding of the lens of the eye), as well as less common dental abnormalities, mental deficiency, cleft palate, and microcephaly (a diminutive-size head and smaller brain capacity) have been found to be related to a rubella infection contracted through the placenta during early development.

The lesions caused by rubella seem to involve the damage of irreplaceable cells at the time of differentiation and early development.

Viral diseases that have a more profound effect on the mother, such as smallpox and polio, often are fatal to the developing fetus, and cause a spontaneous termination of the pregnancy. There has been some evidence that influenza infections during pregnancy have been related to **anencephaly,** a complete deterioration of fetal brain tissue.

Even in the late months of pregnancy, viral diseases can be transmitted across the placenta. Children whose mothers have recently had smallpox, measles, polio, or hepatitis are often born manifesting the symptoms of the disease. Of course, in infants the mortality rate for any viral disease is high.

Transmission of Animal Viruses

Except for the latent viruses, the transmission of animal viruses is the same as that for the transmission of pathogenic bacteria. It is merely a case of the virus (or bacteria) getting to the tissue it is capable of infecting.

There are several ways in which viruses can gain access to the body at large. If this access route also leads them to a susceptible site, infection will take place.

The most obvious of these routes is by direct contact. This includes not only kissing and sexual intercourse, but also contamination of fingers with sputum or fecal material, followed by the inadvertent placing of the fingers into one's mouth or nose, thereby giving the virus access to the internal body.

Indirect contact also serves as a means of transmission. This involves using glasses or utensils that have been contaminated and improperly washed. Indirect contact also involves food and drink prepared by contaminated individuals. Polio, for instance, is transmitted through the feces. It was for this reason that people were warned against going to polluted beaches during the polio epidemics of the 1950s.

The respiratory viruses are most effectively transferred by droplets spread by coughing or sneezing. Many other viruses, including measles, rubella, mumps, smallpox, and polio may also be transmitted in this way.

Other sources of viral infections may be animals or bites of insects, although these are fairly uncommon in the temperate zone. Flies, however, are notorious indirect carriers of infection. Contamination of their appendages is transferred to kitchen counters and food preparations on which they alight.

The success of the transmission of viral diseases is dependent on immunity factors in the host. If the host animal is immune to the virus, either by having built up antibodies in response to a previous infection or by having been induced to do so by appropriate vaccination, the virus cannot successfully infect.

CURES FOR VIRAL DISEASES

The only cure for viral diseases is the natural defensive action of the animal body itself, the buildup of antibodies against the invading virus to check their spread, and the self-destruction of infected cells, thus destroying the viruses contained within them. In some cases, it is deemed advisable to aid the natural defenses with injections of **gamma globulin,** the serum extracted from horse blood that contains a host of antibodies. It is thought that the injection of antibodies into the infected animal may hold the infection in check long enough for the host itself to build up the necessary antibodies. Some physicians, however, consider it unlikely that gamma globulin derived from horses contains the antibodies specific for some human diseases and consider injections of gamma globulin a waste of time and money.

The reason that the infected animal is ultimately left to its own defense mechanisms for combating the infection is simply because of the level of operation of viral infections. Viruses do not manifest a metabolism of their own, but take over the metabolic mechanism of the cell they infect. Therefore, it is impossible to separate the host's metabolism from that of the virus and find a drug that affects only the latter. The formation of viruses involves only the replication of DNA or RNA and the association with the protein. There is no highly specific cell wall, as in bacteria, that permits a drug to inhibit viral formation without affecting basic processes in the host as well.

Often antibiotics are given to patients with viral infections. These antibiotics are not directed toward the viral infection, but rather toward the secondary bacterial infections that might result from a weakening of the system and an associated increased susceptibility to bacterial disease.

SPECULATIONS ON A VIRAL THEME

Many aspects of virus infections are not known. Some have been worked out in bacteriophages, and are not known to be operative in animal viruses because of the inability to effectively translate techniques and experimental procedures. Some are unknown because the effect of the virus is peculiar to the animal system, so there is no way of doing preliminary studies with other, easier-to-handle organisms. This is the case with tumor viruses. No phage has been found that causes a proliferation of bacterial cells in a way that is analogous to a tumor in higher tissues. In plants, cancers such as **crown gall** are *bacterial* in origin. Attempts to bring about tumor formation in tissue culture with animal viruses have been unsuccessful. (Although herpes simplex does cause a proliferation of cells in a tissue culture, animal virologists have been unable to produce a similar effect by infecting tissue culture with viruses that are known to cause tumors in the intact organism.)

It is possible, however, to speculate on some of the interesting features of viral infections.

Why Do Some Viral Diseases Affect Primarily Children?

The most obvious speculation for this question is simply that once a child has the disease, he builds up sufficient antibody protection to insure that he never again will be infected. Therefore, as an adult, he is not susceptible to the disease. Since many people have had the so-called "childhood diseases," there are relatively few susceptible adults, and therefore, the disease remains one associated with childhood.

It is possible that there is more to it than this. The viruses for measles, mumps, and chicken pox may preferentially attack developing tissue, embryonic or early differentiating cells. There would be a profusion of these cells in a developing child, whereas there would be relatively few in an adult. Also, when childhood diseases *are* contacted in adulthood, the effects are often far more severe. Mumps in adult males often attacks the testes, the site of sperm formation where many embryonic cells are present. Testicular mumps infection often leaves the adult male temporarily or permanently sterile. Chicken pox in adults often involves an accompanying pneumonial infection directly attributable to the chicken-pox virus. The cells along the respiratory tract, especially in the lungs, are relatively undifferentiated and have very thin cell membranes. It may be that lacking the preferential embryonic tissue (such as that found in a child) and given access to susceptible embryonic tissue in the adult, the virus will infect the available respiratory-tract cells.

What Is The Significance Of The DNA And RNA Animal Viruses?

As noted in Table III-1, the DNA viruses include herpes simplex and chicken pox, whereas the RNA viruses include the influenza, measles, and mumps viruses. The RNA plant viruses are not capable of infecting (penetrating) a cell unless the cell has been damaged. There is evidence that the RNA animal viruses do not enter the cell, but direct the cell's activities from a site on the cell membrane. It may be that the RNA-based viruses are not able to penetrate and reside within the cell structure.

The DNA viruses, on the other hand, multiply within the animal cell itself. In the case of herpes simplex and chicken pox, the viral multiplication appears to take place in the nucleus. Other DNA viruses, such as the one that causes smallpox, multiply in the cytoplasm of the cells.

But these two types of viruses differ only in their information-carrying system. While it is true that the dual role of RNA in the RNA viruses (for information and for transmission of this information), may represent a less efficient system than that accepted for the central dogma (DNA to RNA to protein), why should this factor affect the ability of the virus to penetrate the cell? The question is left for the reader to speculate about.

CHAPTER 12

CELLULAR DIFFERENTIATION

The development of either a multicellular or a unicellular organism involves two interrelated processes: growth and differentiation. Growth, as related in the previous chapter, refers to an increase in protoplasmic mass. Cellular differentiation involves the change of a cell from an embryonic to a specialized type.

Although the differentiation process is manifested at every level of cell organization, it can, theoretically at least, be traced to changes in the basic cell chemistry. For example, changes may be expressed as an alteration on a metabolic level to complement the function of the specialized or differentiated cell. Such metabolic changes can be traced to alterations of enzyme ratios within the cell, or, for that matter, to the disappearance of previously active enzyme systems and the initiation of new ones. On a chemical level, such changes would involve the preferential synthesis or selective destruction of proteins. The same is true of changes that are observed on a cytological level. The differentiated cell may manifest an alteration of cell organelle ratios, organelle size, appearance of auxiliary structures, and/or allometric growth patterns. Nevertheless, these changes are the result of preferential synthesis and destruction of enzymatic and structural proteins. The differentiation process should, then, be traceable to the most basic level of cellular chemistry, the transcription of DNA to form *m*RNA, and the translation of *m*RNA to form protein.

The genetic basis for differentiation is not a new idea. At the beginning of the twentieth century, many developmental biologists believed that, as the embryonic form of an organism develops, the ensuing cell divisions parceled the information into daughter cells in such a pattern that the genetic information necessary for muscle-cell form and function went into those embryonic cells destined to produce muscle tissue, while the information for nerve-cell form and function went only into those cells destined to produce nerve cells, etc. Only the so-called "germ line," the cells destined to produce the gametes and thus provide genetic continuity through succeeding generations, were believed to contain the total complement of genetic information.

Despite the overwhelming evidence that mitotic divisions that led to the formation of the embryo and eventually, the adult individual, provided each cell with the same chromosome complement, and hence, the same genetic information (see Chap. 8), many biologists held the aforementioned view. This theory was disproven for higher plants by F. C. Stuart, who removed and separated pith cells of an adult plant, and, with proper nutrition, cultured each cell separately into an entirely new plant. In spite of its differentiated character in the adult form, the pith cell contained all the necessary genetic information for the production of cells unlike itself. Therefore, growth and development by mitotic processes *do* provide each cell with the same DNA as every other cell, and the difference between cell types is not attributable to differences in genetic information content.

The alternative idea is that during the differentiating process of each type of cell, certain genes are "turned on" (**activated**), while others are "turned off" (**repressed**). The transcription of genes A, B, E, G, J, and O may produce a series of proteins that cause the cell to become a muscle cell; whereas the transcription of genes A, C, E, I, and M may produce a series of proteins that cause the cell to become an epithelial cell. The problem is, then, how does a given type of cell "turn off" or "on" different genes? Or, for that matter, how does a given type of cell "know" which genes are to be activated or repressed? These questions probe into the very basis of integrated biochemical activities, and when the answers are found, man will understand the dynamic basis of life.

In practical application, it is not possible to trace differentiating systems back through specific chemical changes and to provide a full sequential picture of the events that occur in the differentiation process. At this time it must suffice to present a few glimpses of the phenomenon of cellular differentiation, and hope that the extensive research presently undertaken in this area will some day provide a more profound and comprehensive picture.

EXPRESSION OF CELLULAR POTENTIAL THROUGH DIFFERENTIATION

In order to fully appreciate the complexity of the differentiation process, it is necessary to first recognize those cytological changes that occur over the course of differentiation. These cytological changes have been extensively studied for many types of cells. From these studies has emerged a fascinating account of the physical expressions of cellular potentials that cannot be gleaned from an examination of the final, differentiated cell alone.

Differentiation in a Unicellular Organism

Each type of unicellular organism is a specific differentiated cell, having its own pattern of differentiation. It is impossible to discuss them all, and equally impossible to discuss one and claim that it serves to exemplify all. Therefore, only one example, that of *Acetabularia*, will be discussed, with the understanding that it has been chosen because of its interesting structural features and the biochemical events of the differentiation process that have been partially worked out (see Chap. 9).

Acetabularia is a unicellular green plant that grows one to two inches tall. It is found primarily in tropic and subtropic waters. A single *Acetabularia* plant begins as a diploid spore (see Fig. 12-1). The spore germinates (initiates growth) by producing a filamentous extension that adheres to a rock or pebble on the ocean floor. This initial filament branches (by allometric growth), thus forming several **rhizoids,** or root-like protuberances. While the rhizoids elongate, the spore nucleus enters one of the rhizoids.

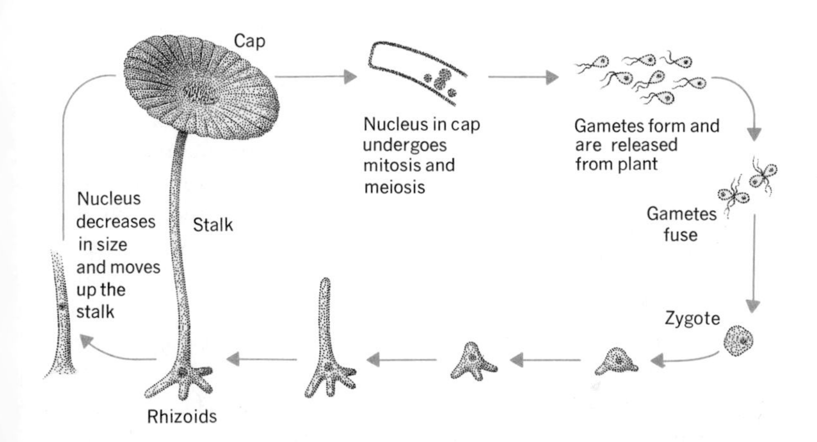

Fig. 12-1 Development of the *Acetabularia* plant.

After the rhizoids are formed and the new plant is firmly anchored to the substratum, an erect filament arises from the spore. This is the **stalk.** Chloroplasts develop in the stalk while the single, large nucleus remains in the rhizoid.

After the first year of growth, a series of branches form at the top of the stem. These branches occur in three sets of radiating filaments. Later the branches and a terminal portion of the stalk die and fall off.

After the second or third year of growth, a mature **cap** is formed at the terminus of the stalk, from one of the three sets of radiating filaments. The other two sets produce two whorles of **paraphases** (sterile hairs).

At this time, the nucleus migrates up to the cap region where it undergoes many mitotic and then meiotic divisions, producing thousands of haploid nuclei. The nuclei migrate through the stem into the filaments of the cap. Each nucleus associates with a chloroplast, some cytoplasm, and an enclosing membrane, to produce units called **protoplasts.** These protoplasts become **biflagellated** (develop two flagella) and swim away from the parental cell upon the deterioration of its cap.

These biflagellate swarmers fuse to form a zygote or spore, which is then capable of producing a new plant.

The life cycle of *Acetabularia,* and indeed, those of all unicellular organisms, is an example of differentiation along a linear pattern. A single spore produces a one-celled plant, which produces spores, etc.

The pattern of differentiation of *Acetabularia* permits many questions to be raised. Hammerling's work (Chapter 9, p. 248) indicates that a cap-forming substance is released by the nucleus, and that the nature of this substance determines the type of cap that is formed. Although the substance has never been isolated and chemically analyzed, there is good reason to believe that it is RNA. If so, are there RNAs for stalk and rhizoid formation? The mechanisms underlying the change in nuclear division from mitotic to meiotic division are also unclear. And how do the nuclei migrate up the stalk? How does each nucleus associate with a chloroplast and initiate the formation of the membrane that eventually separates the protoplasts from one another? The answers to these questions lie on both genetic and functional levels.

Differentiation as Exemplified by Two Different Cells in the Same Multicellular Organism. Except for the gametes, the cells of a multicellular organism have the same chromosomal complement, and hence, identical genetic information. Two cells, identical in information content and similar in cytological appearance, but located in different parts of the animal body, give rise to cells that undergo vastly different differentiation processes to become cells that are totally dissimilar in both structure and function. Such is the case with the **spermatogonia,** destined to produce mature sperm cells, and the **stem cells,** destined to produce red blood cells.

Within the male sex organs are the spermatogonia. These are the embryonic gametogenic cells that are capable, upon division, of producing more spermatogonia or another type of cell (the **primary spermatocyte**) that is destined to undergo meiotic division. The very formation of the primary spermatocyte is a form of functional differentiation. In appearance, the primary spermatocyte is the same as the spermatogonia; yet its next division will be meiotic rather than mitotic, so it is the direct ancestor of the sperm cell (see Fig. 12-2).

Upon meiotic division, the primary spermatocyte gives rise to two haploid secondary spermatocytes. These undergo the second of two meiotic divisions, and four cells called **spermatids** are formed.

The spermatid is not a highly differentiated cell. In fact, it appears very much like the parent primary spermatocyte or the grandparent spermatogonium. Following its formation, however, it begins to exhibit a series of cytological alterations. The cytoplasm around the nucleus begins to contract, and most of the cytoplasm is localized at one end of the cell. The golgi body becomes less membranous and takes on a granular appearance, becoming a very dense region near the nucleus. This is now called the **acrosome.** The centriole takes a position 180° from that of the acrosome, and produces an axial filament. The mitochondria pack around the filament in a well-ordered helical arrangement. Most of the cytoplasm around the area of the filament is sloughed off from the cell. The mature sperm has a **head,** which consists of the acrosome and nucleus; a **neck,** or constriction of the cytoplasm; a **middle piece,** in which the mitochondria, centriole, filament and sheath are found; and a **tail,** composed of the sheath around a spiral fiber.

In the bone marrow and spleen of the same male mammal that

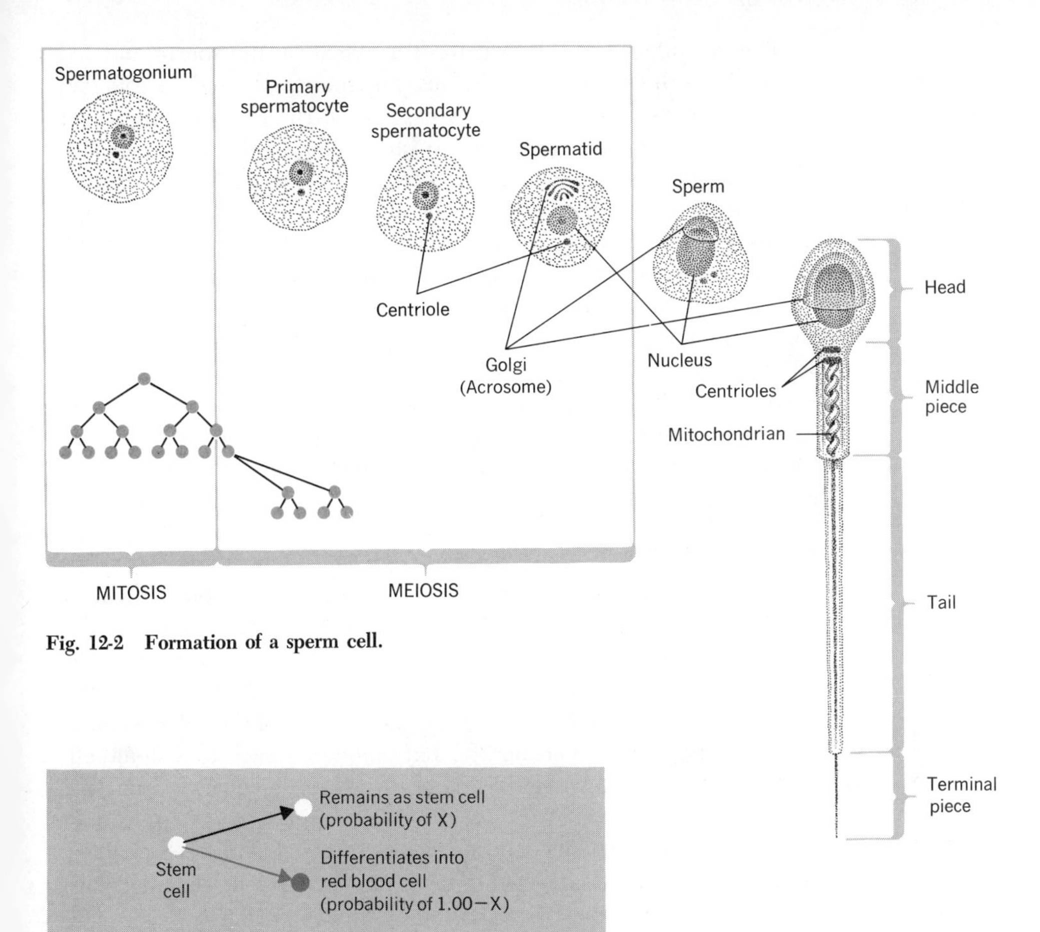

Fig. 12-2 Formation of a sperm cell.

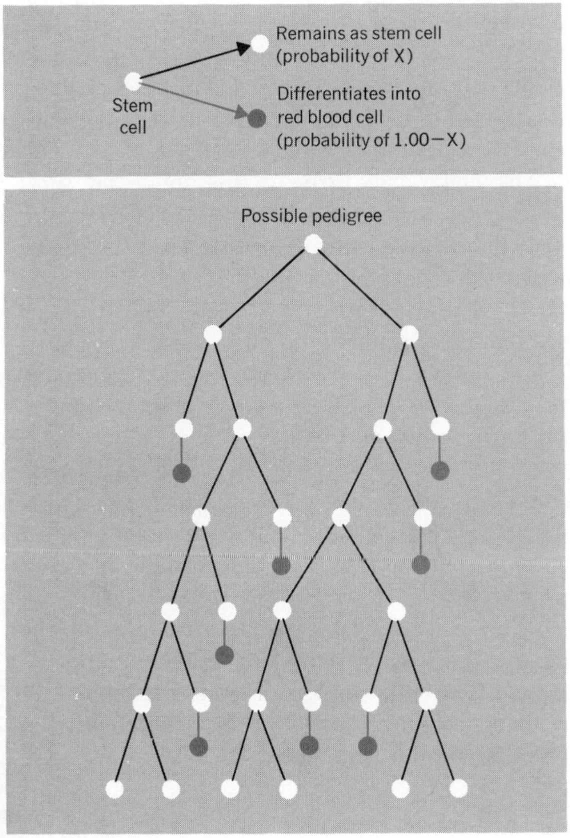

Remains as stem cell (probability of X)
Stem cell
Differentiates into red blood cell (probability of 1.00−X)
Possible pedigree

Fig. 12-3 Possible pedigree of stem cells differentiating into red blood cells.

formed the above-mentioned mature sperm, are the stem cells. Upon division, they give rise either to more stem cells, or to cells that undergo differentiation to form the red blood cells.

When a stem cell is fated to undergo differentiation, the nucleus disintegrates, and all the cytoplasmic contents disappear, except for the *m*RNA and the protein synthetic machinery for hemoglobin. The cell essentially becomes a sac containing the protein hemoglobin, *m*RNAs for hemoglobin, ribosomes, *t*RNA, water, and salts, and it is capable of carrying oxygen until exhaustion of energy sources prohibits further synthesis of hemoglobin (see Fig. 12-3). The average red blood cell lasts for only about 123 days, after which it is destroyed in the liver. The stem cells constantly replenish the red-blood-cell supply.

The advantages of the final form assumed by the sperm and red blood cells for their function is obvious. The sperm cell, with its rounded head and narrow flagellated tail, is capable of the efficient movement necessary to reach the egg. The red blood cell, uncluttered by cytoplasmic organelles or even a nucleus, permits more space for hemoglobin and the oxygen it carries to the cells of the body. But it is a strange and unknown biological phenomenon that permits two cells with the same genetic information to undergo two vastly different differentiation processes, one involving primarily cytological rearrangements, and the other involving the selective destruction of cellular components so that only one protein (hemoglobin), a few species of RNA, and the cell membrane survive. The phenomenon becomes overwhelming when one realizes that these are but two types of cells that are formed in a body that has over one hundred types, each formed from embryonic cells of the same genetic potential.

Dedifferentiation

In the overwhelming number of cases, differentiation is an irreversible process. Once a spermatid has become a mature sperm, the mature sperm cannot revert back to a spermatid. Perhaps this irreversibility is most obvious in the case of the red blood cell, where the destruction of the reproductive factors (the nucleus) of the cell precludes the possibility of the reversion of a red blood cell back to a stem cell.

In certain cases, however, it appears that the change from the embryonic cell type is not as severe, so that the sequence of steps leading away from the embryonic condition can be retraced, and a cell **dedifferentiates.** Dedifferentiation occurs in nature both in normal developmental patterns and under special circumstances.

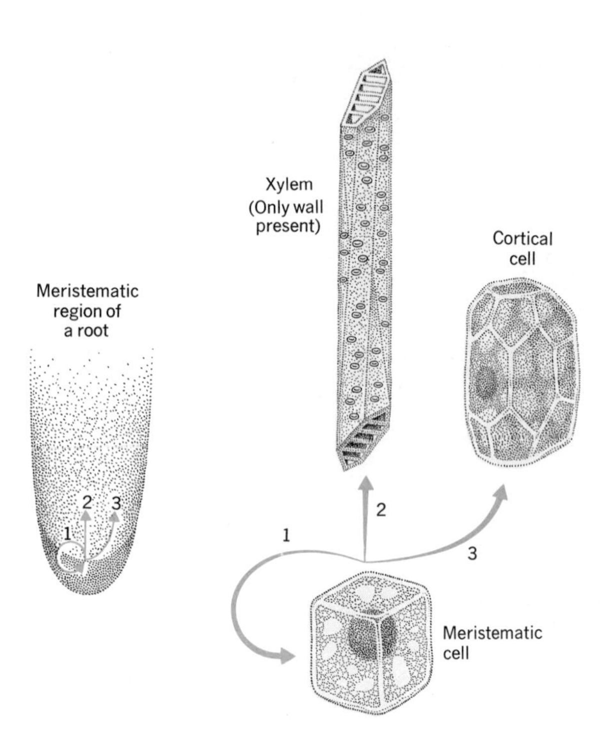

Fig. 12-4 The meristematic cell upon division can form (1) more meristematic cells, or cells that differentiate into (2) vascular cells or (3) cortical cells. The type of cell formed is dependent on its position in relation to the central axis of the plant.

In woody plants, the embryonic tissue of the stem and roots (meristems) gives rise to cells that are fated to become (1) more meristematic cells, (2) cells that differentiate into those that form the xylem and phloem of the vascular tissue, thereby serving to transport minerals, water, and photosynthetic products vertically through the plant, or (3) **cortical cells** (cortex) that serve primarily as fill-in tissue in the stem and roots (see Fig. 12-4). These cortical cells often function as storage depots for the products of photosynthesis. They do not exhibit the degree of differentiation noted in the xylem and phloem cells.

When a woody plant is one or two years old, the cortical cells near the epidermis dedifferentiate, thus becoming embryonic. In doing so, they become a new meristem, called **cork cambium.** The cork cambium gives rise to more cork cambium cells, or cells that differentiate into **cork parenchyma** and **cork cells** (see Fig. 12-5).

The cork cambium, cork cells, and cork parenchyma comprise the outer region of the bark of the woody stem. The dedifferentiation of the cortex cells of the stem is essential for the normal development of the plant. The cells are evidently programmed for this dedifferentiation and subsequent redifferentiation into cork tissue.

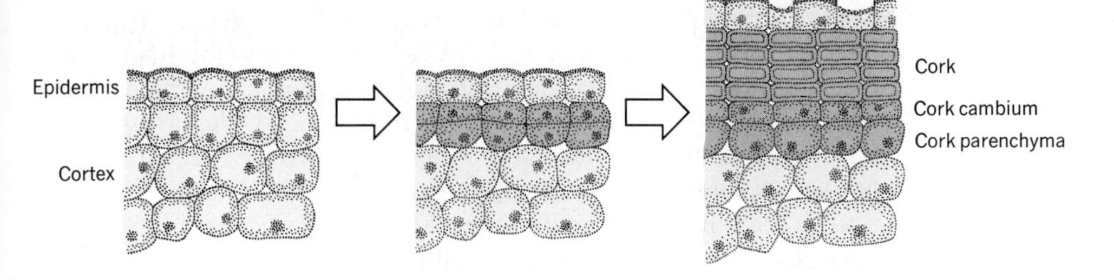

Fig. 12-5 Cortical cells differentiate into cork cambium. Cork cambium cells, upon division, form more cork cambium, or differentiate into cork cells or cork parenchyma, depending on their position with respect to the cambium layer.

In some types of organisms, dedifferentiation occurs as the result of damage to a tissue. When the limbs or tail of salamanders are lost, the exposed bone and muscle tissue dedifferentiates, divides by mitotic divisions, and *redifferentiates* into bone, muscle, nerve, epidermis, etc., as it grows outward to reform the limb. It is not known if dedifferentiated bone forms only bone cells, or if the dedifferentiation process is so complete that general embryonic cells are formed from which any type of tissue arises. Indeed, this process at the organ level, which is called **regeneration,** is a fascinating one, about which relatively little is known. One of the most important unanswered questions that can be raised from a human standpoint is why the regenerative ability is limited to certain lower forms of animals.

In the broadest sense, the bridge between generations can be viewed as dedifferentiation followed by a redifferentiation. Two highly differentiated cells, the egg and the sperm, join to form a zygote, which is the very essence of an embryonic cell. The zygote undergoes mitotic divisions, and ultimately results in the formation of many different types of differentiated cells that make up an entirely new individual.

TRIGGER MECHANISMS OF DIFFERENTIATION

A group of embryonic cells undergoes a series of mitotic divisions, forming more embryonic cells. Some of these cells suddenly undergo various differentiation processes, while others remain in embryonic form.

Obviously, there has been some *internal change* in these cells that has initiated the differentiation process. But there must also be some factor that has *caused* the change to occur. The causative factor *and* the change it has induced constitute a **trigger mechanism** for the differentiation process.

Most of what is known today regarding the chemical dynamics of differentiation concern the trigger mechanisms. Trigger mechanisms appear to be operative on one of three levels of cell chemistry, **tran-**

scription, translation, and **conformation.** That is, trigger mechanisms either effect the transcription of DNA information into the RNA message, the translation of the RNA message into protein, or the conformation of protein from one configuration to another. While it is assumed that in each of these cases, transcription is ultimately effected, it has not been possible, in most cases, to trace the internal change to this level. Nor has it been possible to trace the initial change through the sequence of cellular alterations that ultimately result in the final differentiated form of the cell. It is probably only a minor consolation to the student that researchers in the field of developmental biology also find this paucity of available information rather frustrating.

Trigger Mechanisms on the Transcription Level

For many years, trigger mechanisms on a transcription level were the subject of much speculation. It was thought that certain regions of the DNA molecule were coated, probably with a protein, while others were exposed. Only the exposed regions could be transcribed and hence were active. The coated regions, then, were repressed. The accepted proposal for the activation or repression of genes was the existence of specific repressor proteins that could be removed only by the presence of specific **inducers** (see Fig. 12-6).

The first experimental evidence that supported this proposal was given by Jacob and Monod. The bacterium *E. coli* undergoes a simple form of differentiation on a metabolic level. When the medium in which the bacterium is growing does not contain the disaccharide lactose, the bacterium does not produce the enzymes **β-galactosidase,** which dissociates lactose into two monosaccharides (glucose and galactose) and a **permease** that facilitates the movement of lactose across the cell membrane. If lactose is added to the medium, the bacterium initiates the production of β-galactosidase and permease, thus enabling it to take in substantial amounts of lactose and utilize it as a carbon source. This initiation of these two enzymes presented Jacob and Monod with a simple system for determining the trigger mechanism for this type of differentiation process.

Through a series of complicated genetic studies, Jacob and Monod determined that there were two spatially separated regions of the bacterial chromosome that were involved in determining whether the cell produces β-galactosidase. One was the gene responsible for the synthesis of the β-galactosidase itself, which is called the **operon;** the other was a **regulator gene** that controlled the operation of the operon.

The operon was found to be made up of four separate genes: an **operator** gene (represented *O*), which must be read first, and whose functionality thus determines the functionality of the rest of the operon; the **SG_1** gene, which carries the information for the synthesis of β-galactosidase; and the **SG_2** gene, which carries the information for the synthesis of a permease that permits lactose to enter the bacterial cell more readily and **SG_3** which produces an enzyme the function of which is not understood (see Fig. 12-7). The entire operon (O, SG_1, SG_2, and SG_3) is read as a single unit.

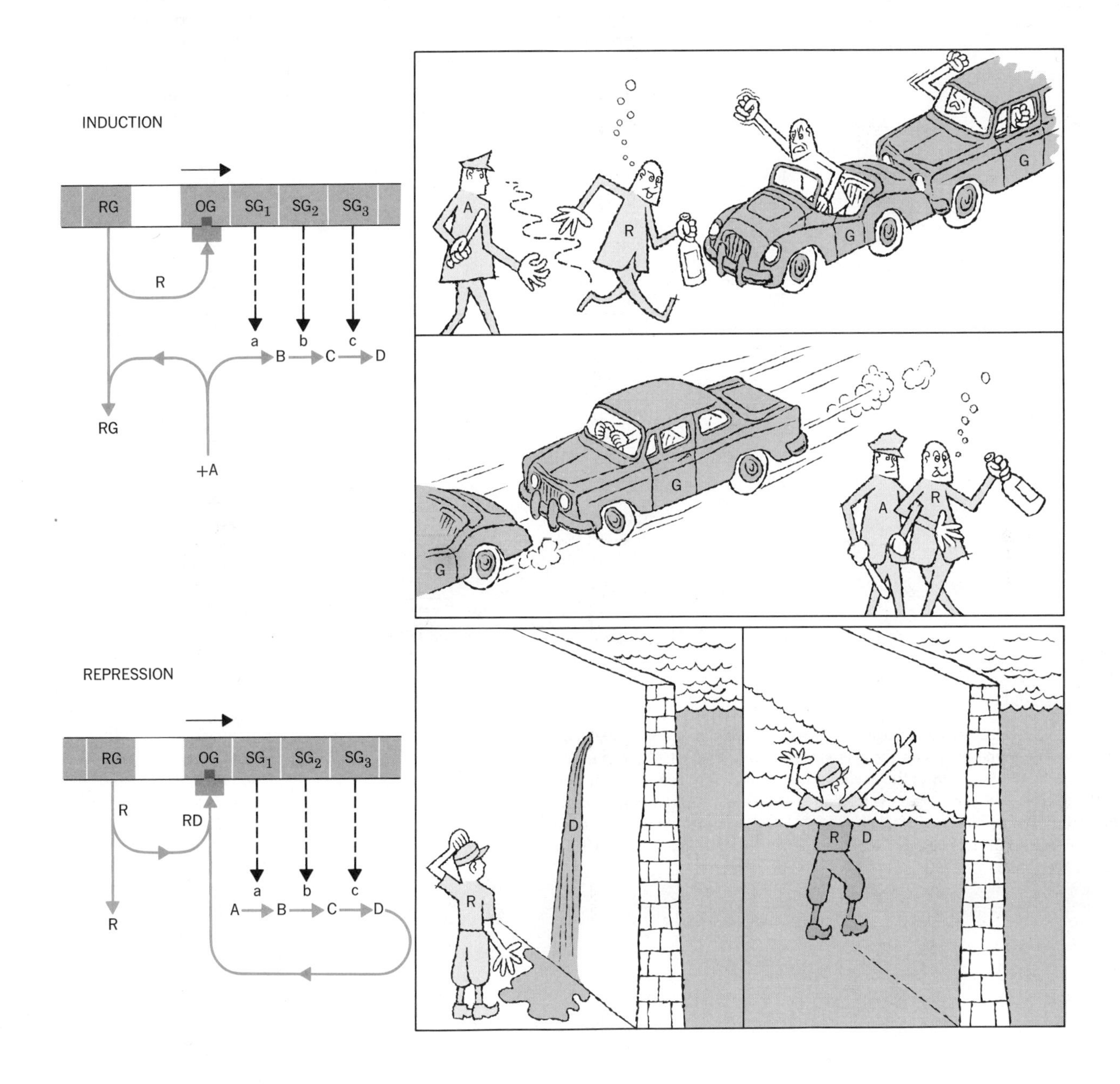

Fig. 12-6 Induction and repression. During induction *A* combines with the repressor substance (R), removing it from its position. Repression occurs when a product of the reaction acts as the gene repressor or stimulates the production of the repressor.

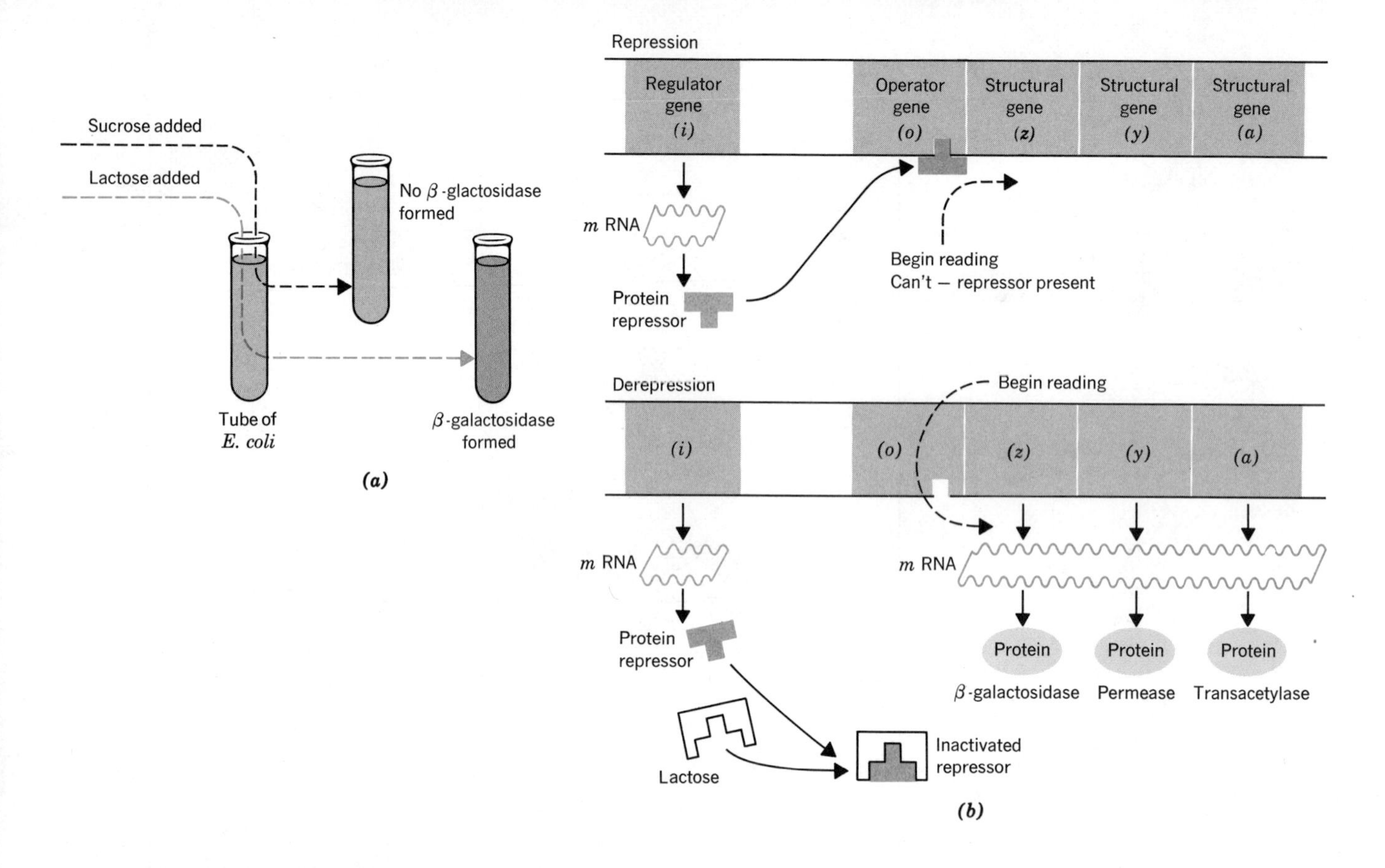

Fig. 12-7 *(a)* β-galactosidase is formed only when the bacteria are in media containing lactose. *(b)* The repressor sits on the operon, preventing the rest of the gene from being read. When lactose is added to the medium, it combines with the repressor, removing it from the operon, and permitting the reading of the genes for β-galactosidase (which aids in the utilization of lactose as a food source), for a permease (which increases the admission of lactose to the cell), and for transacetylase, the function of which is not completely understood.

The regulator gene produces a protein called the repressor substance (represented **RS**). The RS has an affinity for the operator gene of the operon, and (by forming an association with it) masks the DNA, inhibiting it from being read. Since the entire operon must be read as a single unit, and the operating gene must be read first, the masking of the O gene prohibits the reading of the operon (see Fig. 12-8).

When lactose is added to the culture medium, some enters the bacterial cell. The lactose has an affinity for RS, and tends to combine with it. In a sense, the DNA of the operator gene and the lactose compete for the repressor substance. The success of the competitors is based on the number of each type of molecule present in the cell, and since the

Fig. 12-8 Negative and positive feedback.

lactose molecules entering the cell vastly outnumber the single operator gene, the repressor substance is removed from its inhibitory position and becomes bound to lactose. Therefore, lactose itself acts as an inducer for the enzyme that breaks it down.

With the removal of the repressor substance, the operator gene, and hence the whole operon, can be read. β-galactosidase and permease are synthesized, and the cell is now in a state to preferentially permit the entrance of lactose, to utilize it as a carbon source in metabolic processes. Although the regulator gene may continue to produce the repressor substance, it is thought that there is an adequate amount of lactose in the cell to combine with it, thus keeping the operon functional.

When the lactose in the medium is depleted, or if the bacterium is transferred to a medium that does not contain lactose, the repressor substance-lactose-operator equilibrium shifts, and the repressor substance once again attaches to the operator DNA, inhibiting the reading of the operon.

So here is a process by which genes are activated and repressed by the presence of a substance whose breakdown the gene itself controls—a mechanism known as **feedback.**

Feedback is the mechanism by which a substance at a certain concentration level acts to enhance or inhibit syntheses. When syntheses are enhanced, it is called **positive feedback.** When syntheses are inhibited, it is called **negative feedback** (see Fig. 12-8). Both types of feedback mechanisms are important control mechanisms in biological organisms.

The concentration at which the feedback mechanism is triggered is called the **threshold.** The concept of a threshold is also very important in biological systems, and may be viewed either as a probability factor or a cumulative effect. When threshold level is taken as a probability statement, it implies that the greater the number of molecules present, the greater the chance that a given molecule would be in the right place at the right time. The threshold then represents the concentration at which this chance approaches 1.0, or near certainty. When threshold is meant in terms of a cumulative effect, the buildup-effect itself is important. Such is the case with the transmission of nerve impulses. A certain *amount* of electrical charge must be present in order for the synapse (the space between the axon of one nerve cell and the dendrite of another) to be jumped.

When the threshold level of lactose (as a probability statement) is present, the synthesis of β-galactosidase and the permease is initiated and sustained. The β-galactosidase system in *E. coli* is, then, an example of positive feedback. It is tempting to think that the initiation of enzymes in other biological forms operates in much the same manner.

Trigger Mechanisms on the Translation Level

Another level on which trigger mechanisms have been found to operate is that of the translation of RNA to protein. In this case, the DNA information has been successfully transcribed to *m*RNA, but the *m*RNA message is inhibited from being translated into protein. The inducer substance is thought to bring about the exposure of the *m*RNA message, permitting it to be read. This type of system seems to be present in developing animal embryos, and recent evidence indicates that it is also true for seed germination.

After fertilization of an animal egg, the zygote will undergo a series of mitotic divisions. After six or seven divisions (64- or 128-cell stage), a hollow ball is formed that is called the blastula (see Fig. 12-9). The

Fig. 12-9 Development of the zygote to blastula.

blastula is the same size as the original zygote; little more than a parceling up of the cytoplasm of the original zygote appears to have occurred. After this stage has been reached, the cells begin to differentiate, and the hollow ball begins a series of structural changes that are collectively called **gastrulation.** The stage at which differentiation begins is called **gastrula.** The obvious question is, what triggers these undifferentiated blastula cells to initiate the gastrulation process, and at what level does the trigger mechanism operate?

When developing embryos between the zygote and blastula stage are treated with **puromycin,** an antibiotic that inhibits the translation of *m*RNA into protein, their development ceases. This indicates that there is protein-synthesis activity going on in the embryo at this time. When, however, embryos of this stage are treated with **actinomycin D,** which inhibits the transcription of DNA to *m*RNA, development continues to the blastula stage and into the early stages of differentiation that characterize the gastrula stage. This indicates that there is no new *m*RNA made by the developing embryo until the differentiating process has already begun. The *m*RNA that initiates the gastrula stage, then, is always present in the embryo; it is simply not translated.

Further evidence that this is the case is provided by taking embryos between the zygote and blastula stages, and preparing cell extracts from them. When these cell extracts are mixed with ATP and magnesium, new species of proteins are formed. This would seem to show that the *m*RNA for the synthesis of these new proteins is present in a masked form in the developing embryo, and that the extraction procedure disengages the masking factor so that the *m*RNA becomes functional.

Although there is ample evidence to indicate that the level of masking is, in this case, the translation, rather than the transcription level, the factor that induces the *m*RNA to become active is unknown. It is thought that during early development there is a sequential activation of different preformed *m*RNAs.

Trigger Mechanisms on the Conformation Level

The final level on which trigger mechanisms have been shown to act is the protein level itself. In this case, the inducer acts directly on a protein structure, and in doing so, changes the protein to a different structural and functional form. It may well be that the altered protein is capable of acting as a repressor or activator (by either positive or negative feedback mechanisms) on a genetic level, but this is merely speculative.

An example of a trigger mechanism on the conformation level is in the induction of flowers. It has been known for many years that floral induction was primarily dependent on light rather than temperature, water, minerals, etc. It was not until 1957, however, that Hendricks and Borthwick isolated a protein from leaf tissue that underwent a structural change in different qualities of light. When the protein, which for convenience will be called **A,** was exposed to red light (6600 Å), it converted to another protein, which will be called **B** (see Fig. 12-10).

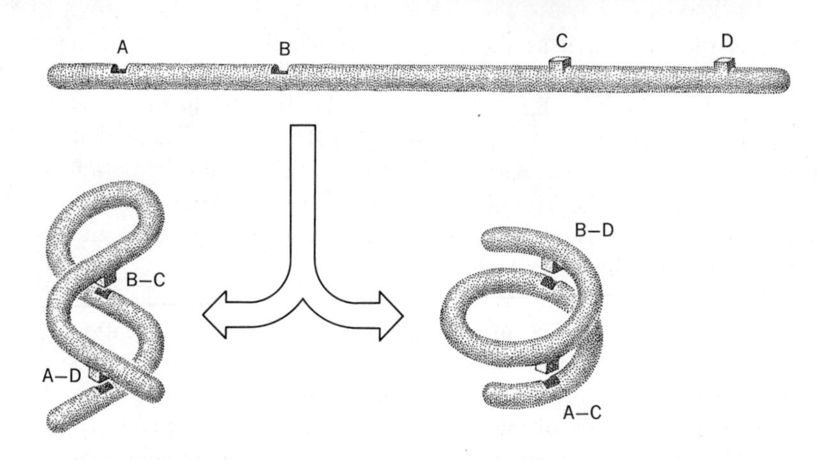

Fig. 12-10 Allosteric stages. The addition of light energy of specific wavelength brings the protein to a new energy level, causing a rearrangement of tertiary or quarternary structure. Light is thought to be absorbed by a chlorophyll-like molecule attached to the protein.

When **B** was exposed to far-red light (7300 Å), it was converted back to **A.** Each protein molecule was arranged in such a way that it absorbed the light of the appropriate wavelength, underwent excitations at specific electronic sites in its structure, and changed its tertiary structure to conform with its new energy level, thereby becoming a different protein, both structurally and functionally. A single protein molecule can refold into different, active **allosteric** states (see Fig. 12-10).

In some plants, the **A** form of the protein (that which was induced by red light) eventually led to the production of the substance **florigen** (see Fig. 12-11). The appearance of florigen signals the beginning of the differentiating process leading to flower formation. It appears, therefore, that the red light induces the **A** protein (on a conformation level), which, in turn, acts as a trigger mechanism for the induction of florigen. Florigen induction may occur at any of the three levels, although it is assumed that it is eventually manifested on the transcription level.

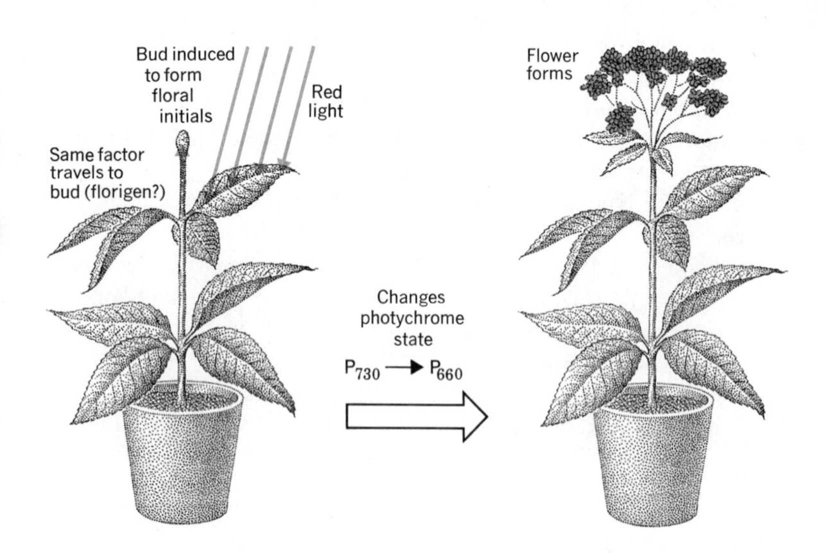

Fig. 12-11 The induction of flowers. In some plants flowering is induced by red light ($P_{730} \longrightarrow P_{660}$) and in others by far red light ($P_{660} \longrightarrow P_{730}$).

NATURE OF THE INDUCING FACTOR

The factors causing the internal change, and thus initiating the differentiation process, are of varied forms. In the example given above, light is an inducing factor for floral induction; the molecule lactose is the inducing agent for β-galactosidase synthesis in *E. coli;* and the inducer for releasing preformed *m*RNA in developing embryos is unknown. There are many other inducing factors that are known to trigger differentiation processes. The cellular mechanisms triggered into operation by these inducing factors are not known.

Hormones as Inducing Factors

A **hormone** is a proteinaceous substance, produced and secreted by a cell, that acts on the basic chemical level of other cells. Although many hormones elicit functional responses from the cells they affect, some elicit developmental responses.

The development of the water mold *Achyla* exemplifies the natural biological utilization of hormones as inducing factors of differentiation. The cells of *Achyla* are long and filamentous, and grow in a tangled mass. Some of the cells penetrate a nutritional source (culture medium in a laboratory, or dead insect or seed in their natural habitat). A colony (growth from a single spore) is either male or female. When male and female colonies grow near one another, and a male and female filament are in close enough proximity for a chemically elicited response, a mating sequence takes place (see Fig. 12-12).

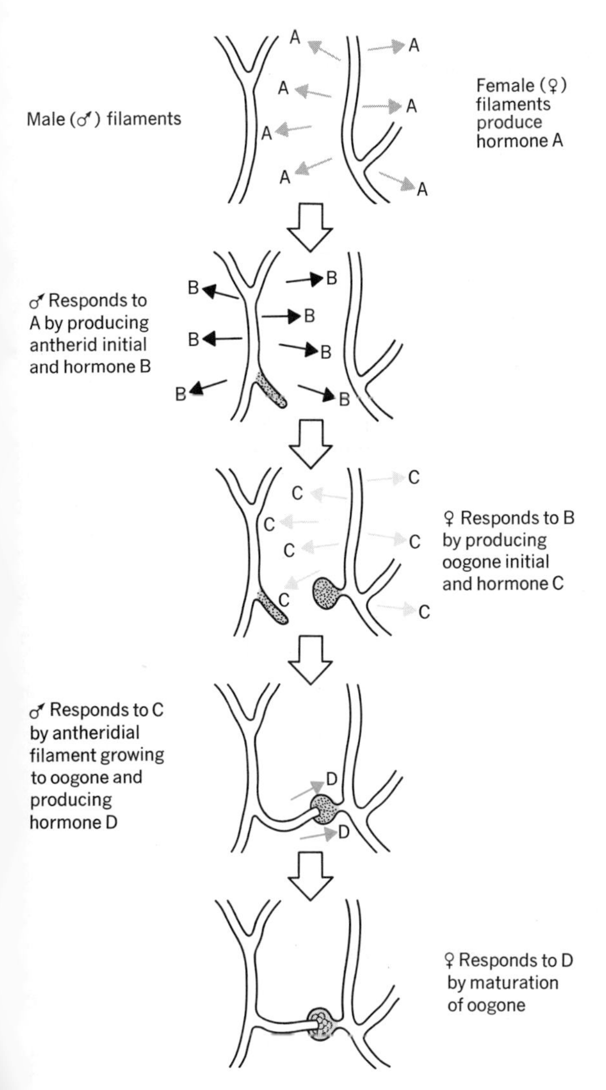

Fig. 12-12 Achylla mating sequence.

The female filament secretes hormone **A.** It is known that hormone **A** is secreted constantly, not just when a male filament is present. The male filament reacts to the secretion of hormone **A** by producing filamentous cytoplasmic extensions, called **antheridial initials.** These contain several haploid nuclei, each of which undergoes mitotic divisions.

The male filament now secretes hormone **B,** which induces the female filament to form a single balloon-shaped cellular extension called the **oogonial initial,** in which there are several haploid nuclei.

The oogonial initial now secretes hormone **C,** which causes the antheridial initials to grow directly toward it and to adhere to it.

The male then secretes hormone **D.** This causes the formation of cell membranes around the haploid nuclei in the oogonial initial, thus converting them into functional female gametes. Finally the antheridia (the male sex organs) put out germ tubes that connect to the female gamete. The male nuclei move through the germ tubes and fertilize the female gametes, thus forming diploid zygotes.

In each of these steps, a different hormone has been detected. Hormone **A** has recently been isolated from several thousand gallons of female cultures. It has been identified as a glycoprotein with a molecular weight of about 120,000, and has been appropriately called **sirenogen.**

As with all of these inducing agents, it is not known how, or (for that matter) on what level of cellular chemistry, the initial change takes place.

Unfavorable Environmental Conditions as Inducing Factors

Many different organisms undergo differentiation leading to spore formation. Even in the healthiest cultures, some of the individual organisms will form spores. These will remain spores for a short time and then revert to vegetative cells, while other vegetative cells become spores. In this way a low percentage (0.00001%) of even the healthiest culture is in spore form. The existence of spores in healthy cultures insures the survival of the population if conditions become unfavorable. When the oxygen is limiting, or there are excessive amounts of carbon dioxide, the number of individuals that form spores markedly increases. Spore formation can be induced in these cultures by simulating the requisite unfavorable conditions.

The small aquatic animal *Hydra* (see Fig. 12-13) forms gametes when it is exposed to crowded conditions. The zygote that results from the joining of the male and female *Hydra* gametes has a protective coating that is more resistant to an adverse environment than are the cells of the typical animal body. The phenomenon of gamete production can be induced by simulating crowded conditions through the introduction of carbon dioxide into the culture.

That unfavorable environmental conditions can act as trigger mechanisms for the production either of spores, or of gametes that fuse to form resistant zygotes, attests to the efficiency of the mechanism.

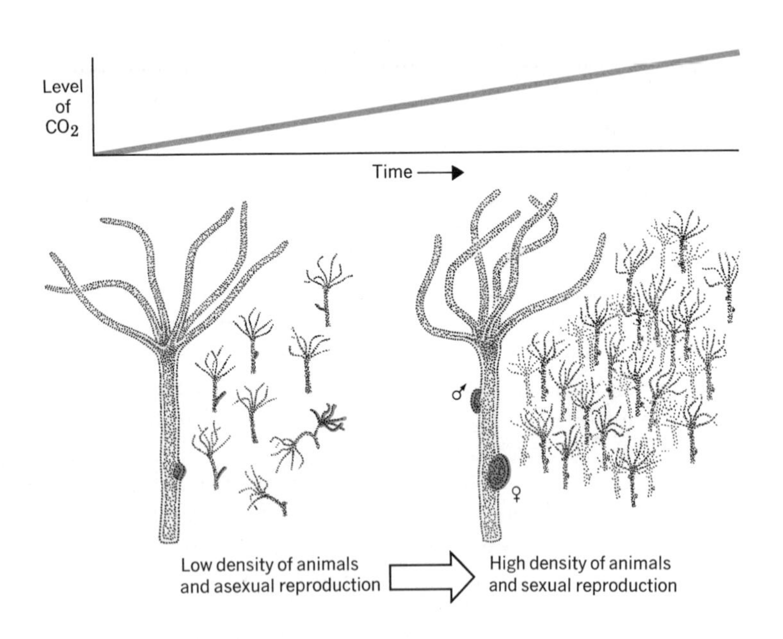

Fig. 12-13 Hydra in crowded conditions.

Presence and Position of Cells as Inducing Factors

There is an overwhelming amount of evidence indicating that the presence of cells in a particular position in the body acts to induce cellular differentiation. Much of this evidence comes from embryological studies.

This same effect, however, has been noted in biological systems that have less complicated potential development patterns than a mammalian, amphibian, or avian embryo. The slime mold, *Dictyostelium discoideum,* has a relatively simple life cycle (see Fig. 12-14). A diploid spore differentiates to form a single amoebalike cell. This cell then undergoes growth and multiplication through simple mitotic divisions, thus forming a number of cells that are called **myxamoeba.** These cells are cytologically and genetically identical.

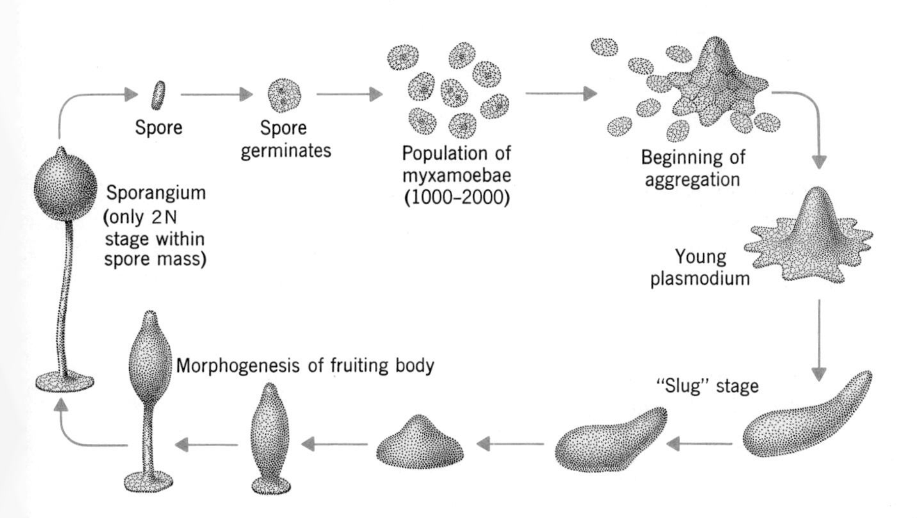

Fig. 12-14 Life cycle of *Dictyostelium discoidum.*

Growth of the population will continue until the supply of bacteria on which they feed is exhausted. The myxamoeba then move toward a central point and form a cellular aggregate. Once the aggregate is formed, it falls on its side, and migrates in a fashion much like a slug. When it comes to rest, it sends up a vertical stalk, at the end of which a fruiting body develops, containing hundreds to thousands of spores. With the formation of the spores, the cells of the stalk die.

Sussman has provided experimental evidence that once the slug is formed, the fate of the cells is fairly determined. When the head and tail segments of the slug are marked by the investigator, it is found that the head region develops into the stalk portion of the **fruiting body,** while the isolated tail region develops into the spores and the **basal disc.** The fate of the cells appears to be determined by the order in which they enter the aggregate. Those cells entering first will form the head portion of the slug, and become programmed for differentiating into stalk cells, while those that enter later are programmed to differentiate into spore or basal-disc cells. The position of the cell in the aggregate is thought to determine the specific set of environmental conditions and chemical influences from surrounding cells, and these program the cell on a transcription level.

If the front end of the slug is separated from the rear end, it continues its migration while the rear end becomes immobile. When the front and rear ends are rejoined, the two will fuse, and migration of the whole slug continues normally. Experiments of this type indicate that there are two cell types in the slug, "leaders" and "followers," so that even this seemingly amorphous blob of cells contains cells of different functional character.

Although it has not been possible to show the environmental and/or chemical influences from one cell to another in the slug or aggregate, it has been possible to show that the aggregation itself is the result of a chemical interaction between cells. There is evidence that in the slime-mold colony, a specially endowed cell differentiates from a myxamoeba, which is called the "initiator cell." When aggregation takes place, it does so by the signal from the initiator cell.

The aggregation of myxamoeba as a chemical response has been demonstrated in the following way. When a stream of water is passed through a myxamoeba population in such a way that any chemical diffusing in the medium is forced only along the direction of the current, it is found that the cells "upstream" from the forming aggregate do not join in aggregate formation (see Fig. 12-15). It appears, therefore, that the **morphogenetic movement** (that movement of the cells as an expression of their differentiating capacities) is indeed a response to a chemical signal given by the initiator cell and the forming aggregate itself.

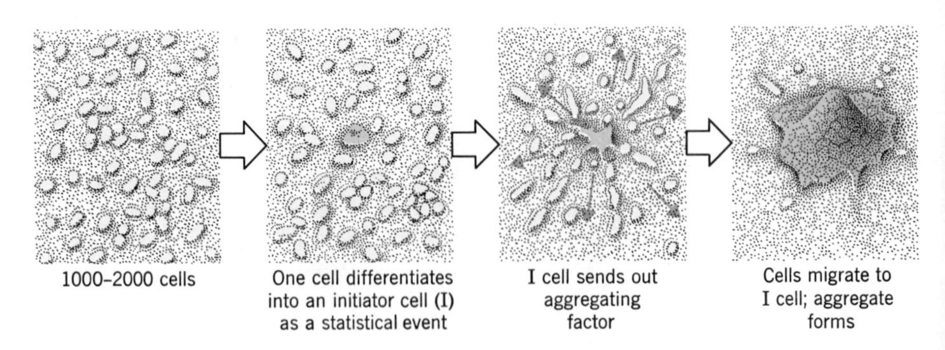

Fig. 12-15 Formation of an aggregate by *Dictyostelium discoidum.*

QUESTIONS

1. How did Stuart disprove the idea that only the cells destined to form the "germ line" retained the full genetic complement?
2. Describe the cytological changes that occur in the development of an *Acetabularia* plant; a sperm cell; a red blood cell.
3. Give an example of dedifferentiation and redifferentiation in a plant.
4. Explain the feedback system operative in the control of β-galactosidase synthesis in *E. coli.*
5. Why is differentiation of the developing gastrula considered an example of trigger mechanisms operative on the translation level?
6. How does a hormone act to induce mating in *Achyla*?
7. How can the presence and position of cells act as inducing factors?

SPECIAL INTEREST CHAPTER IV

CANCER

In the past fifty years, cancer has come close to leading the list of the nemeses of mankind. Cancer, which killed one out of every forty people at the turn of the century, now kills one out of every four. What has caused it to become a more prevalent disease?

In great part it is simply because people are not dying from other diseases. At the turn of the century, many people died of measles, diphtheria, whooping cough, etc.—a long list of diseases that primarily afflicted children. Life expectancy was forty years. Now many of these diseases are controlled by antibiotics and a variety of other drugs. Others are all but eliminated by vaccination. The average life expectancy has climbed to seventy-one years. And cancer is a disease that is more likely to strike during middle age. There are simply more middle-aged people, proportionally, than ever before.

But this is not the only factor. There has been ample evidence to indicate that certain irritating compounds, such as nicotine, act as **carcinogens** (cancer-causing agents). More men and women smoke today than ever before, and hence, more men and women are exposed to the carcinogenic effect of nicotine. The polluted air of the cities contains industrial wastes that have been shown to have a carcinogenic effect. Increasing pollution, therefore, also partially accounts for the increased incidence of cancer.

Microscopic Evaluation of Malignant Tissue

Cancer is a disease at the cellular level of the body. For some unknown reason, the control mechanisms of cells in a given tissue become altered, and the cells undergo divisions, thus forming **neoplastic** growths or **tumors.** Two major types of tumors are distinguished by (1) their morphological characteristics and (2) their tendency to remain at their site of origin. **Benign** tumors usually contain well-differentiated cells, similar to those of the tissue from which they arose. These tumors are always localized, and never establish growths in other parts of the body (**metastasis**). It should be noted, however, that a benign tumor is not

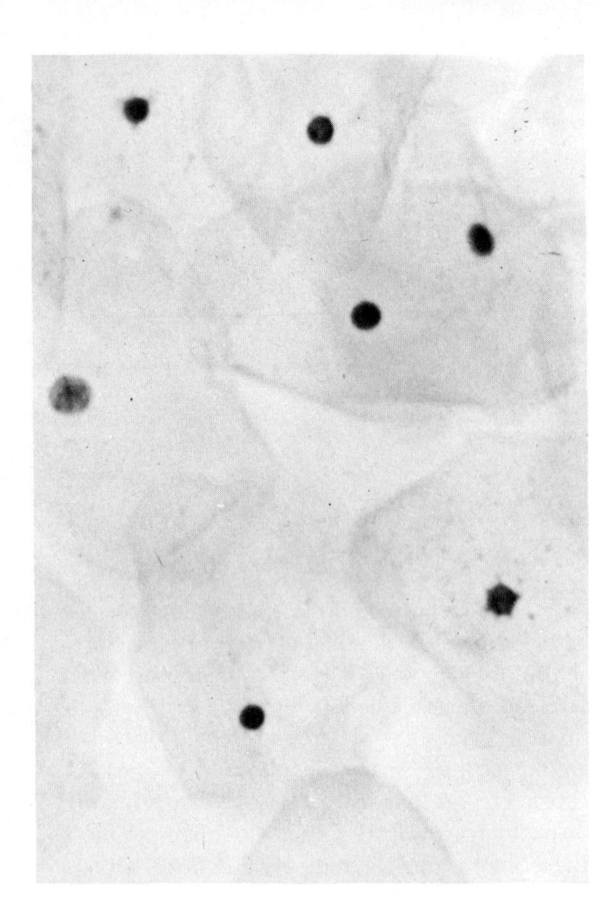

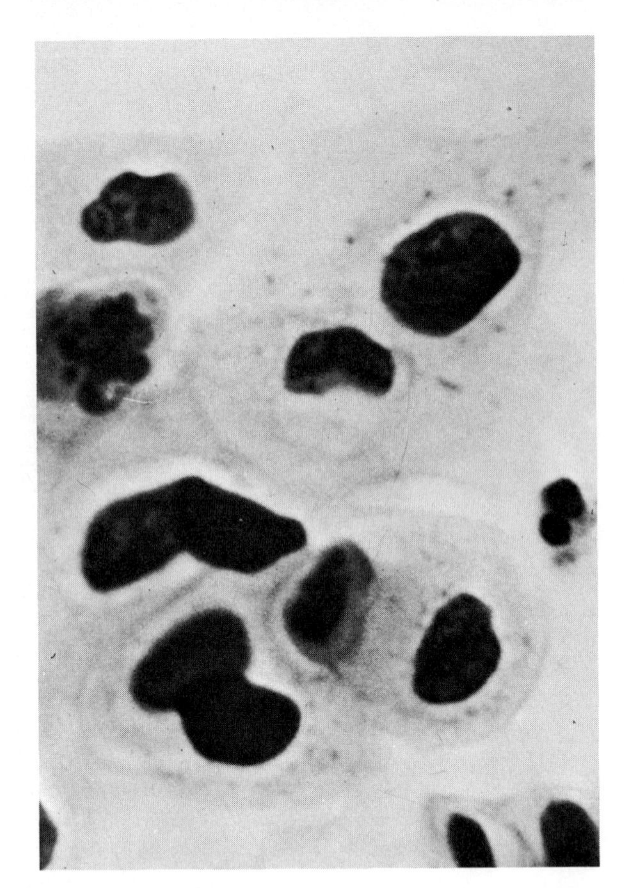

Fig. IV-1 (*a*) **Normal cells of human female uterine cervix and** (*b*) **malignant cells (500×). [American Cancer Society]**

necessarily a *harmless* tumor. Benign tumors may be injurious or lethal if they are located in inoperable regions, such as deep within the brain.

Malignant tumors are neoplastic growths that usually contain embryonic cells that show little or no differentiation (see Fig. IV-1). These cells are often grossly aberrant forms, having large, irregularly shaped nuclei with large nucleoli, and less cytoplasm than the cells typically found in the tissue. The decrease in the cytoplasmic/nuclear volume ratio of cells within a tumor is one of the major criteria in identifying cancerous or malignant cells. (Some malignant growths are exceptionally well-defined, and histologically indistinguishable from the normal tissue, but these are the exceptions rather than the rule.)

Malignant tumors often exhibit less dependence on oxygen than does normal tissue, and even in the presence of oxygen, form lactic acid, the end product of anaerobic glycolysis, rather than carbon dioxide and water. These tumors also manifest self-regulated mitotic cycles, and appear to operate completely independently of the regulatory processes of the organism in which they are found.

The major characteristic of malignant neoplastic growths is their tendency to metastasize. Cells from malignant tumors are apparantly capable of floating away to other tissue sites, and starting new malignant growths. This metastasis makes the surgical removal of advanced malignant tumors almost impossible.

Terminology for Malignancies

Malignancies are categorized according to the type of embryonic tissue from which they arise. Those malignant tumors that arise from tissue of ectodermal or endodermal origin are called **carcinomas,** while those that arise from tissues of mesodermal origin are called **sarcomas.** Some malignant tumors appear to arise from both ectodermal and mesodermal tissues. These are called **mixed malignant tumors.**

Within each of these general categories, the malignancies are named according to the specific tissue affected. For example, a **fibroma** is a benign tumor derived from and composed of fibrous connective tissue, whereas a **fibrosarcoma** is a malignant tumor that originates from fibrous connective tissues.

Since the origin of cancerous growths is not understood, it is not known whether carcinomas can metastasize and become sarcomas in other parts of the body, or vice versa.

Spread of Cancer Throughout the Body

The metastasis, or spread of cancer, throughout the body can occur in several ways. One pattern is by direct invasive growth, the penetration of malignant cells into adjacent tissues, thus starting secondary tumors.

Another common means of metastasis is the spread of malignant cells through the lymphatic system. The tumor cells grow into the lymphatic channels, and dissociate from each other. They are then carried by the lymphatic fluid to the lymph nodes, where they become lodged, and form secondary tumorous growths (see Fig. IV-2).

Metastasis can also occur through the bloodstream. This is a common means for the spread of sarcomas, but many carcinomas may also spread in this fashion. The tumor cells penetrate the blood vessels, and are carried as cell masses to other parts of the body. Tumors that arise in the liver, lung, and bones are often the result of metastasis via the bloodstream (see Fig. IV-3).

Ovarian carcinoma often spreads through the abdominal cavity simply by the detachment and flotation of tumor cells. This is called metastasis by implantation (see Fig. IV-4).

The phenomenon of metastasis is directly related to the character of malignant tissues. Normal animal cells stick together, and in doing so, limit each other's growth by **contact inhibition.** The proteinaceous

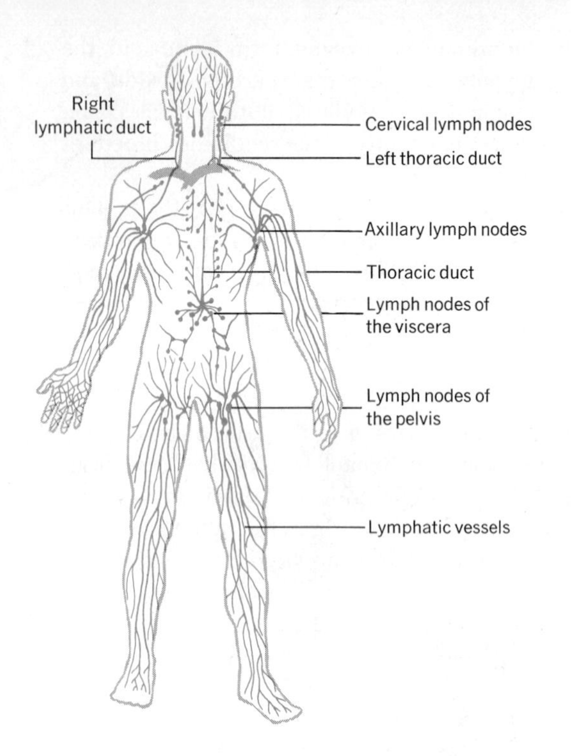

Fig. IV-2 Metastasis through the lymph system.

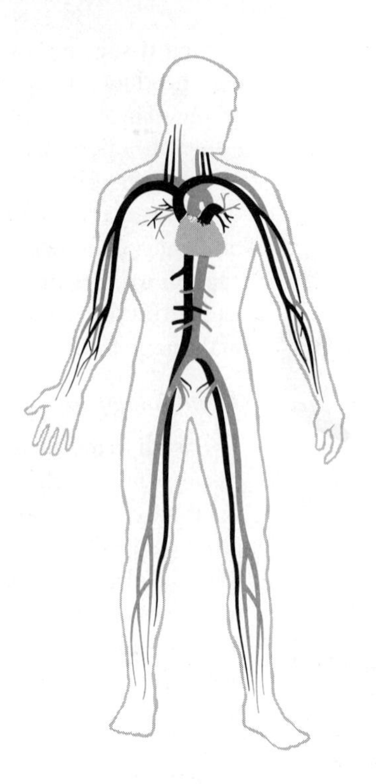

Fig. IV-3 Metastasis through the bloodstream.

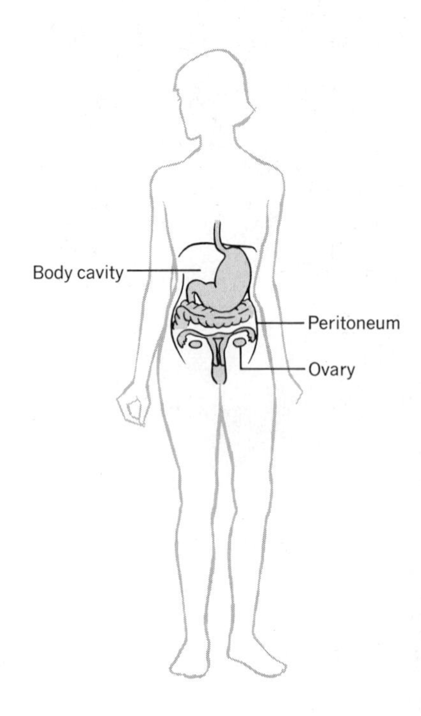

Fig. IV-4 Metastasis through implantation.

"sticking" substance on the surface of these cells causes them to adhere to glass surfaces, a factor that is exploited in tissue-culturing techniques (see p. 318). Often cancer cells do not appear to secrete the "sticking" substance. In tissue culture they do not adhere to the glass of the culture plate. In natural tumorous conditions, they do not exhibit the contact inhibition characteristic of normal cells. Perhaps the loss of this mechanism by which cell growth is held in check leads to the cancerous state. This is not the entire answer, but certainly this feature of cancer cells helps to explain metastasis.

Death by Cancer

There are three basic ways in which cancer can cause the death of an individual.

1. *Obstruction.* Tumors in blood vessels, the digestive tract, the respiratory tract, or any essential lifeline, can form a physical obstruction, causing death. In essence, such a physical obstruction is inhibiting the communication between different parts of the body, thus irreversibly altering the physiological equilibrium.
2. *Pressure.* Tumors, particularly those in an encased area such as the brain, having no room for expansive growth, cause pressure on normal tissue. This

pressure results in the collapsing of blood vessels, thereby shutting off the oxygen supply to normal cells. Death may be directly attributed to the inability of vital cells to function on a metabolic level.

3. *Crowding.* Certain tumors, although not growing in confined areas, crowd the normal tissue cells in such a way that they are ultimately destroyed. In one sense, this is also a form of pressure, and death can result from the imbalance that is created by the loss of normal, functional cells. In another sense, however, the death of normal cells can leave the individual open to bacterial and/or viral infection. The immediate cause of death in many cases of lung cancer is pneumonia.

POSSIBLE CAUSE(S) OF CANCER

Generally, cancer cells are runaway cells, operating without the control mechanisms to stop cell division. The conversion of a normal cell to a malignant form occurs at the level of cellular control, the transcription level.

What initiates this conversion of a normal cell into a malignant form, this turning off of control mechanisms operative in the rest of the tissue? There are three major theories on the causation of cancer. These may not be mutually exclusive, and many scientists are beginning to look upon cancer as a group of diseases in which there may be several origins. It has even been suggested that the causation of cancer may have to be considered as a separate problem for each organ.

The three major hypotheses that are presently in vogue are (1) the irritation hypothesis, (2) the somatic mutation hypothesis, and (3) the microbial (viral) hypothesis.

The Irritation Hypothesis

The irritation hypothesis holds that when cells are exposed to carcinogenic agents, the natural genetic processes are affected in such a way that the control mechanisms are disturbed. In 1912, Yamagiwa and Itchikawa produced carcinomas on the ears of rabbits by painting them with coal tar. The tumorous growth continued even when the tars were removed. Clearly, once a carcinogenic agent induces a cell abnormality, its presence is not necessary for the maintenance of that abnormality. By 1930 the active agent in coal tar was identified as 3,4-benzapyrene. Thus the chemical basis of carcinogenesis was firmly established.

Workers who have had a prolonged contact with coal tar, pitch, soot, or mineral coal often develop skin cancers. Those who work with luminous paints that contain radioactive materials often develop bone cancers. The high incidence of bladder cancer in Egypt is thought to be associated with a disease of the bladder, schistosomiasis, in which the parasite *Schistosoma haematobium* causes an irritation of the bladder lining.* The tissue manifesting the cancerous growth, in each instance, is the one in which the carcinogen is preferentially absorbed.

* Schistosoma infections in Egypt have significantly increased since the construction of the Aswan Dam and the subsequent ecological changes.

When cells in culture are exposed to carcinogenic agents, they do not manifest any overt growth that can be construed as a developing malignancy. This failure to induce malignant growths in tissue culture has, in fact, been one of the delaying factors in finding a causative agent of cancer.

When bacterial or unicellular cultures are treated with carcinogens, there is also no effect similar to malignant growth. Although some carcinogens are also mutagens, and as such, cause certain phenotypic changes in these organisms, the changes are never manifested as aberrant growth patterns suggestive of malignant cell types.

The work that has been done in the field thus far does not rule out the irritation hypothesis, by any means. Investigations have indicated that it is possible for chemical agents to combine with DNA regions, thus repressing their expression, or to combine with repressors, thus allowing the formerly repressed gene to be expressed. It is entirely possible that carcinogens disturb the control factors in cells by these type of actions. Thus the lack of success in experimentally producing cell proliferation may merely indicate that an appropriate system for studying this phenomenon has yet to be found.

The Somatic Mutation Hypothesis

The somatic mutation hypothesis holds that in the course of development, an individual cell undergoes a genetic change that converts it to a malignant form. This cell, and all of its progeny, will manifest the lack of control mechanisms, and a malignant tumor is produced.

The type of mutation that would affect the control mechanisms of the cell would be those that caused the activation of certain normally repressed genes. It could be a genic mutation of the repressed gene itself, or a genic mutation that produces the repressor substance for that site. In the latter case, an aberrant protein or RNA might no longer adequately combine with the gene to repress it, and it would become active.

This hypothesis remains on a purely theoretical level, because there is no way of testing it. When microorganisms undergo mutations, they can exhibit the change by growing on different kinds of media. They may be crossed so that various aspects of the mutation can be studied. When a cell in the animal body undergoes a somatic mutation, it is virtually impossible to study it. Since it is not a gamete, it cannot be crossed to other cells. Therefore, the site of such a mutation cannot be investigated by looking for recombinant types. Also, media for culturing cells is very complex, and the number of cells that take to culturing is few, so that it would be difficult to determine a metabolic change in this way.

The idea of a somatic mutation as the cause of cancer will have to remain an interesting hypothesis until some feasible means for testing is found.

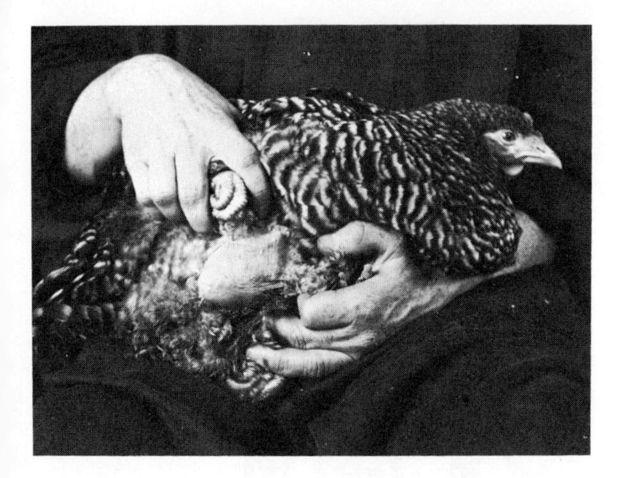

(*a*) The whole tumor

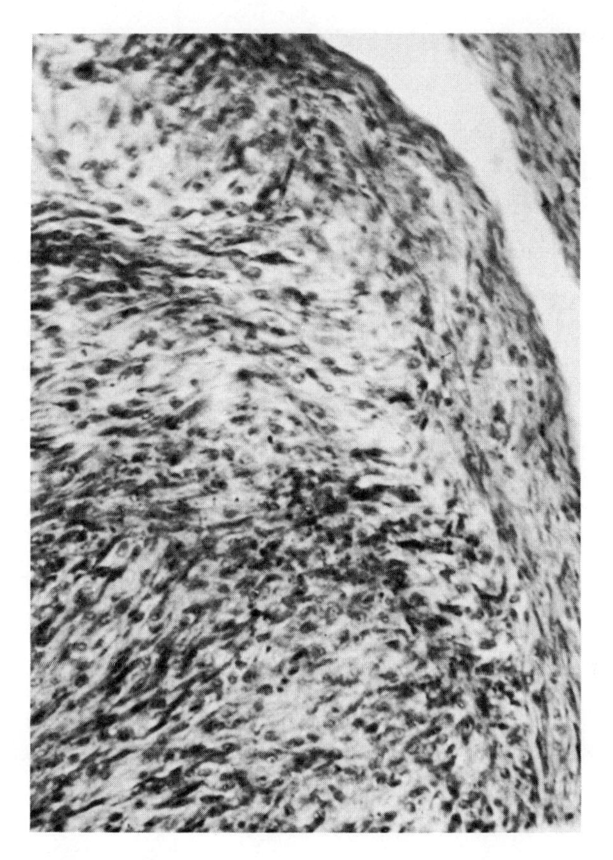

(*b*) Tumor tissue (300×)

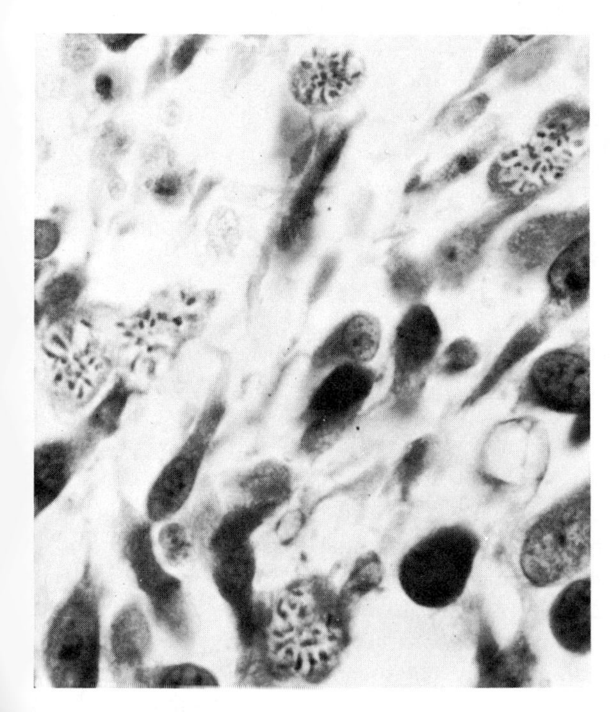

(*c*) Tumor cells (3000×)

Fig. IV-5 Rous sarcoma. [Courtesy of Dr. Peyton Rous]

The Microbial Hypothesis

The microbial hypothesis, which holds that a virus or microbial form is the causative agent for cancer, has gained considerable support.

Rous discovered a sarcoma (appropriately named the **Rous sarcoma**) in chickens. When a cell-free and bacteria-free extract of the tumor is injected into other chickens, they exhibit the growth of similar sarcomas. The Rous-sarcoma virus has been isolated and identified as an RNA virus (see Fig. IV-5).

Tumorous growths on plants, notably crown gall (see Fig. IV-6) have been found to be caused by bacteria. The proliferation of the cells in the crown gall is not unlike that of typical malignant tissue. The cells are relatively undifferentiated, although they do exhibit the cohesiveness typical of plant tissues. Techniques for isolating bacteria are fairly standardized, and since the application of these techniques to human cancers has not yielded a causative agent in the form of a bacterium, it is assumed that this mode of infection is peculiar to plants.

Several tumor-causing viruses have been isolated and identified in animals. Among them is **Bittner's mammary carcinoma virus,** and the

(*a*) Laboratory plant infected with crown gall. [American Cancer Society]

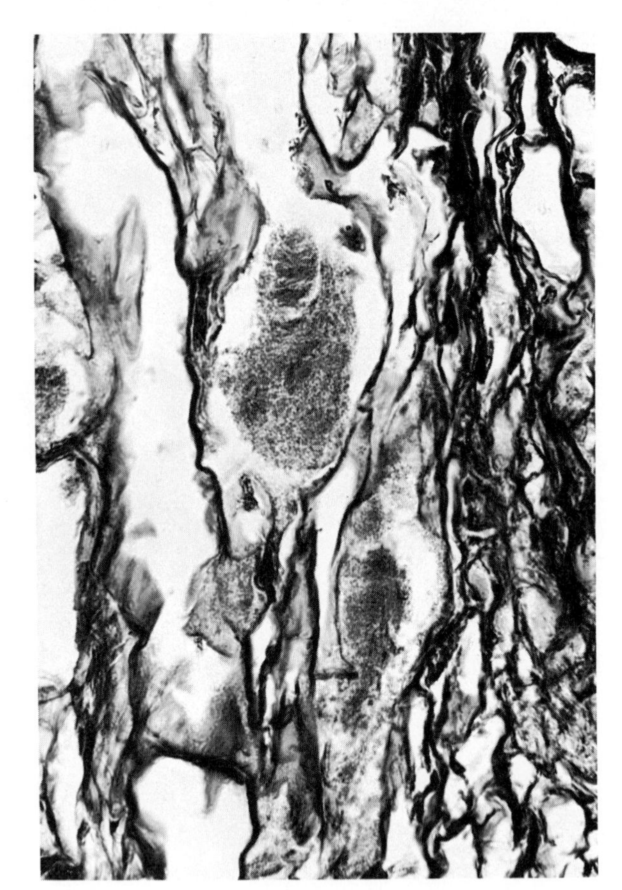

(*b*) Cells of crown gall (3000×). [Carolina Biological Supply House]

Fig. IV-6 Crown gall in plants.

mouse leukemia virus. Indeed, viruses have been isolated from malignant tumors in mice, rats, frogs, hamsters, dogs, monkeys, squirrels, deer, and horses. That man is not included in this list is to many an indication that an insufficient number of human malignant tumors have been studied, and is explained by the impossibility of injecting a prospective cancer-causing virus into a human. The specificity of the viral infection may be such that experimental animals simply will not suffice in this instance.

It is very tempting to think of viruses as causative agents for cancer, for there are so many conjectures that can be made. First of all, the fact that cancer cannot be transmitted from one individual to another (cancer is not an infectious disease) could be explained by a latent virus. Such a latent virus could be coupled with the native DNA, thus traveling through the cell generations. This would also tend to explain the propensity of cancer for some families. The latent virus could simply be carried through either the egg or the sperm.

The carcinogenic agents, then, may do nothing more than activate the virus. The virus then starts to control the cell machinery, and suppresses the control mechanisms for the cessation of cell division.

As tempting as this theory may be, however, there is no experimental evidence that indicates that it is true in human malignancies. In fact, in cell cultures such as HeLa cells (which were taken from a cervical carcinoma of a woman named *He*len *La*ne), no virus has ever been isolated that is capable of infecting other tissues. This may, of course, be due to the specificity of the virus for these cervical cells (benign female cervical cells do not take to culture), or to the fact that such a virus may not be transmissible. Nevertheless, the inability to find a virus as a causative agent in human malignancies does leave the question of viruses as causative agents for cancer very much open to debate.

ATTEMPTS TO CURE CANCER

Clearly any cure for cancer will have to be based on the action of the causative agent on the transcription level of cell activity. The cure for cancer, then, is inexorably tied to the same factors that guide cell differentiation by the activation and repression of genetic sites. And the answers will come only from the concerted effort of both medically oriented and so-called "pure" scientists.

Radiation

X-ray and radium treatments are based on their effects on cellular components (see Special Interest Chapter II). The most important of these effects in the treatment of malignant tumors is that of chromosome breakage, which leads to nonfunctional cells when the chromosome fragments are lost upon division. In the adult, most of the cell types undergo a low rate of division. Only malignant cells are rapidly dividing, and the chromosomal breakage induced by radiation leads to their preferential destruction, although some normal cells are always destroyed as well.

The gamma-ray source is placed as close to the malignancy as possible, to insure both maximum exposure of the cancer, and minimum exposure of normal cells. Radiation may affect most of the malignant cells, but since the occurrence of chromosome breaks is a purely random event, based on probabilities, not all the malignant cells will be affected. When the unaffected cells multiply into a sufficient number, the full effect of the malignacy recurs. Radiation, therefore, may only cause a temporary remission of cancer, although some cures by the irradiation of certain types of cancer may occur.

Radiation treatments are not without auxiliary dangers. Radiation is a mutagenic agent, and so can produce new cancers in heretofore normal cells. Also, excessive exposure to radiation can cause systemic imbalances that could lead to death. The amount of X-radiation to which an individual is exposed must be judiciously watched, lest the cure be his undoing.

Tumors vary in their responsiveness to radioactivity. They have been classified as:

1. *Radiosensitive tumors.* These regress or clinically disappear with a dose of 2500 r or less. Little damage is done to adjacent normal tissue. This group of tumors includes some lymphatic tumors, and chronic leukemia.
2. *Radioresponsive tumors.* These require 2500 to 5000 r for a regression similar to that noted above. The adjacent tissue shows a marked reaction to this dosage, but the injury is not permanent. This group of tumors include some carcinomas of the skin, cervical carcinoma, and carcinoma of the thyroid.
3. *Radioresistant tumors.* These require over 5000 r for any response to be noted. The damage to the normal tissue may equal or exceed that done to the tumor. Such tumors include malignant melanomas (dark, localized growths on the skin that become malignant) and bone sarcomas.

Surgery

When a malignant growth has been discovered shortly after its initiation, the best presently available means of stemming the cancerous tide may be surgical removal. Removal is effected by cutting well into the healthy, normal tissue, and removing part of that as well. Only by removing more than the minimum amount can one be assured of eliminating all malignant cells from that area of the body.

There are several reasons why the success of surgical removal is dependent on early detection. First, new tumors do not crowd normal cells or create pressures or blockages as severely as do well-developed tumors. Severe crowding can lead to the irreversible death of normal cells, and severe blockages or pressures may have gross effects on the equilibrium of the body. In each case, the disturbance of equilibrium may continue after the removal of large tumorous growths. Little or no permanent effect on biological equilibrium is found when small tumors are removed.

Second, the death of normal tissue that results from crowding can greatly reduce the number of cells available to carry out a specific bodily function. Since the surgeon must remove normal as well as malignant tissue, the surgical procedure further reduces the cell number. The reduction may be of sufficient magnitude to seriously hamper the normal function of the body. Again, the fact that small tumors create less crowding, and hence, less death of normal tissue, increases the chance of the body returning to a normal state.

Third, and perhaps most important, is that the early discovery and removal of malignant tumors prevents metastasis. Although it is not impossible for small tumors to metastasize, it is far less likely for them to do so. The removal of a small, newly initiated tumor may well free the body completely of all malignant cells. The removal of a large, well-developed malignant tumor may only free the body from the malignant cells at that particular site. Some of the cells from the cancer may have already spread to different parts of the body, and may be on their way toward the development of secondary growths.

Chemotherapy

The most recent innovations (since 1946) in attempts to cure cancer have been in the field of chemotherapy. Most hospitals now have three units for the treatment of cancer—a surgical unit, a radiation unit, and their most recently acquired chemotherapeutic unit.

There has been some degree of success with the use of drugs, whether alone or in conjunction with X-radiation, for the treatment of some cancers. Among the cancers that have responded to such treatment are the leukemias, the lymphomas, and breast and prostrate cancer. It takes years before one can be certain that a cure has been effected, and since the widespread use of chemotherapeutic agents for malignant growths is fairly recent, insufficient time has elapsed to determine whether these results are truly permanent or only temporary remissions. In other words, all the drugs presently in use are still in the stage of clinical investigation (see Special Interest Chapter I).

All the drugs presently under investigation are **growth-inhibiting compounds.** Clearly, the administration of such drugs could cause widespread cellular damage to normal as well as to malignant cells. Many of the drugs have undesirable side effects traceable to their effect on normal tissue. Yet, when the life of the individual is at stake, the administration of such drugs may be justified.

Different drugs affect different growth processes. One group manifests its effect on the information level. **Alkylating agents, mustard gas,** and a variety of others, destroy the integrity of DNA. **Actinomycin D** inhibits the transcription of DNA to RNA. Other drugs, such as **3-fluorouracil,** are incorporated into the RNA structure, and cause the *m*RNA to be, in part, nonsense.

Another group of drugs includes those that preferentially disturb the equilibrium of malignant cells by affecting phases of the mitotic cycle. It should be noted that any dividing normal cells would be affected in a similar manner. Most of these drugs inhibit the progression of mitosis past metaphase. Certain plant alkaloids, such as **Vinblastine, Vincristine,** and **Colchicine,** are active in this way.

Yet another group of drugs affects cell permeability and mobilization of enzymes. These drugs are primarily steroids, much like those found as hormonal elements in the human body.

Different cancers respond differently to the same drugs. Thus the results of the clinical studies made to date indicate that there is little hope for a universal anticancer agent.

Another area of chemotherapy that is being extensively investigated is the feasibility of cancer vaccines. The development of a single vaccine or series of vaccines assumes a microbial origin for malignancy, a fact that has yet to be demonstrated conclusively. The limited success of the other attempts to cure cancer makes the hope that malignancies are of viral origin, and that a series of vaccines could be developed, the only real chance man seems to have of removing cancer from the top of his list of medical nemeses.

PART FOUR: EVOLUTION

The theme of Part Four is evolution as a dynamic process. Accordingly, the emphasis has been placed on genetics, selection, and adaptation—that is, on processes rather than on products of evolution.

The immediate effects of mutation on cell regulation was discussed in Part Two, The Cell. A discussion of the nature of mutational phenomena and the consequential effects of mutation on a population has been reserved for chapters in this part of the book, so that long-range effects could be related to the evolutionary process involved in the origin of life and the origin of man.

(Opposite) "Bear's Head," Robert Lockhart Jr., photo by W. B. Nickerson

CHAPTER 13

THE THEORY OF EVOLUTION

Although Charles Darwin is credited with formulating the basis of the presently accepted theory of evolution, he was by no means the first man to suggest that life had evolved. Many philosophers and scientists before Darwin sought rational theories of how life and the variety of living forms could have arisen.

Setting the Scene for Darwin

The first post-Renaissance theory of evolution was recorded about 1750 by Buffon, who developed ideas of biological origin that lent a mechanistic quality to the heretofore descriptive science of biology. He proposed that food supply was an important factor in population size; a population would reach a size in direct proportion to the supply of food. Although this may seem to be a rather elementary idea, it must be remembered that Buffon detected this relationship long before the quantitative analysis of natural populations became part of scientific procedure.

Buffon also postulated that structurally similar forms have a common origin. He proposed that monkeys, gorillas, and man had one common origin, while the horse and the donkey had another. He envisioned that small changes in an ancestral type could have resulted in the formation of these similar forms, but he could not conceive of a bridge between structurally diverse groups. His theory of evolution postulated many small branches, each starting from unknown sources.

About this same time, James Hutton (1770), William Smith (1810), and Charles Lyell (1830) collectively contributed to a unified interpretation of the evolution of the inorganic world. Geological evolution involved climatic forces that shaped the world, laying down rock strata in successive layers. Lyell determined that the age of the earth was in the order of millions of years, and therefore, far older than was implied in the Biblical account of special creation (estimated to have occurred about 4000 B.C.). Of biological interest, however, was the occurrence of many types of fossils found in the different strata. Some of the fossils found in the lower strata were of animal types that were similar to many of those found on the earth today; others had no modern counterparts. Moreover, each layer of rock revealed a certain set of fossil types; the more recent the layer, the more complex the body pattern of the fossilized organism.

Around 1770, Erasmus Darwin, grandfather of Charles Darwin, proposed a theory of evolution by the aquisition of hereditarily determined traits. His theory was further developed by Jean Baptiste Lamarck in 1809 into a form known as the **inheritance of acquired characteristics** or **Lamarckian adaptation.** It was a powerful idea, for it provided a *way* in which changes in organisms could come about. It was the first completely comprehensive mechanistic theory that was offered. Furthermore, it was theory that lent itself to predictions and therefore, to testing. According to Lamarckian evolutionary theory, the webbed foot of an aquatic fowl such as a duck would have developed in the following way: The duck would stretch its toes apart to give more push during swimming. The skin between the toes would stretch into a web. This new characteristic would be inherited, and the subsequent generation of ducks would, upon stretching their toes, form a more defined web. Each generation would do the same until the webbed foot seen on ducks today was fully formed. This would then be passed on from generation to generation, essentially unchanged once the perfected state was attained. By the same process, giraffes developed long necks by stretching for the leaves on the higher unpicked branches of trees.

For plants, Lamarck accepted the theory of his compatriot St. Helaire, who developed the notion that plant form is shaped by the combined effects of the environment. Both men presented elegant theories of evolution that encompassed all plants and animals, and mechanistic means by which one form could be converted, over several generations, to another. Lamarck's ideas, as expressed in his book *Philosophie Zoologique,* enjoyed popular acceptance for the next seventy years.

Even though the theory of Lamarckian adaptation was diametrically opposite the literal interpretation of Genesis I, only a few serious objections were made on religious grounds. It must be remembered that Lamarck formulated his theory shortly after the French Revolution. Unlike the "divine right" monarchy that had preceded it, the bourgeois political regime was distinctly non-church-oriented. Royalists in the universities were summarily replaced by freer-thinking bourgeois philosophers. Therefore, even though the Lamarckian theory accounted for the evolution of man from lower forms, it aroused little criticism from the philosophers and influential men in France.

With the popular acceptance also came the inevitable testing of the theory, a testing that went on until the acceptance of Darwinian evolution in the twentieth century. It was certainly obvious that a man who lost his arm in an accident, did not have children with only one arm. Attempts to observe changes in animals in their natural habitats were unsuccessful. Experiments by Weismann in 1890 in which tails of mice were cut off and the tail length of the offspring measured for 20 generations, revealed that on an average, the tails were not shorter. So there was a growing disillusionment with Lamarckian evolutionary theory. It should be noted that the experimental evidence that lent the death blow to Lamarck's theory did not come until 1939. A discussion of this experiment will be given in the next chapter.

Despite the grave questions that arose regarding Lamarckian theory, it enjoyed a resurgence of interest in Bolshevik Russia in the 1930s when some Russian geneticists, such as Lysenko, succeeded in philosophically relating it to Stalinist doctrine. Although most Russian geneticists accept Darwinian concepts, the official party line, as of the late 1960s, was still that of Lysenkoism.

It cannot be overemphasized that even though Lamarck failed to conceive of what is now accepted to be the correct mechanisms of evolution, he did provide a complete and interrelated theory that set a standard for any theory that followed.

By 1844, the idea of the origin of life and the source of biological variation became a matter for serious consideration in England. In that year, Robert Chambers published his book, *The Vestiges of the Natural History of Creation,* in which he presented a modified Lamarckian view of the origin of the organic and inorganic worlds, and carried the evolution of biological form through man. Church-indoctrinated England was not as open to such heretical ideas as bourgeois France, and Chambers incurred the wrath of both clergy and scientists by blatantly dealing with these forbidden subjects. He was bitterly attacked by scientists for his many biological errors, and by the clergy for his overt contradictions to the Bible.

The decade during which Chambers' book enjoyed its success and controversy took much of the edge off the reaction toward Charles Darwin when he published his *Origin of Species* in 1859. The controversy had made the public aware of, and psychologically conditioned to, the idea of evolution. In addition to collecting an overwhelming amount of evidence for evolution, Darwin presented a mechanistic interpretation of evolution to a public that was psychologically prepared to deal intellectually with his cogent arguments for *natural selection* (see Fig. 13-1).

Fig. 13-1 Charles Darwin in 1840, as depicted in a watercolor. [The American Museum of Natural History]

Charles Darwin and Evolutionary Theory

The theory of natural selection, as proposed by Darwin, was a singularly great achievement in biology. It differed from most other ideas in biology in that it related all forms of life, both living and extinct. To develop such a grand design required an enormous amount of experience and a capacity to resolve many paradoxes. Darwin had a keen sense of the meaning of natural selection, beyond that of many evolutionists today, and his explanations for some classical problems rank among the most impressive performances in biological literature.

Darwin's personality was not the type that one associates with a successful man, let alone a great one. While his family held a distinguished reputation (his grandfather Erasmus was a noted scholar, his father, Robert, was an eminent physician, and his mother was a Wedgwood, the family of pottery fame), Charles was a singularly undistinguished individual whose contemporaries viewed him as just a normal boy.

As in many father-son relationships, there was a conflict of interest in the son's professional goal. Charles attempted to follow in his father's livelihood and took up the study of medicine at the University of Edinburgh. His medical interests soon dissipated, and he became something of an academic rover, studying a little science, but finding it too formal for his tastes.

His father stepped in and sent him to Cambridge, in preparation for the ministry (the other acceptable profession in Victorian England). There he spent his energy in card-playing and drinking. He was obviously not ready for clerical commitments.

He did manage, however, to find some constructive interest at Cambridge. His friendship with Professor Henslow stimulated an interest in botany, while his exposure to the geologist Sedgwick stimulated an interest in geological explorations. Through the recommendation of Henslow, Darwin accepted a most inauspicious position as a naturalist without pay aboard the H.M.S. *Beagle.* The ship was to spend five years in exploration, mostly of South America. Darwin was to provide a collection of material representing the natural composition of the areas visited.

Darwin had a rather weak constitution. His unsettled stomach disturbed him unceasingly during the dissection periods at the University of Edinburgh, and almost constantly with seasickness aboard the *Beagle.* In later life, this malady caused him to take many months of uninterrupted rest from his work. Despite his ailment, Darwin was one of the most productive scientists in history. His studies include the *Origin of Species,* the *Descent of Man,* a treatise on plant growth, a monograph of *Cirripideae* (barnacles) and a volume on domestic breeding.

During his five years on the *Beagle,* Darwin recorded almost everything he observed and sent an enormous amount of material back to England. He was a competent observer of plant, animal, and geological specimens, but he was not a professional in any of these disciplines. During the course of his trip, he developed an acute sense of natural balances. Gradually he was transformed from just another observer of nature into an evolutionarily oriented thinker.

He did not consciously develop an evolutionary theory during his sojourn on the *Beagle*, but the seeds of interest were sown. He observed many kinds of natural balances, catastrophic changes, and fossil remains of many communities. He saw nature as a dynamic force, constantly achieving and retaining equilibria, and felt the need for all these factors to be put into a causative pattern.

Upon his return home in 1836, he spent two years writing a book of his experiences, *The Voyage of the Beagle.* In 1838 he came upon an essay by Thomas Malthus entitled *An Essay on the Principle of Populations.* This essay changed Darwin's wisdom of nature into a message of evolution.

Malthus' ideas were of sociological importance. He analyzed the growth of human populations and evaluated the consequences of unchecked growth. He concluded that if humans were to continue to reproduce at their present rate, it would take only a few generations before the population could not be supported by the food supply. This plight would result in the death of a sufficient number of individuals to create a balance with the food supply once again. But it would only be a few generations before the population would once again reach the limited food level. Malthus viewed this as a repetitive phenomenon with no solution.

To Darwin, Malthus' idea assumed different significance. What if, instead of a human population, it were an animal population, living in a restricted area in which the food supply became limiting? A certain number of individuals in the population would certainly die. But Darwin also saw that the factors that determined which individuals would survive were not wholly random. He had long been immensely impressed by the natural variation *within* species. No two individuals of the same species were exactly alike. Juvenile and aged animals for example were not as robust as those in their prime, and therefore, would not be as efficient at capturing their prey. He felt that these differences between organisms of the same population could be the determining factors by which organisms were successful in competing for a limited food supply.

Darwin also realized that the features that determined survival could be based on factors that might be only vaguely related to the direct procurement of food. An animal that was capable of migrating to another habitat and finding the same or a substitute prey on which to feed; an animal that was able to survive within a wide temperature range; or an animal that could grow more rapidly to its adult form, had a better chance for survival than other animals in the population that did not manifest these characteristics. He called those variations present in a natural population, that enabled the individual to survive, **adaptations.**

Darwin by no means excluded plants from his considerations. He recognized that lack of space due to crowding would limit the light supply for plants. Therefore, those that were adapted to grow taller, disperse their seeds further (so that the seeds did not germinate in the same vicinity and become crowded by the parent plant), or function efficiently with less light, would tend to survive. Such adaptations would endow them with a **selective advantage.**

Darwin further recognized the interdependence of all living forms in which plants were the ultimate food source in a food chain. A local habitat maintained an equilibrium between the populations within it. If the plants were crowded and only a few adapted in such a way that they could survive under crowded conditions, the herbivores that fed on these plants would suffer a food shortage. Those herbivores that could either find the plant more successfully, substitute another plant food, or eat the adapted plant form, would survive. All others would die. The carnivorous predators of these herbivores would suffer a similar food shortage, and so on, through the food chain of the community.

Finally, through the interdependence and the natural variation of populations that permitted adaptation and the dynamics of environmental balance, Darwin conceived of a means by which evolution could have taken place. The original form of a plant may have been quite close to the ground. Some of the plants exhibited a variation by growing higher, thereby shading other members of the plant populations and cutting down their exposure to the sunlight. These taller plants produced offsprings of the same height, so that the lower plants were even more limited in their exposure to the sun. Some of the lower plants adapted to require less light for efficient function. Perhaps after thousands of years, there were two species where only one existed before. Both the tall growing plant and the low plant that required less light arose from the original short form. The perpetuation of the two species was based on the fact that they were better adapted to survive in the environment than the plant that had previously dominated the area.

Darwin's understanding of evolution may be generalized as follows. *The change in species by the survival of an organismal type exhibiting a natural variation that gives it an adaptive advantage in an environment, thus leading to a new environmental equilibrium, is evolution by natural selection.* This is often called "survival of the fittest," which is something of a misnomer in that it paints a picture of animals fighting among each other until the fittest survives. The trait on which survival may depend may have nothing whatsoever to do with robustness, agility, or any of the traits one associates with being "fit." Survival can easily depend on coloration, growth potential, flexibility, ability to accept alternative food sources, etc. Modern evolutionists, in fact, consider relative reproductive rates to be the critical factor for survival.

Darwin viewed the fossil evidence as indicative of the different historical periods of the earth. Previously living forms either lacked sufficient variation to give rise to other forms, or possessed such variations and were superseded by descendants better adapted to the environment. Thus they became extinct. He visualized an evolutionary scheme by which the ancestry of all living forms present on the earth today, including man, could be traced back to the simplest living entities known.

Darwin, unlike Lamarck, did not have a satisfying idea of how variations arose on an organismal level, and for many years proponents of Darwinian evolution did not view the two theories as mutually exclusive. Variations arising in a Lamarckian way could be selected for in the manner provided by Darwin.

Darwin himself rejected the amalgamation of the two theories. In his book *Origin of Species,* he dealt at great length with the idea of variation arising by the inheritance of acquired characteristics, but he ultimately found any reconciliation of the two theories untenable. Lamarckian evolutionary theory involves an orientation of characteristic by *individuals,* and is thus independent of the size of the population. Darwinian evolution, on the other hand, is based on the inherent variations among members of a population. The larger the population, the greater the variation. Darwin saw that only through the wide variation present in large populations, and the natural selective pressures that operated on this variation, could a progression of organismal types arise.

After the publication of *Origin of Species,* Darwin received a great deal of criticism. Although some of the criticism was unchecked emotion, which did not require retort, much of it came from fellow naturalists and biologists who could not reconcile certain types of traits with a theory based on natural selection. This latter type of criticism constituted a challenge for Darwin, and one that he met successfully.

Some naturalists had argued that if evolution did occur, it must have been by large-scale changes, rather than the minor modifications postulated by Darwin. For instance, how could the prehensile tail of the South American monkey have evolved? If the original tail was not capable of wrapping around a tree limb, and a minor change was made so that some curling of the tail was possible, it would certainly not be of sufficient degree to hold the monkey, who would fall to the ground and be captured. Therefore, such a minor change would be "selected against" initially, and the monkeys that possessed such a trait would not even reach mating age to perpetuate it. These naturalists felt it was impossible to explain the origin of such a trait by a series of small steps.

Darwin suggested that the initial curling of the tail did not make it prehensile, nor did the animal first possessing it use it for grasping. It was probably used by the animal as an additional hold on the tree, and served to increase its balance. He cited the curling of the tails of mice, which are not prehensile, but appear to increase balance. As more variations were added to make the tail slightly prehensile, it became more of an aid to the feet, and monkeys began to use their feet and tail separately as climbing aids. Thus, as a structural trait was elaborated upon by many small changes, it changed function. Each function endowed the possessor with a selective advantage.

Darwin felt that the emergence of many organs could have come about in a similar manner. The organ manifested an original function that gave the organism possessing it a selective advantage. As the trait developed further, the function of the organ changed. By this argument Darwin explained the presence of such complex organs as the eye, not by one large evolutionary step, but through many adaptive steps, each of which was built into an already existing, more primitive optical system.

He also explained the electric shock of the eel *Torpedo* as a modification of an organ capable of sending electrical impulses, that had a different role before its full elaboration. Recent workers have found that fish related to *Torpedo* have similarly located electric organs that send out impulses as a type of radar system. The evolutionary development of such a radar system may well have given rise to the electric-shock system of *Torpedo.*

Another criticism leveled at Darwin concerned the limitation of adaptation. For instance, why did the giraffe alone evolve a long neck, when there were many other large animals in central Africa that would find it advantageous to pick leaves from the higher branches of trees? Darwin suggested that many other animals did acquire some of the initial steps for this adaptation, but the trait was not compatible with other well-established adaptations. This would comprise an "either-or" situation. An animal would either retain the established adaptations, or he would favor the new adaptations for the longer neck, but never both. Eventually, the adaptations toward a longer neck would disappear because of natural selection.

Modern biology has little to say on this subject because too little is known about the complex physiology of development and the incompatibilities that might arise. Suffice it to say that two types of cattle, beef and dairy, have been developed because it is impossible to select for both high fat content in milk and increased animal weight in a single type. The two traits are simply incompatible.

Alfred Russel Wallace

In 1858, Darwin received a manuscript from Alfred Russel Wallace, who at that time was working in Indonesia. The manuscript presented a theory of evolution by natural selection almost identical to that which Darwin himself was ready to propose. Oddly enough, Wallace's conclusions had also been stimulated by the Malthusian essay and confirmed by his observations of animal and plant life, as well as fossil evidence he had found in the Malayasian archipelago of Indonesia. In his paper to the Linnean Society in 1858, Darwin presented both his and Wallace's paper, giving Wallace full credit for independent work.

When Darwin's book, the *Origin of Species,* was published in 1859, Darwin became the center of much controversy. The fact that this book, written by Darwin alone, exposed the public at large to the theory of natural selection, and that Wallace was still in Indonesia at the time of the controversy and could not champion Darwin's defense by adding his own views, brought the public to associate only Darwin with this theory. Thus the theory is called the Darwinian, rather than the Darwin-Wallace, theory of evolution.

EVIDENCE FOR BIOLOGICAL EVOLUTION

Despite the psychological preparedness of the public for a theory of evolution, the Darwinian concept of natural selection stirred a great controversy. Lamarckian adaptation provided organisms with a control over their destiny. One could visualize animal forms reaching toward those human traits that represent perfection in the animal world, the upright stance, the grasping hand with a pivotal thumb, etc. Darwin, however, presented a picture of organisms at the mercy of a fickle environment, and of a survival based on variations. And, since one cannot predict what environmental factors will create selective pressures, it is impossible to know which variations would be advantageous. Therefore, the term "adaptation" is assigned to an organismal variation only in retrospect.

The lack of man's control over his own evolutionary destiny also disturbed the nineteenth-century man. While the Lamarckian view made it possible to visualize the creation of perfect forms that remained in their perfect state, the Darwinian view held that there was no perfect form. Climatic changes, crowding, mass migrations, etc., could upset the balance of a community, and bring about a total change in the complexion of the habitat. A form that may have been "perfect" for the old environmental equilibrium may not be suited to the new one.

The acceptance of this intellectually satisfying but psychologically disturbing theory rested on the results of the observational and experimental tests devised from its prediction. Darwinian evolution permits two general predictions: (1) if there were an evolution from simple forms to more complex ones, there must be certain structural, developmental, and chemical similarities between different forms of life, and (2) there must be a means by which variation in populations arise, and are transmitted from generation to generation. The remainder of this chapter will deal only with the evidence that has been found to support the idea of a relationship between organisms. The next chapter will deal with the way in which change can initiate and bring about an evolutionary process such as that proposed by Darwin.

Paleontological Evidence

Technically it is not proper to use the evidence that was used in the formulation of a theory to substantiate that theory. In the case of paleontological evidence (the finding of fossils in geological strata), Darwin utilized only the information available in the 1850s, when the science was in an embryonic state. The refinements that have been made in the means of finding and dating paleontological discoveries, in addition to the abundance of discoveries that have been made since Darwin's time, more than justify their use in substantiating the theory.

Geologists envision the Earth's crust as having gone through a series of changes (see Fig. 13-2). These changes involved the alterations of land masses by glacier movements or land shifts that resulted in the rise of mountain ranges, and the leveling of other land. When one land mass came to rest upon another, the tremendous pressure it exerted, and the

Fig. 13-2 Rock strata revealed by erosion from the Colorado River. [U. S. Geological Survey]

lack of available water, caused the conversion of the original land mass into rock strata.

Through the geological ages, then, land mass over land mass formed layers of rock strata, each of which had, at some time, been exposed to the atmospheric environment. The lowest strata solidified some two to three billion years ago. Other evidence indicates that the Earth may have existed as a liquid for about one billion years, and prior to that, a gaseous ball for about one billion years, although this is by no means certain. The age of the Earth is, therefore, between four and five billion years.

For thousands of years scholars had recorded descriptions of fossils and interpreted them as evidence of geological catastrophes. It was thought that these organisms have been trapped between rock strata during a glaciation or general upheaval. Lyell, Hutton, and Smith recognized that many of the strata were subdivided into an extremely large number of thin strata. Many of these substrata contained different combinations of fossils. It became untenable to accept each fossil layer as the result of a geological catastrophe, as these occurrences were far too rare to account for the large number of strata found. The more reasonable assumption was that fossils were the remains of organisms trapped in semiliquid pools that quickly solidified. Since most animals would not be caught in this manner, and plants tend to decay rather than to remain preserved, fossils are thought to represent a small but somewhat representative sample of the earth at any one time in any one locality.

Fossils are produced in many ways. Insects are found in solidified amber, indicating that they had been trapped in plant saps that solidified. Mammals have been found trapped in ice in the Arctic. In most cases, the body is covered quickly after death by falling into quicksand or becoming embedded in drifting sand. The hard parts are preserved and the soft parts are replaced by inorganic salts that serve to preserve the integrity of the animal shape. Fossils may also be the imprint of an animal or plant on soft material that hardened as the animal decayed (see Fig. 13-3). The vast majority of fossils are only fragments of an animal or plant, because of the geologic distortions of the rock.

The age of strata and the fossils they contain is based on radioactive dating. Geologists standardly use uranium dating to estimate the time of solidification of rock. Radioactive uranium decays spontaneously to lead. The half-life (see page 34) of uranium is about 4,500,000,000 years.

Since the uranium of the earth was formed four to five billion years ago when the great pressure of a contracting dust cloud created thermonuclear heat, why should not all uranium measurement give the same age, four to five billion years? The answer lies in the difference between the radioactive decay in a solid and in a liquid. In a liquid, the uranium is diluted, washed away, etc. But when that liquid solidifies, the uranium it contains is not free to move, and undergoes its decay into lead in a highly localized region. The ratio of uranium to its stable product, lead, indicates the time of the solidification of the liquid material and therefore, the time at which it was added to the Earth's crust.

For the determination of the age of fossils, other radioactive materials are analyzed. Radioactive carbon (C^{14}) is a natural radioactive form, produced in the atmosphere from the contact of the naturally occurring C^{12} with UV light. It passes down as $C^{14}O_2$ and enters plants and then animal material. The ratio of C^{14} to C^{12} remains constant during life, because of the constant interaction of biological organisms with the environment. Upon death and fossilization, this ratio decreases as the C^{14} undergoes decay. It is possible therefore, to determine when the fossilized individual lived by comparing its present C^{14} to C^{12} ratio with that usually maintained during life. Radioactive carbon has a half-life of about $5{,}568 \pm 30$ years, and can only be measured up to 25,000 years or about 5 half-lives. This limitation results because the amount of C^{14} in organisms is so small to begin with. Radioactive-carbon dating is excellent for the anthropological studies of early tribal civilizations, but not for the earlier strata.

For most strata, radioactive potassium (potassium^{40}, which decays to argon^{40}) is used. This has a half-life of 1.3 billion years. Also, because of its greater concentration in most rocks, it is a more accurate method of dating fossils than uranium, a relatively rare element. A bed of prehuman fossils in South Africa dated by rock composition gave a result of 500,000 years, whereas the potassium method indicated an age of 1,750,000 years—a most important difference, considering the nature of the fossils.

(*a*) Skeleton of a saber-toothed tiger.

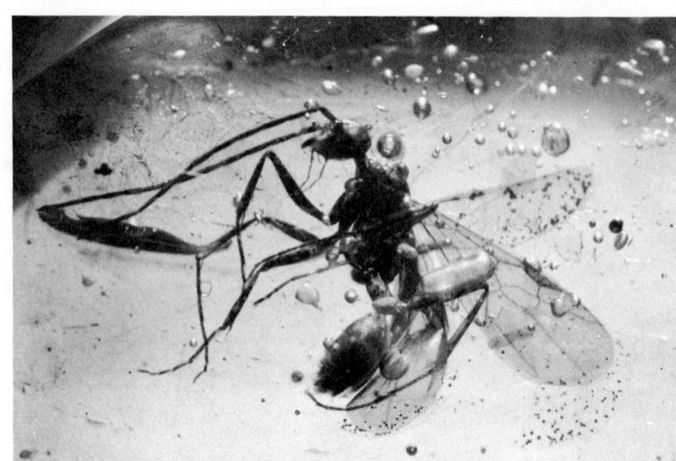

(*b*) Insect trapped in amber.

(*c*) Fossilized clam.

Fig. 13-3 Some fossils. [(*a*) and (*b*), The American Museum of Natural History; (*c*) and (*d*), Steve & Dolores McCutcheon, Alaska Pictorial Service; (*e*), Walter Dawn]

(*d*) **Leaf fossil.**

(*e*) **Fish fossil.**

There is certainly no reason to believe that by pure chance only more primitive forms were entrapped in the older rocks, while more advanced forms escaped entrapment and died and decayed in such a manner that they would not be found. It is far more logical to assume that the fossils found at each stratum are representative of the variety of organisms that existed at that period in history. Table 13-1 presents a chart of the geological time table and the general types of organisms found in the strata during these periods. Notice that there are no fossil records for the Archeozoic period. This may well be due to the lack of preservable structure (such as chitin) in the bodies of organisms present at that time, rather than to an absence of organisms.

TABLE 13-1 GEOLOGICAL TIMETABLE[a]

Millions of Years Ago			Geological Features	Biological Changes	
Era	Period	Epoch		Plants	Animals
Azoic (4500)			Origin of Earth		
Archeozoic (2000)			Mostly igneous and metamorphosed rock	Organic material found in rock-origin of life (?)	
Proterozoic (1200)			Rocks mostly sedimentary	Simple algae	Protozoa bacteria; Sponges
Paleozoic (550)	Cambrian		Mild climate, land flat	Marine algae	Trilobites, sponges
	Ordovician (430)		Much of land submerged	Marine algae	Trilobites, sponges, starfish, corals, armored fish
	Silurian (350)		Same as Ordovician	Land plants	Primitive fish, first land invertebrates
	Devonian		Emergence of much land	Horsetails, ferns	First amphibians and many fish
	Mississippian (310)		Warm, humid climate	Similar to Devonian	Sharks, many amphibians
	Pennsylvanian (260)		Mountain building	Similar to Devonian	Insects and reptiles
	Permian (215)		Mountain building and glaciation	Conifers and cycads	Many reptiles
Mesozoic (200)	Triassic		Warm, great desert areas	Spread of conifers	First dinosaurs and mammallike reptiles
	Jurassic (150)		Large seas in western hemisphere	Angiosperms	Mostly dinosaurs, first birds and mammals
	Cretaceous (125)		Swamps and mountain formation	Many angiosperms	Many insects and birds, primitive mammals
Cenozoic (70)	Tertiary	Paleocene Eocene Oligocene Miocene Pliocene	Formation of Alps and Himalayas	Grass and herbs	Modern mammals and arthropods

[a] This is the reverse of most geologic schemes, which start at the present era and trace backward.

As one proceeds down the geological periods, the increasing complexity of organismal structure with time is quite obvious. Certainly this would tend to substantiate the Darwinian theory of evolution.

Further evidence is offered by the discovery of humanoid bones that date from 10,000,000 years ago. Other skeletal parts that are thought to be those from true humans have been dated at 10 to 50 thousand years ago. So great is the evidence for gradual human evolution that the question is no longer whether man evolved from lower forms, but rather, which of the different types was the first human.

The origin of man will be discussed in a chapter of the same title later in this part.

Evidence from Comparative Anatomy

While paleontological evidence provides the most direct indication of evolution by showing that through the years organismal types became more complex, it only *implies* that the simpler forms gave rise to more complex ones. The comparison of anatomical structure of **extant** (living) organisms provides a more lucid picture of organismal relationships.

One may carry out comparative anatomical studies on levels ranging from body plans to specific bone, muscle, or blood-vessel arrangements within the body. When one considers the body plans of **vertebrate** organisms (those organisms possessing backbones), the similarity in the basic pattern of these diverse animal forms is very apparent. Fish, amphibians, reptiles, birds and mammals all have similar organ systems that perform almost identical functions (see Fig. 13-4).

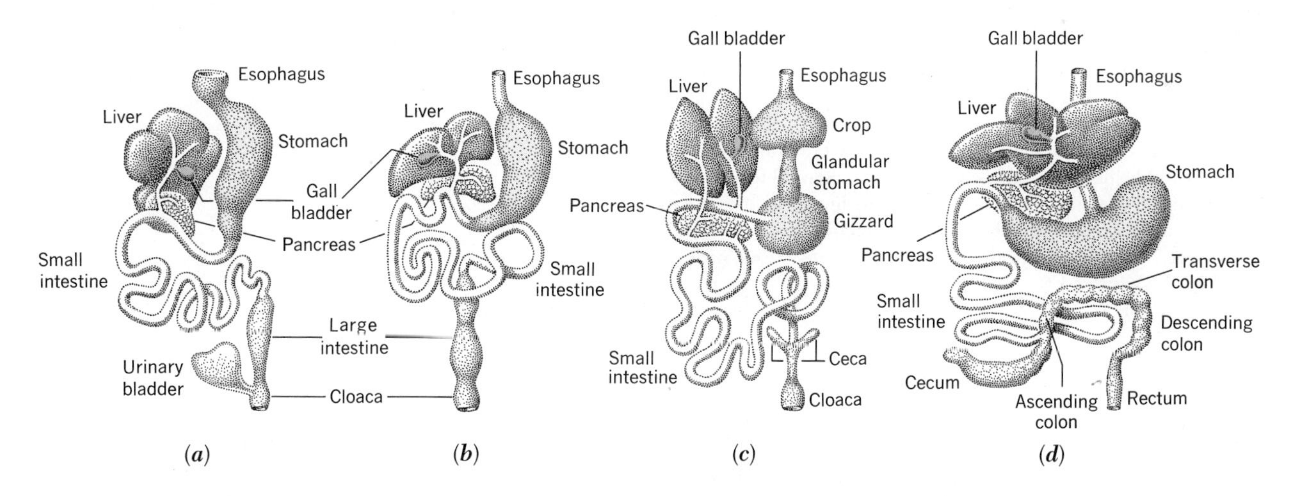

Fig. 13-4 Comparative anatomy of the digestive tract in four-legged animals: (*a*) an amphibian, (*b*) a reptile, (*c*) a bird, and (*d*) a mammal. The basic components are the same, indicating that the sequence of food breakdown has not appreciably changed in the course of evolution.

Further observations have shown that the similarity between corresponding organs in different species is not limited to gross anatomy. The morphology of the tissues and cells is so similar that even an experienced microscopist would have difficulty deciding whether the unlabeled muscle tissue he examines under a microscope came from a frog or a mouse. In addition, these similar organs arise from the same embryological source. When tissues or organs present in individuals of different species (1) arise from the same embryological source and (2) are similar in cell composition and gross anatomical features, they are called **homologous.**

The number of homologous organs in the seemingly diverse vertebrate group is overwhelming. To evolutionists, those organs that have homologous counterparts in simpler organismal types indicate the foundation from which the more complex types elaborated. For example, all vertebrate organisms possess a brain, and the brains of all vertebrates are homologous. The brains of fishes and amphibians are linearly arranged (see Fig. 13-5). In the brains of reptiles, birds, and mammals, certain regions are larger than others, giving the brain a more spherical appearance. The differences are thought to reflect the evolutionary progression of elaboration of the linearly arranged brain.

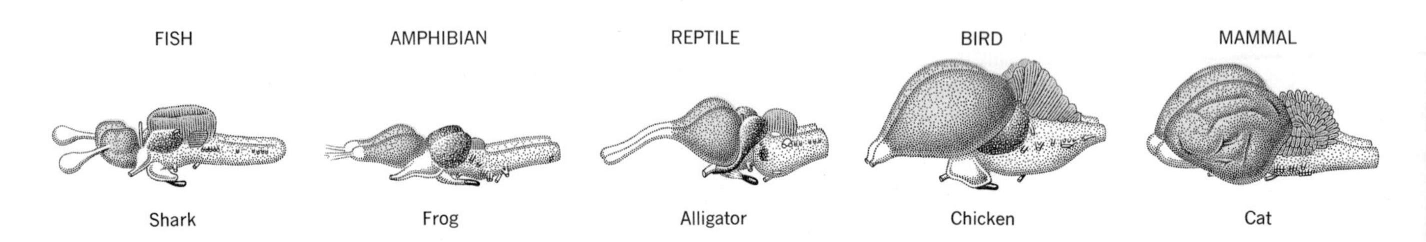

Fig. 13-5 The brains of representative vertebrates show evolutionary progression, especially in the cerebral hemispheres and cerebellum. The cerebral hemispheres are shown lightly stippled; the cerebellum is shown with vertical lines. The increase in the size of these brain areas accompanied changes in their frunction. [From *General Zoology*, by Storer and Usinger, © 1965, McGraw-Hill Book Company. Used with permission of McGraw-Hill Book Company]

It should be noted that the fossil evidence and the comparative anatomical studies of organs other than the brain have led evolutionists to think that the evolution of vertebrates is not linear, but branched. It is presently held that fish gave rise to amphibians, and amphibians to reptiles, which in turn gave rise to both birds and mammals (see Fig. 13-6).

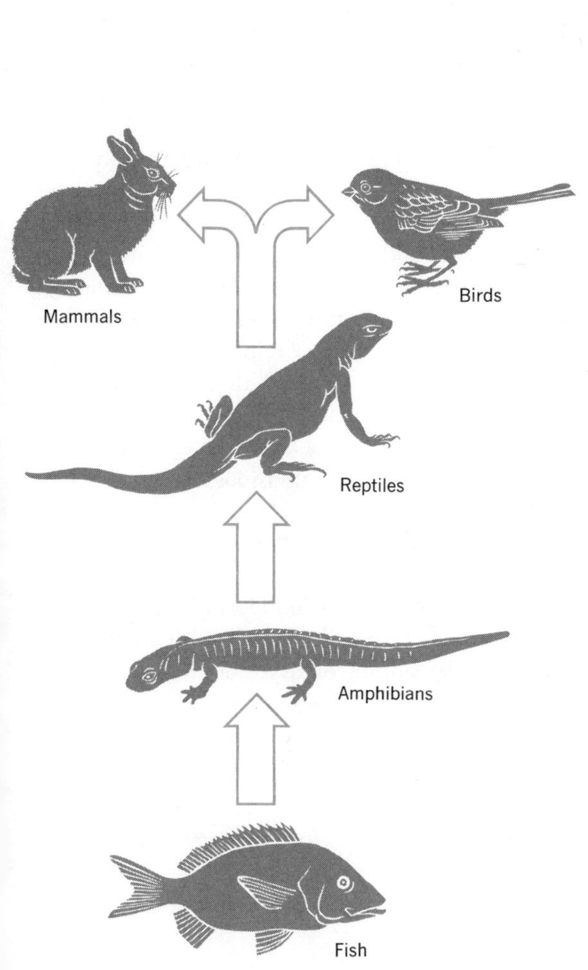

Fig. 13-6 The evolutionary sequence of vertebrates.

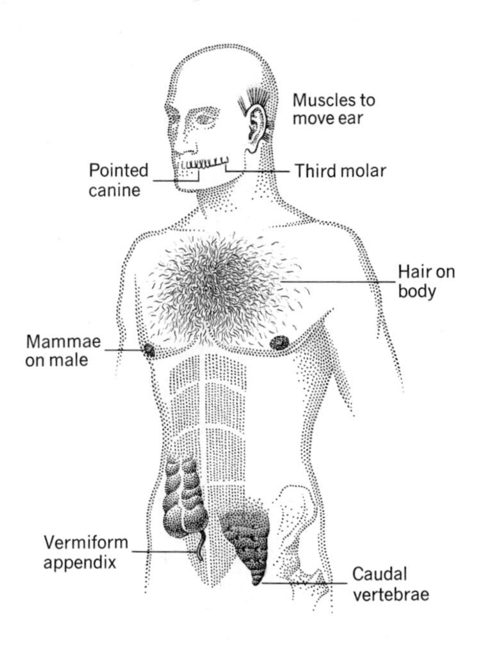

Fig. 13-7 Some vestigial structures in the human body include the vermiform appendix, the caudal vertebrae, the pointed canine, the third molar, muscles that move the ears, hair on the body, and mammae on the male of the species. [From *General Zoology*, by Storer and Usinger, © 1965, McGraw-Hill Book Company. Used with permission of McGraw-Hill Book Company]

Another source of evidence stemming from the studies of comparative anatomy is the presence of organs in various animals that have no apparent function. The typical example of this is the human **coccyx.** The coccyx is found at the base of the spine and consists of four fused vertebrae (the bony units that make up the spine). In structure and embryological development, the coccyx is homologous to the tails found in many vertebrate forms, but in the human it has no function. It does not house any special nerves, nor does it seem to have any element that increases the efficient function of the human (such as balance). It is considered by evolutionists to be a **vestigal** organ, remnants of a tail that decreased in size during the course of evolution. The presence of such vestigal organs (Fig. 13-7) indicates the basic anatomical relationship between higher and lower forms of life.

The overwhelming evidence from comparative anatomical studies of the vertebrates leaves little doubt in the biologist's mind that they are related by common ancestry. Although it has been more difficult, biologists have, through comparative anatomical studies of invertebrates and plants, been able to construct similar schemes of origins for these groups.

Evidence from Comparative Embryology

Certainly one of the most fascinating sources of evidence for the validity of Darwin's theory of evolution has come from studies of comparative embryology. The suspicions of a common ancestry for very diverse sets of organisms from comparative anatomical studies have been supported by embryological data.

According to Darwinian evolution, more complex organisms would have achieved their state by elaborating on the existing developmental patterns of more primitive forms. Hence, one would expect to find that certain relatively simple organisms and more complex ones have many initial developmental steps in common. In fact, the more developmental steps two species have in common, the more closely related they are to a common ancestral form.

Comparative embryological studies revealed that there was one developmental pattern that could be viewed as having undergone a series of branchings. All multicellular animals start their development as a single zygote, and through a series of mitotic divisions, increase in cell number until a blastula is formed (see Fig. 13-8). The developing embryo elaborates upon the blastula stage by forming two fundamental germ layers, **ectoderm** and **endoderm,** in the course of gastrulation (see p. 340). After the differentiation of the ectoderm and endoderm in the gastrula, the third germ layer, **mesoderm,** is formed. (see Fig. 13-9). There are three distinct patterns by which developing embryos of different species produce a mesodermal layer.

Fig. 13-8 The blastula represents an increase in cell number, but not an increase in size.

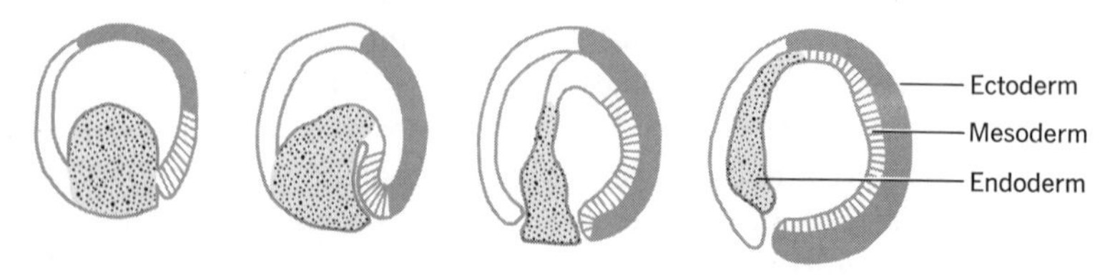

Fig. 13-9 Gastrulation in vertebrate organisms involves the delineation of a mesodermal layer derived from the endodermal layer.

Despite some difference in developmental patterns, comparative embryological studies overwhelmingly support the concept of Darwinian evolution. Two species that follow the same pathway for several steps after the gastrula produce adult forms that are more similar than two species whose pathways diverge after that stage. The different developmental patterns are thought to be indications of different lines of evolution.

Another feature of embryological development that serves to link all multicellular animals is the embryological source of organ systems. Regardless of the way in which the developing animals are programmed for further differentiation of mesoderm, they exhibit similar adult structures derived from the two primary germ layers. The outer covering of all multicellular animals, be it skin, scales, or gelatinous material, is derived from ectoderm. The digestive tract of all multicellular animals is derived from endoderm. The universal features of ectodermally and endodermally derived tissues also indicates the presence of a common ancestral type early in evolutionary history.

The elaboration upon a single developmental pattern is most striking in the development of the mammalian embryo. This embryo passes through a series of stages during which it resembles first an invertebrate form, then a primitive vertebrate form, then a fishlike form with gills, and then an amphibian with limbs, before it finally takes on mammalian characteristics. Even the developing mammalian brain is first linearly laid down, and as certain regions overgrow others, it convolutes to its spherical form.

So impressed by this progression of events was Haeckel that he somewhat overzealously proposed that "ontogeny recapitulates phylogeny," that is, in the course of its embryological development, an organism reviews its evolutionary history. This idea is an oversimplification, because the mammalian embryo never truly resembles either an adult or an embryonic form of a fish, amphibian, or reptile, but merely exhibits some *similar developmental features*. Nevertheless, the resemblances, even if somewhat vague, do support the idea that the developing embryo builds on the developmental patterns of its primitive relatives, an idea predicted by Darwin's theory of evolution.

Evidence from Comparative Cytology

Another type of evidence that indicates that all forms of life are related comes from the cellular level. The very fact that the cell is the unit of structure for all living organisms (except viruses) is thought to reflect the basic relationship among living forms. This relationship is even further emphasized by the fact that it has been possible for biologists to construct a picture of the "generalized" cell from which all other cell types can be inferred.

While comparative anatomical and embryological considerations are somewhat limited to detecting the relationships in multicellular forms, the cytological studies reach into the unicellular forms and serve to show the totality of the relationship among life at all levels of organization.

Evidence from Comparative Biochemistry

Comparative biochemistry, like comparative cytology, indicates a total and encompassing relationship among all living forms, both uni- and multicellular. This relationship is exhibited in the chemical basis of inheritance and cell regulation through the metabolic and structural aspects of cells.

All cells that have been examined thus far (and there is no reason to believe that any exception would be found in the future) have a DNA-RNA-protein information and communication system. All forms contain membranes that are made up of double-layered lipoproteins. All cells (with the exception of a few types of bacteria) utilize the glycolytic pathway. Most bacterial forms, and all uni- and multicellular organisms, have a Krebs cycle and an electron transport system. All are based on ATP as an energy donor. Certainly these factors provide an overwhelming demonstration of the interrelatedness of biological forms. One would hardly expect such complicated structural and functional systems to have developed in identical ways from anything other than a common origin.

Some General Considerations of Evolution

The biologists' conviction that, through a series of changes resulting from natural selective processes, life came to the state known today, is of such magnitude that the entire science of biology has been oriented according to the evolutionary doctrine. Organisms have been reclassified according to proposed evolutionary relationships. Geneticists interpret their results as possible mechanisms of evolution or sources of variation.

So powerful is this idea today, that only a text organization based on evolutionary doctrine would truly represent the science of biology. Therefore, the remainder of this text is based on the concepts of the dynamics of evolutionary processes.

QUESTIONS

1. What observations led Darwin to postulate new ideas regarding the origin of species?
2. Why is the phrase "survival of the fittest" a misnomer?
3. What criticisms were leveled at Darwin, and how did Darwin counter these criticisms?
4. How did paleotological evidence support the Darwinian concept of evolution?
5. What evidence from comparative anatomy, comparative embryology, and comparative biochemistry supports the Darwinian concept of evolution?

CHAPTER 14
SOURCES OF VARIATION

In addition to predicting that morphological and developmental similarities exist among diverse groups of organisms, the Darwinian theory of evolution predicts that the natural variation within a population is perpetuated from generation to generation. Implicit in this prediction is the presence of a paradoxical hereditary mechanism that on one hand provides a stable means for the passage of traits from parent to offspring, and on the other, retains a potential for change. Darwin fervently believed that any organism, at any level of its organization, was mutable.

For more than forty years after the publication of *Origin of Species*, this aspect of Darwin's theory provided a source of argumentative speculation on a philosophical level. Since the formulation of the chromosome theory of inheritance, in the early part of the twentieth century, the concepts of mutation, recombination, and chromosomal aberrations have collectively brought about a comprehensive interpretation of Darwinian evolution on a mechanistic level. The chromosome theory of inheritance provides the stable aspect of the hereditary mechanism, a means by which a full complement of determinants pass from parent to offspring. Mutation, recombination, and chromosomal aberrations are the means by which the potential for change are realized. This chapter will deal with these concepts and how they provide a most overwhelming case for Darwinian evolution and all its implications.

The Meaning of "Sources of Variation"

At first glance, the meaning of "sources of variation" seems to be very simple. It is important to develop the meaning of this statement, however, so that subsequent discussions can be put in their proper perspective.

There are two aspects to a consideration of sources of variation: building blocks, and combinations of those building blocks. Given three genes, *A*, *B*, and *C*, the alleles of which are *a*, *b*, and *c*, there are eight combinations that can exist.

1. Abc
2. aBc
3. abC
4. ABc
5. aBC
6. AbC
7. ABC
8. abc

These eight types would comprise the variation. But the *source* of this variation is (1) the origin of the building blocks themselves, and (2) the origin of their combination. For this reason, mutation (the origin of the building blocks of phenotypic variation), recombination, and chromosome aberrations (the origin of the combination of genetic traits) are all considered sources of variation.

MUTATION AS A SOURCE OF VARIATION

Proponents of the Darwinian theory of evolution greeted DeVries' claim to have found new phenotypes in the evening primrose (1899) with unabashed enthusiasm. Here, they thought, was the first example of man observing evolutionary change, the initiation of new building blocks. Although it was later shown that DeVries had witnessed nothing more than the expression of recessive traits through recombination, rather than actual genic changes, the word he used to describe the event became a part of the genetic vocabulary. That word was **mutation.**

True mutations were first found in *Drosophila melanogaster* by Thomas Hunt Morgan and his students. Using the white-eyed trait, which was a change from the normal red-eyed condition, Morgan, through a series of crosses (see p. 236), showed that its expression was the result of a mutation on the genic level.

In the years that followed, the concept of mutation became more refined and sophisticated.

Mutational Event on a Molecular Level

In its broadest sense, mutation means a hereditary change. It is important from an evolutionary standpoint to make a distinction between **intragenic (point)** and **intergenic** mutations. Point mutations are alterations within a gene. The molecular biologist defines a gene, in most cases, as a sequence of DNA bases that are translated into one polypeptide. A point mutation, then, is a change in the DNA in such a way that it is either translated into a different polypeptide than the one normally formed, or is not translated at all. Intergenic mutations are large modifications, usually involving the deletion or duplication of large regions of DNA. The biochemical imbalances that result from intergenic mutations are often fatal. Biologists feel, therefore, that it is the point mutations that are the primary source of variation for evolution.

The word mutation is standardly used as an abbreviation of point mutation, and will be used as such for the remainder of this text. There are several ways in which a mutation may occur.

1. *Base Substitution:* Base substitution occurs when one base is replaced by another. This can occur in two ways: (a) by the replacement of a base that is removed from the DNA sequence, and (b) by the alteration of a base *within* the sequence (the replacement of an OH radical by an NH_2 group). Both would lead to a change in pairing during replication (see Fig. 14-1).
2. *Duplications:* Duplications are the insertion of one or more bases into the DNA sequence.
3. *Deletions:* Deletions are the removal of one or more bases from the DNA sequence.

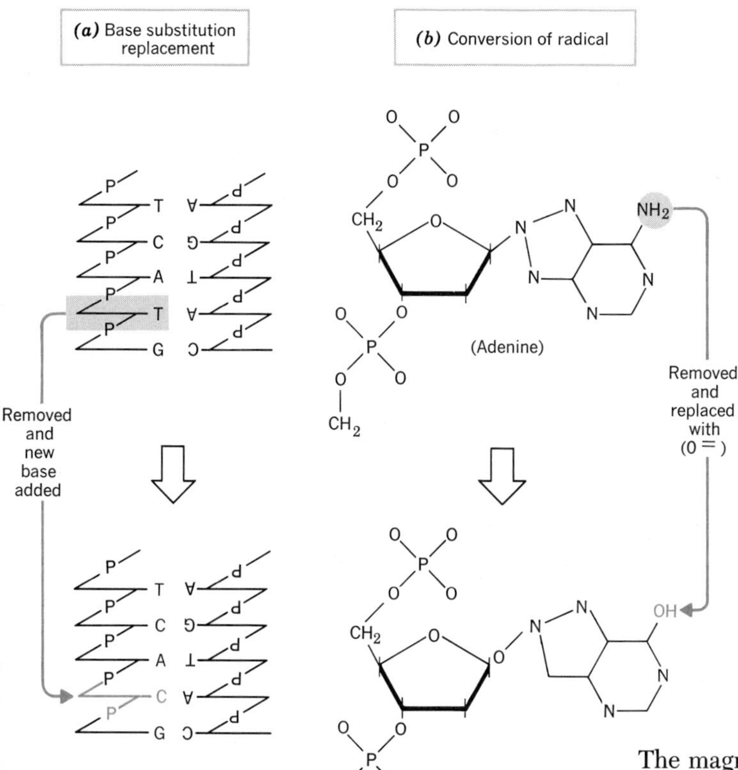

Fig. 14-1 Mutation by base replacement. (*a*) Removal and substitution by a different base. (*b*) Alteration of a base remaining in position.

The magnitude of the change in the polypeptide composition and ultimately the phenotype of the individual possessing a mutation is dependent on (1) whether the change occurs in a critical region of the gene, (2) whether the change throws the reading frame off and thus affects all regions of the gene after the mutation, and (3) the importance of the normal polypeptide in organismal function. If a base substitution, deletion, or duplication affects a triplet that codes for an amino acid essential for the tertiary configuration of the polypeptide, the protein will not fold properly, and will thus not fold into its functional state. If a deletion or duplication includes anything other than a multiple of three bases, the reading frame would be thrown off. If such mutations occurred at the beginning of the reading frame, the entire gene would be read incorrectly, and a "nonsense" polypeptide (having no cellular function) would result. If, on the other hand, the mutation occurred at the end of the reading sequence, it would not result in a grossly aberrant polypeptide.

The severity of the change to the organism ultimately rests on the importance of the normal product for its function. The organism may show little or no sign of phenotypic alteration if the mutant gene is but one of many of the same type. For instance, there are many genes present for the production of the ribosomal RNAs. If but one of these mutates, there are still many others to produce the normal product. Even when redundancy is not a factor, the organism may be able to continue its existence in spite of an aberrant protein. The life and development of a *Drosophila* is hardly dependent on the presence of the enzyme for the synthesis of its normal eye color. But mutations that produce aberrant essential proteins are at best crippling and are usually lethal.

There is yet another aspect of the aberrant protein that is formed by a mutational event. Often a single mutation of a single gene affects the expression of many traits. This multiple expression is called **pleiotropic effect.** It is thought that the normal form of the protein is used as an enzyme in a key developmental or metabolic reaction. The inhibition or slowing of this key reaction affects the many pathways that lead from it. Some forty to fifty simultaneously occurring anatomical and physiological traits in humans have been traced to a single critical reaction in which an aberrant protein is acting as an enzyme. Thus a single mutation can have far-reaching effects; surely this is an important factor in evolution.

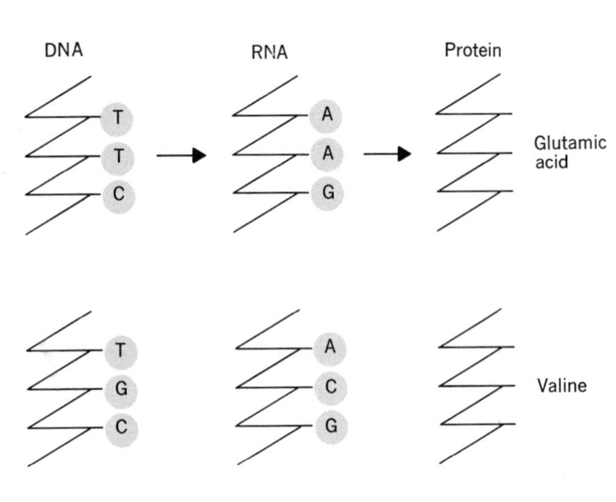

Fig. 14-2 Chain of consequences resulting from an alteration of the DNA sequence in the locus responsible for hemoglobin production.

Point Mutations as Real Events

Point mutations have been confirmed as real events by two lines of evidence. First, the difference between analogous proteins in related species or groups have been found to involve single amino-acid differences. The hemoglobin of normal and sickle-celled anemic humans, for example, differs by one amino acid. Where glutamic acid is present in the former, valine is present in the latter. According to the Khorana code for amino acids, AAG codes for glutamic acid, while ACG codes for valine. The single amino-acid difference, then, can be a single base substitution on the *m*RNA, which can, in turn, result from a single base substitution on the DNA molecule (see Fig. 14-2).

Secondly, mutations in tobacco mosaic virus (TMV) can be induced by nitrous acid, which changes cytosine to uracil. Most of the altered proteins formed in TMV after treatment with nitrous acid were found to be single amino-acid substitutions, which, according to the Khorana code, arose from single base substitutions in the *m*RNA.

Mutations: Lamarckian or Darwinian?

Once it was clear, and firmly accepted, that mutations did indeed occur, it became of the utmost importance to resolve the question of whether a mutational event was "directed" (Lamarckian) or random (Darwinian). It may seem like something akin to beating a dead horse to ask such a question in the late 1930s; yet the discovery of certain phenomena had made asking the question logical, and its answer imperative.

Later work had shown that the mutations that had been discovered by T. H. Morgan and his followers for *Drosophila,* and by other workers for corn, were intergenic rather than point mutations. More and more evolutionists considered only the point mutations to be of evolutionary significance. But these could only be detected in organisms that could be counted in orders of magnitude like 10^9, such as bacteria and viruses. After all, if a point mutation had one chance in a million of occurring, one would have to examine at least one million flies to find it, and even so, there would be no guarantee that the one mutant fly would be in that million. A geneticist, in his lifetime, might be able to count a few million flies, but certainly no more. Viruses and bacteria, on the other hand, can be plated, and examined in such large numbers that an investigator can examine or experimentally select for a given mutational type in 10^9 organisms in a few days. The point mutation that had one in a million chance of occurring would be picked up in the neighborhood of a thousand times in such a population.

$$10^9 \times \frac{1}{10^6} = 10^3$$

Certain phenomena were detected on a bacterial level that were not, and indeed, could not, be detected in *Drosophila.* For example, when viruses are added to sensitive bacteria, most of the bacteria will die from the infection. A few of the bacteria, however, will be resistant and repopulate the culture with resistant cells. The question arose as to whether some of the bacteria had become resistant because of the presence of the viruses (a directed Lamarckian mutation) or whether there were in the natural variation of the bacterial population, some cells that were resistant to viral infection even though they had never been in contact with the virus (random Darwinian mutation). According to the Darwinian interpretation, the viruses acted as a selective agent, producing a pressure under which only those cells that were, by chance, already resistant could survive. The resolution of this problem provided the most critical test of evolutionary theory given to date.

Delbruck and Luria divided a bacterial culture into tubes so that each of the tubes contained approximately the same number of bacteria. Each tube then contained a **subpopulation** of an original large population. The tube cultures with their subpopulations were permitted to grow for several hours. Viruses were then added to each of the tubes in sufficient numbers to infect all the bacteria. After the viral infection was completed, samples from each tube were taken and plated. The plates were incubated, and later the number of resistant cells was tabulated.

According to a Lamarckian view, the mutation to resistance would have occurred at the time the cells were exposed to the viruses. Although one would expect to find different numbers of resistant cells in the different tubes, all would fluctuate only slightly from the average. If one plotted the number of resistant cells against the number of tubes in which that number was found, one would expect a rather narrow bell-shaped curve, similar to that shown in Fig. 14-3.

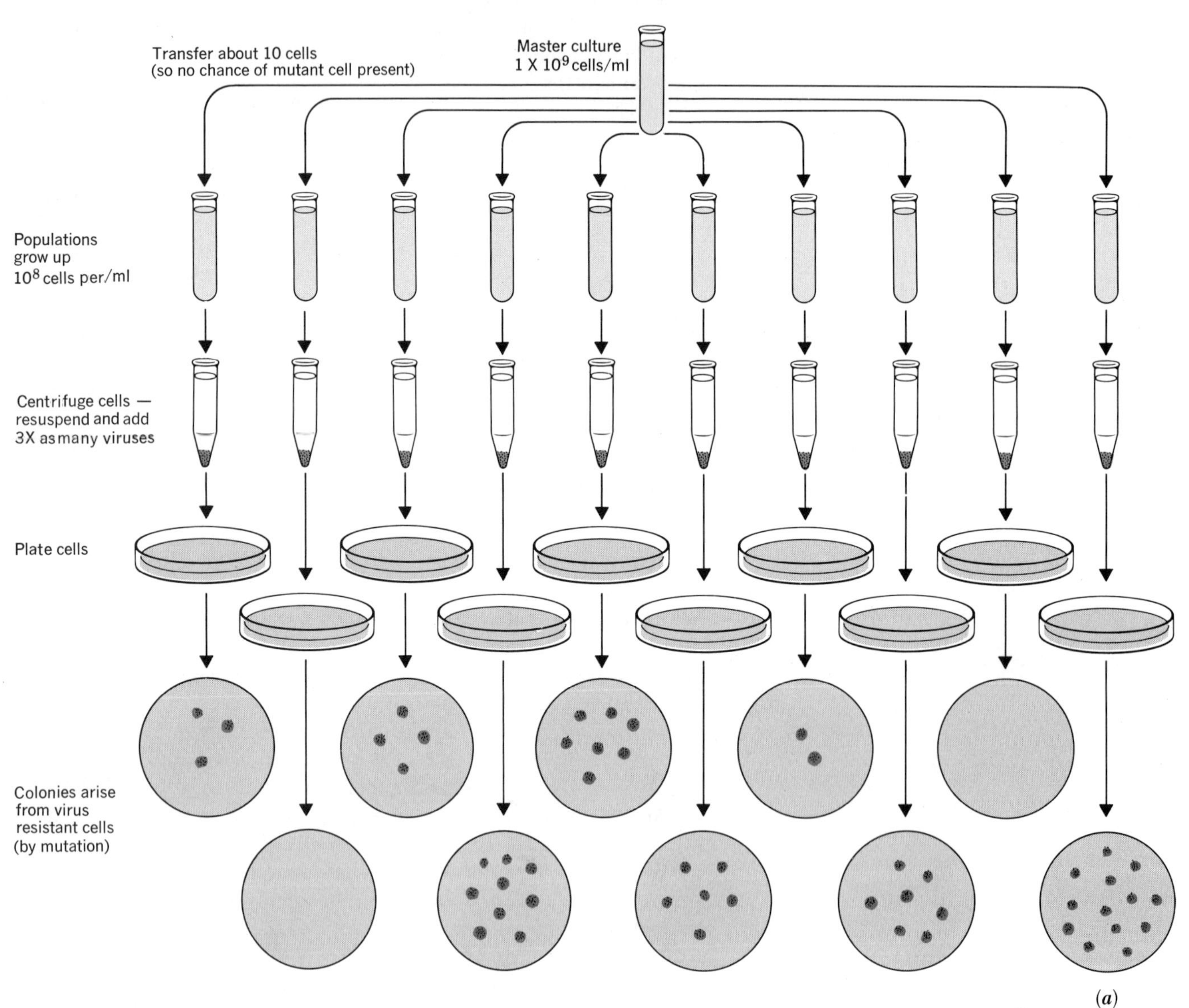

Fig. 14-3 (*a*) Delbruck-Luria experimental setup. (*b*) Expected results based on Lamarckian and Darwinian views of adaptation as seen in a sample of nine clones of cells.

According to a Darwinian view, the mutation to resistance could have occurred at any generation preceding the exposure to the virus. If such a mutation occurred immediately after the formation of the subpopulation, there would have been several hours before the bacteria were exposed to the virus. During this time, the subpopulation was growing, and the mutant forms were passing their resistant trait repeatedly to daughter cells. Therefore, although only one cell had mutated, there would be many resistant cells by the time the bacteria were exposed to the virus. By the same token, a mutation that occurred an hour or two after the subpopulation had been formed would not leave as many resistant cells as the one previously discussed. *The number of resistant cells would be an indication of when the mutation occurred.* One would expect that a tabulation of the numbers of resistant cells in all the tubes would show a wide range of values. If the number of resistant cells were plotted against the number of tubes in which that number was found, a much broader (and possibly asymmetric) bell-shaped curve would be expected than would occur by a Lamarckian interpretation (see Fig. 14-3).

Delbruck and Luria found that there was a great fluctuation in the number of resistant cells from tube to tube, and that when this data was plotted, it generated the broad asymmetrical bell-shaped curve anticipated by a Darwinian interpretation. To insure that this was not simply an acceptable deviation from the Lamarckian curve, they subjected the data to rigorous statistical analysis, and found that the chance that such a fluctuation from the Lamarckian curve would occur was less than 0.1%. Thus, whenever evidence has been sought to confirm Lamarckian evolution, it has not been found, and Lamarckian evolution is viewed as a point of historical interest.

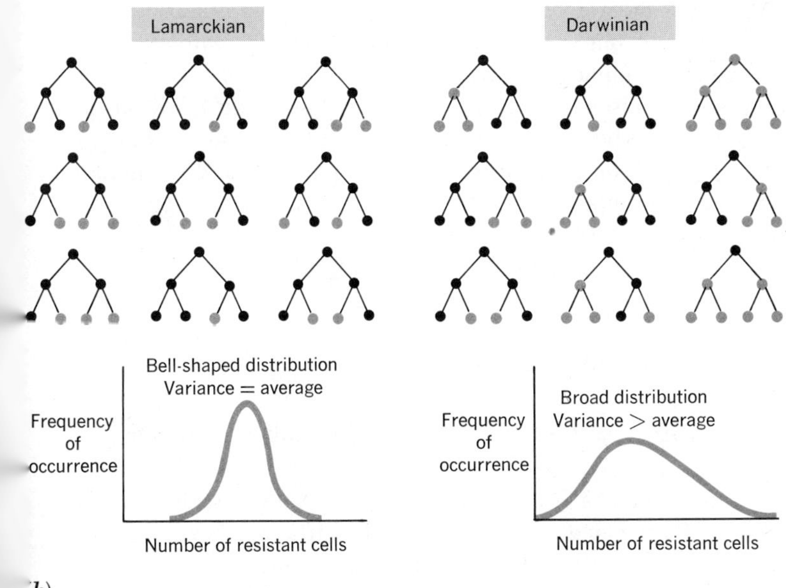

b)

Independence of Mutations

Delbruck and Luria demonstrated that mutation is random with respect to the time of occurrence. But do genes mutate randomly with respect to each other, or do blocks of genes mutate simultaneously?

F. Ryan demonstrated that mutations are independent chronological events in the following experiment. He took a double-mutant strain of *E. coli* that required both histidine (hist$^-$) and leucine (leu$^-$). Both mutant states were capable of undergoing spontaneous reverse mutations, in the course of which the respective genes would become operative and enable the cell to grow in the absence of histidine or leucine.

He placed a few hist$^-$ leu$^-$ cells in each of a series of tubes that contained both histidine and leucine. After the tubes became turbid with bacterial growth, he plated an aliquot from each on duplicate plates, one plate with histidine and no leucine, and the other plate with leucine and no histidine. The frequencies of hist$^+$ and leu$^+$ cells were counted. Few colonies on a plate indicated a late occurrence of a mutation; many colonies suggested an early occurrence of the mutation in the tube. Thus Ryan used the Delbruck-Luria idea that the number of mutant cells indicates the time when the mutation occurred.

If mutational events are interdependent, Ryan expected to find that all the tubes would contain approximately equal numbers of both hist$^+$ and leu$^+$ cells. Instead he found the full range of possible results. Some cultures had many hist$^+$ cells (hist$^-$ $\longrightarrow$ hist$^+$ occurred early in the culture) and few leu$^+$ cells (leu$^-$ $\longrightarrow$ leu$^+$ occurred late). Some cultures had many leu$^+$ but few hist$^+$ cells. Some cultures had approximately equal numbers of both. The experiment demonstrated, therefore, that the genes for histidine and leucine mutate independently of one another.

Frequency of Mutations

Mutations are rare events. The frequency for a mutation of any one gene in man is one in one hundred thousand (10^{-5}). Simple organisms have even lower mutational frequencies, averaging from 10^{-7} to 10^{-9}.

On an organismal level, however, this low value takes on great significance. It has been estimated that there are about one million genes in a human cell. For a human cell in the diploid state ($2N$), therefore, the chance of a mutational event is

$$10^{-5} \times 10^{6} \times 2 = \frac{1}{5}$$

or one in five cells in the human body has a mutation. A human body with its 10^{-15} cells, then, will have

$$\frac{1}{5} \times 10^{-15} = 2 \times 10^{14} \qquad \text{or} \qquad \text{200 trillion mutations}$$

The mutation of a diploid cell (except one that will undergo meiotic division to form gametes) is of relatively little significance from an evolutionary standpoint. Of far greater significance, of course, is the frequency of mutation in the gametic cells. This may be calculated simply by replacing the 2 (which accounted for diploidy) with a 1 (to account for haploidy).

$$10^{-5} \times 10^{6} \times 1 = 1 \times 10^{-1} \quad \text{or} \quad \frac{1}{10}$$

Mutations do not occur simultaneously (Lamarckian), but at each generation (Darwinian), and therefore, as the human develops, mutations will accumulate. A human of 10^{15} cells would have passed through some 48 generations of cell division ($2^{48} = 10^{15}$) in the course of his development. Therefore, the frequency of mutant genes in the gametes formed by the adult human would be

$$48 \times \frac{1}{10} = 4.8 \text{ mutations per gamete}$$

Thus about 4.8 new mutations have occurred during the development of the individual and are present in his or her gametes. The majority of these mutations are small and usually sublethal in effect. They are thought to be deleterious only in large numbers.

The implications of these calculations are far-reaching for all organisms. After ten generations, each human would supposedly bear 40 to 50 mutant genes ($10 \times 4.8 = 48$), and the cumulative buildup would be sufficient to produce deleterious effects. The stockpiling of deleterious genes is offset by the natural selective pressures that tend to eliminate many of these mutant genes in natural populations.

RECOMBINATION AS A SOURCE OF VARIATION

The Mendelian patterns of inheritance and the direct relationship between these patterns and the behavior of chromosomes during meiotic and fertilization processes were viewed with great interest by the proponents of Darwinian evolution. Mendel's single-trait crosses had resulted in F_2 generations in which three fourths of the population manifested a given trait, while one fourth expressed an alternative trait (see p. 231). The F_2 generation could be viewed as exhibiting a variation in regard to this single trait.

The idea was even more encouraging when one considered Mendel's dihybrid crosses. A $\mathbf{P_1}$ generation made up of **AABB** and **aabb** types produced an $\mathbf{F_1}$ generation in which all the members were **AaBb.** The $\mathbf{F_1} \times \mathbf{F_1}$ yielded an $\mathbf{F_2}$ generation in which the ratio was 9 **A-B-** (the dashes indicate that the phenotype would be the same regardless of whether the dominant **A** or recessive **a** genes were present), 3 **A-bb,** 3 **aaB-,** and 1 **aabb.** Four *types* of individuals appeared in the same population. But, what is more, two of these types, the **A-bb** and the **aaB-,** *did not appear in this population before.* They were new variations that had resulted from a **recombination** of the traits. Hence, Mendelian patterns not only indicated how variation could be perpetuated, but also how new variations could arise.

Using Mendelian inheritance as an auspicious start, evolutionary oriented geneticists proceeded to find additional patterns, both Mendelian and non-Mendelian, by which variation could arise. By far the most important contribution of these studies was the growing realization that the genetic material could be completely reshuffled. When a mutation arises, it is a new member of an established genetic community in the individual. A single genetic trait is rarely the object of selective pressures. Selection operates on a *combination of traits* that (if physiologically balanced) might endow the organism possessing them with a selective advantage or (if not physiologically balanced) constitutes a disadvantage. The ultimate test for any mutant trait is in its total genetic environment. The potential reshuffling of the genetic material allows natural selection to test the new genes in all combinations, and assures a maximum opportunity for adaptive traits to remain in the population and for nonadaptive traits to be weeded out.

Evolutionary Significance of the Laws of Mendelian Inheritance

Although the meiotic process and its relation to the patterns of Mendelian inheritance have been discussed in Chapter 9, it is well to review some of the salient features that pertain to the discussion of evolution.

The meiotic process may be outlined in the following way:

1. Meiosis is two divisions, going from one cell to four.
2. DNA synthesis occurs prior to the first division, but *not* prior to the second division.
3. Homologous chromosomes pair at prophase of the first division.
4. Centromeres of sister chromatids do not split at the first division, but do divide and separate at the second division.

The most evident consequence of meiosis is the reduction of the chromosome complement from the diploid to the haploid state. The reduction of chromosome number can be seen to achieve two ends. One is a compensating mechanism for the doubling of the chromosome number during fertilization. Without a meiotic division, fertilization would tend to "stockpile" genetic material. Secondly, by keeping the multiplicity of homologous chromosomes low, meiosis increases the chance that new mutations will be expressed. It is very significant from an evolutionary standpoint that DNA exhibits a uniquely low multiplicity (1 in haploid cells, 2 in diploid cells), while all other cellular components manifest high multiplicities (mitochondria, chloroplasts, etc.). It appears that the expression of new genes and gene combinations has been positively selected for in the course of evolution.

The separation of homologous chromosomes during the first division of meiosis is the cytological counterpart of Mendel's first law of inheritance: segregation. Segregation may seem to be the mechanism of reducing chromosome number, but when allelic genes (**Aa**) are involved, the process accomplishes far more. Segregation provides the separation of allelic genes into different cells, the first step in opening new opportunities of expression. After segregation and meiosis, fertilization will reunite haploid elements back into the diploid state. This process of fertilization is a random one with respect to the genetic content of haploid cells, so if both kinds of gametes (♀ and ♂) possess **A** and **a** genes, the new combinations can be **AA, Aa** and **aa.** The original heterozygous condition of the cells that formed the gametes can be repeated (**Aa**) or changed into two different homozygous states (**AA** and **aa**).

Mendel's second law, that of independent assortment, applies to the behavior of two or more pairs of allelic genes (see Chapter 9). An individual of genotype **AaBb** can produce four types of gametes, **AB, Ab, aB,** and **ab,** in equal frequencies. The relationship of the number of types of gametes to the number of allelic gene pairs is

$$\text{possible gametic genotypes} = 2^n \qquad n = \text{number of allelic pairs}$$

A human that had only one pair of alleles (**Aa**) on each chromosome pair, or a total of 23 allelic pairs, could produce 2^{23} or 5,198,008 different gametes.

The evolutionary significance of independent assortment is in the reshuffling of genes on different chromosomes. New combinations may result in the bringing together of several deleterious genes, which are summarily eliminated by natural selection. In the same way, new combinations may bring about a new physiological balance that endows the individual with a greater selective advantage.

Evolutionary Significance of Non-Mendelian Inheritance Patterns

Both cytological and genetic studies brought about the elucidation of non-Mendelian genetic factors. Once the chromosome theory of inheritance was thoroughly accepted, it became obvious that one could expect recombination only *when two gene pairs under consideration were on different chromosomes.* A chromosome travels through meiotic division as a single unit, and one would expect that two genes that are **linked** on a single chromosome will also travel as a unit. If one crossed two $\mathbf{P_1}$ individuals, **AABB** and **aabb,** and the respective genes were linked, one would expect to find only parental, and not recombinant, types in the $\mathbf{F_2}$ generation (see Fig. 14-4).

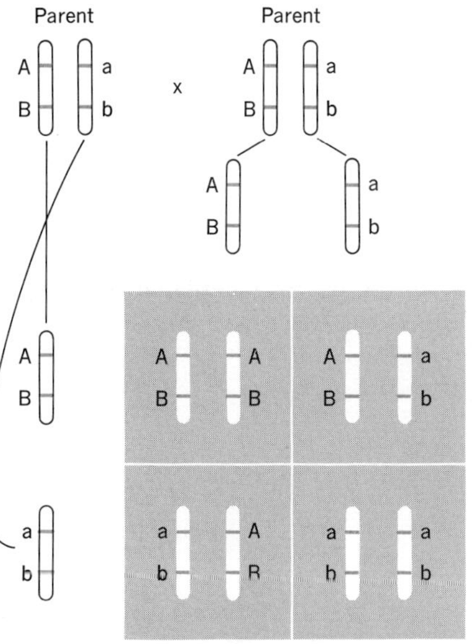

From a practical point of view, pure logic would tell one that if an organism has four chromosome pairs, and five different mutant genes were found for that organism, two of the five must be on the same chromosome. This utilizes the same kind of reasoning as having unpaired white socks and black socks in a draw. A blindfolded person must pull out three socks to be sure of having a pair.

But when many more mutant genes than numbers of chromosomes were found, and the appropriate genetic crosses made, the results were not as clear-cut as geneticists had anticipated. In many crosses parental types were present in far greater numbers in the $\mathbf{F_2}$ generation than would be expected by Mendel's independent assortment; but nevertheless, the recombinant types *were present.* In addition, repeating these crosses again gave a constant ratio of parental to recombinant types, consistently aberrant from that expected by independent assortment.

In 1916, Bridges interpreted this phenomenon by coupling it with the cytological evidence of meiosis that had been noted some twelve

Fig. 14-4 Genetic cross Aa Bb × Aa Bb. If AB and ab are on the same chromosomes, only AB and ab individuals would result.

years earlier. During prophase, the homologous chromosomes are paired. Often, during this pairing, two chromatids seem to become attached to one another at a single point. This point of attachment was called a **chiasma** or chromatid exchange (see Fig. 14-5). Bridges postulated that the chromatids were actually breaking and rejoining in such a way that there was an exchange of genetic material. He termed this genetic phenomenon **crossing-over.** When a cross-over occurred, the genetic factors were recombined. This could account for the appearance of recombinant types even when two genes were on the same chromosome.

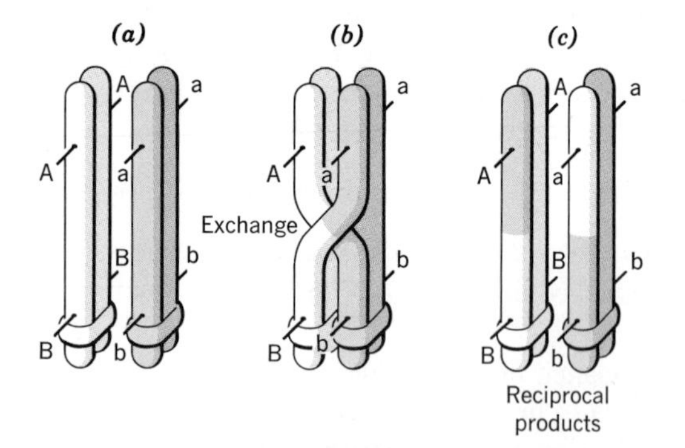

Fig. 14-5 **Schematic view of a crossover. (*a*) Pairing of homologous chromosomes in prophase of meiosis. (*b*) Exchange of chromatids; notice that only two of four strands are involved in exchange. (*c*) Chromosomes resulting from crossover.**

Bridges further saw that crossing-over was a common and frequent event. He reasoned that the farther apart two genes were on a chromosome, the greater is the chance of a cross-over occurring between them. In fact, if two genes were far enough apart on the same chromosome, there would be an equal chance of a cross-over either occurring or not occurring. If an organism possessing two such mutant genes was crossed with one possessing the allelic forms of those genes

$$P_1 \qquad \begin{array}{c|c|c} A & & A \\ B & & B \end{array} \qquad\qquad \begin{array}{c|c|c} a & & a \\ b & & b \end{array}$$

$$F_1 \qquad \begin{array}{c|c|c} A & & a \\ B & & b \end{array}$$

the F_1 would have individuals that were heterozygous for both traits. But, when the F_1 individuals formed gametes by meiotic division, cross-overs would occur. The F_1 generation, then, rather than forming only gametes **AB** and **ab** (the parental genetic arrangement), would form **AB, ab, Ab,** and **aB** in equal numbers.

If such an F_1 individual were backcrossed (see p. 234) to the parental **aabb** type (so that all genetic compositions would be manifested in the phenotypes), the F_2 generation would manifest a 1:1 parental to recombinant ratio. Thus, even though two genes are on the same chromosome, they may be sufficiently far apart to act independently, and give results that one would expect from the law of independent assortment.

But what happens when the genes are closer together? Now the chance of a cross-over occurring between them is less than 50%. If an organism possessing two such genes were crossed with one possessing the allelic forms of those genes

$$P_1 \qquad \begin{array}{c|c|c} A & & A \\ C & & C \end{array} \qquad\qquad \begin{array}{c|c|c} a & & a \\ c & & c \end{array}$$

$$F_1 \qquad \begin{array}{c|c|c} A & & a \\ C & & c \end{array}$$

The F_1 would once again be made up of heterozygous individuals. The gametes from the F_1 individuals would again be of the parental type (**AC** and **ac**) and of the recombinant type (formed by cross-overs, **Ac** and **aC**). But, since the genes are closer together, and there is not an equal chance that a cross-over either will or will not occur, the F_1 individual will have less chance of producing recombinant than parental gametic types.

When such an individual is backcrossed, the F_2 generation will contain both parental and recombinant types of individuals. The ratio of the parental to recombinant types, however, will *not* be 1:1, but will instead by a value that reflects the chance that a cross-over occurred during the formation of the gametes in the F_1 individual.

Bridges detected in this scheme a way in which the spatial relationship between genes could be determined. Thus, when the F_2 of the cross from **AcCc** and **aacc** individuals indicated the presence of 25% recombinant types, it could be viewed* as a 12.5% chance that a cross-over could occur between the **A** and **C** genes. Bridges decided to consider a 0.5% chance of cross-over equal to one cross-over unit. Therefore, genes A and C were 25 cross-over units apart on a chromosome.

But where was *B*? Were the genes on the chromosome arranged

C A B

or

A C B

or, for that matter, was B on a different chromosome? (The results of the cross between **AABB** and **aabb** individuals had produced results that were completely in line with Mendel's law of independent assortment.)

This could be determined by performing a cross between P_1 individuals that possessed **B** and **C** traits.

P_1 B| |B / C| |C   b| |b / c| |c

F_1 B| |b / C| |c

If, when the F_1 individual is backcrossed, the F_2 produces parental to recombinant types in a 1:1 ratio, **B** is either on another chromosome, or lies on the opposite side of **A** from **C** (since **A** is so far away from **B** that there is a 50% chance of a cross-over either occurring or not occurring, then any gene on the other side of A will be even farther away, and would give similar results.) If, however, the F_2 produced parental and recombinant types in a ratio, let us say, of 70:30, then there is a $\frac{30}{2}$% chance of cross-over between genes **C** and **B,** and they are 30 cross-over units apart. The only arrangement they can have on the single chromosome then is

A C B
25 30

And **A** and **B** are 55 cross-over units apart. The only way of determining the position of two genes that are more than 50 cross-over units apart is to find a gene that lies between them, and indirectly calculate the

*Notice from Fig. 14.5 that 1 cross-over gives 2 of 4 (50%) recombinant types; hence, % crossing-over = recombinants/2.

cross-over unit distance by adding the distances between each of the three genes.

Although this has been a digression from the discussion of evolution, the concept of **chromosome mapping,** or the determination of gene placement, has been of such importance to the development of genetic studies that it warrants discussion. Chromosome mapping took on a great significance when bacterial and viral chromosome maps were added to that which Bridges himself established for *Drosophila* (see Fig. 14-6).

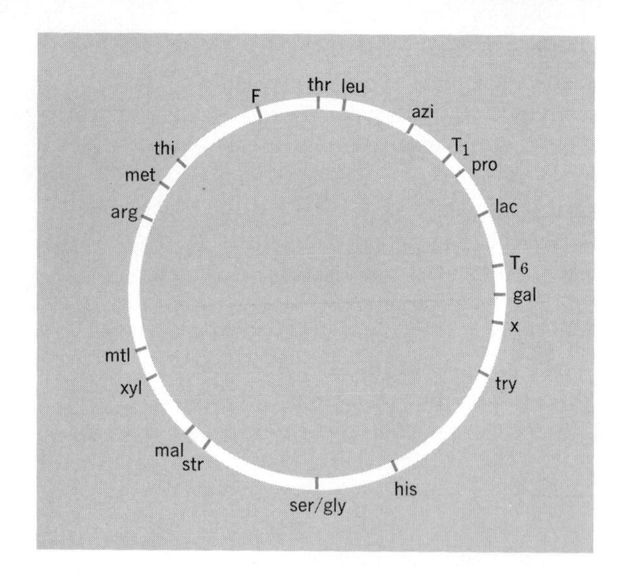

Fig. 14-6 Chromosome map of *E. coli*. Linkage groups indicate that the *E. coli* chromosome is circular. Abbreviations indicate the positions of genes controlling the synthesis of compounds; for example, met = metheonine.

From an evolutionary standpoint, crossing-over has important consequences. It is a means by which genes located on the same chromosome may be recombined with genes of homologous chromosomes. It is a means by which genes existing in a *trans* condition

A	b	
		trans
a	B	

can be brought into a *cis* condition

A	B	
		cis
a	b	

Conversely, it can bring about the separation of linked genes, placing them into many new combinations, some of which may be an advantage in survival. When the processes of segregation, independent assortment, and crossing are combined, one can see that the genetic material is not limited by the kinds of association which can be formed by mutation.

Intragenic Cross-Over

It has recently been found that recombination by crossing-over can even occur *within* a gene. This phenomenon has been described by Benzer in the T-4 bacteriophage. His analysis can be described in several steps. First he produced many mutations of independent origin for the same phenotypic change. The phenotypic expression was for a plaque characteristic where the normal wild type gives a smooth-edged, small plaque and the mutant type (called **r**) gives a rough-edged, larger plaque (see Fig. 14-7).

By mapping the genes for each of the independently produced **r** mutants, Benzer found that they were all localized in the same region of the chromosome. In addition, he found that there were two adjacent genes responsible for the **r** trait. These two genes appeared to have different biochemical roles in the infectious process, but, if either formed an aberrant polypeptide, the same phenotypic expression was manifested. The simplest way of visualizing this is by speculating that both polypeptides formed a single protein. Therefore, if either one were aberrant, the proper protein would not be formed, and the virus would manifest the presence of the nonfunctional protein by a given phenotypic expression.

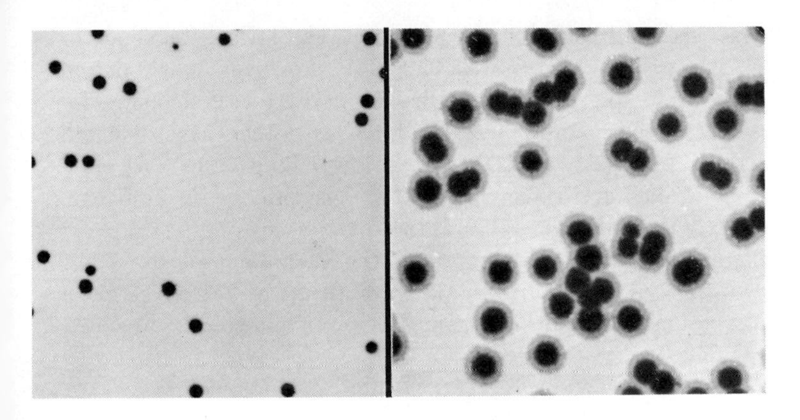

Fig. 14-7 Bacteriophage growth is evidenced by clear areas or plaques on an opaque bacterial growth. The size and texture of the plaque is under genetic control. (*a*) Plaques of T_4; (*b*) plaques of the mutant T_{4r}. [From *Viruses And Molecular Biology* by Dean Fraser, published by The Macmillan Company. Copyright © 1967 by Dean Fraser.]

The two groups of **r** mutants were called simply **A** and **B.** Benzer focused his attention on the **A** group of **r** mutants. It was logical to assume that if a gene were composed of a thousand pairs of nucleotides, an alteration of any might bring about the formation of a nonfunctional protein. Therefore, if two mutants differed in the site of the alteration that led to the mutation, *and* if crossing-over occurred within a gene, it should be possible to form a normal type by "mating" two **r** mutants (see Fig. 14-8).

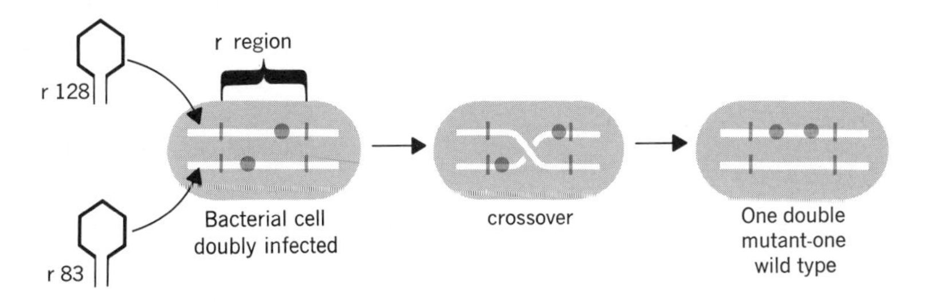

Fig. 14-8 Recombination of nonidentical mutants in some genes leading to a wild-type chromosome.

Benzer indeed found that this was true. The lowest frequency of the cross-over within the gene was .02% (2 cross-overs in 10,000 paired chromosomes), in spite of the fact that he crossed several hundred **r** mutants in all combinations, and could, by his system, have detected lower values. In other words, there is a lower limit as to how close together two mutant sites can be, and still experience an exchange.

Why was there this natural limit to the frequency of crossing-over? Benzer approached this problem from the standpoint of the DNA content of the cell. The entire single chromosome of T-4 phage has a total cross-distance of about 1000 units. His lower value of .02 cross-over units was then

$$\frac{.02}{1000} = 2 \times 10^{-6}$$

of the total distance. The amount of DNA in this single chromosome is about 10^6 nucleotides. If 10^6 nucleotides is equivalent to 1000 cross-over units, then proportionally, 0.02 cross-over units should be equal to

$$\frac{10^6}{1000} = \frac{x}{.02}$$

$$x = 2 \text{ nucleotide pairs}$$

This calculation indicates that a cross-over can occur between two adjacent nucleotides in DNA. Thus, the recombination unit is far smaller than the functional unit. Crossing-over is then more powerful a recombination mechanism than was imagined. Prior to Benzer's work, it was assumed that exchanges only occurred between particular genes. This was, in great part, because only particular genetic systems were studied. From Benzer's work it now seems reasonable to conclude that cross-over can occur between any adjacent nucleotides (by breakage of the phosphate backbone).

If crossing-over takes place within a gene, then new genes can be synthesized! The example of **AB** and **ab** can be extended to the molecular level, where the recombinate types **Ab** and **aB** could be different DNA bases that code for different amino acids (see Fig. 14-9). The first two triplets of the gene pair pictured in Fig. 14-9 code for complementary mRNA and ultimately for the amino acids phenylalanine and glycine. The third triplet of the gene pair differs by *two* nucleotides. The gene represented on the left contains CTA which ultimately codes for asparagine, while the gene on the right codes for lysine. Both asparagine and lysine are diamino monocarboxylic acids, and as such, contain an amino group (NH_3^+) at the terminus of the side group in addition to the amino group involved in the peptide linkage. Hence the side groups of these amino acids would most likely take part in tertiary bonding—*but in the same manner.* Despite the difference in nucleotides, the conformation of the final proteins formed by each of these genes would be extremely similar in form and function.

Upon cross-over between the TA of CTA and the TC of TTC, new codes are formed, so that glutamic acid, rather than asparagine and asparagine rather than lysine are incorporated into the protein. Glutamic acid is a dicarboxylic amino acid, and has a terminal COO^- group which would take part in the tertiary structure in a *very* different manner than asparagine. Thus, the conformation of this protein will be quite different than that of the original protein. A simple cross-over event could produce a type of protein, and have an effect similar to that of a mutation.

It should be mentioned that when two genes produce the same protein (**AA** or **aa**), a cross-over within the gene will not have a mutational effect. Only when a mutant form of the gene differs by two nucleotide bases is there a possibility of an exchange resulting in a new protein.

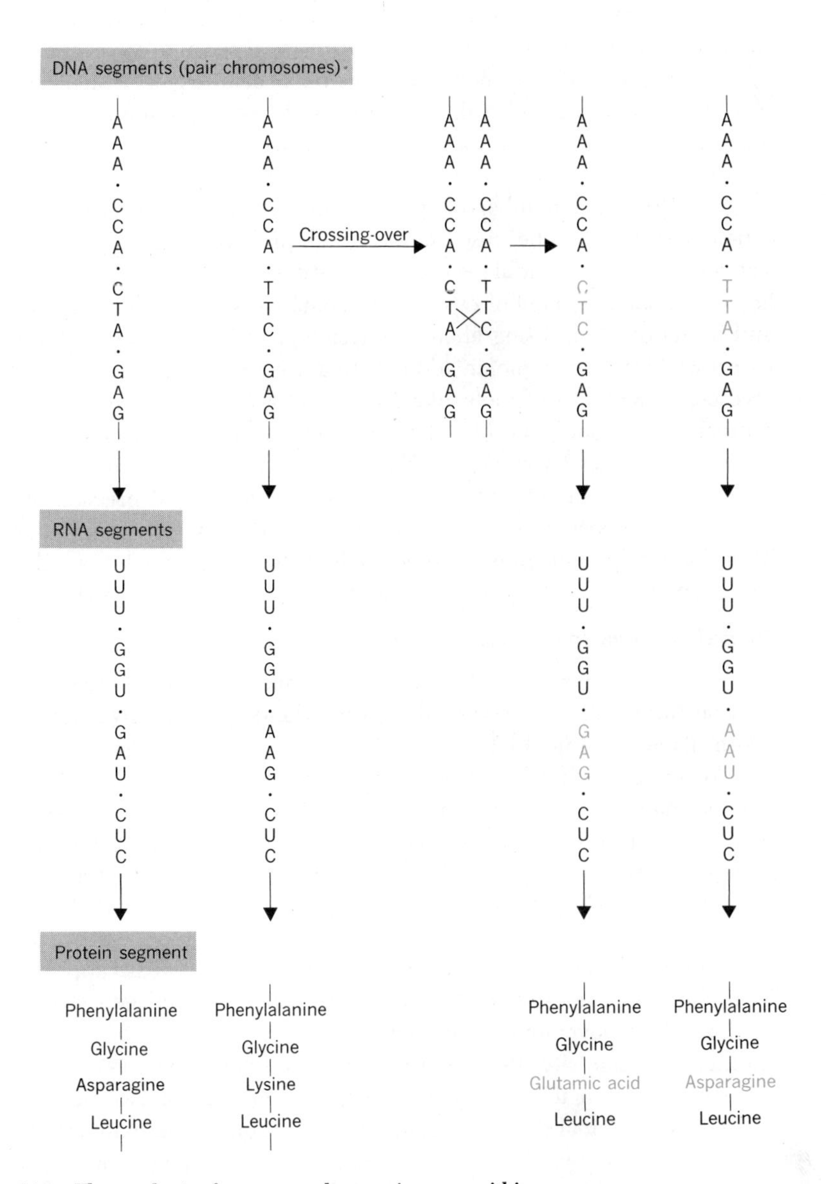

Fig. 14-9 The synthesis of new genes by crossing-over within a gene.

Fig. 14-10 Chromosomal deletion.

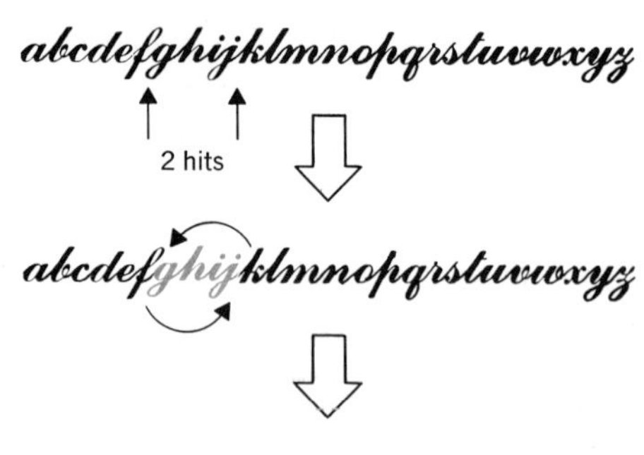

Fig. 14-11 Chromosomal inversion.

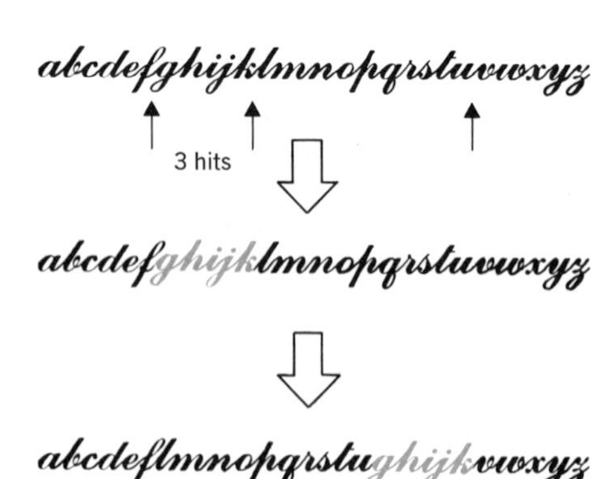

Fig. 14-12 Chromosomal transposition.

CHROMOSOMAL ABERRATIONS AS A SOURCE OF VARIATION

Chromosomal aberrations involve the changes in genetic composition that occur when chromosome breakage and rejoining takes place. There are five major types of such aberrations; Deletions, Inversions, Transpositions, Translocations, and Duplications.

It is impossible to determine the frequency or the range of possible chromosome aberrations. Therefore, one must be satisfied with a qualitative rather than a quantitative discussion.

Deletions

An entire segment of the DNA is broken off and lost during subsequent division. Deletion as a chromosome aberration involves the loss of several genes. It may occur as a terminal deletion, in which there is a single chromosome break, or there may be two chromosome breaks and a rejoining of the chromosome without the middle portion (see Fig. 14-10).

The loss of a number of genes from the genetic complement is usually lethal. This is because there is a high probability that the deleted segment contains essential genetic information. It is possible to conceive, however, that a segment of DNA might contain dispensable information, and by its deletion, bring about a physiological balance impossible in its presence. For example, in addition to a myriad of nonessential genes (eye color, synthesis of molecular forms that are available in the environment, etc.), one gene in a deleted segment may be a repressor for a gene on another chromosome. The removal of the repressor from the genome permits the expression of this other gene. The expression of a formerly repressed gene increases the variation in the population. Therefore, despite the more common deleterious effect of deletions, they can occur in such a way as to be a positive evolutionary force.

Inversions, Transpositions, and Translocations

Two breaks may occur in the same chromosome. When the central portion turns 180° and rejoins the chromosome ends, an inversion has taken place (see Fig. 14-11).

When three breaks occur in a single chromosome, the deleted portion may be inserted into the third break, resulting in a transposition. Thus a block of genes is changing to a new location on the same chromosome (see Fig. 14-12). It should be mentioned that chromosome segments can only join at the site of a break. It appears that the terminal genes of chromosomes are "sealed" and as such are not subject to accepting chromosome segments. New chromosome breaks appear to be highly reactive, and tend to join with other exposed genes. If by their random movements, however, they do not come in contact with another breakage site, they tend to "seal" and become nonreactive.

Translocation involves two breaks, one on each of two chromosomes. The terminal deleted regions of each chromosome switch places and rejoin. The switching of large segments of chromosomes is dependent on random contact. If the two deleted regions, each without a centromere, come in contact, they will join and form an *accentric unit* (denoting the fact that it has no centromere). If the two chromosome segments with centromeres come in contact, they will join to form a *dicentric unit.* In the course of subsequent division, the accentric chromosome is lost, and the dicentric chromosome may be torn in two. Thus a rejoining in any way other than a switching of chromosome segments leads to the death of the cell (see Fig. 14-13).

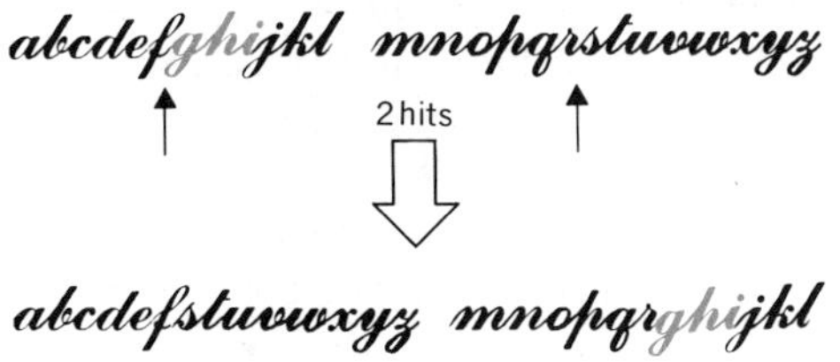

Fig. 14-13 Chromosomal translocation.

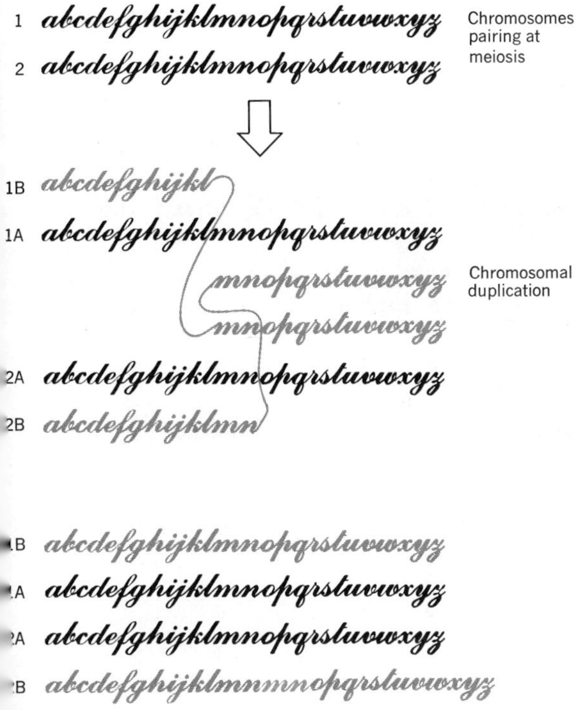

Inversions, transpositions, and translocations all involve a repositioning of a deleted chromosome segment. Usually no genetic material is lost to the cell when these types of chromosomal aberrations occur. The presence of an inverted, transpositioned, or translocated chromosome does present certain problems in meiotic division. These will be discussed in the next chapter.

From an evolutionary standpoint, the most important effect of these chromosomal aberrations is the repositioning of genes. It has been found that genes act differently when placed in different parts of the genome. This is referred to as **position effect,** which can be illustrated in the following way. Three genes **A, B,** and **C** may, by virtue of their position on the chromosome, be read at the same developmental stage of the organism. If **B** is part of a deleted area that rejoins to a region of the chromosome between **M** and **N,** it may be read with **M** and **N.** Because of the different developmental stages during which these genes are read, **B** may express a new trait. Thus variation can arise from a repositioning of the existing genetic material.

Duplication

Duplication of large segments of genetic material involves the insertion of a deleted portion from one chromosome into the site of a single break in another chromosome. Upon meiotic division, the chromosome homologous to the one that has experienced the deletion may go to the poles with the chromosome that has incorporated the deleted region. When fertilization takes place, the genetic information in the deleted region will be present in a **triploid** state. This unnecessary duplicated region will be carried through cell generations by mitotic division (see Fig. 14-14).

Chromosomes pairing at meiosis

Chromosomal duplication

Fig. 14-14 Chromosomal duplication.

Duplication of chromosome segments produces an unnecessary redundancy in the genome—genes whose functions are unnecessary for the stability of the organism. These duplicated segments may be viewed as uncommitted genes, genetic sites where mutational events that carry them far from their normal function can occur without detriment to the organism. They provide the testing sites for evolution, and as such, have great evolutionary significance.

THE PARADOX OF A STABLE BUT MUTABLE HEREDITARY MECHANISM

Darwin's prediction of a stable but totally mutable hereditary mechanism has been adequately demonstrated. The stability of the mechanism lies in the orderly means by which the full complement of genetic material is passed through cell generations by mitotic division, and through organismal generations by meiotic division and fertilization. The stable aspect is thus under organismal control.

The mutable aspect of the system comes under the auspices of the chemical and physical laws that govern all chemical entities. As Darwin predicted, mutability exists at the lowest level of organization. Genes mutate, crossing over can occur between genes, and segments of genes can be transferred from one location to another. Nothing in the chemical composition of the hereditary mechanism is sacred, nothing is so stable that it is not subject to change.

The opposing phenomena of stability and mutability exist side by side. By organismal control, biological entities have found a means by which each cell is provided with at least one gene for every necessary function characteristic of that organism. Through the chemical and physical laws, biological entities evolve.

QUESTIONS

1. Describe the ways in which a mutation may occur.
2. What evidence supports the idea that point mutations are real events?
3. How did the Delbruck-Luria experiment definitively disprove Lamarckian evolution?
4. What is the evolutionary significance of independent assortment?
5. Give the evidence from Benzer's experiment that crossing over can take place within a gene.
6. Describe the five major types of chromosome aberrations, and explain their significance in evolution.
7. What is the significance of recombination in evolution?

SPECIAL INTEREST CHAPTER V

HUMAN HEREDITY

Since antiquity man has been aware of hereditary phenomena in the breeding of date palms, rice, livestock, and work animals. He also saw that he was not different in this respect, and certain family traits had obvious hereditary continuity (see Fig. V-1). Despite man's interest in family lineages, a valid perspective of his genetic status did not emerge until an understanding of heredity had been reached from studies on other organisms.

This conditional development was necessary because human heredity is far more difficult to study than animal and plant heredity. Other organisms can be investigated by first selecting those traits that are easily followed through consecutive generations. Consequently, many traits were not analyzed because the patterns of heredity were unclear, and the problem was discarded. This attitude is not acceptable in human genetics. Imagine a physician having to say that little is known about the inheritance of diabetes because it was too difficult to study, and the research on it was discontinued. Medical research cannot afford this type of luxury.

Another problem in the study of human heredity is concerned with the nature of the creature. Man cannot be manipulated to mate. He produces too few offspring for statistical analysis, and he takes too long to mature and have offspring. He probably stands with the pine trees and elephants as the least desirable research tools for genetic studies.

The third problem in human genetics is that many of man's traits are strongly influenced by the environment. Without his culture, he becomes a savage,* and hence, to measure his full biological stature requires a strong environmental conditioning that masks the underlying hereditary basis. For instance, the existing intelligence tests can be given successfully only to those who are able to read and are conditioned to face abstract challenges. Yet the purpose of such tests is often the measurement of the potential rather than the actual state of man.

*Linnaeus' system of taxonomy included two species of man. Along with men as we know them, were beastlike men inhabiting the forests. It is now known that these men were discarded by their parents, and had to fend for themselves since childhood, apart from the civilized forces of language, abstraction, and emotional suppression.

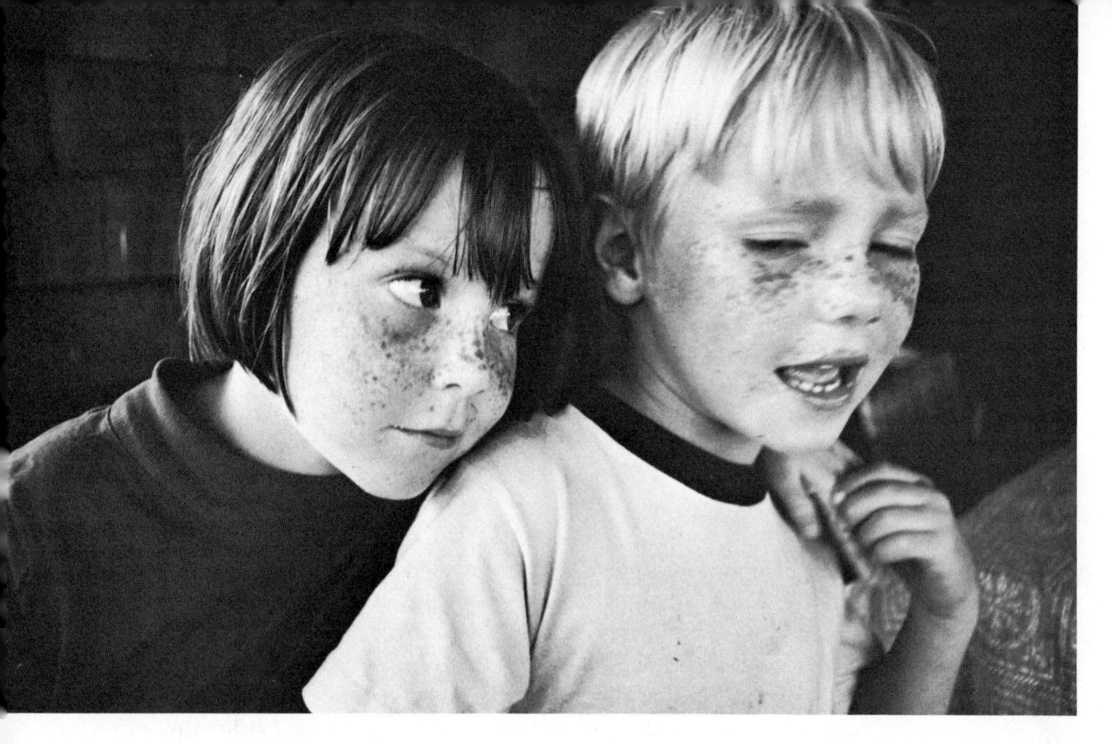

Fig. V-1 Inherited traits in humans: (*above*) Piebald, in which a dominant gene makes the skin produce an uneven pattern of protective melanin pigment when it is exposed to the sun. [Ken Heyman] (*lower left*) Polydactyl, in which an extra finger, in this case a thumb, is present. [Lester V. Bergman] (*lower right*) Brachydactyl, in which fourth finger is longest and fifth is next, in contrast to normal hand where third finger is longest, and fourth or second is next. [American Genetics Association]

(*Facing*) Manifestations of Hapsburg jaw, which is a trait that has appeared through numerous generations of this European ruling family.

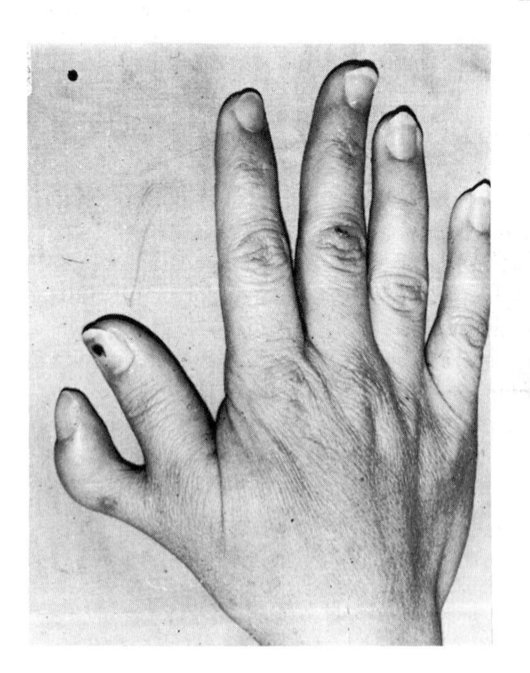

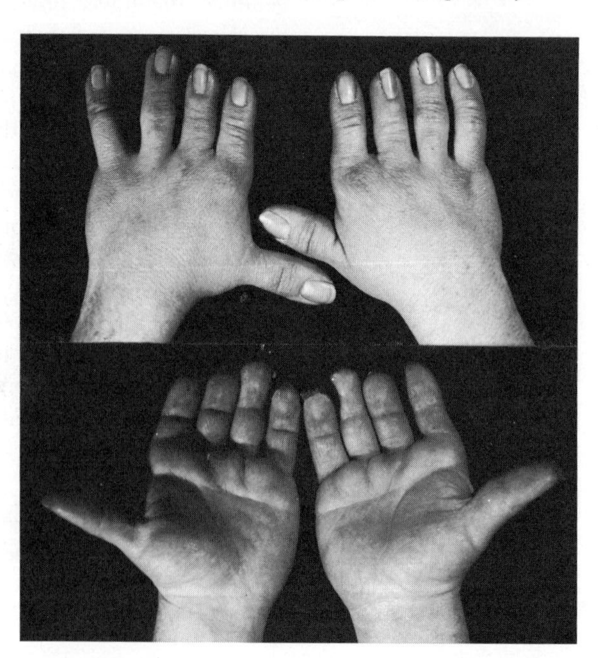

Archduke Ferdinand of Austria (1503-1564)
[Kunsthistorisches Museum, Vienna]

Philip IV of Spain (1605-1665)
[New York Public Library Picture Collection]

Philip III of Spain (1578-1621)
[New York Public Library Picture Collection]

Charles II of Spain (1661-1700)
[New York Public Library Picture Collection]

Hereditary Factors and Patterns of Inheritance

The limitations of man as a research tool, the complexity of certain traits, and a strong environmental influence, make the study of human genetics a formidable task. In cases where a trait (1) is controlled by simple Mendelian recessive or dominant gene, (2) is clearly expressed, and (3) arises at the time of birth, the hereditary basis and the patterns of transmission and expression can be easily determined. In many cases, however, the simple problem of determining whether a trait is inherited is sufficiently difficult to resolve without expecting further conclusions as to the mode of inheritance.

TABLE V-1 COMPARISON OF IDENTICAL AND FRATERNAL TWINS WITH RESPECT TO CERTAIN TRAITS

	Identical Twins		Fraternal Twins	
Trait	Concordant	Discordant	Concordant	Discordant
Blood groups	125	0	60	31
Diabetes	42	23	21	95
Tuberculosis	68	29	45	99
Mental Retardation	74	2	80	138
Schizophrenia	120	54	53	464
Measles	179	10	127	19

The five most commonly employed procedures in human genetics are (1) studies of family pedigrees, (2) Mendelian analyses of groups of families as though they were identical, (3) and (4) two types of statistical methods on twin data, and (5) chromosome studies.

Sometimes a pedigree analysis provides the necessary information for determining both the heritability of a trait and the manner by which it is inherited. Consider the pedigree chart for albinism in Fig. V-2. The recurrent appearance of albino individuals in one family would lead one to strongly suspect that albinism is indeed an inherited trait. Albinism cannot be caused by a gene located on the male-determining **Y** chromosome, because one of the individuals bearing the trait is a female. Similarly, albinism cannot be sex-linked, with the gene on the **X** chromosome, since the fathers pass their X chromosome only to their daughters, and male #9 and his son #14 are albinos. Mendelian inheritance of a dominant gene cannot be an acceptable explanation, because individual #9 has the trait, yet neither of his parents have it. The only remaining common pattern is that based on a Mendelian recessive gene. The pedigree pattern does not exclude this possibility,

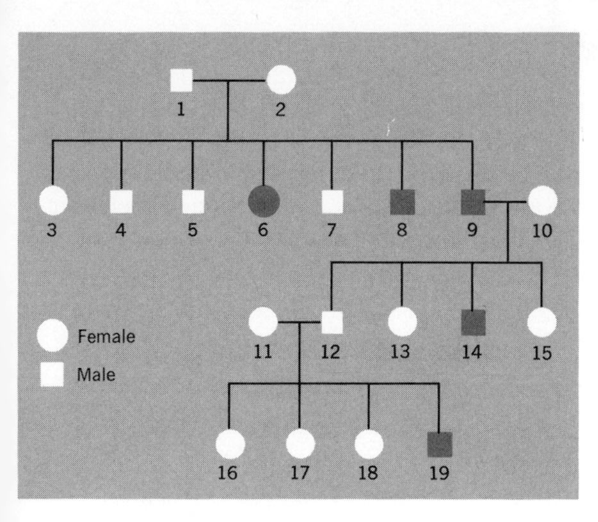

Fig. V-2 Pedigree chart for albinism. Numbers 1 and 2 do not express the trait. Of their children, one female (represented by a circle) and two males (represented by squares) are albino. Albino son (# 9) marries a nonalbino, and they have four non-albino children. A son marries a nonalbino, and of four children, one son is an albino. A pedigree such as this clearly shows that albinism is controlled by a recessive gene.

and hence one could conclude that albinism is *probably* recessive (R?). When the conclusion is supported by deductions from other pedigrees, albinism can be reclassified as almost *definitely* a Mendelian recessive gene present in the homozygous state.

Another type of genetic analysis of human traits is by statistical determination. This particular type of statistical analysis is used when pedigrees have indicated that a trait is heritable, but the pattern of heredity remains elusive. Often in such cases, the expression of the gene is correlated with age. For instance, diabetes is usually not expressed until middle age or later. How does one go about analyzing a family in which many of the children are below the average age at which diabetes might be expressed? One method is to consider a number of families as being genetically similar, and proceeding through a statistical analysis. Consider a number of families in which both parents are diabetic. If the trait is determined by a homozygous recessive condition, then all offspring will be potentially diabetic. If the diabetic condition is brought about by the presence of a single dominant gene (a heterozygous state), then three fourths of the offspring will be diabetic (see Fig. V-3). Diabetes occurs late in life, and with a knowledge of the age of the children, appropriate probabalistic adjustments can be made. In one study it was determined that if diabetes is expressed as a double recessive (*dd*), and the age of the children were taken into consideration, about one fourth of the children should be afflicted at the time of the analysis. Of 173 children tested, 47 were diabetic as compared with the expected number of 42.6. Diabetes is, therefore, considered to be due to a recessive gene present in the homozygous state.

If inherited as a recessive	If inherited as a dominant
Dd X Dd	Dd X Dd
Normal Normal	Diabetic Diabetic
1 DD : 2 Dd : 1dd	1DD : 2Dd : 1dd
¼ offspring diabetic	¾ offspring diabetic

Fig. V-3 Possible modes of inheritance of diabetes.

A more rewarding analysis comes from the comparisons of twins, particularly identical ones. Identical twins come from a common zygote, by separation of the two-celled embryo into two isolated cells, each of which forms an individual. They should, therefore, have identical genetic constitutions (barring the aforementioned possibility of mutations that may occur in either one). On the other hand, fraternal twins come from different fertilizations that have occurred at the same time. In order for fraternal twins to be produced, the mother must undergo a double ovulation. Fraternal twins are no more similar than any other brothers or sisters, and generally have about 50% of their genes in common.

By comparing identical and fraternal twins for several traits to see whether both twins are alike (**concordant**) or not alike (**discordant**), the hereditary basis can be established in some cases (see Table V-1).

The data of Table V-1 indicate that blood groups are hereditary and unaffected by any environmental factors. Diabetes, tuberculosis, mental retardation, and schizophrenia are partly hereditary and partly environmental in cause, or perhaps dependent on age. Measles occurs in equal frequency in both groups, indicating that susceptibility to measles is not inherited and that measles is simply due to an environmental factor.

Many of the important traits of man, such as various aspects of intelligence, are influenced by both hereditary and environmental factors. Valuable advances in the study of human heredity were made by the development of statistical means of determining the degree of heritability. These statistical methods utilize data from studies on identical and fraternal twins.

One statistical procedure is the calculation of the **correlation coefficient** (**r**). The easiest way to illustrate how the correlation coefficient is calculated is to use some *hypothetical* data for some arbitrary feature, such as arm length. Table V-2 gives a set of data for identical and fraternal twins under consideration.

TABLE V-2 HYPOTHETICAL DATA ON ARM LENGTH OF IDENTICAL AND FRATERNAL TWINS

	Identical Twins		Fraternal Twins	
Twin Set	Twin A	Twin B	Twin A	Twin B
1	33.5 in	33.7 in	32.8 in	33.4 in
2	35.1 in	34.8 in	32.7 in	35.2 in
3	22.3 in	22.5 in	33.6 in	34.2 in

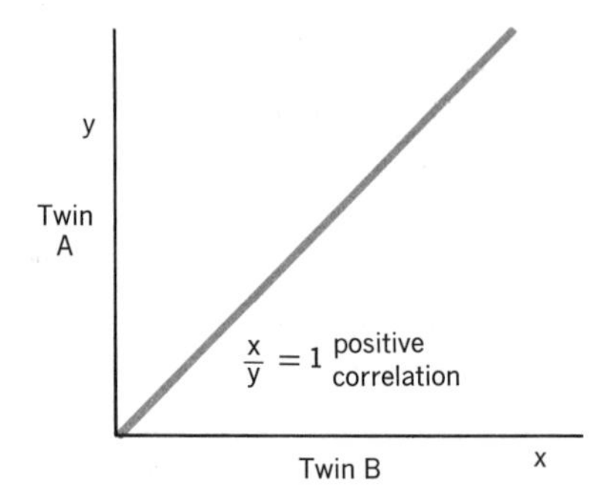

Fig. V-4 Correlation coefficient of an inherited trait.

The specific numbers are unimportant, but the data must be taken from either adults or children of the same sex. What is important is the *relationship* of the numbers. The values for twin A of the sets are used as the value for the abscissa of a graph, and the values for twin B are used as those for the ordinate. For instance, the first identical-twin set would be plotted as $x = 33.5$ and $y = 33.7$. Each pair of twins plotted gives a point on the graph. A line is either drawn from the points, or statistically calculated (see Fig. V-4). The *correlation coefficient is the mathematical expression approximating the slope of the line.*

Both the correlation coefficient for the identical-twin data alone, and the comparative values for the correlation coefficients for both types of twins, are analyzed. A positive slope near the maximum value of 1.0 for the correlation coefficient would indicate that there was close to a one-to-one correlation between the characteristics in both identical twins. Such a value would strongly indicate that the trait was genetically determined. A comparison between the two correlation coefficients of the two types of twins often gives a more conclusive indication of the manner by which the trait comes about. Since identical twins have 100% of their genes in common, while fraternal twins have approximately 50% of their genes in common, one would expect not only that their correlation coefficients for a trait would be different, but also that the value for the identical twins would be positive, and closer to 1.0 than the value

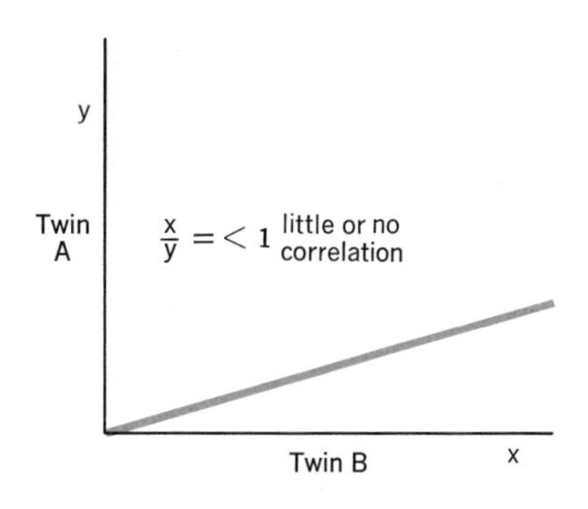

Fig. V-5 Correlation coefficient of a noninherited trait.

for the fraternal twins. If the correlation coefficients of the two types of twins are the same, the chances are small that the particular trait under investigation is inherited (see Fig. V-5).

The degree of **heritability** (**h**) of a trait can be determined by comparing the data on similar traits in the two types of twins by a different statistical procedure. The statistical analysis for degree of heritability is made by considering the **variance** (**V**), that is, a type of average difference between the two groups of twins. If a trait is dependent on genetic rather than environmental factors, identical twins would be more similar and hence express a smaller variance than fraternal twins. A quantitative measure of the degree of heritability is derived from the following formula:

$$h = \frac{V_{\text{fraternal}} - V_{\text{identical}}}{V_{\text{fraternal}}}$$

If, for instance, both groups of twins have the same variance, then the equation becomes $(X - X)/X$ or 0.0. In this case the trait is not inherited. When identical twin groups express almost no variance (ca. 0.0), and fraternal twins have some measurable amount (X), the equation becomes $(X - 0)/X$ or 1.0. A value of this magnitude indicates that the expression of the trait is due entirely to genetic factors. Intermediate values then indicate the degree of heritability (see Table V-3). A list of inherited traits is given in Table V-4.

TABLE V-3 CORRELATION COEFFICIENT (*r*) AND HERITABILITY (*h*) FOR CERTAIN TRAITS IN IDENTICAL AND FRATERNAL TWINS

	Correlation Coefficient			
		Identical Twins		
	Fraternal Twins	Reared Together	Reared Apart	Heritability
Stature	0.64	0.93	0.97	0.81
Sitting height	0.50	0.88	0.96	0.76
Weight	0.63	0.92	0.89	0.78
Binet IQ	0.63	0.88	0.67	0.68
Otis IQ	0.62	0.92	0.73	0.80
Word meaning	0.56	0.86	—	0.68
Arithmetic computation	0.69	0.73	—	0.12
History and literature	0.67	0.82	—	0.45
Tapping speed	0.38	0.69	—	0.50

Biochemical Analyses for Heterozygotes

The determination of the pattern of inheritance for certain traits has found application to practical matters in human genetics. One area of human genetics that has been under intensive study, and has shown great promise in the determination of individual genomes, is biochemical analyses. When the recessive genes governing a trait are present in the

TABLE V-4 HUMAN TRAITS WITH DEFINITE OR SOME HEREDITARY BASIS[a]

Trait	Basis
Achondroplasia dwarfness of the short-limb type	D
Albinism little or no pigment in skin, hair, or eyes	R? D?
Amaurotic idiocy, infantile blindness, motor and mental impairment, death in infancy or childhood	R
Amaurotic idiocy, juvenile as above, but death in childhood or adolescence	R
Cleft palate without harelip	D, v
Congenital club foot	D, v
Deaf-mutism, congenital total deafness	R?, v
Diabetes insipidus excessive urine excretion	R, sl?
Diabetes mellitus low glucose tolerance	R, v
Epilepsy, idiopathic convulsive seizures	R?, v
Criminality	chromosome aberration (?)
Galactosemia galactose not converted to glucose	R
Hemophilia excessive and prolonged bleeding	R, sl
Huntington's chorea progressive muscular spasm, disturbance of speech, dementia	D
Homosexuality	?
Hapsburg's jaw protruding chin, excessively long lower jaw	D
Muscular dystrophy, fascio-scapulo-humeral type progressive wasting of pectoral girdle muscles in childhood	D
Parkinsonism, shaking palsy progressive muscular tremor and rigidity	D, v
Pernicious anemia abnormal red blood cells	R, v
Phenylketonuria phenylpyruvic acid in urine, feeble-mindedness	R
Rickets, vitamin-resistant	D, sl?
Intelligence	multiple factors
Longevity	multiple factors
Manic-depressive psychosis	D(?)
Schizophrenia	R
Skin color	multiple factors
Smoking	?
Mongoloid idiocy	loss of chromosome

[a] D, dominant gene; R, recessive gene; v, variable expression (incomplete penetrance); sl, sex-linked; ?, some evidence of heredibility but uncertain.

homozygous state, an aberrant protein is often formed. In these homozygous individuals the aberrant protein may be detected in the urine or blood. It has recently been found that for certain traits, individuals with a heterozygous condition will have detectable qualities of the aberrant substance in their blood or urine. Although such tests are presently possible for schizophrenia and blood groups, the work undertaken in this area may soon extend to other traits.

Chromosome Aberrations

The cytological study of human cells has made a notable contribution to human genetics. For many years, the chromosome number in man was used solely as a comparative feature between himself and other animal species. In 1892, man's chromosome number was determined to be 46. Some years later, the figure was changed to 48. The accuracy of these counts was hampered because they were made on chance findings of dividing cells in tissues in which the chromosome were spread sufficiently to count.

In 1956 Tjio reported the application of some simple techniques to the study of human cells. He placed cultured cells derived from bone marrow in a colchicine solution that brought the cells to a standstill upon entering metaphase. He then placed the cells in a hypotonic solution, so they would swell, and after that, prepared a squash slide. The chromosomes easily spread out on the slide, and could be readily and accurately counted.

The importance of Tjio's study was not the establishment of the chromosome number, but rather the presentation to biologists of a sophisticated and yet simple means of studying the morphology of individual chromosomes. Cytologists could construct an **ideogram,** a diagram of the morphology of individual chromosomes in human cells (see Fig. V-6). Human chromosomes were identified, and duly numbered arbitrarily, beginning with the longest as number 1. Man has twenty-two pairs of chromosomes that are called **autosomes,** and one pair of **sex chromosomes.** Females always contain two **X** chromosomes as their twenty-third pair. Males contain one **X** and one **Y** chromosome as their twenty-third pair. Thus, man has 46 chromosomes.

In 1959, two teams of medical researchers discovered a relationship between the number of chromosomes and some well-known medical conditions or syndromes. One team discovered that an individual afflicted with **Kleinfelter's syndrome,** a disease characterized by improper development and function of the gonads (egg- and sperm-producing organs), and features of intersexuality, had a chromosome composition of **AAXXY.** The **A** represents one set of autosomal chromosomes (22), and therefore, this individual had 47 chromosomes. The other team discovered that three boys who were mongoloid idiots (**Down's syndrome**) had 47 chromosomes each. Later studies demonstrated that this syndrome involves the addition of an extra 21st autosome. These children were **AAXY** + 21st, or 47.

Once these cases were discovered, it became important to see

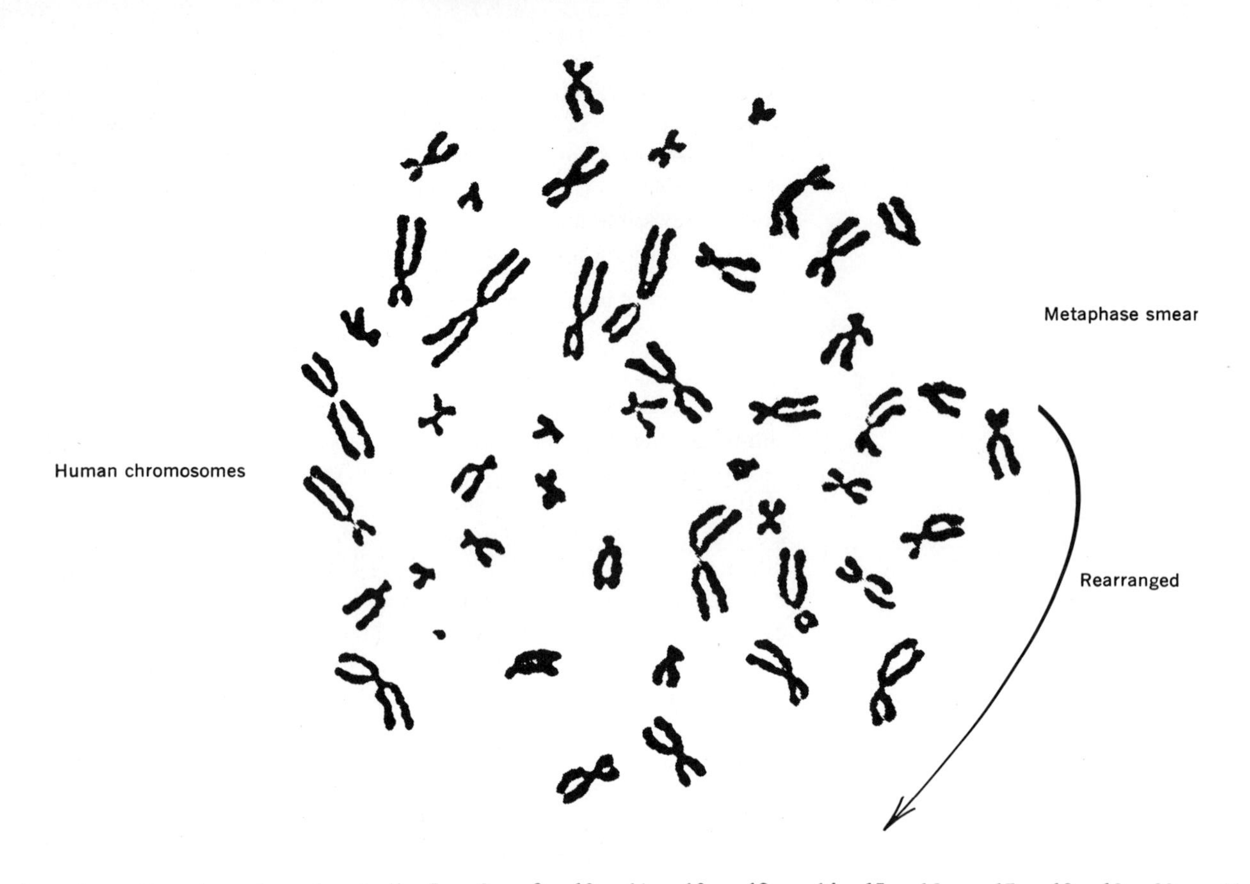

Fig. V-6 Ideogram of human chromosomes and chromosomal aberrations. The photomicrograph can be cut and the chromosome pairs matched and arranged sequentially for easy comparison of normal and abnormal conditions. Intersexuality characteristic of Klinefelter's syndrome is traceable to the presence of an extra X. Individuals manifesting Turner's syndrome (females short in stature often with neck webbing and finger anomalies) are found to have a single X (a genotype of XO). Mongolism or Down's syndrome, seems to be the result of an extra twenty-first chromosome.

whether other chromosome anomalies existed. For example, could not an **AAXXX** individual be possible? Not only was an **AAXXX** person found, but also an **AAXXXX** individual, both of which were fertile but mentally retarded females.

A most valuable variation was found that aided in the elucidation of sex determination in humans. This was an **AAX** (or **AAXO**) individual who was *female.* She had a short, webbed neck and other characteristics that collectively had been known as **Turner's syndrome.** The information available from both Kleinfelter's and Turner's syndromes indicates that the **X** chromosome is responsible for femaleness, and the **Y** chromosome for maleness (see Fig. V-6). Thus, unlike *Drosophila,* in which the determination of sex is in the balance between the female-tending **X** and the male-tending autosomal chromosomes, human sex determination appears to be a function of the number of **X** and **Y** chromosomes. An unbalanced number of **X**'s and **Y**'s results in sex abnormalities. The possible condition **AAY** has not been found, and is assumed to be inviable because the **X** chromosome is necessary for many nonsexual characteristics. This is the case in other organisms, and is assumed to be the same for humans.

As the reports of chromosomal aberrations and associated abnormalities continue to accumulate, several general pictures are gradually emerging. Only two of these generalizations will be mentioned. Two types of leukemia have been linked to anomalies of the 21st chromosome. *Acute* leukemia, which is often associated with mongolism, is thought to be directly related to the trisomic (three instead of two) condition of the 21st chromosome. *Chronic* leukemia is thought to be related to another anomaly of the 21st chromosome. In the case of chronic leukemia, a region of chromosome 21 is missing. The deletion involves perhaps a third of its length, and the shortened chromosome is referred to as a **Philadelphia chromosome.** The presence of a Philadelphia chromosome is thought to cause its cells to multiply faster. Such multiplication is exhibited in the white blood cells, which are present in abnormally large numbers in leukemia patients.

The other general pattern emerging is in regard to the factors leading to mental retardation, which comes from a variety of causes. The additional chromosome 21 causes Down's syndrome, and **AAXXX** and **AAXXXX** females are mentally retarded. Mental retardation can also arise from a trisomic condition of any of the following autosomes: 13, 14, 15, 17, and 18. A reciprocal translocation between a chromosome 13 and 22 also causes mental defects. Man's normal intelligence seems to be based on many different genes, each in appropriate diploid balances, and in the relationship of these genes to other genes in the chromosomes.

One of the more startling results of chromosome studies has been the recent discovery of the **XYY** male (**22AA, 1X, 2Y**). The **XYY** condition appears to be associated with antisocial tendencies, and based on analyses of several thousand individuals, the incidence of the **XYY** condition in prison inmates is remarkably (90%) higher than that found in the typical population (1 in 500). There appears to be a similarity in the types of crimes committed by those convicts with an **XYY** condition. These crimes are generally more brutal in nature, and lack what is regarded as normal compassion associated with social beings. The sociological, psychological, and legal implications of this discovery are far-reaching.

MAN'S GENETIC CONDITION

The knowledge that certain traits in humans are hereditary and the knowledge of the patterns of heredity of these traits have led many geneticists to ponder the state of man's genetic condition. For centuries ethical and legal taboos attempted to discourage inbreeding, that is, mating between closely related individuals. Because of the social structure during past centuries, in which man lived in small communities, it was highly unlikely that one could find an acceptable mate who was not a relative. Therefore, these taboos were extended to cover only individuals that were more closely related than first cousins.

As society became more mobile, outbreeding became the major pattern of mating behavior. Records indicate that before the turn of the century, about 10% of the marriages in the United States were between first cousins. Less than 1% of the marriages today are of this type. Today one makes his or her selection not from a few choices but, especially within college communities, from among twenty or thirty acceptable acquaintances.

As a result of an ever-intensifying outbreeding pattern, many of the deleterious genes once expressed in individuals who bore them in their homozygous recessive condition, and thus died early, or were undesirable mates, now remain underground as hidden recessives. Underground they reside unselected and proliferating through the population. Because of their hidden nature, and the continual outbreeding pattern, more individuals (both numerically and in percentages) in succeeding generations will contain these genes. With every generation, the chance increases for many of these recessive genes to combine and become expressed.

Another factor that has aided in the perpetuation of deleterious genes in the population has been the advances in medical knowledge. Individuals with certain genetic defects often died early. With present-day medical care, many individuals with such defects live to a reproductive age, passing their deleterious genes into the next generation. Of course, medicine is a two-sided coin. The very research that resulted in the perpetuation of deleterious genes will undoubtedly be redirected to alleviate the suffering when these recessive genes reach sufficiently high frequencies to produce many defective individuals.

Yet another factor of grave concern is the number of new deleterious genes that have been added to the population in recent years from the application of atomic energy, especially in bomb testing. Radiation has caused an increased mutation rate in that additional deleterious genes arise in all the populations of the world.

Genetic Counseling

Out of the deep concern of informed individuals for the genetic state of the human species, centers have been established for genetic counseling. These centers have been supported by funds from a number of private and federal foundations, and are staffed by human geneticists.

It is the hope of these concerned individuals that people who know or suspect that a deleterious factor is being perpetuated in their families will seek genetic counseling. At the present time, the majority of counseling cases are parents with a defective child, who want to know whether further children would be similarly afflicted. Some informed couples with some common problem in their backgrounds come to genetic counselors to help them decide whether they should have children.

In most cases, the counselor can inform the couple whether the trait is inherited. With a knowledge of their family backgrounds, he can ascertain the probability of having a child with a certain hereditary defect. Finally, he will offer some advice. In some cases, hospital tests will be made to aid the counselor in determining certain conditions. For instance, Down's syndrome appears to run in some families. It seems that the 21st chromosome has a tendency to undergo an extra replication in these families. When individuals from such families are tested, it is often found that some of their cells contain 47 chromosomes, and others, the normal 46 complement. These individuals are **mosaics,** and although they themselves may be perfectly normal, the chance of producing a gamete with an extra 21st chromosome is greater than in individuals in the general population. A chromosome count would detect the mosaic feature, and a genetics counselor would be able to advise on the basis of the findings.

Biochemical analysis is another type of test that hopefully will be used in the near future when warranted by the family history. The heterozygous condition detected by an abnormally high concentration of certain compounds in the urine or blood provides the genetics counselor with the information as to whether an individual is heterozygous or homozygous recessive.

In about 50% of the cases handled by genetics counselors, the couples do not follow the advice given them.

A list of genetics centers is given in Table V-5. The services are free.

Eugenic Program Solutions

Many thinkers down through the ages have been concerned about the preservation of the more desirable human traits, and have considered the mating pattern of man as a source of future destruction. In the late nineteenth century, this social concern about the genetic state of mankind became embodied in the **eugenics movement.**

The eugenics movement, which at various times has advocated selective human mating, has been greeted with great suspicion by the

TABLE V-5 LOCATION, NAME OF INSTITUTION, AND PRINCIPAL COUNSELOR OF TWENTY-EIGHT GENETICS CENTERS[a]

Location	Institution	Counselor
Berkeley, California	University of California	Dr. Curt Stern
Los Angeles, California	Los Angeles Medical Center	Dr. Stanley Wright
Seattle, Washington	Department of Medicine University of Washington	Dr. Arnold Motulsky
Edmonton, Alberta	Heredity Counseling Service University of Alberta	Dr. Margaret W. Thompson
Tempe, Arizona	Arizona State University	Dr. C. M. Woolf
Austin, Texas	The Genetics Foundation University of Texas	Dr. C. P. Oliver
Norman, Oklahoma	Department of Zoological Science University of Oklahoma	Dr. P. R. David
Winnipeg, Manitoba	Hospital for Sick Children	Dr. Irene Uchida
Minneapolis, Minnesota	University of Minnesota	Dr. S. C. Reed
Minneapolis, Minnesota	Human Genetics Unit State Board of Health	Dr. L. E. Schacht
Rochester, Minnesota	Mayo Clinic	Dr. J. S. Pearson
New Orleans, Louisiana	Genetics Counseling Service Tulane University	Dr. H. W. Kloepfer
Madison, Wisconsin	Department of Medical Genetics University of Wisconsin	Dr. J. F. Crow
Chicago, Illinois	Children's Memorial Hospital	Dr. David Y-Y. Hsia
Ann Arbor, Michigan	The Heredity Clinic	Dr. J. V. Neel
East Lansing, Michigan	Department of Zoology Michigan State University	Dr. J. V. Higgins
Cleveland, Ohio	Department of Biology Western Reserve University	Dr. A. G. Steinberg
Winston-Salem, North Carolina	Bowman Gray School of Medicine	Dr. C. N. Herndon
Toronto, Canada	Hospital for Sick Children	Dr. N. F. Walker
Charlottsville, Virginia	School of Medicine University of Virginia	Dr. R. F. Shaw
Washington, D. C.	Genetics Counseling Research Center	Dr. N. C. Myrianthopoulos
	George Washington University Hospital	Dr. V. A. McKusick
New York, New York	New York State Psychiatric Institute	Dr. F. J. Kallmann
New York, New York	Albert Einstein College of Medicine	Dr. S. G. Waelsch
	Yeshiva University	Dr. Helen Ranney
	Rockefeller University	Dr. A. G. Bearn
Montreal, Quebec	Children's Memorial Hospital	Dr. F. C. Fraser
Providence, Rhode Island	Department of Biology	Dr. C. W. Hagy
Boston, Massachusetts	Department of Immunochemistry Boston University Medical School	Dr. W. C. Boyd

[a] After Reed.

public. To many people, the word "eugenics" has a negative connotation because in the past, attempts were made to relate the science of human heredity to a social movement, despite the fact that few scientifically valid facts were available. Throughout the years, eugenics has been misguided by personal prejudices and pseudoscientific ideas.

With the present state of knowledge of human heredity, and the feasibility of scientifically valid eugenics programs, many feel that eugenics must lose its negative public image. Numerous biologists, including some of the most noted and learned geneticists, feel that a comprehensive eugenics program is the only answer to the problem of the future genetic state of man.

There is, however, little agreement on the type of eugenics program that is required. Some conservative biologists advocate that the primary objective of a eugenics program should be the elimination of deleterious genes from the population, so that future generations would be free from the sufferings brought about by the excessive accumulation of these genes. Their proposal is for the widespread availability of genetic counseling centers, and the *voluntary* acceptance by all members of the population of the importance for counseling before they marry.

This voluntary program would be promulgated by making the public aware of what may be termed a "statistical sense of social responsibility." What is meant by this phrase is that citizens would be educated to realize the importance of their actions with respect to the entire population. If man is to eliminate much of this endless suffering of hereditary ailments, there must be a willingness on everyone's part to participate in the program. When an individual discovers that he or she harbors deleterious genes, he or she must be willing to forego the natural desire to have children. The sacrifice of one or a few individuals would have little effect. Such a eugenics program would only be justified if everyone was willing to participate, regardless of what might be found lurking in the family closet.

The genetic counseling for such a program would have to be greatly expanded. Presently genetics counselors advise on the basis of the type of offspring a given couple may produce, not on the basis of the possible types of grandchildren or great-grandchildren. Thus recessive genes tend to retain their clandestine position. The only way a program of the type outlined above could be truly successful would be if people were willing to forego having children not only for their children's sake, but also for the sake of the generation to which their grandchildren or great-grandchildren might have belonged. The major question is then, could man develop a sense of social responsibility toward his grandchildren when such difficulty has been encountered in his developing a sense of direct responsibility toward his own children (as evidenced by the 50% who gamble on the advice from the genetics counselor)? Can a populace be educated to put their statistical sense of social responsibility before their concern for self-gratification?

Many biologists do not see a widespread genetics-counseling service as the answer to the problem. Some now foresee the solution through a utilization of the information being amassed in the field of molecular biology. Molecular biologists have recently been able to produce synthetic DNA capable of infecting bacterial cells. It may soon be possible to synthesize a DNA that would enter a bacterial cell and act as a lysogenic virus by being incorporated into the host DNA. Such a DNA segment might impart a resistance, as lysogenic viruses do, to the entrance of other viruses. This cell would be forever free of viral infection, and since the synthetic DNA segment is part of the chromosome, the resistance would be inherited. Subsequent research would lead to the successful synthesis of a DNA segment capable of infecting human cells. Could man become forever free of all viral diseases?

Suppose such a synthetic DNA segment could be programmed to become lytic only in cancerous cells. This same segment that imparts a resistance to viral infections would be activated only in the presence of malignant cells, and would destroy the malignancy completely. Man would be free of cancer.

Suppose researchers found the codings to produce such a synthetic DNA segment that carried information for synthesizing certain amino acids that man must now take in with his diet. Upon incorporation, man would become less dependent on his environment. Segments might be synthesized to increase intelligence, physical prowess, mental stability, memory, and on and on. Man could remake himself with the flair of a fashion designer. And because it is the DNA that would be altered, the changes would be immediate and permanent.

At the rate of advance of molecular biology, it is expected that by 1973 man will have made and incorporated a totally synthetic gene into bacteria, and by 1995, the same will be accomplished for man (at least with respect to human-cell cultures). The area of molecular biology is developing so fast that one is too busy keeping abreast of the achievements to try to imagine what genetic gymnastics will eventually be possible.

Should A Solution Even Be Sought?

While the problem of the degenerating genetic state of the human species is indisputable, the solutions offered for the improvement of that state are open to much debate. Widespread genetic counseling is criticized on the basis of sacrifice. Why should this generation sacrifice their self-gratification of having children so that some future generation can be more comfortable? Why should a future generation have priority over this one? A future generation will probably have more information on which to base a program, should one become necessary. A voluntary genetics counseling program to alleviate the suffering of individuals in this generation is all that should be thought of at this time.

Most of the arguments against eugenics programs are leveled against those programs that would grossly alter the human species. One such criticism has been in regard to the choice of "good" and "bad" traits. Those traits that were considered admirable a century ago are far from those that we find the crucial traits for perpetuation. Will not the future generations see still other traits as worthy of perpetuation? If an extreme universal goal is instituted in the near future, subsequent generations may be without the very traits they may find necessary for further improvement of the species, or perhaps even for species survival.

Many other criticisms have been leveled against the consequences of extreme programs. One of the strongest emotionally based arguments involves the procedures that might be involved in maintaining a balanced society, should severe alterations of the species be made by artificial means. Not all individuals could have the more desirable traits. There would have to be some workers talented in given areas, and some unskilled labor. Who is to decide which individuals are to be restricted to subservience by their genetic limitations? The emotional response to the society of Huxley's *Brave New World* is quite strong.

The answer given to the above argument is that an artificial program does not have to imply the creation of a genetically oriented society with caste lines between the workers and intelligentsia. The work would all be done by computerized systems and man, in his new intelligent form, would be part of a homogeneously intelligent population. Even this rebuttal has come under fire. A homogenous population of intelligent beings would greatly reduce the variation in the species. Subsequent selection pressures might obliterate the entire species.

The counterargument against this proposal is that man would control his environment so completely that natural evolutionary forces acting as selection pressures would cease to operate as such. The basic problem, then, is, can man gain such complete control over his environment, and if he can, will he be able to control his enthusiasm to use his knowledge for changing the human species until he is sure of the consequences?

CHAPTER 15

SPECIATION—THE FORMATION OF NEW SPECIES

Most of the two-and-one-half million species that inhabit the Earth today are less than twenty-five million years old. In terms of geologic time, this is very short. Estimates based on fossil evidence indicate that those species existing today are less than 0.01% of all the species that have ever inhabited the Earth. This means that approximately 99.99% of all the species ever present on the Earth are now extinct. It appears then that (1) a species life is short, and (2) the ultimate fate of all species is extinction.

Despite the extinction of approximately 99.99% of all species ever present, the Earth is carpeted with a rich variety of living forms. This is possible because species are not static, immutable groups of organisms. A species maintains a dynamic equilibrium with its internal and external environment.* Shifts in the environment result in modifications within species. The inherent potential for change within a species makes simultaneous species extinction and species formation (*speciation*) possible.

The discussion of speciation centers on the mechanisms through which variation can be translated into species change, rather than on individual cases of speciation. This chapter, then, is a discussion of the modern concept of evolution, the elaboration of, and mechanisms for, the Darwinian concept of speciation.

*Man designates a species by a static identification in the form of a name. But a species name implies no more static a condition than does the name given to an individual human. Joe Smith is a static name. But Joe Smith, the individual, is far from static. He was one type of individual when he was a year old, but quite a different type of individual at ten, twenty, or fifty years of age. A species of bird may have been given a name three hundred years ago. The birds that bear that name today may have a slightly different coloration or flight pattern, or may eat different types of berries or insects than members bearing the same species name three hundred years ago. These alterations may be so slight that they have passed undetected. The analogy of an individual and a species name is meant only to relate to the fact of change, *not* the cause of change. An individual changes in the course of a programmed developmental pattern; a species changes in response to the alterations of the environmental equilibrium.

Dispersive Versus Phyletic Speciation

Phyletic speciation refers to speciation by replacement. A group of organisms that is more adapted to the environment replaces members of the same species that are less adapted. Dispersive speciation refers to speciation by the simultaneous origin of two groups that are significantly different in certain physiological and morphological factors. The terms "dispersive" and "phyletic" are convenient words to describe general patterns. Often, however, individual cases of speciations are not easily pigeonholed into these two categories. For instance, species *A* may give rise to species *B* and *C*. By definition, this would be dispersive speciation. But the same environmental situation once occupied by species A is now occupied by a similar species B. Species C, on the other hand, exists under different environmental conditions (in a different geographical location, at a different level in a pond, growing earlier or later in a season, etc). The replacement of *A* by *B* approximates phyletic speciation, while the genesis of C is more like dispersive speciation.

Another possible occurrence is that while the two groups, B and C, are accumulating adaptations, but before they become totally isolated, hybridization between them might occur. Hybridization would bring about species BC, which could directly replace A in the original environment. Thus events that began as dispersive speciation might change to phyletic speciation (see Fig. 15-1). Dispersive and phyletic speciation, therefore, are not mutually exclusive.

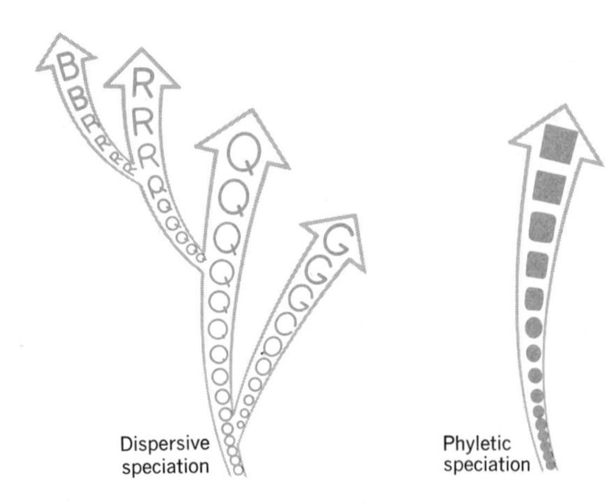

Fig. 15-1 Phyletic and dispersive speciation.

The Concepts of Species and Population

Although a more formal definition of a species will be developed in the chapter on classification of organisms (biosystematics), it is necessary to provide an operational definition for use in the discussions of this chapter. *A species is a group of organisms that have a common morphology, physiology, and behavior, that are found in a given set of environmental circumstances (often in one geographical location), and that are capable of mating with one another to produce viable and fertile offspring.*

A species may exist over a wide geographic range. Because of the topological variation of the environment, it cannot be uniformly distributed over this range, and must be divided into groups (see Fig. 15-2).

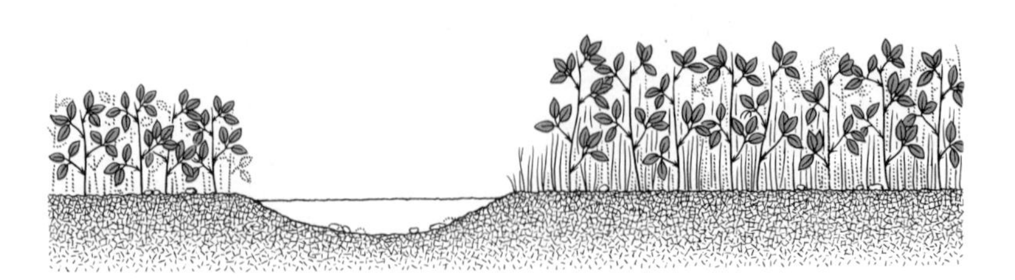
Fig. 15-2 The concept of a population—genetically distinct individuals of the same species exist over a geographic range.

Each group is a **population.** A population is, then, a group that is partially isolated from other groups of the same species. Because of their proximity, members of a population breed more frequently with each other than with other populations of the same species.

Often each population is separated into a number of smaller groups or **demes.** For instance, the fern, *Osmundum rubrum,* will appear in dense patches of several hundred plants. Each of these patches is a deme. The patches may be separated by any distance from a hundred feet to a mile or more, and the sum total of all the demes in one region is a population. The deme and the population, not the entire species, are the units of change in dispersive evolution.

Gene Frequency

The new perspective of the population rather than the species or the individual member of the species as the unit of evolutionary change fostered the development of the field of **population genetics,** which deals with the fate of genes in populations. One of the most important concepts to come from population genetic studies is that of **gene frequency.** Because a population constitutes a breeding unit, the sum total of all the genes of all the members of the population can be thought of as forming a **gene pool.** Members of the next generation derive their genotypes from the gene pool of the parental group (see Fig. 15-3).

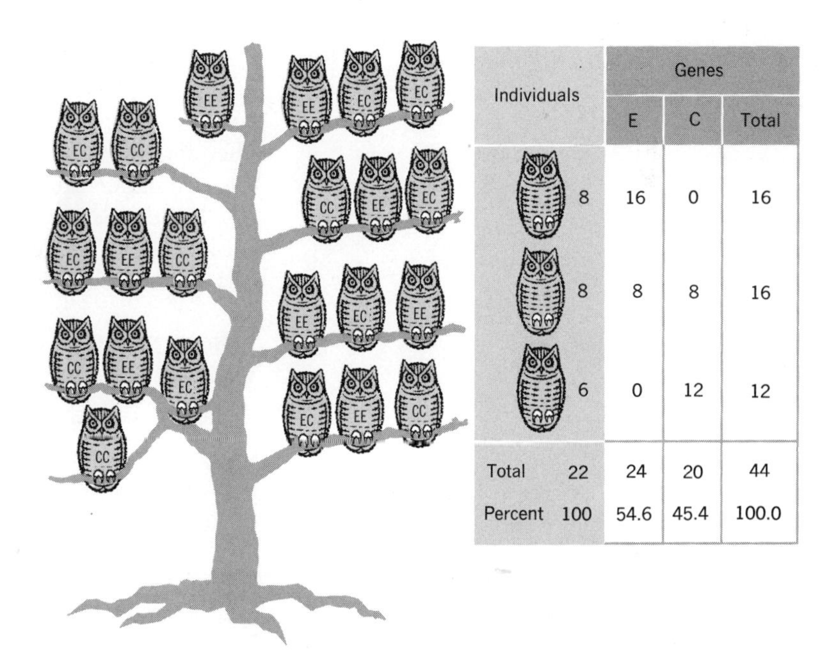

Individuals	Genes		
	E	C	Total
8	16	0	16
8	8	8	16
6	0	12	12
Total 22	24	20	44
Percent 100	54.6	45.4	100.0

Fig. 15-3 A gene pool.

It should be noted that, at least theoretically, a common gene pool exists for all populations of a species. That members of species draw only from the gene pool of their own or proximal populations is only because of their geographical locations.

By viewing the genetics of a population in terms of a gene pool, one can determine the ratio of any gene to its allele(s) within a population. For instance, in an allelic series **A** and **a,** the genotypic combinations of 100 randomly selected individuals might occur as follows:

Genotype	*Number of Individuals*	*Genes Contributed to Gene Pool*
AA	81	162
Aa	18	36
aa	1	2
	100	200

Assuming that these 100 individuals are diploid, there is a total of 200 genes (both **A** or **a**). The gene frequency of this sample can now be determined.

$$\frac{\text{No. of } \mathbf{A} \text{ genes in gene pool}}{\text{Total no. of genes in gene pool for that allelic series}} \times 100 = \text{gene frequency of } \mathbf{A}$$

$$\frac{162 + 18}{200} \times 100 = 90\%$$

$$\frac{\text{No. of } \mathbf{a} \text{ genes in gene pool}}{\text{Total no. of genes in gene pool } (\mathbf{A} \text{ or } \mathbf{a})} \times 100 = \text{gene frequency of } \mathbf{a}$$

$$\frac{2 + 18}{200} \times 100 = 10\%$$

Therefore, 90% of the genes in the gene pool of this allelic series in this population sample for the given trait are **A,** and 10% are **a.**

Notice that in this calculation it is important to be able to distinguish between the homozygous dominant and the heterozygous individuals, as the values for both must be treated separately.

THE SYNTHESIS OF MENDELIAN HEREDITY AND POPULATION BEHAVIOR

Shortly after the rediscovery of Mendel's laws of inheritance in 1901, several workers applied these laws of heredity to natural populations. Some of these workers asserted an erroneous view that dominant genes in a population would tend to take over the gene pool. The basis for this assertion was the Mendelian cross between two heterozygous individuals (**Aa** and **Aa**), which would result in a 3:1 phenotypic ratio of **A**:**a** type individuals. Hardy and Weinberg independently reported their thoughts in 1906 and 1907, respectively, as a mathematical pattern of gene inheritance within a population. Their collective efforts are known as the **Hardy-Weinberg Equilibrium,** a law that raised serious questions with respect to the importance of selection pressures in evolution.

TABLE 15-1 RANDOM MATING POSSIBILITIES OF A POPULATION WITH A GENE POOL 50% A AND 50% a

Crosses	Chance of Cross Occurring = Fraction of One Individual Present in Population × Fraction of Other Individual Present in Population
AA × AA	$\frac{1}{4} \times \frac{1}{4} = \frac{1}{16}$
AA × Aa	$\frac{1}{4} \times \frac{1}{2} = \frac{2}{16}$
AA × aa	$\frac{1}{4} \times \frac{1}{4} = \frac{1}{16}$
Aa × Aa	$\frac{1}{2} \times \frac{1}{2} = \frac{4}{16}$
Aa × AA	$\frac{1}{2} \times \frac{1}{4} = \frac{2}{16}$
Aa × aa	$\frac{1}{2} \times \frac{1}{4} = \frac{2}{16}$
aa × aa	$\frac{1}{4} \times \frac{1}{4} = \frac{1}{16}$
aa × AA	$\frac{1}{4} \times \frac{1}{4} = \frac{1}{16}$
aa × Aa	$\frac{1}{4} \times \frac{1}{2} = \frac{2}{16}$
	$\frac{16}{16} = 1.00$ (the total population)

TABLE 15-2 RESULTS OF MATING DESCRIBED IN TABLE 15-1

Crosses	Chance of Occurrence	Chance of Offspring Being: AA	Aa	aa
AA × AA	$\frac{1}{16}$	$\frac{1}{16}$		
AA × Aa	$\frac{2}{16}$	$\frac{1}{16}$	$\frac{1}{16}$	
AA × aa	$\frac{1}{16}$		$\frac{1}{16}$	
Aa × Aa	$\frac{4}{16}$	$\frac{1}{16}$	$\frac{2}{16}$	$\frac{1}{16}$
Aa × AA	$\frac{2}{16}$	$\frac{1}{16}$	$\frac{1}{16}$	
Aa × aa	$\frac{2}{16}$		$\frac{1}{16}$	$\frac{1}{16}$
aa × aa	$\frac{1}{16}$			$\frac{1}{16}$
aa × AA	$\frac{1}{16}$		$\frac{1}{16}$	
aa × Aa	$\frac{2}{16}$		$\frac{1}{16}$	$\frac{1}{16}$
		$\frac{4}{16}$	$\frac{8}{16}$	$\frac{4}{16}$
	or	$\frac{1}{4}$	$\frac{1}{2}$	$\frac{1}{4}$
	or	25%	50%	25%

The Hardy-Weinberg Equilibrium

The Hardy-Weinberg equilibrium consists of a simple mathematical conclusion of Mendelian inheritance in a population. The first aspect of this interpretation was their insight into the results of a monohybrid Mendelian cross. A cross between two different homozygous parents (**AA** and **aa**) yields an $\mathbf{F_1}$ generation in which all the individuals are heterozygous (**Aa**). If these offspring are mated with each other, the $\mathbf{F_2}$ generation will have three kinds of genotypes, **AA, Aa,** and **aa** (see Fig. 15-3). The ratio of the genotypes will be 1**AA**:2**Aa**:1**aa** and, if one allele is dominant, the phenotypic ratio will be 3**A**:1**a** individuals.

What would happen in the next generation if the individuals resulting from the above cross were to mate randomly? Given such a situation, the following crosses can occur.

female parent	×	*male parent*
AA	×	**AA**
AA	×	**Aa**
AA	×	**aa**
Aa	×	**Aa**
Aa	×	**Aa**
Aa	×	**aa**
aa	×	**aa**
aa	×	**AA**
aa	×	**Aa**

Since one fourth of the population are **AA** individuals, one-half of the population are **Aa** individuals, and one-fourth of the population are **aa** individuals, the matings that will take place are related to the frequency of each type of individual in the population (see Table 15-1). One can now determine the offspring types from the possible crosses in terms of their genotypes. For instance, from the cross **AA** × **Aa,** which has $\frac{4}{16}$ chance of occurring, the offspring produced will be in a ratio of 1**AA**:2**Aa**:1**aa.** Hence, $\frac{1}{16}$ of the offspring will be **AA,** $\frac{2}{16}$ will be **Aa,** and $\frac{1}{16}$ will be **aa** (see Table 15-2). Therefore, on a population level, random mating among these individuals will give rise to a 1:2:1 ratio of the **AA, Aa,** and **aa** genotypes in the next generation. Subsequent generations will manifest exactly the same ratios of individuals. The dominant gene **A** will not become more prevalent in the gene pool!

But real populations never manifest a simple genotypic ratio of 1:2:1, for this implies that each of the two alleles under consideration comprise 50% of the gene pool.

$$\frac{\text{Total gene frequency of } \mathbf{A} \text{ in gene pool}}{\text{Total gene frequency of } \mathbf{A} \textit{ or } \mathbf{a} \text{ in gene pool}} \times 100$$

$$\frac{0.25 + 0.25\ (\mathbf{A} + \mathbf{A} \text{ in } \mathbf{AA} \text{ individuals}) + 0.50\ (\mathbf{A} \text{ present in } \mathbf{Aa} \text{ individuals})}{2.00}$$

$$= 50\% \text{ for the } \mathbf{A} \text{ allele}$$

$$\frac{\text{Total gene frequency of } \mathbf{a} \text{ in gene pool}}{\text{Total gene frequency of } \mathbf{A} \text{ and } \mathbf{a} \text{ in gene pool}} \times 100$$

$$\frac{0.25 + 0.25 + 0.50}{2.00} \times 100 = 50\% \ \mathbf{a} \text{ in gene pool}$$

The above calculations assume that there are 100 individuals and, therefore, 200 genes (200% or 2.00) in the gene pool.

It would be rare that a natural population would have a 1:1 ratio for the gene frequencies of an allelic series. In fact, it is much more common that a ratio of 9:1 or greater exists between the gene frequencies of alleles. What would happen to the gene frequencies in the next generation when one starts with a gene frequency for **A** of 90% and a gene frequency for **a** of 10%?

To cope with this problem, Hardy and Weinberg returned to Mendelian genetics. They realized that the simple relationship between gene and genotypic frequencies follows the binomial theorem.

$$(\mathbf{A} + \mathbf{a})^2 = \mathbf{A}^2 + 2\mathbf{Aa} + \mathbf{a}^2$$

where the exponent 2 is used to indicate the presence of the diploid state. The equation can be made more genetically expressive by modifying it to

$$(\mathbf{A} + \mathbf{a})^2 + \mathbf{AA}:2\mathbf{Aa}:\mathbf{aa}$$

But this is the general equation that assumes $\mathbf{A} = \mathbf{a} = \frac{1}{2}$,

$$\left(\frac{1}{2} + \frac{1}{2}\right)^2 = \frac{1}{4} + \frac{1}{2} + \frac{1}{4}$$

and is only valid when the gene frequencies of both **A** and **a** are 50%. The equation can be generalized for all gene frequencies by multiplying **A** and **a** by appropriate constants, **p** and **q**. The equation now becomes

$$(p\mathbf{A} + q\mathbf{a})^2 = p^2\mathbf{AA}{:}2pq\ \mathbf{Aa}{:}q^2\ \mathbf{aa} \qquad p + q + 1.00$$

When the frequency of the **A** allele is known to be 90%, the frequency of the genotypic classes may be predicted in the following way:

$$\text{since} \quad p + q = 1.00, \quad \text{and} \quad p = 0.9\ (90\% \text{ of } 1.00 = 0.9)$$
$$\text{then} \quad q = 1.00 - 0.9 = 0.10$$

These values may then be added to the equation

$$(0.9\mathbf{A} + 0.1\mathbf{a})^2 = (0.9)^2\ \mathbf{AA}:2(0.9)\,(0.1)\mathbf{Aa}:(0.1)^2\mathbf{aa}$$
$$= 0.81\mathbf{AA}:0.18\mathbf{Aa}:0.01\mathbf{aa}$$

or, in a population sample of 100 individuals, there would be 81 **AA,** 18 **Aa,** and 1 **aa** individuals.

The types of individuals produced in the next generation by such a population can be seen by constructing a composite chart (Table 15-3) listing the possible crosses, the chance of these crosses occurring, and the offspring from these crosses. Even when the **A** allele has a frequency of 0.9 and the **a** allele has a frequency of 0.1, the same frequency ratio will be maintained generation after generation, and the recessive gene will tend to remain in the gene pool.

The Hardy-Weinberg law in its original form, then, states that the gene frequencies of populations will tend to remain stable, generation after generation.*

TABLE 15-3 RESULTS OF A RANDOM MATING IN A POPULATION WITH A GENE POOL 90% A AND 10% a

Crosses	Chances of Cross Occurring	Possible Offspring		
		AA	Aa	aa
AA × AA	0.81 × 0.81 = .656	0.656		
AA × Aa	0.81 × 0.18 = .146	0.073	0.073	
AA × aa	0.81 × 0.01 = .008		0.008	
Aa × Aa	0.18 × 0.18 = .032	0.008	0.016	0.008
Aa × AA	0.18 × 0.81 = .146	0.073	0.073	
Aa × aa	0.18 × 0.01 = .0018		0.0009	0.0009
aa × aa	0.01 × 0.01 = .0001			0.0001
aa × AA	0.01 × 0.81 = .008		0.008	
aa × Aa	0.01 × 0.18 = .0018		0.0009	0.0009
		0.810	0.1800	0.0100
		or 81%	18%	1%

*Another contribution of Hardy and Weinberg, arising from their insight into the 1:2:1 genetic ratio, is a binomial expression, the calculation of gene frequencies when the homozygous dominant individual is indistinguishable from the heterozygous individual (only the homozygous recessive class is known). For instance, what would be the gene frequency of the alleles be if the frequency of the **aa** class is found to be 4%?

$q^2 = 0.04$

$\sqrt{q^2} = q$, therefore $\sqrt{0.04} = q = 0.20$

if $q = 0.20$ and $1 - q = p$, then $1.00 - 0.20 = 0.80 = p$

The gene frequencies are $p = 0.80$; $q = 0.20$

The genotypic frequencies in such a population would be **AA** = 64% (0.8 × 0.8 × 100 = 64%), **Aa** = 32% (0.8 × 0.4 × 100 = 32%), and **aa** = 4% (given).

Thus, one can determine the gene frequencies of two alleles by knowing only the genotypic frequency of the homozygous recessive class.

Fisher, Haldane, and Population Genetics

The Hardy-Weinberg statement that gene frequencies of populations will remain stable generation after generation would appear to be distinctly contrary to the concept of evolution. If a population is to undergo change that leads to the formation of a new species, the gene frequencies must undergo change. If a species maintains the same variation from generation to generation, it could not give rise to new species.

Although both Hardy and Weinberg were aware of the fact that their conclusions were based on certain assumptions, they had never succinctly stated these assumptions as conditions under which the law would be valid. Fisher, and then Haldane, in reviews of the Hardy-Weinberg law, stated that it was imperative that these assumptions be given as conditions. These conditional assumptions are as follows:

1. There is no selective advantage of one allele over the other.
2. The population size is infinitely large and its members are homogenously dispersed.
3. Mutation rates between alleles are equal (**A** ⟶ **a** at the same rate as **a** ⟶ **A**).
4. Migration does not take place.
5. Mating is completely random (panmixis).

The Hardy-Weinberg equilibrium was thus amended and presently states that gene frequencies will remain stable *only when the five conditions mentioned above are met.*

GENE FREQUENCY SHIFTS UNDER NATURAL CONDITIONS

Fisher and Haldane recognized that the Hardy-Weinberg equilibrium rarely operates in natural populations because the five theorical conditions on which it is based are never encountered. They further recognized that the differences between populations, the first step in speciation, must be based on quantitative deviations of those factors contributing to the stability of a species.

By analyzing the five conditions in terms of the effect a change would have on the gene frequency, Fisher developed a series of treatments that show how the separation of populations within a single species can begin.

Consequences of One Allele Having A Selective Advantage Over Another

The almost infinite variety of selective pressures to which the individuals of populations are exposed (food procurement, obtaining mating partners, etc.) ultimately affect the reproductive rate. Hence, differences in reproductive rates indicate the presence of selective pressures in a population.

For instance, if 100 parents of genotypes **AA** or **Aa** can produce 100 offspring (an average of two offspring for each set of parents), but 100 parents of genotype **aa** can produce only 95 offspring (an average of 1.90 offspring for each set of parents), a selection value can be attributed to each class. For both **AA** and **Aa** genotypes, the selection-pressure values would be 0.00; while for the **aa** genotype, the selection-pressure value would be 0.05. The reproductive rates associated with each group would be 1.00 for both the **AA** and **Aa** genotypes, and 0.95 for the **aa** genotype.

Genotype	*Number of Parents*	*Number of Offspring*	*Selection* (s)	*Reproductivity* (k)
AA	100	100	0.00	1.00
Aa	100	100	0.00	1.00
aa	100	95	0.05	0.95

The reproductivity value (k) can be introduced into the binomial equation as follows:

$$(p_1\mathbf{A} + q_1\mathbf{a})^2 = p^2{}_0\mathbf{AA} : 2p_0q_0\mathbf{Aa} : kq^2{}_0\mathbf{aa}$$

The subscripts refer to the generation; so the equation reads that the frequency of genes **A** and **a** in one generation is equal to the genotypic frequencies of the *prior* generation with the reproductivity factor introduced to each. Since k is 1.00 for **AA** and **Aa,** it is not included for these classes in the equation.

Now, given the gene frequency of **A** and **a,** and the genotypic frequency of **aa** equal to 0.95, the fate of the two genes in the next generation can be determined.

$$\begin{aligned}(p_1\mathbf{A} + q_1\mathbf{a})^2 &= (0.9)^2\mathbf{AA} : 2(0.9)(0.1)\mathbf{Aa} : (0.95)(0.1)^2\mathbf{aa} \\ &= 0.81\mathbf{AA} : 0.18\mathbf{Aa} : 0.095\mathbf{aa}\end{aligned}$$

This, of course, does not add up to one (1.00). The necessary adjustments to make the equation total one can be made by finding the value of q.

$$\begin{aligned}q^2 &= 0.095 \\ q &= \sqrt{0.095} = 0.092\end{aligned}$$

The value for p may then be calculated as

$$\begin{aligned}p &= 1 - q \\ &= 1 - 0.092 \\ &= 0.908\end{aligned}$$

The remainder of the equation can be determined by calculating the values for p^2 and pq.

For the purposes of this discussion, however, the gene frequencies p and q are sufficient. Notice that the gene frequency of the **a** gene went from a value of 0.1 to a value of 0.092 in one generation, while that of the **A** gene increased from a value of 0.9 to a value of 0.908. This small selection pressure would be sufficient to almost weed out **a** gene from the population after several thousand generations.

The time factor for natural selective pressures to eliminate genes from a population was greatly underestimated by Darwin. It became obvious to his twentieth-century successors that the failure of many of the early workers to substantiate Darwinian evolution lay in their attempts to view some change over a few generations that actually requires the perspective of thousands. A one-gene gain or extinction requires thousands of years, and for a new species to arise, about a hundred thousand to a million years are necessary. Darwin, when confronted with the problem, gave a rough estimate of only a few thousand years for a new species to arise.

Evolutionists recognize four types of selective pressures, **stabilizing, directive, disruptive,** and **homeostatic. Stabilizing** selection tends to trim away much of the genetic variation that appears by mutation and recombination in an already-successful population. For instance, a population of bacteria may have many mutations for resistence to different drugs and viruses that involve biochemical deficiencies. In the absence of the drug or virus from the bacterial culture, the very biochemical deficiencies that make the resistence possible may be deleterious. This may be expressed by a slower growth rate of the mutant cells. In a fast-growing population, such mutant cells would be selected against. This may be seen most readily by assigning values to such mutants. Imagine that the original population of bacteria was of the order of 10^7, and that 10%, or 10^6, of its members were of some mutant kind. The wild-type bacteria divide every 30 minutes, and the mutant-type bacteria divide every hour (see Table 15-4). We may now follow the progress of such a growing culture. Because of the slower growth rate exhibited by such mutants, a selective situation arises that tends to stabilize the population, slowly eliminating the mutant genes.

TABLE 15-4 SELECTION

Minutes of Growth	Wild Type	Mutant	% Population Mutant
30	1.8×10^7	1×10^6	5.5
60	3.6×10^7	2×10^6	5.5
90	7.2×10^7	2×10^6	2.8
120	1.44×10^8	4×10^6	2.8
150	2.88×10^8	4×10^6	1.4
180	5.76×10^8	8×10^6	1.4
210	1.15×10^9	8×10^6	0.70

When a population of bacteria, such as that outlined above, is exposed to a drug or viruses, the selective force changes to one of **directive** selection. Directive selection is constructive in bringing about evolutionary change. If the presence of the drug or virus is enduring, the change may become fixed by the elimination of all the original wild-type bacteria. If the environment changes back after complete selection, reverse mutations will bring about the original gene, and the relative growth rates of the two types will determine the prevalence of each in the population.

The third type of selection pressure is **disruptive** selection. This is multidirectional and supports the increased variation in a population. For instance, the wings of moths and butterflies are subjected to selection pressures that promote a mimicry of the coloration and patterns of flora or insects in the environment (see Fig. 15-4). But a moth's environment may contain many successful plant and insect forms, the mimicry of any one of which would be advantageous to the moth. Thus there are many forms the coloration and wing patterns may take that would be advantageous in an environment, and the variation of the population is increased.

The fourth type of selection is **homeostatic** selection. This involves the retention of deleterious genes in a population because of either the intervention of nonrandom forces or the selective pressure for the heterozygous individual. The medical advances in the past several centuries have constituted nonrandom forces to which the human population has been exposed. These advances have prolonged the lives of

(*a*) ***Diapheromera fermorata*, the walking stick, blends in so completely with the twig on a tree that it is difficult to see even in the photograph.**

(*b*) **Katydid.**

(*c*) ***Iculullia umbratica*, Shark moth.**

Fig. 15-4 Variation of mimicry. [(*a*), D. Muir—National Audubon Society; (*b*), John H. Gerard—National Audubon Society; (*c*), M. W. F. Tweedie—National Audubon Society]

many individuals with hereditary diseases to mating age, thus retaining deleterious genes in the population.

An example of the natural selective pressure favoring the heterozygous individual may be seen in the hereditary disease, sickle-cell anemia. Certain Negro populations have a gene that causes this disease. Homozygous individuals (**ss**) die at an early age (before sexual maturity) because the hemoglobin protein in the red blood cell is defective and the cells collapse. Individuals homozygous for the wild-type allele (**SS**) will have normal hemoglobin.

The malaria parasite, common to tropic areas, is capable of entering the bloodstream and ultimately the red blood cell, where it multiplies and destroys the cell. Homozygous **SS** individuals are susceptible to malaria. The heterozygous conditions (**Ss**) however, imparts a resistance to the parasite. It is thought that the partially altered hemoglobin (see p. 247) increases the viscosity within the red blood cell, producing an unfavorable environment for the multiplication of the protozoan parasite.

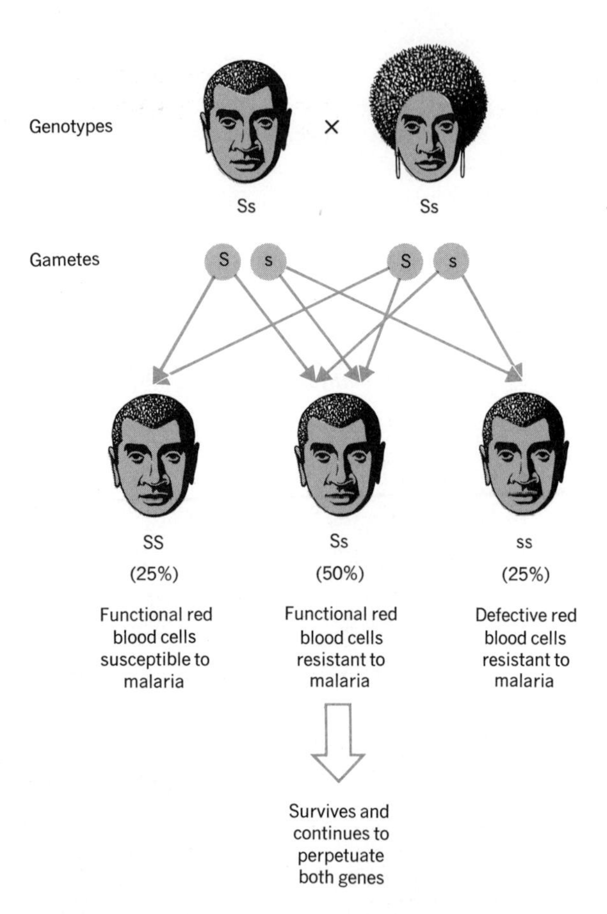

Fig. 15-5 Selection in the sickle-cell anemia condition.

Selection will favor the heterozygous condition, eliminating the other two genotypes at a much higher rate. Most marriages will be between two heterozygous individuals (the **ss** individuals having died before sexual maturity and many of the **SS** individuals having succumbed to malaria), and their offspring will approach the Mendelian ratio of **1SS**:**2Ss**:**1ss** (see Fig. 15-5).

One half of the offspring will be under greater selection pressures, and after ten to fifteen years, during which time the **ss** individuals will be eliminated by sickle-cell anemia, and some of the **SS** individuals by malaria, the heterozygotes can rise far above the initial fifty percent. Here selection maintains a deleterious gene in the population because of the selective advantage of the heterozygous state.

Because of their individual habitats, populations are exposed to different selection pressures. It is conceivable that four different populations of a species could each be exposed to different types of selection pressure (stabilizing, directive, disruptive, and homeostatic). Each specific selection pressure would alter the gene frequencies in the gene pool and lead to slight differences between populations of the same species.

Consequences of a Limited Population Size and Nonrandom Mating

It is impossible to discuss the effects of a limited population size without including a discussion of the consequences of nonrandom mating. These two conditions are inexorably linked together. Random mating, which is the phenomenon by which any individual has the opportunity to mate with any other individual, implies that all possible crosses

AA × **AA**	**Aa** × **Aa**	**aa** × **aa**
AA × **Aa**	**Aa** × **aa**	
AA × **aa**		

occur in frequencies related to the number of each genotype present in the population. This can only be true when a population is infinite in size so that all individuals (regardless of location, age, and mating time) have the same mating opportunities.

In natural situations, where the population size is limited, the last individuals to mate have little selectivity of partners. When an **AA** individual is ready to mate, only **Aa** members of the population may be available. The other **AA** members and **aa** members may have already mated and progressed beyond that stage. So limited population size produces nonrandom mating situations.

Another important aspect of population size that directly relates to a nonrandom mating pattern is the distribution of individuals within the population range. It is impossible to distribute members of the population randomly. Sibs will tend to be reared together and remain in the same locality. Therefore, sibs will tend to mate with one another, and mating becomes nonrandom. This nonrandom mating pattern leads to certain important consequences. When relatives mate, a pattern of partial inbreeding is established. Inbreeding tends to favor homozygosity. This conclusion is illustrated in Fig. 15-6, in which each individual can only mate with itself (**obligate inbreeding**). Close relatives will tend to have similar genotypes and will produce the inbreeding patterns described for obligate inbreeders when they mate together. The result is that more homozygosity will prevail than is expected by the Hardy-Weinberg Law. The consequences of increased homozygosity in severely altering the gene frequencies in the gene pool will be seen more clearly by developing an idea of the fraction of the population members that actually contribute offspring to the next generation.

Generation	Individuals selfed	Percentage		
		AA	Aa	aa
1	Aa × Aa	0	100	0
2	1AA:2Aa:1aa	25	50	25
3	3AA:2Aa:3aa	37.5	25	37.5
4	7AA:2Aa:7aa	43.2	12.5	43.2
5	15AA:2Aa:15aa	46.9	6.25	46.9

Fig. 15-6 The result of obligate inbreeding is increased homozygosity.

Although the average number of offspring per individual may be 1.0, it is unnatural to expect that every parent will have exactly two offspring; instead, some will have none, some one, some two, etc. This kind of distribution is expressed by a formula known as the Poisson distribution

$$\mathbf{P_n = \frac{s^n}{n!} e^{-s}}$$

where **P** is the probability of **n** events occurring when there is an average of **s** number of events for the whole population. The value of **e** is the constant 2.732.

This formula can now be applied to the number of offspring that individuals in the population will give rise to.

Chance of an individual having no offspring

$$P_n = \frac{s^n}{n!} e^{-s} = P_0 = \frac{1^0}{0!}(2.732)^{-1} = \frac{1}{1} \times 0.37 = 0.37 = 37\%$$

Chance of an individual having one offspring

$$P_1 = \frac{s^n}{n!} e^{-s} = P_1 = \frac{1^1}{1!}(2.732)^{-1} = \frac{1}{1} \times 0.37 = 0.37 = 37\%$$

Chance of an individual having two offspring

$$P_2 = \frac{1^2}{2!}(2.732)^{-1} = \frac{1}{2} \times 0.37 = 0.185 = 18.5\%$$

Chance of an individual having three offspring

$$P_3 = \frac{1^3}{3!}(2.732)^{-1} = \frac{1}{6} \times 0.37 = 0.06 = 6\%$$

Chance of an individual having four offspring

$$P_4 = \frac{1^4}{4!}(2.732)^{-1} = \frac{1}{24} \times 0.37 = 0.015 = 1.5\%$$

The equation then states that 37% of the population will have no offspring. If a disproportionate number of individuals in the population are homozygous, it is more reasonable to assume that many homozygous individuals would be included in this 37%. And, if a relatively rare gene is present in such a population, two such genes would most likely be in the homozygous (**aa**) condition. Thus, there is a good chance that two genes could simultaneously be eliminated from the population. If, on the other hand, the rare genes were present in two heterozygous (**Aa**) individuals (which would be favored by random mating), the chance of the **a** gene being retained in the population would double (see Fig. 15-7 and Table 15-5).

TABLE 15-5 CHANCE OF GENE SURVIVAL THROUGH CONSECUTIVE GENERATIONS[a]

Number of replacement copies:	0	1	2	3	4	5	6
Chance for that number:	.37	.37	.18	.06	.015	.004	.0007
	Extinction (37%)	Survival (63%)					

Generation	Extinction	Survival
0		1.00
1	.37	.63
2	.37 + .37(.63) = .60	.63(.63) = .40
3	.60 + .37(.40) = .75	.63(.40) = .25

[a] Assume a gene is replaced by one copy of itself on an average.

Hence, completely independently of selection pressures, limited population size and concomitant nonrandom breeding patterns can bring about alterations in the gene-pool frequencies by (1) fostering the inbreeding patterns that lead to homozygosity and (2) increasing the chance that rare genes will be eliminated from the population because of this homozygosity.

Consequences of Unequal Mutation Rates

The third condition under which a population will tend to remain stable is when the mutation rates of one allele to another are equal. This would insure that despite mutational events, the ratio of the alleles in the gene pool would remain constant. For example, if the mutation

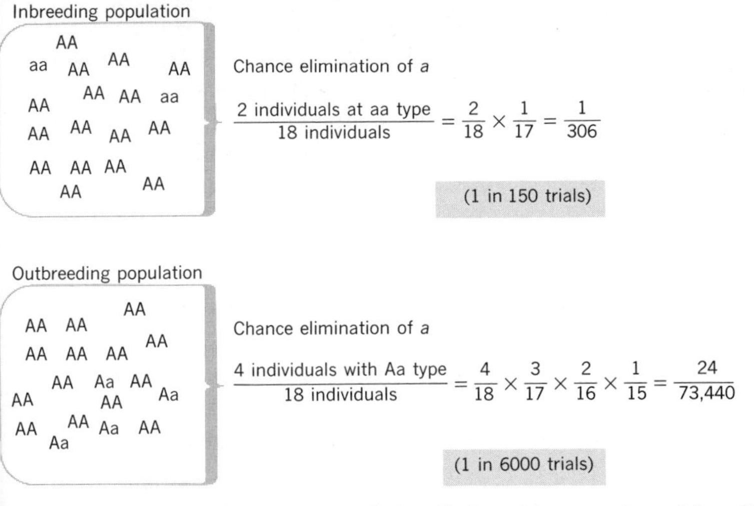

Fig. 15-7 The chance of a mutant gene being eliminated is greater in an inbreeding population.

rate of **A** $\longrightarrow$ **a** is 10^{-6} (one **A** gene in 10^6 mutates spontaneously to the **a** form), and the mutation rate of **a** $\longrightarrow$ **A** is also 10^6, then on the average, for every **A** gene mutating to the **a** allele there will be an **a** gene mutating to the **A** allele. The two events would tend to cancel out one another and the gene-pool frequencies will remain stable.

It has been found that in natural populations the mutations rates are *not* equal. For instance, a deme composed of 50,000 individuals (1,000,000 genes of a given type) might have a gene pool containing 700,000 **A** genes and 300,000 **a** genes. Studies of the next generation might indicate that 70 of the **A** genes had mutated to the **a** allele, while only **3** of the **a** genes had mutated to the **A** allele. The mutation rates will be as follows:

$$\frac{\text{number of mutated genes}}{\text{number of genes from which mutation arose}}$$

$$\frac{70}{700{,}000} = 1 \times 10^{-4}$$

$$\frac{3}{300{,}000} = 1 \times 10^{-5}$$

So the mutation rates of the two alleles differ by a factor of ten.

This difference in mutation rate would cause a small change in the gene pool in the course of one generation. The frequency of the **A** allele would go from 0.70 to 0.69999, while the frequency of the **a** allele would go from 0.30000 to 0.30001.

Consequences of Unequal Migration

Stability of a population or deme demands that it be an isolated genetic unit. Only the individuals present in the population contribute to the genetic makeup of the next generation. Either (1) there must be no introduction of genetic material from members of other populations or demes, or (2) the subgene pool of the individuals migrating to the population must be the same as that of the individuals emigrating from the population.

It would be totally unreasonable to expect that either condition would be met by natural populations. In plant populations, wind- or insect-carried pollen is spread far beyond the bonds of a single deme. In animal populations, the ability to move permits migration between demes. The chance that the pollen entering a given deme contain the same genes in the same ratios as that which has left the deme is highly unlikely.

For example, imagine a deme gene pool made up of 1,000,000 genes, of which 700,000 are **A** and 300,000 are **a.** In the course of a migration to the deme, 40 new **A** genes and 20 new **a** genes are added to the gene pool. In the meantime, 10 **A** genes and 20 **a** genes are lost to the gene pool by emigration. What effect will this have on the ratio of **A** to **a** in the deme gene pool?

New genes:

$$\frac{\text{number of new genes of a given type}}{\text{total number of that type of gene in the gene pool}}$$

$$\frac{40}{10^6} = 0.00004 = 0.004\% \text{ new } \mathbf{A} \text{ genes}$$

$$\frac{20}{10^6} = 0.00002 \text{ or } 0.002\% \text{ new } \mathbf{a} \text{ genes}$$

Loss of genes:

$$\frac{\text{number of lost genes of a given type}}{\text{total number of that type of gene in the gene pool}}$$

$$\frac{10}{10^6} = 0.00001 \text{ or } 0.001\%$$

$$\frac{20}{10^6} = 0.00002 \text{ or } 0.002\%$$

Change in gene pool ratios:

Change in gene pool ratios = original ratio + change

A gene:

$$0.70 + 0.0004 - 0.0001 = 0.70003$$

a gene:

$$0.30 - 0.0002 + 0.0002 = 0.30000$$

Notice that the gene frequencies do not equal 1. This is because in the migration, 60 new genes were added to the gene pool, while only 30 were lost by the emigration. The gene pool is no longer 1,000,000 genes, but 1,000,030 genes.

Migration has an effect similar to mutation on the composition of the gene pool, in that it causes only slight alterations. The same chance factors are also present, and the rare event of a mass migration in one direction could appreciably alter the composition of the gene pool.

RELATIVE IMPORTANCE AND INTERDEPENDENCE OF NATURAL CONDITIONS

It is really quite artificial to simply list the conditions of the Hardy-Weinberg law as separate, independent, and equally important factors. The interdependence of the factors made the simultaneous discussion of population size and nonrandom mating mandatory. Interdependence can also be demonstrated in the following example. A male bird may possess a genotype that gives it a more attractive plumage. Certain female birds will tend to be more attracted to males with this phenotype and preferentially mate with them. The genotype then has a selective advantage *because* it promotes nonrandom mating. Violation of one condition (random mating) has led to the violation of a second condition (no selective advantage). The result of these tandem violations is the formation of a subgroup within the population consisting of males with the noted phenotype and females that are attracted to them.

The interdependence of the Hardy-Weinberg conditions can be most clearly seen in the phenomena of *genetic drift.*

Genetic Drift

Occasionally, a population size will decrease significantly because of food shortage, abundance of predators, migration of individuals, etc. Thus, one violation (unlimited population size) has been perpetrated by others (no migration, no selective pressure, etc.). What is the effect on the gene frequencies in the gene pool when such an event occurs? Imagine 1000 individuals with a distribution of genotypes as follows:

250**AA**:500**Aa**:250**aa**

and the number is reduced to 100 total individuals. The original population had a gene frequency for **A**:**a** of 1:1 (50% **A,** 50% **a**). What will be the frequency of genotypes in the new, small population of 100 individuals? Clearly, it will depend on which individuals of the original population comprise the new population. They might all be of genotype **AA,** or of genotype **Aa,** or **aa,** or some combination thereof. The chance that all one hundred would be of genotype **AA** would be $(0.25)^{100}$ or 10^{-60}, so it is most likely that both genes **A** and **a** would survive.

But the frequencies of the **A** and **a** gene in the new population may be quite deviant from the 1:1 ratio exhibited in the old population. When a small population exhibits a gene frequency deviant from that of the original population, genetic drift has taken place. And genetic drift, which is not caused by violations of the conditions of the Hardy-Weinberg Law, may lead to the violations of other conditions: random mating, new selective pressures, etc.

The particular combinations and specific interdependences of given Hardy-Weinberg conditions are dependent on the way in which the population separates from the parental population. If separation was by simple migration, the new environment may have new selective pressures. In instances where the environment remains the same, a slightly deviant gene frequency might return to its original value by natural selection. If, however, the small population finds itself in a genotypic situation where the new gene frequency has an advantage, the altered frequency will be retained.

The retention of the altered gene frequency can occur even when the new population is exposed to the same environment and thus the same selection pressures as the original population. The reason for this is because the smaller population is more susceptible to previously masked selection pressures that favor the retention of the altered gene frequency (**Founder's Principle**). The original population did not change toward this direction, not because of the absence of the appropriate selection pressure, but rather because of the dominating or stabilizing effect of the large numbers in the gene pool.

This can be seen more clearly if one considers the deviations from the gene frequencies that are expected in populations of different sizes. If the **A** allele has a gene frequency of .50 in one generation, one would expect that even under controlled conditions (Hardy-Weinberg conditions met) there would be some deviation from the exact value of 0.50. The expected deviation can be calculated by means of a statistical procedure called the **standard error.**

$$\text{S.E.} = \sqrt{\frac{pq}{N}}$$

where p and q are the gene frequencies of **A** and **a** respectively, and N is the total number in the population. The standard error may be calculated for populations with 10, 100, and 1000 individuals.

N	S.E.	Frequency Range Expected
10	$\sqrt{\frac{(.5)(.5)}{10}} = 0.16$	$50 \pm 0.16 = 0.34 - 0.66$
100	$\sqrt{\frac{(.5)(.5)}{100}} = 0.05$	$50 \pm 0.05 = 0.45 - 0.55$
1000	$\sqrt{\frac{(.5)(.5)}{1000}} = 0.016$	$50 \pm 0.016 = 0.484 - 0.516$

Therefore, the larger the population, the more stable one would expect the gene frequency to be. And if a population size is greatly reduced by genetic drift, and the gene frequencies are altered, the new gene frequency might have an opportunity to respond to a selective advantage of minor importance in the larger parental population or generation.

Another way in which genetic drift can affect gene frequencies is related to the mutation rate. If a mutational event occurs at the rate of 10^{-6} and the population has 10^8 members, one would expect that about 100 individuals in the population would carry that mutation.

$$\frac{10^8}{10^6} = 10^2$$

If such a population were to undergo a series of events leading to genetic drift, and some of those individuals carrying the mutation survived in the resulting smaller population, the gene frequency of the mutant gene would be greatly increased. For example, the gene frequency in the large population would have been

$$\frac{10^2}{2 \times 10^8 \text{ (diploid)}} = 5 \times 10^{-6} \qquad \text{or} \qquad 0.000005$$

If genetic drift results in a small population composed of only 1000 members, among which is one carrying the mutant gene, the gene frequency of that gene now becomes

$$\frac{1}{2 \times 10^3} = 5 \times 10^{-3} \qquad \text{or} \qquad 0.005$$

which is a significant increase. If, in addition, the gene has a selective advantage, it is more likely to be retained in the smaller population than in the larger one, where stabilizing selection may have played an important role.

Selection Pressure and Mutation Rate

In addition to the interdependence of violations of the Hardy-Weinberg conditions, certain of the violations are more evolutionarily constructive than others. In fact, a classic disagreement existed in the early part of the twentieth century as to the relative importance of selection and mutation. The proper perspective of the roles of both factors was not gained until each was put on a quantitative level.

Selection pressures tend to be of the order of 1%, while mutation rates tend to be of the order of 0.000001%. Clearly, if **A** $\longrightarrow$ **a** at a rate of 0.000001%, but **a** is selected against at a rate of 1%, the **a** gene will never become established in the population. On the other hand, if the selection pressure is 1% for the **a** gene, it will become established in spite of the low mutation rate. The establishment of a mutant gene in a population is dependent on the selection pressures, rather than the mutation rate.

Another important aspect of mutation rate and selection pressure is that mutation rate acts on the genotype, slightly altering the frequencies in the gene pool, while selection pressure operates on the phenotype, and may simultaneously eliminate many genes from the gene pool. Thus, if **A**-type individuals are selected against, both **AA** and **Aa** genotypes will be affected by selection pressures. *It is very significant in evolution that while the gene is the unit of change on a biological level, the in-*

dividual is the unit of change for selection, and the population is the unit of change in evolution.

ACCUMULATING DIFFERENCES—ADAPTATION

Few if any species exhibit the ultimate potential for their genetic material. There are always unexplored means by which organisms can become better suited to their environments. In addition, there may be a multitude of ways a given species could develop to better exploit facets of the environment. Each population of a species may, by chance alterations of the gene frequencies in its gene pool, start on a different road toward this exploitation. Once the particular route is embarked on, the success of the exploitation is dependent on the accumulation of advantageous genetic factors: **adaptation.** Adaptation is, then, the genetically based changes that impart greater survival value.

There are two types of biological adaptations: physiological and genetic. **Physiological adaptation** is the direct response of an individual to an environmental alteration. Examples of this would be the increase in the number of red blood cells that occurs in a body moved to higher altitudes, or the color modification of the chameleon. The capacity for this type of change is built into the organism and occurs in response to the proper stimulus. These are products of evolutionary progress rather than changes that lead to new types of individuals.

Genetic adaptation, on the other hand, is formed by random processes that collectively equip an individual for better survival. It is the successful interplay between the phenotype of the individual and the selective pressures of the total environment. It is the elaboration of genetic traits by mutation coupled with appropriate selective pressures, and the recombination of existing genetic traits to produce a better-adapted individual. Genetic adaptation in phyletic evolution leads to small-scale changes or **microevolution,** and is usually expressed as a change within a species rather than the formation of new species. In dispersive evolution, however, as each population becomes adapted in its own way, it draws away from the others. Adaptations lead to greater and greater differences between the gene pools.

Experimental Demonstrations of Adaptations

In nature, the accumulation of genetic factors that lead to adaptations takes on the order of thousands of generations. It is, therefore, impossible to view this phenomenon under natural conditions. Observational findings of morphological differences between isolated populations of the same species supports the idea of adaptation, but does not prove it.

Genetic adaptation has been most satisfyingly demonstrated with bacterial populations in the laboratory. Unlike the gradual natural process by which adaptation is thought to occur, the experimenter enforces a cataclysmic event that produces a great selective pressure, weeding out all members of the population save those that possess the adaptation.

A population of 10^{10} *E. coli* cells is placed in a broth containing

20 μg of streptomycin per milliliter. After several days the population will be found thriving in this medium. If a few bacteria (10^1) from this new population are inoculated into another flask containing streptomycin broth, comparable growth will be found in only twenty-four hours.

The original inoculum of 10^{10} *E. coli* contained several individuals that had randomly mutated to a streptomycin-resistant form. In a streptomycin medium, this genetically based trait constituted an adaptation. The eventual flourishing population in the later streptomycin medium were the descendants of these adapted cells (and illustrate that one asexually reproductive individual can produce any number of descendants, given the proper conditions).

The addition of streptomycin to a bacterial culture constitutes a cataclysmic event. The adaptation of members of the population is dependent on their capacity to withstand the event, and there appears to be a limit to the degree to which an organism can adapt. If, for instance, the bacteria were subjected to 200 μg/ml instead of 20 μg/ml of streptomycin, all the bacteria would die. There seems to be a limited genetic potential for resistance. This would indicate that a single genic change can lead to only a minor adaptation.

Accumulation of Minor Adaptations

How does an organism undergo the many genic changes necessary to produce major adaptations? It is generally held that major or large-scale adaptations are the results of cumulative genetic events.

The maximum streptomycin level to which at least some members of a 10^{10} population of *Salmonella* bacteria can be adapted is about 5.0 μg/ml. If the descendents of the few resistant cells are then subjected to 10 μg/ml, a few will survive and produce another lush population. The descendents of this population can now be treated with 12 μg/ml, and another selection will take place. Here there is a series of mutations, one building upon the other to eventually produce cells that are adapted to levels of streptomycin almost two and a half times the maximum for nonadaptive types.

It has been calculated that the mutation rate for a single change to streptomycin resistance in Salmonella is 10^{-9}. The chance that all three of these mutations would occur simultaneously would be the product of the three independent mutational events

$$10^{-9} \times 10^{-9} \times 10^{-9} = 10^{-27}$$

This means that one in every 10^{27} cells would be expected to have a triple mutation. This number is probably greater than all the organisms that ever existed.

Here, then, is the key to natural selection—the combination of reproduction and mutation to produce an accumulation of rare events that are impossible as simultaneous events. Such accumulations build and elaborate upon traits within a species undergoing phyletic speciation, and enhance the differences between the parent and daughter species undergoing dispersive evolution.

The evolutionary development of complex adaptive structures (such as legs, eyes, and the many physiological adaptive mechanisms) probably involved more than the action of selective pressures on simple mutations. Undoubtedly, the serial accumulation of adaptive changes made possible the emergence of these complex adaptations.

The transition of minor adaptations to more complex ones has been experimentally demonstrated only for mutation. It is logical to suspect that other sources of variation, such as position effect by chromosome aberration, or reshuffling of traits to new testing positions by recombination, have similar effects toward building complex variations in nature.

Adaptations Under Natural Conditions

The laboratory situation in which populations of *Salmonella* and *E. coli* were exposed to streptomycin was somewhat artificial. In nature, organisms are rarely confronted with an all-or-none situation (streptomycin either in the medium or not). There are, however, a few recorded natural situations in which such an all-or-none situation did arise. One such case is that of so-called **"industrial melanism."**

A comparison of dark- and light-pigmented moths in England now and several centuries ago indicates that in the past century there has been an increased frequency of moths with dark pigmentation. Dark pigmentation has been found to be inherited as a simple Mendelian recessive. The prevalence of the dark-pigmented moths is thought to be an alteration of the gene frequency as an adaptive response to a new environmental situation, the industrial revolution. White or brightly colored moths that once blended into a pastoral environment were glaringly obvious and highly susceptible to their predators when they alighted on sooty buildings. Apparently, the industrial revolution created new selection pressures, resulting in a dramatic shift in gene frequencies and the prevalence of the more protective trait of dark pigmentation (see Fig. 15-8).

Fig. 15-8 Industrial melanism.
[M. W. F. Tweedie—National Audubon Society]

Industrial melanism has been found to occur in over eighty species of moths. Thus the tendency toward darker coloration is not a unique trait, peculiar to one species, but a recurrent trend in many unrelated, reproductively isolated groups.

Although the "all-or-none" situation does arise in nature, it is far more typical for a population to be confronted with a gradient of selective pressures. This is best exemplified by returning to a hypothetical situation with *E. coli* and streptomycin.

An *E. coli* population growing in the soil near a population of *Streptomyces* (a "bacterium" that produces streptomycin) will be confronted with a concentration gradient. At the site of the *Streptomyces* growth, the levels of streptomycin may be as high as 12 μg/ml. A few millimeters from the center of growth, the concentration may be 10 μg/ml, etc. This would create a similar gradient of selective pressures. Those *E. coli* on the periphery of the gradient can undergo minor adaptations that permit them to live there. Some of their offspring may migrate toward higher levels of the gradient, and a few of these will become further adapted. By a continuing process, a new population can infiltrate the entire environment (see Fig. 15-9).

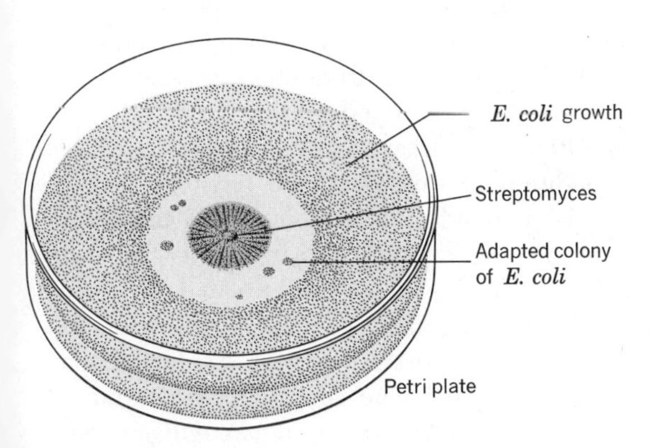

Fig. 15-9 Sequential adaptation of *E. coli* to streptomycin.

Thus, under natural conditions, small adaptive changes can accumulate and lead to new and complex exploitations of the environment. As this sequence occurs, the gene frequencies of the gene pools undergo considerable alterations, and the populations become more divergent morphologically and physiologically.

THE DECISIVE MOVE TOWARD SPECIATION—ISOLATION

The initial alterations in gene frequencies of the gene pool may accumulate through complex adaptations. Finally, by chance alone, these changes include factors that inhibit the interbreeding of two populations. No longer is there a potential for sharing traits; no longer do they contribute to a common gene pool. The elimination of other populations from the gene pool of a group is the final step in speciation. After sexual isolation, a group is committed to an evolutionary development apart from the parental species.

The factors *leading to* the separation of gene pools are called isolation mechanisms. These isolation mechanisms may be of two general types.

Two Groups Theoretically Able to Share the Same Gene Pool

Members of the two groups of the same species can (1) physically mate (copulation in animals, pollen penetration of the stigma in plants), (2) produce viable zygotes, (3) produce viable embryonic forms, and (4) produce viable offspring. Thus laboratory hydrids between the two groups can be obtained.

In nature, however, the two groups are isolated by physical or behavioral factors that inhibit a sharing of their gene pools. Groups that can mate theoretically, but do not, have taken but a few short steps from the parental species.

Geographical Isolation. Two groups may start on the road to sexual isolation by their geographical separation from one another. For a considerable time (thousands of generations) they may be physiologically, morphologically, and genetically capable of producing viable, fertile offspring. Because of the physical separation of their gene pools, adaptive traits accumulated by one group are not shared with the other group. Eventually, by chance alone, their gene pools may incorporate different mutant genes and follow different adaptive routes, so that the two groups either cannot mate for morphological or physiological reasons, or the mating results in inviable or infertile offspring, and speciation is complete.

The classic example of continuing evolution by geographic isolation is that described by Darwin for finches. Darwin found that there were 26 groups of finches among the Galapagos Islands (which lie a few hundred miles west of South America). Only five of these groups were the same as the finches found on the mainland. The other twenty-one were types peculiar to the group of islands. Some of the twenty-one groups interbred quite freely, while others did not. Apparently, each of the groups became isolated by migration. Each group evolved separately from the continental forms as well as from other isolated groups on the islands, forming a series of species and subspecies.

Ecological Isolation. Often two closely related species will thrive in different ecological conditions within the same territory, but no hybrids between them will be found. When members of the representative species are taken into the laboratory, hybrids between them can be obtained. Thus their respective gene pools are isolated physically, but not physiologically. Again, because of the ecological separation that inhibits the sharing of traits, the incorporation of new adaptations may lead to a functional separation of their gene pools, and speciation.

Two subspecies of the mouse (*Peromyscus maniculatis*) inhabit regions that share a common boundary. While laboratory hybrids between these subspecies exist, none have been found in nature. The explanation for this isolation is that one subspecies lives in the forests, and the other on sandy beaches. Apparently members of the two groups rarely encounter each other.

The same situation has been found for another subspecies of mouse in Oregon, as well as for the water snake *Natrix sipedon.* One subspecies of snake is a freshwater race, and the other is a saltwater race.

Seasonal Isolation. Two groups may exist in exactly the same ecological area, but do not interbreed because they become sexually mature at different times of the year or under different conditions. Over ten subspecies of Cypress trees (*Cupressus*) are found in California. They do not interbreed because each race produces pollen at a different time of the year.

In animals it may be the time or conditions for mating that may vary. In the northeastern United States, three species of frog, *Rana pipiens, R. sylvatica,* and *R. clamitans* all mate in the same ponds at different times. *R. sylvatica* begins breeding when the water is 44° F and completes its mating season before the temperature reaches 55° F, the breeding temperature necessary for *R. pipiens*. *R. clamitans* must have temperatures of at least 60° F. By the time this temperature is reached, both *R. sylvatica* and *R. pipiens* are past their breeding stages (see Fig. 15-10).

(*a*) ***Rana pipiens***

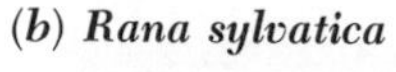

(*b*) ***Rana sylvatica***

(*c*) ***Rana clamitans***

Fig. 15-10 Three species of frogs. The three are isolated by physiological preferences of habitation and mating temperatures. [Hugh Spencer]

Notice that the three frogs are assigned different species names rather than different subspecies as in the previous examples of finches, mice and snakes. This is because the seasonal isolation appears to have progressed beyond the road toward speciation, and has undergone the final, irrevocable step of sexual independence. Hybrids between any of the three species cannot be produced in the laboratory. Although fertilization can be achieved, the laboratory hybrids die in early embryonic stages. There appears to be a basic incompatibility between the genetic and/or cytoplasmic factors in the gametes.

Mating Patterns. Many species of animals, notably birds and insects, embark on mating after highly standardized ritual dances. In some species of birds, plumage of the male stimulates the female sexual interest. So specific are female tastes in such matters that males exhibiting significant deviations from the norm would have little chance in the contest.

Thus, if two groups are geographically or ecologically isolated, and subsequent adaptive changes lead to alterations of their mating patterns, they would become sexually isolated. It should be noted that laboratory hybrids could possibly be formed between two such groups. The isolation by mating patterns is, therefore, based on behavior, rather than on physiological inability to mate.

Two Groups Unable to Share the Same Gene Pool

As adaptations accumulate in two groups separated for physical or behavioral reasons, they may reach a point at which they cannot either physically mate (morphological differences in copulatory organs), or produce viable zygotes, or produce viable embryonic forms, or produce viable, fertile offspring. Laboratory hybrids between these two groups cannot be obtained without manipulation. Thus, by the continuing process of adaptation, isolation mechanisms become speciation mechanisms.

Physiological Isolation. Groups that share the same gene pool, and thus shuffle genetic traits between them, establish certain physiological compatibilities in their mating, fertilization, or developmental processes. When two groups no longer share the same gene pool, although they may theoretically be able to do so, different physiological factors may be developed in each group.

For example, pollen (the male gamete) and stigma (the female organ) in plants have been shown to require specific compatibilities. The success of some interspecies (*interspecific*) tobacco crosses that have been done in the laboratory to produce new varieties are dependent on which parent is the female and whether the pollen can penetrate the entire stigmatal area. Apparently each species has developed characteristic physiological compatibilities among its own members, and resists attempts to crossbreed. The chance of crossbreeding between two such species in nature is negligible.

It has been postulated that sperm can tolerate only narrow ranges of physiological conditions (pH, temperature, etc.) and that these conditions are favorably met in female members of the same species. Deviation from the physiological compatibility would prove lethal to sperm, and few, if any, would survive to fertilize the egg. Therefore, even though members of these two groups could copulate, the mating process would not lead to a viable offspring. There is little proof to substantiate that this factor does arise in the separation of animal populations.

Physiological isolation can occur also as an incompatibility between the embryo and the female parent. Interspecific hybrids between two species of jimson weed (*Datura*) result in the death of the embryo at the eight-celled stage. By removing the embryos (or actually the seeds), and culturing them artificially, viable seeds can be produced that germinate and grow into healthy offspring. As noted, the production of viable offspring requires experimental manipulation, and thus does not occur in nature.

Genetic Isolation. Genetic isolation is usually based on chromosomal incompatibility. This incompatibility between two closely related groups is usually expressed by the production of infertile hybrids. It may arise as the result of individual gene dissimilarities or chromosomal aberrations.

The standard example of individual gene dissimilarities is the horse and donkey. They can mate, and produce mules as functional offspring. Mules, however, are sterile. The horse and donkey each contribute 33 chromosomes to the offspring mule. During meiosis of the gametogenic cells of the mule, however, the chromosomes fail to pair. This failure is thought to be the result of the degree of genic dissimilarity between the horse and donkey chromosomes, so that they do not "recognize" one another as homologues. Since the orderly separation of chromosomes into gametes is necessary for fertilization to take place, the mule is sterile.

Other closely related groups have become genetically isolated because of chromosomal aberrations. Although hybrids between such groups are possible, the hybrids are sterile. If, for instance, the adaptive changes leading to the isolation have included the inversion of a chromosome segment, the meiotic pairing in the gametogenic cells of the hybrid offspring will be inhibited. Since the genes on homologous chromosomes pair only with their allelic forms, the noninverted chromosome must loop into the inverted one (see Fig. 15-11). If a cross-over occurs within the loop, one of the resulting chromosomes would be accentric, and the other would be dicentric. The subsequent loss or tearing apart at anaphase would lead to gametic cells that do not have the full complement of genetic information. Hence they would be either nonfunctional or unable to produce viable zygotes.

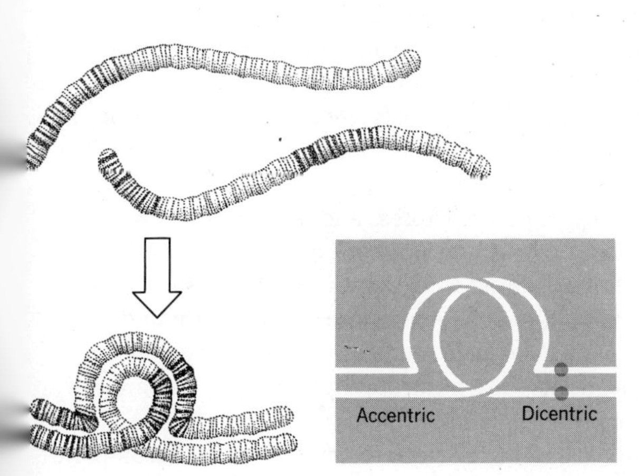

Fig. 15-11 The pairing of chromosomes at meiosis when one of the chromosomes contains an inversion. When a cross-over occurs within the loop, recombinant strands are both non-functional (lower right).

Transpositions and translocations produce similar pairing difficulties in the gametogenic cells of hybrids (see Fig. 15-12). The more deviant the chromosome structure of one group from another, the less chance there is that fertile offspring will be produced by hybridization. Thus, two groups that differ by their chromosome structure are genetically isolated from one another.

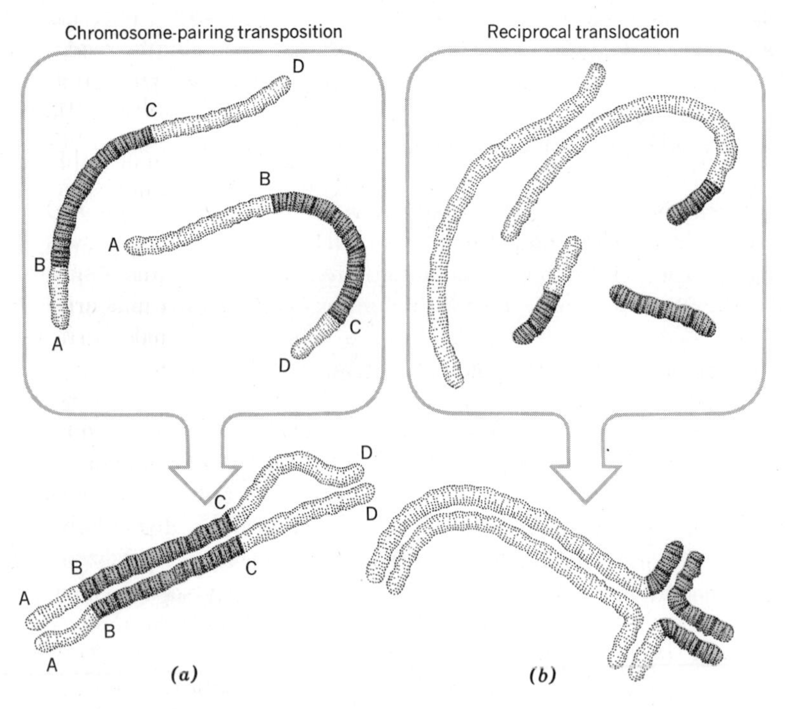

Fig. 15-12 The pairing of chromosomes at meiosis when (*a*) a chromosome contains a transposition, and (*b*) a chromosome contains a translocation.

SPECIATION BY POLYPLOIDY

Throughout this chapter, speciation has been depicted as a gradual process. There are some cases, however, in which a new species can arise in one step. This is called **saltation.**

Species saltation occurs through **polyploidy,** the increase of the number of chromosome sets from diploid ($2N$) to a higher value. The common dandelion, winesap, and Baldwin apples are triploid ($3N$). Oats, sugar cane, cotton, many cultivated roses, and some strains of wheat are tetraploid ($4N$). Some wheat and potatoes are hexaploid ($6N$).

Polyploidy does not create new characteristics, but rather appears to intensify the expression of characteristics present at the diploid level. It appears that these organisms have made a maximum exploitation of the environment through adaptations at the diploid level, and intensify these adaptations through an increased ploidy.

Polyploid individuals can arise in one of two ways: incomplete mitotic division, or hybridization coupled with incomplete mitotic division. Incomplete mitotic division results in the production of a tetraploid (4*N*) cell. Since the cell contains two sets of the same chromosomes, the condition is called **autoploidy.** Gametes from such cells would vary in their ploidy. Some would be diploid; some would be haploid; but most would be gradations in between. The autoploid condition typically leads to gamete inviability, and hence is not thought to be significant in the formation of new species (see Fig. 15-13).

When two closely related but genetically isolated species form a hybrid, the hybrid offspring is infertile. If, however, an incomplete division arises during the development of this hybrid so that tetraploid tissue is formed, this region of the individual can be fertile. The tetraploid hybrid has cells that contain two diploid sets of chromosomes in their nuclei, a condition called **amphidiploidy.** During meiosis each chromosome has a homologue with which to pair. Meiosis of an amphidiploid cell produces diploid gametes. Fertilization between two such gametes will return the amphidiploid condition.

From an evolutionary standpoint, the genesis of tetraploid individuals may mark the origin of a new species. The tetraploid type is reproductively isolated from both parents. Either parental type would contribute haploid gametes, which would join with a diploid gamete to form a nonfertile triploid hybrid. (Triploid plants mentioned earlier all reproduce asexually). The tetraploid strain is also phenotypically different from both parents, often having traits from the parental stock. Crosses between radish (*Raphanus*) (edible root) and cabbage (*Brassica*) (edible leaves) produced a tetraploid hybrid (*Raphanobrassica*) that had the cabbage root and the radish leaves.

Polyploidy has occurred primarily in the plant kingdom. About 60% of existing ferns and seed plants are in a polyploid condition. In general, these 60% tend to be hardier and appear in temperate regions of the world, thus supporting the idea that polyploidy intensifies adaptations rather than initiating them. Polyploidy is almost nonexistent in animals. It is thought that the more closed pattern of development in animals is highly dependent on the diploid state of chromosomes, while the more flexible open pattern of growth in plants permits polyploidization.

EVOLUTIONARY PERSPECTIVE OF ORGANISMS

The most familiar way to describe an organism is as a collection of morphological and physiological characteristics that serve to typify its adaptive state. Another equally valid means of describing an organism is in terms of the various features of the environment that it exploits. For instance, an iris plant can be described as a soft-stemmed plant with a particular type of flower that blooms at a given time of the year, or

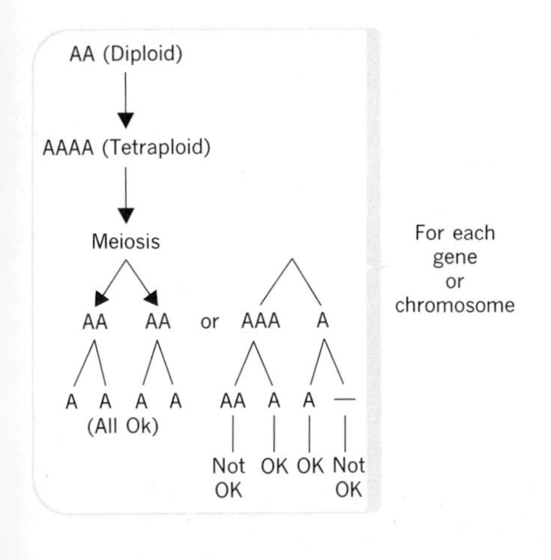

$$\frac{\text{Number of } A \text{ gametes}}{\text{Total gametes}} = \frac{6}{8}$$

For chromosome number of 10

$A_{10}A_{10}$ (2N = 20)

↓

$A_{10}A_{10}A_{10}A_{10}$ (4N = 40)

↓

$$\text{Chance of } A_{10} \text{ gamete} = \frac{6}{8} \times \frac{6}{8} \times \frac{6}{8} \times \frac{6}{8} \times \frac{6}{8} \times \frac{6}{8} \times \frac{6}{8} \times \frac{6}{8} \times \frac{6}{8} \times \frac{6}{8} = \frac{6^{10}}{8^{10}} = 0.056$$

(1 in 20 gametes)

Fig. 15-13 The results of autoploidy.

it can be described in an adaptive perspective as an organism that occupies space during a limited growing season, attracts certain species of bees, etc. The characteristic morphological and physiological traits of the iris constitute its genetically significant phenotype, while the ensemble of environmental exploitations constitute its **adaptive zone.**

The adaptive zone is a collection of environmental features that may be related only by the way in which some organismal type functionally brings them together. As in the example of the iris given above, there appears to be no relationship between a population of bees, the annual rainfall of an area, and light. The iris, however, uses the bees for pollination, the rainfall index for the growth processes, and the nonvisible light for flower induction.

The adaptive zone as a list of environmental factors does not necessarily indicate what form the organism will take. For instance, desert environments have been exploited both by **cacti,** which have juicy photosynthetic stems instead of leaves, and **succulents,** which have thick, juicy leaves with heavy wax coatings (see Fig. 15-14).

Fig. 15-14 Parallel evolution to similar environmental requirements. The cactus, in which water-retaining parts are derived from the stem, and the succulent, in which water-retaining parts are derived from leaves, belong to different families but thrive in common regions. [Russ Kinne—Photo Researchers]

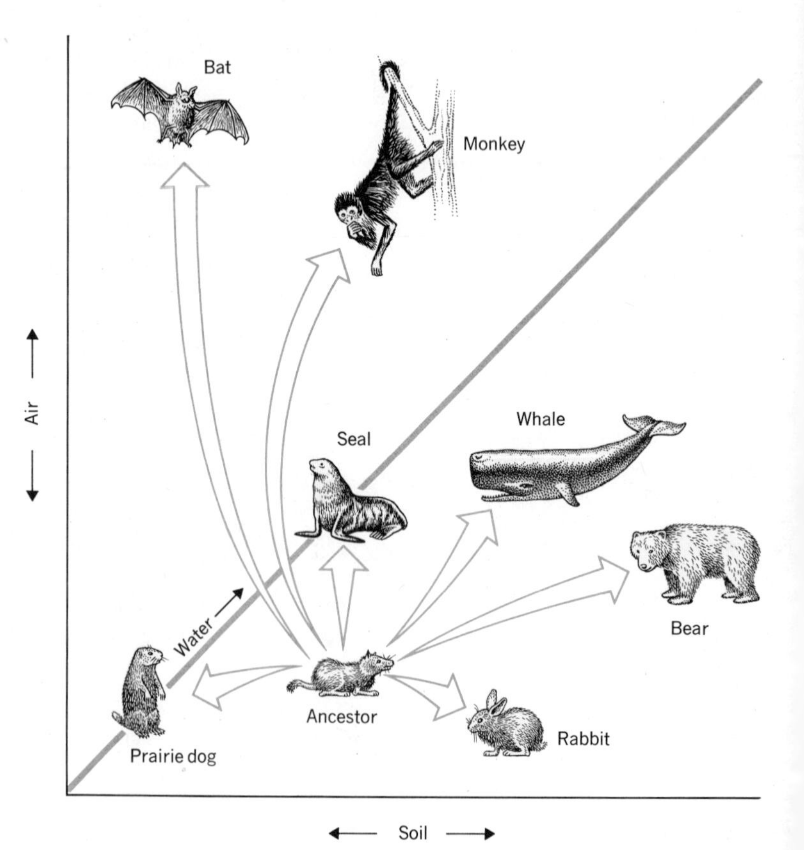

Fig. 15-15 Diagram of adaptive genes (graph), adaptive radiation (arrows), and quantum evolution (mammalian types).

Adaptive Zone and Evolutionary Potential

The potential for evolution is dependent on the range of the adaptive zone. The situation is not unlike that experienced by a human individual in psychological adaptation. If a person is used to experiencing great variation (travel, moving to many different places, meeting many new people), he experiences little anxiety at potential changes. If, however, he is, what is affectionately called, in New England, a "swamp Yankee" (one who has never been more than 20 miles from home), then a move to California would create great anxieties. The former, due to the variation of his experiences, is more adaptable than the latter, who has lacked variation.

The analogy between the psychological and genetic potential of a species for evolution is quite direct. The adaptive zone of a population that has great genetic variation is much larger than one that lacks this variation. Therefore, the former population can survive greater environmental changes than the latter.

When a population is undergoing change, it manifests great variation. When a population has "settled" into an environmental situation, variation is greatly reduced by the effects of stabilizing selection. Such a population is highly adapted to its environment, and even minor environmental changes may lead to the extinction of the entire population. Therefore, evolutionary progression seems to depend on intermediate groups (those in the midst of change) rather than on those that have already adapted to the environment (see Fig. 15-15). Many organisms are considered to have had *common ancestors*, rather than one established species being thought to give rise to another.

QUESTIONS

1. Compare and contrast dispersive and phyletic speciation.
2. Of 200 individuals in a population, 150 are homozygous dominant (AA), 30 are heterozygous (Aa), and 20 are homozygous recessive. What is the frequency of the "A" gene in the gene pool?
3. Under what conditions will the Hardy-Weinberg equilibrium exist?
4. Give examples of disruptive and homeostatic selection.
5. What are the consequences of a limited population size and non-random mating.
6. Why does genetic drift lead to speciation?
7. Compare and contrast physiological and genetic adaptation.
8. Why are sequential adaptations the key to natural selection?
9. How do the following factors lead to a separation of the gene pools: geographical isolation, ecological isolation, seasonal isolation, and difference in mating patterns.
10. Why do physiological isolation and genetic isolation make two groups unable to share the same gene pool?
11. How can species saltation occur?
12. Define an adaptive zone.

CHAPTER 16

THE ORIGIN OF LIFE

The history of evolution must begin with *biopoesis*, the origin of life. But while biopoesis is surely one of the more interesting topics in biology, its very nature demands that any discussion be highly speculative.

Great thinkers down through history have pondered the problem of the origin of life. Such men as Aristotle, Newton, and Harvey accepted some form of **spontaneous generation,** the origin of life through naturalistic, spontaneous processes. These men and their contemporaries accepted the idea of a physiological reproductive process in complex animals and plants, but they also accepted the idea that less complex plant and animal forms sprang to life from nonliving precursors.

SPONTANEOUS GENERATION

So firmly entrenched was the idea of spontaneous generation in philosophy that during the post-Renaissance period when experimentation became the acceptable means of fact-finding, the question that was asked was "What combination of nonliving material gave rise to particular living forms?" rather than "Did spontaneous generation occur?" The early plant nutritionist, von Helmont (1640), recorded his successful experiments in which mice were produced from sweaty shirts and wheat grain. Many experimenters confirmed that decaying meat gave rise to maggots (the larval form of flies).

In the seventeenth century, Francesco Redi performed what is now considered a classical experiment to test the validity of the theory of spontaneous generation. He covered a piece of meat with cloth, and let the meat decay. During the decaying period he noticed that many flies settled on the cloth for short periods of time. After several days he found that there were maggots on the cloth, but on none of the meat underneath (see Fig. 16-1). He concluded that the mature flies had been attracted to the meat and had laid their eggs on the cloth, and postulated that maggots come only from preexisting flies and were not spontaneously generated by any other form of material.

Fig. 16-1 Redi's experimental setup.

While Redi's experiment proved that even the simple forms of life known in the seventeenth century did not arise by spontaneous generation, the discovery of microbes brought the question into the forefront of biological thought once again. After all, maggots are far more complex than those little spherical, rod, or corkscrew-shaped bacterial forms observed under the microscope. Surely these organisms arise by spontaneous generation!

Such thought was supported by the many workers who found that they could obtain microbes for study by preparing meat broths and permitting them to stand for a few days. The meat broth was found to be swarming with microbes.

The Decline and Fall of the Theory of Spontaneous Generation

In the late seventeenth century, Spallanzani tested the theory of spontaneous generation of microbes. He prepared flasks of meat broth, which were boiled for several hours and then sealed. The broth remained clear for months, and when the seals were broken and the broth tested, it was shown to be free of microbes.

Spallanzani's experiments were neither conclusive nor satisfying to many of his contemporaries. Some claimed that by boiling he had driven out a "vital force" necessary for spontaneous generation. When Priestly proclaimed oxygen as a necessary factor for life, it was postulated that the process of boiling and sealing had driven oxygen out of the flask, without which life could not be supported. Many of Spallanzani's successors modified the experimental procedure in attempts to quell the criticisms, but each new technique gave rise to new criticisms. The death blow to the theory of spontaneous generation of microbes was finally dealt by Louis Pasteur in the nineteenth century, some two hundred years after the initial experiments of Spallanzani.

Louis Pasteur devised several experimental means by which the spontaneous generation of microbes was disproved. The simplest and most sophisticated one was with the use of a swan-neck flask. He prepared a meat broth in this flask (see Fig. 16-2), and boiled it for several hours. He then left the flask *unsealed* on a laboratory bench. Previous experiments had shown that the mere boiling of the broth did not

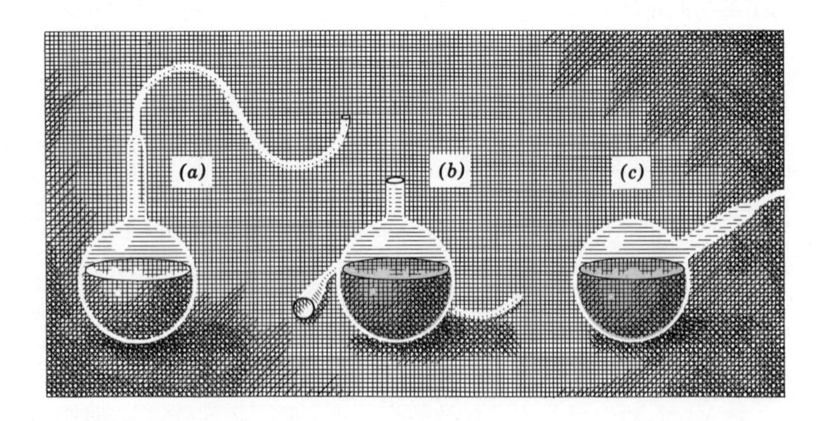

Fig. 16-2 The swan-neck flask of Pasteur. (*a*) While flask remained in place after boiling, no contamination occurred. (*b*) When neck was removed, or (*c*) when neck was wetted by the media, contamination occurred.

diminish its generative capacity, so the experiment could not be criticized on these grounds. The flask was not sealed, and there was a free exchange of air with the environment, so the system did not lack oxygen. Still, the swan-neck flask remained free of microbial contamination for months.

After several months, the flask was tipped, and some of the broth was permitted to flow into the curved neck, after which it was stood upright once again, so this broth flowed back into the vessel. In a few days there were signs of contamination. Pasteur thus showed that microbes are on the dust particles in the air. The swan-neck flask trapped the dust, and hence the microbes, in the curvature of the neck, permitting only the clean and microbe-free air to enter the body of the flask. The microbes, evidently, could not migrate on the dry sides of the flask to the broth and remained on the curved glass surface. When, however, broth washed the surface, some of the entrapped microbes were brought into the body of the flask and proceeded to multiply.

With the end to the theory of spontaneous generation, and the intellectual dissatisfaction with the notion of special creation, there appeared to be a perplexing problem: There were no acceptable ideas on the origin of life left for one to consider.

About the same time that Pasteur was dealing the final death blow to the already dying theory of spontaneous generation, a new idea had come to the fore, natural selection. Almost a century later biologists began to ask: Could not biogenesis have occurred by natural selection on a molecular level? Spontaneous generation, like the phoenix of antiquity, died, and rose again to resume its original position.

SPONTANEOUS GENERATION REVISITED

From 1870 (after Darwin's tenth edition of the *Origin*) to 1940, most areas of science progressed toward a level of maturity where their firmly established principles collectively created a unified picture of the world. Biology had brought life down to the chemical level through physiology and biochemistry. Chemistry had raised the elementary structure of matter to large synthetic molecules. Physics had provided the energy considerations on which reactions were based. Geology had measured the age of Earth as a few billion years, and that of the universe as close to ten billion years. The stage was set for the idea of spontaneous generation to be given a more rigorous inspection. The first form of life was not to be a maggot, or bacterium, or even a virus, but a collection of biologically important molecules, capable of producing organized forms of life through processes of natural selection. The laboratory was the entire surface of the Earth, the time was in terms of several billion years, and the experimental procedure was natural selection.

The Early Earth

According to modern views, the Earth was formed 4.5 to 5 billion years ago by the condensation of an immense dust cloud. The great contracting force pulled the heavier elements (such as iron, uranium, copper, etc.) to the center, while the lighter elements (such as hydrogen and helium) concentrated at the surface. The pressures generated such heat that the core liquified, and many radioactive materials were produced in the several-million-degree heat. The Earth slowly cooled over a period of a million years to the point where small molecules maintained stable states. The lighter elements formed the first atmosphere. Most of these lighter elements eventually escaped, and the surface became nothing but rock exposed to a near vacuum.

Vast quantities of hydrogen and oxygen under great pressure and heat below the surface formed water. The water escaped through fissures of the rocks and formed an atmosphere of vapor. Gradually the temperature of the Earth's surface lowered sufficiently for the condensation of this vapor to occur, and enormous rains poured down. It has been estimated that the rains continued unceasingly for thousands of years.

During these storms, lightening discharged great quantities of electrical energy. The rains wore down the rocks and the mountains were lowered. The water flowed down to the lower levels, eroding surfaces and carrying dissolved minerals with it. The oceans slowly formed, creating great seas of sterile, inorganic media. Subsequent evaporation of the water led to more rain, and more wear on rocks as this water found its way to the sea, bringing with it dissolved salts. The concentration of these inorganic materials began to approach the now prevailing 3% salt content of the sea. Most of the minerals common today were now present. Oxides and sulfates, however, were not formed until free oxygen (a product of photosynthesis) was available in the atmosphere. True soils (which result from a combination of decomposed plants or humus) and oxidized inorganic salts were nonexistent. Volcanic materials (methane, ammonia, hydrogen, helium, etc.) erupted through the earth's mantle. Much of the material became dissolved in the seas.

Such was the chemical state of the prebiological Earth.

Spontaneous Generation on a Molecular Level

The first attempt to describe the origin of life by natural selection on molecules in the pristine seas was made by the Russian biochemist Oparin in 1936. He proposed that since unstable molecular complexes would undergo spontaneous degradation, stable complexes would be favored and tend to accumulate in the seas of the early Earth. This tendency toward stable complexes would result, by chance alone, in the formation of a type of colloidal body well known in physical chemistry as a **coacervate.** A coacervate is a particle (ranging from submicroscopic to microscopic size) composed of two or more colloids. A colloid is an insoluble particle of 0.01 to 0.1 micron in diameter. Its chemical composition might be protein, lipid, nucleic acid, etc. Therefore, the colloidal state refers to the physical rather than the chemical form. These colloids tend to clump together to form the masses.

Oparin further proposed that although these coacervates were not living, they behaved in a manner similar to biological systems (1) by being subjected to natural selection (in terms of a positive selection for stability as expressed through persistent size and constant chemical properties), (2) by being chemically directive (selective accumulation of material), and (3) by reproduction (by fragmentation).

Oparin's ideas created a rash of excitement, but it was not until 1953 that some experimental evidence was gathered to support the hypothesis of spontaneous generation at the molecular level.

Stanley Miller and Harold Urey, in 1953, devised an apparatus that simulated the conditions of the surface of the prebiological Earth, and analyzed the molecular forms that arose. The apparatus (as diagrammed in Fig. 16-3) consisted of exposing water vapor to electric shock of 75,000 volts (to represent the lightening of thunder storms) in an ammonia and methane atmosphere (the atmosphere of the early Earth). After two weeks of treatment, Miller chromatographed the final solution (see p. 68 for the rationale of the chromatographic technique), and found many organic compounds, including both carbohydrates and amino acids. Many of the compounds were important metabolic intermediates of present-day organisms.

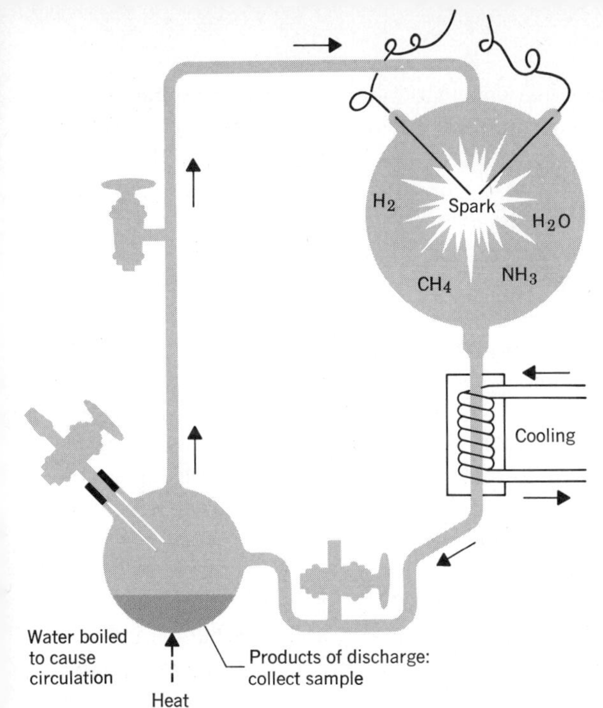

Fig. 16-3 The spark-charge apparatus of Miller. [After Miller]

A number of experimental variations were tried. Those that succeeded in producing biologically active molecules possessed three factors that were homologous to those necessary in the metabolic machinery of biological systems: (1) a high-energy source (electricity, heat, or UV light in simulations of the early Earth, are homologous to high-energy phosphates in metabolic processes), (2) reducing conditions (ammonia as a reducing agent in the early Earth is homologous to NADH or cytochromes in metabolism), and (3) an aquatic medium (the pristine seas of the early Earth are homologous to the internal basis of life in presently existing organisms).

Miller's results were significant in demonstrating that it is possible to pass from the inorganic to a biologically important organic level. But Oparin had postulated large molecular aggregates, and the natural-selection pressures that favor them, rather than simple molecules. Could such large molecules (coacervates or some other association) be produced in the laboratory under conditions that resembled the early Earth?

The challenge was met by Sidney Fox. He and a number of collaborators mixed all twenty amino acids together and heated them on volcanic rock in an oven at 170° C for two hours. The surface of the rock was sprinkled with water (causing it to fizz) and allowed to cool. When the residue was examined under a microscope, millions of small spheres were seen, similar in size and shape to yeast cells (see Fig. 16-4). Fox christened them **proteinoids** or **microspheres** because of their chemical nature. The polypeptides in proteinoids consisted of about eighteen amino-acid residues, but because they were produced spontaneously, each had a different amino-acid sequence.

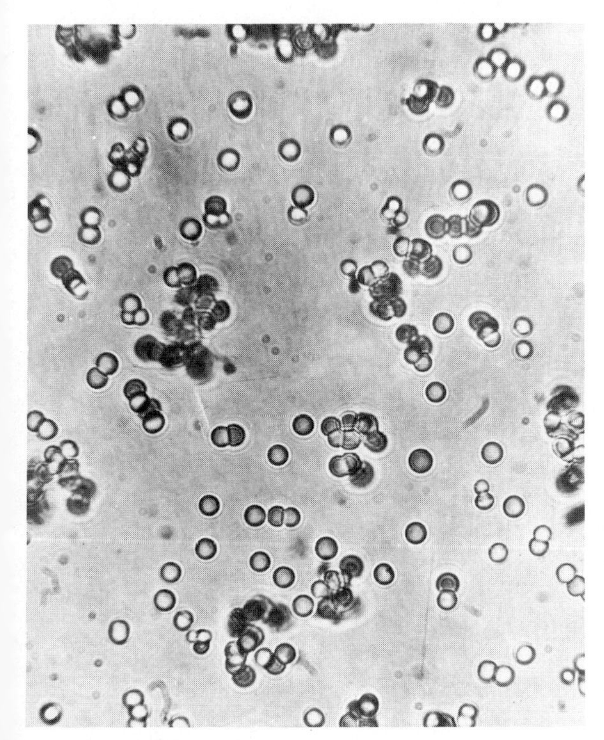

Fig. 16-4 Microspheres resulting from heated mixtures of dry amino acid (2000X). [Courtesy Dr. Sidney Fox, University of Miami.]

There are several reasons why the microsphere may be the large molecular form produced in the pristine seas upon which natural selection operated. First is the fact that microspheres have been produced under conditions that resemble events in the prebiological Earth. Second is the biological activity exhibited by these units. Tests have revealed that these microspheres can carry out a variety of enzymatic reactions, notably oxidations. (Since one microsphere cannot be isolated and tested for its individual enzymatic activity, all microspheres must be tested together.) In addition, it was observed that a small bleb or bubblelike protuberance sometimes grew out of the surface of the sphere, enlarged, and broke away, in a more ordered fashion than Oparin had postulated for molecular aggregates subjected to natural selective forces.

Most curious, however, was the double-layered surface with semi-permeable properties that responded to hypertonic and hypotonic solutions. Thus, the surface of the microspheres is not unlike the cell membranes found in organisms. The double-layered membrane has been thought of as a unique biological structure that could be produced only by specific amino-acid sequences as positioned by genetic regulatory mechanisms. It now appears that the uniqueness of membrane structure has been overestimated, and that the double-layered pattern is spontaneously formed as a chemical property of a host of different proteins with different amino-acid sequences.

THE EOBIONT—THE FIRST LIVING FORM

It had been demonstrated by Miller and Urey that biologically important molecules were not only possible, but probable as geochemical products of the early Earth. Fox has demonstrated that molecular groupings of amino acids into proteinoids could have occurred. Oparin has postulated that the inherent stability of chemical entities would favor the formation of larger, more stable molecules. But the microspheres are not living, and it is a long way from the association of a few amino acids to the formation of the first living entity.

The question is, then, how much elaboration must a collection of molecules, such as a microsphere, have to go through until it becomes a living entity? The answer to this question involves two factors. First, what qualities must a molecular aggregate have in order to be considered living? Second, what is the simplest molecular aggregate that manifests these properties? Therefore, before one can attempt to think in terms of an **eobiont,** the first living entity, one must decide on a set of properties that separate the living from the nonliving.

A Definition of Life

Every biologist at some time tries his hand at defining life, if only to overcome the frustration of spending his life investigating an intangible, philosophical concept. The definition that he assigns usually reflects his own interests, and encompasses those factors that are most important to his area of investigation. What invariably emerges, however, is a set of properties that characterize living systems, rather than a definition of "life." For "life" is, and must always remain, a philosophical rather than a biological concept.

The definition of living systems that will be developed here is arbitrary in that it is meant to provide an operational definition, designed to handle the problems associated with the origin of life.

Although the specific chemical reactions and the functions resulting from those chemical reactions may vary among organisms, certain patterns of organismal activity exist. These patterns can be emp[illegible] to define living systems.

1. *The Capacity for Synthesis:* Living systems convert different energy fo[illegible] into the chemical energy of ATP, which is then used to drive reactions for the synthesis of macromolecules for maintenance and growth.
2. *The Capacity for Self-Regulation:* Living systems contain information content in the form of DNA, which specifies the polymeric syntheses and monomeric degradation that can occur through the regulation of protein synthesis.
3. *The Capacity to Genetically Adapt:* Living systems retain an identity apart from, yet within, their environment while being inherently able to adapt to new environments or improve their efficiency in the same environment by random changes in their information content. In short, living systems undergo evolutionary changes.

Living systems can then be described as *chemical systems capable of elaborate polymeric syntheses, directed in specific patterns by the presence of information (itself in a polymeric form), that is capable of changing to other controlled patterns at more efficient levels of environmental exploitation.* This almost unwieldy description attempts to couple the main features of living systems, self-regulation and natural selection.

The textual material in the remainder of this chapter will be mostly mental exercises. The reason for this approach is simply that nothing is known, and probably nothing ever will be known, about the actual historical events leading to the origin of life. Biopoesis is a very special problem in biology because it is concerned with a particular sequence of events, rather than being like the phenomenon of metabolism or gene recombination, where recurrent activities lead to a steady output of identical products.

The type of analytical method employed here is germane to political history, not science, and we are, therefore, forced to develop the origin of life as a unique sequence of events rather than a scientifically testable one.

The Character of the Eobiont

Given a list of characteristics that set living systems apart from nonliving entities, it is now possible to attempt to find the simplest molecular aggregate that manifests these properties, and assume that this represents the *character* of the eobiont.

The search for the molecular constitution of the ebiont, must begin with a development of ideas regarding its nutritional requirements. Early evolutionists postulated that the first forms of life were photosynthetic. The rationale was simple: photosynthetic organisms required the simplest environment. They could synthesize most of their own molecules and so survive regardless of the environmental contribution. With the realization that organismal complexity is inversely related to its demands on the environment (a photosynthetic organism has more enzyme systems than does a chemoorganotrophic one) and with the proposal that the early seas were almost an organic soup, the perspective changed. It is now believed that the eobiont was heterotrophic, drawing on the environment for all amino acids, nitrogen bases, cofactors, inorganic ions, and energy sources. Such an organism would require very few enzyme systems, since metabolic intermediates for the conversion to energy or polymeric build-up would be available from the environment.

The molecular constitution of such a heterotrophic eobiont can be assumed to have been the simplest combinations of molecular forms that manifested those properties outlined in the definition of living systems. One can postulate the *type* (but never the specific eobiont in two ways. The first is by an examination of the organismal types present on the earth today to determine which would be the most likely candidate. The other is by a theoretical consideration of the minimum chemical composition that an organized living form could possess.

The most obvious place to look for an eobiont-type candidate is in the microbial world. To be considered for candidacy, however, an organism must survive in conditions similar to those of the early Earth. Viruses, although they conform to the definition of life, are dependent on a host cell for many specific polymers that could not have been present in the primordial soup. In many scientists' minds, this consideration has eliminated viruses as candidates for the eobiont.

Many of the bacteria are eliminated because of their structural complexity. There is one group of bacteria that are about one-tenth the size of most bacteria, called the PPLO (pleuropneumonia-like organisms) forms. The PPLO forms have been grown in artificial media, and require no macromolecules or highly unstable metabolic intermediates (see Table 16-1).

TABLE 16-1 GROWTH MEDIUM FOR PPLO ORGANISMS

Na_2HPO_4—KH_2PO_4 (pH 7.8)	0.04 *M*
Na lactate	0.15 *M*
Glucose	5.00 g
Bovine serum fraction C	0.5 g
Albumin	2.0 g
Cholesterol	10.0 μM
Tween	5.0 mg
Glycerol	2.0 mg
Mixed amino acids	4.5 gm
Cysteine	3.5 g
$MgSO_4 \cdot 7H_2O$	50.0 mg
$FeSO_4 \cdot 7H_2O$	0.25 mg
$MnSO_4 \cdot 4H_2O$	0.25 mg
Adenine	5.0 mg
Guanine	5.0 mg
Uracil	5.0 mg
Thymine	2.5 mg
Calcium pantothenate	0.25 mg
Riboflavin	0.25 mg
Pyridoxal HCl	0.05 mg
Nicotinamide	0.5 mg
Thiamine	0.5 mg
Biotin	0.01 mg
DL-lipoic acid	0.01 mg
H_2O	1000 ml

Chemical analysis has shown that the PPLO forms are made up of about 4% DNA, 8% RNA, and about 40 different kinds of enzymes. The chemical analysis does *not* indicate *how many different types of proteins* these organisms are capable of producing. It is possible to calculate the number of different protein types, the *protein potential,* by using the values for the DNA content. The 4% DNA represents about 75,000 nucleotide pairs. Since each protein contains about 300 amino acids, and one amino acid is coded by three bases, the protein potential can be stated

$$\frac{\text{number of nucleotide pairs}}{\text{number of nucleotides to code for one amino acid} \times \text{number of amino acids}} = \text{protein potential}$$

$$\frac{75{,}000}{3 \times 300} = \text{approx. } 80$$

The PPLO forms are capable of producing some 80 proteins, at least 40 of which are the aforementioned enzymes. The value obtained for the protein potential will be used to compare the PPLO type with the type developed through a theoretical consideration.

Determining the simplest form of life by theoretical considerations involves the reduction of a contemporary cell to its lowest metabolic capacity. This is accomplished by replacing enzymatic macromolecules with their simple products, as though they could be supplied by the environment. For instance, the glycolytic breakdown of glucose (a six-carbon molecule) to pyruvate (a three-carbon molecule) would be unnecessary if there were an abundant supply of three-carbon molecules in the medium. A cell in this environment would require almost no enzymes for metabolism. The only enzymic controls that would remain necessary are those for (1) the electron transport system (for generation of ATP), (2) RNA and DNA polymerase (for synthesis of RNA and DNA in growth), and (3) phosphorylating enzymes (for making nucleotides from nitrogen bases) (see Table 16-2).

TABLE 16-2 CELL[a] REDUCED TO LOWEST METABOLIC CAPACITY

Unit of Molecule Function	Class of Molecules	Number of Different Kinds	Multiplicity in Cell	Monomers in Each Kind	Total Monomers
Membrane	Protein	6	500	300	900,000
Phosphorylating	Protein	10	25	300	75,000
Electron Transport System	Protein	10	25	300	75,000
RNA Polymerase	Protein	1	25	300	7,500
DNA Polymerase	Protein	1	25	300	7,500
Ribosomal	Protein	2	25	300	15,000
Messenger	RNA	30	5	1000	150,000
Transfer	RNA	20	100	80	160,000
Ribosomal	RNA	2	50	150	15,000
Genes	DNA	52	2 (double helix)	1000	104,000
					1,509,000

[a] Since the early Earth did not contain oxygen, the metabolic pathways would have been similar to those of present-day anaerobic bacteria.

The calculation of the protein potential of a cell at the lowest level of complexity involves some preliminary estimates. The first preliminary estimate must give the DNA percentage of such a cell. The percentage value of DNA can be obtained from the ratio of the monomers in DNA to the total monomeric content of the cell.

$$\frac{\text{number of monomers in DNA}}{\text{number of monomers in cell}} \times 100 = \%\ \text{DNA}$$

The total monomers of the cell can be obtained by summing the total monomers in each unit of function within the cell, which can be obtained in the following way:

Example: membranes

Number of different kinds of molecules in membranes = 6
Total number of molecules in membranes = 500
Number of monomers in each molecule = 300

$6 \times 500 \times 300 = 900{,}000$ monomers localized in membranes

This type of calculation is the basis for Table 16-2. The protein potential may be obtained by adding the number of different protein molecules in the protein-containing cell units in Table 16-2. Hence the protein potential of this theoretical eobiont is 30.

These approximate calculations indicate that by considering the simplest form of life either from the standpoint of existing organisms or through theoretical manipulations, one is led to the same general estimates. The type of organism postulated by theoretical considerations can and does exist. It is reasonable then, to assume that the eobiont could have been of this degree of organismal complexity. It probably had some, but very few, control mechanisms. Its reproduction was barely beyond the fragmentation method of proteinoids, and its perpetuation was greatly dependent on the continual supply of metabolic precursors from the environment. It was an organism capable of synthesis by the utilization of ATP (which probably arose from the coupling of environmental adenosine and phosphate by phosphorylating enzymes). It could control macromolecular synthesis by a DNA-RNA-protein sequence as found in present forms of life, and it underwent evolutionary advance by natural selection operating on its variation in the eobiont population.

THE ORIGIN OF LIFE

The eobiont, as outlined above, was a complicated little creature. It did not come about by reproduction, but was spontaneously generated through the combinations of monomers available in the environment into polymeric sequences that cumulatively endowed it with the properties of life. But how did the eobiont come about? Was it merely chance, or was it a type of natural selection operating on molecular aggregates? The answer to this requires an insight into the raw statistical probability of life occurring on the Earth.

The Probability of Life

What is the chance of an eobiont simply falling together from the organic ingredients of the environment? Each necessary monomer is one of at least 100 different types present in the early seas. The chance that the proper monomer would be selected from the pool would be $\frac{1}{100}$. Since the model previously proposed for the eobiont was made up of some 1,509,000 monomers, the chance of such an eobiont forming becomes $(\frac{1}{100})^{1,509,000}$

Even if one introduces both a time and a space factor into the calculations, and assumes that an eobiont would be formed if all the proper reactions took place in an hour, in any one cm space on the earth, the number still cannot be reduced to the point where the origin of such an eobiont could have occurred by chance alone.

Another aspect of the statistical analysis of the origin of eobiont involves the genetic basis of life. The initial form of life was not of the type known today, but was instead much less able to control itself and undergo a successful cycle leading to reproduction. Since the control mechanisms were undoubtedly genetically based, one must consider the perpetuation of organismal type through the cycle of reproduction as one does for the survival of mutant genes. It can be shown that at each generation after the arrival of a mutant gene, the chance of its extinction increases![*] Therefore, life would have had to evolve not once, but in the order of 1000 or even 1,000,000 times before a lineage of successful forms arose that were the ancestors of those present on the Earth today.

The small chance of the origin and survival of the first forms of life favors the idea of natural selection operating on a molecular level. This idea is somewhat parallel to the case of adaptation (see p. 441), in which three genes could not mutate simultaneously ($10^{-9} \times 10^{-9} \times 10^{-9} = 10^{-27}$), but could mutate in sequential generations. The origin of life can be viewed in the same way. The coacervate of Oparin, or Fox's microsphere, could acquire further capacities, and selection would favor those manifesting stability and the tendency to fragment. Gradually, enzymes would replace randomly formed proteins, and nucleic acids would appear. Many lineages were probably begun, only to become extinct while others changed into new types and adapted to new environments. Eventually, through this process of natural selection, the eobiont type was formed, not as a single chance event, but as the outcome of a stepwise accumulation of traits, directed by the selection pressures of the total environment. It is thought that life took about one billion years to progress from the stage of small aggregates of molecules to the biological level of the primitive (PPLO) type of cell.

The Origin of Enzymes

On their way toward the eobiont type, molecular aggregates acquired enzyme systems. It is presently thought that enzymes evolved before the DNA code for protein structure. The basis of this belief is the probable abundance of coenzymes (nitrogen-base derivatives, vitamins, and inorganic ions) in the early seas. These coenzymes have an inherent capacity to accelerate chemical reactions even without the protein moiety usually associated with enzymes. In the early environment, these raw coenzymes would have driven a wide variety of reactions, some of which would have been of metabolic importance. Only when these coenzymes became scarce and survival was based on the presence of a system able to incorporate and retain the coenzymes in a stable form, would an information system be important.

Selection would tend to favor specific reactions. The coenzymes in the early seas were far from specific. Some molecular aggregates were capable of surrounding the coenzyme with organic material that limited the shapes of acceptable substrates. The principle of wrapping a coenzyme in organic material can best be illustrated by discussing the effect of heteroconjugate and protein moieties on iron. It should be emphasized

[*] According to the Poisson distribution (see p. 433), there is a 37% chance that an individual will have no offspring. Thus in 37% of the cases where a new gene arises, the new gene is not passed on to the next generation. The offspring of the remaining 63% could carry the new trait. Upon forming the next generation, however, 37% of the individuals will have no offspring. Generation after generation the chance that a new trait is perpetuated cumulatively decreases. There is only a 5% chance that a new gene will survive until the tenth generation.

that this is merely an example of the principle; specific proteins were not formed on the early Earth.

The iron atom can undergo a variety of reactions, only a few of which involve its affinity for oxygen. The presence of the protein of the enzyme catalase, which surrounds the iron atom in a twisted quartet of polypeptides, serves to provide specifically shaped openings that permit only hydrogen peroxide to come in contact with the iron (see Fig. 16-5). In addition to this, the presence of the polypeptides shifts the reducing potential of the iron to a more advantageous state.

It is thought that the wrapping of coenzymes in the early seas had similar effects. Cloaked coenzyme accelerated more specific reactions, and the molecular aggregate to which it belonged was endowed with a selective advantage.

The specific reactions in which the cloaked coenzymes were involved may have led to an increased stability, or a capacity of the molecular aggregate to fragment more readily. Stability would insure the continued presence of the molecular aggregate, the prime prerequisite for further building changes. Fragmentation would assure the multiplication of a successful type, thereby providing more sites on which the building changes that would lead to the eobiont could take place. Thus, selection would favor those enzymes that promoted stability and/or effective fragmentation.

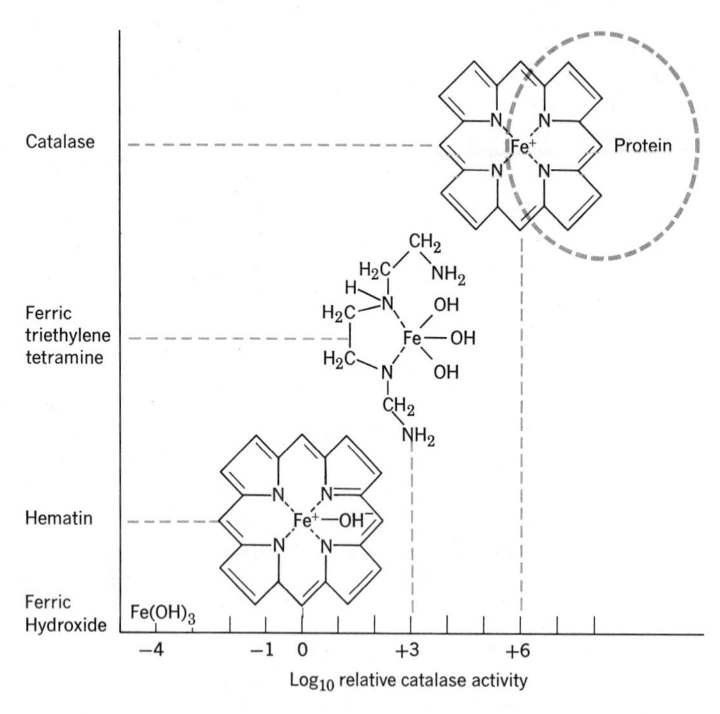

Fig. 16-5 Catalytic activity of several compounds on the decomposition of hydrogen peroxide to water and oxygen (after Wang).

The Origin of DNA

At the very heart of the problem of the origin of life is the emergence of information molecules. In order for the same proteins to be formed without variation, three systems must be present: *information content molecules, decoding molecules* (activating enzymes for each amino acid), and *enzymes for DNA replication.*

At first glance it would appear that such a complex system could have been developed by sequential minor adaptations. The very nature of the components of the system tends to nullify this idea. DNA is not functional, it is informational. A book is useless if there is no one to read it. So is DNA. Information content only takes on meaning when it is decoded and the information is used. If a decoding mechanism (and thus a meaning in terms of selection) were lacking, information molecules would not have been retained to be elaborated upon. The same is true for the decoding mechanism and enzymes for the replication of DNA. Without the informational molecule, they would be useless, and hence would not elicit a positive selective response to their presence. The only conclusion possible is that all three components of the information system arose simultaneously!

But this conclusion is also intellectually dissatisfying. Older theories of biopoesis postulated the random appearance of naked genes, some of which produced enzymes that stimulated DNA replication. The probability that such an enzyme would spontaneously have been coded for by a DNA sequence is infinitesimal. Since there are four nitrogen bases in DNA, and 1000 of these bases make up a gene, the chance that the sequence of bases would code for such an enzyme would be 4^{1000}.

Regardless of how the DNA-RNA-protein sequence came about, once it was established, the eobiont had been formed.

(If the lack of speculation of the origin of the DNA-RNA-protein system leaves one bewildered, he may take some solace in knowing that even scientists that lean toward armchair speculations are no less bewildered.)

The Origin of Cells

Once the eobiont was formed, and the DNA-RNA-protein system established, a variety of control mechanisms became essential. Genes cannot be read and replicated simultaneously, and the genes that control these two functions must be activated and deactivated according to some coordinated scheme. An efficient organism must divide in conjunction with DNA replication. It should regulate the synthesis of proteins selectively, according to the environmental contribution of substrates. The integration of organismal activities into a genetically mediated pattern took the form of the cell.

Thus it is possible to conceive that selective pressures operating on large molecular forms in the pristine seas could initiate a building evolutionary process that some three billion years later carpeted the Earth with a wide variety of living cellular forms.

QUESTIONS

1. Give the experimental evidence that disproved the spontaneous generation of maggots and microorganisms.
2. What conditions of the early Earth provided the necessary factors for the generation of life?
3. Justify the following statement. "Microspheres, while not living entities themselves, had to be formed before a living entity could come about."
4. Characterize the eobiont either according to organisms existing today, or on theoretical grounds.
5. What reasonable speculation is given for the genesis of enzymes?
6. Why does the present concept of the DNA information system make it impossible to speculate satisfactorily on the evolutionary origin of such a system?
7. Comment on the following definition of a cell: A cell is an integration of organismal activities into a genetically mediated pattern.

CHAPTER 17

THE ORIGIN OF MAN

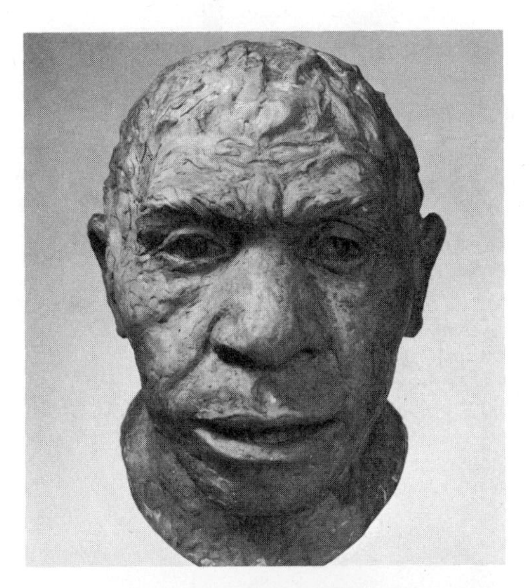

The first remains of a primitive man were found in 1856. They are now known to have been the remains of a Neanderthal man (see Fig. 17-1), but at that time scholars had too little experience to know how to interpret the discovery. During the past century, hundreds of different fragments have been obtained, revealing that a number of kinds of "ape men" and pre-men roamed the earth during the past several million years. At one time it was hoped that the so-called "missing link" could be found, but when many claimed Dubois' discovery of the Java man in 1896 as the "missing link," others wanted to find more links between the three groups—apelike ancestor, Java man, and modern man. As more of these so-called links were unearthed, the assumed linear phyletic origin of man was discarded, and evolutionists began to view man's origin as a complex candelabralike pattern within which many different man-like forms originated.

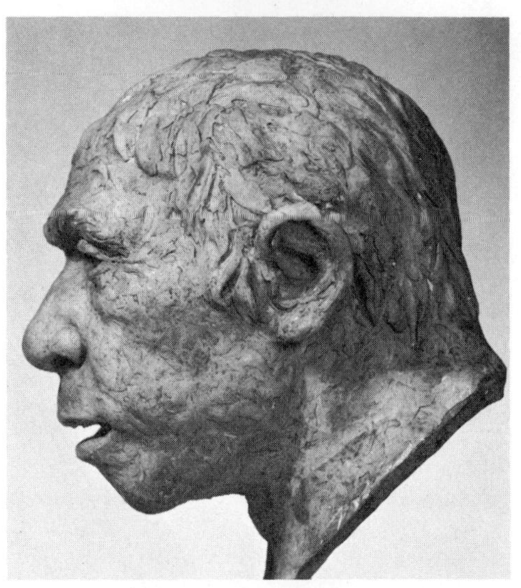

Fig. 17-1 Neanderthal man, or *Homo sapiens neanderthalis.* Although he was small in stature (five and one-half feet), his brain was larger than that of modern man. [The American Museum of Natural History]

Fig. 17-2 Tarsier of Mindanao, Philippines, is geographically restricted, as are the lemurs of Madagascar. What would account for the peculiar distribution of these primitive primates? [San Diego Zoo photo]

Fig. 17-3 The gibbon lives in southeastern Asia and is a higher primate (anthropoid ape). His powerful arms and relatively small size make him a graceful brachiator. [A. W. Ambler—National Audubon Society]

Fig. 17-4 Drawn model of Proconsul. [From *Life: An Introduction to Biology*, 2nd Edition, by George Gaylord Simpson and William S. Beck, © 1957, 1965, by Harcourt Brace Jovanovich, Inc. and reproduced with permission.]

Reconstruction Through the Fossil Record

The beginnings of man can be traced back about seventy million years, when the primate group was represented only by **prosimians** (lemurs and tarsiers). By forty million years ago, most of the prosimians had vanished. Their disappearance coincides with the emergence of many rodents and carnivorous mammals (which were undoubtedly prosimian predators) as well as the large primates. The lemurs have survived mainly on the island of Madagascar, and the tarsiers are now peculiar only to the Philippines (see Fig. 17-2).

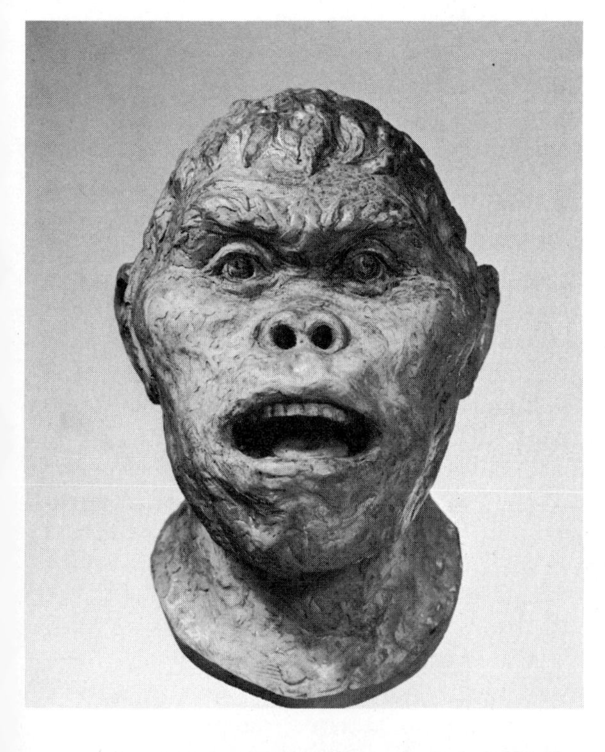

The gibbons can be followed from their earliest appearance 35 million years ago to the present, and their skeletal change indicates that they achieved a **brachiating** habit (long arms for swinging and ground balance) independently of other primate lineages. The first apes do not differ radically from extant ones, and it now seems that the apes are more a group of living fossils than a specialized primate branch of evolution (see Fig. 17-3).

One curious form, **Proconsul** (see Fig. 17-4), of which several hundred remains have been found, had teeth and skull characteristics that are apelike, but it was not brachiating. Rather, it walked on all fours like the Old World monkeys. This creature of 30 million years ago may represent the ancestral stock of the apes, and possibly man.

Proconsul was one of a large number of anthropoid apes collectively known as *Dryopithecus.* They represent a wide variety of forms, some of which were brachiating, others not. This assemblage suggests that the hominoid primates were undergoing extensive evolution during the late Tertiary period.

A period of about 10 million years elapses in the fossil record from the end of the dryopithecines to the appearance of the next major group, the australopithecines. Many remains of members of *Australopithecus* (see Fig. 17-5) have been found in several regions of the world, although the richest finds have been those in South Africa, which seem to date as early as 1,700,000 years ago. The shape of the pelvis is definitely hominoid and not apelike. Therefore, these creatures walked erect. The shape of the femur bone and the arrangement of some of the bones in the foot tend to support this conclusion. Their teeth were also similar to those of humans. The size and shape of the skull were apelike, with a cranial capacity of about 600 cc (modern man has a cranial capacity of 1300 cc).

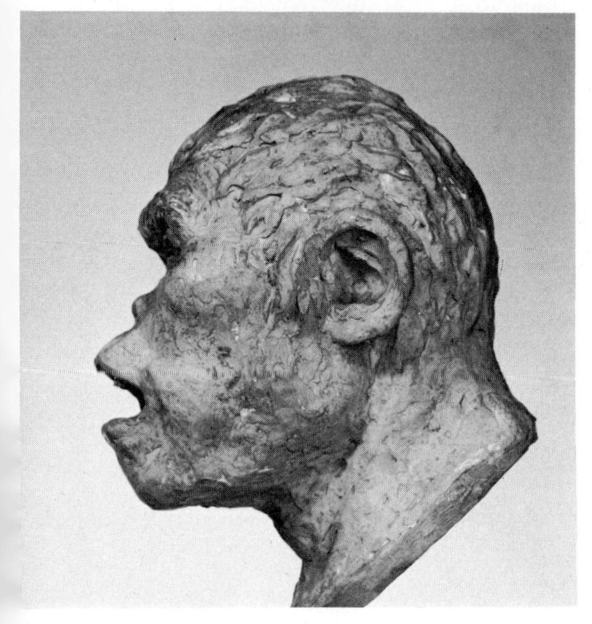

Fig. 17-5 Model reconstruction of *Australopithecus* (southern man) who lived about 1,750,000 years ago. [The American Museum of Natural History]

In one site, the anthropologist L. S. B. Leaky found several small stones with their bones. Since these stones were unlike others found in the stratum, it is thought that they were collected, possibly polished, and used in some manner by these ape-men. The anthropometric variation in the australopithecines over a million-year period indicates that this group experienced considerable evolutionary change. Within this group, mankind probably appeared on the scene, not as an intelligent ape or a manlike creature, but as a genuine man, capable of civilized activities such as burial, tool making, and religion.

A number of decidedly human-type groups appeared in the Mid-Pleistocene period. These groups have been collectively classified as *Homo erectus.* Within this classification is the **Heidelberg man** (550,000 years ago), the **Java man** of Dubois, the **Peking man,** and some of the remains found in South Africa (see Fig. 17-6). The *Homo erectus* individuals were acquainted with the use of fire; they also fashioned weapons and practiced cannibalism. Their brain capacity ranged from 750 to 1200 cc. They had thick skulls, heavy brow ridges, no chin, and large teeth. They existed from 600,000 to 400,000 years ago, and then, for unknown reasons, vanished.

After this period the fossils became more abundant and more similar to modern man. Yet no pattern of evolution seems to emerge. A group of Home sapiens appeared in Europe about 200,000 years ago, and appeared to be replaced by the **Neanderthal man** (H. sapiens *neanderthalensis*), who persisted for about 100,000 years. He was about five feet tall and highly muscular, had a large brain capacity (1450 cc), and practiced most of the ceremonies associated with civilized groups. These men existed in large numbers in geographically separate locations. The skeletal variations found in different parts of Europe indicate that genetically different populations of Neanderthal man arose, which is exactly what would be expected in an evolutionary development of a species.

Within a matter of a few centuries, the Neanderthal man disappeared and was replaced by another H. sapiens sapiens creature, the Cro-Magnon man. This replacement occurred some 35,000 years ago.

Fig. 17-6 Models of (*a*) Java man and (*b*) Peking man. [The American Museum of Natural History]

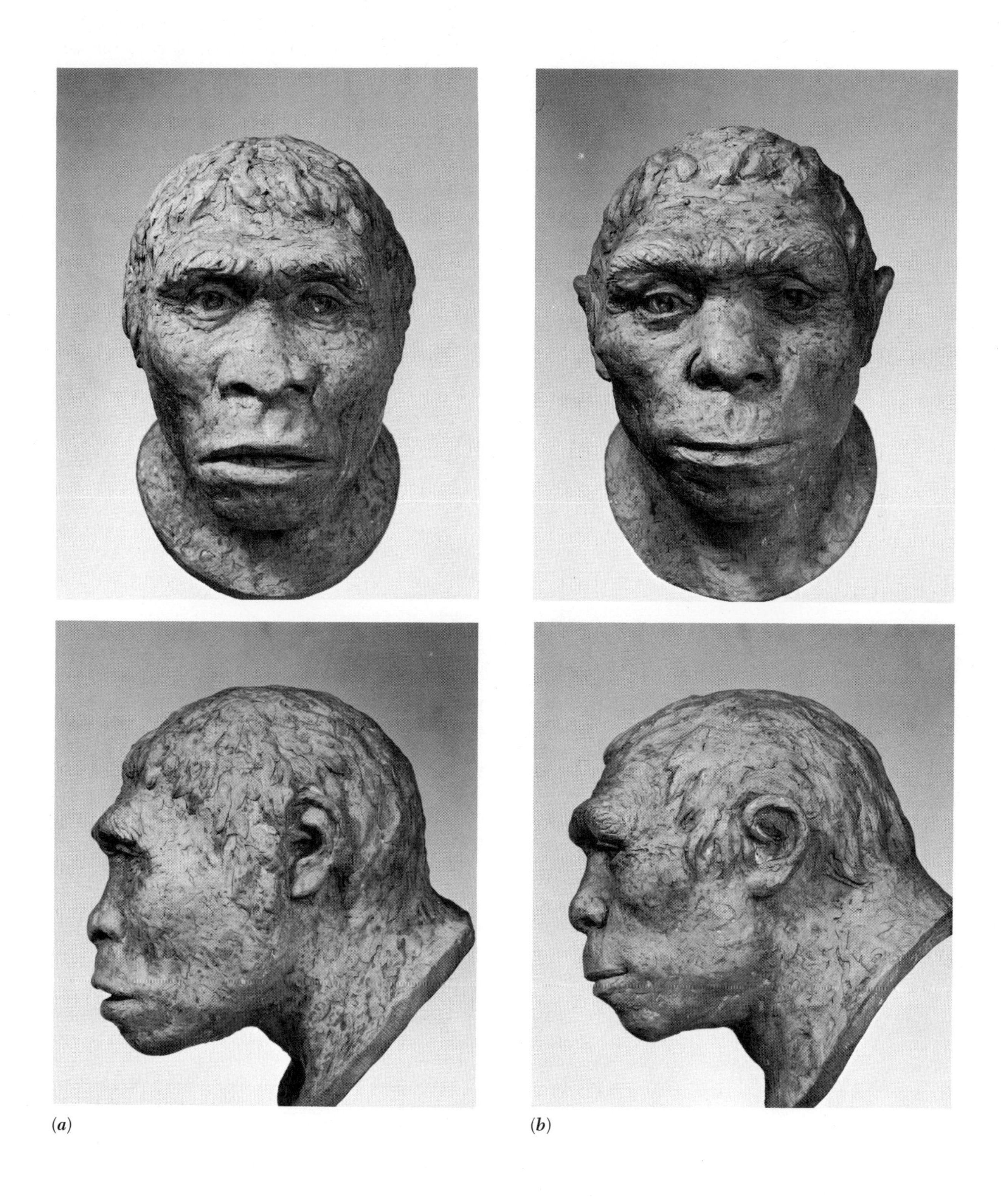

(*a*)

(*b*)

One curious fossil from Mount Carmel in Palestine exhibits the features of both Neanderthal and modern man as though it were the remains of a hybrid type. Whether the Neanderthal group became extinct because of its own inadequacies, was physically exterminated by H. sapiens, or was genetically consumed through hybridization, is unknown. The record does show that modern man appeared, was replaced by the Neanderthal group, and then reappeared.

The fossil record may not be as complete as one would desire, but it does contribute some significant information about man's origin and corrects some erroneous, preconceived ideas. Ten generalizations may be drawn from the fossil evidence that indicate the trends through history of the evolution of man.

1. Manlike tendencies appeared before the modern ape evolved. Hence, man did not evolve from the apes, but both have a common ancestor, probably **Proconsul.**
2. Man's origin was not phyletic speciation. Many variations of pre-man existed at the same time, indicating that many different natural experiments were in progress simultaneously. The notion of a "missing link" suggests a linear evolution, and therefore is erroneous. The answers to man's origin may be found in trends, hybridization between manlike types, and replacement of types.
3. The evolution of man was a rapid process compared to most cases of speciation. Although the actual time necessary to achieve each near-human stage is debatable, the variation of simultaneously existing individuals possessing these semi-human traits indicates that rapid changes were realized at each level of human evolution.
4. The upright position was achieved before the full intellectual capacity of man was realized.
5. Cranial size is not necessarily correlated with mental ability. Neanderthal man had a larger brain than his successor, H. sapiens sapiens.
6. Man today is one species. The variation expressed among the different races is not as great as those that existed at the various time levels during his evolution. Whether modern man's uniformity should be interpreted as the survival of one anthropoid race or as resulting from the convergent evolution of different ancestral groups is not known.
7. Mankind has not undergone any significant biological evolution during the past 35,000 years. This fact should not be interpreted to mean that the evolution of man is complete, because this is a short period on the evolutionary scale. This apparent cessation may be a period of slow change.
8. The fossil record has revealed no major gaps in the evolution of man. Since the search for man's past is but a century old, the quest has been remarkably successful. The number of significant findings discovered over the past twenty years suggests that many more exciting discoveries will be made in the future to provide an even more continuous pattern of human origin.
9. Nothing in the present record indicates how and when speech or religious beliefs originated.
10. The comparative numbers of each type of fossil found indicates that man was not a dominant species until a few thousand years ago, and that during some periods, was probably at the edge of extinction (between *Australopithecus* and the emergence of H. erectus, and again during the Neanderthal dominance.)

HOW DID MAN EVOLVE?

It is one thing to reconstruct a pattern of emergence, such as for the evolution of man, but it is quite a different task to explain why selection pressures favored the development of the particular features exhibited at different levels of evolution. Any explanation of the origin of man from this standpoint is in great part speculative because so little is known about man's ancestors, and, for that matter, about man himself. Some pieces of the puzzle can be put together, but their relative importance and their sequence of occurrence remains problematical. Any hypothesis about biological origin involves three overlapping phases: (1) the nature of the ancestral stock, (2) the kinds of adaptive changes that created a momentum towards the new way of life, and (3) the additional traits that were developed and that polished the product, stabilizing the new adaptive equilibrium.

Nature of the Ancestral Stock

Since man is an animal, many of the evolutionary changes in the entire phylogeny of animals have contributed to his successful establishment on this planet. The emergence of the mammalian form from reptilian stock was marked by several significant changes that established many of the functional features exhibited by man. Among these are:

1. *Viviparity,* the birth of live young (although some reptiles are viviparous and some mammals either lay eggs or have the young remain attached to the mother).
2. The suckling mode of infant nutrition.
3. The presence of hair rather than scales.
4. A homeothermic condition (temperature-controlled).
5. An increase in the size of the neocortex of the brain.
6. A change in the orientation of the skeleton so that the forelimbs are higher off the ground.

These changes to the mammalian character set the stage for the emergence of the primate type, and eventually, man. Viviparity greatly reduced the number of offspring from a single union. Amphibians tend to depend on the law of mass numbers for the survival of the species, by producing thousands of fertilized eggs, of which only two or three ever survive. When the number of offspring was reduced to that which the animal body could carry internally, and the feature was coupled with suckling of the young and prolonged maternal care, the mammalian pattern itself had increased the chance of offspring survival. The homeothermic condition in mammals required an insulation between the animal body and the environment. Hair (or feathers in birds—the other warm-blooded group) act as a more effective insulating material than do the scales of the reptilian types. The increase in the size of the neocortex continues through the more advanced mammalian types, until in man its development is sufficient to release the individual from extensive instinctual behavior. The skeletal orientation proceeds up the mammalian types until in the higher primates and man it manifests a **bipedal** (two-legged) rather than **quadrapedal** (four-legged) stance.

Within the mammalian group, changes occurred leading to the primate form. Comparative anatomical studies have led to the classification of primates into two large groups: the lower primates (tree shrews, lemurs, and tarsiers), and the higher primates (apes, monkeys, gorillas, etc.) (see Table 17-1).

TABLE 17-1 CLASSIFICATION OF THE PRIMATES

Group	Characteristics
Anthropoidae	
Ceboidae—New World Monkeys	Some with long tails Some long tails prehensile (able to grasp)
Cercopitheoidae—Old World Monkeys	Mostly tree dwellers
Hominoidae	
Pongidae	
Pan—Chimpanzee	Tree dwellers
Gorilla—gorilla	Brachiators (able to swing between branches)
Pongo—orangutan	Arms longer than legs
Hylobates } gibbons	Thumb and big toe
Symphalangus } gibbons	No tail
Hominidae	
Homo—man	

Man, monkeys, and apes have a large number of important features in common that set them apart from the lower primates and other mammalian types, marking them as rather unique animals. These features are:

Stereoscopic vision
Color vision
Loss of mobility of external ears
Replacement of a muzzle by a face
Reduction of sense of smell
Loss of tactile hairs (whiskers)
Occurrance of a menstrual cycle
Omnipresent sex drive
Usually single births
Elaborate maternal care

The primate stock from which man arose represents an arboreal mode of life, and the stock is structurally adapted for this. The collar bone, which is larger than in other mammalian forms, affords greater freedom in the movement of the upper limbs. The joint between the collar bone and upper limb allows a wider arc of movement. The five digits associated with the paw of lower mammalian forms is retained, and has developed greater mobility and tactile sensitivity. All these traits allow the early primates to climb trees and grasp branches for swinging.

The ability to see branches more clearly was of adaptive value, while the sense of smell became of secondary value. With the reduction of the nasal muzzle, the eyes, which in lower primates and other mammals are positioned on opposite sides of the head, became more closely set together. This repositioning of the eyes caused the images viewed to overlap, thus giving rise to stereoscopic vision. Since this trait endows one with the ability to determine distance, the eyes of primates became a type of range-finding device. The color vision in the higher primates covers a wider range of the spectrum than in other mammals, thus permitting a greater discernment between objects.

The brain of higher primates is both relatively and absolutely larger than that of other animals. Monkeys have the largest brain relative to their body size in the animal kingdom, but it is smaller on an absolute scale than that of man (see Fig. 17-7). The large absolute size of the brain indicates the presence of many nerve cells, only a fraction of which need be committed to functional control of the body. The increase in brain size led to a coordination of vision and grasping. The ability of the animal to perceive, and have his hands grasp quickly and independently of one another, made the hands exquisite devices for manipulation.

The shape of the skull in higher primates is noticeably different from that of lower primates and other mammalian types. The face and occipital condyles (insertion of the spinal column into the skull) have descended 90°. Were a dog to have this insertion, its nose would point constantly towards the ground (see Fig. 17-8).

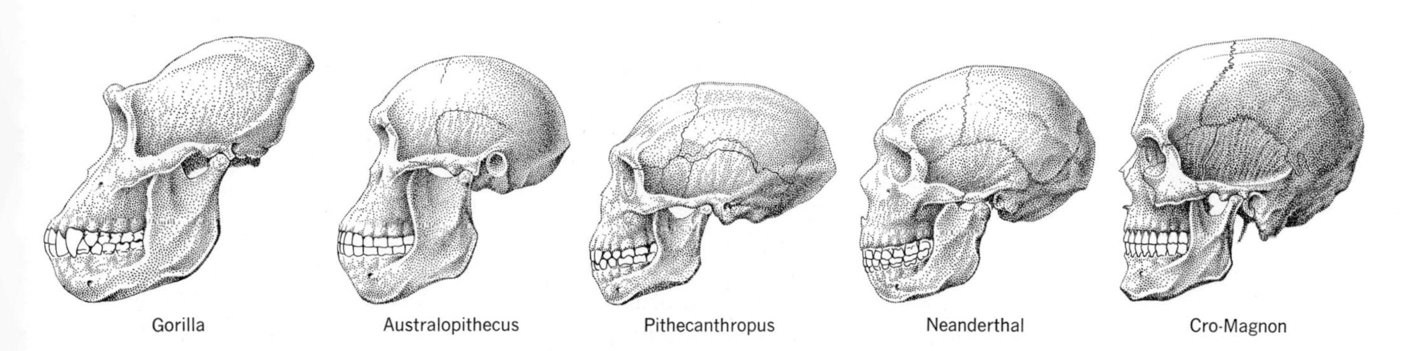

Fig. 17-7 Brain sizes as indicated by the skulls of gorilla, *Australopithecus*, *Pithecanthropus*, Neanderthal man, and Cro-magnon man. [From *Life: An Introduction to Biology*, 2nd Edition, by George Gaylord Simpson and William S. Beck, © 1957, 1965, by Harcourt Brace Jovanovich, Inc. and reproduced with permission.]

(a)

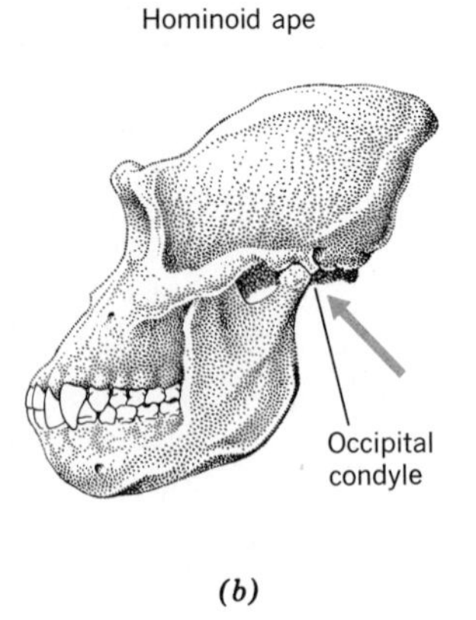

(b)

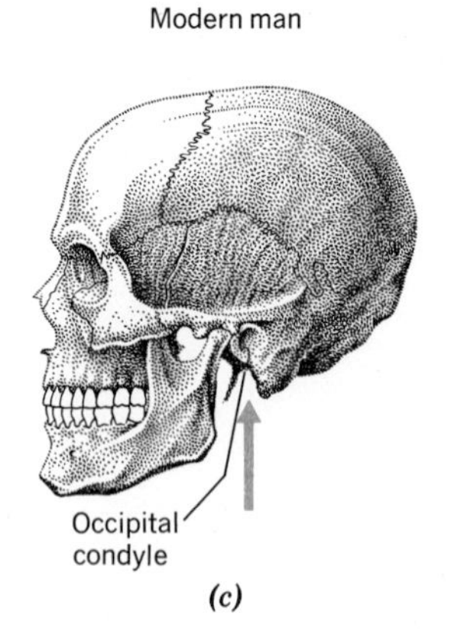

(c)

Fig. 17-8 (*a*) **Head and shoulders of gorilla. Position of head results from the insertion of the occipital condyles as shown in the drawing.** (*b*) **Gorilla skull with posterior occipital condyles of skull inserted on axis of vertebral column; muscles holding head to shoulder.** (*c*) **Human skull with basal occipital condyles inserted into first vertebra; muscles also indicated. [**(*a*)**, Philip Feinberg, Fellow of the New York Microscopical Society]**

The habit of tree dwelling placed restrictions on size, and so the primate stock from which both apes and man arose was rather small, like a monkey. This small size limited his fighting capacity, so he was probably a vegetarian. The ability to swing among trees produced a highly nimble creature who could escape from predators and other dangers quite successfully. He probably slept in trees at night, since he was adapted to daylight by his acute vision. The arboreal way of life imposed great difficulty on rearing the young, so selection pressures favored the birth of only one offspring from a union.

Although the human form may resemble those of the higher primates, man has several unique characteristics of great evolutionary significance. Man differs from other primates in the same way as the primates differ from other mammals: by shifts in development that reshape parts of the body rather than by distinct anatomical features. The anatomical modifications that have occurred during the evolution of man are probably important clues to the natural selective pressures that sustained the adaptive changes. The more obvious differences between man, monkey, and apes are listed in Table 17-2. These differences tend to accentuate (1) the freeing of the upper limbs for grasping objects while the lower ones remain exclusively for locomotion, and (2) the slowing down of development, concomitant with the involvement of the male parent in domestic affairs.

Adaptive Changes and the Creation of a Momentum. Because tree life required nimble hands and a partially erect position, the anatomical features of arboreal primates are similar to those exhibited by man in his terrestrial environment. The full advantage of the marvelously fashioned hands in a manipulative sense could not be realized until the creature descended from the trees and assumed a fully erect position. But while terrestrial life offered the advantages of freedom of the hands and more elaborate care of the young, it also presented problems in means of protection and food gathering. Thus it would appear to be more advantageous for the primates to have remained in trees. But at least two lines of primates came down from trees and assumed a terrestrial existence. The first question, then, is, what brought the primates down from the trees?

It appears that primates were brought down from trees simply because of the demise of forests in many areas of Africa. The glaciations, which started some twenty-five million years ago, brought cascades of ice down through northern Europe, northern America, and northern Asia, converting these areas to Tiagas. Bounded on the north by colder areas, many regions of Africa became plains. Primates in these areas had to adapt to treeless environments or perish. What, then, were some of the adaptations that permitted primates to live in their new treeless environment?

TABLE 17-2 DISTINGUISHING FEATURES OF MONKEY, APE, AND MAN (after Washburn and Avis)

Trait	Monkey	Ape	Man
Locomotion	Quadrupedal	Brachiating	Bipedal
Shoulder	Moderately mobile	Very mobile	Very mobile
Stretching arms	Forward	To side	To side
Arm muscles	Large extensors[a]	Large flexors[b]	Large flexors
Free arm supination[c]	90°	180°	180°
Thumb	Small, but used	Small, but used	Large, much used
Body organs	Adjusted to quadrupedal posture	Adjusted to upright posture	Adjusted to upright posture
Sleep	Sitting	Lying down	Lying down
Growth rate	Slow	Slower	Slowest
Sexual maturity	2–4 years	8 years	14 years
Full growth in males	7 years	12 years	20 years
Infant dependency	1 year	2 years	6–8 years
Female receptivity to sexual intercourse	At ovulation	At ovulation	Continuous
Canine teeth in males	Large	Large	Small
Adult male provides food	Never	Never	Major responsibility
Shelter	None	Temporary nest	Houses

[a] Extensor muscles are those used for reaching.
[b] Flexor muscles are those used for lifting objects.
[c] Supination angle indicates the angle formed between the position of the head and the arm when the arm is at rest.

Several suggestions have been forwarded regarding the sequence of events leading to terrestrial life. Ancestors of man may have become **omnivorous,** able to eat both animal and plants. Animals were caught by dropping rocks or other objects to stun them. The ancestral ape then descended, killed the animal, and returned to the tree tops with his prey. Further development of this skill would permit him to reside for longer periods of time on the ground, and with the hands serving a manipulative function in the terrestrial environment, the momentum toward erect posture was set in motion.

One can also envision that the creatures became group-oriented, and in sufficient numbers, protection against predators was greatly improved. One recent and curious idea is that the primate ancestor developed a body odor detested by other animals. To these other animals the primitive prehuman neither tasted nor smelled good, and so was not prey. The idea that this was so, and might indeed still be true, is supported by the accounts of safari hunters. These hunters have a difficult time in finding some types of animals because these animals can detect man from 800 yards, and take measures to avoid contact. Such an idea might account for the ability of a new and relatively defenseless species to be established in a primitive environment. Possibly the three conditions; (1) emerging omnivorous habits, (2) group orientation, and (3) repugnant odor to potential predators, may have played some role in the descent of man's primate ancestor to the ground.

The descent from arboreal to terrestrial existence did not transform a primate stock into the human species. Indeed, other primate lineages show evidences of similar descents for other reasons. The gorilla lineage probably survived because of the adaptive advantage of brute strength and enormous size. As such a primate increased in size, it was unadapted for tree life and more suitable for terrestrial survival. The chimpanzee, on the other hand, spends some time on the ground, but is primarily a tree dweller. The lineage of the chimpanzee, therefore, does not represent a line of descent to the ground. The descent of man's primate ancestor was not at the expense of his intelligence, as in the case of the gorilla. Rather, his intelligence was the basis for a descent.

The momentum toward modern man began with the existence of this ground-dwelling, apelike creature who made the descent from arboreal life. It seems that it was of adaptive value for the erect posture to become established, before much increase in intelligence took place. (*Australopithecus* was erect, but had only half the cranial capacity of modern man. The size of the brain increased later.)

The increase in brain size and avenues for physical coordination necessitate a dual and simultaneous evolution. Just as the visual centers of the brain increased along with the manipulative features of the hand in the evolution from other mammalian types, other brain-centered control areas had to increase in both size and activity with the development of physical abilities. The two features must go hand in hand. A book is meaningless to one who cannot read. A brain capable of translating visual images to body activity is useless in a body that cannot perform those activities. One who knows how to read is unable to do so without printed material. A body that can perform activities cannot do so without a translation mechanism.

The dual evolution of mental capacity and physical ability intensified the coupling of the eye and the hand. The mind could interpret observed objects and relationships, and provide more opportunities for the hand to manipulate. This was of great adaptive value, for early man was in a hostile environment: left exposed in open areas, too weak to enter physical combat, and a nomad in strange regions. His ability to perceive and use a weapon to kill an adversary or seek shelter was an important feature in his survival.

Larger brains, however, presented two new problems. The first was that bigger brains require larger skulls, and larger skulls created difficulties at birth. The second was the inheritance of intellectual capacity. Patterns of inheritance demand that the *information* one has acquired over a lifetime die with him, and only the *means* by which the information may be gained can be passed on to his offspring. Each new individual must start from scratch in acquiring the knowledge necessary for survival, and this, unfortunately, takes time—time when the unprepared offspring is most open to attack. The solution to these two problems has created characteristics one assumes to be distinctly human traits.

For a human child to be born at the same level of ability as an offspring of a gorilla, or a great ape, would require a gestation time of about eighteen months. Such a child would be nine months old by our reckoning; he could crawl, sit, and respond to negative environments by removing himself from them (crawling away from heat or curling up for warmth). But the head size of such a child would make childbirth almost impossible. It is thought that such selection favored earlier (premature) births and nine months or 266 days from conception became the optimal time for the maximal survival of both mother and child. Premature birth results in a helpless child, requiring constant attention. As a result, the mother becomes shackled to the care of the young, and cannot continue to hunt for food. Food procurement then becomes relegated to the fathers, that is, to those fathers having a genetically endowed proclivity toward domestication. Only the proper combination of premature birth and genetically endowed tendencies toward maternal care and paternal domestication would insure the successful rearing of the young.

Fig. 17-9 Infant gorilla. [San Diego Zoo photo]

Although the above statements are purely speculative, there is some evidence to support the idea of the human as a premature primate form. Only infant primates have the everted lips necessary for suckling, the soft skin, the delicate facial features, and the nearly hairless body (see Fig. 17-9). The human at any age resembles the prenatal or infant primate far more than he does the adult primate types. The retention of the juvenile period is called **neoteny.** The classic example of neoteny is in the case of the amphibian, *Necturus,* commonly called the mud puppy. *Necturus* does not pass through the metamorphosis characteristic of most amphibians whereby the early gill structures are replaced by lungs, but retains the external gills permanently. It attains normal size and reaches sexual maturity. A similar pattern would account for the genesis of some features of the human form: the lack of hair, the retention of the aesthetic facial features, and the protruding lips that are essential for clear speech.

The growth rate of the young human is quite slow compared to that for other primates. Adolescence is far longer (6 to 8 years) than the gestation period, and often several infants at various stages of development have to be reared concurrently. The prolonged helplessness of the young, especially in companionship with both parents, offers a period of time during which knowledge can be passed from one generation to the next. The child appears to pass through periods of intensified interests or tendencies to learn. Periods of tactile curiosity, sound patterns, and a desire for parental acceptance provide the fertile ground for learning. Many scientists believe that these tendencies are built-in behavioral patterns under genetic control. The capacity for learning reaches a peak at fifteen years, after which intelligence does not increase. This corresponds to the time of sexual maturity, and in many societies it is the time at which the already independent child leaves home.

The transmission of parental knowledge to offspring can be accomplished in a number of ways. Drawing can teach the young about different animals, their vulnerable regions and more edible parts. The cave drawings of Dordogne near Bourdeaux, France may have been for teaching rather than aesthetic appreciation (see Fig. 17-10). The young can observe the behavior of the parents, and establish mocking patterns of action. But speech is more efficient than any other means of instruction. Speech is an adaptive trait that is unique to humans, and any physical change that would permit more distinct utterances would be of great value. The everted lips permit such sounds, and would, therefore, have had a selective advantage.

Fig. 17-10 Cave drawings by *Homo sapiens* in the caves of Font-de-Gaume in Dordogne, France, about 20,000 years ago. Were these drawings of religious value, used for hunting instructions, or purely aesthetic? [The American Museum of Natural History]

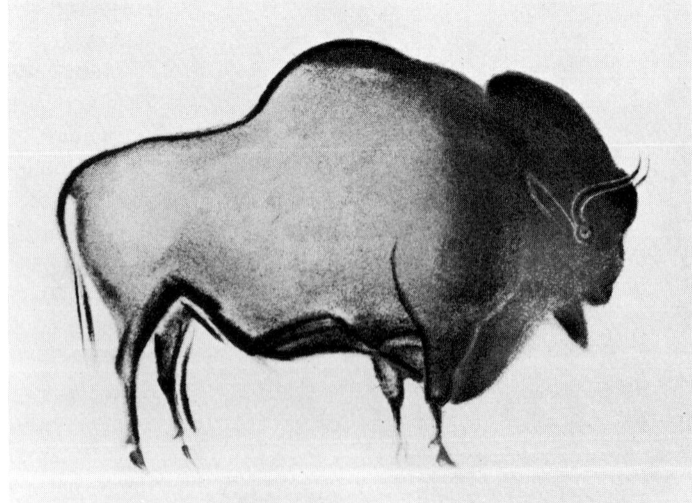

Polishing the Product—The Races of Man

Once a species becomes established, and migration patterns are set up, geographically separated demes and populations arise. The isolation resulting from the ensuing geographical separation leads to the separate evolution of several groups simultaneously. This evolution is not toward a new and different species, but is, rather, directed toward the further adaptation of each group. In a sense, this is the polishing phase of evolution.

This situation is known to exist in many plant and animal species, and leads to the classification categories of subspecies or **races.** It is thought that man followed this typical speciation pattern, and migrated great distances to become spatially isolated, thus evolving into different groups. The characteristics of each group would correspond to their geographic habitat.

Some scientists have given the different human groups a species recognition, but on biological grounds, this is unjustified. A species is composed of interbreeding individuals, and the potential for breeding and producing viable, fertile offspring has been extensively demonstrated between individuals of widely divergent groups. Subspecies or race classification is justified on the evidence that each group (1) manifests consistent and unique structural and physiological characteristics, and (2) is geographically restricted while maintaining a potential for interbreeding with members of different groups.

The problem of how many races of man exist has been a vexing one for anthropologists. Blumenbach (1775) first listed five groups: **Caucasian** (white), **Mongolian** (yellow), **Ethiopian** (black), **Malayan** (brown), and **American** (red). His system was reasonable because it had a manageably small number of groups and did represent different geographical regions; but it was artificial on the grounds that it was based solely on color of skin, a single attribute that tends to clump groups together unnaturally from an evolutionary standpoint. Most later systems have tended to increase the number of races, in some cases up to two hundred. Each scholar has developed his own peculiar listing.

The system of race classification adopted in Table 17-3 is one developed by Coon, Garn, and Birdsell, and modified by Dobzhansky. Although it has the cumbersome number of 34 groups, the criteria used for its construction are biologically valid.

Anyone attempting a categorization of the races of man is confronted with a problem, the constant evolution of his subject. While the *Bushman, Hottentots, Dravidian,* and *Murrayian* groups have undergone little migration or hybridization, the *Neo-Hawaiian, Ladino,* and *North* and *South American Colored* races are newly emerging groups, at best only two centuries old. This listing, then, would have been invalid for the human species two hundred years ago, and will probably be incorrect five hundred years from now, for the historical evidence indicates that the race structure of man has changed every few hundred years. This change is the result of invasion, migration, and exploration.

Theoretically each race should have a unique set of gene frequencies that indicate specific adaptive changes. Negroes and Arabs tend to be tall and slender, an adaptive factor that provides efficient heat dissipation in tropic and subtropic regions. The darker pigmentation of many of the races can be attributed to adaptive traits for camouflage for capturing prey and hiding from their predators, or protective measures against ultraviolet radiation. The Mongoloid group has pads of fat under the cheeks and lining the eyelids (the mongolian fold) that serve to insulate, thus protecting the eyes, nose, and sinuses from the cold. This

TABLE 17-3 THE RACES OF MAN

1. *Northwest European* Scandinavian, Northern German, Northern France, the Low Countries, The United Kingdom, and Ireland.
2. *Northeast European* Poland, Russia, most of Siberia.
3. *Alpine* Central France, South Germany, Switzerland, Northern Italy, and eastward to the shores of the Black Sea.
4. *Mediterranean* From Tangiers to the Dardanelles, Arabia, Turkey, Iran, and Turkomania.
5. *Hindu* India and Pakistan.
6. *Turkic* Turkestan, eastern China.
7. *Tibetan* Tibet.
8. *North Chinese* Northern and central China, Manchuria.
9. *Classic Mongoloid* Siberia, Mongolia, Korea, and Japan.
10. *Eskimo* Arctic America.
11. *Southeast Asiatic* South China to Thailand, Burma, Malaya, and Indonesia.
12. *Ainu* Aboriginal population of northern Japan.
13. *Lapp* Arctic Scandinavia and Finland.
14. *North American Indian* Indigenous population of Canada and U. S.
15. *Central American Indian* From southwestern United States through Central America to Boliva.
16. *South American Indian* Primarily the agricultural people of Peru, Chile, and Bolivia.
17. *Fuegian* Nonagricultural people of southern South America.
18. *East African* East Africa, Ethiopia, and part of the Sudan.
19. *Sudanese* Most of the Sudan.
20. *Forest Negro* West Africa and much of the Congo.
21. *Bushman and Hottentot* The aboriginal inhabitants of South Africa.
22. *Bantu* South Africa and part of East Africa.
23. *African Pigmy* Small-structured people living in the rain forest of equatorial Africa.
24. *Dravidian* The aboriginal populations of Southern India and Ceylon.
25. *Negrito* Small-statured, fuzzy-haired populations scattered from the Philippines to the Andamans, Malaya, and New Guinea.
26. *Melanesian-Papuan* New Guinea to Fiji.
27. *Murrayian* Aboriginal population of Southeastern Australia.
28. *Carpentarian* Aboriginal population of northern and central Australia.
29. *Micronesian* Islands of the Western Pacific.
30. *Polynesian* Islands of the central and eastern Pacific.
31. *Neo-Hawaiian* An emerging population of Hawaii.
32. *Ladino* An emerging population of South and Central America.
33. *North American Colored* The so-called Negro population of North America.
34. *South American Colored* The analogous population of South America.

group appeared in the cold regions of northern Asia, and is well suited for this severe climate. It is thought that the Caucasian races originated in western Europe, where there was less intense ultraviolet light, and so either no pigmentation was developed, or if it was present, it became lost.

Homo sapiens became a successful species about 50,000 years ago. Fossils of Mongolian characteristics dating back 20,000 years indicate that the formation of this race occurred early in man's history.

Since fossils of near-man types have been found in Africa, Java, and China, one theory of man's descent is that he evolved separately from three different ancestral stocks, in different regions. This separate evolution led to the three basic types, *Caucasian, Mongoloid* and *Negro.* Each of these then underwent subspeciation and complicated interbreeding to produce the thirty-four races listed. This, however, is only a hypothesis. The alternative to this hypothesis is, of course, the single origin of man from a single ancestral stock that migrated. Neither argument can be supported by the presently available fossil evidence.

QUESTIONS

1. Review the history of man as revealed through fossil records.
2. What significant changes from the reptilian stock fostered the emergence of the mammalian type of organism?
3. What changes occurred in the primate line when the primates no longer had trees?
4. Give the biological and psychosocial evidence that the human is a premature primate form.

PART FIVE: ECOLOGY AND BIOSYSTEMATICS

The field of ecology is broad enough in scope and sufficiently varied in research approaches to necessitate a selectivity of both material and perspective for the two ecology chapters in this part. The material and perspective chosen were those that complemented and extended the principles developed in the earlier parts of the text, particularly with reference to energy transfers.

The first chapter of Part Five, "Ecology: Levels and Perspectives," is an overview of the field starting with the broadest scope in the ecological realm, the biosphere, and progressing to the lowest level of ecological study, the niche. The discussion of the biosphere includes references to the elemental cycles that have been covered in Chapter 6, so that the perspective of these cycles becomes one of ecological as well as nutritional significance.

The second chapter of Part Five, "Ecology: Energetics and Evolution," extends the concepts of energy requirements up to the population level involving food chains and succession; and succession as a reflection of evolutionary events as life in an ecosystem progresses to energy states at the lowest possible thermodynamic level through increased trophic levels and increased complexity of trophic levels. Thus, this chapter serves both to explain ecological principles and to modify the concept of evolution as a process within an ecosystem that evolves many interdependent populations simultaneously.

Taxonomists have long sought to group organisms in ways that reflect their natural relationships. The "naturalness" of a grouping, of course, depends on the particular point of reference. Many groups considered natural became artificial after the acceptance of the Darwinian theory of evolution. The "Biosystematics" chapter traces the criteria for taxonomic considerations and develops the modern concept of a species.

(Opposite) "Frank & Walter," Robert Lockhart Jr., Kovler Gallery, Chicago

CHAPTER 18

ECOLOGY: LEVELS AND PERSPECTIVES

Ecology is the study of relationships between organisms and their total environment: relationships that are not merely species-environment confrontations, but rather complex interplays between inorganic and organic processes to produce a series of steady-state conditions. The distinction between living and nonliving becomes blurred in the examination of these steady-state systems. Endlessly recycled materials pass through biological and inorganic states. A given oxygen atom can, at one point in time, be part of a carbohydrate molecule, and at some other time, part of a water molecule. In addition, the fate of a given element is determined by the presence or absence of factors in the environment that can act on it. An oxygen atom in the upper atmosphere can absorb UV radiation from the sun, while the same atom in a cell can act as a final hydrogen acceptor in plant or animal respiration.

Past organisms have modified the geochemical and geophysical conditions of the Earth, and thus have provided opportunities for new types of organisms to evolve. Presently adapted organisms provide complex sets of environmental conditions that dictate what types of organisms can coexist in a given environment. Nature draws no lines between life and nonlife, past and present, only between different specific relationships. Ecology is the study of the natural balances produced by these specific relationships.

The scope of ecological relationships is so broad that it has been necessary to compartmentalize these relationships into different areas of study. This compartmentalization has been in the form of ecological levels and perspectives.

The most encompassing level of study is the **biosphere,** the thin layer about the Earth where geochemical and geophysical conditions are conducive to the proliferation of life (and in which organisms, past or present, have existed). It is bounded below by the sterile crust of the Earth, and above by the cold, hostile stratosphere (see Fig. 18-1). The perspective of the biosphere is the universal and fundamental biotic and abiotic features of this spatially defined area. The cyclic pattern of

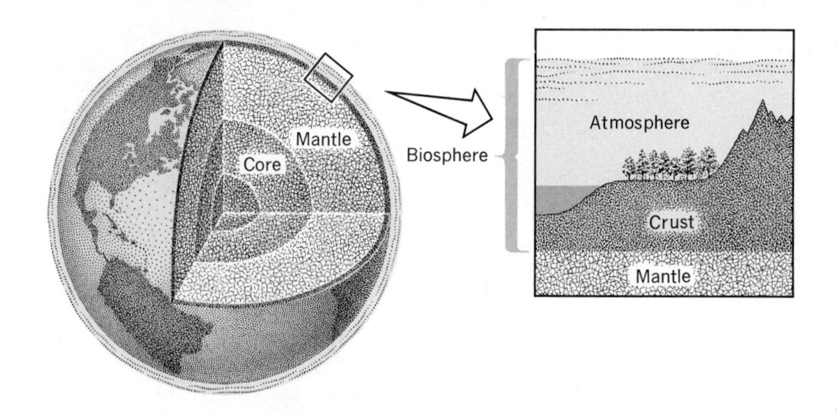

Fig. 18-1 The biosphere.

elements, the properties of elements and common molecules, the geophysical features such as temperature, and the factors that have resulted from geophysical, geochemical, and biological activity, such as soils, are all part of the perspective of the biosphere.

Another level of ecological study is the **biome,** an extensive geographical region wherein a steady-state condition exists among the animals, plants, and microorganisms and the geophysical and geochemical features. Land-based biomes are defined by their rainfall, temperature, and seasonal variation. Aquatic biomes are defined by salinity, water depths, shifts in water depths, temperature, and clarity of the water. The base of reference of a biome, be it terrestial or aquatic, is the plant life. The flora instead of the fauna) of a region is used to characterize the biome because the various factors defining the biome determine what plant life will survive. The flora, as the first step in the food chain (see p. 140), determine which and how many animals can inhabit a region.

The next and more confined level of ecological study is that which examines the **geographical distribution** of biological forms over different land and water masses within the same biome. The perspective of geographical distribution is the evolution of plants and animals in given regions, and certain physiological features of these organisms. Many plants and animals could live in a region of a given biome, but do not, simply because they neither evolved there nor migrated there. Some plant seeds cannot withstand the dryness of deserts or prolonged residence in the sea. Animal migrations are limited by mountain ranges, as well as by deserts and seas. Therefore, despite the similarity of biome conditions on either side of a geographical barrier, the physiological limitations of some animals and plants inhibit their taking advantage of compatible biomes. In general, animals are more limited in regard to migration than plants. The seeds of many plants do withstand transport by ocean currents, and eventually germinate on distant islands where climatic conditions are favorable for growth.

A more confined ecological level is the **ecosystem,** a community of organisms and their abiotic environment that exist under steady-state conditions. The community of organisms within an ecosystem is composed of populations of different species that live together and are interdependent on one another. An aquarium with fish, plants, and bacteria to decompose the wastes of the fish is probably the most familiar example of an ecosystem. In nature, ecosystems are not so well defined. A pond may contain several vertical (different depths) and horizontal (different distances from shore) ecosystems. Each ecosystem of the pond blends into the next.

The most confined level of ecological study is the **niche.** Unlike the common meaning of niche, which refers to a space or location, the scientific reference is to all those factors that characterize the position of a population in a community *except its physical location.* The perspective of a niche, then, is in terms of the highly localized chemical, physical, spatial, and temporal factors required for the survival of a species, as well as those factors that hold its growth in check. Reference to the location of a population within an ecosystem is made by the use of the term **habitat.**

THE BIOSPHERE

In 1913, L. J. Henderson developed the idea that there is a mutual fitness between the environment and the organism. Not only is life highly adapted to survive on Earth, but the Earth's surface is also especially suitable for supporting life. Evolution, as well as the present-day ecological relationships, are possible because of the geophysical and geochemical factors of the biosphere.

One major aspect of the study of ecology is, then, an examination of some of the geophysical and geochemical factors of the biosphere, and the way in which these factors interplay in the environmental-organismal picture. The carbon, nitrogen, and oxygen cycles were discussed in Chapter 6 in relation to the chemical aspects of nutrition. The functional linkage between all forms of life and the inorganic world that results in a constant cycling of these elements is a critical feature of ecological relationships on the biosphere level, and the student should review these cycles and correlate them with the information in this chapter.

The geophysical and geochemical factors of the biosphere will be discussed in this section in regard to their properties and the way in which organisms have adapted to exploit these properties.

Water

One of the major chemical features of the Earth that permitted the origin and perpetuation of living forms is the presence of water. The Earth possesses large quantities of water, both as standing bodies that cover two thirds of the Earth's surface, and as atmospheric vapor, which moistens the land by precipitation. The amount of water present on Earth is perhaps unique among planets.

The evolution of organisms on the Earth is believed to have been inexorably tied to the presence of large amounts of water. Several properties of water contribute to its life-supporting capacity, among which are its stability in a liquid state over a wide temperature range, its effectiveness as a solvent, and its transparancy in both liquid and gaseous states. The 20° to 45° C temperature range for protein activity is well within the 0° to 100° C temperature range at which water is in a liquid state, so that enzyme-mediated reactions proceed in a liquid milieu. Also, chemical reactions are dependent on the dissolution of molecules, so the effectiveness of water as a solvent is a necessary factor for life.

Both the stability of water in a liquid state and some aspects of its effectiveness as a solvent can be traced to features of the hydrogen-bonding capacities of the dipolar water molecule (see Chap. 2). Weak hydrogen bonds are formed between the hydrogen atoms of one water molecule and the oxygen atoms of an adjacent one. The number of molecules involved in the complex depends on the temperature. Large molecular complexes of water exist at near-freezing temperatures, while complexes of four to six molecules are abundant at 20° C. As heat is applied, the hydrogen bonds break. With this breakage, heat is lost. Thus water tends to counterbalance the effects of heat, and in doing so, tends to remain in a liquid state.

The solvency of many types of molecules in water is due to the ability of these molecules to undergo hydrogen bonding with water molecules. Alcohols are commonly soluble in water. The solubility of a protein is dependent on its size and the number of exposed sites that can take part in hydrogen bonding. The solubility of inorganic ions in water is due to the dipolar quality of the water molecule rather than to hydrogen-bonding tendencies (see Chap. 2).

The stability of water in a liquid state contributes to the gross geophysical features of the biosphere in ways that are directly related to the support of life. Much of the solar radiation entering the atmosphere in the form infrared light is absorbed by atmospheric water vapor, as is the infrared radiation radiated from the Earth at night. Thus, water in the atmosphere acts to buffer the temperature of the Earth in much the same way as the glass of a greenhouse prevents heat from leaving that localized area. Accordingly, the absorption of infrared radiation from the Earth at night is termed a *"greenhouse effect"* (see Fig. 18-2).

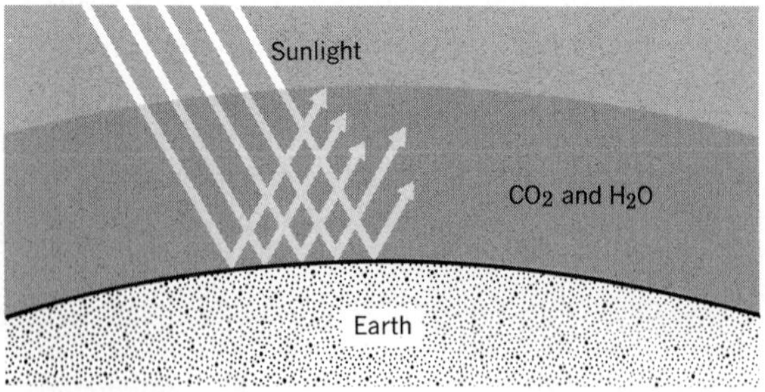

Because of the "greenhouse effect," the land does not overheat during the day, nor does it cool significantly at night; and organisms on the land do not experience sharp temperature changes over short periods of time.

The absorption of infrared light also accounts for the stability of ocean temperatures, a stability that not only affects the oceans themselves, but also the shores and nearby inland regions. These areas have a constant temperature suitable for supporting life.

The distribution of water on land masses can also be traced to the ability of water to absorb large quantities of heat before undergoing a change in state. The vapor given off by the oceans contains consid-

Sunlight
CO_2 and H_2O
Earth

Fig. 18-2 The greenhouse effect is due to heat retention by CO_2 and H_2O.

erable heat. As the vapor absorbs more energy from direct infrared radiation, it rises and moves great distances as cloud formations. When these clouds of water vapor enter a cool region, heat is released, and the water returns to its liquid state in the form of rain.

Other properties of water are also important for life in the biosphere. In both liquid and gaseous states, water is highly transparent to visible light. Solar radiations of the visible spectrum pass readily through atmospheric water vapor and the liquid phase of a plant cell to provide the energy for driving photosynthetic reactions. The transparency of water also permits radiation to penetrate ocean and pond surfaces to supply sufficient energy for submerged aquatic plants.

Gases

In addition to water vapor, the three most important gases that contribute to the character of the biosphere are oxygen, carbon dioxide, and nitrogen. The cycles of all three are discussed in Chapter 6. Some of the physical properties of these gases, and the way in which organisms utilize them within the cycles, is the topic of this section.

Carbon dioxide, present in the atmosphere in trace quantities of about 0.03%, has two properties that contribute to the life-supporting capacity of the biosphere. First, carbon dioxide has a relatively high heat capacity, and therefore can absorb infrared radiation as it enters the atmosphere and as it is radiated from the Earth at night. Thus, carbon dioxide contributes to the "greenhouse effect" in maintaining appropriate temperatures for the survival of life in the biosphere. The "greenhouse effect" is most apparent in different biomes. In a forest, where the lush plant life produces large amounts of carbon dioxide and water, the temperature over a twenty-four hour period varies only slightly. In a desert, where plant life is minimal, the temperatures may vary as much as 80° F over a twenty-four hour period. Some of the larger animals in the Arctic regions have adapted to utilize the "greenhouse effect" for their own survival. Reindeer and musk oxen group together and exhale a dense cloud of water vapor and carbon dioxide. The cloud acts as a temperature buffer, maintaining the animal heat beneath it.

Industrialization in the past century has liberated an enormous quantity of CO_2 by combustion processes. This has resulted in a 12% increase in the carbon dioxide content of the atmosphere. It is believed that the noticeable increase in the average temperature of the Earth over the past century is traceable to the temperature buffering of this additional carbon dioxide. It has been calculated that a doubling of the CO_2 concentration of the atmosphere would raise the average temperature of the Earth's surface by 3.6° C.

The second property of carbon dioxide that contributes to the nature of the biosphere is its high solubility in water to form H_2CO_3.

$$CO_2 + H_2O \longrightarrow H_2CO_3$$

Under standard temperature and pressure, there are about 1.7 liters of CO_2 in every liter of water. This extreme solubility has two important consequences. First, aquatic plants can live below the surface and obtain the carbon dioxide necessary for photosynthetic processes from dissolved CO_2. Even though the high salinity of ocean water lowers the tendency of water to dissolve carbon dioxide, sufficient H_2CO_3 is present, and plants in the sea do not encounter CO_2 deficiencies.

The second consequence of high CO_2 solubility in water is that H_2CO_3 acts as a buffer.

$$H_2CO_3 \rightleftharpoons H^+ + HCO_3^- \rightleftharpoons H^+ + H^+ + CO_3^=$$

Thus dissolved carbon dioxide tends to stabilize the pH of the aqueous environment.

Oxygen is far more prevalent in the atmosphere than carbon dioxide, and comprises about 20% of the atmospheric gases. Oxygen has two interrelated properties that affect the biosphere. First, oxygen is less soluble in water than is carbon dioxide. The oxygen content of fresh water is about 6.5 ml per liter of water at standard temperature and pressure. Sea water, high-temperature water, and rivers and streams with high mineral contents, have even less oxygen. The oxygen supply in artificial aquaria is often the most limiting factor for aquatic life.

The evolution of large aerobic animals, dependent on oxygen as a final hydrogen acceptor in respiration, was possible because of the evolution of iron-based mechanisms by which oxygen could be brought into the aqueous solutions of the animal body. Echinoderms contain no iron-based process, and have an oxygen content similar to sea water. Mollusks (such as clams) have an oxygen-carrying, iron-based, hemoglobin-type pigment, and consequently have a 1 to 2% oxygen level in their blood. Annelid worms have a 2 to 10% blood-oxygen concentration; fish, reptiles, and amphibians have a 5 to 15% blood-oxygen concentration; and mammals have about a 15 to 20% blood-oxygen concentration.

Nitrogen is abundant in the atmosphere, and constitutes about 79% of the total atmospheric gases. Except for its role in the nitrogen cycle, nitrogen has no properties that contribute substantially to the state of the biosphere.

Inorganic Ions

The presence of unique concentrations of inorganic ions in the soil and water environments contributes to the stability of the biosphere. The selective passage of inorganic ions (such as sodium, potassium, magnesium, and calcium) and radicals (such as nitrate and phosphate) into and out of cells, into and out of dissolution in water, and into and out of complex with soil particles, contributes to the steady-state condition between the cell and its environment. The combination and concentrations of inorganic ions and radicals with water trapped in the soil, or the water of an aqueous environment, give character to the environment, and thus determine the types of organisms that can inhabit the region. For example, certain plants (such as cranberries and rhododendrons) thrive in very acid soil, while blueberries can only grow in slightly acid soil, and garden vegetables require neutral or slightly alkaline soils.

The pH of soil environments is directly related to the water-holding capacity of the soil particles. If water passes rapidly through the soil, the inorganic ions and radical dissolved therein will pass downward and become unavailable to growing plants. Calcium, which combines with the OH^- radical of water to form the base $Ca(OH)_2$, can be leached readily from the soil. The leaching of calcium gives rise to highly acid soils. Nitrate ions, the primary source of nitrogen for plants, are particularly susceptible to leaching. The acidity of the soil affects the solubility (and hence, the presence) of other ions, such as iron, manganese, and the phosphates.

The leaching and the washing away of topsoil often brings the dissolved inorganic ions into rivers and streams. The increased organic content of these aqueous environments reduces oxygen dissolution, changes the pH, and generally affects the equilibrial conditions therein.

Soils

Soil is both a component and a product of the biosphere. It is produced by the geochemical, geophysical, and biological factors of the biosphere; and after it is formed, it influences an array of organisms, by both its depth and its quality. This complex combination of inorganic and organic materials serves a dual role, (1) as a substratum for anchorage of plants with roots, and a medium of residence for bacteria, fungi, algae, and a variety of animals, and (2) as a source of nutrients for these living forms.

The inorganic components of soil are formed by the weathering of rock deep beneath the surface. As rock (which is primarily composed of aluminum and silicon compounds) weathers, large separate particles are formed. The weathering of these particles brings about the formation of smaller and smaller particles as one proceeds up toward the surface. Soil particles on or near the surface are characterized according to size.

sand	0.05–2.00 mm
silt	0.002–0.05 mm
clay	less than 0.002 mm

Topsoil, the region in which plant growth takes place, is made up of all three types of soil particles. The proportion of each determines the quality of the soil.

One of the most important features in terms of the quality of soil is its water-holding capacity. The water-holding capacity of a soil is dependent on the proper balance of aeration and the affinity of the soil particles for water. Oxygen and CO_2 exchange is as important for roots as for the upper portion of the plant; most plants grow well only when the soil particles are surrounded by a shell of water and have numerous air-filled spaces. Sandy soils (which contain less than 20% clay and silt particles) are so porous and have so little affinity for water that water tends to pass through rapidly. Because of this drainage, sandy soils are suitable for many types of plant growth. Clay soils, at the other extreme, are not porous, and clay particles have a high affinity for water. Such soils are easily saturated, and plants will literally drown in them because the atmospheric oxygen necessary for root cell respiration is not available. Few plants are adapted to survive in clay soils.

Most species of plants thrive best in soils that contain about equal proportions of the three types of soil particles. Such soils are called *loams* (see Fig. 18-3).

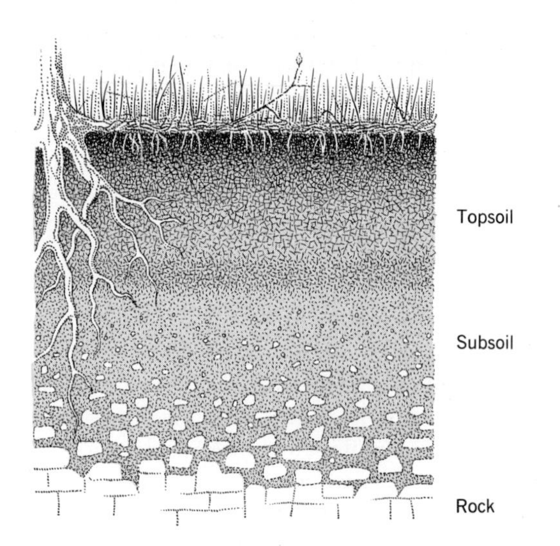

Fig. 18-3 Loam. Topsoil consists of organic debris mixed with sand, silt, and clay. The subsoil is made up of larger rock particles under which lies the soil-forming rock.

While the inorganic content and soil particles are important in characterizing the soil, the organic content also plays an important role. Fertile soil is literally teeming with microorganisms. Studies have indicated that one gram of fertile agricultural soil contains between 15 million and two and one-half billion bacteria,* and about 1,200,000 molds, fungi, algae and protozoans. These organisms, as well as the animals that die and/or are buried in the soil, and the plants growing in a loamy soil that die at the end of the growing season, are decomposed by bacteria present in the soil. It is not surprising, then, that loamy soils are also characterized by large quantities (3 to 10%) of organic material. The decomposition of plant material in the soil has two important consequences. First, the inorganic ions are returned to the soil as material is decomposed. Second, the spongy texture of the decomposing material loosens clay particles and increases soil porosity. **Humus,** the decomposition product of lignin and cellulose, is especially good in this regard.

As mentioned in the previous section, the water-holding capacity of soil is important for maintaining the ion balance as well as for supplying plants with the water they need. Minerals dissolved in water are accessible to roots only when the water is trapped in the topsoil. If water passes through the soil too rapidly, as in sandy soils, the minerals dissolved in the water are also drained away. The structure and composition of soil may be summarized as follows: mineral particles (sand, silt and clay), organic residues (leaves plants and animal bodies that are ultimately broken down to form humus), biological systems (root systems, rodents, insects, worms, etc., as well as microorganisms), water (and the inorganic and organic material dissolved therein), and gases (oxygen, carbon dioxide and nitrogen).

Temperature

A critical geophysical feature of the biosphere is temperature. Over most of the Earth, the temperature fluctuation is not only within the range at which water remains a liquid, but also below the levels where protein and nucleic acid destruction occur. The temperature range that

* The discrepancy between the counts of 15 million and two and one-half billion bacteria in the soil serves to illustrate the technical problem discussed in Chapter 11 (p. 302). Examination of the soil by suitable dilutions and plate counts (viable counts) yielded the result of 15 million bacteria. Direct counts (total counts) yielded the result of two and one-half billion.

occurs in the biosphere has been an important factor in the evolution of a nucleic acid-protein form of life with water serving as a universal medium.

Most types of organisms are adapted to survive within a narrow temperature range. This range is between 10° C and 45° C, and is called the **biokinetic zone.** The most important feature of the biokinetic zone is that it is similar to that for enzyme action. Below 10° C enzyme action is negligible, and above 45° C, destruction of the enzymatic protein molecules occurs.

There are regions of the Earth, however, in which the temperature drops below, or rises above, the limits of the biokinetic zone. Within these regions organisms have adapted in some unique fashion to temperature fluctuations falling outside the biokinetic zone. **Thermophilic** organisms are those that grow at high temperatures, and will, in fact, *not* grow at temperatures within the biokinetic zone. The microorganisms found in the hot springs of Yellowstone Park at a 70° C temperature are thermophiles. The enzymes of these organisms appear to be more stable than comparable ones (electron transport system, DNA replicating enzymes, etc.) of other types of organisms. It is thought that the reason thermophiles do not grow at standard temperatures is because the bonds involved in enzyme reactions are much stronger at lower temperatures. In their natural habitat, the environmental heat contributes part of the energy necessary to dissociate the enzyme-substrate complexes.

Those organisms able not only to survive, but also to grow at temperatures near or below freezing are called **cryophilic** organisms. The cryophilic forms include microorganisms, arctic fish, and invertebrates (animals without backbones). The reason for their resistance to cold is not completely known.

Organisms that grow within the biokinetic zone are called **mesophilic** organisms. Some mesophiles exhibit physiological changes when they are subjected to temperatures either below or above the zone. The adaptation under either condition usually involves elaborate means of protecting the water within the organism. The specific adaptations to cope with adverse temperature conditions vary considerably among different organismal types. Some mammals respond to cold through sleep or **hibernation.** Under similar conditions, some plants produce seeds, or cease growing; some microorganisms produce spores, and some invertebrates form cysts.

Light

The solar energy of light is the key to life on Earth. Photosynthetic organisms utilize light energy for the incorporation of CO_2 into organic structure. Phototrophs are consumed by herbivorous heterotrophs, which are, in turn, consumed by carnivorous heterotrophs. The energy derived from light is converted into chemical bond energy through photosynthesis, and remains as chemical energy through the food chain. Thus light is the ultimate source of energy for all organisms.

Solar radiations that penetrate into the biosphere have wavelengths between 280 mμ and 2,000 mμ. The spectrum of light entering the biosphere has a peak intensity at 480 mμ (see Fig. 18-4). Short wavelengths of 220 mμ have been recorded above the atmosphere with the use of rocket hardware. It appears that these short wavelengths are absorbed, along with most of the UV light (190–320 mμ) by the ozone layer, and thus never enter the biosphere.

In the course of evolution, many organisms have adapted to utilize the different properties and patterns of light. The energy content of light is used as an energy source by green plants. The information content of light (patterns of distribution) is used by both plants and animals. There are two types of information embodied in natural light, indirect and direct. The length of day is correlated with the time of year. Hence, organisms exploit day length as an indirect indication of the prospective rainfall and temperature of that period (or season). Photoperiodism is thus used as a kind of calendar.

In plants, the length of day (or, conversely in some plants, the length of the dark period) triggers the plant into the development of flowers. Plants whose seeds provide dormancy during winter form flowers in the light regime found only in autumn. In other species of plants with seeds that germinate immediately, spring or summer flowering occurs.

Animals also utilize photoperiodism. In birds, the length of day stimulates hormonal production for sexual activities. The rearing of the young demands a plentiful food supply. Mating and rearing of young are correlated with the photoperiod of the season when the environment supports plentiful food, be that food another animal form or a plant.

A variety of means have been developed through adaptive processes to utilize the direct information of light. Simple one-celled organisms often have **phototaxis,** the capacity to detect and move toward the highest light intensity. Some of these organisms have stationary eyespots; others have a sensitive chloroplast that can move in the cell (see Fig. 18-5). The direction from which the optimal intensity of light emanates informs the organism of the light source. The organism or chloroplast becomes positioned in the most advantageous relationship to that source.

Fig. 18-5 Reactions of organisms with light-sensitive organelles when placed in the dark and in the light. (*a*) *Euglena* remain randomly dispersed in a tube kept in the dark but move toward the light source when one is available. (*b*) The chloroplast of filamentous algae expands in volume when the cell is in the dark but becomes oriented so that the greatest surface area is exposed to the light source when one is available.

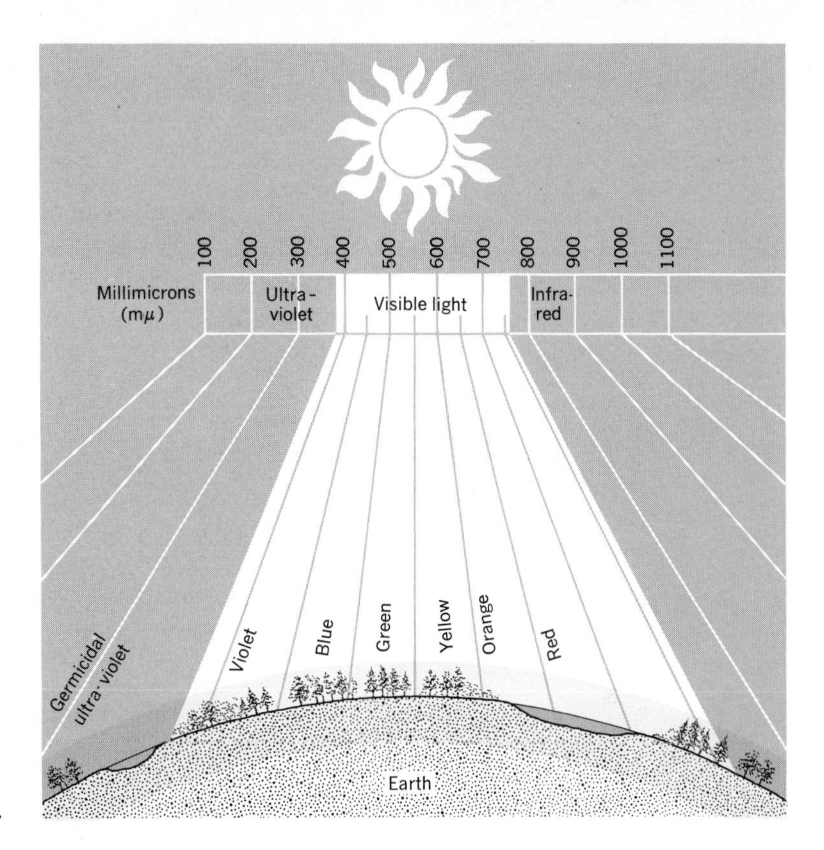

Fig. 18-4 Spectrum of light entering the biosphere.

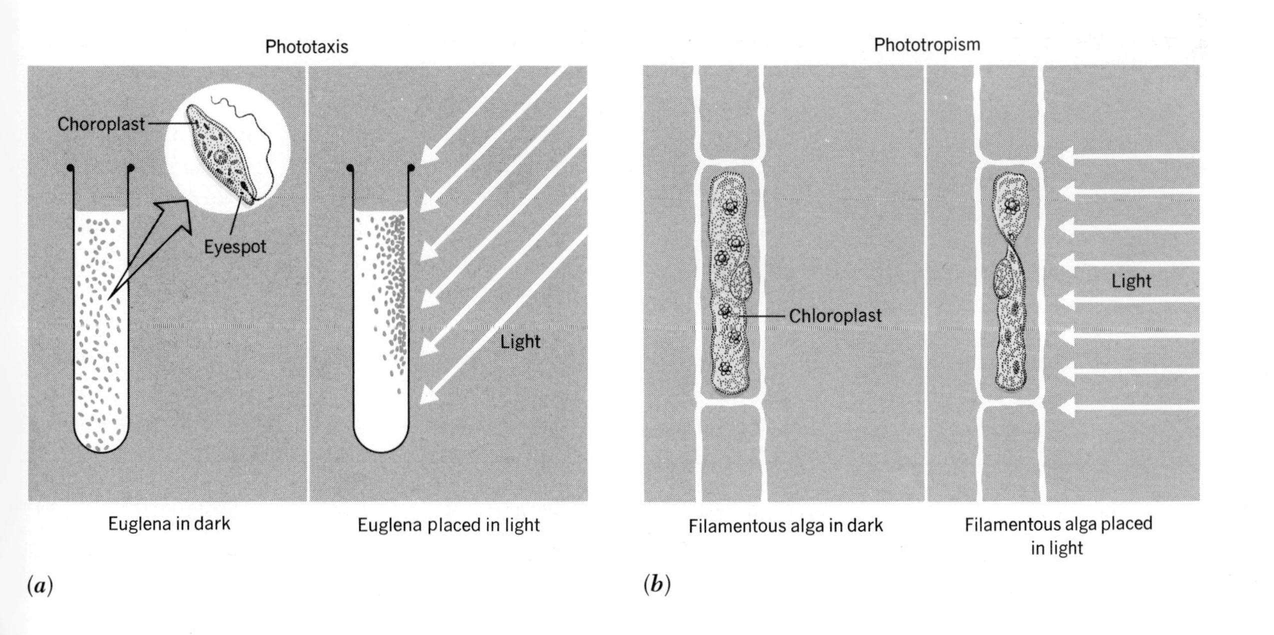

Higher plants and animals often exhibit **phototropism** (bending in response to light) and phototaxis (moving in response to light), respectively. The sunflower follows the sun through the day by appropriate adjustments in the stem. These adjustments in the stem are made in response to the relationship between the stem and the source of light. Bees in search of pollen or in their attempts to escape from an unfamiliar enclosure (such as a car) exhibit positive phototaxis. Both the light source and the solid surface act as positive stimuli for a trapped bee. The instincts that have developed through evolution permit them to exploit light as a useful guide to an escape route, so the insect will cling persistently to the car window because it is both a source of light and a surface upon which the animal can explore. The positive phototaxic tendency in bees is not active, however, when the bee perceives its hive. In entering the darkened hive, the bee exhibits what could almost be termed negative phototaxis.

Higher animals, with their complex organs for both vision and visual interpretation of size, texture, and distance between objects, utilize the information content of reflected light. Some animal forms have developed elaborate means of utilizing reflected light as an information system. In addition to responding to light intensities in a phototaxic manner, bees can determine direction by detecting the plane of polarized light, that is, the parallel light rays reflected off the blue sky. The plane of polarized light changes as the position of the sun changes during the day. Bees couple a facility for the detection of polarized light with a timing device for navigation of their flight. Birds also appear to have a means of detecting polarized light, and may possibly even use the light of stars for nighttime migration.

BIOMES

It is fallacious to present a picture of the biosphere as though all types of plants and animals are randomly dispersed on the surface of the Earth. Climatic conditions differ in various regions of the world. Seasonal variations of temperature, light, and rainfall determine the type of flora that will proliferate in a given land region. Different degrees of salinity or depths in aquatic environments act to limit plant and animal types that can exist there. Therefore, the world is divided into different climatic regions or *biomes*. On land, each biome is identified by the predominant floral types that are present. Aquatic biomes are defined by physical criteria, such as salinity, depth, and distance from shore, and will be discussed later in this section.

The advantage of employing floral criteria for characterizing land biomes is fourfold. First, large plants are more easily recognized than large animals, and areas can be mapped more readily. Second, plants are more homogeneous in coverage and usually exhibit more distinct boundaries than the more migratory, well-separated animals. Third, animals utilize either plants or other animals that in turn utilize plants, as food sources. Therefore, plants usually determine the kinds of animals that will be present. Finally, populations of prevalent plants fluctuate less than those of animals because their requirements are simpler. Therefore, plants are a more fundamental feature of a region than are animals.

Terrestrial Biomes

Ecologists generally divide the world into six major terrestrial biome categories (see Fig. 18-7).

1. Tundra.
2. Taiga or coniferous forest.
3. Temperate deciduous forest.
4. Grassland.
5. Desert.
6. Tropic rain forest.

Labeling a region of the world as a taiga does not imply that the region is uniformly covered by coniferous forest. It means that the most stable steady-state condition that can exist in that region would be that of a taiga.

Although biome categories are generally discussed in terms of wide geographic expanses, it should be noted that there is biome variation even within a given geographical region. For instance, in alpine regions, the high mountainous regions may be more like a tundra, while the valleys have more temperate, deciduous forests (see Fig. 18-6).

A summary of terrestrial biome characteristics is given in Table 18-1.

Aquatic Biomes

Aquatic biomes are either **freshwater** or **marine.** The water pressure, the capacity of water to dissolve oxygen and carbon dioxide, and the freezing points of bodies of water are in great part related to the salinity of the water. The difference in the physical features of these biomes is reflected in the difference in organismal adaptations of the living forms that reside in each of these biomes. A summary of aquatic biome characteristics is given in Table 18-2.

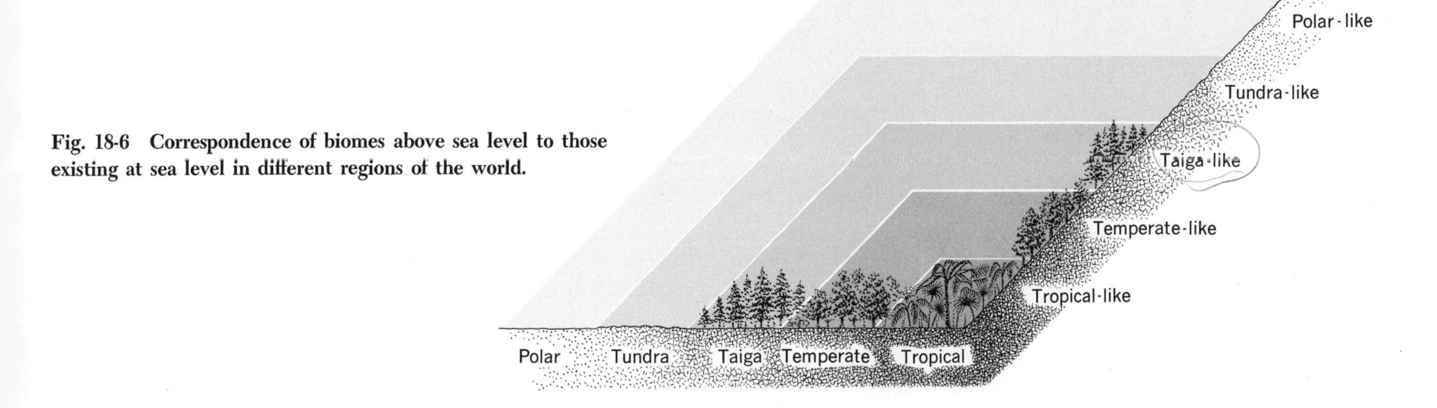

Fig. 18-6 Correspondence of biomes above sea level to those existing at sea level in different regions of the world.

(*a*) tundra

(*c*) deciduous forest

(*b*) taiga

Fig. 18-7 Biomes. [(*a*), Steve and Dolores McCutcheon—Alaska Pictorial Service; (*b*), U. S. Forest Service; (*c*), P. Berger—National Audubon Society; (*d*), FPG; (*e*), Jack Dermid; (*f*), Haas Hannau—Rapho Guillumette]

(*d*) grassland

(*e*) desert

(*f*) tropical rain forest

TABLE 18-1 SUMMARY OF TERRESTRIAL BIOME CHARACTERISTICS

Biome	Example	Temperature	Rainfall Pattern	Examples of: Flora	Examples of: Fauna
Tundra	Arctic, alpine	30° to 70° F permafrost[a]	4 to 7 inches, during growing season	Dwarf trees (willow) Some herbaceous plants Lichens and mosses	Bears Caribou Water fowl Lemmings Reindeer Musk oxen
Taiga	Northern Canada	Below freezing most of year 60° to 70° F during summer	10 to 30 inches a year. Much of the precipitation is snow	Conifers, primarily spruce	Moose Black bears Martins Squirrels Voles Grouse Many insects Birds in summer
Deciduous forest	Eastern United States	Long summers, fairly constant temperatures	20 to 60 inches spread evenly throughout the year	Dominant trees vary in each area: maple, beech, and basswood around Great Lakes; maple, beech, and hemlock in eastern U. S.; shrubs and mosses	All groups of animals: Mammals Birds Reptiles Amphibians
Grasslands	Prairie, pampas, steppe, campo	Similar to that of deciduous forest	10 to 40 inches, mostly in spring	Grass Type dependent on rainfall and soil. Eastern U. S. prairies: big bluestem, Indian grass; Western U. S.: gama and buffalo grass	Buffalo Antelope Small mammals: Mice Gophers Rabbits Prairie dogs Insects Reptiles Amphibians
Desert	Sahara Gobi	Low altitude (hot) High altitude (cold)	1 to 4 inches 1 to 4 inches	Cacti Desert plants that exist most of the time as seeds, germinating and growing quickly after rains	Burrowing animals, reptiles primarily
Tropic rain forest	Near equator	70 to 100° F	at least 80 inches a year	Broadleaf evergreens, epiphytes such as orchids growing on and drawing their nutrients from trees	Many insects Birds Amphibians Rodents

[a] Frozen ground throughout year.

TABLE 18-2 SUMMARY OF AQUATIC BIOME CHARACTERISTICS

Biome	Category	Definition	Example	Zone	Organismal Adaptation
Fresh water	Lotic	Substantial water flow	Rivers, brooks	---	Plants: roots for anchorage in large plants; algae have cellular structures for attaching to rocks or debris Animals: insects and insect larva have suckerlike protuberances, small hooks, suction pads, or gripping claws. larger animals like frogs, snakes, and muskrats swim for short periods of time to avoid displacement by currents fish often show behavioral patterns by moving upstream (salmon, striped bass, and shad) or downstream (eel)
Fresh water	Lentic	Little movement of water	Small lakes, ponds, swamps, bogs, marshes	**Supralittoral:** just above the surface	Plants: grasslike plants; sedges Animals: insects
				Littoral: surface to about six-meter depth	Plants: **hydrophytes** (water-loving plants) both submerged and emergent; algae Animals: insects, snails, worms, amphibians
				Sublittoral: from 6 to 10 meters in depth	Plant: phytoplankton Animal: zooplankton
				Compensation level: about 10 meters in depth	Photosynthetic rate equal to respiratory rate: lowest level at which plants can survive
				Profundal: below 10 meters in depth	Plant: none Animal: fish, crustacean, mollusks
Marine	**Pelagic**	Open sea			
	Photic	From ocean surface (pelagic) to 50 fathoms deep. The 50-fathom level corresponds to the compensation level in freshwater bodies. It is the lowest level at which phytoplankton can grow			Plant: phytoplankton Animal: fish, jelly fish, zooplankton
	Aphotic	From depth of 50 fathoms to ocean bottom (benthic regions)			Plant: none Animal: fish in upper regions
	Benthic	Ocean floor		**Neritic:** ocean floor less than 50 fathoms	Plant: phytoplankton Animal: crustaceans, mollusks, and fish
				Oceanic: oceanic floor greater than 50 fathoms	Plant: none Animal: mostly unknown

GEOGRAPHIC DISTRIBUTION

The primary barriers to animal migration and plant seed dispersal are bodies of water. The distribution of a given species of plant or animal depends on the land mass on which they are found, and the relationship of that land mass to other land masses, as well as the climatic conditions of that region.

Despite migration barriers for animals, and seed dispersal barriers for plants, there are marked similarities between organisms on different continents. Alfred Russel Wallace, in his book *The Geographical Distribution of Animals*, first noted this similarity, and recorded the paradox it presented because of the lack of migration routes. The chance that identical patterns of evolution occurred independently on each continent is remote. How then, did such similar forms come to reside in such removed locations?

Continental Drift

Geological and fossil evidence suggests that the land distribution of the world was not always as it is in its present form. During the past ages, the barriers between land masses were less restricting, and climatic conditions were quite different. For instance, the evidence suggests that Asia, Australia, New Zealand, and several of the Philippine islands were a continuous land expanse until about 100 million years ago. The date given for the separation of these land masses is based on fossil evidence. Marsupials (ancestors of the kangaroo and platypus) migrated from Asia to Australia, but the placental mammals, which appeared in Asia some 40 million years later, did not.

For many years geologists have debated about the nature of the migratory routes. Until quite recently, the most acceptable theory was that wide land bridges existed between the different continents. In the mid-1960s, overwhelming evidence accumulated that it was not the sinking or wearing away of land bridges that brought the world to its present geographic state, but rather the drifting apart of large segments of land that once formed a continuous land mass, a supercontinent, to produce the well-separated continents present today.

The idea that continents have drifted apart was very difficult for geologists to accept. The movement of such large land masses would require tremendous force. From where would such a force come? In the 1930s, the Dutch geophysicist F. A. Vening Meinesz proposed that thermal convections in the Earth's mantle would be of sufficient magnitude to cause continental drift. Over the next thirty years, evidence accumulated that there were thermal convections of just such a magnitude. It is thought that these convections maintain the irregular contour of the Earth by overriding the natural tendency of land moving toward an equilibrium state in which the land of the Earth would be evenly distributed and become covered by water uniform in depth (entropy). Such convection currents would be of a magnitude capable of causing continental drift.

Further evidence for continental drift comes from the profile of the continental shelf that extends from a continent into the sea. Exposure of land to the ocean causes the accumulation of sediment. The amount of sediment accumulation indicates the length of time a given land mass has been directly exposed to the sea. Examination of sediments on the western coast of Africa indicate that the exposure began no earlier than 160 million years ago. Another interesting feature of the western coast of Africa is the different ages of sediments ranging from North to South. Similar data has been found for the eastern coast of South America. It appears that South America and Africa once belonged to a single land mass that broke apart, starting in the north and progressing to the south (see Fig. 18-8), and drifted independently of each other.

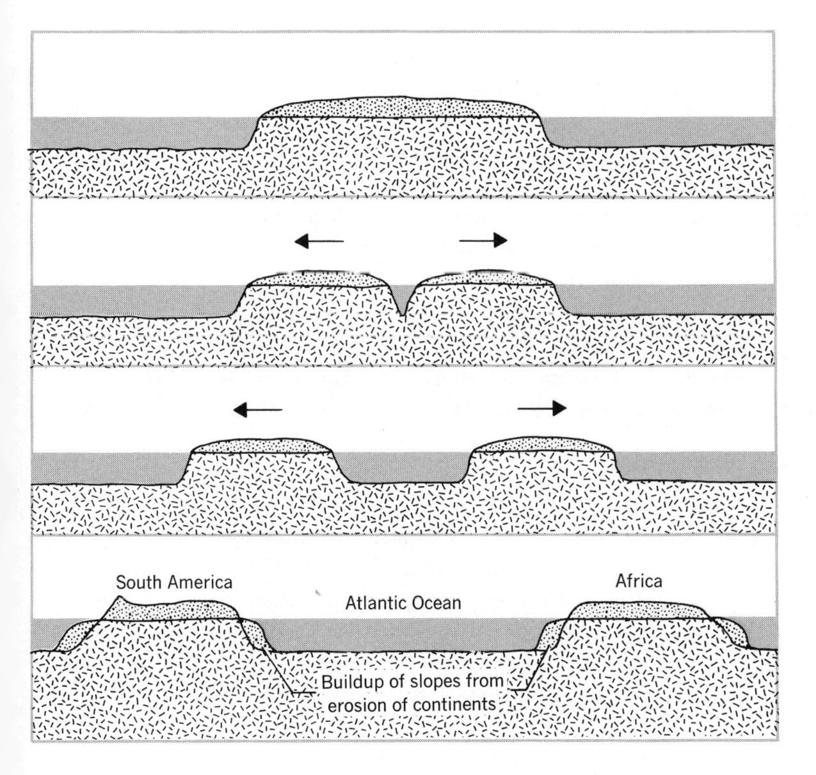

Fig. 18-8 As separation of Gondwanaland occurred to form South America and Africa, sediments built up along the coast. The ages of the sediments indicate how long the continents have been separated.

Still further evidence for continental drift was presented by a study of the history of the magnetic poles. By measuring the magnetic direction in rocks from different continents and of different ages, it is possible to construct a picture of the position of the magnetic pole in relation to a given continent. It was found that the position of the magnetic poles differed both with respect to time and with respect to the continent under investigation. While it may be possible for a magnetic pole to move, it is not possible for the Earth to have had several magnetic poles at the same time. The only acceptable interpretation of the data was that each land mass moved independently in reference to the magnetic poles.

TABLE 18-3 GONDWANALAND AND LAURASIA

Gondwanaland	Laurasia
Antarctica	Arctic
Africa	North America
Australia	Europe
Madagascar	Asia
India	Greenland
South America	Japan
New Zealand	
Philippines	

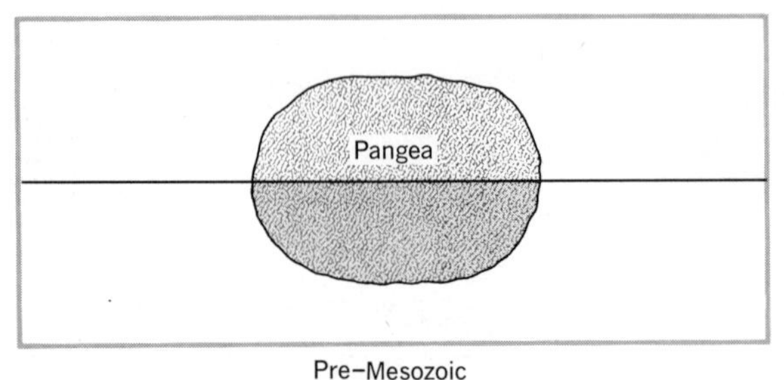

Pre–Mesozoic

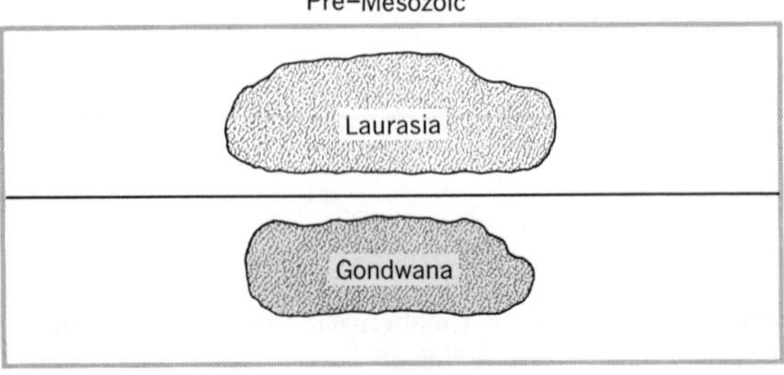

Mesozoic

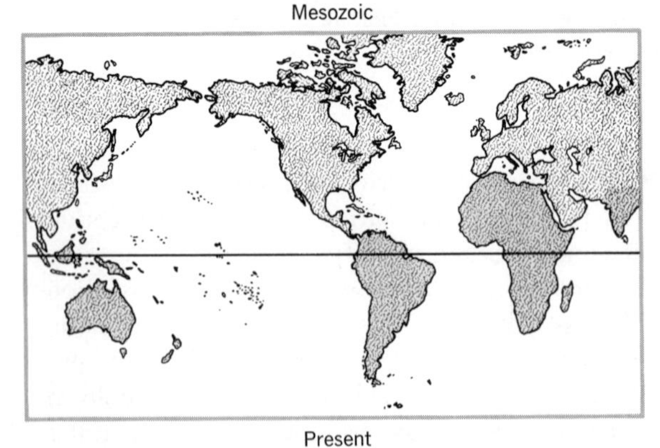

Present

Fig. 18-9 Maps of the world in the Pre-Mesozoic, Mesozoic, and present times.

The evidence accumulated thus far leaves no doubt that the present-day continents once formed a large land mass. At present, the question of whether the major land masses in the northern and southern hemispheres were continuous has not been resolved. Some geologists favor the idea of one continuous supercontinent that has been called *Gondwanaland.* Others favor the idea of two separate land masses, Gondwanaland in the southern hemisphere and *Laurasia* to the north. A listing of the present-day continents and some of the islands that comprised Gondwanaland and Laurasia is given in Table 18-3. The proposed map of the world prior to the Mesozoic era when continental drift was initiated is given in Fig. 18-9.

As the continents drifted and moved to the geographical locations where they are presently found, they were confronted with other land masses. The tremendous pressures exerted by the collision of land masses are believed to have caused the rise of mountain ranges, such as the Himalayas between India and the rest of Asia. (India was part of Gondwanaland. It drifted north to Asia, which was part of Laurasia.)

It is presently believed that all mountain ranges were brought about by the pressure of land mass moving against land mass. If this is true, then the older, predrift mountain ranges such as the Appalachian belt in North America and the Ural mountains in Asia, which are found deep *within* continents, may indicate that the formation of Gondwanaland and Laurasia was from a synthesis of separate smaller blocks. When Gondwanaland and Laurasia broke apart, the older mountain ranges were located within the large land masses that were to remain continents.

Biotic Distribution

The Mesozoic era was the Age of Reptiles, a time when large dinosaurs roamed the lands of both Gondwanaland and Laurasia. As the Mesozoic Era came to an end, and Gondwanaland and Laurasia were breaking up into the present-day continents, the mammals became the predominant fauna. The distinct separation of land masses that created barriers against animal migration did not occur until well into the Cenozoic Era, the Age of Mammals, which began some 70 million years ago.

The similarity of the biota of northern Asia, North America, and Europe indicates that migration between these areas was possible well into the Age of Mammals. The Canadian marten and Russian sable are very similar, and are termed **ecological equivalents** (see Fig. 18-10).

Much of the migration story is unclear. From fossil evidence it appears that the camel family originated in North America, migrated to Europe, and became established as the modern camel in Africa and Asia. Also, the horse seems to have originated in North America, migrated to Asia, and became extinct in America. Elephants apparently appeared first in Egypt, migrated to Europe and America as the mammoth and the mastadon, respectively, and subsequently became extinct. The only two surviving species are the Indian elephant and the African elephant.

(*a*) **Canadian marten**

(*b*) **Russian sable**

Fig. 18-10 [(*a*), **Joseph Van Wormer;** (*b*), **Tass-Sovfoto**]

The times at which the land masses became discontinuous marks the time when geographical barriers inhibited the further migrations of certain plants and animals, and committed these species to separate and distinct evolutions. The separation of South America and Africa appears to have occurred well into the Cenozoic era, when at least the early primates were on the scene. Both South America and Africa contain monkeys, but the differences between the South American or New World monkey and the African or Old World monkey are both numerous and important, and lead to the conclusion that separate and distinct evolutions of these early primates occurred on their respective continents (see Fig. 18-11 and Table 18-4).

(*a*) **Old World monkey**

(*b*) **New World monkey**

Fig. 18-11 [(*a*), A. W. Ambler—National Audubon Society; (*b*), New York Zoological Society photo]

TABLE 18-4 COMPARISON OF OLD WORLD AND NEW WORLD MONKEYS

	Old World Monkeys	New World Monkeys
Position of nostrils	Closely approximated and directed downward	Far apart, and open laterally
Teeth	Three molar teeth	Two molar teeth
Auditory apparatus	Bony auditory meatus	No bony auditory meatus
Blood chemistry	Close to man's	Unlike man's
Thumb position	Opposite other fingers	Not opposite other fingers
Tree life	Arboreal and terrestrial	Many committed to arboreal life
Tails	Never prehensile	Often prehensile
Food storage	Cheek pouches	None

Despite the separation of land masses and the presence of mountain ranges and deserts, many animals and plants did (and, for that matter, still do) migrate. Often this migration is passive on the part of the animal or plant. Seeds of plants are carried great distances by birds or water currents. Insects are blown hundreds of miles by strong winds. Many invertebrates, fungi, seeds of plants, and bacteria are carried on rotting logs into the sea, and subsequently to new shores. Man has intervened, and often inadvertently carries a variety of plant and animal forms to new regions. But man is fairly recent on the biological scene, and is not responsible for older distribution patterns that have been found.

Migration, be it passive or active, brings species into new environments where they are subjected to new selective pressures. If the environment is suitable, or if the members of the species can adapt to the new environment, they will flourish there. The separation of a few members of a species from a large population constitutes genetic drift (see Chap. 15). Thus, as generations pass, a new subspecies or species may arise. A major criterion for dating migrations of various organisms is the species similarity between members found in the original geographical location (as determined by fossil evidence) and those found in regions as the result of migration.

Thus the present biotic distribution is the result of conditions of the Earth at various historical times, and the past and present factors involved in organismal migration. It is, however, impossible to say at the present time just what factors led to specific distribution patterns, or trace the lineage of species in different parts of the world.

ECOSYSTEMS

An ecosystem is a community of organisms and their abiotic environment that exist in a steady-state condition. A decaying log with its termites, fungi, and bacteria is an ecosystem, but so is a forest. The community of organisms within an ecosystem is made up of different species, the populations of which are in different stages of development. As time passes, some members of these populations will reproduce, some will die from accidents and old age, and some will become the food of predators. All the organisms, in all stages of development, contribute to the dynamic equilibrium of the ecosystem.

The abiotic features of the environment also contribute to the steady-state condition of the ecosystem. Climate, availability of minerals, common organic forms, and light, all play roles in determining the type of organisms that will thrive in an ecosystem. The requisites for the life processes of all organisms living within an ecosystem are met by the biotic and abiotic features of that ecosystem. The dynamics of an ecosystem lies within the community relationships that exist among the organisms therein.

Community Relationships

The most direct relationship among members of a community, and one that is confined within the limits of that community, is the supply of energy. As biological entities, the organisms that comprise a community each require a source of energy for maintenance and growth. Photosynthetic organisms obtain this energy directly from light. Heterotrophic organisms obtain energy by feeding on either photosynthetic organisms or other heterotrophic organisms.

In the course of evolution, plants have adapted to obtain maximum exposure to light within the restrictions imposed by their genetic potential and physiological balance. Heterotrophic organisms have evolved means of detecting, perhaps capturing, feeding upon, digesting, and metabolically utilizing the energy derived from their food source. These adaptations have, of course, always reflected the genetic potential and physiological balance of the organism.

Community relationships in terms of food (energy) source may be viewed at several levels within an ecosystem. A common means of viewing food relationships is by categorizing the manner by which the food source is procured.

The complexity of community relationships in terms of manner of obtaining food can be divided into two categories: **exploitation,** in which one organism lives at the expense of another, and **mutualism,** in which organisms benefit from their relationship.

Exploitation is the intervention of one species in the livelihood of another species in a negative fashion. It includes **predation** and **parasitism.** Predation implies that one organism kills and consumes the edible parts of the prey. Parasitism, on the other hand, implies feeding on selected host tissues. Damage imposed by the parasite often leads to death.

Fig. 18-12 Lichen. The growth of lichen directly on rock exemplifies its ability to survive on land lacking rich organic materials. [Mary M. Thacher—Photo Researchers]

The definitions of predation and parasitism are actually artificial conventions imposed on nature by the biologist in an attempt to categorize natural phenomena. These terms are based on extreme examples of different trends. A barn owl capturing field mice is clearly predatory, and a fungus that infects and kills an elm tree is clearly parasitic. But lice and mosquitos on a baboon may cause some discomfort, but little else. Yet the louse is considered a parasite, and the mosquito a predator.

It is generally thought that the situation in which the parasite causes only mild discomfort to the host represents a more highly evolved parasitic relationship. Killing the host is of no advantage to the parasite population. A dead host cannot support a growing parasite, nor aid in its transmission. Therefore, a truly successful parasite is one that causes only mild damage to host tissue.

Another criterion for characterizing highly evolved parasites is the types of hosts they are capable of attacking. A parasite that attacks a series of hosts in a predictible sequence is considered more highly evolved than one that completes its life cycle in a single host. The greater the number of hosts a parasite has, the greater the adaptations it exhibits, and the greater the chance for its survival if one of the hosts becomes temporarily few in number, either through seasonal fluctuations or random changes in host population size.

Mutualism involves those interorganismal relationships in which both members gain some advantage by their association, while a special subcategory, **commensalism,** involves those relationships in which only one species receives some value, and the other is not detrimentally affected.

The most common example of mutualism is the lichen (see Fig. 18-12). A lichen is a type of plant formed by the physical association and physiological cooperation between an alga and a fungus. The phototrophic alga provides organic material for the heterotrophic fungus, while the fungus provides an aqueous environment for the alga.

Many mutualism relationships do not include a lasting physical association. Some insects and flowering plants are considered to have a mutualistic relationship (see Fig. 18-13). The plant benefits from the relationship by the chance transfer of pollen to the stigma of the same or other flowers, thus increasing the opportunity for fertilization and subsequent development of seeds. The insect benefits from the relationship in terms of a source of food (pollen and nectar).

Fig. 18-13 Pollination of a flower by a bumblebee. Notice how thoroughly covered with pollen the bee is. The transfer of this pollen to the stigma of other flowers occurs when the bee moves around within the flower in the course of collecting nectar. [Lynwood Chase—National Audubon Society]

Another level at which community relationships in terms of food (energy) source can be viewed is that of the structural, physiological, and behavioral adaptations of organisms for obtaining their food. Many parasites have chemoreceptors that enable them to detect the presence of an appropriate host, and suckers that enable them to attach to their hosts and thus set up a temporary or permanent residence. Predators have, through the course of evolution, developed a variety of structural, physiological, and behavioral adaptations that enable them to capture their prey. Birds that eat worms seek their prey in early morning, when the water-laden soil induces worms to come to the surface. The beaks of these birds are generally pointed and strong. Their eyes are adapted to detect the movements of worms. Their instinctual behavior pattern drives them to ground level to seek for food. Eagles, on the other hand, feed on small mammals and reptiles. They can detect their prey from hundreds of feet above the ground, and dive directly, grasping their prey with their clawed feet. Their beaks are adapted to enable them to tear the prey apart as they feed (see Fig. 18-14). Members of the cat family generally grasp their prey with their mouths, and use their clawed paws as a means of holding the prey in place. Bears will claw and finally bite their prey to death.

(*a*) **Head of a robin**

(*b*) **Head of an eagle**

Fig. 18-14 [(*a*), George Roos—Photo Researchers; (*b*), Russ Kinne—Photo Researchers]

Adaptations exist in mutualistic relationships as well as in the more violent predator-prey, parasite-host relationships. The alga and fungus that comprise the lichen are each adapted to their way of life. The algal and fungal components can be separated in the laboratory and reared apart. Later mixing of these two components will lead to their reassociation into the **thallus** (plant body) and the differentiation into the reproductive structures characteristic of their association. Mixing any free-living fungus and alga of morphological designs similar to those found in the lichen does not lead to a structural and functional association, indicating that the lichen components have developed a greater potential than that which could be achieved by either separately.

Pollination mechanisms have played an important role in the evolutionary adaptations of flowers, some insects, and some birds. Flowers have adapted to (1) attract insects and birds, (2) permit insects and structurally adapted bird beaks easy access to pollen, and (3) produce nectar. Insects and birds have adapted to (1) respond to color, pattern, and scent of flowers and (2) collect nectar more efficiently. For example, the hummingbird beak is adapted to penetrate the nectaries of flowers; hummingbird eyes are adapted to detect flowers; and those structural and physiological factors that control wing movements are adapted to move in such a way that the bird's body remains relatively stationary, even in flight, as the bird feeds (see Fig. 18-15).

One of the most striking features noticed from an examination of the structural, physiological, and behavioral adaptations is the diversity of food sources within groups of similar evolutionary origin. There are birds that eat worms, others that eat insects, others that eat nectar, still others that eat small mammals and reptiles, etc. There are mammals that eat plants, mammals that eat insects, mammals that eat birds, mammals that eat other mammals, etc. Adaptive radiations at different levels in evolution permitted the expression of structural and physiological adaptations to detect, capture, and feed on many of the different types of flora and fauna of an ecosystem. The presence of suitable prey or hosts in a community represents an opportunity for any group of organisms, regardless of their evolutionary origin, to exploit some food source through appropriate genetically established adaptations. Once a group adapts to capture and eat a prey to infect a host or to exist in a mutualistic relationship with another member of the community, its

Fig. 18-15 Hummingbird feeding from nectaries of flower. [San Diego Zoo photo]

establishment in the community is secure until such time as biotic or abiotic changes occur. Therefore, *the evolution of ecological relationships goes hand in hand with the evolution of species.*

Although the community relationships established through evolutionary processes for obtaining food (and thus the necessary energy for living processes) are most direct and confined within an ecosystem, they are by no means the only basis for community relationships. The functional interrelationship of metabolic processes of all organisms within the ecosystem provides the means by which materials are constantly recycled. An ecosystem is not a closed system; atmospheric elements are brought into terrestrial ecosystems, and soluble materials are brought into aquatic ecosystems. Despite the interaction of the ecosystem with the rest of the biosphere, the majority of recycling material remains within the ecosystem.

The functional interrelationship between metabolic processes of different organisms exists both in the utilization of food material and the deposition of wastes. Herbivores, by definition, are uniquely capable of utilizing plant material. They have bacteria and protozoans in their digestive tract that have the digestive enzymes capable of breaking the cellubiose bonds in cellulose molecules, and thus breaking cellulose down to its glucose monomers. Carnivores and omnivores (including man) do not have comparable symbionts. Hence, cellulose material passes through the carnivores' digestive tract intact.

Metabolic wastes from one organism are used in the life processes of another organism. Oxygen, which can be considered a waste product of photosynthesis, is used by aerobic organisms as a final hydrogen acceptor in their respiratory processes. Carbon dioxide, a waste product of both plant and animal respiration, is used by photosynthetic organisms as a carbon source. Nitrogenous wastes are broken down by a variety of bacteria into nitrites and nitrates that can be utilized by plants. The metabolic processes of plants convert nitrates into plant protein, and thus provide a source of nitrogen (as amino acids) for those organisms that can feed on plants and that require nitrogen in an organic form.

One can also view community relationships in an ecosystem in terms of oxygen consumption of various organismal types; or at various stages in their developmental process; or in terms of mineral composition of the soil and density of plant populations; densities of plant and animal populations. There are a variety of ways of looking at an ecosystem, and a variety of ways in which an experimenter can emphasize one factor, and determine how other factors are related to it.

THE ECOLOGICAL NICHE

The niche of an organism or a species of organism refers to all those factors that characterize the position of the organism within an ecosystem, except its physical location. An ecological niche is a listing of all the chemical, physical, spatial, and temporal factors required for species survival, as well as those factors that hold its growth in check.

Organism-Niche Relationship

As a product of the evolutionary process, an organism is an integrated ensemble of adaptive traits, a complementary organization that has evolved in response to a set of environmental factors. An organism, therefore, exhibits a complementarity to its ecological niche. As previously mentioned, its structural and physiological adaptive structures enable it to detect, perhaps capture, and feed on its food source. Hence, these adaptations reflect the nature of its food. Its pattern of dentation reveals whether the food is animal, vegetable, or both (see Fig. 18-16). The construction of its limbs are indicative of its mode of locomotion, the type of terrain on which locomotion takes place, and possibly, its means of food procurement. Its temperature tolerance reflects the temperature variation within the niche. Pigmentation may complement environmental coloration by mimicry. Reactions to light indicate whether it is **diurnal** and hunts during daylight hours, or **nocturnal** and hunts at night. Each organism is a key that unlocks reservoirs of energy and matter in a unique fashion.

Coexistence

Within an ecosystem, there are many organisms that have the same mineral requirements, use the same food source, or live in the same type of habitat. Such organisms experience a real or potential competition for the biotic and abiotic features of the ecosystem. The relationship that exists between competing organisms is termed **coexistence.**

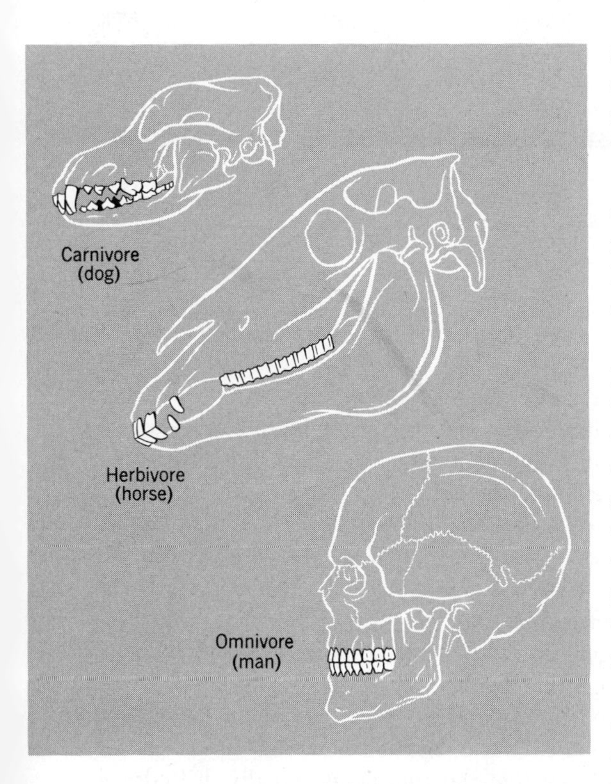

Fig. 18-16 Patterns of dentation for carnivore, herbivore, and omnivore.

Species involved in coexistent relationships may be from very different evolutionary stocks. Eagles and foxes are both predators of small mammals. Birds and insects both make their nests in trees and shrubs. In many cases, however, the coexistent relationship is between two closely related species.

The coexistent relationship between closely related species often pertains to more than just a single competitive area. The degree of competition experienced by the species is directly related to the degree of overlap of their niches. Accordingly, it would be assumed that two closely related species with the same temporal, chemical, physical and spatial requisites for life, could not coexist. This idea was forwarded by Gause, and is known as *Gause's law*. The general proof for this contention is that if two species exploit a given set of environmental conditions in exactly the same way, they would have identical adaptations, and would morphologically be equal. Thus, rather than being two closely related species, they would be of the same species.

In theory, Gause's law is correct. But in nature one finds markedly similar species occurring in what appears to be identical niches, and coexisting in a community. In order to examine the conflict between theory and fact, Park studied two species of flour beetle, *Tribolium*, which appeared to have the same ecological niche. Within the laboratory he used several climatic variables in many replicate trials to test the survival of members of the two species. With hot-wet treatment, only *T. castaneum* survived. With cold-dry treatment, only *T. confusum* survived. Under other climatic conditions, both species survived (see Table 18-5).

Under natural conditions, the fluctuating environment would produce a series of *microniches*, in which certain biotic or abiotic features would exist for short but frequent periods of time. While two species may appear to be occupying the same niche, they are in reality, occupying different microniches within the same ecological niche. Hence, they can readily coexist.

TABLE 18-5 SURVIVAL RATES OF TWO FLOUR-BEETLE SPECIES ACCORDING TO TREATMENT RECEIVED

Treatment	Number of Repeats	Survival Percentage *T. castaneum*	Survival Percentage *T. confusum*
Hot-wet	29	100%	0%
Hot-dry	29	10%	90%
Temperate-wet	28	86%	14%
Temperate-dry	30	13%	87%
Cold-wet	28	29%	71%
Cold-dry	20	0%	100%

QUESTIONS

1. Describe the levels of ecological study.
2. Why is the evolution of organisms considered to be inexorably tied to the presence of large amounts of water on the earth?
3. What is the greenhouse effect?
4. What are the three gases that contribute to the character of the biosphere, and what properties of these gases are directly related to organismal survival?
5. How are soils formed?
6. How does the genetic capacity of certain organisms increase the temperature range that is tolerable to the organism?
7. How is light used by organisms, other than as an energy source?
8. Name and describe the six terrestrial biomes.
9. What is the evidence that the continents actually drifted?
10. Distinguish between exploitation and mutualism.
11. Justify the following statement: "The evolution of evolutionary relationships goes hand in hand with the evolution of species."
12. Define and give examples of coexistence.

CHAPTER 19

ECOLOGY: ENERGETICS AND EVOLUTION

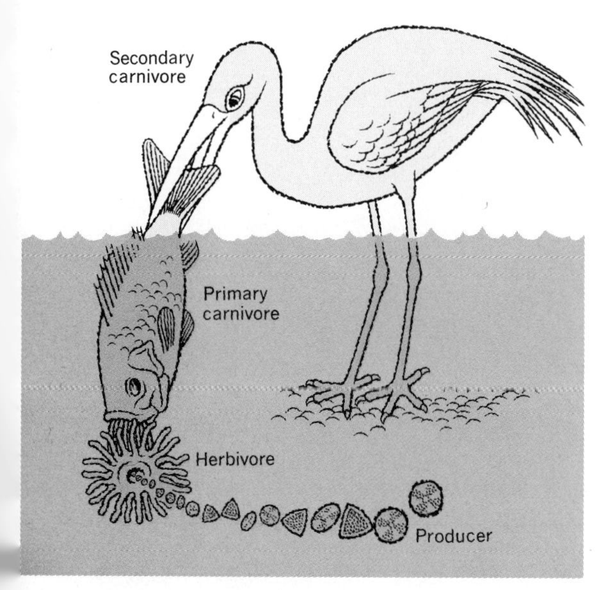

Fig. 19-1 Relation between food chain and carbon cycle.

A prime requisite for the continued existence of an organism is a consistent energy source. The myriad of adaptations that have developed in the course of evolution for capture of prey, mastication, digestion of foodstuffs into usable components, and the capacity to utilize these compounds metabolically, occur as almost endless variations on a basic theme: specific means by which organisms obtain energy.

In a broader sense, mutualism, commensalism, parasitism, and predation are ecological expressions of the aforementioned structural, physiological, and behavioral adaptations. A major feature of community relationships is energy relationships. And, since community relationships exist between members of the biotic community within an ecosystem, a study of the energy relationships among organisms can be effectively made only within this ecological unit.

As previously mentioned, the evolution of species and the evolution of community relationships go hand in hand. Hence, the unit of study of evolution is also the ecosystem.

THE ENERGETICS OF AN ECOSYSTEM

The interdependence of organisms in an ecosystem is based in large part on the food supply. In the sea, a unicellular alga may be eaten by a copepod, which, in turn, is eaten by a herring. In fresh waters, the alga may be eaten by the larva of a mayfly, which is then eaten by the larva of a caddis fly. On land, grass leaves are eaten by a vole, which in turn becomes the food of the weasel. The common pattern for all three examples is:

plant ⟶	**herbivore** ⟶	**carnivore**
alga	copepod	herring
alga	mayfly larva	caddis fly larva
grass	vole	weasel

This linear sequence of predation is termed a food chain (see Chap. 6). If one could follow a single carbon atom, the sequence could be expanded, and recycling of carbon could be detected (see Fig. 19-1).

In one sense, the food chain is an elemental cycle. In another sense, it is an energy chain. As an elemental cycle, carbon is constantly passed from organic to inorganic forms and back again. As an energy chain, it follows the second law of thermodynamics, and therefore, *is unidirectional, always terminating in heat dissipation* (see Chap. 2).

There are many different algal species in the ocean. All of these collectively are the **producers** of the community. There are many different kinds of copepods, crustaceans, and tunicate larvae that eat algae. These are collectively known as the **primary consumers** of the community. The primary consumers or herbivores can be eaten by many kinds of carnivores or **secondary consumers.** The overall picture is not a chain, but rather a **food net** representing many different levels, which are called **trophic levels.** Each trophic level is composed of all the organisms in a community that have a common food source (see Fig. 19-2).

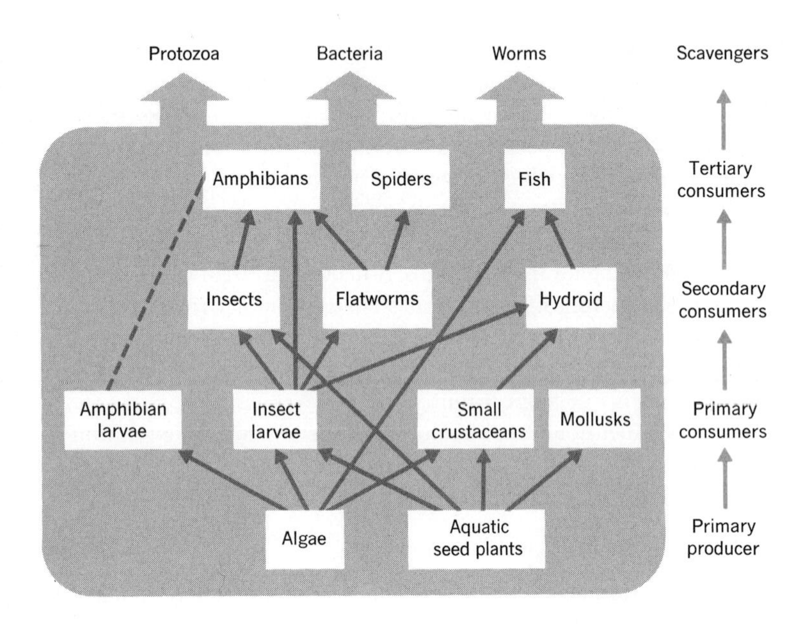

Fig. 19-2 The trophic-level concept.

A community follows a complex conversion of protoplasm. Not only does one organism have a variety of foods available, but also, some organisms change their diet during development (tadpoles are herbivorous, frogs are carnivorous). Many organisms are omnivores, and eat both vegetable and animal matter.

Ecological Pyramids

Several attempts, each unsatisfactory for a different reason, have been made to reduce this maze of community groceries into some simple

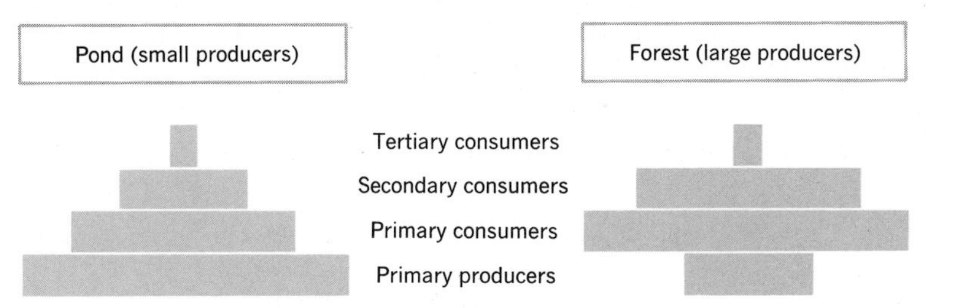

Fig. 19-3 Trophic levels in a pond and a forest. In a pond a large number of producers (phytoplankton) support smaller groups of organisms. In a forest a small number of large producers (trees) support large numbers of primary consumers.

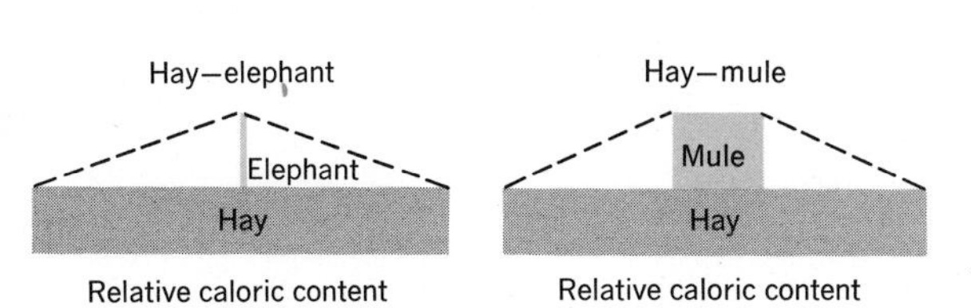

Fig. 19-4 Pyramid of numbers.

representation of the dynamic changes that are going on. Elton, in 1927, recognized the fact that the organisms at the producer trophic level are greater in number than those at the last consumer trophic level, and therefore, there is a sequential reduction of numbers. This relationship can be diagrammed as a **pyramid of numbers** (see Fig. 19-3).

If the width of each level of the pyramid represents the numbers of individuals that support a certain number of predators, as represented by the width of the next level, a quantitative and informative picture is presented. Unfortunately, several common examples produce either an inverted pyramid or an irregularly shaped one. A single plant may have numerous parasites, and all of these may live on the plant at different sites. The situation in which one organism (the plant) supports many organisms (the bacteria or viruses) would produce an inverted pyramid. Also, a ton of hay can support a greater number of mice than elephants, so a new pyramidal shape is formed (see Fig. 19-4).

If, instead of numbers of individuals, the mass of individuals (the **biomass**) is used, the typical pyramid shape is retained even in the cases of parasites or two vastly different species utilizing the same food source. The biomass is usually expressed as grams per cubic meter for aquatic groups, and grams per square meter for terrestrial forms.

Even this method of quantizing the food web leads to some inverted biomass pyramids, and therefore, it is still somewhat artificial. For instance, there is about five times as much biomass of zooplankton (herbivore consumers) as there is of phytoplankton (producers) in the English Channel. These measurements are made on the basis the biomass of both zoo- and phytoplankton collectively available at any one time, or the **standing crop.**

The difficulty in analyzing food nets in this way is quite simple and can be seen in a more common situation: "How can a man survive on a meal that weighs less than himself?" It is erroneous to compare the weight of an animal that required twenty years to grow with the weight of what it consumes in one fifteen-minute period. The comparison should be between the weights over the life of the individual, and how much it consumes during this time.

The total amount of phytoplankton produced over one year in the

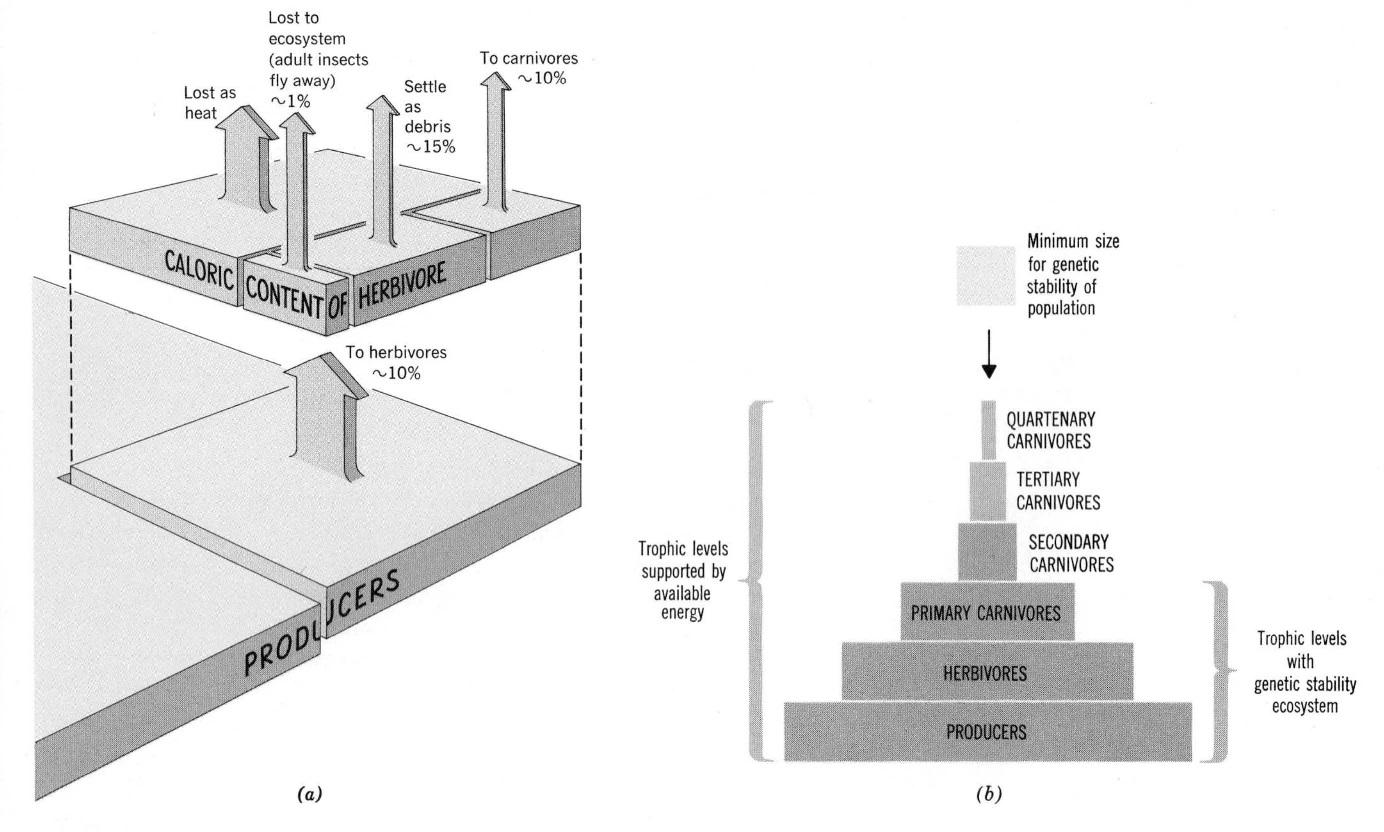

Fig. 19-5 (*a*) **Comparative distribution of energy from producer to primary consumer trophic levels.** (*b*) **Calories per meter^2 required to support minimum genetic size of population.**

English Channel far surpasses the zooplankton. Thus, by using the rate of production of the standing crop, or **productivity,** the pyramid is reinverted, and assumes its typical character.

Finally, the productivity of biomass pyramids has some limitations. The food value of all organisms is not the same. Some organisms have parts that require protoplasmic conversion for their construction, and thus are included in biomass productivity, but are inedible (such as bones, scales, and shells).

To overcome this difficulty, an energy pyramid can be constructed based on the **caloric content of the biomass.** This method, too, has its problems, because it overlooks the difference between the energy that organisms use for their development and their food value to a predator. Organisms do not grow in order that they can become food for others.

In general, a quantitative description of the food chain (or really, the food web) can be expressed in several ways, all of which are inadequate for different reasons. This inadequacy arises because of three factors that are involved in the state of an organism: (1) requirements for growth, (2) the quality of protoplasmic mass available as a food source for others, and (3) the amount of energy expended in obtaining food.

The inability to obtain meaningful quantitative measurements is the death knell to any science. The inadequate results from the data of different ecological pyramids means a new line of approach is necessary. Is there not some means of quantitatively expressing the food web by taking into account the energy expended by an organism in obtaining food, as well as the caloric value of that organism as food?

Lindemann in 1942 developed a mathematical concept for measuring the energy flow through trophic levels. The amount of energy available for trophic level C is a function of the caloric value of the preceding trophic level B (see Fig. 19-5). This caloric value passing from B to C is denoted b'. b' is in turn a function of the energy that trophic level B receives from its food source (b) minus that which it dissipates as respiratory heat (L) in metabolic processes for the maintenance of life through growth, food capture, repair, and reproduction.

$$b' = b - L^{*}$$

For example, consider a group of organisms (trophic level B) dependent on plants. For every 1000 kcal trophic level B receives by consuming the plants, 900 kcal are used for various processes and finally dissipated as heat. The amount incorporated into body structure, and therefore usable for the next trophic level C, is then

$$b'' = b - L$$
$$b' = 1000 - 900$$
$$b' = 100 \text{ kcal}$$

The significant feature of this equation is that it permits a calculation of the ratio b'/b to give a direct comparison of the amount of energy (b') passed on to trophic level C, and the amount of energy (b) consumed by the trophic level under consideration (B).

Slobodkin postulated that the ratio b'/b derived from Lindemann's equation would be significant if the b'/b ratio throughout the food net in an established ecosystem turned out to be a constant. Through the selection of organisms that generate a unilinear sequence of trophic levels

*The equation is actually derived from a partial differential equation for the change in energy in a standing crop over a small period of time.

$$\frac{\delta B}{\delta t} = b - (b' + L)$$

Or, the energy change in a standing crop b, over a small period of time (t), is equal to the energy gained plus the energy lost.

If one assumes that at equilibrial conditions, the rate of change of the standing crop is zero, the equation becomes

$$0 = b - (b' + L)$$

or

$$b' = b - L$$

for laboratory studies, ecological efficiency values have been obtained. For example, Slobodkin measured the ecological efficiency of a unilinear laboratory ecosystem composed of the unicellular alga *Chlamydomonas* (producer); the invertebrate herbivorous animal *Daphnia,* which feeds on the alga; and man, who, by removing *Daphnia* from the system manually, acted as a substitute carnivore. Numbers of *Chlamydomonas* and *Daphnia* were converted to corresponding caloric values. Every four days, 25%, 50%, 75%, and 90% of the *Daphnia* were removed from four identical ecosystems respectively to simulate *Daphnia* predation, and the number of surviving algal cells in each system was determined. The difference between the number of algal cells in these cultures and that in systems where algal growth is not interrupted by predation reflects the eating efficiency of *Daphnia.* The *Chlamydomonas-Daphnia*-Man food chain exhibited a maximum ecological efficiency (about 10%) at 50% predation of *Daphnia* (see Fig. 19-6).

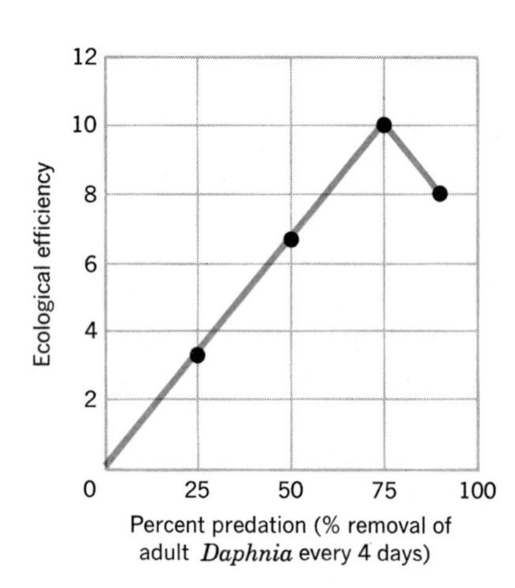

Fig. 19-6 Maximum ecological efficiency for *Chlamydomonas-Daphnia*-man sequence is 10%. This occurs at about 70% predation.

Ecologists have measured the energy value of various steps in the food chains of a variety of communities. Regardless of the types of organisms involved, the ecological efficiency values are about 10% (the ratio of b'/b is $1/10$). This means that 90% of the energy at one trophic level is not available to the subsequent trophic level. It is thought that this efficiency value does indeed reflect some inherent relationship in energy transfer through a community.

The striking consistency of energy efficiency values throughout the food chain, and their relationship to the Lindemann equation, may provide a promising opening to a quantitative study of ecological energetics.

Maintenance of Balance within an Ecosystem

The balance of the ecosystem is dependent on many factors, both biotic and abiotic. Among the most important determinants for the maintenance of ecosystem equilibria is the energy relationships between the different trophic levels. There is a limited quantity of energy available, a limited amount of light energy that is absorbed and converted to a biologically usable form by the producers. The continued existences of all the members of the community at all trophic levels are dependent, each in their turn, on this initial energy content. There can be no greater biomass at any trophic level than what the energy can support. Thus, the maintenance of a balance within an ecosystem is dependent on the controls of population sizes within that ecosystem.

There is much evidence that population size is controlled in natural communities. This evidence is derived from calculations of the **biotic potential** of organisms, that is, the maximum population size under ideal conditions (unlimited food, space, light; absence of predators, parasites, or competitors) compared to the actual state found in communities. The results of such evaluations may be seen in Fig. 19-7.

It should immediately be obvious that the curve for the state of organisms within communities is the sigmoidal one encountered in Chapter 11 on growth. As in the culture tube, the 2^n phase of growth

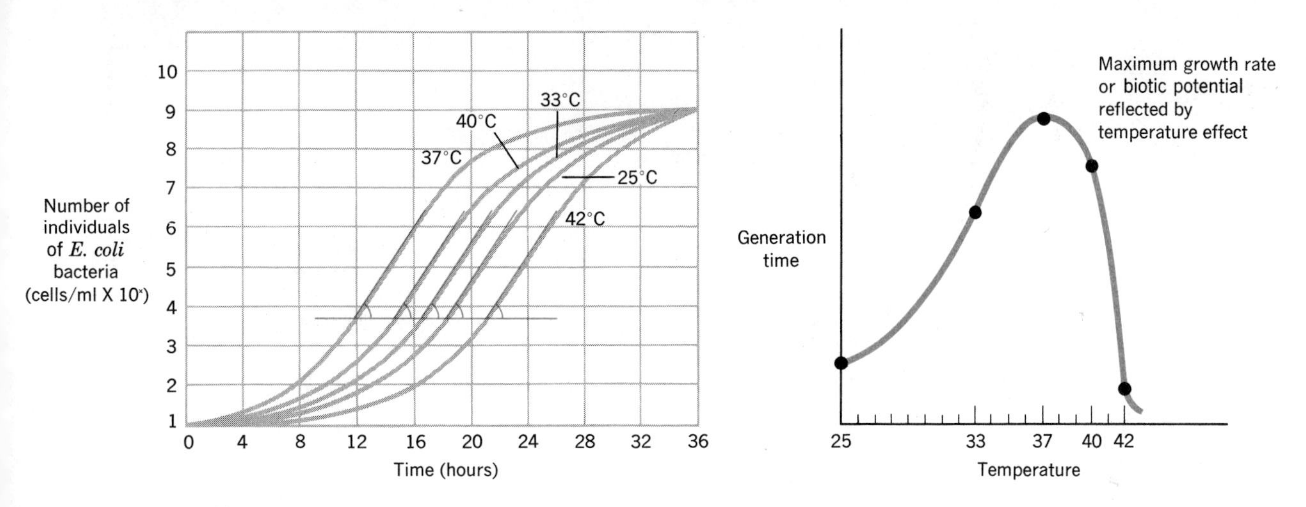

Fig. 19-7 Biotic potential.

is replaced by the stationary phase, during which the birth and death rates are approximately equal when factors within the tube become limiting, so the natural community exerts an environmental resistance toward the overgrowth of populations.

Yet another comparison can be made between the growth curve obtained for bacterial populations in the laboratory and that for the natural populations of any organismal type in a community. The laboratory situation involves an ideal culture medium in which the only limiting factors are the abiotic features of the environment. The curve obtained in these laboratory studies represents the **carrying capacity** of the environment. Natural populations are exposed to a variety of limiting factors. The curve obtained for natural populations fluctuates considerably, and is generally lower than a projected curve for the carrying capacity of the environment (see Fig. 19-8). Not only do organisms fail to attain their full biotic potential, but they also do not remain at the carrying-capacity level of their environments.

The factors that control population size seem to be both external facets of the environment and internal physiological features of the organisms. Among the external environmental features are the predator-prey and parasite-host relationships. An increase in the number of prey is often followed by a population increase of the predator (see Fig. 19-9). It is logical to assume that parasite population size increases in accordance with the increase of that of the host population.

The cause underlying the concomitant increases in predator-prey and parasite-host populations is easily visualized. The more prey that are present in a given community, the greater the chance of a predator finding an abundant food source, growing to sexual maturity, and finally successfully rearing its young.

The opportunity for predator proliferation is increased by the

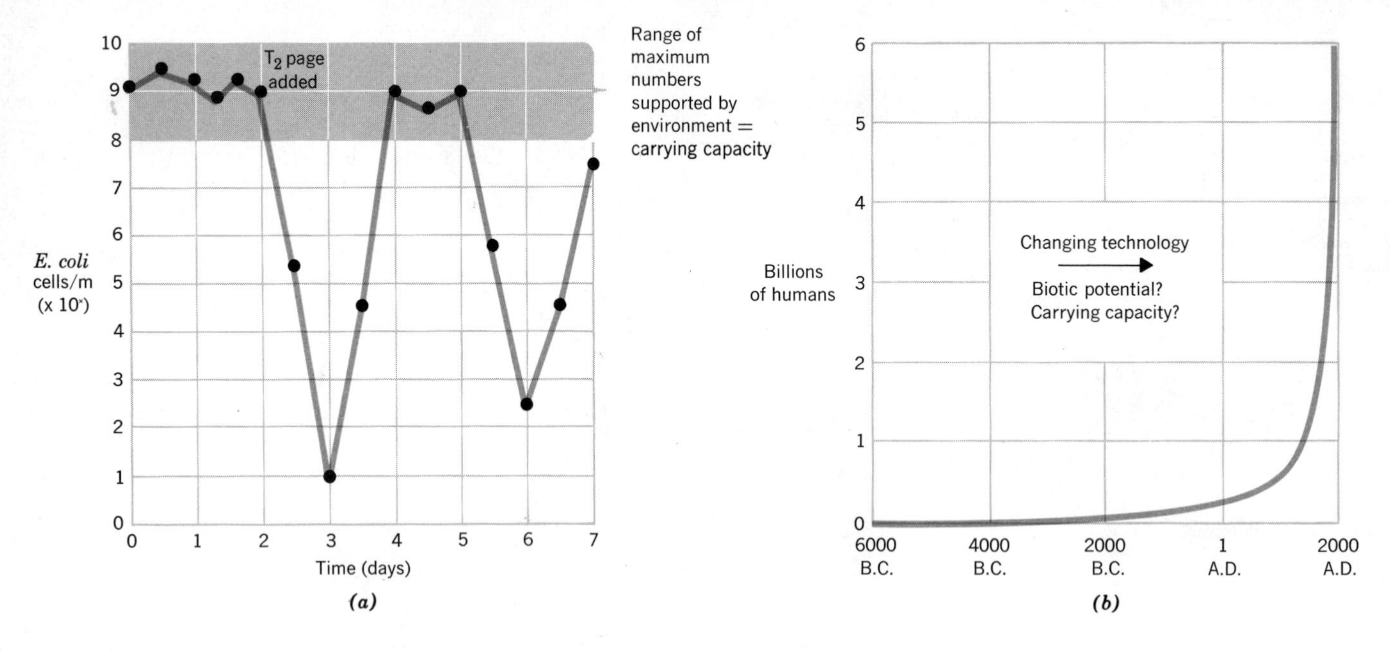

Fig. 19-8 Carrying capacity.

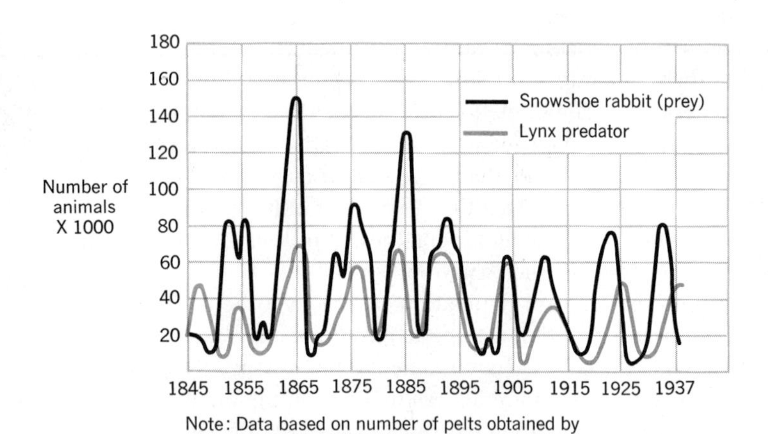

Fig. 19-9 The population levels of the lynx follow the population levels of the snowshoe rabbit.

auxiliary factor of competition, both between members of the prey species and between members of the prey and other species that occupy the same trophic level. As the population of prey increases in number, competition for the products of the preceding trophic level increases. Many of the prey species cannot obtain enough food to sustain their full physiological potential, and so are weakened, and easily found and subdued by their predators.

Evidence is accumulating that seems to indicate that organisms, particularly mammals, have a physiological mechanism for reducing population size under crowded conditions, even when food is not limiting. For example, laboratory mice reared under crowded conditions exhibit a number of curious features. The growth processes of the individual slacken. The length of time required for the young to achieve sexual maturity is lengthened. Even the reproductive inclinations of mature mice diminish. Other factors directly related to the reproductive process are altered. The degree of spermatogenesis in males is markedly lowered; many embryos do not implant normally in the uterus; and females do not produce enough milk to feed their young. These occurrences have been attributed to a change in the physiological state in mice as a response to the crowded conditions.

SUCCESSION AND EVOLUTION

The presence of moist, well-aerated soil, a temperature range of great constancy, long summers, and 20 to 60 inches of rain spread out evenly throughout the year in a region, does not mean that one will find a deciduous forest present. The conditions indicate that a deciduous forest would be the most stable ecosystem or group of ecosystems that could exist in such a region. But it is a long way from the suitable abiotic conditions of temperature and rainfall to the establishment of a deciduous forest. First, primitive forms such as lichens and mosses are established, and by their presence, alter the abiotic features of the environment (pH of the soil, increased nutrients upon their decay, etc.). The altered environment permits the invasion of other forms, such as the grasses, and still other environmental alterations by their establishment. Finally, the flora of a typical deciduous forest is established. This sequential invasion is termed **succession.**

Succession can be viewed as repeating the history of evolutionary events by the invasion of preadapted organisms, rather than by genetic adaptation as occurs during evolution. Primary succession toward a deciduous forest begins with the establishment of lichens. So did the evolution of a deciduous forest.

Before the factual course of succession is considered, two questions should be raised and answered. First, why does succession occur? Given the proper climate, soil, rainfall, and seeds of preadapted plants, why can't the stable deciduous forest be established immediately? Second, how is succession possible? How can a few itinerant types invade an area and compete with the established flora to eventually take over and change the character of the ecosystem? The answer to both questions lies in various adaptive features of plant forms.

A stable deciduous forest cannot be established immediately because of the adaptive features of the trees, shrubs, and mosses that comprise such an ecosystem. The seedlings of trees require shade. They cannot grow in open land. Also, these seedlings grow very slowly, and cannot compete successfully with faster-growing forms. When land lies fallow, all types of seeds fall on it. Yet chiefly the grasses will grow. The shrub and tree seedlings cannot compete with the fast-growing grasses. Thus a deciduous forest could not possibly arise immediately, despite the apparently ideal climatic conditions for one.

Once grasses are established, how do shrubs and finally trees take over such a region? The answer is twofold. First, the establishment of most grass is seasonal. Each year, the competition for space begins anew. And this competition is often not only between established forms and newly invasive ones, but also between seedlings. The seasonal reoccurrence of competition increases the chance that some shrubs will be able to compete in open sunlight and be successful. The commitment of tree seedlings to shady conditions precludes the possibility of their succeeding directly in the competition.

Once a shrub gains success, it grows above the grasses and decreases the intensity of the available light. Shrubs do not die back seasonally, and competition each year is then between established shrubs and grass seedlings. Grasses cannot grow beneath shrubs, but the shady regions under shrubs are suitable for both tree and some shrub seedlings. Thus, the second part of the answer is that the establishment of new forms opens up new ecological niches, and newer forms become established. Once a tree is established, it creates more shade and further inhibits the growth of grass. Eventually, the grasses are expelled from the environment, and trees and shrubs remain.

The climax state of all terrestrial and aquatic biomes is established by sequential replacement stages that comprise different successions. In a prairie (pampas, campo, etc.), the initial steps in succession are essentially the same as those leading to a deciduous forest. The shallowness of the soil, the amount or distribution of rainfall, etc. are not suitable for shrubs to gain a foothold in the region. Hence, grasses represent the climax state of such regions. In tiaga regions, the temperature, rainfall, and wind inhibit deciduous tree seedlings from growing. Therefore, at the stage in succession where trees become established, the conifers become dominant.

It is of little value to discuss the pattern of succession that occurs in each type of biome. A relatively detailed study of succession toward a deciduous forest will be given, with the understanding that it is but an example of the way in which succession can take place.

Primary and Secondary Succession Toward a Deciduous Forest

Succession can begin in an area in which no life is present, in which case it is called **primary succession,** or it can begin in an area that has been so drastically changed that only a few kinds of organisms survive, in which case it is called **secondary succession.** Volcanic eruptions, erosions, sand slides, or excavation by construction work may initiate primary succession, whereas forest fires, disease, or hurricanes are followed by secondary succession.

A typical succession leading to a deciduous-forest climax in a temperate zone will be traced in order to identify the sequential stages and the causes behind the alterations. The first step in the primary succession is the **lichen stage.** The lichens are the first plants to establish a foothold in some barren area, and because of their hardy way of life they are called **pioneer** forms. They can grow on rock surfaces where only microscopic penetration into the substratum is possible. Little water may be available, even though it may rain frequently, because most of the water quickly evaporates or runs off from the impervious rock surface. Weathering causes some of the rock to break down into soil particles. The lichens themselves enhance the breakdown of rock to soil by the action of metabolic acids, or those formed from lichen decay. Gradually, some soil and humus develop, but only as a shallow layer.

The Moss Stage

Mosses begin growing in these shallow depositions of soil, and gradually replace the lichen antecedent. It should be noted that seeds and spores of all kinds of plants enter and germinate, but it is a matter of ecological selection as to which of these can develop under prevailing conditions. Since the mosses are taller than the lichens, they crowd the latter out by capturing most of the light. When the lichens die, the area becomes inhabited by moss.

As the mosses grow and die, the *detritus* (decay material) breaks the rock down even further, and eventually a rich microflora of bacteria and fungi can be established.

The Herb Stage

When a sufficient amount of soil is formed to retain moisture, seedlings of herbaceous plants assume a foothold, and eventually replace the moss in the same way that the moss replaced the lichens. Grass, goldenrod, asters, and stunted woody plants dominate the area. Small mammals, snakes, and a rich variety of insects find suitable niches upon invasion of the area. A nutritionally richer and better-aerated soil layer develops to make the climatic conditions more promising for supporting life.

The Shrub Stage

Shrubs and small trees now replace the grass because (1) the soil is able to support their growth, and (2) upon growth, they are higher, and therefore reduce the amount of sunlight available to lower-growing plants. These taller-growing plants affect the environment by providing better shade over the soil and serving as windbreaks. With the disappearance of the grasses, fewer insects remain. The environment is now suitable for those species of birds that utilize berries as a food source and shrubs for cover. Wet areas become drier and dry areas become moister, so conditions become more **mesic,** by different areas of the region showing little fluctuation from the average water content of the whole region.

The Tree Stage

Trees begin to develop in the moist, shady undercover, and the area becomes dominated by cottonwood, willows, maples, oaks, and birch. These trees form a continuous canopy, leaving some shrubs to survive. The ground once again becomes populated with moss, because there is too little light for grass and too much light for ferns.

As the trees die and fall to the ground, scavengers break them down and a rich humus is formed. There is no dominant type of tree, simply those that could survive under replacement conditions.

With the establishment of trees, the climax of succession to a deciduous forest is reached. Depending on specific climatic conditions, one or a few species of trees will eventually become dominant because they are better adapted to the area (see Table 19-1).

TABLE 19-1 GENERALIZED SCHEME FOR THE SUCCESSION TOWARD A DECIDUOUS FOREST

Stage	Flora	Fauna
Primary		
Lichens	Crustose lichens Fructose lichens	Mites, ants, spiders
Secondary		
Moss	Moss and club moss	Insects and small invertebrates
Herb	Grass, goldenrods, asters, evening primroses	Insects, earthworms, snakes, mice, rabbits, sparrows, deer, skunks, opossums
Shrub	Sumac, aspen, blackberry, red cedar, juniper, mulberry, saplings	Nesting birds, primarily
Tree	Red maple, cottonwood, willows, moss, ferns	Insects, earthworms, salamanders, snakes, turtles, lizards, mice, shrews, fox, raccoons, weasels, skunks
Climax	Oak-hickory or beech-hemlock-maple	A variety of all groups

EVOLUTION OF ECOLOGICAL RELATIONSHIPS

Each adaptation of a species is followed by an alteration of the community relationships and a shifting of the selective pressures within the ecosystem. The intimate relationship among the organisms of a community demands that evolution bring the entire community to a new posture, rather than affecting just one resident species. Thus biological evolution is inexorably tied to the evolution of community relationships.

The Adaptive Value of Succession

Because each species develops in response to its biotic and abiotic environment, one can view all of the typical anatomical and physiological features of a species as adaptations. The answer to the question "what is the adaptive value of *x* structure?" provides the link between the environment and the species. To be sure, the answer to such a question is speculative; but in view of the implications of the Darwinian interpretation of evolution, the answer is most logical.

Relationships in a community have also come about through evolution, and therefore can be viewed as adaptations. It is just as logical to ask "what are the adaptive values of community relationships?" as it is to ask about the adaptive values of anatomical features.

Succession is a sequential turnover of community relationships in which the types of *preadapted* organisms that invade an environment are the same as those that genetically adapted to similar past environments during the course of evolution. Each stage in succession is a product of evolution and has an adaptive value. It is most logical to ask what the adaptive value of each stage is in comparison to the preceding stage. The answer to this question requires logical speculation based on the relationships that exist in an ecosystem.

The adaptive value of each stage in succession, lies in its successful utilization of the minerals, water, etc., but perhaps most important, the *amount of energy available to the community*. A given amount of energy enters a community in the form of sunlight. There are two ways in which energy can be utilized: (1) by increasing the number of trophic levels, and (2) by increasing the size and variation of each trophic level. In order to envision these means of energy utilization as leading to adaptive values, it is necessary to examine the evolution of the stages in succession, with particular emphasis on energy utilization.

Energy Utilization by Increased Number of Trophic Levels

Examine first the effect of increased trophic levels on energy utilization in a community. With the appearance of the lichen stage, only plants and scavenger bacteria that decompose the plants after they die, are present in the community. Thus, there are only two trophic levels.

Shortly after the establishment of the lichens, the mites, ants, and spiders appear. No longer does the death of the lichen have to occur before its energy becomes available. Animals take the place of bacterial scavengers by their ability to directly feed on the plant while it is still living. The presence of simple herbivores fosters the evolution of bacteria capable of decomposing these animals when they die. Thus, three trophic levels exist: (1) the lichen, (2) the spiders, mites, and ants, and (3) the bacteria that decompose these plant and animal forms. The energy captured by the lichens by their photosynthetic process is passed through, and utilized by, a greater number of biological forms. In essence, the presence of any biological form in a community represents a source of energy for any other form that can evolve to feed on it. Sequential feeding increases the number of trophic levels, and hence, the utilization of energy ultimately derived from sunlight by the producers.

Yet another consideration exists in terms of the number of trophic levels. This consideration is the efficiency of energy transfer between levels. The bacteria were probably able to utilize only about 3% of the energy available in the lichen, because they were dependent on natural partial breakdown of the plant material. The spiders, mites, and ants, by direct feeding, could perhaps utilize closer to 10% of the available energy. Hence the evolution of spiders, etc. (active eaters) served to increase the coupling efficiency between trophic levels.

The number of trophic levels in a community tends to increase both in number and in coupling efficiency. Organisms capable of utilizing more energy from the preceding trophic level will replace others whose energy utilization is more limited. The tendency toward an increased number of trophic levels fosters the evolutionary process. A plant that has no predator is exposed only to the selective pressures of the abiotic environment. Once a trophic level is introduced above the producer level, biotic selection pressures come into play, and the plant initiates the spiraling series of events that lead the community to new evolutionary states. Animals evolve to feed on new plant forms, new animals evolve to feed on the new herbivores, etc.

Energy Utilization by Increase in Size and Variation of Trophic Levels

Energy utilization is dependent on the lateral growth of trophic levels as well as the increased number of trophic levels. Since the energy relationships in a community ultimately depend on the photosynthetic forms within that community, the broader and more variant the producer level, the broader and more variant will be the consumer levels. For instance, a field of low plant growth, which is one stage in both the organic evolution and the succession of a deciduous forest, may absorb 40% of the available light, compared to the 10% absorbed by lichens that grow only on rocks. The climax stage, the deciduous forest, may absorb 50% of the available light.

The difference between the amount of light absorbed by the grass-like field and the deciduous forest lies in more efficient mechanisms for light absorption in the community of the latter. A field of low growth has only one level at which light is absorbed. A deciduous forest absorbs about 40% of the light at the canopy level of the trees. The light that passes the tree canopy is absorbed by the shrub leaves below. And the

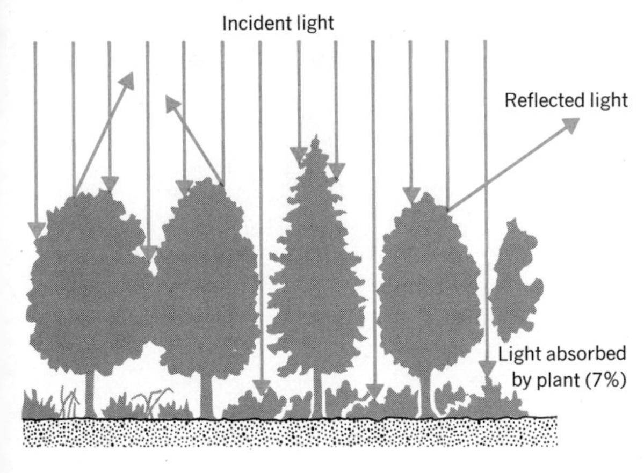

Fig. 19-10 Accessibility of light to low-growing forms in a deciduous forest.

light that filters past the shrub leaves is absorbed by the ferns (see Fig. 19-10). The variation in the producer level permits the absorption of 50% of the available light and also permits variation in subsequent trophic levels. Animals that feed on trees, shrubs, and ferns are all supported in a deciduous forest.

There is one other feature of lateral expansion of trophic levels that must be examined. The stability of the community depends on the presence of several competing species. The presence of competing species acts as a regulator of population size within a trophic level (see Chap. 18), and, at the same time, increases the chance that the trophic level will continue to exist. If a trophic level contains only one or two species, the extinction or expulsion of these species from the community would leave a gap in the trophic sequence that would drastically alter the community. By having a number of species adapting to exploit the same community feature, and thus occupying somewhat similar niches, the community increases in stability.

The evolution of a deciduous forest by sequential replacements in which the trophic levels increased both in number and in variation in suitable environments represents a movement toward the most stable energy conditions that can exist among organisms. Succession is the modern counterpart of evolution, in which preadapted organisms repeat the general sequence of energy relationships once came about through genetic adaptations.

QUESTIONS

1. Distinguish between the carbon cycle and the food chain.
2. Define the following: producers, primary consumers, and secondary consumers.
3. In what ecological situation would the pyramid of numbers be (a) valid, (b) invalid?
4. Define biotic potential and carrying capacity.
5. Describe succession toward a grassland.
6. What is the adaptive value of succession?

SPECIAL INTEREST CHAPTER VI

HUMAN ECOLOGY

Natural selection lost much of its influence when it faced the species *Homo sapiens*. Not only was natural selection incapable of controlling an intelligent creature, but it could not compete with the pace at which man controls other forms of life. Man has accomplished a new feat in evolution: he has overcome ecological restrictions, and in doing so, has realigned other forms of life as well as the physiochemical factors of the environment for his own purposes.

Despite man's ability to overcome ecological restrictions, he remains a biological creature. He still requires water of suitable purity to replenish his daily losses. He remains shackled to a requirement of oxygen fairly uncontaminated with products of combustion. There is still the need for eight amino acids, several vitamins, and a mixture of fatty acids and carbohydrates. Psychologically, he needs space to satisfy his desire for privacy and freedom of movement. Biologically, then, he is no different from his ancestors of 20,000 years ago.

He is very different from his ancestors in the way in which he obtains these necessary factors. Water is purified and transported to his shelter. Food comes from specially bred plants and animals, and is often industrially processed and enriched. Aesthetically pleasing edifices substitute for the disappearing color and scent of nature.

If man is to survive his domestication of the land, he must find solutions to three interrelated problems. First, he must find means to eliminate the waste products from the increased number of individuals on Earth. Second, he must be able to harvest more food in order to maintain the large population. And third, he must find some means of holding the human population size in check; otherwise, solutions to the two previously mentioned problems become insufficient by the time they are achieved.

The problems of food supply and waste elimination are intensifying with the increase in population level. The days of unlimited natural resources has passed. Management is the new perspective to which man must be committed, or he is destined to let nature resolve his problems in a most dreadful fashion.

THE POPULATION AVALANCHE

Since the efficacy of solutions to the problems of food supply and waste removal apply only after a solution to the population explosion has been found, it is logical to first discuss the dynamics of population growth.

The Balances and Checks of Past Populations

There is certainly no reason to suspect that men of prehistoric societies were any different in their reproductive capacities than those for whom records exist. The average woman was able to bear ten to fourteen children during her sexually mature years, just as is the average of women today. Since two parents would have produced some twelve offspring, or six children per individual, the population should have increased by at least fivefold at every generation.

TABLE VI-1 DOUBLING OF THE POPULATION OVER TWENTY CENTURIES

Year (A.D.)	Population (Billions)	Number of Years to Double
1	0.25 (?)[a]	1650 (?)
1650	0.50	200
1850	1.1	80
1930	2.0	45
1975	4.0[a]	35
2010	8.0[a]	?

[a] No population figures are available. The figure was derived by extrapolation.

The population, however, did not show any such increase (see Table VI-1). The reason for the lack of the anticipated fivefold increase in size is simply that while a woman might bear some twelve children, only two or three of them would live to a reproductive age and contribute to the next generation. The high mortality rate can be viewed as the result of man's inability to adequately control his environment. Despite his attempts to domesticate the land through agriculture, and the sea through fishing, he was dependent on natural conditions to yield the necessary food. Two of the three major regulators of population size, **disease** and **famine,** are expressed in such an uncontrolled environment.

But man did gain some control over his environment, and improved his food yield per acre by the introduction of new crops, or by technological innovations. In 1571 the population of China was 64 million. By 1660, 70 years after the introduction of the sweet potato, and 98 years after the introduction of corn, the population had almost doubled to 108 million. After the peanut was introduced into China, the population size increased to 141 million by 1741. Even the intermittent famines that occurred during this period could not offset the tendency toward population increase as the result of new exploitations of the land.

The failure of famine to counteract the population increase based on new food sources is further exemplified by the factors leading to the potato-blight famine in Ireland. At the time the white potato was introduced into Ireland from the Andes in 1754, the human population numbered some three million. By the time the potato blight fungus infected the crop in 1846, the population was eight million. The total increase in population size must include some one-and-three-quarter million who emigrated. Thus in less than 100 years, the population size had trebled. One out of every eight persons alive at the beginning of the famine died during its six years' duration, so the population of Ireland was reduced to seven million, which was twice that of the population before the introduction of the potato.

Man has also gained considerable domination over disease. Vaccination to prevent plagues, chemotherapeutic agents to selectively kill invading microorganisms, and improved surgical techniques to remove diseased tissue without detriment to the individual, have all contributed to a decreased mortality rate.

The effects of such advances on population increases can be seen in Ceylon. From 1871 to 1950, the yearly population increase in Ceylon was 1.7%. In 1950, DDT was introduced to eliminate the mosquitoes, and with them, the malarial parasite they carry. Since the introduction of DDT, the yearly population increase in Ceylon has risen to 2.8%.

But the agricultural advances that permit larger and larger populations to be supported despite famines, and the medical advances that decrease the mortality rate, are not the only factors that have contributed to the population avalanche. Industrialization in the past hundred years has played the important role recently in both supporting large populations and reducing the length and severity of famines. Improved roads, and vehicles (such as automobiles, trucks, and trains) as means for conveying foodstuffs to large agriculturally sterile areas (such as cities) or previously uninhabitable land areas, have permitted large populations to move away from their food source.

The tripling of the population between 1700 and 1850 in England and Wales was not due to more successful exploitation of the land or greater food yield. Neither was it due to control of disease. The population increase was due to the support of larger populations by importing food.

Industrialization has caused a paradoxical inversion of the factors related to population increase. One would expect that the poorer groups would be affected by famine, wars, and disease, and thus their population size would be held in check. But in the past fifty years, the most astounding increases in population size have occurred in just these groups. Although philanthropic organizations have brought modern medical techniques to these areas, the employment of these techniques has not been widely enough dispensed to account for the population increases. Food production in these areas has not increased relative to the population increase. The factor that has made the existence of larger populations possible in these underdeveloped areas has been the technical achievements of the past fifty years. In India the railroad system permits the transportation of food, and thus larger urban areas can be supported. The building of more and better roads in these underdeveloped areas permitted migration away from food sources, and means for food to be brought to new locations. Thus, even in those cultures we condescendingly refer to as "primitive," a woman who bears twelve children can be assured of the survival of about ten.

Another factor related to man's control over his environment comes into play. When famine, disease, and **war** (the third regulator of population size), were the only factors to be considered, the mortality rate was approximately equal to the birth rate. Thus the population size remained somewhat static. Once the mortality rate fell, and a disparity arose between the birth and death rates, two related factors appeared. First, for the first time in the Earth's history, a sizable number of individuals *past* reproductive age continued to live, and in doing so, use the resources of the environment. Second, many of these individuals were women, who, by living past their full reproductive age, could realize their full reproductive potential. The lowered mortality rate not only directly increased the population size, but also permitted the expression of factors that increased the birth rate, and thus, indirectly increased the population size.

The Solution

The discussion of the "solution" to the population avalanche has been succinctly stated by Dorn.*

"The results of human reproduction no longer are solely the concern of the two persons immediately involved, nor of their families, nor even of the nation of which they are citizens. . . . Man has unravelled many of the secrets of nature, and has learned how to control or modify many natural phenomena, but he has not discovered how to evade the consequences of the biological law that no species can multiply without limit.

"There are two ways to control a rapid increase in number—a high mortality or a low fertility. Alone among the millions of biological organisms, man can choose which of these controls shall be used, but there must be one. The choice cannot be avoided; the time for making a free choice will soon pass."

As outlined by Dorn, the solution to the population avalanche is twofold. First there must be a recognition that the problem exists. Second, there must be a willingness to seek a satisfactory solution.

In response to Paul Ehrlich's popularization and indictment of the population explosion in his book *The Population Bomb* (1968), Zero Population Growth Inc. (ZPG), headquartered in Mystic, Connecticut, was established. The ultimate goal of ZPG is to reduce the number of births per year to equal the number of deaths, a goal to be achieved through education and subsequent legislative action.

Voluntary control of reproduction appears, however, to be directly associated with high standards of living rather than with levels of education. Families with low standards of living tend to be large, not necessarily because they are unaware of birth-control methods, but because the only gratification for them in life is to have many children. In addition, many rural families need children to work the land, act as household servants, and care for the parents in old age.

Families with higher living standards view their children from a totally different perspective. Parents are to care for the children. Children are kept dependent on the parents for longer periods of time. Education of the child becomes important. The care, increased dependency, and education of children is costly, and since parents feel that they can adequately support only a few children, they limit their reproductive output voluntarily.

The correlation between standard of living and voluntary birth control has been demonstrated in many parts of the world. In Lebanon,

*Harold F. Dorn, *The Population Dilemma,* Prentice-Hall, 1963.

which is both a Moslem and a Christian country, rapid urbanization during the past twenty years has led to a Western type of family pattern. Concomitant with this Western perspective of the family (care, increased dependency, and education) have been later marriages and birth control by abortion or use of contraceptives. The influence of the urban centers appears to be spreading throughout the country.

A study carried out in India in the 1950s demonstrated that as the standard of living increased, the fertility tended to increase. At a given level of the standard of living, the fertility drastically decreased. The initial increase appears to represent decreasing mortality rates rather than voluntary control at lower levels of poverty (see Fig. VI-1).

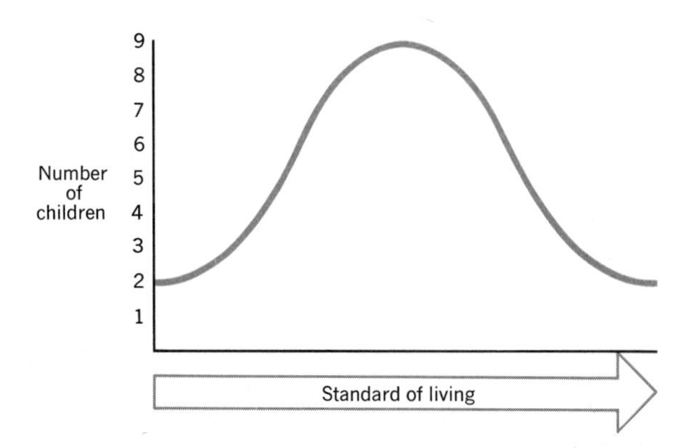

Fig. VI-1 Results of a study in India. Lowest income families have few children because the children die from disease and famine. High income families have few children because of choice.

The fact that birth control is related more to standard of living than to awareness of population increase is evidenced by the fact that some methods of birth control date back to times when there was no reason to be concerned about population increases. Perhaps the earliest record of a means of birth control is the Biblical reference to Onan, who spilled his seed rather than procreating with his dead brother's wife. (The statement that God struck Onan dead for this act is used by the Catholic Church as the Biblical justification for their stand against birth-control methods.)

Ancient cultures developed artificial means of birth control. The Ebeis papyrus, written in Egypt about 1550 B.C., describes a formula of honey and arcacia stem tips that was used as medicated tampons for purposes of birth control. Such a mixture was undoubtedly spermicidal because of the lactic acid that forms. (No mention is made in the papyrus of the infections in the female vaginal tract that undoubtedly resulted from these practices.) Modern history records the use of the condom in the early nineteenth century. Surely its use was not for population control as much as for the fulfillment of Western family values.

Attempts to educate people of lower standards of living in birth-control methods have been frustratingly unsuccessful. If Western man, with his education and freedom from hunger, has not yet reached a level of universal conscience, can he expect one who knows starvation and looks to his children to provide for him, to have such a conscience? Only when the individual sees the need to limit his reproductive output for his own advantage, or for the fulfillment of new values, can full voluntary control of reproduction be expected.

FOOD PRODUCTION

Assuming than man can, by somc acccptablc mcans, rcducc his fertility to the point where each set of parents has no more than two offspring (a one-to-one replacement of individuals or population zero), the time it would take for such a system to be put into worldwide effect would be sufficient for the population to have undergone a substantial increase. The speed with which such an agreement could be made and facilitated would determine the degrees of adjustment that would have to be made in food productivity.

If such events occurred within the next fifty years, it is conceivable that man could, by increasing the efficiency of his exploitations of present farming areas, support the stable, larger population. If such agreement and facilitation are delayed for one hundred years, man may reach the point where no exploitation in terms of efficient use of the land would suffice, and he would have to search for other food sources.

Increasing the Efficiency of Present Land Use

Increased population size has had a dual effect on the land. First, the nutritional needs of the population demand that the land be used more efficiently. Second, the physical presence of the population demands space, and thus many farming areas near large cities have turned into suburbia. It is paradoxical that the very population that requires more food material demands that the land that could produce this material be used for habitation.

The solution to this paradox is to increase the food yield per acre to ever higher levels. The increased efficiency can take two forms. First would be the improvement of the food value, such as increased protein content of grain or butterfat content of milk. Second would be a quantitative increase in productivity.

There are several means by which productivity can be increased, that are now under investigation. One can observe the seedling stage of important crops and conclude that much unused space separates the plants. Mixed crops that can saturate the available space and be harvested at different times to insure the proper development of each crop have yet to be developed.

Another means of increasing the food supply is by increasing the foodstuff yield per plant. Norman Borlang of the Rockefeller Institute was awarded the Nobel Prize in 1970 for his initiation of the so-called "Green Revolution" by developing a new high-yield strain of wheat by crossbreeding Mexican and Japanese strains. This new high-yield strain has been introduced with much success in both Mexico and India.

Productivity can also be increased by changing desert areas to arable land, through irrigation and soil enrichment. Such conversion has been successfully achieved in some regions of Israel, and in some desert areas in southern California, by irrigation with saline water (see Fig. VI-2). For many years scientists had believed that the high salt content in the ocean water was lethal to land plants. A variety of experiments had indicated that this is so. Recently, certain principles of plant growth have come to light that demonstrated that irrigation with saline water might be possible in sandy desert regions. First, water, and those materials that are highly soluble in water (such as sodium and magnesium chloride, the primary constituents of salt water), pass quickly through sand and gravel. Thus the detrimental salt solutions quickly pass the watering levels of the roots, to deeper layers. Secondly, when there is a movement of water upward, toward the roots, the salts tend to come to the surface where they form crystals, rather than remaining in lower layers where they could be detrimental to absorption by decreasing the osmotic pressure. Another factor is that the sodium ion is not absorbed by sand particles, as it is by particles of clay or silt. In fact, the preferen-

Fig. VI-2 A kibbutz in Israel. (*left*) Before reclamation. (*right*) After irrigation with saline water. Tomato plants are flourishing under plastic covers. [Israeli Government Tourist Office]

tial absorption of sodium by clay particles is the main problem in irrigation of soils with saline water. Soils tend to retain the sodium ion, and sand does not.

The failure of the sand to retain the sodium ion seems to be directly related to the feasibility of saline irrigation. Saline water passes by the root system (tiny projectile structures that directly absorb water) so quickly that air pockets form around the root hairs. Condensation occurs within the air pockets, filling them with *fresh water.* Hence the roots have a supply of fresh water from a salinated source (see Fig. VI-3).

Not all plants grow well in desert regions irrigated by saline water. Among those that do, however, are some very important crop plants, such as barley, rye, and sugar cane. The realization of the potential for productivity within the desert lands could well support a substantial increase in population since, at present, most of the 14% of the land

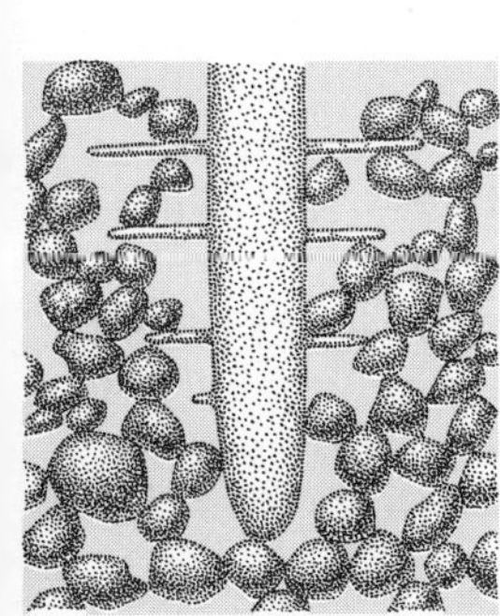

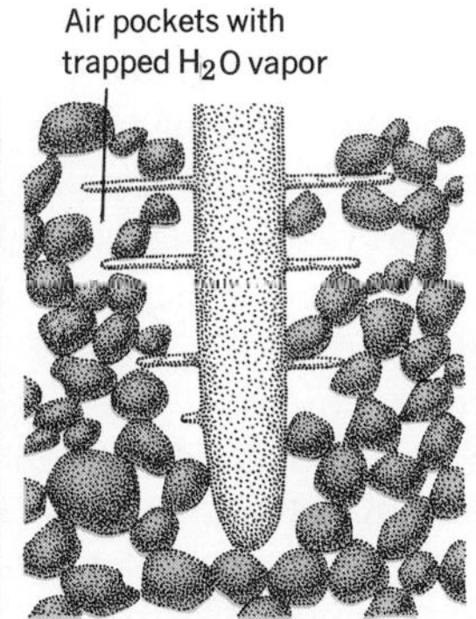

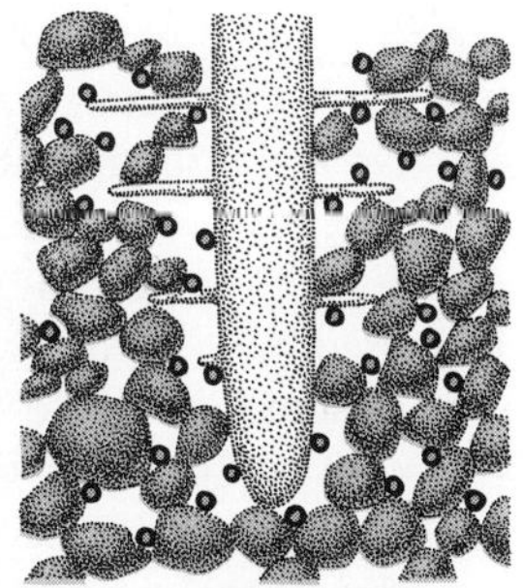

Fig. VI-3 Irrigation with saline water is possible because the salt leaches through the soil while the water vapor is trapped in air pockets in the porous soil. Upon condensation of the water vapor, pure water is available to the plant.

that is desert lies unclaimed.

Productivity is also correlated with the latitude of the region. Tropic lands are two and six times more active in producing plants than cool temperate and tundra regions, respectively. The daily sunshine and year-round growth provide the basis for this difference. Perhaps this is where man should first direct his future efforts.

But man does not live by bread alone. He has a strong appetite for meat, which means that any utopia he attempts to construct should have animals as food sources. How can he increase the total productivity of his animal food?

Man is at the apical trophic level of his ecological community. He is mostly a primary carnivore, and therefore consumes mainly herbivorous animals. The energy he derives from this consumption is about 1% of that available from the plants before passing to the herbivores. If he is to increase the energy he obtains as a carnivore, he must first increase the energy available to the herbivore from plants.

It appears that primary consumers (the herbivores) are not limited in food supply by the available plant food, but by other factors. For instance, one steer and 300 rabbits both weigh about 1300 pounds, and gain 240 pounds from every ton of hay consumed. The difference between the two animals is that it takes only 30 days for the rabbits to gain the 240 pounds, while it takes 120 days for the steer. Rabbits grow four times faster, and therefore are a superior food source for man.

The ratio of the amount of energy incorporated into the body to the amount of energy brought into the body (*the net growth efficiency*) changes during the life of an animal. A chicken grows most rapidly during the first three to four months of its life, during which time it has a net growth efficiency of 35%. After four months, the net growth efficiency slows down to an average of only 4 to 5%. The most efficient means of utilizing chickens as a food source would be to raise them only to the four-month stage. Since hens do not begin laying eggs until they are five to six months old, eggs are also a wasteful means of raising food.

Means to increase the productivity of fish from the ocean are also under investigation. The fishing industry has reached its limit, and is, in fact, probably overfishing its ocean resources in many areas. Improved methods of catching fish have not significantly increased the annual harvest. Studies have shown that fish that have been tagged and released are recaptured with 40% efficiency. This would indicate that there are fewer fish than had previously been imagined.

Greater productivity has been achieved only where the source of plankton, the food of the fish, has been increased. With proper fertilizing of lakes to increase the phytoplankton density, the amount of fish has increased by a factor of three. In some cases it has been found that the fish consume only 1 to 2% of the standing crop of plankton. Much of the rest is consumed by other herbivores, many of which are invertebrate mollusks and crustaceans. If means could be developed to eliminate these predators, thus making the plankton available for the fish exclusively, fish productivity could be vastly increased.

The most obvious means of killing the invertebrate competitors would be through the use of "crustaceacides," or "molluscicides," chemical agents that can be used in sufficient dosages to kill invertebrate plankton-eating organisms without killing fish. Such a program would be similar to that used on land with insecticides. The long-range effects of insecticide use indicate that a thorough understanding of the properties of the chemical agent and the nature of balances within communities would have to be achieved before "crustaceacides" or "molluscicides" could be put into use. The insecticide DDT was freely used in spruce forests to kill the spruce worm. One of the first effects of the spraying was that birds that fed on such worms and insect larva killed by the DDT could not find food, and often died. In addition to affecting the bird populations in the area, DDT caused other effects. Much of the DDT was washed down from the trees by rain, accumulated in the topsoil and was washed into rivers and lakes, where it killed invertebrate larvae. The phytoplankton also incorporated the DDT, and lacking a means of metabolizing and disposing of it, came to accumulate DDT at a level of a few parts per billion. The fish that fed on the phytoplankton further accumulated the chemical agent. Trout were found to have a concentration of four parts per million.

Although the concentration was insufficient to kill the fish in many cases, other detrimental effects occurred. Many salmon were blinded. Fish-eating birds further accumulated DDT in their tissues, sometimes enough to sterilize them, and often enough to kill them.

Examination of the transfer of DDT between trophic levels indicated that this chemical agent is transferred with a 50% level of efficiency. The normal mass transfer between trophic levels is only about 10%. The high transfer of DDT, and the inability of organic tissues to metabolize and excrete the chemical agent, thus fostering its accumulation, have had direct consequences in the biological community. DDT accumulation is also occuring in humans, but as man does not feed exclusively on fish, the concentration has not yet reached dangerous proportions.

These long-range effects have made ecologists acutely aware of the hazards involved in the blatant use of chemical agents to control a single biological form. Before man could tamper with the invertebrates in aquatic environments, he would have to find a chemical agent that (1) does not accumulate in organic tissue, and (2) does not transfer readily between trophic levels. In addition to these factors, he must have a full understanding of the nature of the balance in aquatic communities.

In the final analysis, the means of food production outlined in this discussion are little more than improvements of conventional methods. Since Paleolithic times, man has used biological sources of food, and in doing so, has been part of the energy chain of his environment. In the future, he will extend his farming areas into regions that are presently untamed. He will enrich these regions with chemicals to fortify his exploitations. He will create specially bred animals to take the place of present fauna. But still, he will remain at the apex of the food chain in his environment.

Extensive and intensive food production involves gross rearrangement of the biological composition of the world. As the prairie becomes the wheat field, man will judge the biota of other areas on the basis of which forms he can exploit through further breeding, and which defy exploitation and hence should be relegated to token survival in isolated sanctuaries or zoos. All this will be done to offer the cherished gift of pleasant existence to humans. It is understandable that an individual of the twentieth century might well mourn the passing of nature.

The Search for Other Food Sources

At present there is no sign of a worldwide agreement to lower fertility. Therefore, many responsible governments that can afford the luxury of supporting scientific investigations have provided sizeable grants for finding new sources of food. The hope is, of course, that should the need arise some one hundred years hence, alternative food sources would already be in production.

Several attempts to make synthetic foods have been made. While it is possible to make synthetic materials that contain all the necessary nutrient materials, and the roughage needed by the digestive tract, the human guinea pigs who have tried the synthetic food have pronounced it unsatisfying. Of course, these humans are twentieth-century men who think of food in terms of a large steak covered with mushroom sauce, mashed potatoes, peas, and a salad. Perhaps in some two hundred years when man is used to thinking of food in terms of a bowl of seasoned chemistry, synthetic food may not be so objectionable. Of course, for the sake of the integrity of nature, it would be far wiser for man to make his transition directly from natural to synthetic foods at the time that the food productivity from the land could no longer support his populations. It is perhaps overly optimistic to hope that man will choose to save the biotic communities of the land before he has had his fill.

Another problem of the world's large population today is malnutrition. If population increases are to continue, the malnutrition problem will become even more pronounced. Malnutrition is due to an insufficient intake of the essential amino acids and the necessary vitamins. Attempts to produce synthetic food sources have concentrated on the artificial synthesis and fermentation of the amino acids, with the intention of adding the synthetic material to presently used food material to supplement its nutritional value.

A prime requisite for a synthetic production of food is that it be economical. Since the atomic-driven machines of the future will greatly reduce man's use of petroleum, research is underway to see whether petroleum can be converted inexpensively into nutrient materials. Although some success has occurred in this direction, widespread production is not being carried out.

Another possibility in the future is the design of photosynthetic machines. As man learns nature's secret of carbon fixation, he may be able to use similar methods for driving synthetic food machines. Solar sheets could cover land areas, using the solar energy to drive carbon fixation in factories at the periphery of the field. By this approach man could use the almost limitless source of light energy available to the Earth, even on the rooftops of his futuristic cities.

One of the more unique ideas for a solution has been forwarded by the Nobel laureate, Joshua Lederberg. Lederberg has proposed that some means could be found to incorporate nonpathogenic DNA capable of synthesizing the eight essential amino acids and the multitude of vitamins that comprise the nutritional requirements of man, into human DNA. A human that contained such bacterial DNA would not need protein, or the many foodstuffs that contain vitamins and other nutritive factors. He could synthesize all necessary factors from simple carbohydrates and inorganic nitrates. Food-productivity efforts then could be directed toward sugar-beet growth, rather than the diverse food sources that must now be considered. The scientific area studying the feasibility of this program is called **algeny** (derived from alchemy and genetics).

Many problems must be solved before such a program would be feasible. First, those genes for the synthesis of particular enzymes would have to be isolated. Second, some means would have to be found for them to be incorporated into human DNA. Third, incorporation would have to be made in the gametes or zygote for one generation (after that, the genes would act as part of the human DNA complement, and be passed through generations *in situ*). The present knowledge of enzyme-production sites in DNA, techniques of isolation, tendencies for incorporation, and possible rejection mechanisms of human DNA is so meager that it may be about fifty years before such a program could be put into use.

ELIMINATION OF WASTES

There are three kinds of wastes produced by the human society that contaminate the water and/or the atmosphere of the Earth. The first is direct human contamination of waterways. An increasing population eliminates increasing amounts of human excrement, both feces and urine. The seepage from cesspools and septic tanks into the lakes and inland waterways has left few, if any, unpolluted bodies of fresh water in the northeastern United States. Only the most carefully guarded reservoirs of city water, and judiciously placed wells, are free from polution at this time.

The second kind of wastes arise from man's industrial processes. Factories depend on combustion for the heating of plants and the production of materials. Man operates transportation vehicles that work on combustion principles. Large quantities of carbon dioxide and poisonous carbon monoxide are poured into the atmosphere daily. In the past one hundred and fifty years of industrialization, the amount of carbon dioxide released into the environment has increased the "greenhouse" effect on Earth, raising the temperature of the planet by an average of three degrees Fahrenheit.

Industries are not so selective in their disposal of wastes as to pour them all into the atmosphere. Both biodegradable and nonbiodegradable liquid and solid wastes are dumped into the waterways and oceans, thus further contaminating both fresh and salt water, and reducing fish productivity directly by the presence of poisons and indirectly by altering the biotic conditions and creating ecological imbalances.

Fig. VI-4 Los Angeles smog. [Wide World Photos]

(*a*) A section of the city on a clear and on a smog-filled day.

The third kind of waste produced by human society is radioactive waste. The testing of nuclear bombs contaminates the atmosphere. Seepage from undetonated bombs lost in Spain and Greenland contaminates the land. Seepage from canisters of radioactive wastes from laboratories or industries contaminates the oceans in which they are dumped. As man's use of nuclear power increases, so will the contamination and the problems wrought by that contamination.

Contamination of the Atmosphere

Genetic engineering may someday be able to endow man with the genes necessary to produce the very amino acids and vitamins that he must now take in from an external source. It is highly unlikely, however, that man could be turned into an anaerobic animal, and continue to have the energy necessary for his function as a human. In evolution, the aerobic types gave rise to a wide variety of forms, among which is man. Increased energy derived from an aerobic metabolic process undoubtedly provided certain adaptive advantages. Man will always retain his need for oxygen as a final hydrogen acceptor. Therefore, he must maintain an atmosphere in which the proper proportions of oxygen and carbon dioxide are present if he is to survive.

Yet another factor comes into play. The plants and animals of this planet have evolved in at atmosphere that contained only trace quantities of organic gases, sulfur compounds, nitrogen compounds, etc. The wastes from modern industrial processes have raised the concentration level of these gases in the atmosphere. Not only is man not adapted to utilize these gases in his metabolic processes, but certain of these compounds are also toxic to organic tissue, or tend to accumulate in organic tissue until toxic levels are reached.

The problems of air pollution are not new. Men have faced these

(*b*) **Smog is the result of thermal inversion. Noxious gases are trapped near the ground while fresh clean air remains above.**

problems ever since they banded together in large cities and exercised their ability to make fires. A phenomenon known as *thermal inversion*, in which cold air is trapped under warm air and thus cannot normally disperse the combustion products from large urban areas, was responsible for the London smog in 1661. At this time, John Evelyn wrote:

"Hellish and dismal cloud of sea-coals was so universally mixed with the otherwise wholesome and excellent air, that her inhabitants breathe nothing but an impure thick mist, accompanied with fulgenous and filthy vapour, which renders them obnoxious to a thousand inconveniences, corrupting the lungs, and disordering the entire habit of their bodies; so that cathairs, phthisicks, coughs and consumptions raged more in this one city, than in the whole Earth besides".

And this condition was from the normal combustion of wood and coal.

Since the Industrial Revolution, industrial processes have produced a variety of wastes (see Table VI-2).

TABLE VI-2 AIR CONTAMINANTS[a]

Type	Examples
Solids	Carbon fly ash, ZnO, *$PbCl_2$*
Sulfur compounds	*SO_2*, SO_3, H_2S, mercaptans
Organic	*Aldehydes*, hydrocarbons, tars
Nitrogen compounds	NO_3, NO_2, *NH_3*
Oxygen compounds	O_3 (ozone), *CO*, CO_2
Halogen compounds	*HF*, HCl
Radioactive compounds	Gases, *aerosols*

[a] The most dangerous items are italicized.

Among the most dangerous products are SO_2 (sulfur dioxide), which causes smooth muscle spasms, most often leading to tightening of the bronchioles; ozone, which causes pulmonary fibrosis; particles that cause gastrointestinal disturbances; NO_2, which activates subclinical respiratory infections; and benzapyrene (from the exhaust of diesel engines) which, as previously noted in the special interest chapter on cancer, causes malignancies of the respiratory tract and skin.

The thermal inversions, and the photochemical reactions within the trapped air, greatly increase the deleterious nature of pollution. The typical smogs found in Los Angeles (see Fig. VI-4) are not simple thermal inversions. These smogs contain the trapped olefins (hydrocarbons) and nitrogen oxides that undergo oxidation by photochemical reactions.

$$\text{Olefin} + \text{nitrogen oxide} \longrightarrow \text{nitro-olefins, ozones, and peroxides}$$

The additional ozone content and the production of peroxides substantially increase the pollution.

So many people have minor cases of bronchitis and emphysema that it is difficult to single out those that arose directly from air pollution. Thus it is virtually impossible to effectively demonstrate the effects of industrial wastes on human systems. An interesting study on the incidence of lung cancer in two groups of New Zealanders does give evidence that industrial waste has a deleterious effect. One group was composed of people that were of British extraction, but were born in New Zealand or migrated there before they were thirty; and the other group was made up of people that had immigrated from Great Britain after the age of thirty. Members of both groups smoked about the same. Both groups had similar racial backgrounds (British extraction). Yet the New Zealanders who immigrated from Great Britain (a highly industrialized country, fraught with air pollution) had a 75% greater death rate due to lung cancer than did the native New Zealanders.

The solution to the problem of atmospheric contamination is directly related to the causes. Man must stop contaminating the atmosphere by unnecessary combustion processes. Extreme measures, such as barring all gas-driven automobiles from cities, may have to be placed into effect. Industry must find alternative means of disposing of wastes. Japanese firms have found means of collecting ash particles and forming them into bricks to be used in construction. Means are being devised to convert toxic wastes into nontoxic products that can be metabolized by various types of bacteria. Thus the chemical elements in wastes could proceed through elemental cycles. The solutions are expensive. But the price that human populations may have to pay for the pleasure of enjoying their cars and processed commodities may be far more expensive.

Radioactive wastes emanating from a nuclear explosion have a multitude of effects on the human population. In addition to the deleterious effects of radiation on biological material that was discussed in Special Interest Chapter II, a number of radioactive elements are released. Among these is **strontium**^{90} (Sr^{90}). Sr^{90} is brought down from the atmosphere by rain. Some of the Sr^{90} that falls is absorbed by grass, and is then eaten by cows. Sr^{90} tends to replace calcium. In addition to being incorporated into the skeletal system of the cow, much of the Sr^{90} accumulates in the cow's milk. Cow's milk is an important source of calcium, primarily for children, and the Sr^{90}, instead of the more stable calcium, is incorporated into children's bones and teeth. Sr^{90} has a half-life of 28 days. The high-energy particles emitted from the decay of Sr^{90} is thought to be, in part, responsible for the increased incidence of bone cancer and leukemia in children.

Another radioactive element that is released into the atmosphere from nuclear explosions is iodine^{131}, which is also brought down from the atmosphere by rain. Once at ground or sea level, it is consumed by a variety of plant and animal forms, and proceeds through the food chain. Eventually it is accumulated in the thyroid glands of vertebrate animals, where upon decay it may cause cancer of the thyroid gland.

An awareness of the dangers in contaminating the atmosphere has already led to a ban on atmospheric testing of nuclear weapons. Whether this is a sufficient precaution, only the future will tell.

Contamination of the Water

Contamination of water by direct human products and the products of industry has created a severe problem, one that has no easy solution. Man has evolved to survive in temperate climates. He loses large quantities of water through perspiration and urination, and must take in equally large quantities of water to replace this loss. It is estmated that one man must take in one ton of water a year in food and drink.

Animals and plants that man consumes also require water for growth and sustenance. Although the purity factor in irrigation of plants is not as rigorous as the water required for direct human consumption (see Table VI-3), it nevertheless must be considered. It has been estimated that it takes one half ton of water to produce a pound of plant food.

TABLE VI-3 APPROXIMATE STANDARDS OF WATER PURITY: WATER QUALITY-CONTROL GUIDE

Indicator	Municipal	Swimming	Fish	Livestock and Wildlife	Irrigation
E. coli (cells/100 ml)	1	2400	70	—	5000
Chromium ion (mg/liter)	0.05	—	5–300	5–500	3–17
Cyanide (mg/liter)	0.01	—	0.08	—	—
Phenols (mg/liter)	1.0	—	2700	15,000	—
Sulfates (mg/liter)	250	—	90	250	500

Man's need for water has also extended to his artificial environment, the industries he has created. It takes approximately 365,000 gallons of water to make one ton of rayon; 500,000 gallons of water to process and dye one ton of wool; and a million gallons of water to process 1000 barrels of high-octane gasoline.

The increased population, with its increased need for water for human consumption, plant irrigation, animal feeding, and industry, has raised the water consumption from 41 billion gallons per day in 1900 to 300 billion gallons per day in 1960 in the United States alone. But while man has been increasing his need for pure water, he has been decreasing his supply. Many freshwater lakes and ocean inlets are contaminated with human and industrial wastes.* If population size reaches a point at which it leaves only polluted water in its wake, man would be left with a rather distasteful choice: to die from lack of water, or from disease.

Of course, immunizations against typhoid and some dysenteries are possible. But pathogens can mutate to new pathogenic states, and in doing so, escape the body's defenses. Immunization would, at best, be a temporary solution.

One of the more promising solutions to the problem of water contaminated with human wastes is to support the decomposition of the waste, and hence, encourage the elemental cycles. Two means have been proposed for encouraging the recycling of elemental material in human-waste material. The first is by salting contaminated waterways with populations of decomposing bacteria. The second is by transporting sewage to the ocean, where decomposing bacteria and zooplankton can act on it. Both means are under investigation, but neither is in widespread operation.

Water contaminated by industrial wastes presents a different problem. Many industrial wastes such as mercury, which accumulates in fish as diethyl mercury, are toxic to human tissue. Indeed, if they were not, some industry would have found a means of turning them into a profit, and would not have dumped them in the first place. It is possible to conceive of the development of purification techniques and the construction of factories that utilize these techniques for producing potable water from contaminated sources.

Man may, of course, turn to other sources of water. The most obvious would be the sea. Because of the differences in the osmotic pressure of sea and pure water, and the physiology of the human, man cannot consume sea water. Hence the lament of the ancient mariner, "Water, water everywhere, and not a drop to drink."

Great efforts are in progress by many nations to find feasible ways of desalinating water. While it is possible to distill sea water, such processes are prohibitively expensive for obtaining the necessary amounts. Various attempts have been made to precipitate the salt by the addition of chemical agents. Thus far, none have shown the promise of being economically suitable.

Perhaps the most feasible way to desalinate water is by the use of nuclear-driven distilleries. In the few instances where nuclear energy has been used in industrial processes, severe problems have arisen.

*The water pollution by human excrement is judged on the basis of the number of *E. coli* cells per 100 ml. Although *E. coli* is not a pathogen, it has many requirements similar to the *Salmonella* species that cause dysentery and typhoid. Since *Salmonella* species are difficult to detect, *E. coli* is used as a secondary indicator.

Nuclear energy generates tremendous heat, and it has been difficult to find a means to safely and efficiently dissipate this heat. Attempts to transmit the heat to streams and rivers for dissipation has resulted in raising the temperatures of the water to a level that kills the fish and plants. Some desalination plants are presently in operation, but before desalination of sea water by nuclear energy could become widespread, a solution to the problem of heat dissipation would have to be made.

Assuming that a precipitation method or an inexpensive distilling method could be devised, the industrial process would leave salt as a waste product. What does one do with the salt? If it is returned to the ocean, the salinity of the seas would be increased, and the biotic communities may be destroyed by the imbalances that would be created. If it were buried, the land chosen for a burial site would be eternally unusable, since soil-based terrestrial plants cannot grow in salt water. A means would have to be found to reprocess the waste material to a form that could be deposited safely either in the water or on land.

Another suggestion that has been forwarded is that essential water could be obtained by towing chunks of ice from Antarctica. The expense of such a project would be prohibitive, and it is not really under serious consideration.

Many other ideas involve the retention of presently pure water. A series of six dams have been constructed along the Missouri River that keep the clear, fresh water filled with game fish, from being lost by going to sea. The dams have caused a cessation of spring flooding and the carrying away of the topsoil. Dams, however, are only temporary measures. Still another idea in this direction is digging deep wells and diverting fresh water so it is not lost at sea.

The most promising solution is in terms of reprocessing wastes by converting them into a biodegradable form before they are dumped. Imbalances may be created even by biodegradable wastes. Human excrement is highly biodegradable. Yet biological degradation does not keep pace with production. In addition to converting wastes to a biodegradable form, means must be found to encourage the degradation, and hence, the recycling of waste material. Salting aquatic environments with appropriate bacterial populations, or supplying growth requirements for bacteria already present to restore the balance of the community, might be feasible. Such a solution would mean selecting for, and growing, large populations of appropriate bacteria, along with extensive nutritional tests to ascertain the growth requirements of the bacteria.

While reprocessing may be feasible for industrial wastes, there is no way of reprocessing radioactive wastes. Radioactive wastes enter aquatic environments in two ways. First, some of the radioactive elements from atmospheric testing of nuclear weapons are brought down to the seas, lakes, and rivers, by rain. Second, laboratory wastes are confined in large canisters that are dropped into the ocean. Seepage from these canisters contributes radioactivity to the seas. The effective disposal of radioactive wastes from laboratories remains a problem.

There is one other factor that needs to be considered. If the population continues to increase at its present rate, and industry continues to grow, will there be enough water for human and industrial needs, even if means are found to purify contaminated sources? There is, after all, just so much water and water vapor on the Earth. Some estimates of the maximum population size and industrial use that the water on Earth could support have been unduly pessimistic. Indeed, some of these projected figures have already been exceeded, and water remains available, if not plentiful. Long-range estimates are conflicting. Most agree, however, that for at least the next 30 years, the supply will comply with the demand.

In addition to being concerned with maintaining and increasing the supply of potable water, and water of sufficient purity to be used for industrial processes, man must also have concern for the biotic communities in aquatic environments. While the construction of purification plants for producing potable water from contaminated sources would satisfy the first two concerns, it would leave the body of water contaminated. The contaminants would destroy biotic balances in ecological communities, and reduce the food productivity of fish and the oxygen productivity of phytoplankton, either of which occurrence would be highly detrimental to the human population.

One of the major ecological problems that has arisen in aquatic areas is due to detergents. Detergents contain large quantities of phosphates, which support plant growth. A newly formed lake proceeds from absence of life (**oligotrophy**) through primary and secondary successions to a state in which both plant and animal life thrive (**eutrophy**). In the normal course of events, the lake will eventually enters a phase of deterioration over a period of hundreds of years during which plant life slowly overgrows normal life and fish populations change from game fish (trout, pike, perch, and the like) to scavengers such as carp. Finally, oxygen becomes so limiting that fish die off, and the lake becomes a swamp. In response to the phosphates from both industrial and human wastes, the rate of eutropification by rapid plant growth vastly increases, so that instead of taking hundreds of years for the conversion of a lake to swamp, the feat may be accomplished within a decade.

There was hope that a phosphate substitute, nitrilotriacetic acid (NTA), would stem the effects of detergents, but recent laboratory evidence indicates that NTA causes severe birth defects in animals. The FDA will shortly withdraw those detergents containing NTA from the market.

Yet another major problem in water pollution is that of oil slicks from tanker spillage and offshore wells. In 1968 the oil slick caused by a malfunction and fire in Santa Barbara decimated the wildlife in offshore areas. Thousands of birds perished, their wings covered with oil, and untold numbers of fish suffocated because of the reduced amount of dissolved oxygen in the seawater.

The problem then, is twofold: first, the jeopardizing of ecological balances, and second, the danger of having only polluted water available for human consumption.

Contamination of the Land

Contamination of the land has not always been a problem. Until about a hundred years ago, the garbage of the average family could be dumped on a compost heap, covered, and ignored. Natural decaying processes through the biotic intervention of saprophytic organisms degraded the garbage and aided in the recycling of the chemical elements therein. The garbage contained only a minor amount of nonbiodegradable material to contaminate the land.

Modern society, with its penchant for aluminum cans that defy rust, and for plastic packaging, has created a threat, not only to clean, uncontaminated land, but also to the cycling of elements.

Many communities dump both biodegradable and nonbiodegradable garbage in the same place, making the land perhaps eternally unusable. Many communities separate garbage, dumping biodegradables at one dump site and nonbiodegradables at another, and incinerating paper and plastic. Even this, however, is not the whole answer. There is still a tremendous accumulation of nonbiodegradable material. Part of the answer is in the development of containers that are either biodegradable or subject to rust.

But there seems to be a major problem in "selling" the public to care enough about contamination of the land to do anything about it. A referendum on the ballot of the 1970 election in the state of Washington to prohibit the sale of beverages in "throw-away" containers was soundly defeated.

Man has gained control over his environment, but this control has often been unwisely used. The time has come for him to take stock of his effect on this planet, and to seek, accept, and put into effect solutions for correcting the problems he has created. If man continues on his present road, he is headed for extinction, or at least for a major reduction in his population level in the near future. Whether he will go with the "bang" of thermonuclear war, or with the "whimper" of thirst or starvation would only be a matter of pure speculation.

CHAPTER 20

BIOSYSTEMATICS

The early Earth provided a multiplicity of environmental situations to which living entities could adapt. Through natural selective processes, the Earth became carpeted with a great variety of living forms. Slowly, in the course of evolution, the variety extended to include different levels of structural and physiological complexity.

The mammoth task of grouping organisms in some intrinsically meaningful way is undertaken by taxonomists. But while the goal has always been a natural classification, the criteria by which "natural" was judged has changed.

The criteria used for classification has always been influenced by either prevalent philosophies or methods in other areas of biology. The classification systems from Aristotle (the first philosopher to attempt to put the biological and inorganic world into some semblance of natural order) through Linnaeus (the father of modern taxonomic structure) to Darwin were based on the idea of distinct origins of each organismal group. Both the philosophy and the technological state of society prior to the nineteenth century determined the approach of the taxonomist. According to past philosophies, taxonomists had no need to consider variation within groups. The state of communication before the nineteenth century made it extremely difficult for taxonomists to know whether the organisms they collected were identical to those collected several thousand miles away. Thus, classification was based on collections made of a few specimens within a limited area. In retrospect, this is called the **local communities** approach, or the **museum approach.**

The local communities approach continued long after Darwin espoused his theory of evolution. It was not until the advent of population genetics that taxonomists fully comprehended the implications in the Darwinian interpretation of evolution and realized that natural classification of organisms must be made from a consideration of the variation within a population, the ecological situation in which the organism exists, and the general traits the population exhibits.

The role of taxonomy or **biosystematics,** as the field is now called, is now seen as the presentation of the wealth of biological variation in an orderly manner. It is a compilation of case histories of evolution in action. In more experimental fields, such as biochemistry, investigators design their experiments to reduce variation as much as possible. This enables them to develop a cause-effect relationship for the phenomena under investigation. In the field of biosystematics, however, investigators strive to retain the total variation, so that organisms can be studied as they are encountered in nature. In order to facilitate this approach, taxonomists keep extensive and careful records of their observations. So extensive are these records that a taxonomist is, in great part, a librarian. His large collection of specimens must be labeled, described, identified, and eventually organized into lists and keys. It is only natural to expect that language is an important factor in biosystematics, and rules of language usage have been developed so that there will be some universal, understandable means of communication.

Therefore, before discussing the history of evolution, it is necessary to gain some understanding of (1) the way in which taxonomists view an organism or an organismal type, (2) the basis that has been used in grouping and naming organisms, and (3) the language that is used for organismal classification.

THE CONCEPT OF GENUS AND SPECIES

The most commonly studied groups of organisms are the **species** (plural: species) and **genus** (plural: genera). Organisms that differ from one another by a few traits and that do not interbreed are given different species placement. Many similar species are grouped together under a single genus. Therefore, when two organisms have different generic names, this implies a greater difference between them than if they bore the same generic name and differed only in species rank.

The concept of genus and species has undergone changes that are complementary to the alterations in the approach to classification. In order to gain a historical perspective of this concept, it is appropriate to trace it from antiquity.

The Aristotelian Concept of Genus and Species

Aristotle believed that all organisms were perfect and could be arranged according to their level of organization. This linear sequence is called the **Great Chain of Being,** and represents a tabulation of increasingly more perfect forms.

Aristotle assigned each organismal type a *genos* and an *eidos*. The genos represented the existential state of the organism. A rabbit has a genos of rabbitness. The eidos, on the other hand, gave the organism its individual features. The color, size, awareness, etc., of a rabbit constituted its eidos. The genos was an *immutable pattern* that determined the manner of existence, while the eidos was the somewhat mutable form through which existence takes on individual characteristics.

The Aristotelean concepts of genos and eidos were accepted for 1500 years without modification.

The Renaissance Concept of Genus and Species

Renaissance and post-Renaissance taxonomists developed a Christian interpretation of Aristotle's Great Chain of Being. Each kind of organism was specially created. Therefore, the goal of taxonomy was to ferret out each form of life and identify its position on a natural ascendancy of increasing complexity. The Aristotelian concept of the genus as an immutable existentialist state was retained, as was the concept of a species (eidos) as the descriptive feature of an organism.

The first modern listing of organisms was made by Ray (1680). The father of modern taxonomic structure is, however, considered to be Carl von Linné, the Swedish naturalist who changed his name to Linnaeus. (Latinizing one's name was a common practice among scholars in past centuries.) His important work was his *Systema Naturae* (1735).

Linnaeus is given credit for developing the **binomial nomenclature** that labels each kind of organism by a generic and specific name. Since the role of the species name was the description of specific traits, it often became rather cumbersome. For example, the common honeybee was given the genus name *Apis* and the species name, *pubescens, thorace subgrisso, abdomine fusco, pedibus posticus glabris utrinque margine cilliatus.* The value of the specific name was that it could be expanded as more traits were identified in order to distinguish this group from newly discovered groups. The name became so unwieldly, however, that Linnaeus was forced to use nicknames (the *praenomem trivaile* in Linnaean terms). For the honeybee, the nickname *mellifera* was selected. This trivial name was eventually used so often by others that it became the official species name, and the Linnaean practice of using long descriptive names became obsolete, even within Linnaeus' lifetime.

At the time when Linnaeus was collecting and recording his material, there was little communication between scientists. Most of his published findings were from his own collections. As he primarily collected material from his local region, his listings were far from representative. Within a region each species is a distinctly unique adaptive group. The number of species that expressed the total biological variation in a local community is relatively small. Such a compilation created many large gaps and formed a picture that tended to confirm the idea of special creation. Thus the local community approach to taxonomy reflected the primitive state of the taxonomic art, and led to certain misunderstandings.

But from the cumulative data obtained by the local community approach came the indications of the extent of biological variation and the implications that there was, in reality, a continuum of biological organization. In his later edition of *System Naturae* in the 1750s, Linnaeus included the findings of other taxonomists in Europe. From these additional studies in other local communities, certain overlappings were found. Comparative studies revealed the existence of ecological equivalents, two morphologically similar types that occupy almost the same ecological niche in different communities. About this time, explorers were beginning to return from the New World and Africa with specimens. The gaps between identified species disappeared, and the continuum of biological variation began to be appreciated.

Awareness of the full wealth of biological variation eventually led Darwin to formulate his theory of evolution.

The Evolutionary Interpretation of Groups

Darwin's theory regarding the origin of species provided a completely new interpretation of species relationships. The idea of the *Great Chain of Being* was abandoned and was duly replaced with the idea of common ancestries. In 1866, the anatomist Ernst Haeckel diagrammed phylogenetic trees designed to depict the groups that were most closely related to one another, and the anatomical changes that led to the formation of new groups (see Fig. 20-1). The emphasis in species designation was changed from a listing of trait differences to a perspective of their similarities.

Taxonomists who favored an evolutionary interpretation collected their data in much the same way as their Linnaean predecessors. They examined one or a few members of a species, and from the traits drew conclusions on the species' origin and place on the phylogenetic scale. It became immediately obvious that it was only necessary to slightly rearrange, but not extensively reclassify, the categories of the pre-Darwinian naturalists. Without realizing what they were doing, the early naturalists had separated groups according to evolutionary modifications, using as their criteria traits that were of the greatest evolutionary significance. That their classifications were somewhat valid indicates that the important changes in evolution were obvious to a good observer, whether or not he was an evolutionist.

The importance of the evolutionary interpretation was in the philosophical approach rather than the system of classifications. The characteristics that had been tabulated by early naturalists were now viewed as significant evolutionary changes rather than unique examples of special creation.

Population Analysis—The Contemporary View of a Species

By the turn of the century, a number of disturbing discoveries raised serious questions about the validity of a species as a real group. When a number of individuals of a given species were examined, it was often found that two distinctly contrasting traits existed simultaneously (white and pink flowers, patterns on butterfly wings, etc.). On the other hand, intermediate type between two groups that had been classified as different species were found. Also, some species changed their traits with alterations in the environment.

When geneticists established the genic basis of stable characteristics (ca. 1905), evolution became a genetic problem. But were the traits used by Linnaean and evolutionarily oriented taxonomists to distinguish species genetically based? Plants grown in a valley had different traits

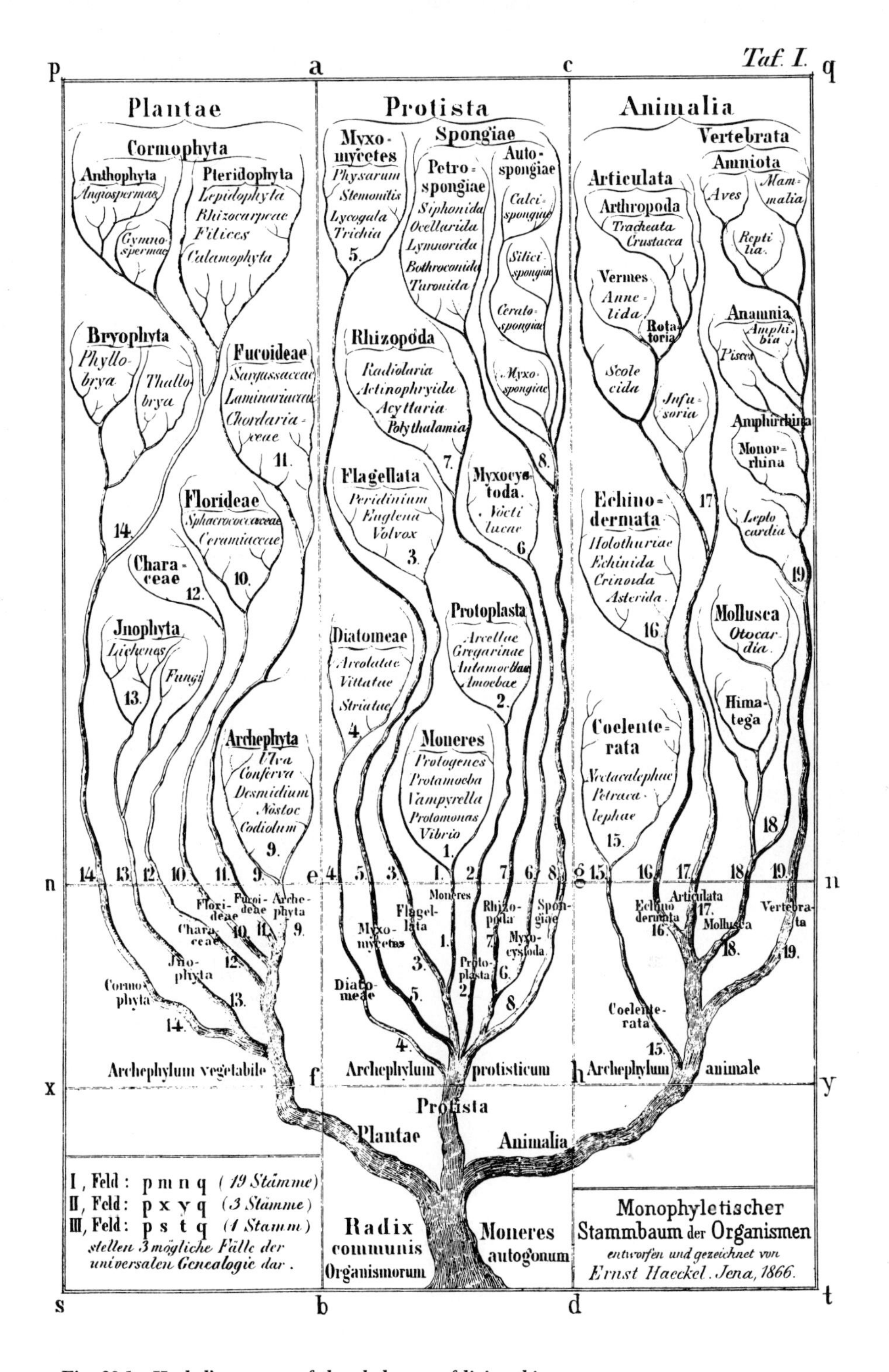

Fig. 20-1 Hackel's concept of the phylogeny of living things.

from those of plants of the same genus that occurred naturally in alpine regions, and on this basis, two species were designated. But when the valley plants were transplanted to higher elevations they exhibited the growth habits of the alpine plants. Were these traits genetically controlled, or were they the result of environmental influences? Was a species description a meaningless ensemble of features?

The resolution of these dilemmas came from the perspectives gained from extensive genetic research. First, the expression of contrasting traits by members of the same species was interpreted as the expression of natural variation, and lent support to the Darwinian view of taxonomy. Second, the presence of intermediate types indicated that the two groups could potentially interbreed, and raised the question as to whether they should be designated separate species. (A group could be in the process of speciation, in which some traits are being replaced by others.) Third, it was found that organisms are most frequently found in the environments to which they are adapted. Placing organisms in environments to which they are not adapted affects gene expression. These organisms in foreign environments may grow and multiply, but they do not thrive and maintain populations for very long. Short-term transplantation experiments are invalid methods of studying evolution.

The perspective gained by genetic experiments altered both the evolutionary and the classification concept of a species. Two isolated populations will evolve independently, accumulating minor adaptive changes. When the changes are insufficient to cause sexual isolation (see Chapter 18), they can be placed in the same species group, regardless of the seemingly important differences in certain traits. Once sexual isolation is effected, however, they must be regarded as separate species. The taxonomic outcome is then related to genetic changes that affect potential interbreeding between populations. The modern concept of a species is, then, *a group of organisms possessing an ensemble of genetically determined traits and variations among individuals that directly or indirectly resulted in its sexual isolation from related groups.* The population is the unit, and all similar populations comprise a species. The understanding of species became one of population dynamics.

To modern biologists, a species is as *real* a group as it was to early taxonomists, but for different reasons. Rather than being a group of organisms with similar characteristics, a species is a breeding aggregate of many populations in various degrees of dissociation. The breeding potential holds the populations of a species, as well as individuals of a population, together, but natural genetic variation tends to isolate individuals. Therefore, a species is a truly functional unit, and as such, is quite real.

The new concept of a species demands the collection of totally different kinds of data than the older concept required. It is no longer considered valid to collect one or a few organisms and describe their traits by the **museum approach.** The modern taxonomist establishes the geographical range of the group, and recognizes the presence of distinct populations. The description of the species is made from collecting many individuals from each population and characterizing them on a statistical basis. The taxonomist usually finds that members of one population have some constant differences from those of other populations. In some cases, the population will exhibit a gradual transition of traits over a geographical range (see Fig. 20-2). This spectrum is called a **cline.**

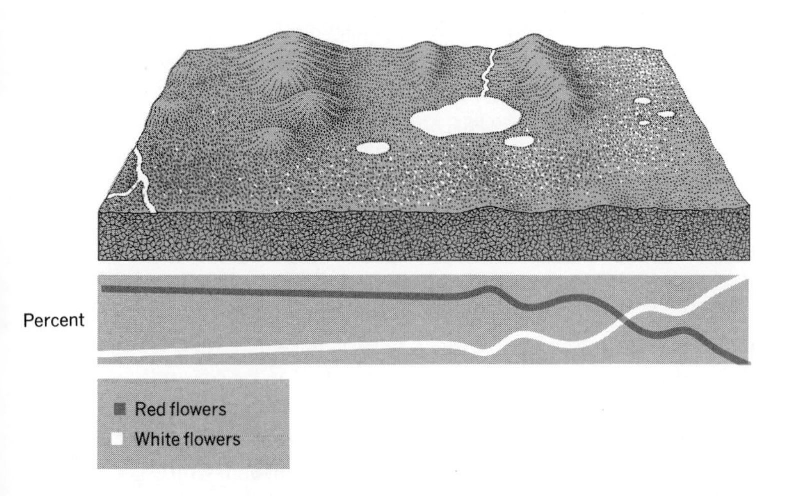

Fig. 20-2 A cline.

The taxonomist may find that these population differences are also expressed in the number or type of chromosomes, and might indicate a subspecies relationship. Or he might find that the difference between populations is manifested only by their growth preference for different climatic conditions. He may bring some of the specimens back to his greenhouse or laboratory, and carry out breeding experiments. In this way, he can identify the breeding structure of the species by (1) testing for interspecific sterility or hybrid inviability and (2) establishing how divergent the sampled populations have become.

From this complex set of data, the taxonomist then attempts to synthesize a picture of the species in terms of the breeding patterns and environmental adaptive characteristics of its populations. By careful analysis, he attempts to gain some insight into the history of the species, and perhaps the adaptive directions toward which it is moving.

The criteria for establishing a species description imposes a gigantic challenge, one that has been met in only a few instances. Because of this difficulty, the vast majority of species have been described by the museum approach. The new species concept, however, has made no major changes regarding the significance of anatomical differences exhibited by different species, so most species groups erected by museum standards are probably correct.

THE HIGHER CATEGORIES

When Linnaeus classified the known organisms of the day, he found that it was not sufficient to place each individual in a species group and all these species groups in various genera. Many of the genera exhibited certain common traits, and Linnaeus was compelled to group them into categories. His acquaintance with the rich variation exhibited by both animals and plants led him to construct two large groups, animals and plants.

Another observation that made further categorizing plausible was the lack of the continuum of variation. Although marked similarities existed among many organismal types, such as cats, primates, bats, etc., the differences were discontinuous, and there were obvious gaps between the groups. In order to cope with both the similarities and the discontinuous variation, Linnaeus constructed a pyramid classification. Several species may be grouped into a single genus; several genera grouped into a single **order;** several orders into a single **class;** and several classes into a single **kingdom.** He recognized only two kingdoms: plants and animals (see Table 20-1). Later taxonomists added the organization levels of **phylum** (between kingdom and order), **family** (between order and genus), and **tribe** (between family and genus), to handle the increasing variation that was uncovered.

TABLE 20-1 LINNAEAN CLASSIFICATION OF THE BIOLOGICAL WORLD[a]

Kingdom: Animalia
- Class: Mammalia
- Class: Aves
- Class: Amphibia
- Class: Pisces
- Class: Insecta
- Class: Vermes

Kingdom: Plantae
- 23 classes of seed plants
- 1 class of all other plants

[a] From *Systema Naturae*, 10th ed., 1758.

The early taxonomists could justify the different kinds of organisms by their value to man. With the advent of the evolutionary interpretation of the group phenomenon, it was necessary to explain *why* the associations among organisms occur. It is impossible to justify the **taxa** (categories) of genera, families, orders, classes, and phyla as natural units in the manner done at the species level. The species is a genetically functional unit. There is a flow of genes among its members, a behavior that makes the species a real group. The higher categories have no genetic interplay between their subgroups, and regardless of the level above the species one considers, no functional explanation can be given to substantiate their erection.

The justification for establishing higher categories lies in their evolutionary history, and is therefore in terms of evolutionary relatedness rather than genetic functionality. Two genera belonging to the same tribe or family are closely related in the sense that at some point in their evolutionary history they were derived from a common ancestral species that is now extinct (see Fig. 20-3).

Natural Versus Artificial Classification

One goal of taxonomists is to create a natural system of classification, a system that truly represents the evolutionary history of the described species. Another goal is to create a uniform classification system. These two goals are sometimes incompatible. Hence it is necessary to be aware of certain artificial aspects in the higher-category classification of organisms.

One artificial aspect of classification is that every species belongs

Fig. 20-3 **Six species in one genus giving rise to two new genera.**

to a genus, family, order, and phylum. While this terminology provides uniformity, it is misleading, because it implies a constant difference between taxonomic levels. One group of organisms cannot be equated with other groups at the same taxonomic level. Consider the classification of three organisms, the colon bacillus, man, and the golden delicious apple (see Table 20-2). There is not the same magnitude of genetic or phenotypic difference between two bacterial species belonging to the genus *Escherichia* and two species belonging to the genus *Pyrus*. So while the taxonomic separations are valid from the standpoint of historical

TABLE 20-2 CLASSIFICATION OF COLON BACILLUS, MAN, AND GOLDEN DELICIOUS APPLE

Kingdom	Procaryota	Metazoa	Metaphyta
Phylum		Cordata	Tracheophyta
Subphylum		Vertebrata	Pteropsida
Class	Schizomyces	Mammalia	Angiospermae
Subclass		Theria	Dictoyledonae
Infraclass		Eutheria	
Order	Eubacteriales	Primates	Rosacales
Suborder		Anthropoidae	
Family	Enterobacteriacaeae	Hominidae	Rosaceae
Tribe			
Genus	Escherichia	Homo	Pyrus
Species	coli	sapiens	malus
Subspecies		sapiens	
Variety			golden delicious
Common name	colon bacillus	man	golden delicious apple

evolution, they are somewhat artificial in equating comparable levels with similar significance of changes.

Another aspect of artificiality arises from the attempt to describe different histories in the same manner. Each of the major taxa represents a unique offshoot in evolutionary history. The number of consecutive higher levels used to define a species depends on the history involved. Some organisms are the products of many significant changes, and are related to many extinct and extant groups. These require the categories of subclass, suborder, etc. to define their evolutionary history. Others have progressed through relatively few steps to achieve their present form, and so require only a few major taxa in order to be classified.

On one hand, classification has to be similar for all organisms so that there can be a uniform means of identifying them. On the other hand, classification must take into consideration the significant changes that have occurred in the evolutionary history of an organism. A purely natural classification for those organisms that have progressed through relatively few evolutionary steps would be simply to omit the unnecessary categories. If there were only one family in an order, it would be unnecessary to designate an order.

The standardly used classification system is a compromise, designed for the uniform treatment of all organisms. Thus, an order is assigned for the sake of uniformity, even though it is unnecessary.

At the higher levels (kingdoms, phyla, and classes) and at the lower levels (species, subspecies, and varieties) the distinctions between categories approach the natural order of evolutionary history. The intermediate levels of classification (family, orders, and tribes) are often arbitrary.

THE MECHANICS OF MODERN CLASSIFICATION

Zoologists, botanists, and bacteriologists have formed their own International Congresses to establish rules of nomenclature. With the exception of the rules for genus and species designations, the rules of the three commissions are different, and there appears to be no prospect of future amalgamation. Since many of these rules are, at best, conventions, the indifference to a unified biological system of classification reveals a professional myopia among scientists.

Regardless of the merits of the individual systems, the lack of uniformity has an effect on the attempt to give the introductory student a simple, universally accepted classification of the major groups of organisms. The schemes given in this text follow those given by most authors of freshman textbooks, with some individual preferences added.

The organismal world is divided into four kingdoms: **Monera, Protista, Metaphyta,** and **Metazoa.** All organisms classified as **Monera** have no true nuclei, chloroplasts, endoplasmic reticulum, or mitochondria. The chromosomes are located in the general cytoplasm, as are the enzyme systems that are usually associated with the particular organelles. The **Protista** kingdom is made up of organisms that possess nuclei, endoplasmic reticulum, mitochondria, and chloroplasts, but lack any true multicellular organization. Organisms of the **Metaphyta** kingdom exhibit true cell types in multicellular associations. They are further characterized by the presence of chlorophyll and cell walls at some time in their life cycle. Organisms without chlorophyll and cell walls that manifest multicellular associations, and are often motile, are classified as **Metazoa.**

Zoologists have chosen to divide the metazoan kingdom into **phyla.** Botanists, on the other hand, have decided to use **divisions** as the major taxon level below the kingdom level. The major groups in each kingdom are shown in Table 20-3.

TABLE 20-3 MAJOR GROUPS IN KINGDOMS

	Common Name
Kingdom: Monera	
Phylum: Schizomycophyta	Bacteria
Phylum: Cyanophyta	Blue-green algae
Kingdom: Protista	
Phylum: Chlorophyta	Green algae
Phylum: Euglenophyta	Euglenoids
Phylum: Charophyta	Stonewarts
Phylum: Phaeophyta	Brown algae
Phylum: Rhodophyta	Red algae
Phylum: Chrysophyta	Diatoms and yellow-brown algae
Phylum: Myxomycophyta	Slime molds
Phylum: Eumycophyta	Fungi
Phylum: Flagellata	Flagellates
Phylum: Sarcodina	Amoebae
Phylum: Ciliata	Ciliates
Phylum: Sporozoa	Sporozoans
Kingdom: Metaphyta	Multicellular plants
Phylum: Bryophyta	Mosses and liverworts
Phylum: Tracheophyta	Vascular plants
Kingdom: Metazoa	Multicellular animals
Phylum: Porifera	Sponges
Phylum: Coelenterata	Jellyfish, corals, hydroids
Phylum: Ctenophora	Comb jellies
Phylum: Platyhelminthes	Flatworms, flukes, tapeworms
Phylum: Aschelminthes	Rotifers, roundworms
Phylum: Nemertea	Proboscis worms
Phylum: Mollusca	Snails, claws, octopuses
Phylum: Annelida	Earthworms, leeches
Phylum: Arthropoda	Insects, lobsters, spiders, millipedes
Phylum: Echinodermata	Starfish, sea urchins, brittle stars
Phylum: Chordata	Tunicates, lamphreys, sharks, rays, bony fish, amphibians, reptiles, birds, mammals

The classification of genus and species must follow universally accepted rules, so that ambiguities are avoided. These rules were established for animal species by the International Congress of Zoologists in 1898, and are accepted in part by both botanists and bacteriologists.

1. The same genus and species names for a plant and an animal should be avoided.
2. A name of a genus should not be used more than once.
3. No names are recognized as official that were used prior to Linnaeus' tenth edition of *Systema Natura* (1758).
4. Scientific names should be either in Latin or latinized and printed in italics.
5. The genus name should be a single word and begin with a capital letter.
6. The species names should be a single or compound word and begin with a small letter.
7. The author of a name is the one who first publishes it in a generally accessible periodical or book along with a description.
8. When a genus is proposed, the type species (first one named) should be indicated.
9. A family name must end in IDAE (this ending is added to the stem of the genus); and a subfamily name must have INAE added to the same stem.

The naming of groups is one of the messier problems in taxonomy. Sometimes a species is named and described a second time. In this case, the **law of priority** recognizes the earlier name. It is common, however, that both names are taken as synonyms. Names are often chosen in moments of unconcern, and names such as *anteromediobasalimagnofasciatipennis* (a species), *Amphionycha knownothing*, *Eudaemonia jehovah*, *Eucosma bandana*, and *Dianachysme* are given.

Although the species name is only a code word, it must be in Latin (albus = white) or latinized (*o'connorii, mexicanus*). The use of Latin today is a convention. It is obvious that a common language is desired, and since that used by the early taxonomists was Latin, it has been retained lest many of the early designations become obsolete. Using the common name is undesirable, since functional language evolves over several centuries and names change. The common sparrow is called Haussperling (German), English sparrow (United States), house sparrow (England), gorrion (Spain), musch (Holland), passera oltramontana (Italy), and moineau domestique (France). If one of these languages were chosen, the names may not persist, because words evolve. Latin, a dead language, will not change, and it can still be learned in most countries.

Classification Keys

An expert can identify a species at a glance; in fact, some have done so in a car moving through a field at twenty miles an hour. Most workers, however, require the use of a **key,** a conventional arrangement of categories by some arbitrarily selected sequence of important characteristics. It is made up of a set of alternatives, each of which either directs the investigator to a new set of alternatives or gives a category name. An example of a key for the different kingdoms would be as follows:

TABLE 20-4 THE GENERA OF THE FAMILY VOLVOCACEAE

1. Colonial envelope flattened	2
1. Colonial envelope globose	3
2. Envelope with an anterior-posterior differentiation	*Platydorina*
2. Envelope without anterior-posterior differentiation	*Gonium*
3. Not over 256 cells in a colony	4
3. Over 500 cells in a colony	*Volvox*
4. All cells in a colony alike in size	5
4. With cells of two distinct sizes	*Pleodorina*
5. Cells forming a sphere	
5. Cells not forming a sphere	*Stephanoon*
6. Cells close together	*Pandorina*
6. Cells not close together	
7. Cells hemispherical	*Volvulira*
7. Cells spherical	*Eudorina*

1. Organisms with no true nuclei but isolated chromosomes, no true chloroplasts, endoplasmic reticulum or mitochondria, the procaryotes . . . . Monera
1. Organisms with nuclei, endoplasmic reticulum, mitochondria, and chloroplasts, the eucaryotes 2
   2. Organisms without true multicellular organization of one or more cells without complex interdependence . . . . Protista
   2. Multicellular organisms 3
3. Multicellular organisms with chlorophyll and cell walls sometime in their life cycle . . . . Metaphyta
3. Organisms without chlorophyll and cell walls, usually motile Metazoa

Thus, by examining the significant characteristics of the organism to be classified, in the order prescribed by the key, the investigator can determine the proper category to which an individual belongs.

Keys are most often used for classifying organisms according to genera or species. An example of this level of classifying can be seen in the key for the genera of the family Volvocaceae, composed of colonial green algal forms (see Table 20-4 and Fig. 20-4).

Although the key, with its associated illustrations and descriptions, is a most accurate and workable system, its use requires considerable

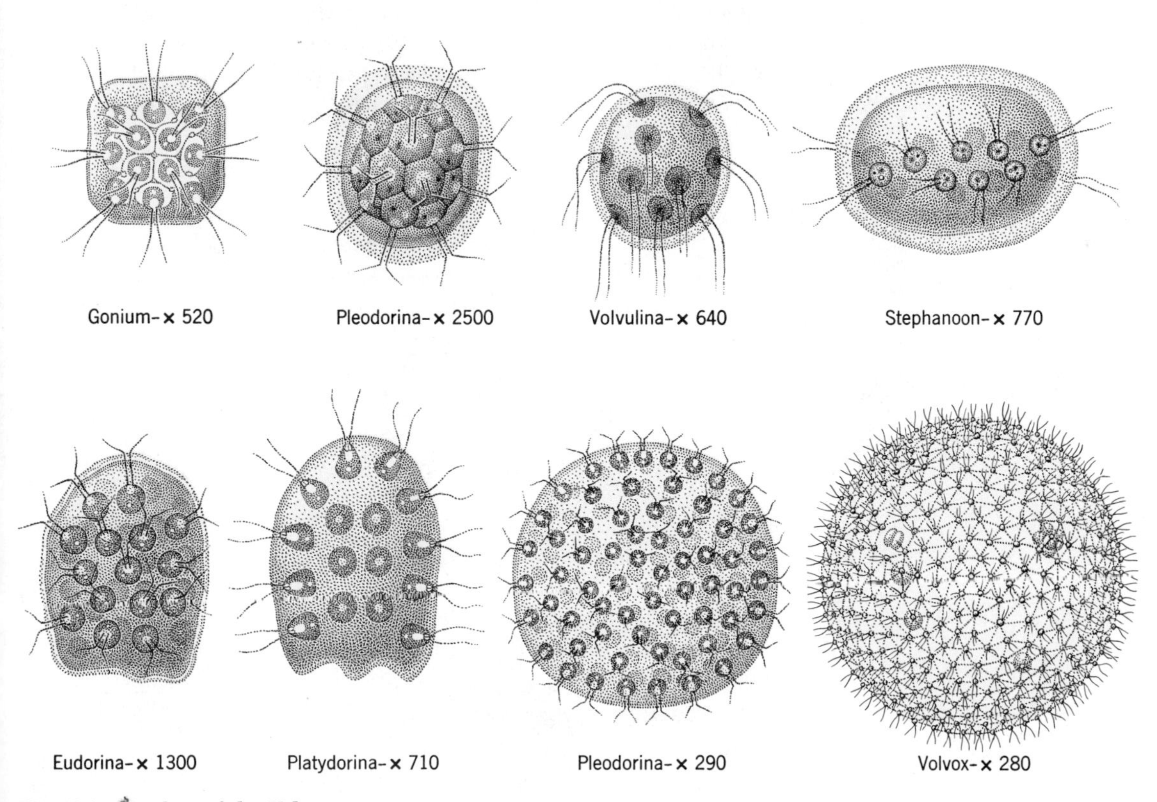

Fig. 20-4 Members of the *Volvocacae.*

skill. One must be familiar with the meaning of the terms and what variation of structure can exist.

The dichotomy of classifying by either the museum or new species-concept approaches is embodied in the typical keys. Based on morphological differences, these keys comprise the very essence of the museum approach. Yet taxonomists who design keys, and investigators who use them, are aware of the evolutionary significance of the different groups, and so impose the perspectives of modern biology on the mechanics of classification.

DESCRIPTION OF GENERA OF THE FAMILY VOLVOCACEAE

1. *Gonium* Mueller, 1773. Colonies have 4, 8, 16, or 32 biflagellate cells embedded in a common gelatinous matrix. All cells capable of forming daughter colonies.
2. *Pamdorina* Bory, 1824. Colonies spherical to ellipsoidal with 4, 8, 16, or 32 biflagellate cells in a homogeneous matrix. Colony with a hollow sphere and cells oppressed against each other. All cells capable of reproduction.
3. *Volvulina* Playfair, 1915. Spherical colonies of 4, 8, 16, or 32 cells separated from each other in a common matrix. Cells hemispherical with the flattened side outward. All cells can multiply.
4. *Stephanoon* Schewiakoff, 1893. Spherical to oblately spherical envelope. There are 8 or 16 cells in two transverse tiers near the equator. Cells of one tier alternate with those in the other. All cells can multiply.
5. *Eudorina* Ehrenberg, 1932. Spherical to obovoid colonies with 16, 32 or 64 cells. Cells separated and near edge of matrix. Cells arranged in transverse tiers, the number of which depends on number of cells. All cells reproductive.
6. *Platydorina* Kofoid, 1899. Flattened horseshoe-shaped colony of either 16 or 32 cells.
7. *Pleodorina* Shaw, 1894. Colonies spherical to broadly ellipsoidal with 32, 64, or 128 cells, well separated from each other. Anterior cells are smaller and incapable of dividing, while posterior cells are large and multiplicative.
8. *Volvox* Linnaeus, 1758. Colonies from 500 to 50,000 cells, spherical to ovoidal. Each cell has separate matrix, hexagonal in outline; 2 to 15 cells are reproductive and usually located in the posterior hemisphere. The great majority of cells are vegetative.

QUESTIONS

1. What is the role of biosystematists?
2. Compare and contrast the Aristotelian concept of genus and species with the Renaissance concept.
3. How does the modern concept of a species differ from the Aristotelian and Renaissance views?
4. Why is a species considered to be a "real" group, and as such, a natural classification?
5. Do higher categories of classification reflect real evolutionary relationships between organisms?

SPECIAL INTEREST CHAPTER VII

EXTRATERRESTRIAL BIOLOGY

For centuries man has gazed up at the heavens and wondered about the possibility of life somewhere else in the universe. Ancient civilizations placed their deities in the heavens, and thought of the stars' patterns and the movement of the sun as reminders of their existence. Modern science-fiction stories about extraterrestrial life have captured many people's fancy, and the controversy over flying saucers has intensified their curiosity.

When scientists as scientists ponder this subject in their more conservative manner, it is formally labeled **exobiology,** the study of extraterrestrial life. Exobiology has been accused of being a scientific anachronism, because it is a science without a subject. No other forms of life have ever been known to visit the earth, either alive or as remains found in meteorites.* No hardware of other civilizations has been obtained, in spite of the numerous claims of observing UFOs (unidentified flying objects). Thus, no unequivocal evidence has been gathered to demonstrate that any other planet has life.

On theoretical grounds, however, there is good reason to entertain the idea.

IS THERE OTHER LIFE IN OUR SOLAR SYSTEM?

The first place to search for extraterrestrial life is on the planets of our own solar system. This is because these planets can be carefully studied with instruments, and will probably be visited within the next decade or two. In fact, they are the only planets *known* to exist at the present time.

Chemical Considerations

An exobiologist uses essentially the same definition of living systems that was developed in a previous chapter. Thus, an entity made up of complex molecular aggregates, incorporating raw materials and energy from the environment to use in complex polymeric syntheses

* Recently a meteorite from Australia has been found to contain 17 different amino acids. Perhaps, they may have been of biological origin from elsewhere in the universe.

for growth and reproduction, would be considered alive. But this definition of a living system is based on the biochemical properties of earth creatures. To what extent is it justified to impose the Earth's biochemistry on Venusians and Marsians?

A science-fiction article (author anonymous) entited "Is Life Possible on the Surface of the Earth?," written by someone from Jupiter who enjoys a rich atmosphere of liquid ammonia, concluded that the answer is "No!" This is because the Earth contains little ammonia, and is too hot for even that small amount of ammonia present to remain liquified. Could exobiologists be committing similar errors in their evaluations of other planets for extraterrestrial life? Could not some form of extraterrestrial life have an organic chemistry founded on such elements as sulfur, silicon, or perhaps some element unknown on earth?

Two arguments negate this premise. First, spectrometric studies of stars and planets, and chemical analysis of meteorites, have shown that there is no element in the solar system that is not present on Earth. Inorganic chemistry is truly a universal discipline. Therefore, another form of life could not be based on an element unknown on Earth. Secondly, the evolution of life on Earth was inexorably bound to the properties of carbon, nitrogen, hydrogen, and oxygen. The properties of carbon permit the formation of large molecular forms necessary for macroscopic structures. No other element manifests this property. Through the exploitation of the hydrogen bond came the crucial linkage in protein configuration, nucleic acid replication, and transcription. In the minds of exobiologists, it is highly unlikely that noncarbon forms of life exist.

Temperature Considerations

The assumption that any extraterrestrial life form would be based on carbon chemistry leads to certain consequences. Immediately, a temperature requirement is established. Macromolecules can form and remain stable only in temperature ranges of 10° to 60° C.

The temperature of a planet is determined by (1) the solar energy it receives from the sun and (2) its capacity to retain the solar energy it receives. Until the recent space probe of Venus, it was impossible to directly measure the temperature of a planet. The most feasible way of estimating these temperatures was in terms of the solar energy the planets received.

This energy value e is directly related to the distance of the planet from the sun. By using the arbitrary value of 1.0 for the Earth, the relative e values are appropriately given terms of percent of Earth's solar energy (see Table VII-1). Venus receives twice as much solar energy and Mars receives half as much solar energy as the Earth. It is most likely that the temperatures of these two planets fall within the 10° to 60° range necessary for the stability of macromolecules.

The space probe by Mariner II detected temperatures far higher (several hundred degrees centigrade) than had been previously calculated for Venus. Apparently the heavy layer of clouds, composed

TABLE VII-1 PLANET AND *e* VALUE

Planet	e	Percent
Mercury	6.68	668
Venus	1.98	198
Earth	1.00	100
Mars	0.433	43
Jupiter	0.37	37
Saturn	0.11	11

chiefly of carbon dioxide, causes the planet to retain much of its solar energy. Prior to this discovery, the apparent temperature, size, position, and density of this planet had made it an excellent candidate for supporting life. This discovery, however, eliminates it from serious consideration.

Mars exhibits a number of interesting features that were thought to indicate the presence of life. Early astronomers described many straight lines on this planet, which might have been the handiwork of intelligent life forms (canals?). Today, with the aid of better telescopes, astronomers have failed to confirm this finding. Apparently the eye, straining at the limits of resolution for the older telescopes, saw "artifacts." The lines in the drawings made by the early astronomers have been judged to have represented formations 15 to 30 miles wide. In theory, nothing less than 50 miles wide could have been seen with the older telescopes.

Mars also has polar caps that exhibit seasonal variations. When the caps recede, the dark regions of the planet seem to expand from the equatorial region, and when the caps enlarge, the dark areas recede toward the equator. These dark areas had been interpreted to be some form of hardy growth that shows a seasonal variation, like lichens.

Measurements of the dark and light regions of Mars with a spectrograph in conjunction with the Mount Palomar telescope reveal some interesting features. The spectrograph measured the amount of absorption of light of 1.5 Å wavelength in both the dark and light regions of the planet. This is the wavelength characteristically absorbed by carbon-hydrogen bonds. The dark areas absorbed far more 1.5 Å wavelength light than did the light regions. Is this reflective of a carbohydrogen material that has seasonal fluctuations, as one would measure for Earth from another planet?

Despite such hopeful findings, a number of facts discredit the possibility of life on Mars. The temperature, although suitable as an average value, fluctuates too widely from day to night to support a carbon-based life. The atmosphere is extremely thin, and is composed mostly of carbon dioxide (twice as much as on Earth), but almost no detectable quantities of oxygen and water. The polar caps, once thought to be ice, have been found to be nitrous oxide, a lethal agent to carbon forms of life. The gravity of Mars is only 38% that of Earth, apparently too small to hold down an atmosphere. In general, Mars seems to have no intelligent life, and does not even appear suitable to support the most elemental types of organisms.

IS THERE OTHER LIFE IN THE UNIVERSE?

The question of other life in the universe would, at first, seem to invite only wild speculation. In the light of known biological and astrophysical facts, however, real estimates can be given that place exobiology, as a science, on firmer grounds than had previously been imagined.

Before embarking on a discussion of other life in the universe, it is necessary to provide some pertinent information on astrophysics.

Astrophysical Background Related to Exobiology

Throughout the universe, the stars are grouped into galaxies. Each galaxy has a characteristic shape and number of stars. Our galaxy, the Milky Way, contains about 10^{11} stars arranged in a flat, pancake-shaped disc (see Fig. VII-1). Many of the stars in each of the galaxies may have planets associated with them, and so many solar systems may exist. The question of whether life exists on a given planet involves (1) the properties of its star and (2) the relationship of the planet to its star.

A star, such as the sun of our solar system, is a large mass of gases held together by gravitational attraction. The brightness of a star is the combined result of gravitation and thermonuclear fusion, converting hydrogen to helium.

Stars progress through a stellar evolution. The first stage involves the formation of the star. During this phase, the star radiates energy (*luminosity*) from the heat created solely by gravitational forces. As stellar evolution progresses, the star radiates increasing amounts of energy. Thus the luminosity of a star in this stage is not constant, and fluctuates too greatly for life to exist on any of the planets that might surround it.

Upon the contraction of the gases in response to the gravitational forces, the temperature rises to at least 4 million degrees centigrade, and thermonuclear reactions occur in the core of the star. A star in this stage is considered to be in the **main-sequence** phase. This phase persists for a long period of time, and is characterized by a constant luminosity and size.

When the hydrogen is exhausted, the luminosity changes, and the star passes into the final, dying phase of stellar evolution. Clearly, then, only the main-sequence phase of a star has a sufficiently prolonged stability for supporting biological evolution on the planets of their solar systems.

TABLE VII-2 CLASSES OF STARS

Class of Star	Temperature (°K)	Color Emitted	Main-Sequence Duration (Years)
Class O	35,000	white	10^7
Class B	23,000	blue-white	↓
Class A	11,000	blue-white to green-white	
Class F	7,500	white	
Class G	5,600	yellowish	
Class K	4,200	yellow-orange	
Class M	3,100	deep orange-red	10^{11}

Fig. VII-1 Spiral shaped galaxies. [(*left*), American Museum of Natural History; (*right*), Mount Wilson and Palomar Observatories]

The life expectancy of the main-sequence phase is of primary importance in considering the capacity of their planets to support life. Three billion years were required for intelligent life to evolve on Earth, and one would expect that it would take a comparable period of time for the same types of events and histories to have occurred on other planets. Evolution, then, must occupy a shorter period of time than the duration of the main-sequence phase.

The duration of this phase is inversely related to the temperature of a star; the hotter the star, the shorter is its main-sequence phase. Stars have been classified according to their temperatures (see Table VII-2). The sequence of letter classifications, OBAFGKM, are remembered by all astronomy students by the sentence, *"Oh, be a fine girl, kiss me."*

Each of these classes is subdivided further by numbers from 0 to 9, thus characterizing stars more specifically. For example, class B0 stars have a temperature of 23,000° K, and class B9 stars have a temperature of 12,000° K.

Type O stars remain in the main-sequence phase for about 10^7 years. Since biological evolution on Earth required 3×10^9 years to produce an intelligent living form, it is clear that such evolution could not occur in solar systems associated with type O or even B stars. Therefore, biological evolution could happen primarily in solar systems belonging to stars from the A to M classes.

In addition to the duration of a star's main-sequence phase, one must also consider the planets that surround the different class stars. If

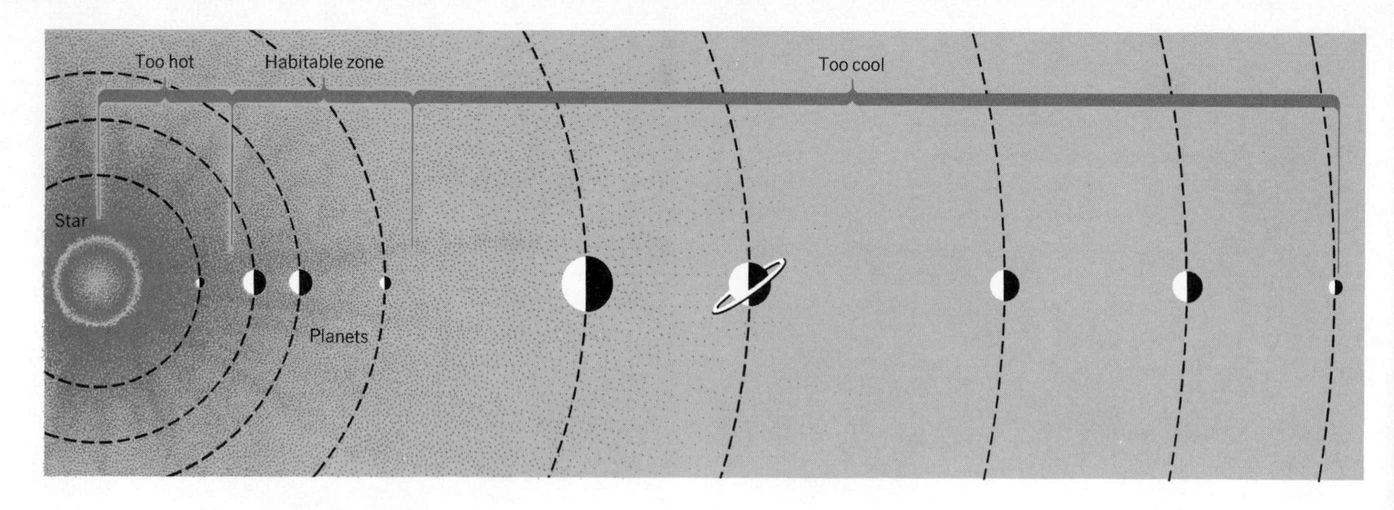

Fig. VII-2 The habitable zone.

a planet is to support life, it must receive a constant amount of radiant energy that is (1) high enough to directly drive biochemical reactions such as photosynthesis, (2) high enough to warm the surface of the planet to temperatures suitable for driving biochemical reactions, and (3) low enough to permit molecular stability. Such a planet must then be positioned an appropriate distance from its sun. The distance range from a star within the admissible temperature ranges is called the **habitable zone** (see Fig. VII-2).

Suns at lower temperature ranges, namely K and M stars, have habitable zones that are smaller by at least a factor of ten (1/10) than the sun of our solar system (which is a G2 star). Since in our solar system three of the nine planets are in the habitable zone, this would mean that there is only $3/9 \times 1/10$ or 3% chance of a planet of a K or M star being found in the habitable zones. Therefore, it is most likely that the solar system of these two types can be eliminated from serious consideration.

From a consideration of main-sequence duration, the O and B stars are eliminated. Considering the size of the habitable zone eliminates K and M stars. Therefore, biological evolution could have taken place in the solar systems of the remaining classes of stars, A, F, and G.

Another factor that must be considered is the relationship of stars to one another. Many stars are components of binary systems, in which two stars are so close that they affect one another's movement. If stars of binary systems have planets, the movement of these planets would also be affected, and would be subject to great changes as one star moves closer to or further away from it. Life would be intolerant of these fluctuations, and close binary systems do not appear as likely prospects for life to exist.

Is there any known nearby star that satisfies these criteria? Before discussing "nearby" stars, it is best to present some ideas of distances in astronomy. The astronomer uses the *light year* (the distance light travels in one year) as a unit of measurement. The speed of light is 186,000 miles per second, and in one year light would travel 186,000 $\times$ 60 (seconds per minute) $\times$ 60 (minutes per hour) $\times$ 24 (hours per day) $\times$ 365 (days per year) or 4.8 trillion miles. The Earth's galaxy, the Milky Way, is about 55,000 light years in diameter, and only 700 light years thick.

The closest star to our solar system is Alpha Centauri, which is 4.3 light years away. But Alpha Centauri is a triple system, and is not likely to have any inhabitable planets associated with its three stars. Of the forty-one other neighboring stars, two, Epsilon Eridana and Rho Certi, are of interest (see Table VII-3). It is not even known whether these two stars have solar systems. If they do, their planets would be the closest ones to Earth on which life *might* be present.

TABLE VII-3 DISTANCES OF STARS FROM EARTH

Star	Distance from Earth in Light Years	Spectral Type	Luminosity
Sun	—	G2	1.00
Epsilon Eridana	10.8	K2	0.34
Rho Certi	11.8	G4	0.38

Are There Planets with Life?

Since we cannot know by direct or indirect observation that there are planets of stars of classes A, F, and G (with the exception of Earth, of course), the question of life on other planets must be considered in terms of a statistical probability. For this calculation, only the stars in our galaxy, the Milky Way, will be considered.

Our galaxy has about 1.6×10^{11} stars. It has been estimated that one out of thirty stars has the desired luminosity and main-sequence life expectancy, and is not a member of a binary system. The number of suitable stars in our galaxy, then, that could have living forms on the planets of their solar systems is

$$\frac{1}{30} \times 1.6 \times 10^{11} = 5.3 \times 10^{9}$$

About 0.8 of these stars probably have planets. So there are

$$0.8 \times 5.3 \times 10^9 = 4.2 \times 10^9$$

number of stars with planets.

Up to this point in the calculation, stars have been the reference points. But life exists on the planets, not the stars. Only the planets that fall within the habitable zone of the stars must be considered for this calculation. It has been calculated that the average number of planets that are in the habitable zone of suitable stars is 1.4, so

$$1.4 \times 4.2 \times 10^9 = 5.9 \times 10^9 \textit{ planets}$$

The chance that a planet with promising conditions for life would actually evolve life can be taken as 0.33. This value is arrived at by examining our own solar system in which only Earth, of the three planets in the habitable zone (Earth, Venus, and Mars), definitely supports life.

$$0.33 \times 5.9 \times 10^9 = 1.8 \times 10^9$$

Or, about two billion planets in our galaxy have had some form of life during some time in the history of the universe.

How many of these planets have had intelligent life? One can assume that intelligent life would be the inevitable result of any biological evolution. Therefore, nearly two billion planets would have had intelligent life.

How many of these planets have intelligent life *now?* The life of an average species is about a million years (10^6), and this figure would probably be true for man. The time for life to evolve intelligent forms capable of developing a technical society is about 3 billion years. Therefore, the fraction of time an intelligent race will be found on a life-yielding planet is

$$\frac{10^6}{3 \times 10^9} = 3 \times 10^{-4}$$

Since 3×10^{-4} of the promising planets would have intelligent life,

$$3 \times 10^{-4} \times 1.8 \times 10^9 = 5.4 \times 10^5$$

there would be about 500,000 planets with intelligent life, present at this time, in our galaxy. On the basis of the volume of the galaxy, and assuming a somewhat equal distribution of planets with intelligent life, they would be about 300 light years away from one another.

COMMUNICATION WITH OTHER PLANETS

Space vehicles (including the type coming from somewhere to us—the UFOs) would have to travel 300 years at the speed of light to reach another planet with intelligent life. Assuming that it were possible to communicate in waves at the speed of light, it would still take 300 years for a message to reach the receiver, and another 300 years for the sender to obtain a response. With such enormous distances, it seems that the *Homo sapiens* on Earth will never have the chance to meet their galactic counterparts.

Even the nearest star, Alpha Centauri, is 4.3 light years away. It would take at least eight years, if it were possible to travel at the speed of light, and 170,000 years at the 17,000 mph of our present space craft, to travel one way. If man is to ever visit planets outside of our solar system, it would clearly be necessary for his craft to be capable of travel at speeds near that of light.

The Feasibility of Space Craft Speeds Near that of Light

Traveling at or near the speed of light, a round trip to a system 12 light years away would take 24 years. Such a space craft would be powered by propulsion, that is, some form of fuel will be burned and exhausted, and the rate of the fuel exhaust will determine the rate of acceleration. The best fuel presently known is the nuclear fusion by which hydrogen is converted to helium at 100 percent efficiency. This type of fuel will give an exhaust velocity $\frac{1}{8}$ the speed of light (c), and for every ton of payload, a billion tons of fuel is needed. It would be impossible, for a rocket to take off with that much fuel, so one might look for an even better fuel.

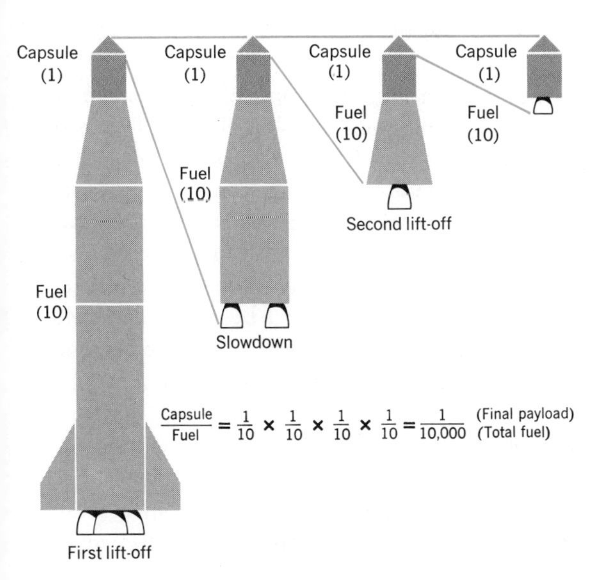

Fig. VII-3 Justification for the use of exponential notation in calculating fuel necessary for rockets.

The ultimate energy source is the interaction of matter and antimatter. This propellant, if it were even feasible, would be converted only into energy (gamma rays) and would rapidly give an exhaust velocity of about c itself. For a craft to go at $0.99\,c$ with this fuel, would require about 14 times as much fuel as payload, certainly a reasonable ratio. But since the fuel will have to be burned four times (to start, stop, return start, and return stop), the ratio of payload to fuel would have to be $1:14^4$ or $1:40{,}000$. The exponential form is used because the initial payload must include the fuel for the other trips (see Fig. VII-3). A payload of ten tons requires 200,000 tons of matter and 200,000 tons of antimatter. The energy to get such a space craft off the ground would result in an emission of 10^{18} watts as gamma rays. So great is this amount of energy that enormous destruction would result. Space flight is impractical if the Earth is to survive the takeoff.

Calling All Stars

If space travels to inhabited or uninhabited planets is unfeasible, there is still another means to meet other forms of intelligent life—through communication. Successful communication involves three problems: (1) the nature of the signal, (2) the kinds of symbols that will be used, and (3) the information that will be communicated.

Signals can be sent by electromagnetic radiation, but since there is an enormously wide range of wavelengths, which will be the one used? This problem is akin to two friends agreeing to meet on a certain day in New York City, without arranging where the meeting is to take place. Assuming that both friends are unfamiliar with the city, how would they go about finding each other? Both would think in terms of the parts of New York the other would be most familiar with, and therefore, most likely to go. Thus the possibilities could be reduced to two or three places.

The uncertainty of which wavelength to use would be a mutual one between any two planets wishing to communicate. Fortunately, the problem can be solved. First, we are limited to the radiation we can receive because of the dense atmosphere around the Earth. There are two openings: one is the "optical window" at about 10^{-5} to 10^{-4} cm (the range used by light astronomers), and the other is the "radio window" at about 10 to 1,000 cm. This latter window is used in the new science of radioastronomy.

But at which frequency in the radio window would signals most likely be transmitted from or received by Earth? There is one obvious frequency, 1420 mc (megacycles). This frequency is universal "sound" because it radiates from the vast amount of neutral hydrogen atoms (one per cm^3) in space. Any planet wishing to communicate would have this knowledge of astrochemistry.

So the first problem can be solved by use of signals of electromagnetic radiation at a frequency of 1420 mc (or 21 centimeters in wavelength).

A system for the transmission and reception of electromagnetic signals over a hydrogen line is not only possible, but also highly feasible from a financial standpoint. Receivers have been built for under a thousand dollars that can detect signals emanating from a one-milliwatt source at this frequency ten light years away. More complex receivers can detect radio waves originating from as far as 100 light years away, and so a considerable portion of the Milky Way is under surveillance. The cost of sending messages from Earth would be about one dollar an hour. Comparing the cost and distances involved with those necessary for space travel, it appears that radioastronomy, rather than any other means, will discover extraterrestrial life.

The next problem concerns the type of symbols that will be used to send information. One can only hope that the choice by radioastronomers follows a universally accepted logic, so that other planets would most likely use similar symbols. Several suggestions have been forwarded, and the most promising is the construction of a visual picture through the binary system of information theory (see Chapter 10). This assumes that any form of intelligent life has some means of seeing objects, a reasonable assumption in view of the success of the various mechanisms that have evolved on Earth for absorbing light. The binary system is the simplest, and corresponds to the universal language of mathematics.

Since the binary system is arbitrary, and pictures are two-dimensional, the linear message sent by spacing signals must contain the key to its interpretation—a two-dimensional interpretation.

Then the problem is what message should be sent. Communication involves a common knowledge of the meanings of the symbols and some common experience between the sender and receiver. Intelligent forms of life capable of sending and receiving radio messages must have some knowledge of mathematics and chemistry. With these two disciplines, plus visual experience, it is possible to describe who one is, what is one's place in a solar system, what is one's chemical composition, what kind of creature one is, and what one is capable of doing in a technical sense.

An ingenious message has been devised by F. Drake, a facsimile of which (as constructed by B. Oliver) is presented in Fig. VII-4. The message is made up of a series of signals, represented by 1, and pauses, represented by 0. The first 1 represents the beginning of the message. When Drake first conceived this program, he sent the message (by mail) to many of his colleagues. Several of these recipients were able to correctly decipher and interpret the message. This success indicates that it would, indeed, be possible for intelligent beings on other planets to gain significant meaning from such a message, or, for that matter, for us to interpret such a message sent from another planet.

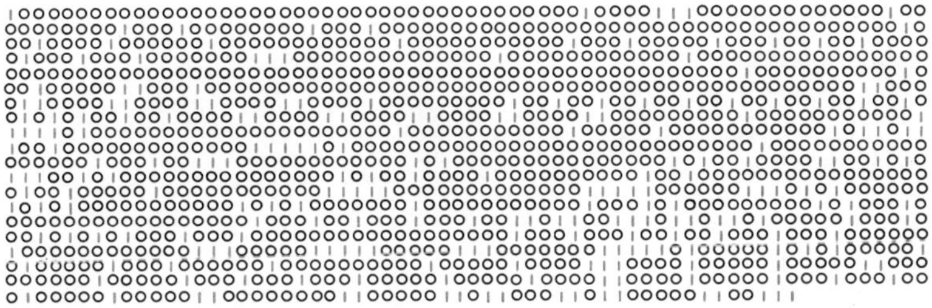

Fig. VII-4 Drake's hypothetical message from another planet.

This message is composed of 1271 1 or 0 symbols. The number 1271 is the product of two prime numbers, 31 and 41. This suggests that if the message is arranged by either 31 or 41 units across a two-dimensional block with either 41 or 31 rows, respectively. When the message is arranged into 41 rows and 31 columns, the 1's form the pictorial pattern illustrated in Fig. VII-5.

Careful viewing of this two-dimensional picture reveals a number of ideas. It should be noted that Drake constructed this picture as if it were a message we received from another planet. Therefore, the information it contains pertains to a solar system other than ours. In the top left hand corner is a sun. In the columns under the sun, the planets of the solar system are numbered by the binary system. A biped is pointing to the fourth planet, indicating that it is their site of residence. The two erect bipeds, with different sex characteristics, and the small individual between them reveals a bisexual mode of reproduction. The female points to the binary number for eleven (1011) at the top of the bracket on the right hand side, to indicate the height of the average individual. But eleven what? What unit of measurement is employed? Since the message is coming over a frequency of 1420 mc/sec, which has a wavelength of 21 centimeters, the height must be 11 $\times$ 21 centimeters or about 7 feet.

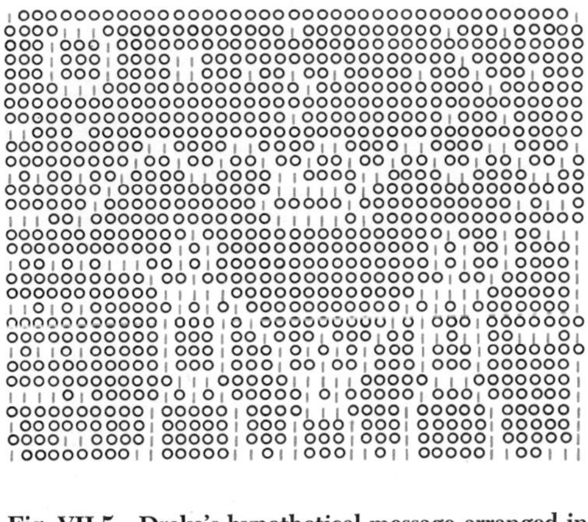

Fig. VII-5 Drake's hypothetical message arranged in 31 rows, 41 characters long. To emphasize the message, the *1*'s are colored.

Above the bipeds is a wavy line, under which is a fish, representing a water-based environment. Since the line is next to the third planet, it implies that this third planet may have a water-based environment.

Above the wavy line is a series of three pictures, giving the particle configurations of hydrogen, carbon and oxygen, the main chemical constituents of their living systems.

In conclusion, it appears that the only way man will ever communicate with his galactic equivalents is by radio messages. Within this medium, one can logically solve the problems of the nature of the signal, the kind of symbol, and the information that would be transmitted.

The two assumptions are, of course, that intelligent beings on other planets would (1) follow the same reasoning procedures as exobiologists on Earth, and (2) want to communicate. It is possible that other, more advanced societies have communicated with so many other planets that they are hardly interested in communicating with someone from the stellar boondocks.

GLOSSARY

α particles Subatomic particles located in the nucleus of an atom.

Absorption The phenomenon by which matter or energy passively enters a system (for example, absorption of nutrients from the digestive tract of animals into the blood stream, and absorption of light by chlorophyll).

Absorption spectrum The relative amounts of energy absorbed by a molecule at different wavelengths of light.

Acetabularia A single-celled alga growing in tropical marine waters and attaining a height of approximately one inch.

Achyla A water mold.

Acid A chemical substance that acts as a proton (H^+) donor.

Acid group The organic group —COOH that tends to act as a proton donor, and in doing so is converted to $—COO^-$.

Acridine A dye used as a mutagen and as an inhibitor of the division of the kinetosome.

Acrosome The vesicle from the Golgi body once it takes on a granular, dense appearance near the nucleus of the sperm cell.

Actinomycin D An inhibitor of the translation of DNA to RNA.

Action spectrum The relative activity of a system at different wavelengths of light.

Active site The region of an apoenzyme in direct physical contact with the substrate.

Active transport The movement of molecules through a membrane at the expense of energy.

Adaptation (genetic) The phenomenon by which organisms with appropriate genes as part of the natural variation in a population are able to survive under certain environmental or physiological conditions.

Adaptation (physiological) The mechanisms within an organism that permit continued existence over a range of environmental conditions by activating or repressing organismal functions (for example, an increase in heartbeat accomplishes faster pumping and hence increased movement of blood during exercise).

Adaptive radiation The exploitation of many ecological conditions by organisms of a new biological level of organization (for instance, the wealth of variety of biological organisms that arose after the evolution of reptiles—snakes, lizards, dinosaurs, turtles, etc.).

Adaptive zone The range of ways in which a species exploits its environment.

Addiction The physical or psychological dependence on a drug so that life is unbearable without its constant presence in the body.

Adenine The purine that is part of the structure of ATP (adenosine triphosphate), DNA, and RNA.

Adenosine diphosphate (ADP) The molecule resulting from the donation of one high-energy phosphate group from ATP in a coupled reaction.

Adenosine monophosphate (AMP) The molecule resulting from the donation of two high-energy phosphate groups (a pyrophosphate) from ATP in a coupled reaction.

Adenosine triphosphate (ATP) The primary donor of high-energy phosphate groups in metabolic-coupled reactions.

Aerobic respiration The stepwise series of reactions in the degradation of glucose that requires the presence of oxygen.

Agar A substance used as a surface on which or within which microorganisms are grown for research or culture purposes. Agar is inert and has no nutritional value.

Algeny An area of biological study that is examining the possibility and feasibility of endowing man with genes to synthesize all metabolic requirements except glucose.

Aliquot A measured portion of a whole (for example, "A series of 50 ml aliquots was taken from the culture at five-minute intervals").

Alleles Genes, contrasting or differing in expression, that control the same trait, that segregate at meiosis, and that exist as different genes on the same site of homologous chromosomes.

Allometric growth The phenomenon by which shape is generated by unequal growth rates in different regions of an organism.

Allosteric states The response of protein structure of different energy levels. A protein may have a characteristic shape at each of several energy levels.

Amination The chemical reaction in which an amine group becomes part of an organic molecule.

Amine group The nitrogen group associated with organic molecules that has the tendency to accept protons ($—NH_2 + H^+ \longrightarrow —NH_3^+$).

Amino acid A member of a class of organic molecules containing some twenty forms, each having the characteristic amine and acid groups.

```
H2N  H  COOH
   \ | /
     C
     |
     R
```

Ammonia The hydrogenated form of an amine group (NH_4), toxic to most organisms.

Amoeba A single-celled protozoan characterized by the projection of false feet (pseudopodia) by the alteration of sol and gel states of the cytoplasm.

Anaerobic respiration The stepwise series of reactions in the breakdown of glucose that occurs in the absence of oxygen.

Analgesic A drug administered to diminish pain sensation.

Anaphase The stage during the mitotic cycle when sister chromatids separate and proceed to the cell poles.

Anaphase I The stage in meiosis when homologous chromosomes separate and proceed to the poles.

Anaphase II The stage in meiosis when sister chromatids of the now haploid cells separate and proceed to the poles.

Antheridial initial The first cell formed in the development of the male sex organ (antheridium) in ferns, mosses, and fungi.

Antheroceros A liverwort in which the egg cells always contain one chloroplast.

Antibody The chemical substance formed in response to, and that combines with, an antigen.

Antibiotic A substance produced by a microorganism, that is toxic to other microorganisms.

Antigen A chemical substance, most frequently protein in nature, that elicits the formation of antibodies.

Antihistamine A drug that counteracts the effects of histamine.

Arboreal existence A mode of living primarily in trees.

Artifact A structure observed through the microscope that is a product of optic rather than of biological phenomena.

Asexual reproduction Growth accomplished by mitotic divisions resulting in a group of genetically identical cells.

Ascorbic acid Vitamin C, a deficiency of which causes the disease scurvy.

Atom The smallest unit into which an element may be divided.

Atomic number The number of protons or electrons possessed by an atom.

Atomic weight The combined weight of the protons and neutrons of an atom, usually given in grams per liter to achieve the gram molecular weight.

Attenuation The process by which the virulence of a virus is diminished.

Australopithecus A pre-man type dating from 1,700,000 years ago, fossils of which have been found in many parts of the world.

Autocoid A chemical of medicinal value produced within an organism.

Autograft A graft from one part to another of the same body.

Autoploidy The state in which a cell or organism contains two or more sets of the same chromosomes.

Autoradiograph A photograph of a cellular component in which the parts of that component that contains radioactive material appear as dark spots.

Autoradiography The experimental technique in which a photographic emulsion is placed over a specimen containing radioactive material. The emission of atomic particles from the specimen leaves exposed areas or spots on the emulsion. Photography of the specimen and superimposition of the developed emulsion produce an autoradiograph.

Autosome A chromosome that is not responsible for the conferring of sex upon the individual that possesses it.

Autotroph An organism that does not require an organic carbon source.

Auxotroph An organism that requires a vitamin, not as a carbon source, but as a source of a coenzyme that it is not capable of synthesizing.

Average The value arrived at by summing a group of values and dividing by the number of values in the group (mean).

β-particles Subatomic particles in the atomic nucleus.

Backcross A genetic cross in which an F_1 individual is crossed to a homozygous recessive individual like that in the P_1 generation.

B complex A group of vitamins including some of the most important coenzymes, thiamine, cobalamine, pyridoxine, etc.

Bacteria A group of organisms all of which have procaryotic cell types.

Bacteriophage Viruses that parasitize bacteria. Each bacteriophage is highly specific, and is therefore capable of parasitizing only one species.

Balanced growth The normal growth of an organism during which each tissue has a growth rate complementary to all other tissues.

Barbiturates A class of drugs used as analgesics, that act to uncouple the oxidative-phosphorylation process.

Basal body The body located at the base of a cilium.

Base A chemical substance that has the tendency to accept protons (H^+). In inorganic chemistry, most bases contain a hydroxyl (OH^-) group. In organic chemistry, the amine group serves as the proton acceptor, and the molecule containing it is regarded as a base.

Base deletion The removal of a nitrogen base, a purine or pyrimidine, from the DNA structure.

Base substitution The alteration of the structure of a purine or pyrimidine by removal or addition of side groups that convert the molecule to a different purine or pyrimidine, respectively.

Benign Referring to a nonmalignant tumor.

Beriberi A disease caused by insufficient vitamin B_1.

Biflagellate A cell characterized by the presence of two flagella.

Binary system A two-symbol combination (1 and 0, + and −, yes and no, etc.) used in information systems.

Biochemistry The study of the chemical aspects of living organisms.

Biogenesis The theory that holds that living forms come only from preexisting living forms.

Biokinetic zone The temperature range (5° to 60° C) over which organisms can survive.

Biology The study of living organisms and their processes.

Biomass The density per unit volume of a group of organisms in a community.

Biome The ecological perspective in which knowledge of the flora of a region permits prediction of the climate and animal life to be found there.

Biotin A vitamin that acts as a coenzyme for carboxylation reactions.

Biophysics The study of the physical behavior of molecules in biological systems.

Biosphere The region, extending below and above the surface of the Earth, in which organisms can survive.

Biosynthesis The production of complex structures from simple building blocks through enzymatically mediated reactions.

Biosystematics The methods and criteria used for organismal classification.

Biotic community All the organisms living in, and contributing to, the environment in a deliniated geographical region.

Biotic potential The number of organism of different species that a deliniated geographical region can support under ideal conditions.

Bipedal The mode of walking on two legs.

Bit of information The average number of questions that must be asked in a given information system in order to determine a given amount of information.

Blastula The stage in the development of the embryo of some multicellular organisms in which a hollow ball of about 250 to 500 cells exist.

Botany The study of plants.

Brachiating habit The pattern of arboreal existence in which the animal swings from tree limb to tree limb.

Buffer The combination of a weak acid and its salt. The buffer is characterized by a variety of possible equilibrium states depending on the amount of H^+ and OH^- present in solution.

Calorie The amount of energy required to raise one cubic centimeter of water from 14.5° C to 15.5° C.

Cambium A lateral layer of embryonic cells that give rise to xylem and phloem in vascular plants.

Cancer A disease of unknown origin characterized by a disturbance in the growth rate of the affected tissue.

Carbon The element that is the basis for organic structure.

Carbon dioxide A waste product of metabolism of living organisms produced by a rearrangement of the bonds of a free acid group to form $O{=}C{=}O$.

Carbon cycle The elemental cycle in which carbon passes through the metabolic processes of living organisms, always returning to an arbitrary starting point.

Carcinogen A cancer-causing agent.

Carcinoma A malignant tumor of endodermal or ectodermal origin.

Carnivore An organism that eats only other animals.

Carotene A pigment found in the chloroplast that operates in conjunction with chlorophyll in the photosynthetic process.

Carrying capacity The total population of a species that an ecosystem can support.

Cell The simplest independent system manifesting biological processes.

Cell division Nuclear division operating in conjunction with cytokinesis.

Cell membrane (unit membrane) The protein-lipid-protein structure surrounding the cell.

Cell plate The cytoplasmic figure formed during telophase of plant-cell mitosis at the site where a new cellulose partition will be laid down between the two new cells.

Cell theory The theory expressed by Scheiden and Schwann that all living organisms are composed of cells.

Cellulose The insoluble glucose polymer of which the cell wall is composed.

Cell wall The portion of the plant cell external to the cell membrane, mostly made of cellulose.

Central dogma Name given to the sequence of events leading to protein synthesis in the cell.

Centrifugation The experimental procedure that utilizes the physical principle that in a system revolving rapidly around a fixed point, heavier material will tend to go toward the periphery of the generated circle.

Centriole The cell organelle found near the nucleus of cells that seems to be associated with the origin of the spindle.

Centromere The part of a chromosome to which the spindle fibers are attached in nuclear divisions.

Chain reaction The chemical phenomenon in which subatomic particles emitted from atomic nuclei collide with other atomic nuclei and cause the further emission of subatomic particles.

Chemical bond The physical association of two atoms.

Chemical reaction The physical events leading to the formation of products from one or more reactants under specific energy conditions.

Chemolithotroph An organism that requires organic energy sources and inorganic electron donors.

Chemoorganotroph An organism that requires organic energy sources and organic electron donors.

Chiasma The cytological phenomenon in which nonsister chromatids of homologous chromosomes appear joined at homologous regions along their length.

Chloramphenicol A protein-synthesis inhibitor.

Choline A vitamin that acts as a coenzyme in fat metabolism and transmethylation.

Chlorophyll A plant pigment existing in many forms, chlorophyll *a*, *b*, *c*, *d*, and *e*, each capable of being raised to a higher energy level when exposed to appropriate wavelengths of light.

Chloroplast The organelle in plant cells where the entire photosynthetic process takes place.

Chloroplast fraction The portion of cellular material obtained by differential centrifugation that contains only the chloroplasts.

Chondriosome Name used for mitochondria before 1945.

Chromatid The linear unit of the chromosome containing at least one double helix of DNA.

Chromatin Passé term used to describe the dense stained material in chromosomes, now identified as nucleoprotein.

Chromatography An experimental technique employing the specific migration properties of substances in a compatible solvent.

Chromosome The universal carrier of hereditary material, genetically recognized as a linkage group, and found in procaryotes, the nuclei of eucaryotic cells, chloroplasts, and mitochondria.

Chromosome map A depiction of the relative distances between genes according to crossover units.

Chromosome complement The chromosomes characteristic to a given species.

Cilium The hairlike organelle for locomotion in eucaryotic cells containing the 9:2 arrangement of fibrils. See FLAGELLUM.

Cleavage Early cell division in an embryo.

Clone A population of cells derived from a single cell, and hence, genetically identical.

Coacervate theory Theory expressed by Oparin that the origin of life was preceeded by the formation of mixed colloidal units.

Coexistence The relationship existing between two species that compete for biotic or abiotic features of the environment.

Colloid The state of two chemical substances that are immiscible in one another.

Colony The localized population of bacteria derived from a single cell grown on an agar surface.

Commensalism The relationship between organisms in which one organism derives some advantage from the relationship while the other does not.

Community The total sum of populations of different species living in an ecosystem.

Community approach A technique for determining the classification of an organism in the phylogenetic scheme according to ecological and genetic criteria.

Comparative anatomy An area of study within biology that focuses on the similarities and dissimilarities of structures of organisms at different phylogenetic levels.

Comparative biochemistry An area of study within biochemistry that deals with the similarities and dissimilarities of metabolic processes of organisms at different phylogenetic levels.

Comparative embryology An area of study within embryology that examines the similarities and dissimilarities of embryological development in organisms at different phylogenetic levels.

Comparative cytology An area of study within cytology concerned with the similarities and dissimilarities of cell structure in organisms at different phylogenetic levels.

Complete medium Culture fluid in which microorganisms or cell cultures are grown that contains all the nutrients required for growth.

Compound A collection of identical molecules.

Competitive inhibition The phenomenon in which two substrates compete for the same site on the enzyme.

Conformation The relationship of parts, or the shape of a structure.

Conidia The haploid spores of *Neurospora*.

Conservation of energy The essence of the first law of thermodynamics. See FIRST LAW OF THERMODYNAMICS.

Conservation of matter The physical law stating that matter can be neither created nor destroyed.

Conservative mode of DNA replication A proposed means by which DNA could replicate, this mode involves a retention of the helical structure while new bases, attracted to the existing strands, form a new helix.

Contact inhibition The cessation of growth when a cell is surrounded by other cells.

Continental drift The theory, presently widely accepted, that there was originally one large land mass that broke into two continents, Laurasia and Gondwanaland, and subsequently divided and drifted to form the land masses of the present Earth.

Cork The outer layer of woody plants made up of dead cells.

Correlation coefficient An approach to human genetics that uses identical and fraternal twin data to determine the heritability of traits.

Coupled reaction A chemical reaction in which the energy from an exergonic reaction drives the endergonic reaction.

Covalent bond A chemical bond in which an electron is shared by two atoms.

Cristae The ultramicroscopic membraneous shelves within mitochondria.
Crossing over The genetic phenomenon in which linked genes recombine.
Cyanocobalamine Vitamin B_{12} active as a coenzyme in methyl transfers.
Cyclic phase of light reaction Series of reactions in which the electron in the triplet state is returned to chlorophyll *a*.
Cytochromes A series of iron-based porphrin molecules active in the electron transport system.
Cytokinesis The division of the cytoplasmic portion of a cell into two daughter cells.
Cytoplasm General term for the viscous-liquid phase inside a cell.
Cytosine A pryimidine; one of the four nucleotide bases found in DNA and RNA.

Dark reaction The series of reactions in the photosynthetic process for which light is not necessary.
Deamination The removal of an NH_2 group from an organic amine.
Death phase That portion of the growth curve that reflects a culture in which the death rate exceeds the growth rate.
Dedifferentiation A phenomenon in regeneration in which differentiated cells return to an embryonic state.
Degradation A series of reactions in which molecules are broken down to shorter carbon chains.
Deme A subunit of a population.
Denitrification The process by which various forms of nitrogenous compounds return to the atmosphere as N_2.
De novo To arise from nothingness.
Density gradient centrifugation Separation of the components of a homogenate by centrifugation in a gradient of solute in which each component is pulled to the level equal in density to itself.
Deoxyribose A ribose molecule of the formula $C_5H_{10}O_4$, while ribose typically has the formula $C_5H_{10}O_5$.
Deoxyribonucleic acid (DNA) The hereditary material.
Desert A terrestrial biome characterized by low rainfall, poor soil, temperature extremes, and scarcity of living forms.
Detritus Organic debris.
Differentiation The processes involved in the alteration of cell shape and organelle ratios in a specialized cell.
Digestion The process by which chemolithotrophs and chemoorganotrophs break down polymers to monomeric forms so that they may be taken into the body proper for use.
Digestive system The system in multicellular animals responsible for digestion.
Digram A two-unit symbol read as a single symbol (such as st and ch).
Diploidy The condition in which a nucleus possesses two sets of chromosomes.
Dipolarity The condition in which two sides of a structure have opposite charges.
Directive selective pressure The type of selection in which the only members of a species to survive are those that manifest change in a single direction.
Disaccharide Two monosaccharides joined by an ester linkage.
Disruptive selective pressure A type of selection in which the species adapts to several environments simultaneously.
Dispersive speciation Two or more species arising from a single species.
Disulfide bridge The joining of the two sulfur atoms in terminal positions of two cysteine molecules to form a double bond (—S═S—).
DNA See DEOXYRIBONUCLEIC ACID.
Dominance The phenomenon in which the trait controlled by one gene in a heterozygous individual is expressed in the phenotype of that individual.
Double helix Two strands, each coiled in a helical fashion, wrapped around one another.
Drosophila melanogaster The fruit fly used in genetic research.
Drug A chemical substance, foreign to the body, that alters the chemical and physiological process of the body.
Dryopithecus A group that includes a large number of anthropoid (apelike) types.
Dry weight The weighing of a chemical or biological material after the removal of water.
Dynamic equilibrium A balance in which the products are constantly converted to reactants and reactants to products in such a way that the number of molecules in any one state at any one time remains the same.

e value The percent of the Earth's solar energy present on other planets.
E_0 values A series of values reflecting the redox potential of molecules. Any molecule will accept protons from any other molecule with a more positive e_0 value.
Eobiont The first living entity.
Ecological efficiency The percentage of the caloric content available in a given trophic level that is used by the next trophic level.
Ecological equivalents Two species from the same ancestral stock, geographically isolated but evolved in similar environments, that have similar adaptive features.
Ecological isolation Two species capable of mating and producing viable offspring but which do not because they live in different ecosystems.
Ecological pyramid A means of depicting the relationship of certain features of trophic levels in which trophic levels are represented by rectangles and placed in an appropriate sequence.
Ecology The study of organisms in their biotic and abiotic environments.
Ecosystem The biotic and abiotic features of an environment shared by a community of organisms.
Ectoderm The outermost cell layer in the embryos of triploblastic organisms that differentiates into skin and the nervous system.
Electron The particle carrying a negative charge that is found, 90% of the time, in the volume assigned as an orbital.
Electron microscope A microscope based on the principles of the passage of electron beams through a vacuum.

Electromagnetic radiation The emission of particles that have wave properties and respond to an electromagnetic field.
Electron transport system (ETS) A series of reactions in which an electron is passed sequentially to several molecules and ultimately to oxygen to form water, during which sufficient energy is generated for oxidative phosphorylation.
Element A substance that cannot be destroyed or decomposed by available levels of heat or electrical energy.
Elemental cycle The passage of elements through the metabolic processes of living organisms leading to the regeneration of any given form of that element.
Endergonic reaction A chemical reaction that requires energy in order to proceed.
Endoderm The innermost cell layer in the embryo of triploblastic organisms that develops into the digestive tract and accessory organs.
Endoplasmic reticulum The series of ultramicroscopic shelflike membranous projections within a cell.
Energy The ability to do work.
Energy acceptor The molecules in a coupled reaction that use the energy released in the exergonic reaction.
Energy donor The molecules in a coupled reaction that yield the energy to drive the endergonic reaction.
Energy of activation The energy that must be put into some exergonic systems before the systems will proceed spontaneously.
Entropy Heat that cannot be translated into work; a tendency toward randomness.
Enzyme A molecule, usually made up of a protein and a heteroconjugate, that acts as a catalyst in metabolic reactions.
Enzyme-mediated reaction A reaction that proceeds in biological systems in conjunction with an enzyme.
Enzyme-substrate complex The molecule formed by the physical association of the enzyme and one or more substrates as intermediates in enzyme-mediated reactions.
Epigenesis The idea in embryology that the entire organism develops from an originally undifferentiated mass of cells.
Esophagus The part of the digestive tract that food enters after the pharynx and leaves upon entrance into the stomach.
Euglena A unicellular, flagellated organism, classified as a green alga, that has chloroplasts and an eyespot.
Evolution The process by which a species gives rise through natural selection to other species more adapted to the existing environment.
Excretion The elimination of metabolic, primarily nitrogenous, wastes.
Exergonic reaction A chemical reaction that upon completion yields greater energy than that which was put into the system.
Exobiology The study of extraterrestrial organisms.
Experimental cytology An approach to the study of cells that utilizes the microscope in conjunction with biochemical and biophysical experimental techniques.
Exponent A numerical value written to the right and slightly above another numerical value to indicate the number of times the number is to be multiplied by itself.
Exploitation The ecological relationship in which one type of organism gains at the expense of another.
Extinct No longer present in the world population.
Extrapolation The calculation of a value beyond a given series of values, often based on graphic representation of the results and an extension of the trend represented.
Extraterrestrial Outside the limits of the Earth or its atmosphere.

F_1 generation First filial generation, the offspring of a genetic cross that is the first in what may be planned as a series of genetic crosses.
F_2 generation Second filial generation, the offspring of a genetic cross involving a member of the F_1 generation.
Flavin adenine dinucleotide (FAD) A coenzyme active in fat metabolism and the first step of the electron transport system.
Fatty acid A linear molecule containing a terminal acid group and a hydrogenated carbon chain. Three fatty acids and one glycerol molecule react to form a molecule of fat.
Fermentation The metabolic breakdown of glucose to produce ethyl alcohol or lactic acid.
Ferridoxin A molecule found in the chloroplast that accepts the triplet-state electron from chlorophyll *a*; this molecule has the lowest E_0 value in biological systems.
Fertilization The union of two gametes to form the zygote.
Fixation Preparation of cellular material for microscopic examination by killing it in such a way that the material retains its structural integrity.
First law of thermodynamics The physical law that states that energy can be transformed from one state to another, but neither created nor destroyed.
First-order reaction A chemical reaction that proceeds at a rate that can be mathematically expressed as $y = x$.
Flagellum The hairlike projection that characterizes some procaryotic cells and that is revealed by electron micrographs to be composed of a single fibril.
Florigen A plant hormone that is triggered by phytochrome and induces flowering.
Fluoride An element that replaces hydroxyl groups in teeth and bone.
Fluoridation The addition of fluoride to a water supply.
Folic acid Formed by the combination of para-aminobenzoic acid and a pterin moiety; the hydrogenated form, tetrahydrofolic acid, is active as a coenzyme in formate transfers.
Food chain The sequence in which producers are consumed by herbivores, herbivores by primary carnivores, etc.
Food net A modification of the food-chain concept that takes into account the variety of organisms, predators and parasites, that prey on other organisms.
Food vacuole A membrane-bound sac in certain protozoans that is isotonic with the cell and contains food materials and enzymes that digest the food.
Fossils Remains of organisms or parts of organisms as either impressions or preservations.
Free energy The energy resulting from a chemical reaction, that is available for work.
Fresh water Nonmarine water, such as that found in rivers, most lakes, and ponds.

G_1 Stage The growth period or gap that follows the entrance of a nucleus into interphase after telophase.
G_2 Stage The growth period or gap that follows the DNA synthesis period (S stage) during interphase.

Galactose A monosaccharide that in some organisms is readily converted to glucose.

Gastrula The stage in embryological development in which the three primary germ layers (ectoderm, endoderm, and mesoderm) first make their appearance.

Gastrulation The differentiation process that brings the embryo from the blastula to the gastrula stage.

Gametes The haploid cells, formed by meiosis in diploid organisms, that join to form the zygote.

Gametogenic cells The diploid cells that undergo meiosis to form the gametes.

Gause's law The principle that two species with similar ecological requirements cannot coexist.

Gene The unit of heredity that controls a trait or series of related traits; the unit of DNA encoded for the production of a polypeptide.

Gene frequency The percentage of a given gene in the gene pool.

Gene pool The sum total of all the allelic genes in a population.

Generation time The duration required for a population of cells to double in number; the time for one mitotic cycle.

Genetic code The combination of nucleotide bases that determines the placement of a given amino acid in a protein chain.

Genetic death The failure of a cell or individual to reproduce.

Genetic drift The phenomenon of a small, nonrepresentative part of the population surviving and leading to a significant change in the gene frequency.

Genetic isolation The inability of two similar species to produce viable F_2 because the parental chromosomes do not pair at prophase I of meiosis in the F_1 individual.

Genetics The study of heredity and variation.

Genotype The genetic makeup of an individual; the sum total of all genes.

Geographical distribution The localization of organisms in particular geographical regions.

Geographical isolation The inability of organisms to interbreed because they are separated by physical barriers such as oceans or mountain ranges.

Glucose A six-carbon sugar, the breakdown of which constitutes the respiratory process.

Glycogen A polymer composed of glucose monomers that acts as a storage molecule in most animals.

Glycolysis The series of reactions through which one molecule of glucose is converted into two molecules of pyruvate.

Golgi body or **apparatus** Membranous shelves oriented in layers associated with the endoplasmic reticulum and apparently involved in secretory cellular processes.

Gondwanaland One of the two major land masses that existed in the Mesozoic era according to the theory of continental drift.

Granum Stacks of disc-shaped units within the chloroplast that contain chlorophyll.

Grassland A terrestrial biome characterized by moderate rainfall and temperature; also, a stage in the succession sequence toward a deciduous forest.

Greenhouse effect The phenomenon in which carbon dioxide and water in the atmosphere absorb and retain heat energy, thus maintaining a relatively narrow temperature range in certain geographical regions.

Ground state The condition of a molecule when at its lowest energy level.

Growth An increase in active protoplasmic mass.

Growth curve A graphic representation of the pattern of growth of a population of cells or organisms.

Guanine A purine associated with deoxyribonucleic acid and ribonucleic acid.

Habitable zone The range distance from a star within which suitable quantities of energy exist to sustain life.

Habitat The location of a population within an ecosystem.

Half-life The duration of time required for half the number of atoms of a given radioactive sample to decay to a more stable form.

Hallucinatory drug A chemical substance that affects brain processes in ways that result in visual distortions.

Haploid organism Organisms in which the majority of cells in the life cycle contains one set of chromosomes.

Haploidy The condition in which one set of chromosomes exists in a cell nucleus.

Hardy-Weinberg equilibrium A set of five conditions that, when fulfilled, lead to the stability of gene frequency.

Heidelberg man Classified as *Homo erectus,* this type of man lived approximately 550,000 years ago.

Helix A three-dimensional configuration that revolves around an axis in such a way that any point along the figure is the same distance from the central axis as any other point. Not synonymous with a spiral.

Heme The iron-containing porphrin molecule associated with globin proteins in the hemoglobin molecule.

Hemoglobin The oxygen-carrying molecule in the blood of many vertebrate and invertebrate animals.

Herbivore Animal that consumes only living plant material.

Heredity The study of traits that are passed from generation to generation.

Hereditary patterns The relationships between members of a family exhibiting a given trait or set of traits.

Heteroconjugate A molecule composed of two or more monomers belonging to different classes.

Heterotroph An organism that requires an organic carbon source.

Heterozygous A state characterized by the possession of two alleles (Aa).

Hibernation A phenomenon in some animals in which the metabolic processes are slowed, so that the animal goes into a long period of sleep.

High-energy bond The energy difference between the original molecule and the two resultant molecules in a reaction.

High-energy phosphate A phosphate group with low resonance joined to another phosphate group. The resumption of resonance when this group becomes free is associated with the release of considerable energy.

Histamine An autocoid that causes a swelling of the membranous lining of the respiratory tract.

Histogram A graphical representation of data in which rectangles are used to illustrate quantities (for example, a bar graph).

Histoincompatability The destructive reaction of in situ tissue against any foreign protein that comes in contact with it.

Holotroph An organism that is capable of ingesting entire organisms.

Homeostasis The maintenance of a dynamic equilibrium between the biological processes at all levels of biological organization of an organism through the presence of control devices.
Homeostatic selective pressure The retention of deleterious genes, often because of the selective advantage of the heterozygous individual over either homozygous state.
Homo erectus An early species of man that became extinct 400,000 years ago.
Homograft A tissue graft between members of the same species.
Homologous chromosomes Chromosomes that contain genes controlling the same traits (alleles) at the same loci; chromosomes that pair at prophase of meiosis.
Homologues Synonymous with HOMOLOGOUS CHROMOSOMES.
Homozygous The state in which the same allele is present more than once (SS, ss).
Hormone A secretion of a ductless gland that proceeds into the bloodstream and acts upon a distant organ or set of organs.
Humus Decayed organic matter in soils.
Hybrid A heterozygous individual resulting from a cross between two homozygous individuals (Aa, the F_1 individual from the cross AA $\times$ aa).
Hydra A freshwater species of animal that grows about one-quarter inch high and that contains ectoderm, endoderm, and a primitive third layer, mesoglea.
Hydrogen bond A weak chemical bond formed by the sharing of a hydrogen atom by two molecules.
Hydrolases A class of enzymes that mediate hydrolytic reactions.
Hydrolysis The chemical breakdown of a large molecule into two smaller molecules by the addition of water.
Hydrolytic reaction Synonymous with HYDROLYSIS.
Hydrophilic groups Parts of molecules that possess electrical charge or hydroxyl groups that form weak ionic associations or hydrogen bonds respectively with water molecules. Literally, water-loving.
Hydrophobic groups Parts of molecules that do not possess groups compatible with water, and thus tend to remain unassociated with water. Literally, water-fearing or -hating.
Hypothesis A tentative idea suggested to explain some data.
Hypertonic condition A condition in which the water concentration is less than that within the cell.
Hypersensitivity The allergic reaction associated with the entrance of certain foreign material into a body.
Hypotonic condition A condition in which the water concentration is greater than that within the cell.

Immunity The presence of appropriate antibodies that were produced as the result of past exposure to either the disease organism or a protein similar to that of the disease organism (vaccine).
Independent assortment The phenomenon in which one member of a pair of homologous chromosomes has an equal chance of going to the pole with either member of another pair of homologous chromosomes during meiosis.
Induced mutation A change in the genotype caused by the introduction of a mutagen (mustard gas, UV light, X ray, etc.) in the course of an experiment.
Information content The amount of information possessed by a system.
Information theory Study of the measurement and properties of messages and codes.
Inositol Classified as a member of the B complex of vitamins, the function of this compound is unknown.
Insulin A hormone of protein structure produced by the pancreas and active in glucose-glycogen metabolism.
Intergenic crossover A crossover between two linked genes controlling different traits.
Interphase The time during the mitotic cycle when the chromosomes are uncoiled and metabolic processes are directed toward growth through synthesis.
Interspecific Between species.
Intragenic crossover A crossover within a gene or cistron.
Inversion A chromosome anomaly in which a chromosome breaks at two points, and the internal region rotates 180° and rejoins the end regions of the chromosome.
In vitro Biological processes taking place outside an organism.
In vivo Biological processes taking place within an organism.
Ion An atom or group of atoms possessing an electrical charge.
Ionic bond The chemical bond between two atoms of opposite charge.
Isomerase A class of enzymes that convert organic compounds from D- to L- or from L- to D- forms.
Isomerism The phenomenon in which organic molecules have the same empirical formulae, but differ in the planar orientation of groups.
Isotonic condition The situation in which the concentration of water inside and outside the cell is equal.
Isotope The form of an element having one or more neutrons than that characteristic for the element.
Interpolation The calculation of a value within a given series of values, but which does not constitute a preexisting point on the graph.

Java man A primitive man classified as *Homo erectus.*

Karyokinesis Division of the nucleus.
Kinetic energy Energy in motion.
Kinetosome The basal body of a cilium.
Krebs cycle A series of reactions during which a series of dehydrogenations take place and two acid groups are removed from molecules and eliminated as carbon dioxide for each turn of the cycle.

Lag phase The part of the growth curve reflecting the time when organisms are adjusting to the environment and preparing for synthetic processes of growth.
Lamarckian adaptation The theory of adaptation forwarded by Lamarck, holding that heritable traits are purposefully acquired.
Lateral redundancy Repetition of the same message in different parts of a series of messages.
Laurasia Name given to one of the two land masses believed to have existed during the Mesozoic era, according to the theory of continental drift.
Law of mass action The chemical law stating that the rate of a reaction is proportional to the concentration of the reactants.

Leaf The vegetative organ of a plant, usually with a large surface area to volume ratio, in which most of the photosynthetic activity occurs.

Lentic Referring to aquatic biome in which there is little movement of water.

Lethal mutation The alteration of a gene responsible for the production of an essential polypeptide. The lethality of a mutation depends on several factors. For example, if the capacity to synthesize an amino acid is lost, but the organism is supplied with that amino acid, the mutation is only potentially lethal.

Leucoplasts Colorless plastids in plant cells that are thought to act as storage depots for starch.

Light A form of energy that manifests both quantum and wave properties.

Light reaction of photosynthesis The series of reactions of the photosynthetic process for which light is necessary.

Light year The distance light travels in one year—the speed of light (186,000 ft/sec) × the number of seconds in a minute (60) × the number of minutes in an hour (60) × the number of hours in a day (24) × the number of days in a year (365).

Lignase A class of enzymes that catalyze the condensation of two molecules.

Limiting factor A reactant that limits the amount of product that can be formed because it is present in small quantities.

Limiting reaction When two or more reactions occur in tandem, the rate of the slowest determines the maximum rate at which the last reaction can take place.

Linear redundancy A number of identical messages in a one-dimensional array.

Lithotroph An organism that uses an inorganic electron donor.

Lipid An organic chemical composed of a molecule of glycerol bonded to three fatty acid chains.

Log phase The portion of the growth curve reflecting when the population is increasing at an exponential rate.

Lotic Referring to an aquatic environment in which the water is rapidly moving.

Low-energy phosphate A phosphate group undergoing resonance, the detachment of which is associated with an output of little energy.

Luminosity The radiation of light energy.

Lyase A class of enzymes that catalyze the addition of groups to double-bonded carbon units or the removal of groups so that double-bonded carbon units are formed.

Lysergic acid diethylamide (LSD) A hallucinatogenic drug.

Lysis The breakage of a cell by the bursting of its wall or membrane.

Lysosome An organelle containing ribonuclease and other destructive enzymes. The rupture of lysosomes causes cell deaths.

Macromolecular complex A multibonded association of a number of large molecules visibly discernible with the electron microscope.

Magnification The enlargement of an image to a size greater than the natural size of the object, through the use of lenses.

Main-sequence phase of a star A star manifesting a temperature greater than 4,000,000° C.

Malignancy A cancerous growth.

Malthusian theory The idea that human population growth would undergo cyclic phases of growth and starvation, as the population size could not be supported by the food supply.

Mammals Vertebrate organisms that feed their young with milk produced in the mammary glands of the female.

Marihuana A drug derived from hemp that when sniffed or smoked produces a euphoric effect.

Marine Referring to bodies of salt water.

Marsupials A subclass of the mammalian group, characterized by a typical premature birth of the young, and the continuance of their development in the pouch of the mother. The mammary glands are within the pouch.

Mass Size and density per unit volume.

Mating patterns The rituals that precede copulation, very specific to given species, particularly among birds and insects.

Mating type Designation made for species in which no morphologically distinct male or female gametes exist.

Matter Anything that manifests mass.

Me-too drug A drug that differs in nonessential side groups from a drug that has already been proven and marketed.

Meiosis The type of nuclear division that takes place in a diploid cell, resulting in the formation of four haploid cells.

Mendelian laws of heredity The laws of heredity including that for segregation of alleles and independent assortment of pairs of alleles as proposed by Mendel.

Membrane A macromolecular complex composed of protein, carbohydrate, and lipid, found as a component of several cell organelles.

Meristematic cells The embryonic cells of a plant that upon division give rise to cells that either remain embryonic or elongate and ultimately differentiate.

Messenger RNA (*m*RNA) A strand of RNA synthesized in the nucleus of a cell, directly complementary to a portion of the DNA, that migrates into the cytoplasm where it becomes associated with the ribosones; the encoded determinant for a sequence of amino acids.

Metabolic pathway A series of stepwise reactions leading to the degradation or biosynthesis of a given molecule.

Metabolism The sum total of all the chemical reactions in a biological system.

Metaphase The stage of mitosis in which the chromosomes line up on the metaphase plate.

Metaphase I The stage of meiosis in which bivalents line up on the metaphase plate.

Metaphase II The stage of meiosis in which the chromosomes line up on the metaphase plate at the second division.

Metaphase plate The narrow region, indistinguishable from the rest of the cellular material, within which the chromosomes reside during metaphase.

Metastasis The spreading of cancer to different parts of the body.

Microevolution The small-scale changes resulting from genetic adaptation.

Microscope An instrument used by histologists and cytologists to magnify the images of biological material so that the component parts are discernible.

Mitochondria The organelles in which the reactions associated with aerobic respiration take place.

Microniche The environmental localization suitable for a few individuals to survive.

Microspheres Proteinaceous bodies formed by a random combination of component proteins by an experimental technique that simulates the early environment of the Earth.

Microsomal fraction A portion of cellular material obtained by differential centrifugation, containing the ribosomes and fragments of endoplasmic reticulum.

Microtubules Bar-shaped tubules of helically coiled fibers that constitute the spindle fiber.

Microvilli Invaginations of the cellular or unit membrane.

Mitotic cycle The series of stages that transpire when an embryonic cell undergoes nuclear and cytoplasmic division.

Mitosis Synonymous with MITOTIC CYCLE.

Molecule The smallest chemical entity that manifests the property of a pure substance.

Monomer A molecule that can exist either by itself or in association with like molecules to form a polymer; the building blocks for biosynthesis.

Monosomes Units composed of RNA in the cytoplasm, the ribosome.

Monosaccharide A monomer of a carbohydrate.

Morphogenesis The pattern of changes during the emergence of biological forms.

Mutant An organism or gene manifesting a genetic change.

Mutation A change in the genetic information of a cell; an alteration of DNA.

Multicellular Referring to groups of cells, organized as tissues, that contribute to the function of the organism as a whole, and that regulate one another.

Multinucleate Condition in which a cell has more than one nucleus.

Mutualism The relationship between organisms in which both derive benefit from the association.

Narcotic Any addicting drug that produces euphoric states.

Natural selection The heart of Darwin's theory of evolution; it holds that within the natural variation of a population, those able to live more successfully in a given environment will survive and reproduce. The idea has been modified so that the modern concept of natural selection involves the idea of the survival of the type within the species that produces more offspring than other types.

Neanderthal man A primative man classified as *Homo sapiens neanderthalis* that lived until about 100,000 years ago.

Negative feedback A mechanism in a self-regulatory system, whereby a change in one direction is converted into a command for the cessation of additional change.

Neoplastic growth The growth of new tissue, such as that in a tumor.

Neurospora A fungus used in genetic research.

Neutron A particle having weight, but no charge, and found in atomic nuclei.

Niacin A member of the B complex, nicotinamide is converted to NAD by the addition of adenine diphosphate.

Niche The biotic and abiotic position of an organism within an ecosystem.

Nitrogen cycle The passage of the element nitrogen through the metabolic processes of all organisms on Earth, leading to a regeneration of any given form.

Nitrogen-fixing bacteria Bacteria capable of converting atmospheric nitrogen to nitrate form.

Noncompetitive inhibition The type of inhibition that occurs when a molecule present in a system forms a strong physical association with an enzyme, thus preventing the enzyme from taking part in the normal reaction.

Noncyclic reaction of photosynthesis The part of the light reaction in which the electron lost by chlorophyll *b* is replaced by the electron gained from the photolysis of water.

Non-Mendelian genetics Genetic phenomena that do not proceed according to Mendel's laws of heredity, such as the recombination of linked genes in proportions that reflect their distance apart. Also, cytoplasmic heredity.

Nonpolar groups Side groups of molecules that bear no charge or hydroxyl groups, and hence do not combine with water.

Nuclear division Division of the nucleus by either mitosis or meiosis, which may or may not be accompanied by cytokinesis.

Nuclear membrane (**nuclear envelope**) A macromolecular complex exhibiting pores about 50 Å in diameter, and that surrounds the nucleus in the cell.

Nucleolar-organizer region A region on one chromosome that exhibits a high concentration of RNA during interphase, and is responsible for the regeneration of the nucleolus.

Nucleic acid Polymers of nucleotides, either ribonucleic acids or deoxyribonucleic acids.

Nucleolus An organelle in the nucleus of a cell composed primarily of RNA and involved in the synthesis of ribosomes.

Nucleoplasm The protoplasmic portion of the nucleus. A semi-antiquated term.

Nucleotide A heteroconjugate composed of a nitrogenous base, a sugar, and a phosphate group.

Nucleotide pool The nucleotides available for DNA and *m*RNA synthesis in the nucleus of a cell.

Nucleus (**atomic**) The area of an atom where the protons and neutrons are located.

Nucleus (**cellular**) An organelle that contains the hereditary information and controls the metabolic processes through its control of protein synthesis.

Nutritional classification Classification of organisms based on their nutritional requirements and the metabolic reactions underlying these requirements.

Nutritional requirements The molecular forms certain organisms must take in from the environment in order to continue their existence. The necessity of taking in organized carbon structures implies the inability of the organism to synthesize a given molecule.

Objective lens The lens of a microscope into which the image of the specimen first passes. Objective lenses of different magnification powers are usually found on a revolving nosepiece so that appropriate ones may be lined up with the body tube.

Obligate inbreeding The phenomenon resulting from a situation in which mating occurs only between members of a single population, or in the case of some plants, where self-fertilization becomes the mode, and the plant can breed only with itself.

Ocular The eyepiece of a microscope.

Oogonial initial The cell from which the oogonia, or egg-producing cells, of the plant will develop.

Operon An integrated cluster of regulator and structural genes.

Organ A group of tissues that together perform a necessary function.

Organelle A body within a cell composed of two or more macromolecular complexes.

Organotroph An organism that uses organic compounds as electron donors.

Osmosis The passage of water from an area of greater concentration to an area of lesser concentration.

Ovum The female reproductive cell containing the haploid number of chromosomes.

Oxidative phosphorylation The conversion of ADP to ATP by the addition of an inorganic phosphate group from the medium operating in conjuction with and deriving energy from the electron transport system.

Oxidation-reduction reactions Chemical reactions involving the transfer of electrons from one molecule to another.

Oxidizing agent The molecule in an oxidation-reduction reaction that acts as the electron donor.

Oxidoreductase A class of enzymes that mediate oxidation-reduction reactions.

Oxygen An element of biological importance that acts as a final hydrogen acceptor in the electron transport system.

Oxygen cycle The passage of the element oxygen through the metabolic pathways of all organisms, ultimately leading to the regeneration of any given form.

P_1 generation The first parental generation in what may be a series of genetic crosses.

Paleontology The study of extinct organisms by reconstruction of their adaptions from deductions based on their fossil remains.

Palisade cells Columnar photosynthetic cells found in the leaves of vascular plants.

Pantothenic acid Classified with the B complex of vitamins, this undergoes reaction to combine with a dinucleotide to become Coenzyme A.

Paper chromatography An experimental technique utilizing the differential migration of solutes in a compatible solvent on a filter-paper surface.

Paraphases Sterile hairs.

Parasitism A relationship between two organisms in which one derives advantage while the other is harmed. Parasitism usually refers to a smaller or biologically lower-level species living at the expense of a larger or biologically higher species.

Para-aminobenzoic acid (PABA) A vitamin that is converted to folic acid and ultimately to tetrahydrafolic acid, and that is active in formate transfers.

Peking man A primitive man classified as *Homo erectus*.

Pellegra A disease caused by a deficiency of niacin.

Pellet The product of centrifugation, it contains the heavier material that is forced to the bottom of the centrifuge tube.

Penicillin An antibiotic produced by the mold *Penicillium* and toxic to a number of pathogenic and nonpathogenic bacteria.

Peptide bond A bond between the amine group of one amino acid and the acid group of another.

Permease A class of enzymes that increases the permeability of the cell membrane to a particular molecule.

Pernicious anemia A disease associated with vitamin B_{12} deficiency.

pH The negative logarithm of the hydrogen-ion concentration used as an indication of the acidity or alkalinity of a solution. pH 7 is neutral. Values below 7 indicate an acidic environment; above 7, an alkaline environment.

Pharmocology The study of the effect of drugs on living organisms.

Phenotype The genetically controlled traits exhibited by an individual.

Phloem Long, narrow cells in the vascular plants, used for the transport of food.

Phospholipids Polar lipids containing phosphate groups.

Phosphorescence The light emissions associated with the return of the pi electron to its normal orbit.

Photolysis Literally, light-splitting. Refers specifically to the breakage of the water molecule during photosynthesis by the energy derived from the excitation of chlorophyll *b*.

Photon A unit of light energy.

Photolithotroph An organism that uses light as an energy source, and an inorganic molecule as an electron donor.

Photoorganotroph An organism that uses light as an energy source, and an organic compound as an electron donor.

Photoreactivation The phenomenon in which the effects of ultraviolet light as a mutagen are largely reversed by exposure of the biological material to visible light.

Photosynthesis The series of reactions that take place in organisms possessing chlorophyll that result in the synthesis of glucose.

Phototaxis The movement of an organism in response to the presence of light. Negative phototaxis is the movement of an organism away from the light source, while positive phototaxis is the movement of an organism toward the source of light.

Photosynthetic pigment A colored compound that transmits its triplet-state electron to a series of reactions that ultimately result in the production of ATP.

Photosynthetic phosphorylation The conversion of ADP to ATP by the addition of an inorganic phosphate from the medium, operating in conjunction with the light reaction of photosynthesis. Photosynthetic phosphorylation is the key to the conversion of light energy to chemical energy.

Phototroph An organism that utilizes light as an energy source.

Phototropism The bending of a part of an organism in response to light. Negative phototropism refers to a bending away from light, while positive phototropism refers to a bending toward the light source.

Phytoplankton The algae in an aquatic body, taken collectively as a photosynthetic mass.

Phyletic speciation The attainment of a slight modification in a species through selective processes.

Physiological isolation A situation in which two species are unable to produce viable offspring because of physiological differences in their sex organs.

Phytols The side groups associated with the magnesium porphrin of chlorophyll.

***Pi* electron** An electron associated with a molecule rather than a single atomic nucleus.

Pinocytic vesicle An ultramicroscopic vacuolelike structure apparently formed by a pinching off of the microvilli, and thought to be responsible for the entrance of certain material into the cell.

Pinocytosis The process by which material is brought into the cell in pinocytic vesicles.

Placebo A dummy medication; a pill containing dextrose.

Plant A multicellular living organism nutritionally classified as a photolithotroph.

Plastid A body within a plant cell, usually classified according to color.

Pleuralpneumonia group A group of microorganisms manifesting simple nutritional requirements, composed of monomers, and thought to be similar to the eobiont.

Point mutation A genetic change in a single gene.

Poison A substance that inhibits the function of a biological system, possibly terminating its continued existence. Noncompetitive inhibitors are poisons.

Polar groups Side groups of organic molecules that contain the electrical charge of hydroxyl groups and therefore combine readily with water.

Poles of the spindle The termini from which the spindle fibers radiate, thought to be formed by the products of the division of the centriole.

Polymer A linear or branched molecule made up of monomer subunits.

Polypeptide A sequence of amino acids.

Polyploidy A nuclear condition in which more than two sets of chromosomes are present.

Polysaccharide A polymer made up of monosaccharides, usually six-carbon sugars, and most typically of glucose units. The most frequently found polysaccharides in the biological world are glycogen and starch.

Polysome A linear arrangement of monosomes on a series of ribosomes; these are the *m*RNA units.

Populations A group of organisms of the same species living in an ecosystem.

Population genetics An approach to genetic study by examining populations and deducing or mathematically computing gene frequency and patterns of heredity.

Porphrin A molecule composed of nitrogen and carbon units arranged in a complex ring, and usually associated with a metal ion. The porphrin of heme is associated with iron; that of chlorophyll, with magnesium.

Potential energy Energy available for work.

Predation The organismal relationship in which one organism feeds on another.

Prey The organism in predation that is fed upon.

Primary consumers The trophic level that feeds on the producers. Primary consumers include microorganisms that feed on live plant material as well as herbivorous animals.

Primary carnivores Animals that eat only herbivorous animals.

Primary structure of protein The sequence of amino acids in a protein.

Primary succession Occurs when succession begins in an area where no life is present.

Pristine seas The oceans of the prebiological Earth.

Proconsul An apelike but not brachiating pre-man type that lived 30 million years ago.

Producers Members of the trophic level responsible for the production of organic material from inorganic precursors. Plants are considered the producers in a terrestrial environment; phytoplankton are the producers in an aquatic environment.

Product The molecular form or forms resulting from a chemical reaction.

Productivity A value assigned to the producer level to indicate the change in caloric value available to the next trophic level over a period of time.

Prosimians Initial group of primates to evolve, now represented only by the lemurs and tarsiers.

Prophase The stage in mitosis when the chromosomes condense (coil) and sister chromatids become visible. The stage in meiosis when homologous chromosomes pair, in addition to condensing.

Protein One or more polypeptides arranged in a three-dimensional configuration to give a specific shape and reactive potential.

Proteinase A class of enzymes that break peptide bonds.

Protein synthesis The series of events, generally called the central dogma, resulting in the formation of a protein.

Proton A positively charged particle in the nucleus of an atom.

Protoplasm Antiquated term for living material of the cell.

Protoplast A bacterial cell without a cell wall.

Puromycin An antibiotic that inhibits the translation of *m*RNA into protein.

Pyramid of numbers A means of depicting trophic levels to illustrate the relative number of organisms in the trophic levels.

Pyrodoxine Vitamin B_6, which undergoes reaction with ATP to form pyridoxal phosphate, active in the amination of keto acids to form amino acids.

Pyruvate The molecule that is the terminus of glycolysis. Several pathways are possible from pyruvate, depending on the enzymatic capacities of the organism.

Q_{10} A numerical value representing the number of times faster the rate of a reaction proceeds at a temperature 10° C higher.

Quadrapedal Walking on four legs.

Quantosomes Particles found in the chloroplast associated with the lamellae.

Quantum The tiny energy packet in which light travels. Synonymous with photons.

Quantum theory of light The model that views light as being composed of quanta that are given off by any light emitter, and that travel intact through space.

Quarternary structure of proteins The structure generated by the linkage of two or more polypeptide units to form the final protein.

Radiation damage The physical harm to biological systems perpetrated by electromagnetic radiation.
Radiation sickness The physical symptoms associated with radiation damage to the body.
Radioactive isotope An isotope that gives off subatomic particles and in doing so, is converted to either the element of which it is an isotope or a different element.
Random mating A situation in which any member of a population has an equal chance of mating with any other member of the population.
Range The distribution of a group of numerical values describing specifically the lowest and highest values.
Reactant The molecule that enters into a chemical reaction.
Recessive factor The gene controlling a trait that is expressed in the phenotype only when the same hereditary material is present on both homologous chromosomes.
Recombination New associations of old genes.
Redifferentiation Part of the regeneration process, dedifferentiation brings cells back to an embryonic state, and regeneration alters cell destinies.
Redox potential An E_0 value assigned to different molecules according to the readiness with which they will give up electrons.
Redox reaction See OXIDATION-REDUCTION REACTION.
Reducing agent A molecule that accepts electrons in an oxidation-reduction reaction.
Redundancy The repeat of messages expressed elsewhere in the information system (as in a glossary).
Regeneration The growth phenomenon of some animal forms in which an amputated portion of the body is replaced by dedifferentiation, growth, and redifferentiation processes.
Regulator The gene that controls the reading of other genes.
Repression The turning off of a system.
Reproduction The process involving the creation of a new individual by sexual or asexual processes of the preceding generation.
Resolution The distance between two points when they are seen as two points rather than as one.
Resonance The movement of one or more electrons from orbital to orbital among a group of atoms.
Riboflavin Vitamin B_2; this is converted to FMA and FAD in biological systems.
Ribonuclease A class of enzymes that break down RNA.
Ribosomes (monosomes) The RNA tables associated with the endoplasmic reticulum upon which the *m*RNA resides when it is read.
Ribosomal fraction The portion of cellular material, derived from differential centrifugation, that contains the ribosomes.
RNA (ribonucleic acid) A single-stranded nucleotide chain. There are three or more types of RNA, all of which are associated with the events outlined in the central dogma.
Roentgen A measured unit of high-energy radiation; one roentgen equals the amount of energy absorbed by one gram of biological material that dissipates 83 ergs.
Root The plant organ that serves for support and sometimes for storage, that is responsible for the intake of water and inorganic ions.

S stage The period during interphase when DNA synthesis takes place.
Salicylates A group of acids, the most well known of which is acetylsalicylic acid (aspirin), used as mild analgesics.
Sarcoma A malignant tumor of mesodermal origin.
Saltation Speciation in a single large jump, occurring over two or three generations.
Scientific method An approach to scientific inquiry involving the formation of a hypothesis and a means of testing the hypothesis under controlled conditions.
Seasonal isolation The inability of two species to produce a viable offspring because of a difference in the time of year when the male and female are fertile.
Secondary carnivore An animal that eats only animals that eat only animals.
Secondary consumer Any organism that feeds on primary consumers.
Secondary structure of proteins The α-helical structure generated by the angles of the bonds of the peptide linkage, and the juxtaposition of O and OH groups leading to the formation of hydrogen bonds.
Secondary succession Succession in an area that already has primitive forms of life.
Segregation The separation of alleles during meiosis.
Selective advantage The advantage for survival existing for an organism that is by some heritable trait better suited for the environment than an organism with another trait.
Self-assembly Specific associations of molecules to form cellular structures by spontaneous processes rather than from an enzyme-mediated system.
Semiconservative replication The multiplicative pattern proposed by Watson and Crick for their model of DNA; this pattern involves an unwinding of the existing strands of DNA and the attraction of complementary bases.
Semipermeable Term applied to a membrane that allows some substances to pass through, and prohibits the passage of others.
Sex chromosomes The chromosomes in an organism involved in the determination of sex.
Sickle-cell anemia A genetic disease in which the homozygous recessive individual dies, while the heterozygous individual has a selective advantage because of the inability of the malarial parasite to enter sickled red blood cells.
Sirenogen A sex hormone produced by a water mold.
Sister chromatids Two chromosomes originating from the duplication of one chromosome.
Smog A word derived from the words "smoke" and "fog" to describe the atmospheric condition during a thermal inversion when combustion products are trapped close to the earth.
Soil A combination of rock particles of different sizes, and organic debris in which and on which certain animals and plants reside.
Soluble RNA (*t*RNA) 4S RNA.

Specimen A biological entity under observation.
Spectrum The color bars created by breaking white light into its component wavelengths. The range of energy content in electromagnetic radiation.
Sperm The motile gamete formed by the male of a species.
Spermatogonia The cells that upon meiotic division will give rise to the four spermatids, that will in turn, differentiate into sperm.
Spindle fibers An ensemble of fibers composed of microtubules, important in the separation of chromosomes during nuclear division.
Spontaneous generation The creation of biological entities from nonbiological sources.
Spontaneous mutation A change in the genetic information by natural processes, that is, without deliberate introduction of a mutagen.
Stabilizing selective pressure Selective pressure that favors the exclusion of genes that would increase the adaptive zone of the species.
Standing crop The energy content of a population available to feeders any one time.
Star A heavenly body generating tremendous heat as the result of thermonuclear explosions at the surface, and sometimes surrounded by planets in set orbits.
Starch A polymer made up of glucose molecules produced by some plant cells.
Starch grains Bodies that stain blue-black with iodine in some plant cells; these seem to be storage depots for starch.
Stationary phase The part of the growth curve that reflects the time when the rate of growth equals that of death.
Steady state Synonymous with HOMEOSTASIS.
Stem The vegetative organ of vascular plants that supports the leaves, and contain the vascular tissue for the transport of materials to and from the leaves and roots.
Stepwise reactions A series of reactions, each accomplishing a single discrete change and requiring energy within the limits of the system, that characterize metabolism.
Sterioisomerism The phenomenon in which different molecules of the same structure cause a change in the orientation of polarized light either to the left (levorotary) or to the right (dextrorotatory).
Stochastic model A model of a system taking into consideration the probabilities involved in the occurrence of any facet of that system.
Stomach An organ of the digestive system in which food is ground into chyme and protein digestion begins.
Streptomycin An antibiotic produced by bacteria of the *Streptomyces* genus.
Stroma The "fluid" phase of the chloroplast.
Strontium^{90} The element that tends to replace Ca^{++} in plant and animal tissues.
Substrate The reactant of an enzyme-mediated reaction.
Substrate-level phosphorylation A high-energy phosphate such as that found on phospho-enol pyruvate, used to convert ADP to ATP.
Succession The series of organismal replacements that occur in a given geographical location and lead to a stable climax form.
Sulfanilamide An analogue of para-aminobenzoic acid.
Supernatant The phase resulting from a centrifugation that is in a liquid state.
Svedberg unit The measurement for ultracentrifugations that takes into consideration the drift of minute particles according to their size, shape, and density.
Synthesis Literally, building. In metabolism, the production of larger, more complex types of molecules from simple precursors.

Taxonomy Synonymous with BIOSYSTEMATICS.
Taxonomic key A series of statements concerning morphological features arranged in such a way that by following the appropriate descriptions, the investigator is brought to the various classification levels for the organism he is trying to identify.
Temperate deciduous forest The typical forest found in temperate zones; this represents the final stage in succession.
Temperature A universally agreed-on measure of the heat content of a system.
Tertiary structure of proteins The conformation generated by the bonding of the side groups of amino acids (for instance, disulfide bridge or ionic bond).
Tetracyclines A group of broad-spectrum antibiotics produced by bacteria.
Thallus Plant body.
Thermodynamics The study of energy transformations.
Thermophyllic organisms Those organisms that thrive and perhaps only exist at temperatures above 60° C.
Thiamine Vitamin B_1, the lack of which causes the disease beriberi; and which combines with a pyrophosphate to form TTP, a coenzyme active in both the glycolytic pathway and the Krebs cycle.
Threshold The lowest level of intensity necessary for a stimulus to elicit a response.
Tiaga A biome characteristic of cold, sub-arctic regions consisting primarily of conifer forests.
Tissue A population of cells in a multicellular organism having a common function.
Total cell count The number of live and dead cells in a culture.
Trace elements Elements required for biological systems in concentrations less than 10^{-5} M.
Transcription The lining up of RNA bases along the DNA gene, according to a sequence dictated by the sequence of nucleotide bases of the reading strand of the DNA.
Transferases A class of enzymes that mediate reactions in which transfers of groups are involved.
Transfer RNA Synonymous with SOLUBLE RNA.
Transformation principle The factor or molecule responsible for the transformation of one genetic type of cell to another type; found to be DNA.
Translation The process in which *t*RNA units, with their appropriate amino acids in tow, line up on the *m*RNA in an order dictated by the *m*RNA units.
Translocation (1) A phenomenon in which two chromosomes are broken and the free portion of one joins with the other, while the free portion of the other joins with the first. The genome is complete only when the two chromosomes bearing the translocations remain together after meiotic division. (2) The movement of food within a plant.

Transmethylation The transfer of a methyl group from one organic molecule to another.
Transposition A phenomenon in which a chromosome breaks in three places, and the segment rejoins in a different part of the chromosome. (e.g. *abc defghi jklm opqrstu* because of the breaks and rejoining becomes *abcjklmdefghiopqrstu*.)
Trigram A three-symbol unit read as a single symbol (such as str, ing, and ion).
Triplet-state electron The electron that, upon the addition of energy to a molecule, is raised to a higher energy level.
Tritiated thymidine Thymine prepared with radioactive hydrogen.
Tropic rain forest The final stage of succession in tropic regions composed primarily of giant fern trees with aerial roots, and broadleaf conifers.
Trophic level A composite of all the organisms that feed on and derive all their energy from another composite of all organisms that derive their food in the same manner.
Trypanosomes Protozoans classified as Mastigophora, that cause sleeping sickness.
Tumors Localized groups of cells resulting from the resumption of cell division by some differentiated cells of a given tissue.
Tundra A biome representing the final stage of succession in polar or polarlike regions.

Ultraviolet light Electromagnetic radiation ranging from 1800 Å to 3300 Å.
Uracil A nitrogen base (pyrimidine) found only in ribotide form, and hence only in RNA.
Unicellular Referring to organisms in which all biological processes proceed within a single cell.
Unit membrane Synonymous with CELL MEMBRANE.

Vaccine A preparation of dead or attenuated viruses that elicit an antibody response that will be effective if the true virulent form of the virus enters the body.
Valence The combining power of an atom; the valence reflects the number of electrons that must be shared with another atom to achieve electrical neutrality.
Variance (σ^2) A numerical calculation describing the extent to which data can deviate from the mean and still be part of the phenomenon under observation.
Vestigial organ A structure in a degenerate state and of no known function, that has as yet not been eliminated from a species.
Viable cell count A count of only the live cells in a population.
Virus A minute unit of biological function, composed of nucleic acid and protein, carrying out no metabolism itself, but able to control the metabolism of the cell it infects.
Vitamin A molecular form that is either a coenzyme or is readily converted into a coenzyme; that cannot be synthesized by the organism; and that therefore becomes part of the organism's nutritional requirements.
Water The universal solvent, due in part to its dipolar quality.
Wavelength The distance between the crests (or for that matter, any corresponding points) along a wave.
Wave theory of light The idea that asserts that light manifests properties associated with wave mechanics, similar to those waves on the surface of a body of water.
Wet weight The weighing of material in which the water has been only partially removed.
Work The movement of an object over a distance.

X-rays Electromagnetic radiation containing sufficient energy to cause chromosome breaks.
Xenograft A tissue graft between members of different species.
Xylem The conductive tissue in plants responsible for the movement of water and minerals; these cells are hollow tubes that have died in the process of differentiation.

Zero-order reaction A reaction that proceeds at a rate that can be represented by the general equation $y = a$.
Zoology The area of biology concerned exclusively with the study of animals.
Zygote The diploid cell formed as the result of the fertilization of two haploid gametes. In diploid organisms, the formation of the zygote represents a return to the typical ploidy level of the species. In haploid organisms, the formation of the zygote initiates a period of atypical ploidy, terminated by the meiotic division of the zygote to form four haploid cells.

INDEX

*Boldface numbers indicate tables or figures.